The Abel Prize 2013–2017

Niels Henrik Abel 1802–1829
The only contemporary portrait of Abel, painted by Johan Gørbitz in 1826

Helge Holden • Ragni Piene
Editors

The Abel Prize 2013–2017

ABEL
PRISEN

Springer

Editors

Helge Holden
Department of Mathematical Sciences
NTNU Norwegian University of Science and Technology
Trondheim, Norway

Ragni Piene
Department of Mathematics
University of Oslo
Oslo, Norway

Photo of Pierre Deligne on part title page by Heiko Junge/NTB Scanpix
Photo of Pierre Deligne on first page of his Curriculum Vitae by Cliff Moore Institute for Advanced Study
Photo of Yakov G. Sinai on part title page by Håkon Mosvold Larsen/NTB Scanpix
Photo of Yakov G. Sinai on first page of his Curriculum Vitae by Harald Hanche-Olsen
Photo of John F. Nash and Louis Nirenberg on part title page by Berit Roald/NTB Scanpix.
Photo of Louis Nirenberg on the first page of his Curriculum Vitae by Harald Hanche-Olsen
Photo of Sir Andrew J. Wiles on part title page by Audun Braastad.
Photo of Sir Andrew J. Wiles on first page of his Curriculum Vitae by Harald Hanche-Olsen
Photo of Yves Meyer on part title page by Thomas Brun/NTB Scanpix.
Photo of Yves Meyer on the first page of his Curriculum Vitae by Harald Hanche-Olsen

Additional material to this book can be downloaded from http://extras.springer.com.

ISSN 2661-829X ISSN 2661-8303 (electronic)
The Abel Prize
ISBN 978-3-319-99027-9 ISBN 978-3-319-99028-6 (eBook)
https://doi.org/10.1007/978-3-319-99028-6

Mathematics Subject Classification (2010): 00-XX

This Springer imprint is published by the registered company Springer Nature Switzerland AG
The registered company address is: Gewerbestrasse 11, 6330 Cham, Switzerland

Au reste il me parait que si l'on veut faire des progres dans les mathematiques il faut étudier les maitres et non pas les ecoliers. Niels Henrik Abel†

† "Finally, it appears to me that if one wants to make progress in mathematics, one should study the masters, not their students." In: "Memoires de Mathématiques par N. H. Abel", Paris, August 9, 1826, in the margin of p. 79. Original (Ms.fol. 351 A) in The National Library of Norway. Reprinted with permission.

Preface

This book constitutes the third volume[1] in a series on the Abel Laureates, covering the period 2013–2017.

We keep the same structure as that of the previous volumes. There is one chapter per year. Each chapter starts with the full citation from the Abel Committee, followed by an autobiographical piece by the laureate. Then comes an article on the scientific accomplishments of the laureate. In the first chapter, L. Illusie writes on Pierre Deligne, while in the second chapter, the team, C. Boldrighini, L. Bunimovich, F. Cellarosi, B. Gurevich, K. Khanin, D. Li, Y. Pesin, N. Simányi, and D. Szász, led by K. Khanin, presents the work of Yakov G. Sinai. In the third chapter, C. De Lellis writes on the work of John Nash, Jr. and R. Kohn on the work of Louis Nirenberg. The work of Andrew Wiles is presented by C. Skinner in the fourth chapter, and in the last chapter, A. Cohen writes on the work of Yves Meyer.

Tragically, John Nash, Jr. and his wife Alicia died in an automobile accident on their way home to Princeton after the Abel Prize events in Oslo. Nash had prior to the Abel ceremony agreed to write his autobiographical piece, but this was not to be. Sylvia Nasar, the author of the bestselling biography[2] of John Nash, kindly volunteered to write a brief biography for this volume. In addition, we reproduce with the kind permission of the Nobel Foundation, the short autobiography that Nash wrote on the occasion of receiving, in 1994, The Sveriges Riksbank Prize in Economic Sciences in Memory of Alfred Nobel.

Each chapter contains a complete bibliography and a curriculum vitae, as well as photos—old and new.

The last chapter is meant to give, through a collection of photos, an idea of all the activities that take place in connection with the Abel Prize, especially those that involve children and youth. For in the Statutes of the prize it says:

---

[1] H. Holden, R. Piene (eds.): *The Abel Prize 2003–2007. The First Five Years*, Springer, Heidelberg, 2010, and H. Holden, R. Piene (eds.): *The Abel Prize 2008–2012*, Springer, Heidelberg, 2014.

[2] S. Nasar: *A Beautiful Mind*, Simon and Schuster, New York, 1998.

> The main objective of the Abel Prize is to recognize pioneering scientific achievements in mathematics. The Prize shall also help boost the status of the field of mathematics in society and stimulate children and youth to become interested in mathematics.

These other activities thus include mathematics competitions—the Niels Henrik Abel competition for high school students and the UngeAbel (previously KappAbel) competition for class teams of elementary school pupils—and the Bernt Michael Holmboe Memorial Prize, an annual prize awarded in connection with the Abel Prize ceremony, to a teacher or a group of teachers, who have done extraordinary efforts in mathematics teaching in Norway.

The Abel Board also supports annual international conferences, the Abel Symposia. It also supports mathematics in the developing world, by a yearly donation to the International Mathematical Union. This included support for the Ramanujan Prize in the years 2005–2012, and, from 2013 on, the Abel Visiting Scholar program administered by IMU's Commission for Developing Countries.

The back matter contains updates regarding publications and curriculum vitae for all laureates, as well as the full prize citations for the years 2003–2012. Finally, we list the members of the Abel Committee and the Abel Board for this period.

The annual interview of the Abel Laureates, aired on Norwegian national TV, can be streamed from the Springer website. Transcripts of the interviews have been published, and publication details can be found in the back matter.

We would like to express our gratitude to the laureates for collaborating with us on this project, especially for providing the autobiographical pieces and the photos. We would like to thank the mathematicians who agreed to write about the scientific work of the laureates, and thus are helping us in making the laureates' work known to a broader audience.

Thanks go Marius Thaule for his LaTeX expertise and the preparation of the bibliographies as well as copyediting the manuscripts.

The technical preparation of the manuscript was financed by the Abel Board.

Trondheim, Norway
Oslo, Norway
June 6, 2018

Helge Holden
Ragni Piene

Contents

Part I
2013 Pierre Deligne

"for seminal contributions to algebraic geometry and for their transformative impact on number theory, representation theory, and related fields"

ABEL
PRISEN

Citation

The Norwegian Academy of Science and Letters has decided to award the Abel Prize for 2013 to **Pierre Deligne**, Institute for Advanced Study, Princeton, New Jersey, USA,

> for seminal contributions to algebraic geometry and for their transformative impact on number theory, representation theory, and related fields

Geometric objects such as lines, circles and spheres can be described by simple algebraic equations. The resulting fundamental connection between geometry and algebra led to the development of algebraic geometry, in which geometric methods are used to study solutions of polynomial equations, and, conversely, algebraic techniques are applied to analyze geometric objects. Over time, algebraic geometry has undergone several transformations and expansions, and has become a central subject with deep connections to almost every area of mathematics. Pierre Deligne played a crucial role in many of these developments. Deligne's best known achievement is his spectacular solution of the last and deepest of the Weil conjectures, namely the analogue of the Riemann hypothesis for algebraic varieties over a finite field. Weil envisioned that the proof of these conjectures would require methods from algebraic topology. In this spirit, Grothendieck and his school developed the theory of l-adic cohomology, which would then become a basic tool in Deligne's proof. Deligne's brilliant work is a real tour de force and sheds new light on the cohomology of algebraic varieties. The Weil conjectures have many important applications in number theory, including the solution of the Ramanujan–Petersson conjecture and the estimation of exponential sums.

In a series of papers, Deligne showed that the cohomology of singular, non-compact varieties possesses a mixed Hodge structure that generalized the classical Hodge theory. The theory of mixed Hodge structures is now a basic and powerful tool in algebraic geometry and has yielded a deeper understanding of cohomology. It was also used by Cattani, Deligne and Kaplan to prove an algebraicity theorem that provides strong evidence for the Hodge conjecture.

With Beilinson, Bernstein and Gabber, Deligne made definitive contributions to the theory of perverse sheaves. This theory plays an important role in the recent proof of the fundamental lemma by Ngo. It was also used by Deligne himself to greatly clarify the nature of the Riemann–Hilbert correspondence, which extends Hilbert's 21st problem to higher dimensions. Deligne and Lusztig used l-adic cohomology to construct linear representations for general finite groups of Lie type. With Mumford, Deligne introduced the notion of an algebraic stack to prove that the moduli space of stable curves is compact. These and many other contributions have had a profound impact on algebraic geometry and related fields. Deligne's powerful concepts, ideas, results and methods continue to influence the development of algebraic geometry, as well as mathematics as a whole.

Mathematical Autobiography

Pierre Deligne

In what follows, I dwell on some major influences on my mathematical education. The account Luc Illusie gives of my work is much more systematic. I would like to begin by thanking him for it.

I was born 1 month after the liberation of Brussels. My mother often told me how a providential school of herring saved Belgium from starvation, and how Holland had it much worse in the winter of 1944/1945.

My siblings are 7 and 11 years older than me. My parents highly valued education, and we were the first generation in the family to attend university. I enjoyed my brother's explanations of mathematical facts he had just learned. Looking at the thermometer made negative numbers easy to grasp, but that $(-1) \times (-1)$ is $+1$ was another matter. Of course, my brother was saying "is", not "is better defined to be because...". Much later, I was very surprised that historians did not use a year 0—presumably because chronologies preceded the taming of negative numbers. When he was in high school, my brother showed me how to solve second degree equations. In his college textbook, I read about the degree three case.

I have been extremely lucky, both with the people I met, who helped me, and that the time of my youth was a time for the creation of tools, my inclination. At 14, I met Mr. Nijs, who was a high school teacher. He saw my interest in mathematics, and took the risky, but fortunate decision to give me Bourbaki's *Éléments de Mathématique,* starting with the four chapters on Set Theory. I cared that in mathematics "true" meant true, not just arguable, and here at last was an

---

Electronic Supplementary Material The online version of this chapter (https://doi.org/10.1007/978-3-319-99028-6_1) contains supplementary material, which is available to authorized users.

P. Deligne (✉)
Institute for Advanced Study, Princeton, NJ, USA
e-mail: deligne@math.ias.edu

H. Holden, R. Piene (eds.), *The Abel Prize 2013–2017*,
https://doi.org/10.1007/978-3-319-99028-6_1

Me at age 7. Taken in September 1952. I am in a "louveteau" (younger boy scout) uniform. (Photo: private)

idealization of what a proof was, serving as an anchor. A welcome contrast to the sloppy beginning of Euclidean geometry learned in school. I also read A. Heyting's *Intuitionism, an introduction.* From intuitionism (and later from E. Bishop's less dogmatic *Constructive Analysis*), I learned that non effectivity of proofs is usually due to the use of excluded middle (a statement or its negation is true) rather than to the use of the axiom of choice.[1] In writing up proofs, I continue to try avoiding proofs by contradiction. I find that when not too costly, a construction gives a better understanding.

A second piece of luck was that J. Tits was then at Brussels University (ULB). While still in high school, I could attend his course on Lie groups, as well as the seminar he was organizing with F. Bingen and L. Waelbroeck. At one of his lectures, he defined the center of a group, stated it is an invariant subgroup, started proving it, and then stopped, saying: "in fact this is obvious. As I could define the center, it is stable by any automorphism, a fortiori by inner automorphisms." This is how I realized the power of "transport of structures", the principle that when we have two sets S_1 and S_2 with some structures s_1 and s_2, and an isomorphism $f\colon (S_1, s_1) \xrightarrow{\sim}$

[1]For a nice description of what is involved here, I refer to the recent article by A. Bauer: Five stages of accepting constructive mathematics, Bull. AMS **54** 3 (2017) 481–498.

(S_2, s_2), anything done on (S_1, s_1) can be transported to (S_2, s_2), and that this is especially useful when $(S_1, s_1) = (S_2, s_2)$, but f is not the identity. This is an analogue in mathematics to Curie's principle: "Lorsque certaines causes produisent certains effets, les éléments de symétrie des causes doivent se retrouver dans les effets produits."[2]

The mental hygiene needed to apply transport of structures is natural, especially so when using a language, unlike Russian, which distinguishes between "a" and "the". It is well explained in chapter "Curriculum Vitae for Pierre R. Deligne" of Bourbaki's Set Theory (original French edition, not the second edition "revue et diminuée"). Unfortunately, the categorical analogue, asking that no distinction be made between equivalent categories, remains rules of thumb, such that "equality makes sense between morphisms, not between objects." An equality sign between objects usually means an isomorphism has been constructed, and compatibilities between such isomorphisms have to be taken care of.

I cherish a piece of advice Tits gave me: "Do what you like". When I was 20, he told me it was time to go to Paris, made it possible and introduced me to Grothendieck. The next 2 years (the second as a "pensionnaire étranger" at École Normale Supérieure), I mainly attended Grothendieck's seminar at IHÉS and Serre's lectures at Collège de France, with the rest of each week needed to understand the lectures I had listened to and to fill gaps in my education. At the end of each year, I would return to Brussels to pass exams at the University. This was possible thanks to the European system, where only mathematical courses, plus some physics, were required. The American system, with its distributional requirements, would have been suffocating. I was also helped by fortunate previous readings, made possible by browsing the open stacks of the library of the ULB department of mathematics. The shelving by alphabetical order of authors encouraged serendipity. Two books which took me a long effort to digest, but proved very useful, were de Rham's *Variétés Différentiables,* and Godement's *Théorie des faisceaux.*

The next year (1966/1967) was lost to military service. (Belgium was still occupying parts of Germany at that time.)

As Grothendieck wrote in *Récoltes et Semailles,* he was building "houses" where mathematical ideas would not be cramped. He had around him some of the best young French mathematicians who, inspired by him, were helping at that task. He asked me to write exposés XVII and XVIII of SGA4, respectively about cohomology with compact support and duality in etale cohomology. Doing so, I learned how to write, as well as the subject matter. My first draft was returned to me with two injunctions: "Proofs should be complete" and "False statements are not allowed." The second seems obvious, but is not when it concerns signs in homological algebra.

[2]Translation: When certain causes produce certain effects, the elements of symmetry of these causes must be found in the produced effects.

Orsay thesis. February 1972. (Photo: private)

I admire how Grothendieck was able, so many times, to develop a framework in which difficulties of proofs dissolved. This remained for me an ideal, which I rarely approached.

His philosophy of motives has been a guiding thread in many of my works, including some for which this is not immediately apparent. I mean here the philosophy, not Grothendieck's precise definition of a category of pure motives over a field k. This precise definition is reasonable only if one assumes the so-called "standard conjectures", for which the evidence is meager.

Let us consider algebraic varieties over a field k. We have for them many cohomology theories (with coefficient fields of characteristic zero) which seem to repeat the same story in different languages: Betti (for $k \hookrightarrow \mathbb{C}$), de Rham (for smooth varieties over k of characteristic zero), crystalline, ℓ-adic (for ℓ a prime invertible in k). The philosophy of motives tells the following.

(A) Each of these theories factors through a motivic theory H^*_{mot}, with values in the category of motives over k. This category is a $\mathbb{Q}$-linear abelian category, in which the Hom groups are finite dimensional. The theory h^* is deduced from H^*_{mot} by applying a *realization functor* real$\{h^*\}$: an exact functor from motives to the abelian category in which h^* takes values. Of course, natural isomorphisms, exact sequences, spectral sequences,... making sense across theories are images of the same in the category of motives, and relations between theories, such as comparison isomorphisms, are induced by relations among the realization functors.

Models: for smooth projective varieties, $\mathrm{Pic}^0(X)$ (viewed as an object of the category of abelian varieties taken up to isogeny) plays the role of a motivic H^1: all h^1 are deduced from it by applying suitable functors. For H^0, we have the more elementary model of rational representations of $\mathrm{Gal}(\bar{k}/k)$ (Artin motives).

(B) The category of motives has a tensor product, compatible with the various realization functors, and giving rise to a motivic Künneth formula. This tensor product turns the category of motives into a tannakian category over $\mathbb{Q}$. Tannakian categories were invented by Grothendieck for this purpose, and are akin to categories of representations of algebraic groups. The case of Artin motives, where the group is $\mathrm{Gal}(\bar{k}/k)$, led to the terminology "motivic Galois group."

Here are applications I made of the motivic philosophy.

Definition of Mixed Hodge Structures

I try to remain aware of what I don't understand, and of "discrepancies". One discrepancy which occupied me a great deal at the time is that while the eigenvalues of the Frobenius acting on ℓ-adic cohomology are ℓ-adic numbers, the Weil conjecture is about their complex absolute values. The discrepancy relevant for mixed Hodge theory is between the scope of applicability of the Hodge versus the ℓ-adic theories. Let X be a smooth projective variety over, say, a number field $k \subset \mathbb{C}$. Hodge theory gives a Hodge structure of weight n on $H^n_B(X) := H^n(X(\mathbb{C}), \mathbb{Q})$, that is a decomposition $H^n_B(X) \otimes \mathbb{C} = \bigoplus_{p+q=n} H^{p,q}$, with $\overline{H^{pq}} = H^{qp}$. For each ℓ, etale cohomology gives an action of $\mathrm{Gal}(\bar{k}/k)$ on $H^n_\ell(X) = H^n(X(\mathbb{C}), \mathbb{Q}) \otimes \mathbb{Q}_\ell$, turning H^n_ℓ into an ℓ-adic representation of weight n (weight n refers to the complex absolute values of eigenvalues of Frobenius elements, and was at that time conjectural).

Etale cohomology continues to provide an action of Galois on $H^*_\ell(X)$ when X is not supposed to be projective and smooth. Further, spectral sequences of geometric origin abut an increasing *weight filtration* W such that each $\mathrm{Gr}^W_p(H^n_\ell(X))$ is a subquotient of some $H^p_\ell(Y)$, with Y projective and smooth. The motivic philosophy suggests that the weight filtration is motivic, that is comes from a filtration W of $H^n_{\mathrm{mot}}(X)$, and that $\mathrm{Gr}^W_p H^n_{\mathrm{mot}}(X)$ is pure of weight p. Applying the Betti realization functor, we would get on $H^n_B(X) = H^n(X(\mathbb{C}), \mathbb{Q})$ a weight filtration W and for each p a Hodge structure of weight p on $\mathrm{Gr}^W_p H_B(X)$.

In the ℓ-adic case, one does not just have a weight filtration and a pure structure on the successive quotients, but an ambient abelian category of ℓ-adic representations of Galois. The motivic philosophy forces the question: "What is the Hodge analogue?" The solution appears when one admits the primacy of the Hodge filtration over the Hodge decomposition: $H^n_B(X)$ carries a mixed Hodge structure, given by a filtration F of $H^n_B(X) \otimes \mathbb{C}$ inducing the Hodge filtrations of the pure subquotients $\mathrm{Gr}^W_p H^n_B(X)$.

For projective varieties, an extension of an abelian variety by a torus (taken up to isogeny) plays the role of a motivic H^1: to X, one attaches the quotient of $\mathrm{Pic}^0(X)$ by its unipotent radical. For X over $\mathbb{C}$, this motivic H^1 is determined by $H^1_B(X)$, with its mixed Hodge structure. For general varieties, one should similarly consider 1-motives (up to isogeny).

Definition of Shimura Varieties

Period mapping domains are moduli spaces of Hodge structures on a fixed vector space V, compatible with structures on V expressible in the language of multilinear algebra. Hermitian symmetric spaces correspond to the special case where Griffiths transversality (at first order, F^p moves within F^{p-1}) is satisfied. This makes it natural to think of their arithmetic quotients as moduli spaces of motives M, endowed with structures s expressible using the tensor product. A way to express such an (M, s) is: a functor, compatible with $\otimes$, from the category of representations of a reductive group G over $\mathbb{Q}$, to the category of motives. Level structures should be given as well. Conditions have to be imposed, and fields of definition should be subfields of $\mathbb{C}$ above which they make sense. Shimura emphasized an algebra with involution giving rise to the (classical) group G, and case-by-case characterized his canonical model by properties of CM points. Emphasizing G, as motives suggested, allowed for uniform definitions, where the properties of CM points appeared as functoriality for a morphism $G_1 \to G_2$, with G_1 a torus, $G_2 = G$.

Morphisms Between Motives

Grothendieck's definition of the category of pure motives is reasonable only provided that there are "enough" algebraic cycles. On this question, almost no progress has been made since the 1960s. I have made attempts to find substitutes for algebraic cycles, with some success only in situations closely related to abelian varieties and where monodromy groups are "large". [D20] and [D21] concern cohomology groups H for which one can construct injections to $H^1(A)$ (resp $H^2(A)$), for A an abelian variety, with the same good properties as if they were induced by a motivic map in the sense of Grothendieck. In [D48], I show that Hodge cycles on abelian varieties enjoy many of the properties of algebraic cycles.

Conjectures on Critical Zeta Values

Motives give rise to zeta functions $\zeta(M, s)$. The value at an integer n depends only on the Tate twist $M(n)$ of M. For n "critical", $\zeta(M, s)$ was in many cases expressible as a rational multiple of "periods". If to make a conjecture one insists on using only the de Rham and Betti realizations of $M(n)$, with the natural structures they carry, one is quickly led to the conjecture I made.

Later, Beilinson understood I was simply taking the volume of an Ext^1-group in the category of mixed Hodge structures, and that for general integral values of s, this Ext^1-group should be taken modulo a motivic Ext^1.

Me making pancakes on an open fire in Ormaille. Around 1979. (Photo: C. Tate)

Relations Between Multizeta Values

Here we leave the categories of pure motives on which Grothendieck was concentrating, to consider iterated extensions of Tate motives. Over number fields, such categories of mixed Tate motives can actually be defined, and the size of a motivic Galois group imposes linear relations between multizeta values.

When I was in high school, I had no idea one could get paid for doing mathematics. My father would have liked me to become an engineer. I was planning to become a high school teacher, and do mathematics as a hobby. That I could earn a living by doing what I liked best came as a pleasant surprise. I should add that the situation then was much better than it is at present for young people. Many jobs were available thanks to the expansion of higher education.

The IHÉS (Institut des Hautes Études Scientifiques) had been created by Motchane in 1958. He was inspired by the example of the IAS (Institute for Advanced Study), and asked advice from Oppenheimer, its director. France, however, had no tradition of philanthropy for the sciences, and Motchane succeeded against great odds. He took good advice, convinced Dieudonné and (at Dieudonné's instigation) Grothendieck to accept his risky offer, and managed to convince first industrialists, and later governments to give money for his creation to survive, sometimes tenuously. IHÉS became my paradise.

Daughter Natalia in basket, son Alyosha on my back. (Photo: private)

In 1985, I moved to another paradise, the IAS. My self-imposed obligation to give each year a seminar was becoming heavy, and I did not feel it to be wise to spend all my life in one place. I was also attracted by the beauty of open spaces in the US, and by the presence in Princeton of Langlands and Harish-Chandra (who alas passed away shortly before my arrival).

In 1996/1997, there was at the IAS a yearlong effort to understand what string theorists were doing. One of the motivations was that they were able to make wholly unexpected predictions—which so far have always turned out to be correct—even in very classical parts of algebraic geometry. My aim that year was to learn the rules for making such predictions. I failed. The stumbling blocks were not the ones I expected: an absence of proof is a challenge, an absence of definition is for me deadly. I felt expelled from Cantor's paradise to the world of Euler, where formulae are assumed to have meaning, with no distinction between defining and computing.

At the banquet for my 61st anniversary (Fall 2005). Left: Mozzochi, behind him Langlands. Clockwise around my head: Esnault, Messing, Beilinson. Right: half face of Luc Illusie, and behind him Nicholas Katz. (Photo: private)

As Euler would say "Let us compute $\sum(-1)^n n!$",[3] physicists would say "Let us compute the path integral related to such or such lagrangian".

I still would very much like to understand why this formalism led to so many correct predictions.

[3]De seriebus divergentibus, Opera Omnia I **14** 585–617.

Pierre Deligne: A Poet of Arithmetic Geometry

Luc Illusie

Dix choses soupçonnées seulement, dont aucune (la conjecture de Hodge disons) n'entraîne conviction, mais qui mutuellement s'éclairent et se complètent et semblent concourir à une même harmonie encore mystérieuse, acquièrent dans cette harmonie force de vision. Alors même que toutes les dix finiraient par se révéler fausses, le travail qui a abouti à cette vision provisoire n'a pas été fait en vain, et l'harmonie qu'il nous a fait entrevoir et qu'il nous a permis de pénétrer tant soit peu n'est pas une illusion, mais une réalité, nous appelant à la connaître.

— A. Grothendieck, *Récoltes et Semailles*, Deuxième partie, I B 4 1.

Grothendieck's philosophy of motives permeates Deligne's work. No one has made the multiple voices of arithmetic geometry sing in harmony better than Deligne. Almost every one of his articles echoes or corresponds to another one, sometimes far away. I have tried to make this counterpoint perceptible.

The plan roughly follows a tentative chronological order—awkward and artificial as it is to establish such an order, since Deligne was often working on several distinct themes at the same time. Despite the interaction between the various parts, I think that each main section can be read independently. An important part of Deligne's work consists in his conjectures. I recap them in Sect. 10 and discuss those that had not appeared in the previous sections. In Sect. 11, I list Deligne's expository articles.

This report is by no means comprehensive. The contributions that I have only briefly mentioned or not discussed at all are numerous, and each of them would have deserved a careful analysis.

References to articles in the list of publications of Deligne are given in the form [D***,****].

L. Illusie (✉)
Université Paris-Sud, Orsay Cedex, France
e-mail: luc.illusie@math.u-psud.fr

H. Holden, R. Piene (eds.), *The Abel Prize 2013–2017*,
https://doi.org/10.1007/978-3-319-99028-6_2

1 Foundational Work: Topology, Homological Algebra, Étale Cohomology

Deligne's first contributions were inspired by questions related to the new territories that Grothendieck was exploring: sites and topoi, derived categories, étale cohomology. He did not just solve riddles, but, with a view towards geometric applications, built solid foundations for new techniques which were to become of standard use.

Deligne's foundational work is not limited to the topics discussed in this section. See Sects. 2.1, 2.3, 4.2 "Homological Algebra Infrastructure", Sect. 4.3 "Axiomatization of Shimura Varieties", Sect. 5.8 "t-Structures", 7.4, and 9.1 for other important basic contributions.

1.1 General Topology

A site is a category equipped with a Grothendieck topology. A topos is a category equivalent to the category of sheaves on a site. A point t of a topos T is a functor $F \mapsto F_t$ from T to the category of sets commuting with arbitrary inductive limits and with finite projective limits. The empty topos, i.e., the one object, one map category of sheaves on the empty space has no point. Deligne gave the first example of a non-empty topos having no point ([2], IV, 7.4): the topos of sheaves on the site defined by the category of Lebesgue measurable subsets of the segment $[0, 1]$, up to measure zero sets, with maps deduced from inclusions, and the topology defined by covering families consisting of countable unions (up to measure zero sets).

In the positive direction, Deligne gave a convenient sufficient condition for a topos T to have *enough points*, i.e., a conservative family of points: if T is *locally coherent*, i.e., is locally defined by a site having finite projective limits and in which any covering family $(U_i \to U)_{i \in I}$ has a finite sub-covering, then T has enough points ([3], VI 9). Topoi arising from certain topologies on schemes, such as the fpqc topology or Voevodsky's h-topology, are easily seen to be locally coherent, though the existence of enough points is not clear. It was later observed that Deligne's theorem is equivalent to Gödel's completeness theorem on first order logic ([137], p. 243).

Though topoi without points can be considered as pathological, for a number of basic results in the theory, the hypothesis of the existence of enough points looked artificial, and it was a challenge to do without it, for example, to prove stability of flatness under inverse images. Deligne solved this question by an elegant extension of D. Lazard's theorem on flat modules, involving a new technique of *local inductive limits* ([3], V, 8.2.12).

In the early 1980s Deligne constructed *oriented products* of topoi, with an application to a theory of nearby cycles over bases of any dimension, see Sect. 7.4.

1.2 Spectral Sequences

Degeneration and Decomposition in the Derived Category

Let $\mathscr{A}$ be an abelian category. An object K of the bounded derived category $D^b(\mathscr{A})$ is called *decomposable* if there is an isomorphism $K \simeq \oplus H^i(K)[-i]$ in $D^b(\mathscr{A})$. If K is decomposable, then, for any cohomological functor T from $D^b(\mathscr{A})$ to an abelian category $\mathscr{B}$, the spectral sequence ([254], III 4.4.6)

$$E_2^{pq} = T(H^q(K)[p]) \Rightarrow T(K[p+q])$$

trivially degenerates at E_2. In [D3, 1968] Deligne proves that the converse holds, and derives from this useful criteria of decomposability (*loc. cit.*, 1.5, 1.11):

Theorem 1 *Let $K \in D^b(\mathscr{A})$.*

(a) *If there exist an integer n and a morphism $u : K \to K[2]$ in $D^b(\mathscr{A})$ such that, for all $i \geq 0$, u^i induces an isomorphism $H^{n-i}(K) \xrightarrow{\sim} H^{n+i}(K)$, then K is decomposable.*
(b) *If there exist endomorphisms π_i of K in $D^b(\mathscr{A})$ such that $H^j(\pi_i) = \delta_{ij}$, then K is decomposable.*

He applies this to get degeneration results for Leray spectral sequences. Let $f : X \to Y$ be a proper and smooth morphism of schemes, purely of relative dimension n, with Y connected and X having a relatively ample invertible sheaf $\mathcal{O}(1)$.

(i) Assume Y separated and of finite type over $\mathbf{C}$, and let $f^{\mathrm{an}} : X^{\mathrm{an}} \to Y^{\mathrm{an}}$ denote the induced morphism on the associated complex analytic spaces. Then $Rf_*^{\mathrm{an}}\mathbf{Q}$ is decomposable in $D^b(Y^{\mathrm{an}}, \mathbf{Q})$, and, in particular, the Leray spectral sequence

$$E_2^{pq} = H^p(Y^{\mathrm{an}}, R^q f_*^{\mathrm{an}}\mathbf{Q}) \Rightarrow H^{p+q}(X^{\mathrm{an}}, \mathbf{Q}) \qquad (1)$$

degenerates at E_2.[1]

This follows from Theorem 1 (a) applied to the endomorphism of degree 2 of $Rf_*^{\mathrm{an}}\mathbf{Q}$ defined by the Chern class $u \in H^2(X^{an}, \mathbf{Q})$ of $\mathcal{O}(1)$, in view of the hard Lefschetz theorem on one fiber of f. If Y is smooth over $\mathbf{C}$ and f is assumed only *proper* and smooth (no existence of a relatively ample line bundle is demanded), then the conclusions of (i) still hold ([D16, 1971], 4.1.1). This time, this follows from Theorem 1 (b) applied to the endomorphisms of $Rf_*^{an}\mathbf{Q}$ defined by liftings to $X \times_Y X$ of Künneth components of the cohomology class of the diagonal of a fiber of $X \times_Y X$.

(ii) Assume Y separated and of finite type over an algebraically closed field k, and let ℓ be a prime number invertible in k. Assume that the hard Lefschetz

[1]As Serre observed (*loc. cit.*, 2.10), when Poincaré duality is available on the base, this degeneration can also be proved by an extension of Blanchard's method in [33].

theorem holds for $H^*(X_y, \mathbf{Q}_\ell)$ for one geometric fiber X_y of f, i.e., if $v \in H^2(X_y, \mathbf{Q}_\ell(1))$ is the image of the Chern class of $\mathcal{O}(1)$, then, for all $i \geq 0$, $v^i : H^{n-i}(X_y, \mathbf{Q}_\ell) \to H^{n+i}(X_y, \mathbf{Q}_\ell)(i)$ is an isomorphism. Then, similarly, using a variant of (a) allowing for Tate twists ([D3, 1968], 1.10), $Rf_*\mathbf{Q}_\ell$ is decomposable, and in particular, the Leray spectral sequence

$$E_2^{pq} = H^p(Y, R^q f_*\mathbf{Q}_\ell) \Rightarrow H^{p+q}(X, \mathbf{Q}_\ell) \tag{2}$$

degenerates at E_2.

At the time, the hard Lefschetz theorem was known only in special cases (e.g., varieties liftable to characteristic zero). It was later proved in general by Deligne [D46, 1980] (see Sect. 5.6 "First Applications", Hard Lefschetz theorem). Actually, according to Deligne ([D3, 1968], 2.9), Grothendieck, using a weight argument, had conjectured the degeneration of (2) for Y proper and smooth over k. His argument was the following. After standard reductions, we may assume that k is the algebraic closure of a finite field $\mathbf{F}_q$, and that $f : X \to Y$ comes by extension of scalars from a proper and smooth $f_0 : X_0 \to Y_0$. By the Weil conjectures (in the generalized form proved by Deligne in [D46, 1980], see Theorem 22), for any (i, j), the lisse sheaf $R^j f_{0*}\mathbf{Q}_\ell$ is pure of weight j and $H^i(Y_0, R^j f_{0*}\mathbf{Q}_\ell)$ is pure of weight $i+j$, hence all differentials of (2) must vanish, as their sources and targets have different weights. This doesn't prove the decomposability of $Rf_*\mathbf{Q}_\ell$, but, assuming Y to be only smooth over k, Deligne later found another argument (also based on Theorem 22), namely that $H^i(Y_0, R^j f_{0*}\mathbf{Q}_\ell)$ is mixed of weights $\geq i + j$, showing the desired decomposability. But all these weight arguments assume Y smooth over k.

In ([D3, 1968], 5.5) Deligne also gave a complement to this decomposition theorem for relative Hodge cohomology in characteristic zero.

Deligne will return to this topic several times:

- in [D53, 1982], with the so-called *decomposition theorem* (see Sect. 5.8 "The Purity and Decomposition Theorems")
- in [D65, 1987], with decompositions of the de Rham complex in characteristic $p > 0$ under certain lifting and dimension assumptions (see Sect. 4.6)
- in [D75,1994], where he revisits the above decomposition criteria in the framework of triangulated categories endowed with a t-structure, and constructs distinguished decompositions (see [49] for variants).

Décalage of Filtrations

In the early 1960s it had been observed that a spectral sequence could sometimes appear under different disguises: starting at E_1, or starting at E_2, with E_2 equal to the previous E_1 up to a certain renumbering. A typical example is provided by the spectral sequences arising from a bicomplex. Let $(M^{\bullet,\bullet}, d', d'')$ be a bicomplex of an abelian category A, concentrated in a quadrant $i \geq a$, $j \geq b$, and let $K := \mathbf{s}M^{\bullet,\bullet}$ be the associated simple complex, with $K^n = \oplus_{p+q=n} M^{p,q}$, $d = d' + d''$. The filtration on K induced by the naive truncation of $M^{\bullet,\bullet}$ relative to the first degree gives rise to a spectral sequence

$$'E_1^{p,q} = H''^q(M^{p,\bullet}) \Rightarrow H^{p+q}(K)$$

with E_2 term $'E_2^{p,q} = H'^p H''^q(M^{\bullet,\bullet})$, where H' (resp. H'') denotes cohomology relative to d' (resp. d''). On the other hand, the filtration W on K induced by the filtration of $M^{\bullet,\bullet}$ defined by the canonical truncations with respect to d'' gives rise to a spectral sequence

$${}_W E_1^{p,q} = H'^{2p+q} H''^{-p}(M^{\bullet,\bullet}) \Rightarrow H^{p+q}(K).$$

One has the coincidence

$${}_W E_1^{p,q} = {}'E_2^{2p+q,-p}, \tag{3}$$

with d_2 for $'E$ corresponding to d_1 for ${}_W E$. Deligne realized that this was a special case of a phenomenon produced by what he called *décalage* (shift) of filtrations. If $F = (F^p)_{p\in\mathbf{Z}}$ is a decreasing filtration on a complex (L, d) of A, Deligne defines the *filtration décalée* $\mathrm{Dec}(F)$ on L by

$$\mathrm{Dec}(F)^p L^n = F^{p+n} L^n \cap d^{-1}(F^{p+n+1} L^{n+1}).$$

The spectral sequences of L filtered by F and $\mathrm{Dec}(F)$ are related in the following way. The obvious homomorphism of complexes

$$\begin{aligned}(E_0^{p,q}(L, \mathrm{Dec}(F)), d) &= (\mathrm{gr}^p_{\mathrm{Dec}(F)} L^{p+q}, d)\\ &\to (E_1^{2p+q,-p}(L, F), d_1) = (H^{p+q}(\mathrm{gr}_F^{2p+q} L, d), d_1)\end{aligned}$$

is a quasi-isomorphism, and for $r \geq 1$, induces isomorphisms of complexes

$$(E_r^{p,q}(L, \mathrm{Dec}(F)), d_r) \xrightarrow{\sim} (E_{r+1}^{2p+q,-p}(L, F), d_{r+1}) \tag{4}$$

([D16, 1971], 1.3.3, 1.3.4). When one takes for F the filtration on K induced by the naive truncation of $M^{\bullet,\bullet}$ relative to the first degree, one has $\mathrm{Dec}(F) = W$, which explains (3).

Deligne devised this mechanism of décalage in 1965. It turned out to be a crucial technical tool in his construction of mixed Hodge structures on smooth schemes over $\mathbf{C}$, see Sect. 4.2 "Homological Algebra Infrastructure". Since then, décalage was used occasionally in de Rham or crystalline cohomology in positive characteristic, see, e.g., ([206], 7.2.1, [207], 2.26).

1.3 Cohomological Descent

Let $(X, \mathscr{O})$ be a ringed space, and let $\mathscr{U} = (U_i)_{i \in I}$ be an open covering. Giving a sheaf $\mathscr{F}$ of $\mathscr{O}$-modules on X is equivalent to giving a family of sheaves of $\mathscr{O}$-modules $\mathscr{F}_i$ on U_i and gluing data $g_{ij} : \mathscr{F}_i|U_{ij} \xrightarrow{\sim} \mathscr{F}_j|U_{ij}$ on the intersections $U_{ij} = U_i \cap U_j$, satisfying the usual cocycle condition. This no longer holds in general for objects of the derived category $D(X, \mathscr{O})$. Indeed, a cohomology class $u \in H^n(X, \mathscr{F})$ is a morphism $u : \mathscr{O} \to \mathscr{F}[n]$ in $D(X, \mathscr{O})$, and for $n > 0$, u locally vanishes. In 1965 Deligne conceived a theory by-passing this difficulty, later called *cohomological descent*, that worked in a much greater level of generality: for ringed sites or topoi, and *hypercoverings*, a generalization due to Cartier and Verdier of the notion of covering family. A full account was written up by Saint-Donat in ([3], Vbis). See ([D29, 1974], 5) for an introduction, and [128] for an overview.

The following example, discussed by Deligne in ([D29, 1974], 5.3), is of crucial importance for mixed Hodge theory (see Sect. 4.2 "Mixed Hodge Theory", The general case).

Let $X_\bullet$ be a simplicial topological space. A *sheaf* $\mathscr{F}^\bullet$ on $X_\bullet$ is the data of a family of sheaves $\mathscr{F}^n$ on X_n and of X_f-maps $\mathscr{F}^\bullet(f) : \mathscr{F}^n \to \mathscr{F}^m$ for each non-decreasing map $f : [n] \to [m]$, where $[n]$ denotes the ordered set $\{0, \cdots, n\}$, $X_f : X_n \to X_m$, and by an X_f-map one means an element of $\mathrm{Hom}(X_f^*\mathscr{F}^n, \mathscr{F}^m) \simeq \mathrm{Hom}(\mathscr{F}^n, X_{f*}\mathscr{F}^m)$, the maps $\mathscr{F}^\bullet(f)$ having to satisfy the condition $\mathscr{F}^\bullet(gf) = \mathscr{F}^\bullet(g)\mathscr{F}^\bullet(f)$. With morphisms defined in the obvious way, sheaves on $X_\bullet$ form a topos $\widetilde{X_\bullet}$, which was first defined by Deligne. It was later called the *total topos* of $X_\bullet$, and studied in great generality in ([3], VI 7.4).

An augmentation $a : X_\bullet \to S$ defines a morphism from $\widetilde{X_\bullet}$ to the topos of sheaves on S, hence a pair of adjoint functors (a^*, a_*), which extend to a pair of adjoint functors (a^*, Ra_*) between the corresponding derived categories $D^+(\widetilde{X_\bullet}, \mathbf{Z})$ and $D^+(S, \mathbf{Z})$. If S is a point, one writes $R\Gamma(\widetilde{X_\bullet}, -)$ (or $R\Gamma(X_\bullet, -)$ for Ra_*. For $M \in D^+(X_\bullet, \mathbf{Z})$, $H^*(X_\bullet, M) = H^*R\Gamma(X_\bullet, M)$ is the abutment of a spectral sequence

$$E_1^{pq} = H^q(X_p, M_p) \Rightarrow H^{p+q}(X_\bullet, M). \tag{5}$$

For any $K \in D^+(S, \mathbf{Z})$, we have an adjunction morphism

$$K \to Ra_*a^*K. \tag{6}$$

A key result of the theory of cohomological descent is the following theorem ([D29, 1974], 5.3.5), ([3], Vbis, 3.3.3, 4.1.6):

Theorem 2 *Assume that a is a proper hypercovering, which means that, for each $n \geq -1$, the canonical map*

$$(\varphi_n)_{n+1} : X_{n+1} \to (\mathrm{cosk}_n\ \mathrm{sk}_n X_\bullet)_{n+1} \tag{7}$$

(induced by the adjunction map $\varphi_n : Id \to \mathrm{cosk}_n \mathrm{sk}_n$*) is proper and surjective (with the convention that* $(\varphi_{-1})_0 = a_0 : X_0 \to S$*). Then* (6) *is an isomorphism.*

Here, as usual, sk_n denotes the n-th skeleton functor, associating with a simplicial space $X_\bullet$ over S the underlying n-th truncated simplicial space (restriction to the category of ordered sets $[m]$ with $m \leq n$), and cosk_n, the n-th coskeleton functor, which is its right adjoint. For example, $\mathrm{sk}_0 X = X_0$, and, for a space Y over S, $\mathrm{cosk}_0 Y$ is the simplicial space $[n] \mapsto (Y/S)^{[n]}$. The space $(\mathrm{cosk}_n \ \mathrm{sk}_n X_\bullet)_{n+1}$ appearing in (7) consists of the maps from the n-th skeleton of the standard simplex Δ_{n+1} (a "simplical n-th sphere") to $X_\bullet$. Its construction involves a finite projective limit, deduced from the gluing of n-th faces along the $(n-1)$-th skeleton of Δ_{n+1}.

The isomorphism (7) induces an isomorphism

$$H^*(S, K) \xrightarrow{\sim} H^*(X_\bullet, a^* K),$$

thanks to which $H^*(S, K)$ can be analyzed through the spectral sequence (5), which reads

$$E_1^{p,q} = H^q(X_p, a_p^* K) \Rightarrow H^{p+q}(S, K),$$

a generalization of the Čech spectral sequence for a locally finite covering of S by closed subsets.

Note that by ([3], *loc. cit.*) (6) would still be an isomorphism if, instead of being a proper hypercovering, a was assumed to be a hypercovering for the topology on the category of S-spaces generated by usual open coverings and proper surjective maps (an analogue of the Voevodsky topology on schemes). In particular, if $b : S_h \to S$ is the corresponding morphism of sites, the adjunction map $K \to Rb_* b^* K$ is an isomorphism (cf. ([246], 10.2) for a similar result on schemes). This observation was used by Beilinson in his proof of the p-adic de Rham comparison theorem [28].

1.4 Duality and Finiteness Theorems in Étale Cohomology

Global Duality

Poincaré duality in étale cohomology for quasi-projective morphisms $f : X \to Y$ and coefficients $\Lambda = \mathbf{Z}/n\mathbf{Z}$ with n invertible on Y, in the form of the construction of a pair of adjoint functors $(Rf_!, Rf^!)$ between the derived categories $D^+(X, \Lambda)$ and $D^+(Y, \Lambda)$, was established by Grothendieck in 1963 and he talked about it in his seminar SGA 4 [4] in 1964. A sketch was given by Verdier in [252]. Grothendieck proposed to Deligne, who had not attended SGA 4, to write up a more comprehensive version. This resulted in the exposés [D4, D5, 1969]. Not only did

Deligne generalize the set-up and fill in details, but he made a number of original contributions.

Derived Functors

A large part of [D4, 1969] is devoted to foundations for the theory of derived functors. At the time, Verdier's thesis had not been published.[2] Hartshorne's treatment in [113] was insufficient for the needs of the étale global duality formalism, and plagued by a number of sign mistakes. Deligne clarified the sign conventions for multiple complexes, and introduced a more flexible notion of right (resp. left) derived functor of a functor F, with values in categories of ind- (resp. pro-) objects of derived categories. For example, if A and B are abelian categories, and $F : K^+(A) \to K^+(B)$ is an exact functor, then Deligne defines $RF(K)$ as the ind-object "colim"$_{s:K\to K'}F(K')$ of $D^+(B)$, where s runs through the filtering category of classes up to homotopy of quasi-isomorphisms in $K^+(A)$; F is said to be *derivable* at K if $RF(K)$ is essentially constant. This new viewpoint was especially useful for the construction of generalized Künneth isomorphisms ([D4, 1969], 5.4), and turned out to be essential for the definition of derived functors of non additive functors ([D4, 1969], 5.5.5) (see also [120]).

Diagram Compatibilities

By an ingenious argument of fibered and cofibered categories Deligne solved Artin's perplexity in ([4], XII 4) about the coincidence of the base change maps defined in the two natural ways ([D4, 1969], 2.1.3). His method was later exploited to prove the diagram compatibilities involved in the Lefschetz–Verdier trace formula ([5], III).

The Functors $Rf_!$ and $Rf^!$

For $\Lambda = \mathbf{Z}/n\mathbf{Z}$ and $f : X \to Y$ *compactifiable*, i.e., of the form $f = gj$ with g proper and j an open immersion, Grothendieck defined the direct image with proper support functor $Rf_! : D^+(X, \Lambda) \to D^+(Y, \Lambda)$ by $Rf_! = Rg_* j_!$, where j is the extension by zero functor. This definition was forced by the requirement of transitivity, and the proper base change theorem guaranteed independence of the choice of the compactification. However, a rigorous treatment demanded the verification of a number of compatibilities, that Deligne neatly axiomatized in a gluing lemma ([D5, 1969], 3.3) that can be used in other contexts (it was recently revisited by Liu and Zheng [176]).

[2]The published version [254] doesn't treat derived functors either.

For n invertible on Y, Grothendieck defined $Rf^! : D^+(Y, \Lambda) \to D^+(X, \Lambda)$ for f *smoothable*, i.e., of the form $f = gi$, with g smooth and i a closed immersion, by $Rf^! = Ri^! g^*[2d](d)$, where d is the relative dimension of g, and $Ri^!$ is the derived functor of the functor $F \mapsto \mathscr{H}^0_X(F)|X$. Independence of the factorization was guaranteed by the relative purity theorem, and the main bulk of the global duality theorem rested on the definition of a trace morphism $\mathrm{Tr} : Rg_! g^*[d](d) \to \mathrm{Id}$ making $Rf^! = g^*[2d](d)$ right adjoint to $Rg_!$. This is the approach explained by Verdier in [252]. Deligne chose a different path, enabling him to get rid of the assumption of the existence of such a factorization. Imitating Verdier's method to prove Poincaré duality for topological spaces, he showed that, for f compactifiable, $Rf_!$ admits a right adjoint $Rf^!$, thus shifting the core of the proof of the global duality theorem to the calculation of $Rf^!$ for f smooth, i.e., recovering the formula $Rf^! = f^*[2d](d)$.

He also realized that the same method could be used in the quite different context of global duality for coherent sheaves on locally noetherian schemes, provided that a suitable direct image with proper support functor $Rf_!$ could be defined. This is what he does in the appendix [D2, 1966] to Hartshorne's seminar [113].

Picard Stacks and Geometric Class Field Theory

Whatever the method used to prove global duality in étale cohomology, at the end of the day the key point is Poincaré duality on curves. For X a smooth connected curve over an algebraically closed field k and n invertible in k, the fact that cup-product followed by the trace isomorphism $\mathrm{Tr} : H^2_c(X, \mu_n) \xrightarrow{\sim} \mathbf{Z}/n\mathbf{Z}$ is a perfect duality between $H^1_c(X, \mu_n)$ and $H^1(X, \mathbf{Z}/n\mathbf{Z})$ is a by-product of geometric class field theory: if $X = \overline{X} - D$, where $\overline{X}$ is proper, smooth, connected and D a reduced divisor, then, given a closed point x_0 of X, the Abel–Jacobi map $X \to \mathrm{Pic}^0_D(\overline{X})$, $x \mapsto \mathscr{O}(x - x_0)$, induces an isomorphism

$$\mathrm{Hom}(\mathrm{Pic}^0_D(\overline{X})_n, \mathbf{Z}/n\mathbf{Z}) \xrightarrow{\sim} H^1(X, \mathbf{Z}/n\mathbf{Z}), \tag{8}$$

where $\mathrm{Pic}_D(\overline{X}) := H^1(\overline{X}, {}_D\mathbf{G}_m)$, ${}_D\mathbf{G}_m$ is the sheaf of sections of $\mathbf{G}_m$ congruent to 1 mod D, and $(-)_n$ denotes the kernel of the multiplication by n (so that $H^1_c(X, \mu_n) = \mathrm{Pic}^0_D(\overline{X})_n$). See ([D39, 1977], Arcata VI 2.3, Dualité 3.4) for a short, self-contained proof.

In ([D5, 1969], 1.5.2, 1.5.14) Deligne gives a generalization of this, where X/k is replaced by a smooth relative curve X/S, S a base scheme, and $\mathbf{Z}/n\mathbf{Z}$ by a certain complex of abelian sheaves on the big fppf site of S. First, in the *proper* case, he has the following general theorem, that he calls *formule des coefficients universels*:

Theorem 3 *Let $f : X \to S$ be a projective and smooth curve, K a complex of abelian sheaves on S_{fppf}, locally isomorphic, in the derived category, to a complex of the form $[G^{-1} \to G^0]$, where G^i is of one of the following types: smooth of finite presentation, inverse image of a torsion sheaf on the small étale site of S, affine and equal to the kernel of an epimorphism of smooth groups, defined by a flat,* *quasi-coherent sheaf on S. Then there is a natural isomorphism*

$$\tau_{\leq 0}R\mathcal{H}om(\tau_{\leq 0}Rf_*(\mathbf{G}_m[1]), K) \xrightarrow{\sim} \tau_{\leq 0}Rf_*f^*K. \tag{9}$$

Applied to $K = \mathbf{G}_m[1]$, this gives a refined version of the self-duality (with value in $\mathbf{G}_m[1]$) of the Jacobian $\underline{\mathrm{Pic}}^0_{X/S}$. In the *non proper* case, for $f : X \to S$ a smooth curve of the form $X = \overline{X} - D$, with $\overline{X}/S$ proper of relative dimension 1 and $D \subset \overline{X}$ a closed subscheme finite over S, (8) is refined to an isomorphism $\mathcal{E}xt^1(\underline{Pic}_{D,\overline{X}/S}, G) \xrightarrow{\sim} R^1f_*G$, for G a torsion sheaf annihilated by an integer invertible on S.

To prove Theorem 3, Deligne first reformulates it in terms of *Picard stacks*, a sheaf-theoretic generalization of Grothendieck's notion of Picard category [107].[3] Then he uses a technique of *integration of torsors*, consisting in the construction of *symbols* generalizing those of geometric class field theory ([235], III). For a group G of the type described in Theorem 3, these symbols associate to an invertible sheaf $\mathcal{L}$ on a projective smooth curve X over S and a G_X-torsor K on X a G-torsor

$$\langle \mathcal{L}, K] \tag{10}$$

on S ([D5, 1969], 1.3.10), which depends functorially on $\mathcal{L}$ and K, additively on $\mathcal{L}$, and whose formation is compatible with any base change. For $G = \mathbf{G}_m$, and $\mathcal{M}$ the line bundle corresponding to a $\mathbf{G}_m$-torsor M, the line bundle associated with $\langle \mathcal{L}, M]$ is denoted $\langle \mathcal{L}, \mathcal{M}\rangle$. For D a (relative) Cartier divisor on X, $\langle \mathcal{O}(D), \mathcal{M}\rangle$ is the norm $N_{D/S}(\mathcal{M})$, and, for D and E effective, $\langle \mathcal{O}(D), \mathcal{O}(E)\rangle = \det Rf_*(\mathcal{O}_D \otimes^L \mathcal{O}_E)$. When $S = \mathrm{Spec}(k)$, k an algebraically closed field, functoriality of the construction yields, for rational functions f, g on X, the classical *product formula* $\prod_{x\in X(k)}\langle f, g\rangle_x = 1$, where

$$\langle f, g\rangle_x = (-1)^{v(f)v(g)}(g^{v(f)}/f^{v(g)}), \tag{11}$$

v denoting the valuation at x.

Deligne revisited these questions in [D68, 1987] and [D73, 1991]. In [D68, 1987], which is an amplification of a letter to Quillen (and earlier private notes[4]), Deligne uses the symbols (10) to write a *Grothendieck–Riemann–Roch formula without denominators* for a projective smooth curve $f : X \to S$ and a line bundle $\mathcal{L}$ on X, in the form of a canonical, base change compatible isomorphism

$$\det(Rf_*\mathcal{L})^{\otimes 12} \xrightarrow{\sim} \langle \omega, \omega\rangle \otimes \langle \mathcal{L}, \mathcal{L} \otimes \omega^{-1}\rangle^{\otimes 6}, \tag{12}$$

[3]Picard stacks appear in deformation theory: in [61] Deligne sketched a method to use them to give an alternate proof of the theorems of ([121],VII) on deformations of torsors and group schemes—a program that has not yet been carried out. See [44, 208, 255] for recent developments.

[4]*Le déterminant d'une courbe*, undated.

where $\omega = \Omega^1_{X/S}$. Actually, in (*loc. cit.*, 9, 10, 11), a more general isomorphism is constructed for vector bundles. In the case of a projective smooth curve $X/\mathbf{C}$ and a vector bundle, both endowed with hermitian metrics, an enriched isomorphism is defined, taking into account the Ray–Singer analytic torsion. In [D73, 1991], he gives analytic variants and refinements of the above product formula, see Sect. 4.5 "Link with the Tame Symbol".

Symmetric Künneth Formula

The proper base change theorem in étale cohomology implies a very general Künneth formula for cohomology with proper support: given a base scheme S and a finite family $(u_i : X'_i \to X_i)_{i \in I}$ of compactifiable morphisms of S-schemes, $\Lambda = \mathbf{Z}/n\mathbf{Z}$, and a family of objects K_i of $D^-(X'_i, \Lambda)$, the natural map

$$\otimes^L_{\text{ext}} Ru_{i!}K_i \to Ru_!(\otimes^L_{\text{ext}} K_i),$$

where $u = \prod u_i : \prod_S X'_i \to \prod_S X_i$ and $\otimes_{\text{ext}}$ denotes an external tensor product, is an isomorphism ([D4, 1969], (5.4.4.1)). In particular, taking $I = \{1, \cdots, n\}$, $X_i = X$, $X'_i = X'$, $u_i = f$, and $K_i = K$ for each i, we have an isomorphism

$$(Rf_!K)^{(\otimes^L_{\text{ext}})^n} \xrightarrow{\sim} R(f^n_S)_!(K^{(\otimes^L_{\text{ext}})^n}). \tag{13}$$

In ([D4, 1969], 5.5.21) Deligne proves a formula which looks like being deduced from (13) by taking invariants under the symmetric group S_n: for S quasi-compact and quasi-separated, and f quasi-projective, and $K \in D^b(X', \Lambda)$, of tor-amplitude in an interval $[0, a]$, then there is defined a *symmetric Künneth map*

$$L\Gamma^n_{\text{ext}}(Rf_!K) \to R\mathrm{Sym}^n_S(f)_! L\Gamma^n_{\text{ext}}K, \tag{14}$$

which is an isomorphism. Here $L\Gamma^n_{\text{ext}}$ is an external variant of the derived functor of the (non additive) functor Γ^n (n-th component of the divided power algebra Γ over Λ), and $\mathrm{Sym}^n_S(f) : \mathrm{Sym}^n_S(X') \to \mathrm{Sym}^n_S(X)$ is the morphism induced by f^n_S on the symmetric power $\mathrm{Sym}^n_S(X') = X'^n/S_n$. Despite its appearance, (14) is not a formal consequence of (13). Its proof is by dévissage, and for n invertible on S, by reduction to the case of curves and a transcendental argument. Deligne will use (14) several times:

(a) in ([D39, 1977], Fonctions L modulo ℓ^n et modulo p), to give an alternate proof – and a generalization – of Katz's congruence formula SGA 7 ([7], XXII) for the zeta function of a proper scheme over $\mathbf{F}_q$;
(b) the proof of the functional equation of Grothendieck's L-functions on a curve (see Sect. 6.3 "The Case of Function Fields");
(c) in [74], to prove the product formula for local constants (cf. (149)) in the tame case, using a strategy developed in his letter to Serre [D30, 1974].

Finiteness

For $\Lambda = \mathbf{Z}/n\mathbf{Z}$, let $D^b_c(X, \Lambda)$ denote the full subcategory of $D^b(X, \Lambda)$ consisting of complexes with constructible cohomology sheaves. For a compactifiable morphism $f : X \to Y$, the preservation of D^b_c by $Rf_!$ is an easy corollary of the proper base change theorem and the structure of the cohomology of curves, and was proved by Artin and Grothendieck in 1963 ([4], XIV). However, the preservation of D^b_c under Rf_*, assuming n invertible on Y (otherwise there are simple counter-examples), was a wide open problem in the 1960s, even when Y is the spectrum of a field k (except for X/k smooth and locally constant coefficients, by Poincaré duality). In the early 1970s Deligne made a breakthrough, by simultaneously proving several basic finiteness theorems ([D39, 1977], Th. finitude):

- **Generic constructibility** If S is a noetherian scheme over which n is invertible, and $f : X \to Y$ a morphism of S-schemes of finite type, then, for $K \in D^b_c(X, \Lambda)$, there exists a dense open subscheme U of S such that $Rf_*K|Y_U$ belongs to $D^b_c(Y_U, \Lambda)$ and is of formation compatible with any base change $S' \to U \subset S$ (where $Y_U := Y \times_S U$).
- **Finiteness and biduality over regular bases of dimension** ≤ 1 If S is a regular noetherian scheme of dimension ≤ 1 over which n is invertible, and $f : X \to Y$ a morphism of S-schemes of finite type, then Rf_* sends $D^b_c(X, \Lambda)$ to $D^b_c(Y, \Lambda)$.

  Let $D_{ctf}(X, \Lambda)$ denote the full subcategory of $D^b_c(X, \Lambda)$ consisting of complexes K of finite tor-dimension. Then, for $K \in D_{ctf}(X, \Lambda)$, DK belongs to $D_{ctf}(X, \Lambda)$, and the biduality map $K \to DDK$ is an isomorphism, where $D := R\mathscr{H}om(-, a^!\mathbf{Z}/n\mathbf{Z})$, $a : X \to S$ being the structural map.
- **Constructibility of nearby cycles** Let S be a strictly local trait, i.e., the spectrum of a strictly henselian discrete valuation ring, with closed point s and generic point η, and let X be an S-scheme of finite type. Then, for $K \in D^b_c(X_\eta, \Lambda)$, the complex of nearby cycles $R\Psi_\eta K$ belongs to $D^b_c(X_s, \Lambda)$, and its formation is compatible with base change by any surjective morphism $S' \to S$ of strictly local traits.

Proofs of the basic results of SGA 4 [4] proceed by dévissage and reduction to relative curves. To establish the above theorems Deligne used an ingenious new method, later called the *global to local method*. Roughly speaking, the principle is the following. In order to prove that a certain canonical map u (like the biduality map, or the base change map for $R\Psi$) is an isomorphism, one uses induction on the relative dimension. Cutting by a finite number of pencils one constrains the cone C of u to be concentrated on a union Σ of a finite number of geometric points. One then concludes by a global argument, knowing that $R\Gamma(\Sigma, C) = 0$, hence $C_x = 0$ for all $x \in \Sigma$.

Since then the global to local method has been successfully applied to various problems: construction of du Bois complexes (see Sect. 4.2 "The du Bois Complex"), Gabber's theorems on the compatibility of $R\Psi$ with duality and external tensor product ([125], 4.2, 4.7), ([24], 5.1) to mention only a couple of them.

Given a regular noetherian base S of dimension ≤ 1 and an integer n invertible on S, Deligne's theorems imply the existence of a Grothendieck formalism of *six operations* in $D_{ctf}(-, \Lambda)$ over S-schemes of finite type. A variant of this formalism for ℓ-adic coefficients ($\mathbf{Z}_\ell$, $\mathbf{Q}_\ell$, $\overline{\mathbf{Q}}_\ell$) for a prime ℓ invertible on S was constructed by Deligne [D46, 1980] (for S satisfying certain restrictive hypotheses), and later improved and generalized by several authors (see Sect. 5.6 "Mixed Sheaves, Statement of the Main Theorem", (b)). Over bases S of higher dimension, and for torsion coefficients, an extension of the above formalism, under (necessary) assumptions of quasi-excellency, has recently been obtained by Gabber [131], using new tools.

2 Algebraic Stacks

2.1 Deligne–Mumford Stacks

The notion of *stack*—a fibered category over a site in which objects as well as morphisms can be glued—and of *gerbe*—a stack in groupoids in which fibers are locally non-empty and any two objects of a fiber are locally isomorphic—are due to Grothendieck, and were used by Giraud to develop a theory of non-abelian cohomology [100]. However, the purpose was purely topological (and cohomological). Deligne showed that one could do algebraic geometry with them. The motivation was to build a geometric framework that could incorporate the automorphism groups preventing moduli problems to be represented by schemes. In [198] Mumford had defined *moduli topologies* $\mathscr{M}$ whose "open subsets" were families of curves of genus g and "intersections" involved isomorphism schemes between families. The terminology was misleading as those topologies were not Grothendieck topologies on a category, nor objects of $\mathscr{M}$ sheaves on a site. Nonetheless they gave a hint to what should be the right notion to introduce.

Let $\mathscr{S}$ be the category of schemes, and let $\mathscr{S}_{\text{et}}$ be the corresponding étale site. In [D8, 1969] Deligne and Mumford define an *algebraic stack*, later called *Deligne–Mumford stack*, as a stack in groupoids $\mathscr{X}$ over $\mathscr{S}_{\text{et}}$ such that the diagonal $\mathscr{X} \to \mathscr{X} \times \mathscr{X}$ is representable by schemes and there exists a surjective étale morphism $X \to \mathscr{X}$ where X is a scheme. They then extend to these new objects classical results of algebraic geometry, such as Chow's lemma or the valuative criterion for properness, critical in one of the two proofs of their main theorem.

Closely related objects are:

- **orbifolds**, which are topological or differentiable analogues of Deligne–Mumford stacks, in the case when an open and dense subset is an ordinary space; introduced by Satake in 1956 under the name of V-varieties, they play an important role in differential geometry (e.g., Thurston's work) and string theory;
- **algebraic spaces**, introduced just before Deligne–Mumford stacks by Artin and Knutson [157], which are Deligne–Mumford stacks with trivial inertia groups.

Shortly afterwards Artin proposed a generalization of the notion of Deligne–Mumford stack, now called *Artin stack*, with "étale" replaced by "smooth" for the surjective map $X \to \mathscr{X}$, and gave powerful criteria for a stack in groupoids (on $\mathscr{S}_{\text{et}}$ or its variant $\mathscr{S}_{\text{fppf}}$) to be representable by an Artin stack [14, 15]. While Deligne–Mumford stacks are well adapted to the study of moduli of curves (see below), Artin stacks are needed for other types of moduli problems, such as moduli of vector bundles, and those appearing in the theory of stable maps and Gromov-Witten invariants, and in the Langlands program (Langlands correspondence over function fields, geometric Langlands correspondence).

2.2 Moduli of Curves of Genus ≥ 2

Let g be an integer ≥ 2. If k is an algebraically closed field, a *stable curve of genus* g over k is a proper, reduced, connected, 1-dimensional k-scheme C such that C is smooth except for ordinary double points, $\dim H^0(\mathscr{O}_C) = g$, and any smooth rational component of C meets the other components in at least three points. If S is a scheme, a *stable curve of genus* g over S is a proper, flat S-scheme whose geometric fibers are stable curves of genus g. Let $\mathscr{M}_g$ be the fibered category over $\mathscr{S}$, whose fiber at S is the groupoid of S-stable curves of genus g and S-isomorphisms. It is shown in [D8, 1969] that $\mathscr{M}_g$ is a Deligne–Mumford stack over $\mathscr{S}_{\text{et}}$ (or, for short, over Spec $\mathbf{Z}$), in which the diagonal map $\mathscr{M}_g \to \mathscr{M}_g \times \mathscr{M}_g$ is finite and unramified. Let $\mathscr{M}_g^0 \subset \mathscr{M}_g$ be the open substack such that $\mathscr{M}_g^0(S)$ consists of *smooth* (stable) curves over S. The two main results of *loc. cit.* are:

Theorem 4 *The stack $\mathscr{M}_g$ is proper and smooth over* Spec $\mathbf{Z}$, *and $\mathscr{M}_g - \mathscr{M}_g^0$ is a divisor with normal crossings relative to* Spec $\mathbf{Z}$.

For a smooth stable curve $f : X \to S$ of genus g over a scheme S over which an integer $n \geq 1$ is invertible, a *Jacobi structure of level n* on X is defined as a homogeneous symplectic isomorphism between $R^1 f_*(\mathbf{Z}/n\mathbf{Z})$ and $(\mathbf{Z}/n\mathbf{Z})^{2g}$.

Theorem 5 *For any integer $n \geq 1$, let $\mu : {}_n\mathscr{M}_g^0 \to$ Spec $\mathbf{Z}[1/n, e^{2\pi i/n}]$ be the stack*[5] *classifying smooth stable curves of genus g endowed with a Jacobi structure of level n. Then the geometric fibers of μ are normal and irreducible.*

In particular, for any algebraically closed field k, $(\mathscr{M}_g^0)_k$, and the coarse moduli quotient $(M_g^0)_k = (H_g^0)_k/\text{PGL}(5g-6)$, where $(H_g^0)_k$ is the scheme of tri-canonical smooth stable curves of genus g over k (a dense open subscheme of a certain Hilbert scheme), are irreducible.

A key ingredient in the proof of Theorem 4 is the stable reduction theorem for curves, deduced in *loc. cit.* from Grothendieck's semistable reduction theorem for abelian varieties (independent proofs were found later, see, e.g., [16, 223]). The

[5] A scheme, for $n \geq 3$, by Serre's rigidity lemma.

proof of Theorem 5 uses the fact that the result is known for the fiber at the standard complex place of Spec $\mathbf{Z}[1/n, e^{2\pi i/n}]$ by Teichmüller theory.

It was later proved by Knudsen and Mumford—and, independently, by Mumford using another method—that coarse moduli spaces of stable curves are projective [154–156, 199].

2.3 *Moduli of Elliptic Curves*

Classically, a *modular curve* X is the quotient $\mathscr{H}/\Gamma$ of the upper half-plane $\mathscr{H} = \{z \in \mathbf{C}, \operatorname{Im} z > 0\}$ by a congruence subgroup Γ of $\mathrm{SL}(2, \mathbf{Z})$. This is the complement of a finite number of points (the *cusps*) in a compact Riemann surface, hence an algebraic curve over $\mathbf{C}$. The interpretation of X as a (coarse) moduli space for elliptic curves endowed with a so-called *level* structure and its relation with modular forms has given rise to a huge literature on the geometry and arithmetic of these curves, starting with the pioneering works of Igusa and Eichler–Shimura. In [D24, 1973] Deligne and Rapoport give a comprehensive account of the state of the art in 1972. Their monograph also contains new constructions and results. I will only briefly mention some of these pertaining to the compactification of modular curves and their reduction modulo p.

Generalized Elliptic Curves and Compactifications

Given a base scheme S, a *generalized elliptic curve* over S is defined as a proper and flat scheme $p : C \to S$, whose every geometric fiber is either a proper, smooth, connected curve of genus 1 or a Néron n-gon ($n \geq 1$, together with a commutative group scheme structure on the subscheme C^{reg} of smooth points of C and an extension of this action to C rotating the graphs of the n-gons ([D24, 1973], II 1.12). For $n \geq 1$ fixed, invertible on S, a *level n structure* on C is an isomorphism $C_n^{\mathrm{reg}} \xrightarrow{\sim} (\mathbf{Z}/n\mathbf{Z})_S^2$ (compatible with the action of C^{reg} on C and of $\mathbf{Z}/n\mathbf{Z}$ on $(\mathbf{Z}/n\mathbf{Z})^2$ by translation on the second factor), where the subscript n denotes the kernel of the multiplication by n. If S is a scheme over which n is invertible, let $\mathscr{M}_n[1/n](S)$ denote the groupoid of generalized elliptic curves over S with level n structure (morphisms being S-isomorphisms). One of the main results of [D24, 1973] is:

Theorem 6 *$\mathscr{M}_n[1/n]$ is a proper and smooth Deligne–Mumford stack of relative dimension 1 over* Spec $\mathbf{Z}[1/n]$*, and the complement of the open substack $\mathscr{M}_n^0[1/n]$ such that $\mathscr{M}_n^0[1/n](S)$ consists of elliptic curves over S is finite and étale over* Spec $\mathbf{Z}[1/n]$*. For $n \geq 3$, $\mathscr{M}_n[1/n]$ is a projective and smooth scheme over* Spec $\mathbf{Z}[1/n]$.

The e_n-pairing $\Lambda^2 E_n \xrightarrow{\sim} \mu_{nS}$ for E in $\mathscr{M}_n^0[1/n](S)$ defines a morphism $\mathscr{M}_n^0[1/n] \to \operatorname{Spec} \mathbf{Z}[\zeta_n, 1/n]$, and the coarse moduli space of $\mathscr{M}_n^0[1/n] \otimes_{\mathbf{Z}[\zeta_n, 1/n]} \mathbf{C}$ is the (affine) modular curve $\mathscr{H}/\Gamma$ corresponding to the principal congruence subgroup $\Gamma = \Gamma(n)$ consisting of matrices congruent to the identity matrix mod n.

Reduction mod *p*

The above results are extended to level H structures for congruence subgroups Γ_H inverse images of subgroups H of $\mathrm{GL}(2, \mathbf{Z}/n\mathbf{Z})$, such as $\Gamma_0(n) = \Gamma_H$ for $H = \left(\begin{smallmatrix} * & * \\ 0 & * \end{smallmatrix}\right)$. The reduction modulo a prime number p of the corresponding modular stacks $\mathscr{M}_H[1/n]$ is examined. For p dividing n, a model $\mathscr{M}_H$ of $\mathscr{M}_H[1/n]$ over $\operatorname{Spec} \mathbf{Z}$ is needed: Deligne and Rapoport define $\mathscr{M}_H$ as the *normalization* of $\mathscr{M}_1$ in $\mathscr{M}_H^0[1/n]$. The stack $\mathscr{M}_{\Gamma_0(p)}$ has a modular interpretation: it classifies pairs $(C/S, A)$ of a generalized elliptic curve over S and a rank p locally free subgroup A meeting each irreducible component of any geometric fiber of C. They prove a refinement of the Eichler–Shimura congruence formula ([D24, 1973], V 1.16):

Theorem 7 *The stack $\mathscr{M}_{\Gamma_0(p)}$ is regular, proper and flat over* Spec **Z**, *of relative dimension* 1, *smooth outside the supersingular points of characteristic p, and with semistable reduction at these points; $\mathscr{M}_{\Gamma_0(p)} \otimes \mathbf{F}_p$ is the union of two irreducible components crossing transversally at the supersingular points. Moreover, the (open) coarse moduli space $M^0_{\Gamma_0(p)}$ is the spectrum of the normalization of $\mathbf{Z}[j, j']/(\Phi_p(j, j'))$, where $\Phi_p(j, j')$ is the modular equation, a polynomial congruent to $(j - j'^p)(j' - j^p)$ modulo p.*

For more general groups H the definition of $\mathscr{M}_H$ as a normalization made it difficult to study its reduction mod p. Drinfeld's notion of *full level N structures*, providing a simple modular interpretation of $\mathscr{M}_H$, solved the problem. See Katz–Mazur's treatise [148] for a systematic exposition of this theory.

3 Differential Equations, de Rham Cohomology

3.1 The Canonical Extension and Hilbert's 21st Problem

The Curve Case

Let X be a projective, smooth, connected curve over $\mathbf{C}$, Y a (possibly empty) finite subset of closed points, and $U = X - Y$. An *(algebraic) differential equation* on U is the datum of a vector bundle E on U equipped with a connection $\nabla : E \to E \otimes \Omega^1_U$. Let $y \in Y$. The connection ∇ is said to have (at most) *regular singular points along* Y if the following condition is satisfied:

(reg) there exists a vector bundle $\overline{E}$ over X extending E such that, for any point y in Y, given a local parameter t of X at y, and $D = t\partial/\partial t$, the (additive) endomorphism $\nabla(D)$ of E near y leaves $\overline{E}$ stable, i.e., the entries of its matrix in a local basis (e_i) of $\overline{E}$ are sections of $\mathcal{O}_X$ (in other words, if $\nabla e_i = \sum a_{ij} e_j$, the differential forms a_{ij}'s have poles of order at most one at y).

Given a smooth connected curve U over **C**, the smooth projective model $X/\mathbf{C}$ such that $X - U$ is finite is unique, and (reg) depends only on (E, ∇). Moreover, it is a classical result of Fuchs, re-interpreted by Deligne, that (reg) is equivalent to a *moderate growth condition* along Y on the solutions of the associated analytic differential equation ∇^{an} on the corresponding Riemann surface U^{an}, i.e., near each point y of Y, the solutions are $O(|t|^{-N})$ in fixed sectors, see ([D11, 1970], II 1.19). Solutions of ∇^{an} form a locally constant sheaf of finite dimensional **C**-vector spaces

$$E^{\nabla} := \text{Ker}(\nabla^{\text{an}} : E^{\text{an}} \to E^{\text{an}} \otimes \Omega^1_{U^{\text{an}}})$$

on U^{an}. Given a point x_0 of U, this local system is determined by its stalk $E^{\nabla}_{x_0}$ at x_0 and its monodromy representation

$$\rho(\nabla) : \pi_1(U^{\text{an}}, x_0) \to \text{GL}(E^{\nabla}_{x_0}). \tag{15}$$

Hilbert's 21'st problem was the following: given a finite dimensional **C**-vector space V, and a representation

$$\rho : \pi_1(U^{\text{an}}, x_0) \to \text{GL}(V),$$

can one find an algebraic differential equation (E, ∇) on U, with at most regular singular points along Y, such that $\rho = \rho(\nabla)$? Deligne positively answered the question, and in fact solved a more general problem in higher dimension.

Higher Dimension: The Riemann–Hilbert Correspondence

Let X be a smooth scheme over **C**, $Y \subset X$ a normal crossings divisor, $j : U \hookrightarrow X$ the complementary open immersion. Differential forms on U with logarithmic poles along Y (a generalization of differentials of the 3rd kind on curves) briefly appear in ([117], Lemma 17). However, their formal definition, and that of the corresponding de Rham complex, are due to Deligne. In a letter to Atiyah [60], Deligne defines $\Omega^\bullet_X\langle Y\rangle$ as the subcomplex of $j_*\Omega^\bullet_U$, where $\Omega^\bullet_U$ is the (algebraic) de Rham complex of $U/\mathbf{C}$, consisting of forms ω having a pole of order ≤ 1 along Y and such that $d\omega$ enjoys the same property. This complex, later called *de Rham complex of X with logarithmic (or log) poles along Y*, and now usually denoted $\Omega^\bullet_X(\log Y)$, was to play a fundamental role in Deligne's mixed Hodge theory and spur many important developments in algebraic and arithmetic geometry. Its components $\Omega^p_X(\log Y) = \Lambda^p \Omega^1_X(\log Y)$ are locally free of finite type: if étale locally near a point of Y, $(t_1, \cdots, t_n)$ are sections of $\mathcal{O}_X$ such that $(dt_1, \cdots, dt_n)$ form a basis of Ω^1_X and Y is defined by the equation $t_1 \cdots t_r = 0$, then $\Omega^1_X(\log Y) = \oplus_{1 \leq i \leq r} \mathcal{O} \frac{dt_i}{t_i} \oplus \oplus_{i > r} \mathcal{O} dt_i$; the dual of $\Omega^1_X(\log Y)$ is the subsheaf $\mathcal{D}er_Y(X/\mathbf{C})$ of the tangent bundle $\mathcal{D}er(X/\mathbf{C}) = (\Omega^1_X)^\vee$ consisting of vector fields tangent to each branch of Y. If E is a vector bundle on X, a *connection on E with log poles along Y* is an additive map $\nabla : E \to E \otimes \Omega^1_X(\log Y)$ satisfying the Leibniz rule; ∇ is said to be *integrable* if $\nabla^2 : E \to E \otimes \Omega^2_X(\log Y)$ is zero. If $\nabla : E \to E \otimes \Omega^1_U$ is an integrable connection on a vector bundle E on U, ∇ is said to have *regular singular points* (or to be *regular*) *along* Y if there exists a vector bundle $\overline{E}$ on X extending E, and a connection $\overline{\nabla}$ with log poles along Y on $\overline{E}$ extending ∇. This is a local condition at points x of Y which can be expressed in terms of sectorial moderate growth of the entries of a fundamental matrix solution of ∇ along each branch of Y passing through x.

Fix now a smooth scheme U over $\mathbf{C}$, separated and of finite type. By Nagata's compactification theorem,[6] there exists a dense open immersion $j : U \hookrightarrow X$ with $X/\mathbf{C}$ proper, and by Hironaka one may further require that X is smooth and $Y = X - U$ is a divisor with normal crossings. Let's call such a compactification a *good* compactification. If U is of dimension >1, a good compactification of U is not unique, but any two good compactifications $j_1 : U \hookrightarrow X_1$, $j_2 : U \hookrightarrow X_2$ are dominated by a third one, i.e., there exists a good compactification $j : U \hookrightarrow X$ mapping to j_1 and j_2. Given an algebraic differential equation on U, i.e., a vector bundle E on U equipped with an integrable connection ∇, one says that ∇ is *regular at infinity* if for one good compactification X (or, which can be shown to be equivalent, for any good compactification X) of U, ∇ is regular along $Y = X - U$. Basic examples are the relative de Rham cohomology group $E = \mathcal{H}^n_{\mathrm{dR}}(Z/U) := R^n f_* \Omega^\bullet_{Z/U}$, for $f : Z \to U$ proper and smooth, equipped with its Gauss–Manin connection ∇ (the regularity was proved by Deligne ([D11, 1970], II 7.9), and, independently, by Griffiths and Katz, see [142]).

Let $\mathcal{M}_{\mathrm{reg}}(U)$ denote the category of vector bundles on U equipped with an integrable connection which is regular at infinity, and let $\mathcal{L}(U)$ denote the category of locally constant sheaves of finite dimensional $\mathbf{C}$-vector spaces on U^{an}. As in the case of curves, we have a functor

$$\mathcal{M}_{\mathrm{reg}}(U) \to \mathcal{L}(U) \tag{16}$$

associating with (E, ∇) the local system $E^\nabla := \mathrm{Ker}(\nabla^{\mathrm{an}} : E^{\mathrm{an}} \to E^{\mathrm{an}} \otimes \Omega^1_{U^{\mathrm{an}}})$. In ([D11, 1970], II 5.9) Deligne proves the following theorem:

Theorem 8 *The functor* (16) *is an equivalence of categories.*

[6] Nagata's original proof is obscure to today's readers. A modern presentation was given by Deligne in [D112, 2010].

The equivalence (16) was later called *Riemann–Hilbert correspondence*. The statement of Theorem 8 is global, but the proof relies on a local analytic construction. No longer assuming X proper, and supposing Y is a union of smooth branches Y_i $(1 \leq i \leq r)$, given an *analytic* differential equation (M, ∇) on U^{an}, Deligne shows that there exists a unique (analytic) extension $(\overline{M}, \overline{\nabla})$ of M on X^{an} having log poles along Y^{an}, and such that for each k, if $R_k := \mathrm{Res}_{Y_k}(\overline{\nabla}) \in \mathrm{End}(\overline{M} \otimes \mathscr{O}_{Y_k})$ is the residue of $\overline{\nabla}$ along Y_k, the eigenvalues of R_k at each point of Y_k have real parts in the interval $[0, 1)$. This extension is called the *canonical extension* of ∇. Identifying $\mathscr{L}(U)$ with the category of analytic differential equations on U^{an}, the canonical extension, combined with the GAGA functor, yields a quasi-inverse to (16).

Given an algebraic differential equation (E, ∇) on U, regular at infinity, an extension $(\overline{E}, \overline{\nabla})$ on X with log poles along Y is not unique, see [88] for a discussion of this. Theorem 8 has given rise to several variants and generalizations—the first ones by Deligne himself, see Sect. 3.3 "Discontinuous Crystals"—the most important one being the Riemann–Hilbert correspondence between regular holonomic $\mathscr{D}$-modules and perverse sheaves [139, 140, 191].

3.2 *Betti–de Rham Comparison Theorems*

Let U be a smooth scheme over $\mathbf{C}$. The resolution of the constant sheaf $\mathbf{C}$ on U^{an} by the holomorphic de Rham complex $\Omega^\bullet_{U^{\mathrm{an}}}$ (*Poincaré lemma*) induces an isomorphism

$$H^*(U^{\mathrm{an}}, \mathbf{C}) \xrightarrow{\sim} H^*(U^{\mathrm{an}}, \Omega^\bullet_{U^{\mathrm{an}}}).$$

On the other hand, Grothendieck proved in [106] that the GAGA comparison map

$$H^*(U, \Omega^\bullet_U) \to H^*(U^{\mathrm{an}}, \Omega^\bullet_{U^{\mathrm{an}}}) \tag{17}$$

is an isomorphism, so that the Betti cohomology of U^{an} can be calculated purely algebraically as the algebraic de Rham cohomology of U. For $U/\mathbf{C}$ proper, this is an immediate consequence of Serre's classical GAGA theorem. In the general case, which is easily reduced to the case where X is separated, or even affine, this follows from the existence of a good compactification $j : U \to X$ (Sect. 3.1 "Higher Dimension: The Riemann–Hilbert Correspondence") and a local calculation ([117], Lemma 17). In [D16, 1971] Deligne gives refinements of this result, of local nature, involving certain canonical filtrations.

Let us assume, for simplicity, that Y has strict normal crossings, i.e., is a sum of smooth divisors Y_i $(1 \leq i \leq r)$ crossing transversally.[7] Consider the inclusion

$$\Omega^\bullet_{X^{\mathrm{an}}}(\log Y^{\mathrm{an}}) \hookrightarrow j^{\mathrm{an}}_* \Omega^\bullet_{U^{\mathrm{an}}}. \tag{18}$$

[7] One can achieve this by a sequence of blow-ups ([138], 7.2).

As $j_*^{\mathrm{an}}\Omega^\bullet_{U^{\mathrm{an}}} \xrightarrow{\sim} Rj_*^{\mathrm{an}}\Omega^\bullet_{U^{\mathrm{an}}}$ since j is affine, by the Poincaré lemma the right hand side calculates $Rj_*^{\mathrm{an}}\mathbf{C}$. Deligne proves ([D16, 1971], 3.1.8) that (18) is a quasi-isomorphism, and even a *filtered quasi-isomorphism*, with respect to increasing filtrations W and τ on the left and right hand sides respectively, i.e., induces quasi-isomorphisms on the associated graded complexes $\mathrm{gr}^W \to \mathrm{gr}^\tau$. The filtration τ is the filtration by the canonical truncations $\tau_{\leq i}$. For $n \geq 0$, $W_n\Omega^p_X(\log Y)$ (resp. $W_n\Omega^p_{X^{\mathrm{an}}}(\log Y^{\mathrm{an}})$) is the subsheaf of $\Omega^p_X(\log Y)$ (resp. $\Omega^p_{X^{\mathrm{an}}}(\log Y^{\mathrm{an}})$) additively generated by local sections of the form $a \wedge d\log f_1 \wedge \cdots \wedge d\log f_m$, with $m \leq n$, a in Ω^{p-n}_X (resp. $\Omega^{p-n}_{X^{\mathrm{an}}}$), and f_i in $j_*\mathcal{O}^*_U$ (resp. $j_*^{\mathrm{an}}\mathcal{O}^*_{U^{\mathrm{an}}}$). The associated graded complex is calculated by the *Poincaré residue* isomorphism:

$$\mathrm{Res} : \mathrm{gr}^W_n \Omega^\bullet_X(\log Y) \xrightarrow{\sim} \oplus\Omega^\bullet_{Y_I}[-n]$$

(resp.

$$\mathrm{Res} : \mathrm{gr}^W_n \Omega^\bullet_{X^{\mathrm{an}}}(\log Y^{\mathrm{an}}) \xrightarrow{\sim} \oplus\Omega^\bullet_{Y_I^{\mathrm{an}}}[-n]),$$

where I runs through the subsets $1 \leq i_1 < \cdots < i_n \leq r$ of $\{1, \cdots, r\}$ with n elements, and $Y_I := Y_{i_1} \cap \cdots \cap Y_{i_n}$. The statement that (18) is a filtered quasi-isomorphism with respect to W and τ follows from this and the calculation of $\mathrm{gr}^\tau_n Rj_*^{\mathrm{an}}\mathbf{C}[n] = R^n j_*^{\mathrm{an}}\mathbf{C}[-n]$:

$$R^n j_*^{\mathrm{an}}\mathbf{C} \xrightarrow{\sim} \Lambda^n R^1 j_*^{\mathrm{an}}\mathbf{C} \xrightarrow{\sim} \oplus\mathbf{C}_{Y_I^{\mathrm{an}}}.$$

This result is at the core of the construction of a mixed Hodge structure on $H^*(U^{\mathrm{an}}, \mathbf{Z})$ (see Sect. 4.2 "Mixed Hodge Theory"). The algebraic analogue of (18),

$$\Omega^\bullet_X(\log Y) \hookrightarrow j_*\Omega^\bullet_U, \tag{19}$$

is also a quasi-isomorphism, but it is no longer the case that it is a filtered quasi-isomorphism with respect to W and τ. Instead, Deligne proves in ([D11, 1970], II 3.13) that it is a filtered quasi-isomorphism with respect to the (decreasing) filtration F on the left hand side given by the naive truncations (the so-called *Hodge filtration*) and a (decreasing) filtration P on the right hand side ([D11, 1970], II 3.12), already introduced in a letter to Atiyah [60] in the more general framework of complexes of differential operators, called the *filtration by the order of the pole* (for Y smooth, $P^n j_*\mathcal{O}_U = \mathcal{O}_X(-nY)$, $P^n j_*\Omega^p_U = P^{n-p} j_*\mathcal{O}_U \otimes \Omega^p_X$). By the classical GAGA theorem, (18) and (19) yield the isomorphism (17). The quasi-isomorphisms (18) and (19) also imply that the inclusion

$$j_*\Omega^\bullet_U \otimes \mathcal{O}_{U^{\mathrm{an}}} \hookrightarrow j_*^{\mathrm{an}}\Omega^\bullet_{U^{\mathrm{an}}} \tag{20}$$

is a quasi-isomorphism. In ([D11, 1970], II 3, 6) Deligne generalizes the fact that (19), (20) are quasi-isomorphisms to algebraic differential equations (E, ∇)

on U, extended to $(\overline{E}, \overline{\nabla})$ on X, $\overline{\nabla}$ having log poles along Y, hence getting a comparison isomorphism

$$H^*(U, \Omega^\bullet_U(E)) \xrightarrow{\sim} H^*(U^{\mathrm{an}}, \Omega^\bullet_{U^{\mathrm{an}}}(E^{\mathrm{an}})). \tag{21}$$

generalizing (17). He will consider again the filtration by the order of the pole in ([D29, 1974], 9.2) (in relation with a theorem of Griffiths on Hodge theory of smooth hypersurfaces) and in [D70, 1990] (for its extension to the singular case).

3.3 Crystalline Cohomology

Discontinuous Crystals

In his exposé [108] in [1] Grothendieck introduced crystalline sites, crystals and crystalline cohomology, both in characteristic zero and in characteristic $p > 0$. In positive characteristic the theory was extensively developed by Berthelot in his thesis [30]. The case of characteristic zero retained less attention. In his seminar at the IHÉS in 1970 [62] Deligne generalized both the Riemann–Hilbert correspondence (16) and the comparison isomorphism (21) to certain crystals on possibly singular schemes over $\mathbf{C}$. Let X be a scheme separated and of finite type over $\mathbf{C}$, and let X^{an} be the associated analytic space. The category $\mathscr{L}(U)$ of (16) is replaced by the category $\mathscr{C}ons(X)$ of *algebraically constructible* sheaves V of $\mathbf{C}$-vector spaces, i.e., for which there exists a finite partition of X into locally closed subschemes X_i (for the Zariski topology) such that $V|(X_i)^{\mathrm{an}}$ is locally constant with finite dimensional fibers. Let X_{cris} denote the crystalline site of $X/\mathbf{C}$, consisting of $\mathbf{C}$-nilpotent thickenings $U \hookrightarrow \overline{U}$ of open subschemes U of X, and $\mathscr{O}_{X_{\mathrm{cris}}}$ the sheaf of rings on X_{cris}, $(U \hookrightarrow \overline{U}) \mapsto \Gamma(\overline{U}, \mathscr{O}_{\overline{U}})$. The category $\mathscr{M}_{\mathrm{reg}}(U)$ is replaced by a category $\mathscr{C}r(X)$ consisting of *pro-coherent crystals* of $\mathscr{O}_{X_{\mathrm{cris}}}$-modules M having the property that there exists a partition of X into *smooth* locally closed subschemes X_i of X such that $M|(X_i)_{\mathrm{cris}}$ is given by a vector bundle E_i on X_i equipped with an integrable connection ∇_i which is regular at infinity. Then Deligne defines a functor

$$\mathscr{C}ris(X) \to \mathscr{C}ons(X) \tag{22}$$

generalizing (16), which he proves to be an equivalence of categories. The reason for the introduction of pro-objects lies in the need of realizing the functor extension by zero for constructible sheaves on the crystalline level, which he does by the techniques he had developed in [D2, 1966] for Grothendieck global duality in the context of coherent sheaves (Sect. 1.4 "Global Duality", the functors $Rf_!$ and $Rf^!$). If V is the constructible sheaf associated with an object M of $\mathscr{C}ris(X)$ by (22), Deligne constructs a canonical isomorphism

$$H^*(X_{\mathrm{cris}}, M) \xrightarrow{\sim} H^*(X^{\mathrm{an}}, V) \tag{23}$$

generalizing (21). For $M = \mathscr{O}_{\text{cris}}$, V is the constant sheaf of value $\mathbf{C}$ on X^{an}, and the isomorphism (23) solves a conjecture of Grothendieck. Moreover, if X is embedded as a closed subscheme of a smooth scheme Z, and $\hat{Z} = \varinjlim Z_n$ the completion of Z along X, then the left hand side of (23) is the completed de Rham cohomology

$$H^*(X_{\text{cris}}, \mathscr{O}_{X_{\text{cris}}}) \xrightarrow{\sim} H^*(\hat{Z}, \Omega^\bullet_{\hat{Z}}) := \varprojlim_n H^*(Z_n, \Omega^\bullet_{Z_n}).$$

In this particular case, (23) was re-discovered by Hartshorne in [114]. It was quite recently revisited by Bhatt [31], who proved relative variants, using derived de Rham complexes. However, it seems that the relation between the crystalline approach and that of $\mathscr{D}$-modules, briefly mentioned at the end of Sect. 3.1 "Higher Dimension: The Riemann–Hilbert Correspondence", is not yet well understood.

Liftings of K3 Surfaces, Canonical Coordinates

Let X_0 be a K3 surface over an algebraically closed field k of characteristic $p > 0$, and let $T = T_{X_0/k} \simeq \Omega^1_{X_0/k}$ be its tangent bundle. Rudakov and Shafarevich [217] and, by other methods, Nygaard [202] and Lang–Nygaard [161] proved that $H^0(X_0, T) = 0$. This implies that the formal versal deformation S of X_0 is universal and formally smooth of dimension 20 over the Witt ring $W(k)$. Using a crystalline Chern class argument, Deligne [D49, 1981] deduced that, given a non-trivial line bundle $\mathscr{L}_0$ on X_0, the formal scheme $\Sigma(\mathscr{L}_0)$ pro-representing the deformations of $(X_0, \mathscr{L}_0)$ is cut out in S by one equation f not divisible by p, which, by Grothendieck's existence theorem, implies that X_0, together with a polarization, can be lifted to a polarized K3 surface X over a finite extension of the field of fractions of $W(k)$.

In [D50, 1981] Deligne examines more closely the structure of S and $\Sigma(\mathscr{L}_0)$ when X_0 is *ordinary*, i.e., its crystalline cohomology $H^*(X_0/W)$ ($W = W(k)$) is an ordinary F-crystal, which means that its Hodge and Newton polygons coincide, or equivalently that the absolute Frobenius F on $H^2(X_0, \mathscr{O})$ is non-zero. Assuming moreover $p > 2$, he proves that $S = \operatorname{Spf} A$ has a structure of formal torus of rank 20 over W, and that one can choose *canonical coordinates* $q_i \in 1 + \mathbf{m}A$, $1 \leq i \leq 20$, $\mathbf{m}$ the maximal ideal of A, such that $A = W[[q_1 - 1, \cdots, q_{20} - 1]]$, and a distinguished basis $(a, b_1, \cdots, b_{20}, c)$ of $H = H^2(X_0/W)$ in which, for the lifting φ of Frobenius to A given by $\varphi(q_i) = q_i^p$, the Gauss–Manin connection ∇ on H and its Frobenius endomorphism are given by simple formulas (in particular, $Fa = a$, $Fb_i = pb_i$, $Fc = p^2c$). This follows from a general structure theorem for ordinary F-crystals of level ≤ 1 (based on a lemma of Dwork), and a Dieudonné type theory for infinitesimal liftings of a K3. In particular, $q_i = 1$ defines a lifting X_{can} of X_0 to W, called the *canonical lifting*, by analogy with the Serre–Tate canonical lifting of an ordinary abelian variety. Moreover, the first crystalline Chern class of $\mathscr{L}_0$ corresponds to a character $\chi = (x_1, \cdots, x_{20}) \in \mathbf{Z}_p^{20}$ $(\xrightarrow{\sim} \operatorname{Hom}(\widehat{\mathbf{G}^{20}_{m\,\mathbf{Z}_p}}, \widehat{\mathbf{G}_{m\mathbf{Z}_p}}))$

of S, and the equation f above defining $\Sigma(\mathscr{L}_0)$ is $\prod q_i^{x_i}$, i.e., $\Sigma(\mathscr{L}_0) = \operatorname{Ker}\chi$. The structure theorem mentioned above applies to the crystalline H^1 of ordinary abelian varieties, and yields *canonical coordinates* q_{ij} $(1 \le i, j \le g)$ on the corresponding formal moduli space. Shortly afterwards it was shown by Katz [145] that these coincide with the parameters defined by Serre–Tate using the equivalence between liftings of an abelian variety and those of its p-divisible group.

The above theory for K3's could be carried out with minor adjustments assuming only k perfect. The restriction $p > 2$ posed more serious problems. It was removed by Nygaard [203]. Given an ordinary K3 surface X_0 over a perfect field k of characteristic $p > 0$, Nygaard at the same time gave a functorial description, à la Serre–Tate, of the group structure on S, as pro-representing the functor of liftings of the Artin–Mazur p-divisible group $\Psi_{X_0/k}$ (enlarged formal Brauer group), and, using the Kuga–Satake–Deligne abelian variety associated with X_0 (see Sect. 5.3 "K3 Surfaces"), proved, for k finite, the Tate conjecture for X_0. After partial results by several authors [51, 52, 190, 204, 249], the Tate conjecture has been established by Madapusi Pera [187] for all K3's over finitely generated fields of characteristic not equal to 2.

The de Rham–Witt Complex

Let X be a proper and smooth scheme of dimension d over a perfect field k of characteristic $p > 0$, and let $W = W(k)$ be the Witt ring on k. In [34], assuming $p > 2$ and $d < p$, Bloch constructed a projective system of complexes $C_n^\bullet = (C_n^0 \to \cdots \to C_n^d)$ $(n \ge 1)$ on the Zariski site of X, where C_n^q is the sheaf of typical curves $TC_n\mathscr{K}_{q+1}$ on the symbolic part of Quillen's K-group K_{q+1}, together with operators $F : C_n^q \to C_{n-1}^q$, $V : C_n^q \to C$, enjoying remarkable properties, in particular:

(i) construction of a projective system of isomorphisms $H^*(X, C_n^\bullet) \xrightarrow{\sim} H^*(X/W_n)$, where $H^*(X/W_n)$ is Berthelot's crystalline cohomology, hence, by applying $\varprojlim_n$, an isomorphism $H^*(X, C^\bullet) \xrightarrow{\sim} H^*(X/W)$, where the action of the absolute Frobenius on $H^*(X/W)$ is deduced from the endomorphism of $C^\bullet := \varprojlim_n C_n^q$ given by $p^q F$ on C^q,
(ii) degeneration at E_1 modulo torsion of the spectral sequence

$$E_1^{ij} = H^j(X, C^i) \Rightarrow H^{i+j}(X, C^\bullet)(\xrightarrow{\sim} H^{i+j}(X/W)),$$

called the *slope spectral sequence*, with $H^j(X, C^i)/(\text{torsion})$ being finitely generated over W, and, together with F and V being the Cartier module of a p-divisible group, in such a way that $(H^j(X, C^i) \otimes \mathbf{Q}, p^i F)$ calculates the part of slope in $[i, i+1)$ of $H^{i+j}(X/W) \otimes \mathbf{Q}$.

In [66], revisiting work of Lubkin on bounded Witt vectors, Deligne sketched a purely differential geometric construction of Bloch's complex that could work without any restriction of dimension or characteristic. His program was carried out in [122], where the new complex was called the *de Rham–Witt complex*, and proved to coincide with that of Bloch in its range of definition. Generalizations, refinements, and applications of (i) and (ii) were given. This theory has generated many developments up to now.

3.4 Irregular Connections

In the mid 1970s Deligne got interested in irregular connections. In a letter to Katz of Dec. 1, 1976 ([D107, 2007], p. 15) he wrote: "Je collectionne les analogies entre conducteur de Swan et irrégularité d'un système différentiel (au sens de Malgrange, Gérard, Levelt)." The prototype of this is the analogy between the irregular $\mathscr{D}$-module $(\mathscr{O}_X, \nabla)$ on the affine line $X = \mathrm{Spec}(\mathbf{C}[x])$, with $\nabla(1) = -dx$, whose solutions on X^{an} are ce^x, $c \in \mathbf{C}$, and which has irregularity 1 at infinity, on the one hand, and the Artin–Schreier sheaf $\mathscr{L}_\psi$ on $\mathrm{Spec}(\mathbf{F}_q[t])$ (see (111)), which has Swan conductor 1 at infinity, on the other hand.

In his letter, Deligne sketched a proof of a semicontinuity result for the irregularity similar to the one he proved for the Swan conductor (see (42)). He continued to think about this topic and had an extensive correspondence with Malgrange and Ramis, published in [D107, 2007]. In (*loc. cit.*, p.1), he mentions four characteristic $p > 0$ phenomena, and states problems that they suggest on holonomic $\mathscr{D}$-modules with not necessarily regular singularities:

(a) construction of a *Betti structure* (generalizing the classical Stokes structure in dimension 1), periods;
(b) definition of nearby cycles;
(c) (real) Hodge filtration and slopes;
(d) global and local epsilon factors.

He proposes a solution to (b) in a letter to Malgrange (*loc. cit.*, pp. 37, 167). He also suggests an analogue of Laumon's stationary phase principle (134); the construction of $R\Psi$ uses his notion of *tensor product* of abelian categories (see Sect. 9.1), where the universal cover of the punctured disc is "replaced" by the category of finite dimensional $\mathbf{C}((t))$-vector spaces with a connection. He addresses (a) and (c) in (*loc. cit.*, Théorie de Hodge irrégulière), first written in March, 1984, and revised in August, 2006. In particular, as regard to (c), he defines a filtration on a twisted de Rham complex on a curve, and shows degeneration at E_1 of the corresponding spectral sequence (*loc. cit.*, p. 123). This is generalized to higher dimension in ([91], th. 1.2.2). A complete answer to (a) (in arbitrary dimension) was given by T. Mochizuki [193].

Problem (d), i.e., finding an analogue of the product formula for the global constant of the functional equation of L-functions (see Sect. 6.3), was part of the subject of the seminar he ran at the IHÉS in 1984. The seminar was unfinished. Laumon took faithful notes, that remained unpublished. The subject was revisited in the early 2000s by Beilinson, Bloch, and Esnault [25, 39–41]. There, they also discuss (a), give a solution to (d) in the de Rham context, and in a joint unpublished manuscript with Deligne [26], give a solution to (d) in the Betti context, following the line of proof proposed by him in his seminar. A full treatment of (d) is given by Beilinson in [27].

The general problem in the background is the construction of a Riemann–Hilbert correspondence for (not necessarily regular) holonomic $\mathscr{D}$-modules extending that of the regular case (cf. Sect. 3.1 "Higher Dimension: The Riemann–Hilbert Correspondence"). It has been actively studied during the past 30 years. In the 1980s a solution in dimension 1 was known to the contributors of [D107, 2007], though it seems difficult to give a precise reference. In arbitrary dimension, d'Agnolo and Kashiwara [58] have recently constructed a fully faithful functor, compatible with the six operations, from the derived category of cohomologically holonomic complexes of $\mathscr{D}$-modules to a certain derived category of $\mathbf{R}$-constructible enhanced ind-sheaves. A criterion for detecting objects of the essential image by restriction to curves is given by T. Mochizuki in [195].

3.5 Monodromy of the Hypergeometric Equation, Lattices

Let X be a connected, smooth scheme, separated and of finite type over $\mathbf{C}$, $(\mathscr{V}, \nabla)$ a vector bundle on X with an integrable connection, $V = \mathscr{V}^{\nabla}$ the local system on X^{an} of its horizontal sections. If $x_0 \in X^{\mathrm{an}}$ is a base-point, V corresponds to a homomorphism (the *monodromy representation*, cf. (15))

$$\rho : \pi_1(X^{\mathrm{an}}, x_0) \to \mathrm{GL}(V_{x_0}). \tag{24}$$

When $(\mathscr{V}^{\mathrm{an}}, \nabla^{\mathrm{an}})$ underlies a polarizable variation of $\mathbf{Q}$-Hodge structures on X^{an}, then $V = V_{\mathbf{Q}} \otimes \mathbf{C}$ for a $\mathbf{Q}$-local system $V_{\mathbf{Q}}$, ρ factors through $\mathrm{GL}((V_{\mathbf{Q}})_{x_0})$, and, by a theorem of Deligne and Griffiths–Schmid, the identity component G^0 of the Zariski closure G (in $\mathrm{GL}((V_{\mathbf{Q}})_{x_0})$) of the image Γ of ρ is semisimple (see Sect. 4.2 "The Fixed Part and Semisimplicity Theorems"). This is the case, for example, when $(\mathscr{V}, \nabla)$ is defined by a relative de Rham cohomology group $\mathscr{H}^n_{\mathrm{dR}}(Z/X)$, for $Z \to X$ proper and smooth, equipped with its Gauss–Manin connection, in which case ∇ is regular (Sect. 3.1 "Higher Dimension: The Riemann–Hilbert Correspondence"). Apart from this, and from a related finiteness theorem due to Deligne (Theorem 16), little seems to be known in general on the monodromy representations (24). In the geometric situation just mentioned, the determination of $\Gamma \subset G$ is already, in each case, a difficult problem.

Basic questions, for example, are to decide whether the image Γ of ρ in its Zariski closure in $\mathrm{GL}(V_{x_0})$ (or a suitable real algebraic subquotient G of it) is *discrete*, and in this case, if it is a *lattice* in $G(\mathbf{R})$, i.e., Γ has finite covolume, and, if so, if it is *arithmetic*, i.e., roughly speaking, commensurable[8] to $G(\mathcal{O})$, for a ring of integers $\mathcal{O}$ of a totally real number field (see ([D63, 1986], 12) for a precise definition).

A classical laboratory for these questions is the *hypergeometric differential equation* on $\mathbf{P}^1_{\mathbf{C}}$

$$x(x-1)y'' + (c-(a+b+1)x)y' - aby = 0, \tag{25}$$

for a, b, c in $\mathbf{C}$, which has regular singular points at 0, 1, ∞. Its study goes back to Euler, and it has been the subject of extensive work for over 200 years, with an enormous amplification during the past 40 years. In [D63, 1986], Deligne and Mostow revisit (25) and higher dimensional analogues.

The starting point is Schwarz's study of the monodromy representation of (25). Here $X = \mathbf{P}^1_{\mathbf{C}} - \{0, 1, \infty\})$, $\mathcal{V}$ is of rank 2, with a basis (e_1, e_2) such that

$$\nabla(\partial_x)(e_1) = e_2,\ \ \nabla(\partial_x)(e_2) = \frac{abe_1}{x(1-x)} + \frac{((a+b+1)x-c)e_2}{x(1-x)}.$$

For x_0 a base-point in X, $W := V_{x_0}$ is of dimension 2, with the hypergeometric function

$$F(a,b,c;x) = \sum_{n\geq 0} \frac{(a,n)(b,n)}{(c,n)} \frac{x^n}{n!} \tag{26}$$

(c not an integer ≤ 0) as a distinguished solution (defined by (26) for $|x| < 1$ and by analytic continuation outside, using its classical integral representation). Instead of (24), Schwarz considered the projective representation

$$\rho : \pi_1(X, x_0) \to \mathrm{Aut}(\mathbf{P}(W))(= \mathrm{PGL}_2(\mathbf{C})) \tag{27}$$

defined by the local system of lines in $V^\vee$, $x \mapsto w(x) = w_2(x)/w_1(x) \in \mathbf{P}(V_x)$, where (w_1, w_2) is a basis of W (viewed as multivalued functions on X, or a single valued function $\widetilde{w}$ on its universal cover). Schwarz gave criteria on (a, b, c) for $\rho(\Gamma)$ to be finite, and, more generally, for its discreteness in $\mathrm{PGL}_2(\mathbf{C})$. Picard extended this to a 2-variable analogue of $F(a, b, c; x)$, but his proof contained fatal errors. In [D63, 1986], Deligne and Mostow correct it, and generalize it to the d-variable case ($d \geq 1$). Namely, given rational numbers μ_i $(0 \leq i \leq d+1)$, none of which is an integer, they consider the function of $(x_2, \cdots, x_{d+1})$ given by the

[8]Lattices Γ_1 and Γ_2 are called commensurable if $\Gamma_1 \cap \Gamma_2$ is of finite index in Γ_1 and in Γ_2.

integral representation

$$F(x_2, \cdots, x_{d+1}) = \int_1^{\infty} u^{-\mu_0}(u-1)^{-\mu_1} \prod_{2 \leq i \leq d+1} (u - x_i)^{-\mu_i} du \tag{28}$$

(generalizing that for $F(a, b, c; x)$ for $d = 1$, $\mu_0 = c - a$, $\mu_1 = 1 + b - c$, $\mu_2 = a$, up to a product of constant gamma factors), where each x_i is different from 0, 1, ∞, and the x_i's are pairwise distinct. Let ω be the differential form on $\mathbf{P}^1_{\mathbf{C}} - S$ under the integral sign on the right hand side of (28), where $S = \{0, 1, \infty, x_2, \cdots, x_{d+1}\}$. It is shown in (*loc. cit.*, 3) that by integrating ω on suitable cycles on $\mathbf{P}^1_{\mathbf{C}} - S$, one obtains a projective local system $\mathbf{P}(\mathscr{H})$ of rank d on the open subspace Q of $(\mathbf{P}^1_{\mathbf{C}})^d$ consisting of points $(t_i)_{1 \leq i \leq d}$ such that $t_i \neq 0, 1, \infty$ and $t_i \neq t_j$ for $i \neq j$, i.e., a map (of Schwarz type)

$$\widetilde{w} : \widetilde{Q} \to \mathbf{P}^d_{\mathbf{C}}, \tag{29}$$

hence a monodromy representation (for o a base-point in Q)

$$\rho : \pi_1(Q, o) \to \mathrm{PGL}_d(\mathbf{C}) \tag{30}$$

($\mathscr{H}$ is the local system of horizontal sections of a relative $\mathscr{H}^1_{\mathrm{dR}}$ for a relative curve C over Q and a regular singular rank 1 $(\mathscr{L}, \nabla)$ on C). One of the main results of *loc. cit.* is the following theorem:

Theorem 9 *Let μ_∞ be the order of the pole of ω at ∞. Assume that $0 < \mu_i < 1$ for all i, including $i = \infty$, and that $(1 - \mu_i - \mu_j)^{-1}$ is an integer for all $i \neq j$ such that $\mu_i + \mu_j < 1$. Then the monodromy group Γ image of ρ (30) is discrete, and, in fact, is a lattice in* PU$(1, d)$, *i.e., has finite covolume.*

Then Deligne and Mostow investigate when Γ is arithmetic. In the last sections, they give a criterion for arithmeticity, and provide examples of non-arithmetic lattices for $d = 2, 3$ (and, in fact, $\mathbf{Q}$-algebraic families of them). In [197], Mostow had announced the first such example for $d = 2$ (with a different construction, using complex reflections).

A natural continuation of this work is the study of *commensurability* between lattices in PU$(1, n)$. This is the subject of [D74, 1993], where Deligne and Mostow examine several categories of lattices: (a) those coming from reflections (like in [197]), (b) those arising as monodromy groups of hypergeometric local systems (as above), (c) the lattices Γ such that the quotient by Γ of a hyperbolic complex ball in $PU(1, n)$ is a certain orbifold. They discuss commensurability in each category, and between lattices of different ones.

4 Hodge Theory

4.1 Hodge I

Hodge theory has a long history, going back to Abel, Riemann, Picard, and others (see, e.g., P. Griffiths's talk *Abel to Deligne*, IAS, 14 October, 2013 for a survey). The Hodge decomposition of $H^n(X, \mathbf{C})$, for X a compact Kähler variety, led to the notion of *pure Hodge structure*, whose systematic study (variations, moduli, relations with hermitian symmetric domains and Shimura varieties, Mumford–Tate groups) began in the late 1960s.

For $n \in \mathbf{Z}$, a *pure Hodge structure H of weight n* is the data of a finitely generated $\mathbf{Z}$-module $H_{\mathbf{Z}}$ and a decomposition of the $\mathbf{C}$-vector space $H_{\mathbf{C}} = H_{\mathbf{Z}} \otimes \mathbf{C}$ into $\oplus_{p+q=n} H^{pq}$, with $H^{qp} = \overline{H^{pq}}$, or, equivalently, a finite decreasing filtration F of $H_{\mathbf{C}}$ n-opposed to its complex conjugate $\overline{F}$, i.e., satisfying $F^p \oplus \overline{F}^q = H_{\mathbf{C}}$ for $p + q = n + 1$, with $H^{p,q} = F^p \cap \overline{F^q}$ for $p + q = n$. For example, the *Hodge structure of Tate* $\mathbf{Z}(1)$ is the Hodge structure of rank 1 and weight -2, purely of bidegree $(-1, -1)$, with integral lattice $2\pi i \mathbf{Z} \subset \mathbf{C}$ (and for $n \in \mathbf{Z}$, $\mathbf{Z}(n)$ is its n-th tensor power). The notion of pure $\mathbf{Q}$- (resp. $\mathbf{R}$-) Hodge structure of weight n is defined similarly, with $\mathbf{Z}$ replaced by $\mathbf{Q}$ (resp. $\mathbf{R}$).

In his talk at the Nice ICM [D15, 1971] Deligne introduced a generalization of this notion, which he called *mixed Hodge structure*. A mixed Hodge structure H consists of the following data:

(a) a $\mathbf{Z}$-module $H_{\mathbf{Z}}$ of finite type (the *integral lattice*);
(b) a finite increasing filtration W of $H_{\mathbf{Q}} := \mathbf{Q} \otimes_{\mathbf{Z}} H_{\mathbf{Z}}$ (the *weight filtration*);
(c) a finite decreasing filtration F of $H_{\mathbf{C}} := \mathbf{C} \otimes_{\mathbf{Z}} H_{\mathbf{Z}}$ (the *Hodge filtration*).

These data are subject to the condition that, for each $n \in \mathbf{Z}$, $\mathrm{gr}_n^W H_{\mathbf{Q}}$, with the filtration induced by F on $\mathbf{C} \otimes_{\mathbf{Q}} \mathrm{gr}_n^W H_{\mathbf{Q}}$, is a pure $\mathbf{Q}$-Hodge structure of weight n. The numbers

$$h^{pq} = \dim_{\mathbf{C}} H^{pq}, \tag{31}$$

where $H^{pq} = \mathrm{gr}_F^p \mathrm{gr}_{\overline{F}}^q \mathrm{gr}_{p+q}^W H_{\mathbf{C}} = (\mathrm{gr}_{p+q}^W H_{\mathbf{C}})^{p,q}$ are called the *Hodge numbers* of H.

With the obvious definition of morphisms, it is proved in ([D16, 1971], 2.3.5) that mixed Hodge structures form an abelian category, in which morphisms are strictly compatible with the weight and Hodge filtrations.

In [D15, 1971], Deligne sketched a program of construction of mixed Hodge structures on the cohomology of complex algebraic varieties, generalizing classical Hodge theory for smooth projective ones. He carried it out in [D16, 1971] and [D29, 1974]. The idea that Betti cohomology groups of arbitrary complex algebraic varieties should carry such a structure was suggested by Grothendieck's—at the time—conjectural theory of weights in the ℓ-adic cohomology of algebraic varieties over finite fields (coming from the Weil conjectures), and the link between Betti cohomology and ℓ-adic cohomologies provided by Grothendieck's conjectural theory of motives. Deligne's definition and study of mixed Hodge structures made it possible to formulate (and sometimes prove) precise conjectures concerning *weight filtrations* in the two contexts. This is the subject of his report [D15, 1971].

First, Deligne discusses the problem of the existence of a *weight filtration* in the ℓ-adic setting. Let X_0 be a normal, integral scheme of finite type over $\mathbf{Z}$, with generic point η, and geometric point $\overline{\eta}$ over η. Let ℓ be a prime number invertible on X_0. A lisse $\mathbf{Q}_\ell$-sheaf $\mathscr{H}$ on X_0 corresponds to a continuous representation $\rho : \pi_1(X_0, \overline{\eta}) \to \mathrm{GL}(H)$, where H is a finite dimensional $\mathbf{Q}_\ell$-vector space. Given such a sheaf $\mathscr{H}$, Deligne conjectures (*loc. cit.*, 2.1) that, if $\mathscr{H}$ "comes from algebraic geometry", then $\mathscr{H}$ admits a unique increasing filtration W by lisse sheaves, such that each $\mathrm{gr}_i^W \mathscr{H}$ is "punctually pure of weight i" (see Sect. 5.6 "Mixed Sheaves, Statement of the Main Theorem"). By "comes from algebraic geometry", one can for example ask that $\mathscr{H} = R^i f_{0!}\mathbf{Q}_\ell$ (or $\mathscr{H} = R^i f_{0*}\mathbf{Q}_\ell$) for $f_0 : Y_0 \to X_0$ separated and of finite type, demanding that $\mathscr{H}$ is lisse. For $X_0/\mathbf{F}_q$, Deligne said that, assuming (i) resolution of singularities, (ii) the Weil conjectures, one could "in many cases" define a conjectural filtration W. A few years later, he proved (ii), and though (i) is still open today, he constructed W unconditionally in ([D46, 1980], 3.4.1) (see Sect. 5.6 "First Applications", The weight filtration). Nowadays, de Jong's alterations serve as a good substitute for resolution, and could be used (in conjunction with the Weil conjectures) to construct W. In the simple case of the complement of a divisor with normal crossings in a projective and smooth scheme over $\mathbf{C}$, Deligne explains how this filtration arises as the abutment of a certain spectral sequence, and what is the (non conjectural) analogue (involving mixed Hodge structures) that one obtains in Hodge theory, which is the main theme of [D16, 1971].

In the second part, Deligne makes a parallel between the ℓ-adic cohomology of families over a trait on the one hand, and Hodge theory of families over a complex disc on the other hand. The conjectures he formulated (or suggested) there, concerning weights and monodromy, the so-called *weight-monodromy conjectures*, turned out to be a focus of interest in both ℓ-adic cohomology and Hodge theory (see Theorem 25, and Sect. 10).

4.2 *Hodge II and Hodge III*

Homological Algebra Infrastructure

As Deligne explains in [D15, 1971], thanks to Hironaka's resolution of singularities, the desired weight filtrations on the Betti cohomology of complex algebraic varieties can be defined as abutment filtrations of spectral sequences whose initial terms are the cohomology of projective smooth ones. Showing that their combinations with the Hodge filtrations appearing on these abutments give rise to mixed Hodge structures relies on new tools of homological algebra: (i) *filtered derived categories* (introduced in ([120], V) in the case of finite filtrations), and used systematically in [D29, 1974] (ii) *décalage of filtrations*, mentioned in Sect. 1.2 "Décalage of Filtrations".

This décalage is used in the proof of a technical result on which all geometric constructions of mixed Hodge structures ultimately rest, the so-called *lemma of two filtrations* ([D16, 1971], 1.3.16), ([D29, 1974], 7.2). Given a complex K of an abelian category, equipped with two filtrations W and F, F being biregular, i.e., inducing on each component a finite filtration, this lemma provides a handy criterion to ensure that the three natural filtrations[9] cut out by F on the E_r terms of the spectral sequence of K filtered by W coincide (and on $E_\infty^{p,q} = \mathrm{gr}^p{}_W H^{p+q}(K)$ coincide with the filtration induced by F on the right hand side).

Deligne associates with complex algebraic varieties finer objects than mixed Hodge structures, namely *mixed Hodge complexes*. Their definition involves filtered and bi-filtered derived categories, whose definition we recall first (*loc. cit.*, 7.1).

Let $\mathscr{A}$ be an abelian category. The category $K^+F(\mathscr{A})$ is the category of bounded below filtered complexes (K, F) (with F biregular) and homotopy classes of maps preserving the filtration. The *filtered derived category* $D^+F(\mathscr{A})$ is the triangulated category deduced from $K^+F(\mathscr{A})$ by inverting the filtered quasi-isomorphisms, i.e., morphisms u such that $\mathrm{gr}_F(u)$ is a quasi-isomorphism.

The category $K^+F_2(\mathscr{A})$ is the category of bounded below bi-filtered complexes (K, F, W) (with F and W biregular) and homotopy classes of maps preserving F and W. The *bi-filtered derived category* $D^+F(\mathscr{A})$ is the triangulated category deduced from $K^+F_2(\mathscr{A})$ by inverting the bi-filtered quasi-isomorphisms, i.e., morphisms u such that $\mathrm{gr}_F\mathrm{gr}_W(u)$ is a quasi-isomorphism.

Let $n \in \mathbf{Z}$. A *Hodge complex of weight n* is a triple $(K_\mathbf{Z}, (K_\mathbf{C}, F), \alpha)$, where $K_\mathbf{Z}$ is an object of $D^+(\mathbf{Z})$ with finitely generated cohomology groups, $(K_\mathbf{C}, F)$ an object of $D^+F(\mathbf{C})$, and $\alpha : K_\mathbf{C} \xrightarrow{\sim} K_\mathbf{Z} \otimes \mathbf{C}$ an isomorphism of $D^+(\mathbf{C})$ such that the following conditions are satisfied:

(i) the differential of $K_\mathbf{C}$ is strictly compatible with the filtration, in other words ([D16, 1971], 1.3.2), the spectral sequence of $(K_\mathbf{C}, F)$ degenerates at E_1;
(ii) F induces on each $H^k(K_\mathbf{C})(\xrightarrow{\sim} H^k(K_\mathbf{Z}) \otimes \mathbf{C})$ a filtration which is $(n+k)$-opposed to its complex conjugate, i.e., such that $H^k(K_\mathbf{C}) = \oplus_{p+q=n+k}(F^p \cap \overline{F}^q)H^k(K_\mathbf{C})$, in other words, defines a pure $\mathbf{Q}$-Hodge structure of weight $n+k$ on $H^k(K_\mathbf{Q})$.

For A a noetherian subring of $\mathbf{C}$ such that $A \otimes \mathbf{Q}$ is a field (e.g., $\mathbf{Q}$, $\mathbf{R}$), one similarly defines an A-Hodge complex of weight n, by replacing $K_\mathbf{Z}$ by K_A, with K_A in $D^+(A)$, with cohomology groups finitely generated over A.

[9]They come from the three descriptions of E_r^{pq}: as a subobject of a quotient of K^{p+q}, as a quotient of a subobject of K^{p+q}, as a quotient of a subobject of E_{r-1}^{pq}.

An *A-mixed Hodge complex* is a tuple $K = (K_A, (K_{A\otimes\mathbf{Q}}, W), (K_{\mathbf{C}}, W, F), \beta, \alpha)$, where K_A is an object of $D^+(A)$ with finitely generated cohomology modules, $(K_{A\otimes\mathbf{Q}}, W)$ an object of $D^+F(\mathbf{Q})$ (W being increasing), $(K_{\mathbf{C}}, W, F)$ an object of $D^+F_2(\mathbf{C})$, $\beta : K_{A\otimes\mathbf{Q}} \xrightarrow{\sim} K_A \otimes \mathbf{Q}$ an isomorphism in $D^+(A \otimes \mathbf{Q})$, $\alpha : (K_{\mathbf{C}}, W) \xrightarrow{\sim} (K_{A\otimes\mathbf{Q}}, W) \otimes \mathbf{C}$ in $D^+F(\mathbf{C})$ such that for all $n \in \mathbf{Z}$, the triple

$$(\mathrm{gr}_n^W K_{A\otimes\mathbf{Q}}, (\mathrm{gr}_n^W K_{\mathbf{C}}, F), \mathrm{gr}_n^W \alpha : \mathrm{gr}_n^W K_{\mathbf{C}} \xrightarrow{\sim} \mathrm{gr}_n^W K_{A\otimes\mathbf{Q}} \otimes \mathbf{C})$$

is an $A \otimes \mathbf{Q}$-Hodge complex of weight n.

For $A = \mathbf{Z}$, one simply says mixed Hodge complex. An A-mixed Hodge complex K such that $\mathrm{gr}_W^i K_{A\otimes\mathbf{Q}} = 0$ for $i \neq n$ can be viewed, by forgetting W, as an A-Hodge complex of weight n. The link between mixed Hodge complexes and mixed Hodge structures is provided by (a) of the following theorem (*loc. cit.*, 8.1.9) (whose other statements yield the basic degeneration results for the spectral sequences arising from geometric situations):

Theorem 10 *Let K be a mixed Hodge complex.*

(a) *For each $k \in \mathbf{Z}$, the shifted filtration $W[k]$ ($W[k]_p = W_{p-k}$) on $H^k(K_{\mathbf{Q}})$ and the filtration F of $H^k(K_{\mathbf{C}}) = H^k(K_{\mathbf{Q}}) \otimes \mathbf{C}$ define a mixed Hodge structure.*
(b) *The spectral sequence of $(K_{\mathbf{Q}}, W)$ (weight spectral sequence) degenerates at E_2.*
(c) *The spectral sequence of $(K_{\mathbf{C}}, F)$ (Hodge spectral sequence) degenerates at E_1.*
(d) *For each $p \in \mathbf{Z}$, the spectral sequence of $\mathrm{gr}_F^p K_{\mathbf{C}}$ filtered by W degenerates at E_2.*

Morphisms of mixed Hodge complexes are defined in the obvious way. It follows from Theorem 10 (and the fact that morphisms of Hodge structures are strictly compatible with the Hodge filtrations) that a morphism $u : K \to L$ of Hodge complexes such that the underlying morphism $u_{\mathbf{Z}} : K_{\mathbf{Z}} \to L_{\mathbf{Z}}$ is an isomorphism of $D(\mathbf{Z})$ is an isomorphism. This does not extend to morphisms of mixed Hodge complexes. However, if $u : K \to L$ is a morphism of mixed Hodge complexes such that $u_{\mathbf{Z}} : K_{\mathbf{Z}} \to L_{\mathbf{Z}}$ is an isomorphism, then the morphism deduced by décalage of W,

$$(K_{\mathbf{Z}}, (K_{\mathbf{Q}}, \mathrm{Dec}(W)), (K_{\mathbf{C}}, \mathrm{Dec}(W), F)) \to (L_{\mathbf{Z}}, (L_{\mathbf{Q}}, \mathrm{Dec}(W)), (L_{\mathbf{C}}, \mathrm{Dec}(W), F))$$

is an isomorphism (the décalage comes from (d) and (4)). In particular, $H^n(u) : H^n(K) \to H^n(L)$ is an isomorphism of mixed Hodge structures.

Mixed Hodge Theory

The Smooth Case

If X is a projective, smooth scheme over $\mathbf{C}$, the associated complex analytic variety X^{an} is Kähler, hence, for each $n \in \mathbf{Z}$, $H^n(X, \mathbf{Z})$[10] comes equipped with a pure Hodge structure of weight n. The filtration F on $H^n(X, \mathbf{C})$ doesn't depend on the Kähler structure, as it is the abutment filtration of the Hodge to de Rham spectral sequence

$$E_1^{pq} = H^q(X, \Omega_X^p) \Rightarrow H^{p+q}(X, \mathbf{C})(\xrightarrow{\sim} H^{p+q}(X, \Omega_X^\bullet)), \tag{32}$$

which degenerates at E_1. In ([D3, 1968], 5.3), Deligne shows that if X is only assumed *proper* and smooth, the same degeneration holds for the similar spectral sequence (32) (an algebraic proof of this was later given in [D65, 1987], see Sect. 4.6), and the abutment filtration F again provides $H^n(X, \mathbf{Z})$ with a pure Hodge structure of weight n.

Let X be a separated and smooth scheme of finite type over $\mathbf{C}$. As we have seen in Sect. 3.1 "Higher Dimension: The Riemann–Hilbert Correspondence", by Nagata's compactification theorem, followed by Hironaka's resolution, one can find a dense open embedding $j : X \hookrightarrow \overline{X}$, with $\overline{X}/\mathbf{C}$ proper and smooth, and $D = \overline{X} - X$ a strict normal crossings divisor. Recall (Sect. 3.2) that the inclusion (18) (of analytic complexes)

$$\Omega_{\overline{X}}^\bullet(\log D) \hookrightarrow j_*\Omega_X^\bullet$$

is a filtered quasi-isomorphism, where the left (resp. right) hand side is filtered by W (resp. τ). By the Poincaré lemma, $j_*\Omega_X^\bullet \xrightarrow{\sim} Rj_*\mathbf{C}$, so we get a filtered complex $(Rj_*\mathbf{Q}, \tau)$ with $Rj_*\mathbf{Q} \xrightarrow{\sim} \mathbf{Q} \otimes Rj_*\mathbf{Z}$, and a bifiltered complex $(\Omega_{\overline{X}}^\bullet(\log D), W, F)$ (where F is the naive filtration), with a filtered isomorphism $\mathbf{C} \otimes (Rj_*\mathbf{Q}, \tau) \xrightarrow{\sim} (\Omega_{\overline{X}}^\bullet(\log D), W)$. Applying $R\Gamma(\overline{X}, -)$ we get an object $R\Gamma(X, \mathbf{Z})$ of $D^+(\mathbf{Z})$, an object $(R\Gamma(X, \mathbf{Q}), W)$ of $D^+F(\mathbf{Q})$ (W being induced by τ on $Rj_*\mathbf{Q}$) with an isomorphism $\mathbf{Q} \otimes R\Gamma(X, \mathbf{Z}) \xrightarrow{\sim} R\Gamma(X, \mathbf{Q})$, and an object $R\Gamma(\overline{X}, \Omega_{\overline{X}}^\bullet(\log D), W, F)$ of $D^+F_2(\mathbf{C})$ with a filtered isomorphism $\mathbf{C} \otimes (R\Gamma(X, \mathbf{Q}), \tau) \xrightarrow{\sim} R\Gamma(\overline{X}, \Omega_{\overline{X}}^\bullet(\log D), W)$. Deligne proves the following result ([D29, 1974], 8.1.7, 8.1.8):

[10] In this section, sheaves on and cohomology groups of schemes of finite type over $\mathbf{C}$ are taken with respect to the classical topology (on the associated complex analytic spaces), and we omit the superscript "an" for brevity.

Theorem 11 *The triple*

$$(R\Gamma(X,\mathbf{Z}), W, F) = (R\Gamma(X,\mathbf{Z}), (R\Gamma(X,\mathbf{Q}), W), (R\Gamma(\overline{X}, \Omega^\bullet_{\overline{X}}(\log D), W, F)), \tag{33}$$

endowed with the above isomorphisms, is a mixed Hodge complex.

In particular, by Theorem 10, for each $n \in \mathbf{Z}$,

$$(H^n(X,\mathbf{Z}), (H^n(X,\mathbf{Q}), W[n]), (H^n(\overline{X}, \Omega^\bullet_{\overline{X}}(\log D), W[n], F)) \tag{34}$$

is a mixed Hodge structure. As compactifications $j : X \hookrightarrow \overline{X}$ as above form a connected category, it follows from the remark after Theorem 10 that this mixed Hodge structure $((H^n(X,\mathbf{Z}), W, F), j)$ is independent of the compactification j: the structures associated with various j''s are related by a transitive system of isomorphisms. As morphisms $f : X \to Y$ can be embedded in morphisms of compactifications, it depends functorially on X. The same holds for the complexes $(R\Gamma(X,\mathbf{Z}), \mathrm{Dec}(W), F)$ deduced from (33) by décalage of the filtration W (*loc. cit.* 8.1.16).

The General Case

Let X be a scheme (or algebraic space) separated and of finite type over $\mathbf{C}$. Using Hironaka's resolution, Deligne shows that by a step by step construction (axiomatized in ([3], Vbis, 5.1) and recalled in (*loc. cit.*, 6.2.5)) one can construct a commutative diagram of simplicial $\mathbf{C}$-schemes

$$\begin{array}{ccc} Y_\bullet & \xrightarrow{j_\bullet} & \overline{Y}_\bullet \\ {\scriptstyle a}\downarrow & & \downarrow \\ X & \longrightarrow & \mathrm{Spec}\,\mathbf{C}, \end{array} \tag{35}$$

where a is a proper hypercovering (for the classical topology), $\overline{Y}_\bullet$ a simplicial $\mathbf{C}$-scheme which is proper and smooth in each degree, and $j_\bullet$ a map which is in each degree n a dense open immersion such that the complement $D_n = \overline{Y}_n - Y_n$ is a strict normal crossing divisor. The constructions of the smooth case yield a triple of (filtered) complexes on $\overline{Y}_\bullet$:

$$(Rj_{\bullet *}\mathbf{Z}, (Rj_{\bullet *}\mathbf{Q}, \tau), (\Omega^\bullet_{\overline{Y}_\bullet}(\log D_\bullet), W, F)), \tag{36}$$

with isomorphisms $\mathbf{Q} \otimes Rj_{\bullet *}\mathbf{Z} \xrightarrow{\sim} Rj_{\bullet *}\mathbf{Q}$, $\mathbf{C} \otimes (Rj_{\bullet *}\mathbf{Q}, \tau) \xrightarrow{\sim} (\Omega^{\bullet}_{Y_\bullet}(\log D_\bullet), W)$. By applying $R\Gamma(\overline{Y}_\bullet, -)$ to (36) we get a triple of (filtered) complexes:

$$(R\Gamma(\overline{Y}_\bullet, Rj_{\bullet *}\mathbf{Z}), (R\Gamma(\overline{Y}_\bullet, Rj_{\bullet *}\mathbf{Q}), \delta\tau), (R\Gamma(\overline{Y}_\bullet, \Omega^{\bullet}_{Y_\bullet}(\log D_\bullet), \delta W, F)), \tag{37}$$

with isomorphisms $\mathbf{Q}\otimes R\Gamma(Y_\bullet, \mathbf{Z}) \xrightarrow{\sim} R\Gamma(\overline{Y}_\bullet, Rj_{\bullet *}\mathbf{Q})$, $\mathbf{C}\otimes(R\Gamma(\overline{Y}_\bullet, Rj_{\bullet *}\mathbf{Q}), \delta\tau)$ $\xrightarrow{\sim} (R\Gamma(\overline{Y}_\bullet, \Omega^{\bullet}_{Y_\bullet}(\log D_\bullet), \delta W)$. Here the filtrations $\delta\tau$ and δW are obtained by a diagonal process: if M is a complex on $\overline{Y}_\bullet$, $R\Gamma(\overline{Y}_\bullet, M)$ is the total complex of a bicomplex $K^{\bullet,\bullet}$, whose second degree corresponds to the simplicial degree: $K^{\bullet,q}$ calculates $R\Gamma(\overline{Y}_q, M_q)$; if M is filtered by an increasing filtration W, the filtration δW on $\mathbf{s}K^{\bullet,\bullet}$ is given by $(\delta W)_n(\mathbf{s}K) = \oplus_{p,q} W_{n+q}(K^{p,q})$, with associated graded $\mathrm{gr}_n^{\delta W}(\mathbf{s}K) = \oplus_q \mathrm{gr}_{n+q}^W(K^{\bullet,q})[-q]$.

By Theorem 2, the adjunction map

$$\mathbf{Z} \to Ra_*\mathbf{Z}$$

is an isomorphism, hence $R\Gamma(\overline{Y}_\bullet, Rj_{\bullet *}\mathbf{Z}) = R\Gamma(X, \mathbf{Z})$, so that (37) can be rewritten

$$(R\Gamma(X, \mathbf{Z}), (R\Gamma(X, \mathbf{Q}), W), (R\Gamma(\overline{Y}_\bullet, \Omega^{\bullet}_{Y_\bullet}(\log D_\bullet), \delta W, F)), \tag{38}$$

(where the filtration W on $R\Gamma(X, \mathbf{Q})$ is induced by the diagonal filtration δW of $(R\Gamma(\overline{Y}_\bullet, Rj_{\bullet *}\mathbf{Q}))$. Deligne proves the following generalization of Theorem 11:

Theorem 12 *The triple* $(R\Gamma(X, \mathbf{Z}), W, F)$ *defined by* (38) *(and the above isomorphisms) is a mixed Hodge complex.*

Again, by Theorem 10, for each $n \in \mathbf{Z}$,

$$(H^n(X, \mathbf{Z}), (H^n(X, \mathbf{Q}), W[n]), (H^n(\overline{X}_\bullet, \Omega^{\bullet}_{\overline{X}_\bullet}(\log D_\bullet), W[n], F)) \tag{39}$$

is a mixed Hodge structure, and, as before, it follows from the remark after Theorem 10 that, up to a transitive system of isomorphisms, it does not depend on the choice of the diagram (35) (and a similar statement holds for the complexes $(R\Gamma(X, \mathbf{Z}), \mathrm{Dec}(W), F))$. Moreover, Deligne shows that this mixed Hodge structure is functorial in X: a morphism $f : X_1 \to X_2$ induces a morphism of mixed Hodge structures $f^* : (H^n(X_2, \mathbf{Z}), W, F) \to (H^n(X_1, \mathbf{Z}), W, F)$ (which is automatically strictly compatible with the filtrations W and F).

Concerning the Hodge numbers $h^{pq} = h^{pq}(H^n(X, \mathbf{Z}))$ (31), Deligne proves ([D29, 1974], 8.2.4) that they are concentrated in the square $[0, n] \times [0, n]$ (and even in the smaller square $[n - N, n - N]$ if $n \geq N = \dim(X)$), an that in addition: if X is *proper* (resp. *smooth*) they are concentrated on or under (resp. above) the diagonal $p + q = n$. When X is smooth, the bottom (i.e., smallest weight) part

$W_n H^n(X, \mathbf{Q})$ of the weight filtration is the image of $H^n(\overline{X}, \mathbf{Q})$ in $H^n(X, \mathbf{Q})$ for *any* smooth compactification $X \hookrightarrow \overline{X}$.

Simplicial Variants

Let $X_\bullet$ be a simplicial scheme (or algebraic space) over $\mathbf{C}$, whose components X_n are separated and of finite type. Similar constructions equip the cohomology groups $H^n(X_\bullet, \mathbf{Z})$ with mixed Hodge structures $H^n(X_\bullet, \mathbf{Z}), W, F)$, functorial in $X_\bullet$. The spectral sequence (5)

$$E_1^{pq} = H^q(X_p, \mathbf{Z}) \Rightarrow H^{p+q}(X_\bullet, \mathbf{Z})$$

is a spectral sequence of mixed Hodge structures. Deligne gives two applications of this.

The first one concerns *relative cohomology*. If $f : Y \to X$ is a continuous map between topological spaces, the relative cohomology complex of f (or X mod Y) could be defined as $C[-1]$, where C is a cone in $D^+(\mathbf{Z})$ of $f^* : R\Gamma(X, \mathbf{Z}) \to R\Gamma(Y, \mathbf{Z})$. However, this definition is not functorial, and, when f underlies a morphism of separated schemes of finite type over $\mathbf{C}$, doesn't yield a definition of mixed Hodge structures on the groups $H^n(C)$ making the relative cohomology exact sequence an exact sequence of mixed Hodge structures. Instead, Deligne defines the *cone of f*, $C(f)$, as the simplicial scheme which is the push-out of the diagram (of simplicial schemes)

$$\begin{array}{ccc} Y \times (\{0\}, \{1\}) & \longrightarrow & X \sqcup \operatorname{Spec} \mathbf{C} \\ \downarrow & & \downarrow \\ X \sqcup (Y \times \Delta(1)) & \longrightarrow & C(f) \end{array} \tag{40}$$

(where the left vertical arrow is defined by the inclusion of $\Delta(0) = (\{0\}, \{1\})$ into $\Delta(1)$,[11] and the top horizontal arrow sends $(y, 0)$ (resp. $(y, 1)$) to $f(y)$ (resp. $\operatorname{Spec} \mathbf{C}$).[12] Thus the cohomology groups

$$H^n(X \bmod Y, \mathbf{Z}) := H^n(C(f), \mathbf{Z})$$

[11] $\Delta(n)$ is the simplicial set $[p] \mapsto \operatorname{Hom}([p], [n])$.

[12] The n-th component of $C(f)$ is the disjoint union of X_n, Y_i for $i < n$, and $\operatorname{Spec} \mathbf{C}$.

come naturally equipped with a mixed Hodge structure, and one checks (*loc. cit.* *8.3.9*) that the exact sequence of relative cohomology

$$\cdots \to H^n(X \bmod Y, \mathbf{Z}) \to H^n(X, \mathbf{Z}) \to H^n(Y, \mathbf{Z}) \to \cdots$$

is an exact sequence of mixed Hodge structures. This formalism is used in the proof of the above assertions on the Hodge numbers. it could also be applied to define mixed Hodge structures on compactly supported cohomology groups. However, this is not discussed in *loc. cit.*, and would not suffice to construct a good theory of duality. A formalism of *six operations* in Hodge theory was later provided by M. Saito's theory of *mixed Hodge modules* [221, 222]. See Sect. 4.5 for its bearing on Deligne cohomology groups.

The second application is to algebraic groups. The cohomology of Lie groups and their classifying spaces was extensively studied by Borel in the 1950s. Let G be a linear algebraic group over $\mathbf{C}$. The Betti cohomology of its topological classifying space BG can be calculated as the cohomology of the simplicial scheme (where $[p]$ is the ordered set $(0, \cdots, p)$)

$$B_\bullet G = ([p] \mapsto G^{[p]}/G)$$

(sometimes called the *nerve* of G); in stack theoretic language, this is $\mathrm{cosk}_0(\mathrm{Spec}\,\mathbf{C} \to [\mathrm{Spec}\,\mathbf{C}/G])$. The corresponding spectral sequence (cf. (5)),

$$E_1^{pq} = H^q(G^p, \mathbf{Z}) \Rightarrow H^{p+q}(B_\bullet G, \mathbf{Z}),$$

sometimes called the *Eilenberg–Moore spectral sequence*, underlies a spectral sequence of mixed Hodge structures. Using it and the splitting principle, Deligne proves (*loc. cit.* 9.1.1, 9.1.5):

Theorem 13 *Let G be a linear algebraic group over* $\mathbf{C}$.

(a) $H^{2n-1}(B_\bullet G, \mathbf{Q}) = 0$, *and* $H^{2n}(B_\bullet G, \mathbf{Q}) \otimes \mathbf{C}$ *is purely of type* (n, n).
(b) *If G is connected, the primitive part P^* of the Hopf algebra $H^*(G, \mathbf{Q})$ is a mixed sub-Hodge structure, $P^{2i} = 0$, P^{2i-1} is purely of type (i, i) and $H^*(G, \mathbf{Q}) = \Lambda^* P^*$ as mixed Hodge structures.*

Deligne mentions that similar results hold in ℓ-adic cohomology. He gives some details in ([D39, 1977], Sommes trigonométriques, 8.2).

The Fixed Part and Semisimplicity Theorems

While Deligne was developing his theory of mixed Hodge structures, variations of (pure) Hodge structures and their local and global monodromies were being studied by Griffiths and Schmid by analytic methods. Mixed Hodge theory enabled him to prove key results on the global monodromy of variations of Hodge structures of geometric origin.

The main one is the so-called *fixed part theorem*. Let S be a smooth, separated scheme over $\mathbf{C}$, and $f : X \to S$ a proper and smooth morphism. Then, by Deligne's criteria in [D3, 1968], the Leray spectral sequence of f,

$$H^p(S, R^q f_*\mathbf{Q}) \Rightarrow H^{p+q}(X, \mathbf{Q})$$

degenerates at E_2 (cf. (1)) (we write here f for f^{an}). In particular, for all n, the edge homomorphism

$$H^n(X, \mathbf{Q}) \to H^0(S, R^n f_*\mathbf{Q})$$

is surjective. The fixed part theorem is the following statement ([D16, 1971], 4.1.1):

Theorem 14 *With the above notation, let $\overline{X}$ be a smooth compactification of X. Then the induced morphism*

$$H^n(\overline{X}, \mathbf{Q}) \to H^0(S, R^n f_*\mathbf{Q}) \tag{41}$$

is surjective.

In general, the restriction map $H^n(\overline{X}, \mathbf{Q}) \to H^n(X, \mathbf{Q})$ is far from being surjective: its image is the bottom layer $W_n H^n(X, \mathbf{Q})$ of the weight filtration of $H^n(X, \mathbf{Q})$.

Theorem 14 has several remarkable consequences. Here is one which plays a crucial role in the next theorem:

Corollary 1 *Under the assumptions of Theorem* 14, *suppose S connected. Let $(R^n f_*\mathbf{Q})^0$ be the largest constant sub-local system of $R^n f_*\mathbf{Q}$ (thus, for $s \in S$, the restriction map $H^0(S, R^n f_*\mathbf{Q}) \to (R^n f_*\mathbf{Q})^0_s$ is an isomorphism). Then $(R^n f_*\mathbf{Q})^0_s$ underlies a sub-Hodge structure of $H^n(X_s, \mathbf{Q})$, inducing on $H^0(S, R^n f_*\mathbf{Q})$ a Hodge structure which is independent of s.*

Actually, the conclusion holds assuming only S reduced and separated (and connected). In particular, a global section a of $R^n f_*\mathbf{C}$ on S is of Hodge type (p, q) at one point s, then a is of type (p, q) everywhere.

Deligne mentions in a footnote to *loc. cit.* that one can deduce from results of Griffiths and Schmid a generalization of Corollary 1, with $R^n f_*\mathbf{Q}$ replaced by a *polarizable variation of (pure) Hodge structures* on the (smooth scheme) S.

In *loc. cit.* Deligne proves a general semisimplicity theorem for representations of the fundamental group of a good connected topological space S associated with continuous variations of pure $\mathbf{Q}$-Hodge structures on S satisfying a number of properties, verified for example in the case of "algebraic" variations (by Corollary 1) or variations à la Griffiths–Schmid as above. He derives from it the following consequence ([D16, 1971], 4.2.9):

Theorem 15 *Let S be as in Corollary 1, $s \in S$, and $n \in \mathbf{Z}$. Let $f : X \to S$ be a morphism, with $X/\mathbf{C}$ separated and of finite type, such that $R^n f_* \mathbf{Q}$ is a local system on S. Let G be the Zariski closure of the image of the representation $\rho : \pi_1(S, s) \to \mathrm{Aut}(R^n f_* \mathbf{Q})_s$, and G^0 the identity component. Then:*

(a) *The radical of G^0 is unipotent.*
(b) *If f is proper and smooth, ρ is semisimple (hence G^0 is semisimple).*

Part (a) will later resonate in ℓ-adic cohomology (see Sect. 5.6 "Ingredients of the Proof", Theorem 23).

In turn, the Weil conjectures resonate in Hodge theory: Deligne will show that they imply that the weight filtration of $H^n(X, \mathbf{Q})$ (36) is a *discrete invariant*, i.e., is invariant under algebraic deformation of X (see Theorem 39).

The semisimplicity theorem, combined with distance decreasing properties of Griffiths period maps, implies the following striking finiteness result ([D66, 1987], 0.5):

Theorem 16 *Let S be a smooth, connected, scheme over* **C**, *and N a nonnegative integer. There exists only a finite number of isomorphism classes of local systems on S of* **Q**-*vector spaces of rank N that are direct summands of local systems underlying a polarizable variation of Hodge structures.*

For $f : X \to S$ proper and smooth, with S as above, consider the monodromy representation ρ as in Theorem 15. Theorem 16 implies ([D66, 1987], 0.1):

Corollary 2 *Fix (S, s) and the integer $N \geq 0$. For variable n and (proper and smooth) $f : X \to S$, the associated monodromy representations ρ which are of dimension N form a finite number of isomorphism classes.*

1-Motives

While Grothendieck's conjectural theory of motives inspired Deligne's construction of mixed Hodge theory, in turn, mixed Hodge theory suggested an (even more remote) theory of *mixed motives*. Though such a theory (or even a precise formulation of it) seems to be still out of reach today, interesting pieces could be defined and studied unconditionally, namely, (i) *1-motives*, and (ii) *mixed Tate motives* over a number field. In the 1980s Deligne proposed a conjectural formalism for (ii), which was later constructed as a by-product of Voevodsky's theory (see Sect. 9.2 "Mixed Tate Motives", (c)). He developed (i) at the end of [D29, 1974].

If A is a complex abelian variety, the homology group $H_{\mathbf{Z}} = H_1(A, \mathbf{Z})$ is the kernel of the (surjective) exponential map $\mathrm{Lie}(A) \to A$. Let F be the (one step) filtration on $H_{\mathbf{C}} = H_{\mathbf{Z}} \otimes \mathbf{C}$ defined by the kernel of the (surjective) homomorphism $H_1(A, \mathbf{Z}) \otimes \mathbf{C} \to \mathrm{Lie}(A)$. Then $(H_{\mathbf{Z}}, F)$ is a polarizable pure Hodge structure of type $((-1, 0), (0, -1))$, and it has been known since Riemann that this construction defines an equivalence between the category of complex abelian varieties and that of polarizable pure Hodge structures of type $((-1, 0), (0, -1))$.

In ([D29, 1974], 10.1) Deligne gives a similar geometric interpretation for certain mixed Hodge structures of weight between -2 and 0. He first defines the beautiful geometric notion of 1-motive.

A 1*-motive M over a scheme S* consists of the following data: an abelian scheme A and a torus T over S, an extension G of A by T, and a morphism $u : X \to G$, where X is a group scheme over S, which étale locally is the constant group scheme defined by a finitely generated and free $\mathbf{Z}$-module.

Let $M = (X, A, T, G, u)$ be a 1-motive over $\mathbf{C}$. Deligne constructs a mixed Hodge structure $T(M)$ of type $t = ((0,0), (-1,0), (0,-1), (-1,-1))$, called the *Hodge realization* of M, whose integral lattice $T(M)_{\mathbf{Z}}$ is torsion free, and isomorphisms of pure Hodge structures $H_1(T,\mathbf{Z}) \xrightarrow{\sim} \mathrm{gr}^W_{-2}T(M)_{\mathbf{Z}}$, $H_1(A,\mathbf{Z}) \xrightarrow{\sim} \mathrm{gr}^W_{-1}T(M)_{\mathbf{Z}}$, $X \xrightarrow{\sim} \mathrm{gr}^W_0 T(M)_{\mathbf{Z}}$, where $T(M)_{\mathbf{Z}}$ is endowed with the filtration W induced by the filtration W on $T(M)_{\mathbf{Q}}$. He shows that $M \mapsto T(M)$ is an equivalence from the category of 1-motives over $\mathbf{C}$ to that of mixed Hodge structures H of type t such that $H_{\mathbf{Z}}$ is torsion free and $\mathrm{gr}^W_{-1}H$ is polarizable.

For 1-motives over an algebraically closed field k, Deligne defines similar realizations in the ℓ-adic and de Rham contexts. In addition, using Grothendieck's formalism of bi-extensions, he constructs a self-duality $M \mapsto M^*$ of the category of 1-motives over k, which, when $k = \mathbf{C}$ induces on the Hodge realization $T(M) \mapsto \mathrm{Hom}(T(M), \mathbf{Z}(1))$ (where $\mathbf{Z}(1)$ is the Hodge structure of Tate (Sect. 4.1)).

Let $X/\mathbf{C}$ be separated and of finite type, of dimension $\leq N$. For $n \geq 0$, $H^n(X, \mathbf{Z})$ has a mixed Hodge structure $\mathscr{H}$, whose Hodge numbers h^{pq} are concentrated in the square $[0,n] \times [0,n]$ (12). From $\mathscr{H}$ one can deduce 1-motives, that Deligne denotes by I and II_n: I (resp. II_n) is the largest mixed sub-Hodge structure (resp. quotient Hodge structure) of $(\mathscr{H}_{\mathbf{Z}}/\mathrm{torsion})(1)$ (resp. $(\mathscr{H}_{\mathbf{Z}}/\mathrm{torsion})(n)$) which is purely of type $((-1,-1), (-1,0), (0,-1), (0,0))$. He makes the following conjecture ([D29, 1974], 10.4.1):

Conjecture 1 The 1-motives I and II_n (for $n \leq N$) and II_N (for $N \geq n$) admit a purely algebraic description.

In (*loc. cit.*, 10.3) he proves it for curves. In a slightly different form, the conjecture was proven (independently) by Ramachandran [212], and Barbieri-Viale, Rosenschon, M. Saito [18].

The notion of 1-motive has given rise to many developments and generated a huge literature, see [19] for a recent survey.

The du Bois Complex

Let X be a quasi-projective scheme over $\mathbf{C}$. In ([D29, 1974], 9.3) Deligne constructs a complex K of coherent sheaves on X, concentrated in nonnegative degrees, with differential given by differential operators of order ≤ 1 such that $\mathbf{C} \to K^{\mathrm{an}}$ is a quasi-isomorphism, and factors into $\mathbf{C} \to \Omega^{\bullet}_{X^{\mathrm{an}}} \to K$, where the first arrow is the natural augmentation. In particular, $H^*(X^{\mathrm{an}}, \mathbf{C})$ appears as a direct summand of the analytic de Rham cohomology $H^*(X^{\mathrm{an}}, \Omega^\bullet_{X^{\mathrm{an}}})$ (a result previously proved by Bloom and Herrera by other methods).

In his letter [64], he proposes a strong refinement of this. He makes the following conjecture:

Conjecture 2 Let $\mathscr{X}$ be a complex analytic space. Let $\varepsilon : \mathscr{Y}_\bullet \to \mathscr{X}$ be a proper hypercovering, with $\mathscr{Y}_n$ smooth over $\mathbf{C}$ for all n. Consider the total complex $R\varepsilon_*\Omega^\bullet_{\mathscr{Y}_\bullet/\mathbf{C}}$, filtered by the Hodge filtration $\Omega^{\geq p}_{\mathscr{Y}_\bullet/\mathbf{C}}$ on $\Omega^\bullet_{\mathscr{Y}_\bullet/\mathbf{C}}$, an object of the bounded below derived category of filtered complexes of sheaves of $\mathscr{O}_{\mathscr{X}}$-modules, the filtration being biregular, with differential given by differential operators of order ≤ 1, and $\mathscr{O}_{\mathscr{X}}$-linear associated graded. Then, in this category, $R\varepsilon_*\Omega^\bullet_{\mathscr{Y}_\bullet/\mathbf{C}}$ is independent of the choice of ε, namely there should exist a transitive system of isomorphisms between these objects when ε varies. Denote this object by $\underline{\Omega}^\bullet_{\mathscr{X}}$. Then, in particular, for all p, $\underline{\Omega}^p_{\mathscr{X}} := \mathrm{gr}^p\underline{\Omega}^\bullet_{\mathscr{X}}$ is a well defined object of $D^+(\mathscr{X}, \mathscr{O}_{\mathscr{X}})$. Moreover, for $\mathscr{X} = X^{\mathrm{an}}$ for $X/\mathbf{C}$ a projective scheme, $\underline{\Omega}^\bullet_{\mathscr{X}}$ should be the analytification of a similar object $\underline{\Omega}^\bullet_X$ on the Zariski site of X, and the spectral sequence

$$E_1^{pq} = H^q(X, \underline{\Omega}^p_X) \Rightarrow H^{p+q}(X, \underline{\Omega}^\bullet_X)$$

should degenerate at E_1 and abut to the Hodge filtration of (the mixed Hodge structure) of $H^*(X^{\mathrm{an}}, \mathbf{C})$.

The analytic conjecture is still open. In the algebraic case, in a second letter [68], Deligne explained how, in the projective case, one could prove the desired independence by the global to local argument, that he had used to prove finiteness of étale cohomology (see Sect. 1.4 "Finiteness"). Details and generalizations were written up by du Bois in his thesis [85]. The complex $\underline{\Omega}^\bullet_X$ was later called the *du Bois complex*, and singularities for which $\mathscr{O}_X \to \mathrm{gr}^0\underline{\Omega}^\bullet_X (= R\varepsilon_*\mathscr{O}_{Y_\bullet})$ is an isomorphism, *du Bois singularities*. They are important in birational geometry. They were studied by various authors after du Bois (Steenbrink, Ishida, Kollár, etc.), see [233] for recent applications.

Hodge Theory and Rational Homotopy

Let M be a connected CW-complex (or simplicial set). By different methods Quillen [211] and Sullivan [245] (see ([D31, 1975], 3) for the details of the construction) attached to M anticommutative $\mathbf{Q}$-differential graded algebras $\Omega^*(M)$ which capture its rational homotopy type. Furthermore, Sullivan introduced the notion of *minimal model* of such an object, namely an anticommutative $\mathbf{Q}$-differential graded algebra $\mathscr{M} = \oplus_{n\geq 0}\mathscr{M}^n$ quasi-isomorphic to $\Omega^*(M)$, with $\mathscr{M}^0 = \mathbf{Q}$, $\mathscr{M}$ free in the graded sense, generated by its indecomposable elements, and with $d\mathscr{M} \subset (\mathscr{M}^{>0})^2$. Such a minimal model is unique up to isomorphism. It has the property that $H^*(\mathscr{M}) = H^*(M, \mathbf{Q})$, and, if M is simply connected, then the dual of

$\pi_*(M) \otimes \mathbf{Q}$ is $\mathcal{M}^{>0}/(\mathcal{M}^{>0})^2$. When M underlies a Kähler, or algebraic, structure, inputs from Hodge theory or from the Weil conjectures produce constraints on $\mathcal{M}$.

In [D31, 1975] Deligne et al. prove that if M is a compact Kähler manifold, then $\mathcal{M} \otimes \mathbf{R}$ is a minimal model of the cohomology algebra $H^*(M, \mathbf{R})$ (with zero differential).[13] In particular, if in addition M is simply connected, the whole real Postnikov tower of M can be reconstructed from $H^*(M, \mathbf{R})$. See [196] for a survey.

When M underlies a scheme X separated and of finite type over $\mathbf{C}$, Deligne uses the Weil conjectures to show that $\mathcal{M} \otimes \mathbf{Q}_\ell$ comes equipped with a rich structure (weight filtration with natural splittings) (see Sect. 4.2 "First Applications", $\mathbf{Q}_\ell$-homotopy type).

4.3 *Shimura Varieties*

In the 1960s Shimura, in connection with his work on complex multiplication, studied quotients of hermitian symmetric domains by discrete congruence subgroups. By a theorem of Baily–Borel, such quotients turn out to be algebraic. Sometimes they are moduli spaces for abelian varieties with additional structures. In a number of cases Shimura showed that they can be defined over a number field.

In his Bourbaki report [D13, 1971] Deligne defined a large class of such objects, which he later called *Shimura varieties* in his Corvallis survey ([D42, 1979], 2.3). These two texts have become standard references for the foundations of the theory, in which he introduced new angles and approaches that proved to be seminal.

Axiomatization of Shimura Varieties

Deligne emphasized (and popularized) the use of the real torus (sometimes called, nowadays, the *Deligne torus*)

$$\mathbf{S} := \prod_{\mathbf{C}/\mathbf{R}} \mathbf{G}_m, \tag{42}$$

Weil restriction of $\mathbf{G}_m$ from $\mathbf{C}$ to $\mathbf{R}$, to express a *real Hodge structure* V, defined by a finite dimensional $\mathbf{R}$-vector space $V_\mathbf{R}$ and a bi-grading V^{pq} of $V_\mathbf{C} = V_\mathbf{R} \otimes \mathbf{C}$ satisfying $V^{pq} = \overline{V}^{qp}$, as an action of $\mathbf{S}$ on $V_\mathbf{R}$, i.e., a homomorphism of real algebraic groups $h : \mathbf{S} \to \mathrm{GL}(V)$: V^{pq} is the summand where $h(z)$, for $z \in \mathbf{C}^* = \mathbf{S}(\mathbf{R})$, acts by $z^{-p}\bar{z}^{-q}$ (with the conventions of ([D42, 1979], 1.1.1.1)). The weight decomposition $V = \oplus_{n \in \mathbf{Z}} V^n$, where $V^n = \oplus_{p+q=n} V^{pq}$ can be read on the *weight homomorphism* $w_h : \mathbf{G}_m \to \mathrm{GL}(V)$, which is the restriction to $\mathbf{G}_m \subset \mathbf{S}$ (corresponding to $\mathbf{R}^* \subset \mathbf{C}^*$) of h^{-1}: V^n is the summand where $w_h(\lambda)$ is $x \to \lambda^n x$.

[13] In the terminology of (*loc. cit.*, p. 260), $\mathcal{M} \otimes \mathbf{R}$ is a *formal consequence* of $H^*(M, \mathbf{R})$.

In ([D13, 1971], 1.5) and ([D42, 1979], 2.1.1), Deligne considers the following object (later called *Shimura datum*):

$$(G, X), \tag{43}$$

where G is a reductive group over $\mathbf{Q}$, and X a $G(\mathbf{R})$-conjugacy class of morphisms $h : \mathbf{S} \to G_{\mathbf{R}}$ of algebraic groups over $\mathbf{R}$, satisfying the following conditions (i)–(iii):

(i) For $h \in X$, $\mathrm{Lie}(G_{\mathbf{R}})$, endowed with the Hodge structure defined by the composition $\mathbf{S} \to G_{\mathbf{R}} \to \mathrm{GL}(\mathrm{Lie}(G_{\mathbf{R}}))$ where the second map is the adjoint representation, is purely of type $\{(-1, 1), (0, 0), (1, -1)\}$.
(ii) The involution $\mathrm{int}(h(i))$ of the adjoint group $G^{\mathrm{ad}}_{\mathbf{R}}$ is a Cartan involution.[14]
(iii) The adjoint group of G has no factor G' defined over $\mathbf{Q}$ into which h projects trivially.

Axioms (i) and (ii) imply that X can be described in the following two ways:

(a) X is a finite disjoint union of hermitian symmetric domains (= hermitian symmetric spaces with negative curvature, i.e., with no compact nor euclidian factor),
(b) at least in the case where the restriction of $h \in X$ to $\mathbf{G}_m$ is defined over $\mathbf{Q}$, X is a parameter space for $G(\mathbf{Q})$-equivariant $\mathbf{Q}$-variations of polarizable Hodge structures associated with representations of G.

More precisely, for (b), Deligne proves that X has a unique complex structure such that, for each representation $\rho : G_{\mathbf{R}} \to \mathrm{GL}(V)$ (V a finite dimensional $\mathbf{R}$-vector space), the Hodge filtration F_h of $V_{\mathbf{C}}$ induced by $\rho h : \mathbf{S} \to \mathrm{GL}(V)$ varies holomorphically with h and satisfies Griffiths transversality $\nabla F^i_h \subset F^{i-1}_h \otimes \Omega^1_X$. Condition (ii) has a Hodge theoretic interpretation. By (i), for $h \in X$, the image of the restriction w_h of h^{-1} to $\mathbf{G}_m$ lies in the center of G, in particular $C = h(i)^2$ is central, so $\mathrm{int}(h(i))$ is an involution; by an elementary key lemma in ([D20, 1972], 2.8), condition (ii) is equivalent to requiring that for all representations $\rho : G \to GL(V)$ (or for one faithful representation ρ), V is C-polarizable, i.e., admits a G-invariant bilinear form ψ such that $\psi(x, Cy)$ is symmetric and positive definite – which is equivalent to the polarizability, in the usual sense, of the homogeneous components of V. Concerning (a), Deligne proves a converse: any hermitian symmetric domain is a connected component of an X as above. As for (iii), it is seen to be equivalent to saying that G^{ad} has no factor G' such that $G'(\mathbf{R})$ is compact, and by the strong approximation theorem, it ensures that $\overline{G}(\mathbf{Q})$ is dense in $\overline{G}(\mathbf{A}^f)$, where $\overline{G}$ is the universal cover of the derived group of G, and $\mathbf{A}^f = \widehat{\mathbf{Z}} \otimes \mathbf{Q}$ is the ring of finite adeles.

[14] I.e., an involution σ such that the real form $(G^{\mathrm{ad}}_{\mathbf{R}})^{\sigma}$ of $G^{\mathrm{ad}}_{\mathbf{R}}$ relative to the complex conjugation $g \mapsto \sigma(\overline{g})$ is *compact*, in the sense that $(G^{\mathrm{ad}}_{\mathbf{R}})^{\sigma}(\mathbf{R})$ is compact.

Deligne defines a *Shimura variety* (relative to a Shimura datum (G, X)) as a quotient

$$_K M_{\mathbf{C}}(G, X) := G(\mathbf{Q})\backslash(X \times (G(\mathbf{A}^f)/K)), \tag{44}$$

where K is a compact open subgroup of $G(\mathbf{A}^f)$. Such a variety is a finite disjoint union of quotients of connected components of X by arithmetic subgroups of $G(\mathbf{R})$. It is a complex analytic space, which (by Baily–Borel) has a natural structure of quasi-projective scheme over $\mathbf{C}$, unique for K small enough. For variable K, the varieties $_K M_{\mathbf{C}}(G, X)$ form a projective system, with finite transition morphisms, and Deligne considers its projective limit

$$M_{\mathbf{C}}(G, X) := \varprojlim_K {}_K M_{\mathbf{C}}(G, X), \tag{45}$$

a $\mathbf{C}$-scheme equipped with a natural right action of $G(\mathbf{A}^f)$, such that $_K M_{\mathbf{C}}(G, X) = M_{\mathbf{C}}(G, X)/K$. It is this action which makes Shimura varieties especially interesting in view of the Langlands program.

The simplest example of such a structure is the tower of modular curves $M_n^0(\mathbf{C}) = \mathscr{H}/\Gamma(n)$ (cf. Theorem 6). It corresponds to the Shimura datum (G, X) where $G = \mathrm{GL}_2$, $X = \mathbf{C} - \mathbf{R} = \mathscr{H} \cup -\mathscr{H}$ the conjugacy class of the canonical inclusion $h_0 : \mathbf{S} \hookrightarrow G$ $(x + iy \mapsto \begin{pmatrix} x & y \\ -y & x \end{pmatrix})$, i.e., the homogeneous space $\mathrm{GL}_2(\mathbf{R})/\mathbf{R}^*\mathrm{SO}_2(\mathbf{R})$, orbit of i in $\mathbf{C}$ under the natural action of $\mathrm{GL}_2(\mathbf{R})$. We have $M_n^0(\mathbf{C}) = {}_{K_n} M_{\mathbf{C}}(G, X)$ for $K_n = \mathrm{Ker}(G(\widehat{\mathbf{Z}}) \to G(\mathbf{Z}/n))$ ([D24, 1973], 5.3). The projective limit $M_{\mathbf{C}}(G, X) = \varprojlim_n M_n^0(\mathbf{C})$, denoted M_∞ in ([D6, 1969], 3.7), has an action of $G(\mathbf{A}^f)$, a fact which, according to Deligne, was first noticed by Shafarevich.

A basic generalization is the Shimura datum $(G = \mathrm{GSp}(2n), X)$, where X is the unique conjugacy class of homomorphisms $h : \mathbf{S} \to \mathrm{GSp}(2n)_{\mathbf{R}}$ satisfying condition (i) (or, simply, such that h_w sends λ to the homothecy or ratio λ^{-1}); X is the union $S^{\pm}$ of two Siegel upper half spaces. The corresponding Shimura varieties are moduli spaces for principally polarized abelian varieties with level structure.

Canonical Models

In this set-up Deligne tackles the question of the existence of models of Shimura varieties over number fields. He essentially follows Shimura's method, but in order to do so, he develops foundational preliminaries. They comprise the following notions: *reflex field*, *special (or CM) points*, *canonical models*.

- The *reflex field* $E(G, X) \subset \mathbf{C}$ of a Shimura datum (G, X) is the field of definition of the composite morphism $\mu_h : \mathbf{G}_{m\mathbf{C}} \hookrightarrow \mathbf{S}_{\mathbf{C}}(= (\mathbf{G}_m^2)_{\mathbf{C}}) \xrightarrow{h} G_{\mathbf{C}}$, for $h \in X$, where the first map is $(z \mapsto (z^{-1}, 1))$. When G^{ad} is simple, $E(G, X)$ is either a totally real field, or a totally imaginary quadratic extension of a totally real field ([D13, 1971], 3.8).

- A point $h \in X$ is called *special* (or *of CM type*) if there is a torus $T \subset G$ (over $\mathbf{Q}$) such that $h : \mathbf{S} \to G_{\mathbf{R}}$ factors through $T_{\mathbf{R}}$. A point x of ${}_K M_{\mathbf{C}}(G, X)$ or $M_{\mathbf{C}}(G, X)$, class of $(h, g) \in X \times G(\mathbf{A}^f)$, is called *special* if h is special; the $G(\mathbf{Q})$-conjugacy class of h (which depends only on x) is called the *type* of x, and the corresponding reflex field is denoted $E(x)$. In the example of modular curves, a point $x \in X = \mathbf{C} - \mathbf{R}$ is special if and only if x generates an imaginary quadratic extension E of $\mathbf{Q}$ (and $E = E(x)$). Using the reciprocity isomorphisms for the tori T, Deligne defines an action of $\mathrm{Gal}(\overline{\mathbf{Q}}/E(\tau))$ on the set $M(\tau)$ of special points of a given type τ, commuting with the right action of $G(\mathbf{A}^f)$.
- A *canonical model of* $M_{\mathbf{C}}(G, X)$ is a scheme $M(G, X)$ over $E(G, X)$, equipped with an action of $G(\mathbf{A}^f)$, and an equivariant isomorphism $M(G, X) \otimes_{E(G,X)} \mathbf{C} \xrightarrow{\sim} M_{\mathbf{C}}(G, X)$, such that the special points are algebraic and for each type τ the action of $\mathrm{Gal}(\overline{\mathbf{Q}}/E(\tau))$ on $M(\tau)$ is induced by that of $\mathrm{Gal}(\overline{\mathbf{Q}}/E(G, X))$ on $M_{\mathbf{C}}(G, X)$. Deligne proves that a canonical model, if it exists, is unique up to a unique isomorphism ([D13, 1971], 5.5).

In the case of Shimura varieties corresponding to moduli of abelian varieties with additional structures (later called *of PEL type*, for "polarization", "endomorphism", "level"), and especially in the case of Siegel modular spaces, the modular interpretation yields such canonical models. In the general case, the existence of canonical models is a difficult problem. To construct canonical models, Deligne follows a method due to Shimura, which relies on the following criterion ([D42, 1979], 3.1):

Proposition 1 *If there exists an embedding* $G \hookrightarrow \mathrm{CSp}(2n)$ *sending* X *to the Siegel double-space* $S^{\pm}$, *then* $M_{\mathbf{C}}(G, X)$ *admits a canonical model.*

In [D42, 1979] Deligne mentions, at the end of the introduction, that his theorem on absolute Hodge cycles (see Sect. 4.4) leads to a simpler proof than his earlier one in [D13, 1971]. It doesn't seem, however, that it has been published.

As for the construction, very roughly speaking, the idea is to define $M(G, X)$ as the closure of the set of special points in the canonical model of a Siegel modular variety. Deligne shows that the existence of a canonical model for a datum (G, X) depends only on the derived and adjoint groups of G and a connected component of X, and that such a model exists for G $\mathbf{Q}$-simple adjoint of type A, B, or C, or for certain types D.

4.4 Absolute Hodge Cycles

Let X be a smooth projective $\mathbf{C}$-scheme. Recall that $H^{2n}(X^{\mathrm{an}}, \mathbf{Q}(n))$ is a pure $\mathbf{Q}$-Hodge structure of weight 0 ($\mathbf{Z}(n)$ is the pure Hodge structure of weight $-2n$, purely of type $(-n, -n)$, with integral lattice $(2\pi i)^n\mathbf{Z}$). The rational Betti cohomology class is a homomorphism

$$cl : CH^n(X) \otimes \mathbf{Q} \to H^{2n}(X^{\mathrm{an}}, \mathbf{Q}(n))^{0,0} \tag{46}$$

where $CH^n(X)$ is the Chow group of codimension n cocycles, and the group on the right hand side, called the group of *rational Hodge cycles* (of degree $2n$), is the intersection, in $H^{2n}(X^{an}, \mathbf{C})$, of $H^{2n}(X^{\mathrm{an}}, \mathbf{Q}(n))$ with the part of bidegree $(0, 0)$ of $H^{2n}(X^{\mathrm{an}}, \mathbf{Q}(n)) \otimes \mathbf{C}$, i.e., the part of type (n, n) of $H^{2n}(X^{\mathrm{an}}, \mathbf{Q}) \otimes \mathbf{C}$. The *Hodge conjecture* asserts that (46) is surjective.

In the late 1970s, Deligne discovered that, in the case of an abelian variety A, rational Hodge cycles enjoy many of the (motivic) properties they would have if we knew the Hodge conjecture. In particular, strikingly, the property for a cohomology class in $H^{2n}(A^{\mathrm{an}}, \mathbf{C})$ of being a rational Hodge cycle is invariant under automorphisms of $\mathbf{C}$. In order to formulate his theorem, he introduced new objects, which he called *absolute Hodge cycles*, possessing all the good properties of cohomology classes of algebraic cycles. This notion turned out to change the perspective one had so far on Grothendieck's theory of motives, by suggesting unconditional approximations of it (as suggested by Deligne in ([D43, 1979], 0.9), and developed by several authors, see especially [13]).

Let k be an algebraically closed field of characteristic zero, which, for simplicity, will be assumed to admit an embedding into $\mathbf{C}$. Let X be a proper and smooth scheme over k. Let $\mathbf{A}^f = (\prod_\ell \mathbf{Z}_\ell) \otimes \mathbf{Q}$ be the ring of finite adeles of $\mathbf{Q}$. One can make a single package of de Rham cohomology and all ℓ-adic cohomology groups of X, in the following way. For m, r in $\mathbf{Z}$, let

$$H^m_{\mathbf{A}}(X)(r) := H^m_{dR}(X/k)(r) \times H^m(X)(r), \tag{47}$$

where $H^m(X)(r) := (\prod_\ell H^m(X, \mathbf{Z}_\ell(r))) \otimes \mathbf{Q}$ (restricted product of the $H^m(X, \mathbf{Q}_\ell(r))$ relative to the $\mathbf{Z}_\ell(r)$), and $H^m_{dR}(X/k)(r)$ is $H^m_{dR}(X/k) := H^m(X, \Omega^\bullet_{X/k})$, with Hodge filtration $F^i(H^m_{dR}(X/k)(r)) := F^{i+r} H^m_{dR}(X/k)$. This is a (finitely generated and free) $k \times \mathbf{A}^f$-module. Let $\sigma : k \hookrightarrow \mathbf{C}$ be an embedding and $\sigma X := X \otimes_{(k,\sigma)} \mathbf{C}$. We have a comparison isomorphism

$$\sigma^* : (H^m_{dR}(X/k)(r) \otimes_{(k,\sigma)} \mathbf{C}) \times H^m(X)(r) \xrightarrow{\sim} H^m_B(\sigma X)(r) \otimes (\mathbf{C} \times \mathbf{A}^f), \tag{48}$$

where, for a proper smooth scheme $Y/\mathbf{C}$, $H^m_B(Y)(r) := H^m(Y^{\mathrm{an}}, \mathbf{Q}(r))$. Deligne makes the following definitions.

Let $n \in \mathbf{Z}$. An element $t = (t_{dR}, t_{et}) \in H^{2n}_{\mathbf{A}}(X)(n)$ is called a *Hodge cycle relative to* σ if its first component $t_{dR} \in H^{2n}_{dR}(X)(n)$ lies in $F^0 H^{2n}_{dR}(X)(n) = F^n H^{2n}_{dR}(X/k)$, and the image of t by σ^* lies in $H^{2n}_B(\sigma X)(n)$ (diagonally embedded in $H^{2n}_B(\sigma X)(n) \otimes (\mathbf{C} \times \mathbf{A}^f)$); for $k = \mathbf{C}$ and $\sigma = Id$, it means that t is the image of a rational Hodge cycle (cf. (46)).

One says that t is an *absolute Hodge cycle* if t is a Hodge cycle relative to *any* embedding $\sigma : k \hookrightarrow \mathbf{C}$. There is an obvious generalization of these definitions to elements of Tate twisted tensor products of tensor powers of $H_{\mathbf{A}}^{m_\alpha}(X_\alpha)$ and $H_{\mathbf{A}}^{n_\alpha}(X_\alpha)^\vee$ for (X_α) a family of smooth projective schemes over k, and in particular for *cohomological correspondences*.

A codimension n cycle Z in X has a class $cl(Z)$ in $H_{\mathbf{A}}^{2n}(X)(n)$ which is an absolute Hodge cycle. But there are other easy examples, such as the Künneth components of the cohomology class of the diagonal in $X \times X$ (though it is one the standard conjectures that they are algebraic, see Sect. 5.2). A beautiful example, which may have been the source of Deligne's inspiration, is Deligne's cohomological correspondence between a K3 surface and its associated Kuga–Satake variety (see Sect. 5.3).

The main result is the following theorem ([D47, 1981], 3, T. 1), ([D52, 1982], I 2.11)):

Theorem 17 *Let k be an algebraically closed field embeddable in $\mathbf{C}$, and let X/k be an abelian variety. If $t \in H_{\mathbf{A}}^{2n}(X)(n)$ is a Hodge cycle relative to one embedding $\sigma : k \hookrightarrow \mathbf{C}$, then t is an absolute Hodge cycle.*

The idea of the proof is to reduce, by a deformation argument, to a case where Hodge cycles can be proved to be absolutely Hodge, namely when X is of CM-type. In order to do so, Deligne applies two key results, which he calls *principles A and B*.

Principle A is a statement of *Tannakian* nature. It says that if we are given a finite family (t_i) of (possibly Tate twisted) $\mathbf{Q}$-valued Betti cohomology classes (over $\mathbf{C}$) which are absolute Hodge cycles, then any Betti cohomology class fixed by the *Mumford–Tate* (Tannakian) group defined by the (t_i)'s is again an absolute Hodge cycle (see ([D52, 1982], 3.8) for a precise statement).

Principle B is a *deformation* statement. It says that, given $f : X \to S$ proper and smooth, with S connected and smooth over $\mathbf{C}$, a horizontal global section t of $R^{2p} f_* \Omega^\bullet_{X/S} \times (R^{2p} f_* \widehat{\mathbf{Z}}(p) \otimes \mathbf{Q})$ which is horizontal for the Gauss–Manin connection and whose de Rham cohomology component lies in F^0, then, if at one closed point s of S, t_s is an absolute Hodge cycle, then, for all s', $t_{s'}$ is an absolute Hodge cycle ([D52, 1982], 2.12).

The proof of these principles is not difficult. The main bulk of the proof of Theorem 17 consists in: (a) proving that it holds for abelian varieties of CM type (b) constructing a deformation space S having the following property: the given (A, t_0) is the fiber at one point s of a pair of an abelian scheme X/S and a horizontal class t as in principle B, such that there exists a point s_1 at which X_{s_1} is of CM type. The space S is a certain Shimura variety. A slightly different approach, with simplifications due to André and Voisin, is given in Charles-Schnell's notes [53].

For $k = \mathbf{C}$, Deligne makes the following conjecture (weaker than the Hodge conjecture)[15] ([D43, 1979], 0.10):

Conjecture 3 Every Hodge cycle is absolute Hodge.

This conjecture is still wide open today. See [53] for a recent discussion.

[15]He calls it "hope".

4.5 Deligne Cohomology

In the early 1970s, in an unpublished work, Deligne introduced certain variants of the de Rham complex of a complex manifold, which turned out to play an important role in various questions pertaining to mixed Hodge theory and the arithmetic of L-functions.

Let X be a complex manifold, A a subring of $\mathbf{C}$ as in Sect. 4.2 "Homological Algebra Infrastructure", and $n \in \mathbf{Z}$. Deligne defined the complex

$$A_X(n)_{\mathscr{D}} = (0 \to A_X(n) \to \mathscr{O}_X \to \Omega_X^1 \to \cdots \to \Omega_X^{n-1} \to 0) \tag{49}$$

(where $A(n) := (2\pi i\mathbf{Z})^{\otimes n} \otimes A$). There is a natural quasi-isomorphism

$$\mathrm{Cone}(A_X(n) \oplus \Omega_X^{\geq n} \overset{(1,-1)}{\to} \Omega_X^{\bullet})[-1] \to A_X(n)_{\mathscr{D}}, \tag{50}$$

in other words, $A_X(n)_{\mathscr{D}}$ can be thought of as a derived category kernel of the above map $(1, -1)$.

Let now X be a projective and smooth scheme over $\mathbf{C}$. Let us write $A_X(n)_{\mathscr{D}}$ for $A_{X^{\mathrm{an}}}(n)_{\mathscr{D}}$. Deligne considered the cohomology groups

$$H^i(X, A_X(n)_{\mathscr{D}}) \tag{51}$$

(calculated for the classical topology on X^{an}). They are now called *Deligne* (or *Deligne–Beilinson*) (or sometimes, *absolute Hodge*) cohomology groups. They contain deep information, as examplified below.

Link with the Hodge Conjecture and Intermediate Jacobians

The long exact sequences deduced from (50) yield in particular short exact sequences

$$0 \to J_n(X)^0 \to J_n(X) \to H^{2n}(X, \mathbf{Z}(n)) \cap H^{n,n} \to 0, \tag{52}$$

where $H^{2n}(X, \mathbf{Z}(n)) \cap H^{n,n}$ denotes, with an abuse of notation, the group of integral Hodge classes of degree $2n$ which modulo torsion are of type (n, n) (cf. Sect. 4.4), $J_n(X) := H^{2n}(X, \mathbf{Z}_X(n)_{\mathscr{D}})$, and

$$J_n(X)^0 = H^{2n-1}(X, \mathbf{C})/(\alpha(H^{2n-1}(X, \mathbf{Z}(n))) + F^n H^{2n-1}(X, \mathbf{C})) \tag{53}$$

is the *Griffiths intermediate Jacobian* (α denoting the map induced by the inclusion $\mathbf{Z}(n) \hookrightarrow \mathbf{C}$). For $n = 1$, $J_1(X) = \mathrm{Pic}(X)$. Let $Z^n(X)$ denote the group of codimension n cycles on X. Deligne constructed (see, e.g., ([256], 12.3.3)) a morphism (the *Deligne cycle class*)

$$c_{\mathscr{D}} : Z^n(X) \to J_n(X), \tag{54}$$

lifting the classical morphism $c : Z^n(X) \to H^{2n}(X, \mathbf{Z}(n)) \cap H^{n,n}$. It factors through the Chow group $\mathrm{CH}^n(X)$, and, on the subgroup $\mathrm{CH}^n(X)_{\mathrm{hom}}$ of cycles homologically equivalent to zero, induces the *Griffiths Abel–Jacobi map*

$$\mathrm{CH}^n(X)_{\mathrm{hom}} \to J_n(X)^0.$$

The morphism $c_{\mathscr{D}}$ is compatible with the intersection product on Chow groups, and on a product on Deligne cohomology deduced from a certain associative pairing $A_X(m)_{\mathscr{D}} \otimes A_X(n)_{\mathscr{D}} \to A_X(m+n)_{\mathscr{D}}$. While the image of $c \otimes \mathbf{Q}$ is expected to be the whole group of rational Hodge classes (Hodge conjecture), there is as yet no conjecture predicting which subgroup of $J_n(X)$ is the image of $c_{\mathscr{D}}$ (see [D104, 2006] for a brief discussion).

Link with the Tame Symbol

Let X be a complex analytic manifold. The product $\mathbf{Z}_X(1)_{\mathscr{D}} \otimes \mathbf{Z}_X(1)_{\mathscr{D}} \to \mathbf{Z}_X(2)_{\mathscr{D}}$ can be rewritten as a morphism in $D(X, \mathbf{Z})$

$$\mathscr{O}_X^* \otimes_{\mathbf{Z}}^L \mathscr{O}_X^* \to [\mathscr{O}_X^* \overset{\mathrm{dlog}}{\to} \Omega_X^1](1), \tag{55}$$

where the complex in the right hand side is placed in degrees $(-1, 0)$. This morphism, and its generalization with $\mathscr{O}^*$ replaced by a commutative complex analytic group G, plays a central role in [D73,1991]. For $G = \mathscr{O}^*$ and X of dimension 1 (a Riemann surface), the right hand side is quasi-isomorphic to $\mathbf{C}^*[1]$. If $f : \mathbf{Z} \to \mathscr{O}^*$, $g : \mathbf{Z} \to \mathscr{O}^*$ are invertible holomorphic functions on an open subset U of X, then the composition of $f \otimes^L g \to \mathscr{O}^* \otimes^L \mathscr{O}^*$ with (55) gives an element of $H^1(U, \mathbf{C}^*)$, i.e., the class of a $\mathbf{C}^*$-torsor (f, g) on U. Deligne shows that, for $U = X - \{x\}$, x a point on X, if f and g are meromorphic at x, the image in $\mathbf{C}^*$ by the residue map $H^1(X - \{x\}, \mathbf{C}^*) \to \mathbf{C}^*$ is the tame symbol $\langle f, g \rangle_x = (-1)^{v(f)v(g)}(g^{v(f)}/f^{v(g)})$ with v the valuation at x (cf. (11)). The fact that (55) (and its generalization mentioned above) come from an actual pairing of complexes enables to define torsors, not just isomorphism classes of them. Classical formulas on symbols then translate into new phenomena, that Deligne studies in detail.

Link with Mixed Hodge Structures and Regulators

Beilinson showed [21] that the Hodge complex $(R\Gamma(X, \mathbf{Z}), (R\Gamma(X, \mathbf{Z}) \otimes \mathbf{C}, F))$ (of weight zero) constructed by Deligne (cf. Theorem 11) underlies a finer object, namely an object $R\underline{\Gamma}(X, \mathbf{Z})$ of a certain derived category $D^b_{\mathscr{H}p}$ of polarizable mixed Hodge complexes, which he proved to be equivalent (via a realization functor, see Sect. 5.8 "t-Structures") to $D^b(\text{MHS})$, where MHS denotes the (abelian) category of polarizable mixed Hodge structures on Spec($\mathbf{C}$). The same holds with $\mathbf{Z}$ replaced by $A(n)$ (and MHS by the category MHS_A of polarizable mixed A-Hodge structures), and the miracle (*loc. cit.*) is that one can re-write Deligne cohomology groups as

$$H^i(X, A_X(n)_{\mathscr{D}}) = \text{Hom}^i_{D^b(\text{MHS})}(A, R\underline{\Gamma}(X, A(n))). \tag{56}$$

(then (52) comes from a Leray spectral sequence, as MHS has cohomological dimension 1). More generally, Beilinson defined $R\underline{\Gamma}(X, A(n))$ in $D^b(\text{MHS}_A)$ assuming only $X/\mathbf{C}$ separated and of finite type, using simplicial techniques à la Hodge III (Sect. 4.2 "Mixed Hodge Theory"), thus obtaining a definition of Deligne cohomology by (56), as well as another construction of the Deligne cohomology class (54). An alternate approach today is to consider the constant sheaf $A(n)_X$ on X as an object of M. Saito's category $D^b(\text{MHM}_A(X))$ (the bounded derived category of *mixed A-Hodge modules* on X), apply Ra_* ($a : X \to \text{Spec}(\mathbf{C})$) and use the equivalence between $D^b(\text{MHM}_A(\text{Spec}(\mathbf{C}))$ and $D^b(\text{MHS}_A)$.

Let now X be a projective and smooth scheme over $\mathbf{Q}$, and $X_{\mathbf{C}} = X \otimes \mathbf{C}$. The description (56) of Deligne cohomology as an extension group fits with the conjectural description of the so-called *motivic cohomology* $H^i_{\mathscr{M}}(X, \mathbf{Z}(n))$ as an extension group $\text{Hom}^i_{D^b(\mathscr{M}\mathscr{M})}(\mathbf{Z}, Ra_*\mathbf{Z}(n)_X)$ in a derived category of mixed motives over $\mathbf{Q}$, as suggested by Deligne in his letter to Soulé ([76], A. Motifs). In any case, Deligne's real cohomology groups of $X_{\mathbf{C}}$ appear as targets for Beilinson's regulator maps (see [241], 3.3) for precise statements.

4.6 *Liftings mod p^2 and Hodge Degeneration*

In [D65, 1987] an elementary algebraic proof of the Hodge degeneration and Kodaira–Akizuki–Nakano vanishing theorems is given. By a usual spreading out argument, these theorems are deduced from a decomposition theorem in positive characteristic. Before stating it, recall that, if k is a field of characteristic $p > 0$, X/k a smooth scheme, $F : X \to X'$ the relative Frobenius, where X' is deduced from X by base change by the Frobenius endomorphism of k, we have the *Cartier isomorphism*

$$C^{-1} : \oplus\Omega^i_{X'} \xrightarrow{\sim} \oplus\mathscr{H}^i(F_*\Omega^\bullet_X), \tag{57}$$

which is $\mathcal{O}_{X'}$-linear, graded, multiplicative, coincides with the natural inclusion in degree zero, and sends a 1-form dx to the class of $x^{p-1}dx$ in $\mathcal{H}^1$. The main result of *loc. cit.* is the following:

Theorem 18 *Assume k perfect. Let $W(k)$ be the ring of Witt vectors on k, $W_n(k) = W(k)/p^n W(k)$, and let X/k be a smooth scheme. With the above notation, with any smooth lifting of X to $W_2(k)$ there is associated an isomorphism*

$$\oplus_{i<p} \Omega^i_{X'} \xrightarrow{\sim} \tau_{<p} F_* \Omega^\bullet_X \tag{58}$$

in $D^b(X', \mathcal{O}_{X'})$, inducing C^{-1} in each degree i (where $\tau_{<p} = \tau_{\leq p-1}$ is the canonical truncation).

In particular, if X is of dimension $< p$, (58) is a decomposition of the de Rham complex $F_*\Omega^\bullet_X$ (in the sense of Sect. 1.2 "Degeneration and Decomposition in the Derived Category"). It is shown in *loc. cit.* that such a decomposition still holds if X is of dimension p, provided that X is furthermore assumed to be proper. Whether this extends to dimension $> p$ is still an open question in general.

The proof relies on the following simple observation, due to Mazur: if $F : X \to X'$ lifts to $\widetilde{F} : \widetilde{X} \to \widetilde{X}'$ (such a lifting exists only locally), then C^{-1} lifts to a morphism $\varphi_{\widetilde{F}} : \Omega^i_{X'} \to Z^i F_* \Omega^\bullet_X$ (where $Z^i = \mathrm{Ker}(d)$), which is multiplicative, and sends the image in $\Omega^1_{X'}$ of a 1-form ω on X to $\frac{1}{p}\widetilde{F}^*\widetilde{\omega}$, where $\widetilde{F}^* : \Omega^1_{\widetilde{X}'} \to p\widetilde{F}_*\Omega^1_{\widetilde{X}}$ is the morphism induced by $\widetilde{F}$, and $\widetilde{\omega}$ lifts ω.

For X proper and smooth of dimension $\leq p$, and liftable to $W_2(k)$, the degeneration at E_1 of the Hodge to de Rham spectral sequence $E_1^{ij} = H^j(X, \Omega^i_{X/k}) \Rightarrow H^{i+j}(X, \Omega^\bullet_{X/k})$ follows from the decomposability of $F_*\Omega^\bullet_{X/k}$ by a simple dimension count. A more subtle argument (due to Raynaud) is needed for the Kodaira-Akizuki-Nakano vanishing theorem under the same assumptions.

Several variants and generalizations (general bases, log poles) are discussed in *loc. cit.* In the recent years the method has been often imitated. See for example [207] for a "mod p^2" proof of an analytic theorem of Barranikov and Kontsevich (and [91] for generalizations).

4.7 The Hodge Locus

Let $f : X \to S$ be a projective and smooth morphism, with S separated and of finite type over $\mathbf{C}$. Let $s \in S(\mathbf{C})$, $p \in \mathbf{Z}$, and $u \in H^{2p}(X_s, \mathbf{Z}) \cap H^{p,p}$ a Hodge class. The locus T_u where, in a simply connected neighbourhood U of s in S^{an}, u remains of type (p, p) (as a constant section of the local system $R^{2p} f_*\mathbf{Z}$ trivialized on U) is a closed analytic subspace of U. It had been observed long ago (probably in the 1960s) that the Hodge conjecture implies that the germ of T_u at s is algebraic. In [D80, 1995], Deligne, Cattani and Kaplan prove this consequence unconditionally. In fact, they prove a stronger result, for polarized variations of Hodge structures:

Theorem 19 *Assume $S/\mathbf{C}$ smooth. Let $(\mathscr{V}_{\mathbf{Z}}, \mathscr{V} = \mathscr{V}_{\mathbf{Z}} \otimes \mathscr{O}_{S^{\mathrm{an}}}, F^\bullet, \nabla, Q)$ be a polarized variation of Hodge structures on S of weight* 0*. Fix an integer $K \geq 0$. Let*

$$S^{(K)} := \{(s \in S, u \in \mathscr{V}_s) | u \in (\mathscr{V}_{\mathbf{Z}})_s \cap F^0\mathscr{V}_s, Q(u, u) \leq K\}. \tag{59}$$

Then $S^{(K)}$ is finite over S (hence algebraic).

This implies algebraicity of the components of the so-called *Hodge locus*:

Theorem 20 *Assume $S/\mathbf{C}$ smooth, quasi-projective. The components of the analytic subset (the Hodge locus)*

$$R^{2p} f_*\mathbf{Z} \cap F^p \mathscr{H}^{2p}_{\mathrm{dR}}(X^{\mathrm{an}}/S^{\mathrm{an}}) \tag{60}$$

of the (analytic) fibre bundle $\mathscr{H}^{2p}_{\mathrm{dR}}(X^{\mathrm{an}}/S^{\mathrm{an}}) = R^{2p} f_*\mathbf{Z} \otimes \mathscr{O}^{\mathrm{an}}_S$ are finite covers of algebraic subsets of S.

The core of the proof of Theorem 19 is a delicate extension theorem for integral Hodge classes in a polarized variation of Hodge structures over a product of punctured discs, making heavy use of Schmid's nilpotent orbit and SL(2) orbit theorem.

These results are so far the strongest evidence towards the Hodge conjecture. See [50, 257] for comments and complements.

5 The Weil Conjectures

Deligne's contribution to the proof of the Weil conjectures is his most famous achievement. It broke down what seemed to be an impassable barrier. With the refinements, generalizations, and applications he derived, it changed the face of arithmetic geometry.

In this section, all schemes are assumed to be noetherian and separated unless otherwise stated.

5.1 The Zeta Function of a Variety Over a Finite Field

Basic Definitions

We fix a prime number p and denote by $\mathbf{F}$ an algebraic closure of $\mathbf{F}_p$. If q is a power of p, we denote by $\mathbf{F}_q$ the subfield of $\mathbf{F}$ with q elements. Following Deligne's conventions in [Weil I] and [Weil II],[16] when q is fixed, we denote by a subscript 0 objects over $\mathbf{F}_q$, and remove it to denote the object deduced by base change to $\mathbf{F}$.

Let q be a power of p, and let X_0 be a scheme of finite type over $\mathbf{F}_q$. Let me recall the definition of the zeta function of X_0, see, e.g., ([Weil I], 1). This is the formal series

$$Z(X_0/\mathbf{F}_q, t) := \prod_{x \in |X_0|} (1 - t^{\deg(x)})^{-1}, \tag{61}$$

where $|T|$ denotes the set of closed points of a scheme T, and, for $x \in |X_0|$, $\deg(x) := [k(x) : \mathbf{F}_q]$. We omit $/\mathbf{F}_q$ when no confusion can arise. The Galois group $\mathrm{Gal}(\mathbf{F}/\mathbf{F}_q)$ is topologically generated by the Frobenius substitution σ, $\sigma(a) = a^q$. It acts on the set $X_0(\mathbf{F})$ of points of X_0 with value in $\mathbf{F}$ (which is also the set of closed (or rational) points of $X = X_0 \otimes_{\mathbf{F}_q} \mathbf{F}$). This action of σ on $X_0(\mathbf{F})$ is the same as that induced by the $\mathbf{F}_q$-endomorphism F of X_0 (identity on the underlying space, and raising to the q-th power on $\mathcal{O}_X$). Closed points of degree d of X_0 correspond bijectively to orbits of F on $X_0(\mathbf{F}) = |X|$ of cardinality d. More generally, $X_0(\mathbf{F}_{q^n})$ is the set $|X|^{F^n}$ of fixed points of $|X|$ under F^n:

$$X_0(\mathbf{F}_{q^n}) = |X|^{F^n}. \tag{62}$$

In particular, $\#X(\mathbf{F}_{q^n}) = \sum_{x \in |X_0|, \deg(x)|n} \deg(x)$, which implies that $Z(X_0, t)$ can be re-written

$$Z(X_0, t) = \exp(\sum_{n \geq 1} \#X_0(\mathbf{F}_{q^n}) \frac{t^n}{n}). \tag{63}$$

Statement of the Weil Conjectures

In [260] Weil made the following celebrated conjectures. Suppose that X_0 is projective, smooth, of dimension d. Then:

(a) (*rationality*) $Z(X_0, t)$ belongs to $\mathbf{Q}(t)$;
(b) (*functional equation*) $Z(X_0, t)$ satisfies an equation of the form

$$Z(X_0, t) = \pm q^{-d\frac{\chi(X)}{2}} t^{-\chi(X)} Z(X_0, \frac{1}{q^d t}), \tag{64}$$

where $\chi(X)$ is the Euler–Poincaré characteristic of X, defined as the self-intersection number of X in $X \times_{\mathbf{F}} X$;

[16] [Weil I] = [D27, 1974], [Weil II] = [D46, 1980].

(c) (*product decomposition and weights*)

$$Z(X_0, t) = \frac{P_1(t) \cdots P_{2d-1}(t)}{P_0(t) \cdots P_{2d}(t)}, \tag{65}$$

with $P_i(t)$ a polynomial with coefficients in $\mathbf{Z}$, of the form

$$P_i(t) = \prod_{1 \leq j \leq b_i} (1 - \alpha_{ij} t), \tag{66}$$

where the α_{ij}'s are algebraic integers, all of whose conjugates are of absolute value $q^{i/2}$, i.e., are *q-Weil integers of weight i*.[17] Moreover, for X geometrically connected, $P_0(t) = 1 - t$, $P_{2d}(t) = 1 - q^d t$.

Statement (c) was traditionally called the *Riemann hypothesis for varieties over finite fields*. Weil proved (a), (b), (c) for curves and abelian varieties [258, 259].

The polynomials P_i, and in particular their degrees b_i, are uniquely determined by X_0. In particular, (65) implies $\chi(X) = \sum (-1)^i b_i$. In [260] Weil observed this analogy of $b_i = b_i(X)$ with a Betti number and conjectured that, for a projective smooth variety $\mathcal{X}$ over a number field K, given an embedding of K into $\mathbf{C}$, for each i, the topological Betti number $b_i(\mathcal{X} \otimes_K \mathbf{C}) := \dim H^i(\mathcal{X} \otimes_K \mathbf{C}, \mathbf{Q})$ should be equal to the numbers b_i corresponding to the reductions of $\mathcal{X}$ at places of good reduction. Elaborating on this, in [261] Weil conjectured the existence of a cohomology theory with coefficients in a field Q of characteristic zero, functorially assigning to every projective, smooth variety Y over an algebraically closed field k a finite dimensional Q-algebra $H^*(Y) = \oplus H^i(Y)$, satisfying a Künneth isomorphism and Poincaré duality, and such that for $k = \mathbf{F}$ and $Y = X = X_0 \otimes \mathbf{F}$ as above, the cardinality of $X_0(\mathbf{F}_{q^n}) = |X|^{F^n}$ could be calculated by a Lefschetz fixed point trace formula

$$\# X_0(\mathbf{F}_{q^n}) = \sum (-1)^i \mathrm{Tr}(F^{n*}, H^i(X)). \tag{67}$$

Furthermore, the alternating sum of the dimensions of $H^i(Y)$ should equal $\chi(Y)$, and, for $k = \mathbf{C}$, these dimensions should coincide with the topological Betti numbers of $Y(\mathbf{C})$. By (63) the fixed point formula (67) would yield a product decomposition for $Z(X_0, t)$ of the form

$$Z(X_0, t) = \prod_{0 \leq i \leq 2d} \det(1 - F^* t, H^i(X))^{(-1)^{i+1}}, \tag{68}$$

[17]For $w \in \mathbf{Z}$, a *q-Weil number* (resp. *q-Weil integer*) *of weight w* is an algebraic number (resp. algebraic integer) all of whose conjugates are of absolute value $q^{w/2}$.

which would imply (a), and (b) by Poincaré duality, $\alpha \mapsto q^d/\alpha$ giving a bijection from the set of reciprocal roots of $\det(1-F^*t, H^i(X))$, i.e., the eigenvalues of F^* on $H^i(X)$, to that of $\det(1-F^*t, H^{2d-i}(X))$. However, (68) would not a priori solve (c), as there would remain to prove that the polynomial $\det(1-F^*t, H^i(X))^{(-1)^{i+1}}$ has coefficients in $\mathbf{Q}$ and its inverse roots would be q-Weil integers of weight i. Note that Serre observed that such a cohomology theory could not have coefficients in $\mathbf{Q}$. More precisely, he showed that, given an algebraically closed field k of positive characteristic, there could not exist a contravariant functor H^1 on the category of smooth projective schemes over k with value in finite dimensional vector spaces over a subfield Q of $\mathbf{R}$, compatible with products, and such that $H^1(E)$ be of dimension 2 for E an elliptic curve over K (see ([4], IX, Introduction)).

Grothendieck's Trace Formula

As the search for a "Weil cohomology" was actively pursued, Dwork's proof [86] of the rationality of $Z(X,t)$ for *any* X of finite type over k, using methods of p-adic analysis and no cohomology theory, came as a big surprise. However, it gave no insight into (c), nor even into (b). A few years later, the formalism of étale cohomology constructed by Artin, Grothendieck, and Verdier provided Weil cohomologies, namely the ℓ-adic cohomologies, one for each prime number ℓ invertible on the schemes under consideration. Through the use of constructible coefficients and derived categories this formalism displayed a power and flexibility that had not been dreamed of by Weil. Concerning *torsion* coefficients, I recalled the main points in Sect. 1.4 "Global Duality". The case of ℓ-adic coefficients raised new questions, which were fully solved only much later (see the end of Sect. 1.4 "Finiteness"). But the results in SGA 4 [4] and SGA 5 [5] sufficed to establish (a), (b), and a decomposition of the form (67): finiteness of $\mathbf{Q}_\ell$-cohohomology with compact support of schemes separated and of finite type over an algebraically closed field was known, and for such an $X_0/\mathbf{F}_q$ (and $\ell \neq p$), with the notation of (62), Grothendieck had proved the trace formula

$$\#X_0(\mathbf{F}_{q^n}) = \sum(-1)^i \mathrm{Tr}(F^{n*}, H^i_c(X, \mathbf{Q}_\ell)), \tag{69}$$

giving a decomposition

$$Z(X_0, t) = \prod_{0 \le i \le 2d} \det(1 - F^*t, H^i_c(X, \mathbf{Q}_\ell))^{(-1)^{i+1}}, \tag{70}$$

for $\dim X_0 = d$, in particular recovering Dwork's theorem. Moreover, Poincaré duality had been established, thus giving a functional equation (64) for $X_0/\mathbf{F}_q$ *proper* and smooth. In fact, Grothendieck proved a much more general formula than (69), with $\mathbf{Q}_\ell$ replaced by any (constructible) $\mathbf{Q}_\ell$-sheaf $\mathscr{F}_0$, namely, if $\mathscr{F}$ is the inverse image of $\mathscr{F}_0$ on X,

$$\sum_{x\in X_0(\mathbf{F}_{q^n})} \mathrm{Tr}(F^{*n}, \mathscr{F}_x) = \sum_i (-1)^i \mathrm{Tr}(F^{*n}, H^i_c(X, \mathscr{F})), \tag{71}$$

which implies

$$L(X_0, \mathscr{F}_0, t) = \prod_{0\le i\le 2d} \det(1 - F^*t, H^i_c(X, \mathscr{F}))^{(-1)^{i+1}}. \tag{72}$$

Here the L function on the left hand side (sometimes denoted $Z(X_0, \mathscr{F}_0, t)$) is defined by

$$L(X_0, \mathscr{F}_0, t) := \prod_{x\in|X_0|} \det(1 - F_x t^{\deg(x)}, \mathscr{F}_0)^{-1}, \tag{73}$$

with the notation

$$\det(1 - F_x t^{\deg(x)}, \mathscr{F}_0) := \det(1 - F_x t^{\deg(x)}, \mathscr{F}_{\overline{x}}), \tag{74}$$

where $\overline{x} \in |X|$ denotes a geometric point over x and F_x the endomorphism of the stalk $\mathscr{F}_{\overline{x}}$ induced by $F^{*\deg(x)}$ (which leaves $\overline{x}$ fixed) (the right hand side does not depend on the choice of $\overline{x}$). For $\mathscr{F}_0$ the constant sheaf $\mathbf{Q}_\ell$, $L(X_0, \mathscr{F}_0, t) = Z(X_0, t)$. Formula (72) for X_0 a curve was to be a basic tool in Deligne's proof of the Weil conjectures.

5.2 A False Good Plan: Grothendieck's Standard Conjectures

At this point, what remained to be proved of (c) was the following statements (for $X_0/\mathbf{F}_q$ projective, smooth, $\ell \neq p$ and all i):

(I) **Integrality and independence of** ℓ: *The polynomial*

$$P_{i,\ell}(t) := \det(1 - F^*t, H^i(X, \mathbf{Q}_\ell)) \in \mathbf{Q}_\ell[t]$$

has coefficients in $\mathbf{Z}$ *and is independent of* ℓ.

(W) **Weights of Frobenius:** *The eigenvalues of* F^* *on* $H^i(X, \mathbf{Q}_\ell)$ *are* q*-Weil numbers of weight* i.

Actually, in view of (70), (W) implies (I) by a lemma of Fatou ([Weil I], 1.7).

Inspired by Serre's proof of analogues of the Weil conjectures for Kähler varieties ([236], ([240], 45)), Grothendieck [109] made certain conjectures on algebraic cycles, which he called *standard conjectures*, and thanks to which Serre's arguments could be transposed into the context of varieties over finite fields, thus yielding a proof of (I) and (W). The conjectures were the following ([109], see also Kleiman's exposition [152, 153]).

Let k be an algebraically closed field of characteristic $p \geq 0$ and ℓ a prime number $\neq p$. Let X be a projective, smooth, irreducible scheme over k, purely of dimension d, $L \in H^2(X, \mathbf{Q}_\ell(1))$ the class of an ample line bundle on X. Then:

(i) **Hard Lefschetz:** For $i \leq d$ the cup-product map

$$\cdot L^{d-i} : H^i(X, \mathbf{Q}_\ell) \to H^{2d-i}(X, \mathbf{Q}_\ell(d-i)) \tag{75}$$

is an isomorphism.

(ii) **Algebraicity of Λ:** Fix an isomorphism $\mathbf{Q}_\ell \simeq \mathbf{Q}_\ell(1)$. Assume that (i) holds. Let $\Lambda : H^i(X, \mathbf{Q}_\ell) \to H^{i-2}(X, \mathbf{Q}_\ell)$ be the operator defined by the following commutative square (where the rows are isomorphisms)

$$\begin{array}{ccc} H^i(X, \mathbf{Q}_\ell) & \xrightarrow{L^{d-i}} & H^{2d-i}(X, \mathbf{Q}_\ell) \\ \Lambda \downarrow & & \downarrow L \\ H^{i-2}(X, \mathbf{Q}_\ell) & \xrightarrow{L^{d-i+2}} & H^{2d-i+2}(X, \mathbf{Q}_\ell) \end{array}$$

for $0 \leq i \leq d$ (and the similar diagram for $\Lambda : H^{2d-i+2}(X, \mathbf{Q}_\ell) \to H^{2d-i}(X, \mathbf{Q}_\ell)$, with vertical arrows interchanged). Then there exists an algebraic cycle $\lambda \in CH^{d-1}(X \times X) \otimes \mathbf{Q}$, *independent of* ℓ, such that, for all $a \in H^*(X, \mathbf{Q}_\ell)$, one has

$$\Lambda a = \mathrm{pr}_{2*}(\mathrm{pr}_1^* a.[\lambda]),$$

where $[\lambda] \in H^{2d-2}(X \times X, \mathbf{Q}_\ell)$ is the cohomology class of λ, and $\mathrm{pr}_{2*} : H^*(X \times X, \mathbf{Q}_\ell) \to H^{*-2d}(X \times X, \mathbf{Q}_\ell)$ is the Gysin homomorphism.

(iii) **Hodge positivity:** Assume (i) holds. For $j \leq d$, let

$$P^j(X, \mathbf{Q}_\ell) := \mathrm{Ker}(L^{d-j+1} : H^j(X, \mathbf{Q}_\ell) \to H^{2d-j+2}(X, \mathbf{Q}_\ell))$$

be the *primitive part* of $H^j(X, \mathbf{Q}_\ell)$, and let $A^i(X)$ denote the $\mathbf{Q}$-vector subspace of $H^{2i}(X, \mathbf{Q}_\ell)$ generated by the cohomology classes of elements of the Chow group $CH^i(X)$. Then, for $i \leq d/2$, the $\mathbf{Q}$-valued symmetric bilinear form on $A^i(X) \cap P^{2i}(X, \mathbf{Q}_\ell)$,

$$(x, y) \mapsto (-1)^i \mathrm{Tr}(L^{d-2i} x \cdot y),$$

where $\mathrm{Tr} : H^{2d}(X, \mathbf{Q}_\ell) \to \mathbf{Q}_\ell$ is the trace map, is positive definite.

Elementary arguments show that (i) and (ii) imply the integrality and independence conjecture (I) above. On the other hand, arguments similar to those of Serre [236] show that the validity for all X of (i), (ii) and (iii) together implies the weight conjecture (W) (for all X). In addition, (i), (ii), (iii) imply the coincidence of homological and numerical equivalence for algebraic cycles, another famous conjecture. See [109] (and [152, 153] for more details).

Establishing the standard conjectures seemed to be a royal path towards (I) and (W). As they would yield much more, being the foundation of his theory of motives, Grothendieck in [109] considered their proof as "the most urgent task in algebraic geometry". Unfortunately, (ii) and (iii) proved to be intractable, and they are in fact still widely open today. For comments on the difficulty of constructing interesting algebraic cycles, see [D76, 1994]. Ignoring the conventional wisdom of the time, Deligne proved conjectures (I) and (W) by a totally different method, and, eventually, deduced from their proof the hard Lefschetz conjecture (i) (see Theorem 29). But he first established special cases, exploiting the motivic interplay between ℓ-adic cohomology of varieties over finite fields and Hodge theory when working over schemes of finite type over $\mathbf{Z}$. Though these special cases were of no utility for the general one, the methods he developed there had a lasting impact.

5.3 *Partial Results Using Hodge Theory*

With the notation of Sect. 5.2, assume that $\dim X_0 = 2$. Then, thanks to (68), (I) holds, as it holds for $i = 1$ or 3 as a corollary of the case of abelian varieties, and for the same reason, (W) holds for $i \neq 2$. It thus remains to show that the eigenvalues of F^* on $H^2(X, \mathbf{Q}_\ell)$ have all their complex conjugates of absolute value q.

K3 Surfaces

In [D20, 1972], Deligne proves this, i.e., (W) for $i = 2$, for X_0 a *liftable* K3 surface. Actually, as we have seen in Sect. 3.3 "Liftings of K3 Surfaces, Canonical Coordinates", Deligne later showed that this hypothesis of liftability is always satisfied [D49, 1981]. His proof is inspired by Grothendieck's philosophy of motives, which suggests a comparison between $H^2(X)$ and $\mathrm{End}(H^1(A))$ for a certain abelian variety A, via a construction due to Kuga–Satake over $\mathbf{C}$ [158]. More precisely, Deligne shows that, up to enlarging the finite field $\mathbf{F}_q$, there exist a complete discrete valuation ring V of mixed characteristic, with residue field $\mathbf{F}_q$, and fraction field K, an abelian scheme $\mathscr{A}$ over V, with complex multiplication by the even part $C = C^+(L_{\mathbf{Z}})$ of the Clifford algebra of the underlying lattice of a certain polarized Hodge structure of weight zero, type $((-1, 1), (0, 0), (1, -1))$, and rank 21, and a $\mathrm{Gal}(\mathbf{F}/\mathbf{F}_q)$-equivariant isomorphism

$$C^+(P^2(X, \mathbf{Z}_\ell)(1), \psi) \xrightarrow{\sim} \mathrm{End}_C(H^1(A, \mathbf{Z}_\ell)), \tag{76}$$

where $A = \mathscr{A}_{\mathbf{F}}$, ψ is the symmetric bilinear form on $H^2(X, \mathbf{Z}_\ell(1))$ defined by the cup-product, and P^2 denotes the primitive part, i.e., the orthogonal of the class $\xi \in H^2(X, \mathbf{Z}_\ell(1))$ of an ample line bundle on X. By Weil, the eigenvalues of F on the right hand side of (76) are q-Weil numbers of weight zero, hence also those of F on $\Lambda^2(P^2(1)) \subset C^+(P^2(1))$, and finally, also those of F on $P^2(1)$ as $\dim P^2 = 21 > 2$, which proves (W) (as $F^*\xi = \xi$).

A key ingredient for (76) is Kuga–Satake's construction of an abelian variety B associated with a (polarized) K3 Y over $\mathbf{C}$. For such a Y, the primitive part $P_{\mathbf{Z}} = P^2(Y, \mathbf{Z}(1))$ underlies a polarized Hodge structure of weight zero and type ((-1,1),(0,0),(1,-1)), with $h^{-1,1} = 1$ and $h^{0,0} = 19$. The action of the Deligne torus $\mathbf{S}$ on $P_{\mathbf{R}}$ (giving the Hodge decomposition of $P_{\mathbf{C}}$) lifts to the group $\mathrm{CSpin}(P_{\mathbf{R}})$ acting by left translations on the even part $C_{\mathbf{R}} = C^+(P_{\mathbf{R}})$ of the Clifford algebra $C(P_{\mathbf{R}})$. The corresponding (real) Hodge structure on $C_{\mathbf{R}}$ is purely of type ((0,1),(1,0)) and polarizable. Together with the lattice $C_{\mathbf{Z}} = C^+(P_{\mathbf{Z}})$, it defines an abelian variety B (see Sect. 4.2 "1-Motives"), such that

$$C_{\mathbf{Z}} = H^1(B, \mathbf{Z}).$$

The abelian variety B has complex multiplication by $C_{\mathbf{Z}}$, and (tautologically) left multiplication defines an isomorphism of algebras

$$u : C_{\mathbf{Z}} \xrightarrow{\sim} \mathrm{End}_{C_{\mathbf{Z}}}(H^1(B, \mathbf{Z}))$$

(where $H^1(B, \mathbf{Z})$ is considered as a right $C_{\mathbf{Z}}$-module), equivariant under the action of CSpin, acting by conjugation on the left hand side and left multiplication on H^1. Lifting $X_0/\mathbf{F}_q$ to $Y/\mathbf{C}$ (via some embedding of K into $\mathbf{C}$) yields such a pair (B, u), but it's unclear whether B would descend to a finite extension of K (with good reduction over V), and $u \otimes \mathbf{Z}_\ell$ to an isomorphism of type (76). This is nonetheless the case. To show it Deligne constructs—via an algebraicity theorem of Borel (complementing the Baily–Borel theorem on quotients of hermitian symmetric domains by torsion free arithmetic subgroups)—a variant (and refinement) of u with parameters, as an isomorphism of families of polarized Hodge structures over a formal moduli space of K3's. As mentioned in Sect. 3.3 "Liftings of K3 Surfaces, Canonical Coordinates", his construction has been extensively used since then in all questions pertaining to the Tate conjecture for K3 surfaces. The "motivic injection" $H^2(X) \hookrightarrow H^1(A) \otimes H^1(A)$ cryptically mentioned by Deligne in 1.3 of [D20, 1972] can be realized by an *absolute Hodge cycle* (Sect. 4.4), cf. [53].

Complete Intersections of Hodge Level ≤ 1

At the same time, Deligne used a similar argument in [D21, 1972] to prove (W) for complete intersections $X_0/\mathbf{F}_q$ of odd dimension $n = 2m + 1$, and of Hodge level ≤ 1, i.e., such that $H^j(X_0, \Omega^i_{X_0/\mathbf{F}_q}) = 0$ for $i + j = n$ and $|j - i| > 1$. Here the

Kuga–Satake variety is replaced by the intermediate Jacobian

$$J(Y) = H^n(Y, \mathbf{Z})\backslash H^n(Y, \mathbf{C})/H^{m+1,m}(Y)$$

(for Y a complete intersection of dimension n and level ≤ 1 over $\mathbf{C}$).

5.4 *Integrality and Independence of ℓ*

Despite their ingenuity and their beauty, the above results offered no clue towards the general case of (Sect. 5.2, (W)). Meanwhile, Deligne made progress on two related issues: integrality of eigenvalues of Frobenius, and rationality and independence of ℓ, i.e., (Sect. 5.2, (I)).

Integrality

In ([D19, 1972], XXI, Appendice) Deligne proved a weaker result than (Sect. 5.2, (I)), namely that the reciprocal zeroes of $P_{i,\ell}(t)$, which a priori are only ℓ-adic units, are *algebraic integers*. But he did it in greater generality, for coefficients, and with q-divisibility refinements. More precisely, he proved the following theorem:

Theorem 21 *Let $X_0/\mathbf{F}_q$ be separated and of finite type, of dimension $\leq n$, and let $\mathscr{F}_0$ be a (constructible) $\overline{\mathbf{Q}}_\ell$-sheaf on X_0. Assume that, for any closed point x_0 of X_0, the eigenvalues of F_{x_0} on $\mathscr{F}_x$ (with the notation of* (73)*) are algebraic integers. Then, for all i, the eigenvalues of F on $H^i_c(X, \mathscr{F})$ are algebraic integers, and for $i \geq n$, they are divisible by q^{i-n}.*

The proof proceeds by dévissage and fibration into curves. Assuming resolution of singularities, Deligne also shows that the eigenvalues of F on $H^i(X, \mathscr{F})$ are algebraic integers (and even proves a relative variant of this for $f_0 : X_0 \to Y_0$ and Rf_{0*}). The restrictive hypothesis could later be lifted, using de Jong's alterations. Further results, pertaining to the number of rational points over finite fields, or variants of the integrality theorem over local fields were obtained in the 2000s by Esnault, Deligne–Esnault [D103, 2006], Esnault–Katz (see [127] for a survey) and Zheng [264].

Independence of ℓ

Shortly afterwards Deligne made a breakthrough on (Sect. 5.2, (I)). Namely, he proved it *assuming $p > 2$ and that $X_0/\mathbf{F}_q$ lifts to a projective scheme in characteristic zero*. Deligne did not write up his theorem, but it was the subject of an exposé by Verdier at the 1972–1973 Bourbaki seminar [253]. Though the result itself couln't be of any help for the general case of (I) (not to speak of (W)), its proof already contained some of the key ingredients that Deligne was to use in [Weil I].

The strategy, which will be that of [Weil I], is to proceed by induction on the dimension of X_0, using *the monodromy of Lefschetz pencils*. Assume that (I) has been shown in dimension $\leq n$, and let $X_0/\mathbf{F}_q$ be of pure dimension $n+1$. After a possible finite extension of $\mathbf{F}_q$, let $X_0 \hookrightarrow P_0 = P^N_{\mathbf{F}_q}$ be a closed embedding, and let $(X_0)_t = X_0 \cap (H_0)_t)_{t \in D_0}$ (D_0 a line in $P_0^\vee$) be a Lefschetz pencil of hyperplane sections of X_0. Recall that this means the following. Let $\Delta_0 = (D_0)^\vee \subset P_0$ be the *axis* of the pencil. Then Δ_0 is transverse to X_0, and if $\widetilde{X_0}$ is the blow-up of $X_0 \cap \Delta_0$ in X_0, and $f_0 : \widetilde{X_0} \to D_0$ the canonical projection, induced by $\mathrm{pr}_2 : X_0 \times P_0^\vee \to P_0^\vee$, with fiber $(X_0)_t$ at $t \in D_0$, then there exists a finite closed subscheme S_0 of D_0 such that $f|U_0 := D_0 - S_0$ is smooth, and the geometric fibers of f_0 above points of S_0 have one single singular point, which is ordinary quadratic.

By the weak Lefschetz theorem, for any geometric point $\overline{t}$ in U above a closed point $t \in U_0$, the restriction map

$$H^i(X, \mathbf{Q}_\ell) \to H^i(X_{\overline{t}}, \mathbf{Q}_\ell) \tag{77}$$

is an isomorphism for $i < n$ and injective for $i = n$. The induction assumption, combined with Poincaré duality, thus reduces to proving (I) for $i = n$ (or, because of Theorem 21, that $\det(1 - FT, H^n(X, \mathbf{Q}_\ell))$ is in $\mathbf{Q}[T]$ and independent of ℓ). Moreover, the above injection for $i = n$ shows that, for all such point $\overline{t}$ above t, and F_t as in (73),

(*) $\det(1 - F^{\deg(t)}T, H^n(X, \mathbf{Q}_\ell))$ *divides* $\det(1 - F_t T, H^n(X_{\overline{t}}, \mathbf{Q}_\ell))$.

Let $\mathscr{G}$ be the set of polynomials $g(T) = \prod(1 - \alpha T)$, where α are ℓ-adic units, such that, for all closed point t of U_0, $\prod(1 - \alpha^{\deg(t)}T)$ divides $\det(1 - F_t T, H^n(X_{\overline{t}}, \mathbf{Q}_\ell))$. By (*), $\det(1 - FT, H^n(X, \mathbf{Q}_\ell))$ belongs to $\mathscr{G}$. Belonging to $\mathscr{G}$ does not imply any rationality or independence of ℓ condition. However, since $\det(1 - F_t T, H^n(X_t, \mathbf{Q}_\ell))$ is in $\mathbf{Z}[T]$ and independent of ℓ, so is the lcm G of the polynomials g in $\mathscr{G}$, as the family of such g is defined over $\mathbf{Q}$, i.e., if g is in $\mathscr{G}$ and σ is an automorphism of $\overline{\mathbf{Q}}_\ell$, then g^σ is in $\mathscr{G}$. Let's call G the *Deligne gcd* of the polynomials $\det(1 - F_t T, H^n(X_t, \mathbf{Q}_\ell))$. By definition, $\det(1 - FT, H^n(X_{\overline{k}}, \mathbf{Q}_\ell))$ divides G. The miraculous property, that Deligne proved in [253], is that if the given Lefschetz pencil has sufficiently big monodromy, then in fact,

$$\det(1 - FT, H^n(X, \mathbf{Q}_\ell)) = G. \tag{78}$$

Here it is assumed that $p > 2$ and that X_0 lifts in characteristic zero as a smooth projective scheme. In ([Weil II], 4.5) Deligne proved the same result without these restrictions, and in a slightly stronger form (Sect. 5.6 "First Applications", Theorem 31).

Let me say a few words about the proof of this gcd theorem, under the assumptions made at the beginning of Sect. 5.4 "Independence of ℓ". For a fixed geometric point $\overline{u}$ in U above u, the choice of paths from $\overline{u}$ to geometric points

$\bar{s}$ above closed points $s \in S_0$ defines a finite family of *vanishing cycles* ($\delta_{\bar{s}} \in H^n(X_{\bar{t}}, \mathbf{Q}_\ell)(m)$), for $n = 2m$ or $2m+1$, all conjugate up to sign[18] under the geometric fundamental group $\pi_1^{\mathrm{geom}}(U_0) := \pi_1(U, \bar{u})$. Let

$$E \subset H^n(X_{\bar{u}}, \mathbf{Q}_\ell) \tag{79}$$

be the $\mathbf{Q}_\ell$-linear subspace generated by the $\lambda\delta_{\bar{s}}$ for $\lambda \in \mathbf{Q}_\ell(-m)$. This subspace is called the *vanishing subspace* of $H^n(X_{\bar{u}}, \mathbf{Q}_\ell)$. It is stable under the action of $\pi_1^{\mathrm{geom}}(U_0)$, and is an absolutely irreducible representation of it. Let's write H^n for $H^n(X_{\bar{u}}, \mathbf{Q}_\ell)$. Cup-product defines a non-degenerate bilinear form

$$\langle x, y\rangle := \mathrm{Tr}(xy) : H^n \otimes H^n \to \mathbf{Q}_\ell(-n), \tag{80}$$

symmetric for n even, alternating for n odd, compatible with the action of $\pi_1^{\mathrm{geom}}(U_0)$ (and even that of $\pi_1(U_0, \bar{u})$). The Picard–Lefschetz formula implies that

$$E^\perp = (H^n)^{\pi_1^{\mathrm{geom}}(U_0)}. \tag{81}$$

This doesn't use the lifting assumption on X_0. This assumption, however, was to be used in a critical way in the proof of the gcd theorem. First of all, it implies that the hard Lefschetz theorem holds for X_0 and its hyperplane sections, which in turn implies that the restriction of $\langle, \rangle$ to E is non-degenerate:

$$E \cap E^\perp = 0, \tag{82}$$

and gives an orthogonal decomposition

$$H^n = H^n(X, \mathbf{Q}_\ell) \oplus E, \tag{83}$$

in other words,

$$H^n(X, \mathbf{Q}_\ell) = H^n(X_{\bar{t}}, \mathbf{Q}_\ell)^{\pi_1^{\mathrm{geom}}(U_0)}, \tag{84}$$

where the left hand side is considered as a subspace of $H^n(X_{\bar{t}}, \mathbf{Q}_\ell)$ by (77). The action of $\pi_1^{\mathrm{geom}}(U_0)$ on H^n induces a representation

$$\rho : \pi_1^{\mathrm{geom}}(U_0) \to \mathrm{GL}(E), \tag{85}$$

whose image is contained in $\mathrm{Sp}(E)$ (resp. $\mathrm{O}(E)$) for n odd (resp. even). By a theorem of Kazhdan–Margulis,[19] for n odd, $\rho(\pi_1^{\mathrm{geom}}(U_0))$ is open in $\mathrm{Sp}(E)$. For n

[18] The hypothesis $p > 2$ enables to apply the results of Katz in ([7], XVIII) – actually $p > 2$ *or* n odd would suffice for this reference; see also the comment after Theorem 31.

[19] According to Katz [143], privately communicated to P. Deligne in 1971.

even, $\rho(\pi_1^{\mathrm{geom}}(U_0))$ may be finite. Deligne shows that there exists an integer $d_0 \geq 1$ such that for any $d \geq d_0$ and any Lefschetz pencil of hyperplane sections of X_0 relative to the d-th multiple of the given embedding $X_0 \to P_0$, then $\rho(\pi_1^{\mathrm{geom}}(U_0))$ is open in $\mathrm{O}(E)$. The proof uses a lifting to characteristic zero, and a transcendental argument over $\mathbf{C}$, based on formulas of A'Campo and Thom–Sebastiani [234] on the monodromy of certain isolated singularities. Once the openness of the image of $\pi_1^{\mathrm{geom}}(U_0)$ in $\mathrm{Sp}(E)$ (or $\mathrm{O}(E)$) is achieved, then Deligne proves (78) by a rather involved argument using the Chebotarev density theorem. This argument will again be crucial in [Weil I] (see Sect. 5.5, Step 3).

5.5 *Weil I*

Combined with a new idea coming from the theory of modular forms, the techniques of monodromy of Lefschetz pencils finally enabled Deligne to prove *the* Weil conjecture (Sect. 5.2, (W)) for any projective, smooth $X_0/\mathbf{F}_q$ [Weil I]. Let me sketch the main points in the proof.

Step 1: *It suffices to show that, for any $X_0/\mathbf{F}_q$, projective, smooth, geometrically connected of* even *dimension $d+1$, any eigenvalue α of F^* on $H^{d+1}(X, \mathbf{Q}_\ell)$ is an algebraic number, all of whose conjugates $\sigma\alpha$ satisfy*

$$q^{\frac{d+1}{2}-\frac{1}{2}} \leq |\sigma\alpha| \leq q^{\frac{d+1}{2}+\frac{1}{2}}. \tag{86}$$

By induction on the dimension of X_0, using Poincaré duality and the weak Lefschetz theorem, one is reduced to proving that, for X_0 geometrically connected and of dimension n, the eigenvalues α of F^* on $H^n(X, \mathbf{Q}_\ell)$ are algebraic numbers all of whose conjugates $\sigma\alpha$ satisfy $|\sigma\alpha| = q^{n/2}$. If m is a positive, even integer, then, by Künneth, α^m is an eigenvalue of F^* on $H^{nm}(X^m, \mathbf{Q}_\ell)$. Applying (86) to X_0^m with $d+1 = nm$, and letting m tend to infinity yields the desired result, i.e., $|\sigma\alpha| = q^{n/2}$.

Step 2: *Reduction to an estimate on the eigenvalues of F^* on $H_c^1(U, \mathscr{E}/(\mathscr{E} \cap \mathscr{E}^\perp))$.*
Let $X_0/\mathbf{F}_q$ be as in Step 1. As in Sect. 5.4 "Independence of ℓ", after possibly making a finite extension of $\mathbf{F}_q$, choose a Lefschetz pencil (over $\mathbf{F}_q$) of hyperplane sections $(X_0)_t = X_0 \cap (H_0)_t)_{t \in D_0}$ of X_0, satisfying a few additional rationality conditions (exceptional set $S_0 \subset D_0$ consisting of rational points, geometric point $\overline{u}$ over u with $u \in U_0 = D_0 - S_0$ rational over $\mathbf{F}_q$, X_u smooth admitting a smooth hyperplane section over $\mathbf{F}_q$, vanishing cycles $\delta_{\overline{s}}$ defined over $\mathbf{F}_q$). We keep the notation of Sect. 5.4 "Independence of ℓ". As the relative dimension d of $f_0 : \tilde{X}_0 \to D_0$ is odd, much of the monodromy theory described in Sect. 5.4 "Independence of ℓ" (with $n = d$) is still valid. Local monodromies are tame, the vanishing cycles are conjugate up to sign, and, with E defined as above, (81) holds. A notable difference, however, is that hard Lefschetz is no longer known, so that a priori

$E \cap E^{\perp}$ might be non-zero, and has to be taken care of. In particular, the orthogonal decomposition (83) might fail. A good point, nevertheless, is that the representation

$$\rho : \pi_1^{\mathrm{geom}}(U_0) \to \mathrm{GL}(E/(E \cap E^{\perp})), \tag{87}$$

analogous to (85), is absolutely irreducible, and, as above, its image is open in $\mathrm{Sp}(E/(E \cap E^{\perp}))$. The representation $H^d = H^d(X_{\overline{u}}, \mathbf{Q}_\ell)$ is the stalk at $\overline{u}$ of the lisse sheaf $R^n f_{0*}\mathbf{Q}_\ell|U$ (more generally, all the sheaves $R^i f_{0*}\mathbf{Q}_\ell|U$ are lisse, (80) is induced by a non-degenerate, alternate pairing

$$\langle , \rangle : (R^d f_{0*}\mathbf{Q}_\ell|U) \otimes (R^d f_{0*}\mathbf{Q}_\ell|U) \to \mathbf{Q}_\ell(-d), \tag{88}$$

and, because of the rationality assumptions on the pencil, E, $E^{\perp}$ and $E \cap E^{\perp}$ are the stalks at $\overline{u}$ of lisse subsheaves $\mathscr{E}_0$, $\mathscr{E}_0^{\perp}$, and $\mathscr{E}_0 \cap \mathscr{E}_0^{\perp}$ of $R^d f_{0*}\mathbf{Q}_\ell|U$. Since $H^{d+1}(X, \mathbf{Q}_\ell) \to H^{d+1}(\widetilde{X}, \mathbf{Q}_\ell)$ is injective, it suffices to check (86) with X_0 replaced by $\widetilde{X}_0$. A careful analysis of the Leray spectral sequence of $f : \widetilde{X} \to D$ via the Picard–Lefschetz theory shows that it suffices to prove the following assertion:

(A1) *Any eigenvalue of* F^* *on* $H^1(D, j_*(\mathscr{E}/(\mathscr{E} \cap \mathscr{E}^{\perp})))$, *where* $j : U \hookrightarrow D$, *is an algebraic number, all of whose conjugates* $\sigma\alpha$ *satisfy*

$$q^{\frac{d+1}{2}-\frac{1}{2}} \le |\sigma\alpha| \le q^{\frac{d+1}{2}+\frac{1}{2}}. \tag{89}$$

To prove (A1) it suffices to prove

(A2) *Any eigenvalue of* F^* *on* $H_c^1(U, \mathscr{E}/\mathscr{E} \cap \mathscr{E}^{\perp})$ *is an algebraic number, all of whose conjugates* $\sigma\alpha$ *satisfy*

$$|\sigma\alpha| \le q^{\frac{d+1}{2}+\frac{1}{2}}. \tag{90}$$

To see this, Deligne observes that, for any lisse $\mathbf{Q}_\ell$-sheaf $\mathscr{L}_0$ on an open subscheme $j_0 : U_0 \hookrightarrow C_0$ of a proper, smooth curve $C_0/\mathbf{F}_q$, then, if D denotes the dualizing functor $R\mathscr{H}om(-, \mathbf{Q}_\ell)[2](1)$ on C_0, then

$$D(j_{0*}\mathscr{L}_0) = j_{0*}\mathscr{L}_0^{\vee}[2](1), \tag{91}$$

so that $H^i(C, j_*\mathscr{L})$ is dual to $H^{2-i}(C, j_*\mathscr{L}^{\vee})(1)$ by Poincaré duality ([D39], 1977, Dualité). This property is a particular case of the self-duality of an intermediate extension: here $j_{0*}\mathscr{L}_0[1]$ is the intermediate extension $j_{0!*}$ of the perverse sheaf $\mathscr{L}_0[1]$. It was first noted by Deligne in ([D26, 1973], 10.8), as being the reason for the simple form of the constant of the functional equation of L-functions (see Sect. 6.3). As $H^1(D, j_*(\mathscr{E}/(\mathscr{E} \cap \mathscr{E}^{\perp}))$ is a quotient of $H_c^1(U, \mathscr{E}/\mathscr{E} \cap \mathscr{E}^{\perp})$, (90) gives half of the inequalities (89), hence the other half by duality.

Let $\mathscr{F} := \mathscr{E}/(\mathscr{E} \cap \mathscr{E}^{\perp})$. To prove (A2) one may assume U_0 affine, and one wants to use Grothendieck's cohomological expression for $L(U, \mathscr{F}, t)$. As $H_c^0(U, \mathscr{F}) = 0$

(U affine), and $H^2_c(U, \mathscr{F}) = 0$ by (81), Grothendieck's formula reads

$$L(U, \mathscr{F}, t) = \det(1 - F^*t, H^1_c(U, \mathscr{F})) \tag{92}$$

To exploit (92) one needs information on the local factors $\det(1 - F_x t^{\deg(x)}, \mathscr{F}_0)^{-1}$ of L, for x a closed point of U_0. This is provided by the next two steps.

Step 3: *For every closed point x of U_0, $\det(1 - F_x t, \mathscr{F}_0) \in \mathbf{Q}[t]$.*

The proof follows the same lines as that of the gcd formula (78). But it requires more work, as the orthogonal decomposition (83) is no longer available. It is replaced by the filtration $\mathscr{E}_0 \cap \mathscr{E}_0^\perp \subset \mathscr{E}_0 \subset \mathscr{H}_0^d := R^d f_{0*}\mathbf{Q}_\ell$. The starting point is the observation that the sheaves $\mathscr{E}_0 \cap \mathscr{E}_0^\perp$, $\mathscr{H}_0^d/\mathscr{E}_0$, as well as the sheaves $R^i f_{0*}\mathbf{Q}_\ell$ for $i \neq d$ are geometrically constant, hence that, if $\mathscr{G}_0$ is any of them, there exist ℓ-adic units α_i in $\mathbf{Q}_\ell^*$ such that, for any closed point x of U_0, $\det(1 - F_x t, \mathscr{G}_0) = \prod(1 - \alpha_i^{\deg(x)} t)$.

Step 4: *The lisse sheaf $\mathscr{F}_0 = \mathscr{E}_0/(\mathscr{E}_0 \cap \mathscr{E}_0^\perp)$ on U_0 is pure of weight d, i.e., for any closed point x of U_0, the eigenvalues of F_x on $\mathscr{F}_0$ are $q^{\deg(x)}$-Weil numbers of weight d.*

This is the core of the proof, the place where a new idea enters. Deligne says at the beginning of section 3 in [Weil I] that the above result was "catalyzed" by the reading of a paper of Rankin [213]. For the genesis of this idea, see [143] and [163].

The main tool is again Grothendieck's formula (72), applied to U_0, which may be and is assumed affine, and the lisse sheaf $\mathscr{F}_0^{\otimes 2n}$ over U_0, for an integer $n \geq 1$. One has $H^0_c(U, \mathscr{F}^{\otimes 2n}) = 0$ (U affine), and

$$H^2_c(U, \mathscr{F}^{\otimes 2n}) = (\mathscr{F}_{\overline{u}}^{\otimes 2n})_{\pi_1^{\mathrm{geom}}(U_0)}(-1) = (\mathscr{F}_{\overline{u}}^{\otimes 2n})_{\mathrm{Sp}(E/(E\cap E^\perp))}(-1)$$

as the image of $\pi_1^{\mathrm{geom}}(U_0)$ is open in $\mathrm{Sp}(E/(E \cap E^\perp))$, where $\overline{u}$ is a geometric point in U. By a theorem of H. Weyl on co-invariants of the symplectic group, this yields $H^2_c(U, \mathscr{F}^{\otimes 2n}) = \mathbf{Q}_\ell(-nd - 1)^N$ for a certain integer $N \geq 1$. Therefore Grothendieck's formula reads

$$L(U_0, \mathscr{F}_0^{\otimes 2n}, t) = \frac{\det(1 - F^*t, H^1_c(U, \mathscr{F}^{\otimes 2n}))}{(1 - q^{nd+1}t)^N}. \tag{93}$$

Observing that each factor $P_x(t) := \det(1 - F_x t^{\deg(x)}, \mathscr{F}_0^{\otimes 2n})^{-1}$ is a formal series with rational (by Step 3) and nonnegative coefficients, one deduces from (93) that the poles of $P_x(t)$ are of absolute value at least q^{-nd-1}, and hence that every conjugate α of an eigenvalue of F_x on $\mathscr{F}_0^{\otimes 2n}$ is of absolute value $\leq q^{\deg(x)(\frac{d}{2}+\frac{1}{2n})}$, which by letting n tend to infinity yields the inequality $|\alpha| \leq q^{\deg(x)\frac{d}{2}}$, and hence $|\alpha| = q^{\deg(x)\frac{d}{2}}$ by duality.

The arguments in the proof yield a general statement of purity ([Weil I], 3.2) for lisse sheaves equipped with a perfect symplectic pairing satisfying certain rationality and monodromy conditions. This statement, however, doesn't seem to have had any other applications. But it suggested to Deligne a generalization, which was of critical use in [Weil II], see (Sect. 5.6 "Ingredients of the Proof", Purity criterion).

Step 5 (final step): *Proof of (A2).* By Step 3 the right hand side of (92) has rational coefficients, hence any eigenvalue α of F^* on $H^1_c(U, \mathscr{E}/(\mathscr{E} \cap \mathscr{E}^\perp))$ is an algebraic number. To show that any conjugate of α has absolute value $\leq q^{\frac{d+1}{2}+\frac{1}{2}}$ it suffices to show that the left hand side is absolutely convergent for $|t| < q^{-(\frac{d+1}{2}+\frac{1}{2})}$, which follows easily from Step 4.

5.6 Weil II

In [Weil II] Deligne gave an alternative proof of (Sect. 5.2 (W)), with far reaching generalizations. Grothendieck's conjectural philosophy of weights over finite fields had inspired Deligne's construction of mixed Hodge theory. Now that the Weil conjectures were proven, mixed Hodge theory in turn inspired Deligne's work in [Weil II] and [D53, 1982].

The convention made at the beginning of Sect. 5 is still in force. In addition, we fix a prime number ℓ invertible on all schemes to be considered, and denote by $\overline{\mathbf{Q}}_\ell$ an algebraic closure of $\mathbf{Q}_\ell$.

Mixed Sheaves, Statement of the Main Theorem

Deligne made the following basic definition ([Weil II], 1.2.2). Let X_0 be a scheme of finite type over $\mathbf{Z}$, and $\mathscr{F}_0$ a (constructible)[20] $\overline{\mathbf{Q}}_\ell$-sheaf on X_0. One says that $\mathscr{F}_0$ is *punctually pure* if there exists $n \in \mathbf{Z}$, called the *weight* of $\mathscr{F}_0$, such that, for any closed point x of X_0, with residue characteristic p and $[k(x) : \mathbf{F}_p] = n_x$, the eigenvalues of F^{n_x} on the stalk $(\mathscr{F}_0)_{\overline{x}}$ of $\mathscr{F}_0$ at a geometric point $\overline{x}$ over x where F is the geometric Frobenius $a \mapsto a^{1/p}$ of $k(\overline{x})$, i.e., on $\mathscr{F}_{\overline{x}}$ in the notation of (73), are p^{n_x}-Weil numbers of weight n. One says that $\mathscr{F}_0$ is *mixed* if $\mathscr{F}_0$ admits a finite filtration whose successive quotients are punctually pure. The weights of those quotients which are nonzero are called the (punctual) *weights* of $\mathscr{F}_0$. The category of mixed sheaves is stable by sub-objects, quotients, extensions, tensor products, and inverse images.

Given a mixed $\overline{\mathbf{Q}}_\ell$-sheaf $\mathscr{F}_0$ on X_0, one can ask whether it admits a "better" filtration with successive quotients pointwise pure, hopefully functorial in $\mathscr{F}_0$, and which relation should then hold between the weights of the nonzero quotients. It is shown in (*loc. cit*, 3.4.1), as a consequence of the main theorem of *loc. cit.*, that for

[20] We will consider only constructible $\overline{\mathbf{Q}}_\ell$-sheaves, and omit "constructible" in the sequel.

lisse sheaves, there exists indeed such a functorial filtration, called the *filtration by the punctual weights*, see Sect. 5.6 "First Applications".

The main result of *loc. cit.* is the following theorem (*loc. cit.*, 3.3.1):

Theorem 22 *Let $f_0 : X_0 \to S_0$ be a morphism between schemes of finite type over $\mathbf{Z}$, and $\mathscr{F}_0$ be a mixed $\overline{\mathbf{Q}}_\ell$-sheaf on X_0 of weight $\leq n$. Then, for all $i \in \mathbf{Z}$, $R^i f_! \mathscr{F}_0$ on S_0 is mixed of weight $\leq n + i$.*

In particular, if X_0 is of finite type over $\mathbf{F}_q$, $H^i_c(X, \mathbf{Q}_\ell)$ (where $X = X_0 \otimes_{\mathbf{F}_q} \mathbf{F}$) is mixed of weight $\leq i$. By Poincaré duality, if $X_0/\mathbf{F}_q$ is smooth, $H^i(X, \mathbf{Q}_\ell)$ is mixed of weight $\geq i$, which echoes the analogous results in Hodge theory ([D29, 1974], 8.2.4) (see Sect. 4.2 "Mixed Hodge Theory", The general case). As a corollary, the Weil conjectures (Sect. 5.2, (I) (W)) hold for any *proper* and smooth $X_0/\mathbf{F}_q$:

Corollary 3 *If $X_0/\mathbf{F}_q$ is proper and smooth, then, for all i, the polynomial* $\det(1 - Ft, H^i(X, \mathbf{Q}_\ell))$ *belongs to $\mathbf{Z}[t]$, and its reciprocal roots are q-Weil integers of weight i.*

In fact, Deligne proves refinements and generalizations of Theorem 22, of the following two types:

(a) **Weil sheaves**. Let $W(\mathbf{F}/\mathbf{F}_q)$ be the *Weil group*, i.e., the subgroup of $\mathrm{Gal}(\mathbf{F}/\mathbf{F}_q)$ generated by the geometric Frobenius $F : a \mapsto a^{1/q}$. A *Weil sheaf* $\mathscr{F}_0$ on X_0 is a $\overline{\mathbf{Q}}_\ell$-sheaf $\mathscr{F}$ on X, together with an action of $W(\mathbf{F}/\mathbf{F}_q)$ lifting the action of $W(\mathbf{F}/\mathbf{F}_q)$ on $X = X_0 \otimes_{\mathbf{F}_q} \mathbf{F}$. By pull-back to X, $\overline{\mathbf{Q}}_\ell$-sheaves define Weil sheaves, but the category of Weil sheaves is larger. For example, a Weil sheaf on $\mathrm{Spec}\, \mathbf{F}_q$ corresponds to an element $u \in \overline{\mathbf{Q}}_\ell^*$; it is a $\overline{\mathbf{Q}}_\ell$-sheaf if and only if u is an ℓ-adic unit. If X_0 is normal, geometrically connected, any *lisse irreducible* Weil sheaf is deduced by torsion from a $\overline{\mathbf{Q}}_\ell$-sheaf $\mathscr{G}_0$, i.e., there exists $b \in \overline{\mathbf{Q}}_\ell^*$ such that $\mathscr{F}_0 = \mathscr{G}_0 \otimes \overline{\mathbf{Q}}_\ell^{(b)}$, where $\overline{\mathbf{Q}}_\ell^{(b)}$ is a rank 1 Weil sheaf on $\mathrm{Spec}\, \mathbf{F}_q$ on which the geometric Frobenius F acts by multiplication by b ([Weil II], 1.3.14). This implies that Grothendieck's formula (72) extends to Weil sheaves.

The definitions of *punctually pure* and *mixed* readily extend to Weil sheaves. Theorem 22 (for $X_0/\mathbf{F}_q$) holds with $\overline{\mathbf{Q}}_\ell$*-sheaf* replaced by *Weil sheaf.*

As elements α of $\overline{\mathbf{Q}}_\ell^*$ which are q-Weil numbers of weight n are characterized by the fact that for any isomorphism $\iota : \overline{\mathbf{Q}}_\ell^* \xrightarrow{\sim} \mathbf{C}$, $\iota\alpha$ has absolute value $q^{n/2}$, Deligne finds it convenient to work separately for each ι. One defines the ι*-weight* of α (rel. to q) as $2\log_q|\iota\alpha|$, and the corresponding notions of (punctually) ι*-pure* and ι*-mixed* for Weil sheaves on X_0. These ι-weights are real numbers, which are not necessarily integers. We will sometimes say "weights" instead of "ι-weight" when no confusion can arise.

(b) **Six operations**. The construction of a triangulated category $D^b_c(X, \overline{\mathbf{Q}}_\ell)$ stable under Grothendieck's six operations ($\otimes$, $\mathrm{R}\mathscr{H}om$, f^*, $Rf^!$, Rf_*, $Rf_!$) raised problems that had not been tackled in SGA 5 [5], nor in the unpublished thesis of Jouanolou. In ([Weil II], 1.1.2) Deligne proposes a definition which works well for schemes X separated and of finite type over a field satisfying certain finiteness conditions (verified for example if it is either finite or algebraically closed): for the integer ring R of a finite extension E_λ of $\mathbf{Q}_\ell$, with maximal ideal m, $D_c^b(X, R)$ is defined as the 2-inverse limit of the categories $D_{ctf}^b(X, R/m^n)$ ($n \geq 1$) (consisting of finite tor-dimension complexes K_n of R/m^n-sheaves with constructible cohomology, where transition maps are given by $\otimes_{R/m^{n+1}}^L R/m^n$), $D_c^b(X, E_\lambda) := E_\lambda \otimes D_c^b(X, R)$, and $D_c^b(X, \overline{\mathbf{Q}}_\ell)$ is the 2-inductive limit of the $D_c^b(X, E_\lambda)$ for $E_\lambda \subset \overline{\mathbf{Q}}_\ell$. More flexible definitions were later given independently by Gabber (unpublished) and Ekedahl [87]. Quite recently, a new and seemingly better approach was developed by Bhatt and Scholze [32].

Let X_0 be a scheme of finite type over $\mathbf{F}_q$. An object K_0 of $D_c^b(X_0, \overline{\mathbf{Q}}_\ell)$ is called *mixed* if, for all $i \in \mathbf{Z}$, $\mathscr{H}^i K_0$ is mixed. The full subcategory

$$D_m^b(X_0, \overline{\mathbf{Q}}_\ell) \subset D_c^b(X_0, \overline{\mathbf{Q}}_\ell) \tag{94}$$

consisting of mixed complexes is a subtriangulated category, and it follows from Theorem 22 that, for schemes separated and of finite type over $\mathbf{F}_q$, it is stable under the six operations.[21] Given an integer w, a mixed complex K_0 over such a scheme $a_0 : X_0 \to \operatorname{Spec} \mathbf{F}_q$ is said to be of *weight* $\leq w$ if, for all $i \in \mathbf{Z}$, the pointwise weights of $\mathscr{H}^i K_0$ are $\leq w + i$. It is said to be of *weight* $\geq w$ if DK_0 is of weight $\leq -w$; here D is the dualizing functor $D(-) = R\mathscr{H}om(-, Ra_0^! \overline{\mathbf{Q}}_\ell)$. It is said to be *pure of weight* w if it is both of weight $\leq w$ and $\geq w$. If one denotes by $D_{\leq w}^b$ (resp. $D_{\geq w}^b$, resp. D_w^b) the full subcategory of D_m^b consisting of complexes of weight $\leq w$ (resp. $\geq w$, resp. w), Theorem 22, for a morphism $f_0 : X_0 \to S_0$ of schemes of finite type over $\mathbf{F}_q$ can be reformulated by saying that $Rf_{0!}$ sends $D_{\leq w}^b(X_0, \overline{\mathbf{Q}}_\ell)$ into $D_{\leq w}^b(S_0, \overline{\mathbf{Q}}_\ell)$. Dually, Rf_{0*} respects $D_{\geq w}^b$. In particular, if f_0 is *proper*, then Rf_{0*} respects purity, more precisely, sends D_w^b into D_w^b. This formalism was to acquire its full force with the introduction of *perverse t-structures* in [D53, 1982], see Sect. 5.8.

Ingredients of the Proof

- **Purity criterion for real sheaves**
  The basic estimate (Sect. 5.5, Step 4), obtained by the so-called *Rankin–Selberg method* relay on two facts: (i) The coefficients of the polynomial $\det(1 - F_x t, \mathscr{F}_0)$ belong to $\mathbf{Q}$; (ii) $H_c^2(U, \mathscr{F}^{\otimes 2n}) = \mathbf{Q}_\ell(-nd-1)^N$, a consequence of a theorem of H. Weyl on co-invariants of the symplectic group, and the openness of the image of the geometric fundamental group into $\mathrm{Sp}(E/(E \cap E^\perp))$.
  In (i), the fact that the coefficients were supposed rational was not essential: *real* instead of rational would suffice. For (ii), it turns out that, in the absence of any assumption on the geometric monodromy group, general results on the global

[21] See ([Weil II], 6.1.11). For schemes over Spec $\mathbf{Z}$, it seems that only generic variants are available ([Weil II], 6.1.2).

monodromy of lisse Weil sheaves coming from class field and algebraic groups theory still enable to get enough control on the factors of the L-function.

The starting point is that, for a smooth curve $X_0/\mathbf{F}_q$, the Weil group $W(\mathbf{F}/\mathbf{F}_q)$ acts very strongly on the prime to p abelian quotient of the geometric fundamental group $\pi_1(X, \overline{x})$ of X_0: the largest quotient of it on which the action is trivial is finite. This is an easy consequence of class field theory. Deligne proves a generalization for $X_0/\mathbf{F}_q$ normal, geometrically connected (the image of $\pi_1(X, \overline{x})$ in the quotient of $W(X_0, \overline{x})$ by the closure of its derived group is an extension of a finite group of order prime to p by a pro-p-group) (a generalization which is in fact not needed).

An important corollary is that if $\mathscr{L}_0$ is a lisse sheaf of rank 1 on X_0, corresponding to a character $\chi : W(X_0, \overline{x}) \to \overline{\mathbf{Q}}_\ell^*$, then there exists $c \in \overline{\mathbf{Q}}_\ell^*$ and a character of finite order ε such that $\chi(g) = c^{\deg(g)}\varepsilon(g)$. In particular, $\mathscr{L}_0$ is ι-pure, of weight the weight of c. Given a lisse sheaf $\mathscr{F}_0$ on X_0, Deligne defines *determinantal weights* of $\mathscr{F}_0$ as the numbers $w(\Lambda^d \mathscr{G}_0)/d$, for a constituent[22] $\mathscr{G}_0$ of $\mathscr{F}_0$ of rank d, where w denotes the ι-weight relative to q. These numbers play a crucial role in the purity theorem below.

From this Deligne deduces the following key result, which he calls *Grothendieck's global monodromy theorem*[23] ([Weil II], 1.3.8), and which echoes Theorem 15:

Theorem 23 *Let $\mathscr{F}_0$ be a lisse Weil sheaf on a scheme X_0 of finite type, normal and geometrically connected over $\mathbf{F}_q$. Let $\overline{x}$ be a geometric point of X, let*

$$G^0 = G^0(\mathscr{F}_0) \tag{95}$$

be the Zariski closure of the image of $\pi_1(X, \overline{x})$ in $\mathrm{GL}(\mathscr{F}_{\overline{x}})$, and let G^{00} be its identity component. Then the radical of G^{00} is unipotent.

In particular, if $\mathscr{F}_0$ is semisimple, so that G^{00} is reductive, then G^{00} is *semisimple*. Moreover, if G is the extension of $\mathbf{Z}$ by G^0 defined by push-out of the Weil group $W(X_0, \overline{x})$ by $\pi_1(X, \overline{x}) \to G^0$,

$$0 \to G^0 \to G \stackrel{\deg}{\to} \mathbf{Z} \to 0,$$

then the center of G has an image of finite index in $\mathbf{Z}$ by the degree map. This corollary enables to describe the behavior of determinantal weights under tensor operations. The upshot is the following theorem, which generalizes ([Weil I], 3.2), and plays a crucial role in the proof of the curve case of Theorem 22 for pure sheaves (see Sect. 5.6 "Ingredients of the Proof", Squaring of a curve):

[22] I.e., a simple quotient in a Jordan–Hölder filtration by lisse subsheaves.

[23] Grothendieck proved this theorem at a talk he gave during the SGA 7 seminar, on March 26, 1968.

Theorem 24 *Let $X_0/\mathbf{F}_q$ be a smooth, geometrically connected curve, and let $\mathscr{F}_0$ be a lisse ι-real Weil sheaf on X_0. Then its constituents are ι-pure.*

Here ι is an isomorphism $\overline{\mathbf{Q}}_\ell \xrightarrow{\sim} \mathbf{C}$, "$\iota$-real" means that, for any closed point $x \in X_0$, $\iota\det(1 - F_x t, \mathscr{F}_0) \in \mathbf{R}[t]$, and "$\iota$-pure" means that for any such constituent G_0, there is $w \in \mathbf{R}$ such that the eigenvalues α of F_x on $\mathscr{G}_0$ are of ι-weight $w_{N(x)}(\alpha)$ equal to w, where $N(x) = q^{\deg(x)}$, with the notation of (Sect. 5.6 "Mixed Sheaves, Statement of the Main Theorem", (b)).

- **Weight monodromy theorem**

Let $X_0/\mathbf{F}_q$ be as in Theorem 24, and let $j_0 : U_0 \hookrightarrow X_0$ be the complement of a finite closed subscheme S_0 of X_0. As in (Sect. 5.5, Step 2, (A1)), sheaves of the form $j_{0*}\mathscr{F}_0$, for $\mathscr{F}_0$ a lisse Weil sheaf on U_0 are of special interest, in view of the duality between $j_{0*}\mathscr{F}_0$ and $j_{0*}\mathscr{F}_0^\vee(1)[2]$ (91). For $\mathscr{F}_0$ pure, the weights of $j_{0*}\mathscr{F}_0$ decrease at points of S_0:

Lemma 1 *If $\mathscr{F}_0$ is a lisse ι-pure (Weil) sheaf of ι-weight β on $U_0 = X_0 - S_0$, then for any closed point s of S_0, and any eigenvalue α of F_s on $j_{0*}\mathscr{F}_0$, one has $w_{N(s)}(\alpha) \leq \beta$ (where $w_{N(s)}$ means a ι-weight).*

Actually, Deligne gave formulas for $\beta - w_{N(s)}(\alpha)$ in terms of the local monodromy of $\mathscr{F}_0$ near s. Let $\overline{s}$ be a geometric point over s, $X_{0(s)}$ (resp. $X_{0(\overline{s})}$) the henselization (resp. strict henselization) of X_0 at s (resp. $\overline{s}$), and $\overline{\eta}$ a geometric point over the generic point η of $X_{0(s)}$. Let $W(\overline{\eta}/\eta)$ be the Weil group, inverse image of $W(\overline{s}/s)$ in $\mathrm{Gal}(\overline{\eta}/\eta)$. Then $V := \mathscr{F}_{\overline{\eta}}$ is an ℓ-adic representation ρ of $W(\overline{\eta}/\eta)$. By *Grothendieck's local monodromy theorem* ([237], Appendix), there exists an open subgroup I_1 of the inertia group $I \subset W(\overline{\eta}/\eta)$ and a nilpotent morphism $N : V(1) \to V$, called *logarithm of the monodromy*, such that $\rho(g) = \exp(Nt_\ell(g))$ for all $g \in I_1$, where $t_\ell : I \to \mathbf{Z}_\ell(1)$ is the ℓ-primary component of the tame character. It follows that the ι-weights (rel. to $N(s)$) of the eigenvalues of a lifting $\widetilde{F}$ in $W(\overline{\eta}/\eta)$ of the geometric Frobenius $F_s \in \mathrm{Gal}(\overline{s}/s)$ do not depend on the choice of $\widetilde{F}$. They are called the *ι-weights* of $\mathscr{F}_{0\eta}$. When they are integers, there exists a unique finite increasing filtration W of V, called the *weight filtration*, which is stable under N and such that $\mathrm{gr}^i_W(V)$ is ι-pure of weight i. On the other hand, whether the ι-weights of $\mathscr{F}_{0\eta}$ are integers or not, by general nonsense on nilpotent endomorphisms in an abelian category, there exists a unique finite increasing filtration M of V, called the *monodromy filtration*, such that $NM_iV(1) \subset M_{i-2}V$ and N^i induces an isomorphism $\mathrm{gr}^M_i V(i) \xrightarrow{\sim} \mathrm{gr}^M_{-i} V$.

Deligne deduced from Lemma 1 the following theorem ([Weil II], 1.8.4), now called *weight monodromy theorem*:

Theorem 25 *If $\mathscr{F}_0$ is ι-pure of weight β, then, for all $j \in \mathbf{Z}$, $\mathrm{gr}^M_j \mathscr{F}_{0\eta}$ is ι-pure of weight $\beta + j$. In particular, if β is an integer, then $\mathscr{F}_{0\eta}$ is ι-mixed with integral weights, and its monodromy filtration M coincides with its weight filtration up to shift: $M_j = W_{\beta+j}$.*

Once Theorem 22 is proven, Theorem 25 can be applied to $\mathscr{F}_0 = R^n f_{0*}\mathbf{Q}_\ell$, for $f_0 : Y_0 \to X_0$ proper and smooth, which is lisse and punctually pure of weight n. Then $\mathscr{F}_{0,\overline{\eta}} = H^n(Y_{\overline{\eta}}, \mathbf{Q}_\ell)$, and the monodromy operator N induces isomorphisms

$$N^i : \mathrm{gr}_W^{n+i} H^n(Y_{\overline{\eta}}, \mathbf{Q}_\ell) \xrightarrow{\sim} \mathrm{gr}_W^{n-i} H^n(Y_{\overline{\eta}}, \mathbf{Q}_\ell), \tag{96}$$

where W is the weight filtration.

A slightly more general statement is given in ([Weil II], 1.8.5). This theorem was inspired by the theory (due to Deligne and Schmid) of variations of Hodge structures on the punctured disc (see Sect. 4.1). The mixed characteristic analogue of Theorem 25 is still an open conjecture (see Sect. 10). The proof of Theorem 25 relies on Grothendieck's trace formula, and the determination of the monodromy filtration of a tensor product, using the Jacobson–Morosov theorem.

- **Hadamard–de la Vallée-Poussin method**

It follows from Lemma 1 and Grothendieck's formula (72) that under the assumption of Lemma 1, for any eigenvalue α of F on $H^1(X, j_*\mathscr{F})$, one has

$$w_q(\alpha) \leq \beta + 2. \tag{97}$$

The desired formula is $w_q(\alpha) = \beta + 1$, or (equivalently, using duality), $w_q(\alpha) \leq \beta + 1$. The variant of the Rankin–Selberg method used in (Sect. 5.6 "Ingredients of the Proof", Squaring of a curve) requires to know that, in the situation of Theorem 24, if $\mathscr{F}_0$ is ι-pure, then the polynomials $\iota\det(1 - Ft, H^1(X, \mathscr{F}))$ have real coefficients. The estimate (97) does not suffice for this, as it does not exclude cancellation between the numerator and the denominator in Grothendieck's formula (72). However, using a method inspired to him by Mertens's proof of the Hadamard–de La Vallée-Poussin theorem asserting that the Riemann zeta function $\zeta(s)$ does not vanish on the line $\mathscr{R}s = 1$, Deligne proved the following refinement of (97):

$$w_q(\alpha) < \beta + 2, \tag{98}$$

which indeed suffices to prove

$$\iota\det(1 - Ft, H^1(X, \mathscr{F})) \in \mathbf{R}[t]. \tag{99}$$

The Riemann zeta function is replaced by the function $s \mapsto L(X_0, \mathscr{F}_0, q^{-s})$. The proof is a piece of analysis on continuous linear complex representations of a certain real Lie group $G_\mathbf{R}$, extension of $\mathbf{Z}$ by a compact group $G^0_\mathbf{R}$, defined by the global monodromy group of $\mathscr{F}_0$ (see (95)) via an isomorphism $\iota : \overline{\mathbf{Q}}_\ell \xrightarrow{\sim} \mathbf{C}$, equipped with a set of distinguished conjugacy classes in $G^0_\mathbf{R}$ (playing the role of prime numbers) defined by the semisimple parts of the Frobenius F_x at closed points x of X_0.

- **Squaring of a curve**

More or less standard dévissages reduce the proof of Theorem 22 to the following special case ([Weil II], 3.2.3):

Lemma 2 *Let X_0 be a projective and smooth curve over $\mathbf{F}_q$, $j : U_0 \hookrightarrow X_0$ a dense open subscheme, and $\mathscr{F}_0$ a lisse, pointwise ι-pure sheaf on U_0 of weight β. Then any eigenvalue α of F on $H^i(X, j_*\mathscr{F})$ satisfies $w_q(\alpha) = \beta + i$.*

In [Weil I] Deligne had used high cartesian powers of the given X_0 (Sect. 5.5, Step 1), Lefschetz pencils of arbitrary odd relative dimension (Sect. 5.5, Step 2), and high tensor powers of $\mathscr{F}$ (Sect. 5.5, Step 4). Quite strikingly, the proof of Lemma 2 uses only $X_0 \times X_0$ and $\mathscr{F}_0 \otimes_{\text{ext}} \mathscr{F}_0$ over it, where $\otimes_{ext}$ denotes an external tensor product $\text{pr}_1^* \otimes \text{pr}_2^*$. The main point is to prove, by induction on the integer $k \geq 0$, that if $\beta = 0$, then any eigenvalue α of F on $H_c^1(U, \mathscr{F})$ satisfies $w_q(\alpha) \leq 1 + 2^{-k}$ (starting with (97) for $k = 0$). For this, Deligne uses a good pencil of hyperplane sections of $Z_0 := X_0 \times X_0$, $Z_0 \stackrel{\pi_0}{\leftarrow} \widetilde{Z}_0 \stackrel{f_0}{\rightarrow} D_0$, $((Z_0)_t = Z_0 \cap H_t)$, $t \in D_0 \stackrel{\sim}{\rightarrow} \mathbf{P}^1_{\mathbf{F}_q}$ (where π_0 is the blow-up of the axis). The proof combines an analysis of the vanishing cycles at the points of non acyclicity of $(f_0, \mathscr{G}_0)$, where $\mathscr{F}_0 \otimes_{\text{ext}} \mathscr{F}_0$, and Theorem 24 and (99) applied to the ι-real sheaf $\mathscr{G}_0 \oplus \mathscr{G}_0^\vee$.

A few years later, another proof of Theorem 22 was given by Laumon [171]. It uses neither the Hadamard–de la Vallée-Poussin method nor that of squaring, but instead relies on deep properties of Deligne's ℓ-adic Fourier transform (see Sect. 5.7). However, Deligne's purity criterion (Theorem 24) and the weight monodromy theorem (Theorem 25) remain crucial ingredients.

First Applications

We briefly discuss here the applications of Theorem 22 contained in [Weil II], with the exception of *estimates of exponential sums*, to which we devote Sect. 5.6 "Exponential Sums".

- **The weight filtration**

By definition, if $\mathscr{F}_0$ is a mixed $\overline{\mathbf{Q}}_\ell$-sheaf on a scheme X_0 of finite type over $\mathbf{F}_q$, $\mathscr{F}_0$ admits a finite filtration whose graded pieces are punctually pure (Sect. 5.6 "Mixed Sheaves, Statement of the Main Theorem"). Thanks to Theorem 22, more can be said when $\mathscr{F}_0$ is *lisse*. Deligne shows that in this case there exists a *unique* increasing filtration of $\mathscr{F}_0$ by lisse sheaves,

$$\cdots \subset W_{i-1}\mathscr{F}_0 \subset W_i\mathscr{F}_0 \subset \cdots,$$

called the *weight filtration*, such that each $\text{gr}_i^W \mathscr{F}_0$ is punctually pure of weight i. Moreover, if $\mathscr{G}_0$ is a second mixed lisse sheaf, and $u_0 : \mathscr{F}_0 \rightarrow \mathscr{G}_0$ is a morphism, then u_0 is *strictly compatible* with the weight filtrations. This positively answers a conjecture he made in his report ([D15, 1971], 2.1), see Sect. 4.1. Later,

Deligne proved in [D53, 1982] a better statement, for *mixed perverse sheaves*, see Theorem 37.

- **The Ramanujan–Petersson conjecture**

This is the most spectacular application of the Weil conjectures. Serre calls it a theorem of Deligne–Deligne, as it relies on the construction of ℓ-adic representations associated with modular forms, made earlier by Deligne (see Sect. 6). The result is the following (see, e.g., ([D28, 1974], Sect. 1) for the basic definitions and notation on modular forms):

Theorem 26 *Let N be an integer ≥ 1, k an integer ≥ 0, $\varepsilon : (\mathbf{Z}/N\mathbf{Z})^* \to \mathbf{C}^*$ a character, $f = \sum_{n\geq 1} a_n q^n$ ($q = e^{2\pi i z}$) a modular form on $\Gamma_0(N)$ of weight $k+2$ and character ε, which is cuspidal and primitive. Then, for all prime p not dividing N, one has*

$$|a_p| \leq 2p^{\frac{k+1}{2}}. \tag{100}$$

In particular, for $N = 1$, $k = 10$, $f = \Delta = q\prod_{n\geq 1}(1-q^n)^{24} = \sum_{n\geq 1}\tau(n)q^n$, we have, for all primes p,

$$|\tau(p)| \leq 2p^{11/2}, \tag{101}$$

as conjectured by Ramanujan.

Deligne stated Theorem 26 in ([Weil I], 8.2), leaving it to the reader to fill in the details of the proof using his Bourbaki exposé [D6, 1969] to construct a suitable ℓ-adic representation $\rho_{f,\ell}$ attached to f. In ([Weil II], 3.7.1), he gives an elegant argument for the main point (5.1) of [D6, 1969], using Theorem 22. Namely, if S_0 is a smooth curve over $\mathbf{F}_q$ and $f_0 : E_0 \to S_0$ an elliptic curve, then $R^1 f_{0*}\mathbf{Q}_\ell$ is lisse and pure of weight 1, hence $\mathrm{Sym}^k R^1 f_{0*}\mathbf{Q}_\ell$ is pure of weight k, and therefore, by Theorem 22 (or Lemma 2),

$$\mathrm{Im}(H^1_c(S, \mathrm{Sym}^k R^1 f_*\mathbf{Q}_\ell) \to H^1(S, \mathrm{Sym}^k R^1 f_*\mathbf{Q}_\ell)), \tag{102}$$

which can be identified with $H^1(\overline{S}, j_*\mathrm{Sym}^k R^1 f_*\mathbf{Q}_\ell)$, where $j_0 : S_0 \hookrightarrow \overline{S}_0$ is a proper smooth compactification of S_0, is pure of weight $k+1$. This avoids the delicate compactification lemma 5.4 of [D6, 1969]. See (Sect. 6.1, Theorem 40) for the definition of $\rho_{f,\ell}$.

The analogue of Theorem 26 for $k = -1$ was proven by Deligne and Serre in [D28, 1974]. The proof relies again on the construction of $\rho_{f,\ell}$ for $k \geq 0$.

- **Equidistribution of angles of Frobenius**

As a by-product of the Hadamard–de la Vallée-Poussin method, Deligne obtains general equidistribution results for the conjugacy classes in the compact group $G^0_{\mathbf{R}}$ (Sect. 5.6 "Ingredients of the Proof", Hadamard–de la Vallée Poussin method) of the semisimple parts of the Frobenius elements F_x at closed points x of X_0

([Weil II], 3.5.3). In particular, he proves a general Sato–Tate equidistribution theorem in equal characteristic:

Theorem 27 *Let $f_0 : E_0 \to X_0$ be an elliptic curve, with $X_0/\mathbf{F}_q$ of finite type, normal, geometrically connected, of dimension N, such that the modular invariant j of E/X is not constant. For $x \in X_0(\mathbf{F}_{q^n})$, let $\theta(x) \in [0, \pi]$ be the number such that the eigenvalues of F_x on $(R^1 f_* \mathbf{Q}_\ell)_x = H^1(E_x, \mathbf{Q}_\ell)$ are $q^{\frac{n}{2}} e^{\pm i\theta(x)}$. For $\theta \in [0, \pi]$, let $\delta[\theta]$ denote the Dirac measure at θ. Then, when n tends to infinity, the measure $\frac{1}{q^{nN}} \sum_{x \in X_0(\mathbf{F}_{q^n})} \delta[\theta(x)]$ vaguely converges towards the measure $\frac{2}{\pi} \sin^2\theta d\theta$.*

The original Sato–Tate conjecture for elliptic curves E over $\mathbf{Q}$ with no complex multiplication ([248], p. 106) was recently established by Clozel, Harris, Shepherd–Barron, and Taylor (assuming E has multiplicative reduction at one prime) [54, 112, 250]. For a discussion of motivic variants, see [239].

- **Semisimplicity**

The next result is an analogue of the semisimplicity theorem for variations of Hodge structures ([D16, 1971], 4.2.6) (cf. Theorem 15 (b)):

Theorem 28 *Let $\mathscr{F}_0$ be a lisse, punctually ι-pure (Weil) sheaf on a scheme X_0 of finite type over $\mathbf{F}_q$ and geometrically normal, then the pull-back $\mathscr{F}$ of $\mathscr{F}_0$ on X is semisimple.*

In view of Grothendieck's global monodromy theorem (Theorem 23), this implies:

Corollary 4 *Under the assumptions of Theorem* 28, *the identity component of the global geometric monodromy group* (95) *of $\mathscr{F}_0$ is semisimple.*

On the other hand, by a specialization argument Deligne deduces the following corollary, which generalizes ([D16, 1971], 4.2.9):

Corollary 5 *If S is a normal scheme of finite type over an algebraically closed field of characteristic $\neq \ell$, $f : X \to S$ a proper and smooth morphism, then the sheaves $R^i f_* \mathbf{Q}_\ell$ are semisimple.*

- **Hard Lefschetz theorem**

Let k be an algebraically closed field of characteristic $p \neq \ell$, and let X/k be a projective and smooth scheme, purely of dimension d. Then the hard Lefschetz conjecture (Sect. 5.2, (i)) holds:

Theorem 29 *If $L \in H^2(X, \mathbf{Q}_\ell(1))$ is the class of an ample line bundle $\mathscr{O}(1)$ on X, then, for all $i \in \mathbf{Z}$, the cup-product map* (75) *is an isomorphism.*

The proof exploits the consequences of Corollary 5 on the monodromy of Lefschetz pencils. Up to replacing $\mathscr{O}(1)$ by a power, we may assume that it defines an embedding in $\mathbf{P}^N$ such that L is the class of a general member Y of a Lefschetz pencil of hyperplane sections $(X_t = X \cap H_t)_{t \in D}$ of X (cf. Sect. 5.4 "Independence of ℓ" and 5.5, Step 2). We know by the weak Lefschetz theorem that, if $n = d - 1$,

$H^n(X, \mathbf{Q}_\ell)$ injects into $H^n(Y, \mathbf{Q}_\ell)$. Let $U = D - S$ be the complement of the set of points t at which X_t is singular, and u a rational point of U, such that $Y = X_u$. It follows from the Picard–Lefschetz formula and the inductive assumption that hard Lefschetz holds in dimension n, hence for Y, so that

$$H^n(X, \mathbf{Q}_\ell) = H^n(Y, \mathbf{Q}_\ell)^{\pi_1(U,u)}. \tag{103}$$

As $H^n(Y, \mathbf{Q}_\ell)$ is a semisimple representation of $\pi_1(U, u)$ by Corollary 5, the restriction to $H^n(Y, \mathbf{Q}_\ell)^{\pi_1(U,u)}$ of the cup-product pairing (80) is non-degenerate, and by known arguments ([7], XVIII) this implies that hard Lefschetz holds for X.

As a corollary, (103) holds for *any* X, and as the right hand side of (103) is the orthogonal of the vanishing subspace $E \subset H^n(Y, \mathbf{Q}_\ell)$ (79), formulas (82) and (83) hold, i.e., $E \cap E^\perp = 0$, and we have an orthogonal decomposition

$$H^n(Y, \mathbf{Q}_\ell) = H^n(X, \mathbf{Q}_\ell) \oplus E. \tag{104}$$

- **Local invariant cycle theorem**

This the following result ([Weil II], 3.6.1):

Theorem 30 *Let S be the henselization at a rational point of a smooth curve over an algebraically closed field k of characteristic $p \geq 0$, and let $f : X \to S$ be a proper morphism. Let s be the closed point of S, and let $\overline{\eta}$ be a geometric generic point. Assume that X is essentially smooth over k, and $X_{\overline{\eta}}$ is smooth. Then, the specialization morphism*

$$\mathrm{sp}^* : H^*(X_s, \mathbf{Q}_\ell) \to H^*(X_{\overline{\eta}}, \mathbf{Q}_\ell)^{\mathrm{Gal}(\overline{\eta}/\eta)} \tag{105}$$

is surjective.

Here sp^* is the composition $H^*(X_s, \mathbf{Q}_\ell) \xrightarrow{\sim} H^*(X, \mathbf{Q}_\ell) \to H^*(X_{\overline{\eta}}, \mathbf{Q}_\ell)$, where the first isomorphism is the inverse of the proper base change isomorphism, and the second one the restriction, whose image is contained in $H^*(X_{\overline{\eta}}, \mathbf{Q}_\ell)^{\mathrm{Gal}(\overline{\eta}/\eta)}$.

After reduction to the case where f comes by base change and localization from a proper $f_0 : X_0 \to Y_0$, where Y_0 is a smooth curve over $\mathbf{F}_q$, with smooth geometric generic fiber and $X_0/\mathbf{F}_q$ is smooth, the proof uses a weight argument, based on Theorem 22 and the weight monodromy theorem (Theorem 25).

The analytic analogue of Theorem 30 for a projective morphism $f : X \to D$ over a disc, with $X/\mathbf{C}$ smooth and f smooth outside $0 \in D$ (cf. ([Weil II], 3.6.4)) was shown by Steenbrink [242] to follow from Deligne's weight argument, based here on mixed Hodge theory, and in particular on the weight monodromy theorem over $\mathbf{C}$.[24] When the inertia I acts unipotently through $t_\ell : I \to \mathbf{Z}_\ell(1)$, so that $H^*(X_{\overline{\eta}}, \mathbf{Q}_\ell)^{\mathrm{Gal}(\overline{\eta}/\eta)} = \mathrm{Ker}\, N$, the surjection (105) is refined into long exact

[24] Whose proof in *loc. cit.* was incorrect, see Sect. 10, The weight monodromy conjecture.

sequences (*Clemens–Schmid exact sequences*), see ([266], 7.8):

$$\to H^{2N-i}(X_s)^\vee(-N) \to H^i(X_s) \overset{\mathrm{sp}}{\to} H^i(X_{\overline{\eta}})$$

$$\overset{N}{\to} H^i(X_{\overline{\eta}})(-1) \to H^{2N-i-2}(X_s)^\vee(-N) \to, \qquad (106)$$

where $H^*(-) = H^*(-, \mathbf{Q}_\ell)$, and X is supposed to be purely of dimension N.

In ([Weil II], 6.2.9) Deligne gives a generalization of Theorem 30, where the hypotheses of smoothness (resp. essential smoothness) of X_η (resp X/k) are dropped, and the constant sheaf $\mathbf{Q}_\ell$ is replaced by a *potentially pure* complex $K \in D^b(X, \mathbf{Q}_\ell)$.

- **gcd theorem**

The hard Lefschetz theorem, and (as a corollary) the orthogonal decomposition (83), having been established, Deligne could prove the gcd formula (78) unconditionally, and with no assumption on the geometric monodromy group:

Theorem 31 *Let $X_0 \subset \mathbf{P}^N_{\mathbf{F}_q}$ be a projective and smooth scheme of pure dimension $n+1$ over $\mathbf{F}_q$, equipped with a Lefschetz pencil $((X_0)_t = X_0 \cap (H_0)_t)_{t \in D_0}$ (defined over $\mathbf{F}_q$) of hyperplane sections of X_0, whose axis is supposed to be sufficiently general if $p = 2$ and n is even. Let S_0 be the set of closed points t in D_0 at which X_{0t} is singular, and let $G(T) \in \overline{\mathbf{Q}}_\ell[T]$ be the lcm of the polynomials $\prod(1 - \alpha_i T) \in \overline{\mathbf{Q}}_\ell[T]$ such that for all $t \in (D_0 - S_0)(\mathbf{F}_{q^r})$, $\prod(1 - \alpha_i^r T)$ divides $\det(1 - FT, H^n(X_t, \mathbf{Q}_\ell))$ (in $\overline{\mathbf{Q}}_\ell[T]$). Then (78) holds, i.e.,*

$$G(T) = \det(1 - FT, H^n(X, \mathbf{Q}_\ell)),$$

(and $G(T)$ is in $\mathbf{Z}[T]$ and independent of ℓ).

The proof uses the determination of the global geometric monodromy group (95) of the sheaf $\mathcal{E}_0$ (cf. Sect. 5.5, Step 2) on $D_0 - S_0$, which is either finite, or open in the automorphism group of $E = \mathcal{E}_t \subset H^n(X_t, \mathbf{Q}_\ell)$ equipped with its intersection form. This determination, in turn, relies on the conjugacy of the vanishing cycles up to sign. It is to prove this conjugacy that Deligne uses the restrictive hypothesis on the pencil for $p = 2$ and n even. This restriction was later shown by Gabber and Orgogozo [210] to be superfluous.

Katz and Messing [149] deduced from Theorem 31 that for any projective and smooth[25] $X_0/\mathbf{F}_q$, and any $i \in \mathbf{Z}$, then

$$\det(1 - Ft, H^*(X, \mathbf{Q}_\ell)) = \det(1 - Ft, H^i(X_0/W(\mathbf{F}_q))), \qquad (107)$$

[25] Suh [243] showed that (107) holds more generally assuming only $X_0/\mathbf{F}_q$ proper and smooth, and used it to prove the evenness of odd degree ℓ-adic Betti numbers of X. Generalizations of this last result to intersection cohomology are given by Sun-Zheng [247].

where on the right hand side $H^i(X_0/W(\mathbf{F}_q))$ denotes the i-th crystalline cohomology group of X_0 and F the r-th power of the absolute Frobenius endomorphism, for $q = p^r$.

- **$\mathbf{Q}_\ell$-homotopy type**

Let X be a scheme separated and of finite type over an algebraically closed field k and let ℓ be a prime number different from the characteristic of k. In ([Weil II], 5) Deligne defines a $\mathbf{Q}_\ell$-dga (differential graded algebra) $A(X)$ (or rather an object of the corresponding derived category), depending functorially on X, with the following properties:

(i) $H^*(A(X)) = H^*(X, \mathbf{Q}_\ell)$;
(ii) if $k = \mathbf{C}$, $A(X) = \mathbf{Q}_\ell \otimes \mathscr{M}$, where $\mathscr{M} = \mathscr{M}(X(\mathbf{C}))$ is a Sullivan minimal model of the rational homotopy type of the topological space $X(\mathbf{C})$ (see Sect. 4.2 "Hodge Theory and Rational Homotopy"). Using Theorem 22, he constructs *gradings* $\mathscr{W}_j$ of $\mathscr{M}$ compatible with the weight filtration W of $H^i(X(\mathbf{C}), \mathbf{Q})$ ($W_n = \oplus_{j \le n} \mathscr{W}_j$), as announced in [D34, 1975]. These gradings are such that the corresponding action of $\mathbf{G}_{m\mathbf{Q}}$ on $\mathscr{M}$ is by automorphisms of its dga structure. For X normal, the existence of these gradings implies a theorem of Morgan on the pro-unipotent completion of $\pi_1(X)$, and for X proper and smooth, that $\mathscr{M} \otimes \mathbf{Q}_\ell$ is a minimal model of $H^*(X, \mathbf{Q}_\ell)$ (with zero differential) (in particular, all Massey products are zero).

The definition of $A(X)$ relies on a construction of Grothendieck and Miller. A simpler approach is provided by the pro-étale theory of Bhatt–Scholze, which directly produces a $\mathbf{Q}_\ell$-dga $R\Gamma(X, \mathbf{Q}_\ell)$ [32].

Exponential Sums

Estimates of exponential sums were one of the first applications of the Weil conjectures, and one of those which have been the most extensively studied.

- **The method**

The main tool is the cohomological interpretation of exponential sums via Grothendieck's trace formula (71) applied to certain ℓ-adic sheaves associated with them, combined with the bounds on the eigenvalues of Frobenius given by the Weil conjecture (Theorem 22).

The starting point is the so-called *dictionary between functions and sheaves* on schemes over finite fields, introduced by Deligne in a letter to D. Kazhdan [73] (see [171], 1.1, [124], 1). As in Sect. 5.6 "Mixed Sheaves, Statement of the Main Theorem", let X_0 be a scheme separated and of finite type over $\mathbf{F}_q$ and $\mathscr{F}_0$ a $\overline{\mathbf{Q}}_\ell$-sheaf on X_0. This sheaf defines a function

$$t_{\mathscr{F}_0} : X_0(\mathbf{F}_q) (= X^F) \to \overline{\mathbf{Q}}_\ell, \ x \mapsto \mathrm{Tr}(F, \mathscr{F}_x). \tag{108}$$

More generally, for $\mathscr{F}_0 \in D_c^b(X_0, \overline{\mathbf{Q}}_\ell)$, one defines

$$t_{\mathscr{F}_0} : X_0(\mathbf{F}_q) \to \overline{\mathbf{Q}}_\ell, \ x \mapsto \sum_i (-1)^i t_{\mathscr{H}^i(\mathscr{F}_0)}. \tag{109}$$

The law $\mathscr{F}_0 \mapsto t_{\mathscr{F}_0}$ transforms $\otimes$ into product, inverse image into inverse image, and, by Grothendieck's trace formula and proper base change theorem, $Rg_{0!}$ into g_{0*} (for $g_0 : X_0 \to Y_0$).

Let $\psi_0 : \mathbf{F}_p \to \mathbf{C}^*$ be a nontrivial character, e.g., $\psi_0(x) = e^{2\pi i x/p}$. Choosing an isomorphism $\iota : \overline{\mathbf{Q}}_\ell \xrightarrow{\sim} \mathbf{C}$, we can view ψ_0 as a character with values in $\overline{\mathbf{Q}}_\ell^*$. We denote by $\psi : \mathbf{F}_q \to \overline{\mathbf{Q}}_\ell^*$ the character deduced from ψ_0 by composition with $\mathrm{Tr}_{\mathbf{F}_q/\mathbf{F}_p}$. Let A_0 be the affine line over $\mathbf{F}_q$. The Artin–Schreier $\mathbf{F}_q$-torsor

$$0 \to \mathbf{F}_q \to A_0 \overset{x \mapsto x^q - x}{\to} A_0 \to 0 \tag{110}$$

defines, by pushout by ψ^{-1}, a lisse $\overline{\mathbf{Q}}_\ell$-sheaf $\mathscr{L}_\psi$ of rank 1 on A_0, such that

$$t_{\mathscr{L}_\psi} = \psi : A_0(\mathbf{F}_q) = \mathbf{F}_q \to \overline{\mathbf{Q}}_\ell^*. \tag{111}$$

If $f_0 : X_0 \to A_0$ is an $\mathbf{F}_q$-morphism, one defines

$$\mathscr{L}_\psi(f_0) := f_0^* \mathscr{L}_\psi \tag{112}$$

By the compatibility $t_{Rg_{0!}} = g_{0*}$ and the Leray spectral sequence of f_0, we have

$$\sum_{x \in X_0(\mathbf{F}_q)} \psi(f_0(x)) = \sum_i (-1)^i \mathrm{Tr}(F^*, H_c^i(X, \mathscr{L}_\psi(f))). \tag{113}$$

([D39, 1977], Sommes trigonométriques, 2.3). When $H_c^i(X, \mathscr{L}_\psi(f))$ happens to be concentrated in a single degree, and one is able to calculate both its dimension and its weights, then Theorem 22 gives a bound on the absolute value of the exponential sum in the left hand side of (113). The following theorem (([Weil I], 8.4), ([Weil II], 3.7)), suggested to Deligne by Bombieri, is a typical example:

Theorem 32 *Let $Q \in \mathbf{F}_q[X_1, \cdots, X_n]$ be a polynomial of degree d prime to p, whose homogeneous component Q_d of degree d defines a smooth hypersurface in $\mathbf{P}^{n-1}_{\mathbf{F}_q}$. Consider Q as an $\mathbf{F}_q$-morphism $A_0^n \to A_0$. Then:*

(i) *$H_c^i(A^n, \mathscr{L}_\psi(Q)) = 0$ for $i \neq n$, $H_c^n(A^n, \mathscr{L}_\psi(Q))$ is of dimension $(d-1)^n$ and is pure of weight n, and therefore, by* (113)*,*
(ii) $|\sum_{x \in \mathbf{F_q}^n} \psi(Q(x))| \leq (d-1)^n q^{\frac{n}{2}}$.

A generalization of Theorem 32 is given by Katz in ([144], 5.1.2) (see also [166]).

- **Gauss, Jacobi, and Kloosterman sums**

The Artin–Schreier torsor (110) is a particular case of the *Lang torsor*. Given a connected algebraic group $G_0/\mathbf{F}_q$, the Lang torsor $\mathscr{L}$ is the $G_0(\mathbf{F}_q)$-torsor on G_0 given by

$$G_0 \to G_0, \quad x \mapsto Fx.x^{-1} \tag{114}$$

Push-outs of the Lang torsor by ℓ-adic representations of $G_0(\mathbf{F}_q)$ define ℓ-adic sheaves on G_0. In particular, for G_0 commutative, a character $\chi : G_0(\mathbf{F}_q) \to \overline{\mathbf{Q}}_\ell^*$ defines, by push-out by χ^{-1}, a lisse $\overline{\mathbf{Q}}_\ell$-sheaf of rank 1 $\mathscr{F}(\chi)$ on G_0, hence, for an $\mathbf{F}_q$-morphism $X_0 \to G_0$, a sheaf

$$\mathscr{F}(\chi, f_0) := f_0^* \mathscr{F}(\chi) \tag{115}$$

on X_0. At a point $x \in X^F$, the endomorphism F_x^* of $\mathscr{F}(\chi, f_0)_x$ is given by multiplication by $\chi(f(x))$:

$$(F_x^* : \mathscr{F}(\chi, f_0)_x \to \mathscr{F}(\chi, f_0)_x) = \chi(f(x)), \tag{116}$$

so that by Grothendieck's trace formula one gets the following generalization of (113):

$$\sum_{x \in X_0(\mathbf{F}_q)} \chi(f_0(x)) = \sum_i (-1)^i \mathrm{Tr}(F^*, H_c^*(X, \mathscr{F}(\chi, f_0))). \tag{117}$$

For $G_0 = \mathbf{G}_m$, and n prime to p, consider the *Kummer torsor* $\mathscr{K}_n$ on $\mathbf{G}_m$,

$$1 \to \mu_n \to \mathbf{G}_m \overset{x \mapsto x^n}{\to} \mathbf{G}_m \to 1; \tag{118}$$

the Lang torsor (114) is a particular case of it: we have $\mathscr{L} = \mathscr{K}_{q-1}$. In ([D39, 1977], Sommes trigonométriques) Deligne gives applications of (117) to standard identities between classical exponential sums, and old and new estimates of them. Here is a brief sample.

(i) *Gauss sums.* With Deligne's convention in (*loc. cit.* 4.1), the Gauss sum associated with $\psi : \mathbf{F}_q \to \overline{\mathbf{Q}}_\ell^*$ (as above) and the multiplicative character $\chi : \mathbf{F}_q^* \to \overline{\mathbf{Q}}_\ell^*$ is

$$\tau(\chi, \psi) = -\sum_{x \in \mathbf{F}_q^*} \psi(x)\chi^{-1}(x). \tag{119}$$

Let $\mathscr{F}(\psi\chi^{-1})$ denote the pull-back on $\mathbf{G}_m$ by the diagonal inclusion $\mathbf{G}_m \to \mathbf{G}_a \times \mathbf{G}_m$ of $\mathscr{F}(\psi \times \chi^{-1} : \mathbf{F}_q \times \mathbf{F}_q^* \to \overline{\mathbf{Q}}_\ell^*)$ on $\mathbf{G}_a \times \mathbf{G}_m$. This is a pure lisse sheaf of rank 1 and weight zero. One has $H_c^i((\mathbf{G}_m)_{\overline{k}}, \mathscr{F}(\psi\chi^{-1})) = 0$ for $i \neq 1$, $H_c^1((\mathbf{G}_m)_{\overline{k}}, \mathscr{F}(\psi\chi^{-1}))$ is 1-dimensional, and

$$\tau(\chi, \psi) = \mathrm{Tr}(F^*, H_c^1((\mathbf{G}_m)_{\overline{k}}, \mathscr{F}(\psi\chi^{-1}))). \tag{120}$$

If χ is nontrivial, then $H_c^* \xrightarrow{\sim} H^*$, so that H_c^1 is pure of weight 1, which yields the classical formula $|\tau(\chi, \psi)| = \sqrt{q}$ (which, as Deligne observes, also reflects Poincaré duality).

(ii) *Jacobi sums.* For $1 \leq i \leq n$, let $\chi_i : \mathbf{F}_q^* \to \overline{\mathbf{Q}}_\ell^*$ be a nontrivial character, such that the product of the χ_i's is nontrivial, and let $\chi : (\mathbf{F}_q^*)^n \to \overline{\mathbf{Q}}_\ell^*$, $\chi(x) = \prod \chi_i(x_i)$. Deligne (*loc. cit.* (4.14.2)) defines the associated Jacobi sum as

$$J(\chi) = (-1)^{n-1} \sum_{x_1, \cdots, x_n \in \mathbf{F}_q^*, \sum x_i = -1} \chi_1^{-1}(x_1) \cdots \chi_n^{-1}(x_n). \tag{121}$$

Let $G_0 = \mathbf{G}_m^{n+1}$ and $\mathscr{F}(\underline{\chi})$ the rank 1 lisse sheaf on G_0 associated with the multiple Kummer torsor $G_0 \to G_0, x \mapsto x^{q-1}$ and $\underline{\chi} = ((\chi_1 \cdots \chi_n)^{-1}, \chi_1, \cdots, \chi_n)$. Then (*loc. cit.*, 4.16), one has $H_c^i(X, \mathscr{F}(\underline{\chi}^{-1})) = 0$ for $i \neq n-1$, $H_c^{n-1}(X, \mathscr{F}(\underline{\chi}^{-1}))$ is of dimension 1 and pure of weight $n-1$, where $X_0 = ((x_0 + \cdots + x_n = 0) \cap \mathbf{G}_m^{n+1})/\mathbf{G}_m$, and

$$J(\chi) = \mathrm{Tr}(F^*, H_c^{n-1}(X, \mathscr{F}(\underline{\chi}^{-1}))). \tag{122}$$

In particular, $|J(\chi)| = q^{\frac{n-1}{2}}$. The classical formula expressing a Jacobi sum in terms of Gauss sums has a nice cohomological expression (*loc. cit.*, 4.17), which is a key ingredient in the proof. This interpretation is also at the source of a cohomological proof of a generalized form of Weil's theorem on the existence, over number fields, of algebraic Hecke characters associated with Jacobi sums (*loc. cit.*, 6.5).

(iii) *Kloosterman sums.* Given an integer $n \geq 1$, and $a \in \mathbf{F}_q$,

$$K_{n,a} := \sum_{x_1 \cdots x_n = a} \psi(x_1 + \cdots + x_n), \tag{123}$$

where ψ is as above, is called a *generalized Kloosterman sum.* Let $(V_a)_0 \subset A_0^n = \mathbf{A}_{\mathbf{F}_q}^n$ be the hypersurface of equation $x_1 \cdots x_n = a$, and $\pi : A_0^n \to A_0$ (resp. $\sigma : A_0^n \to A_0$) be the map defined by the product (resp. sum) of coordinates. Thus $(V_a)_0$ is the fiber of π at a. Analyzing the Leray spectral sequence of $\pi : A^n \to A$ for $\mathscr{L}_\psi(\sigma)$ (112), Deligne (*loc. cit.*, 7.4) shows that, if H_c^* denotes $H_c^*(V_a, \mathscr{L}_\psi(\sigma)|V_a)$, then $H_c^i = 0$ for $i \neq n-1$, and, for $a \neq 0$,

H_c^{n-1} is of dimension n and pure of weight $n-1$. By (113), we have

$$K_a = (-1)^{n-1}\mathrm{Tr}(F^*, H_c^{n-1}(V_a, \mathscr{L}_\psi(\sigma)|V_a)), \tag{124}$$

and, for $a \neq 0$,

$$|K_a| \leq nq^{\frac{n-1}{2}}. \tag{125}$$

The rank n lisse sheaf

$$\mathscr{K}l_{\psi,n} := R^{n-1}\pi_{0!}\mathscr{L}_\psi(\sigma) \tag{126}$$

on $\mathbf{G}_m$, with fiber $H_c^{n-1}(V_a, \mathscr{L}_\psi(\sigma)|V_a)$ at the point a, was called *Kloosterman sheaf* and studied by several authors, first by Katz [146], who determined its global geometric monodromy group (95) and gave applications to equidistribution properties of the angles of Kloosterman sums (123), in the spirit of ([Weil II], 3.5.3) (cf. Theorem 27). For recent developments, see [115].

Deligne's exposé *Sommes trigonométriques* in [D39, 1977] was the beginning of a long series of studies on exponential sums, by Katz, Laumon and others, in which the cohomological methods initiated there were reinforced by further tools (also initiated by Deligne), such as the ℓ-adic Fourier transform or geometric convolutions. See [173] for a survey of results up to 1999.

5.7 *The ℓ-Adic Fourier Transform*

Definition and First Properties

In the letter to D. Kazhdan mentioned above [73], Deligne introduced operations on complexes of sheaves on schemes over finite fields lifting operations on functions given by convolution or transformation by kernels. In particular, for $\mathbf{G}_a = (\mathbf{G}_a)_{\mathbf{F}_q}$, and $\psi : \mathbf{F}_q \to \overline{\mathbf{Q}}_\ell^*$ as above, he defined the *Fourier transform*

$$\mathscr{F}_{\psi,0} : D_c^b(\mathbf{G}_a, \overline{\mathbf{Q}}_\ell) \to D_c^b(\mathbf{G}_a, \overline{\mathbf{Q}}_\ell) \tag{127}$$

by $K \to R\mathrm{pr}'_!(\mathscr{L}_\psi(xy) \otimes \mathrm{pr}^*K)$, where pr (resp. pr$'$) is the first (resp. second) projection, and $\mathscr{L}_\psi(xy)$ is the sheaf (112) for $f_0 : \mathbf{G}_a^2 \to \mathbf{G}_a$, $(x, y) \mapsto xy$. For a function $f : \mathbf{F}_q \to \overline{\mathbf{Q}}_\ell$, let $\widehat{f} : \mathbf{F}_q \to \overline{\mathbf{Q}}_\ell$ denote its Fourier transform, defined by $\widehat{f}(y) = \sum_{x\in\mathbf{F}_q} f(x)\psi(xy)$. The proper base change theorem and Grothendieck's trace formula imply that, for $K \in D_c^b(\mathbf{G}_a, \overline{\mathbf{Q}}_\ell)$,

$$t_{\mathscr{F}_{\psi,0}(K)} = \widehat{t_K}, \tag{128}$$

with the notation of (109). Deligne gives a number of basic properties of $\mathscr{F}_\psi$ which, via (128) imply standard properties of the Fourier transform on functions: involutivity, exchange of product and convolution, Plancherel formula.

Deligne proposed generalizations where $\mathbf{G}_a$ is replaced by a unipotent group. The case of a vector space has been extensively studied [147, 171].

Let k be a perfect field of characteristic $p > 0$, $\overline{k}$ an algebraic closure of k, and, for q a power of p, $\mathbf{F}_q$ the subfield of $\overline{k}$ with q elements. Given an r-dimensional k-vector space E, Laumon defines

$$\mathscr{F}_\psi : D_c^b(E, \overline{\mathbf{Q}}_\ell) \to D_c^b(E', \overline{\mathbf{Q}}_\ell), \tag{129}$$

$$\mathscr{F}_\psi(K) := R\mathrm{pr}'_!(\mathscr{L}_\psi(\langle , \rangle) \otimes \mathrm{pr}^* K)[r], \tag{130}$$

where E' is the dual of E, pr, pr', are the canonical projections, $\langle , \rangle : E \times E' \to \mathbf{G}_a$ the canonical pairing, and $\mathscr{L}_\psi$ the pull-back by $\langle , \rangle$ of the Artin–Schreier sheaf $\mathscr{L}_\psi$ on $\mathbf{A}^1_{\mathbf{F}_p}$ corresponding to the character $\psi = \psi_0 : \mathbf{F}_p \to \overline{\mathbf{Q}}_\ell$ (so that, if $k = \mathbf{F}_p$, $\mathscr{F}_{\psi,0} = \mathscr{F}_\psi[-1]$ with the notation of (127)). The standard properties mentioned above are proved in [171] (in a slightly more general framework).

Laumon's Contribution and Applications

Deligne's construction remained dormant until, in the late 1970s, Verdier made the following observation. Instead of (129), one can consider the functor

$$\mathscr{F}_{\psi,*} : D_c^b(E, \overline{\mathbf{Q}}_\ell) \to D_c^b(E', \overline{\mathbf{Q}}_\ell), \tag{131}$$

defined by

$$\mathscr{F}_{\psi,*}(K) := R\mathrm{pr}'_*(\mathscr{L}_\psi(\langle , \rangle) \otimes \mathrm{pr}^* K)[r]. \tag{132}$$

Forgetting supports gives a natural map

$$\mathscr{F}_\psi(K) \to \mathscr{F}_{\psi,*}(K). \tag{133}$$

Verdier observed that, surprisingly, (133) is an isomorphism. In other words, the Fourier transform commutes with duality: $\mathscr{F}_\psi DK = D\mathscr{F}_{\psi^{-1}} DK(r)$, where $D = R\mathscr{H}om(-, a^!\overline{\mathbf{Q}}_\ell)$ is the Grothendieck–Verdier dualizing functor on an $\mathbf{F}_q$-scheme $a : X \to \operatorname{Spec} \mathbf{F_q}$. Verdier's unpublished argument was global (and possibly incomplete). Laumon's proof in ([147], 2.1.3, 2.4.4) is local, and shows more. After reduction to E of dimension $r = 1$, and E' compactified into a projective line $P' = E' \cup \infty'$, Laumon proves that, with Grothendieck's notation for vanishing cycles,

$$R\Phi_{\mathrm{pr}'}(j_!\mathscr{L}_\psi(\langle , \rangle) \otimes \mathrm{pr}^* K)_{(\overline{x}, \overline{\infty'})} = 0, \tag{134}$$

where $j : E\times E' \hookrightarrow E\times P'$ is the inclusion, $\mathrm{pr}' : E\times P' \to P'$ is the projection, and $(\overline{x}, \overline{\infty'})$ is a geometric point of $E\times P'$ above ∞', such that K has lisse cohomology sheaves at $\overline{x}$: in other words, $(\mathrm{pr}', j_!\mathscr{L}_\psi(\langle,\rangle)\otimes \mathrm{pr}^*K)$ is locally acyclic above ∞' at points where K is lisse. This property is an analogue of the classical *stationary phase principle*.

Laumon in [171] exploited this ℓ-adic stationary phase principle (134) and the isomorphism (133) to prove Deligne's conjecture on the global constant of the functional equations of L-functions (Sect. 6.3 "The Case of Function Fields", (149)), and to give an alternate proof of Deligne's main theorem (22) bypassing the use of the Hadamard–de la Vallée-Poussin method. The local additive convolution product defined by Deligne in his seminar [74] is a key ingredient, and is extensively studied in [171]. It follows from the isomorphism (133) that $\mathscr{F}_\psi$ transforms perverse sheaves into perverse sheaves. Applications of this to *uniform* estimates of exponential sums are given by Katz and Laumon in [147]. At about the same time, other applications of the ℓ-adic Fourier transform were found by Brylinski [48] (Lefschetz theory for intersection cohomology, Radon transform), and Lusztig ([179–184] (theory of character sheaves)).

5.8 *Perverse Sheaves*

In the monograph [D53, 1982] Deligne and his co-authors construct a general formalism of truncation in triangulated categories, which they apply to develop a theory of intersection cohomology in the étale setting, in positive characteristic. This work was inspired by the definition and study of perverse and intersection cohomology groups for certain singular stratified spaces by Goresky and MacPherson on the one hand, and the Riemann–Hilbert correspondence for regular holonomic $\mathscr{D}$-modules on smooth $\mathbf{C}$-schemes on the other hand. Over finite fields, combined with Deligne's results in [Weil II] discussed above, it led to *purity and decomposition theorems*.

t-Structures

Let $\mathscr{D}$ be a triangulated category. A *t-structure* on $\mathscr{D}$ ([D53, 1982], 1.3.1) is the datum of a pair of strictly full subcategories $(\mathscr{D}^{\leq 0}, \mathscr{D}^{\geq 0})$ such that, if for $n\in\mathbf{Z}$, $\mathscr{D}^{\leq n} := \mathscr{D}^{\leq 0}[-n]$, $\mathscr{D}^{\geq n} := \mathscr{D}^{\geq 0}[-n]$, one has:

(i) $\mathrm{Hom}(X,Y) = 0$ if $X\in\mathscr{D}^{\leq 0}$, $Y\in\mathscr{D}^{\geq 1}$,
(ii) $\mathscr{D}^{\leq 0}\subset\mathscr{D}^{\leq 1}$ and $\mathscr{D}^{\geq 0}\supset\mathscr{D}^{\geq 1}$,
(iii) for all $X\in\mathscr{D}$, there exists an exact triangle $A\to X\to B\to$ with $A\in\mathscr{D}^{\leq 0}$ and $B\in\mathscr{D}^{\geq 1}$.

A *t-category* is a triangulated category equipped with a t-structure. A typical example is provided by the derived category $\mathscr{D} = D(\mathscr{A})$ of an abelian category $\mathscr{A}$, with the *standard* t-structure where $\mathscr{D}^{\leq 0}$ (resp. $\mathscr{D}^{\geq 0}$) is the subcategory consisting of complexes K such that $H^iK = 0$ for $i > 0$ (resp. $i < 0$). In this case, the intersection $\mathscr{D}^{\leq 0} \cap \mathscr{D}^{\geq 0}$ is $\mathscr{A}$. It is shown in *loc. cit.* that in any t-category $\mathscr{D}$, the intersection $\mathscr{C} := \mathscr{D}^{\leq 0} \cap \mathscr{D}^{\geq 0}$, called the *heart* of $\mathscr{D}$, is an abelian subcategory of $\mathscr{D}$, stable by extensions, in which short exact sequences come from exact triangles by forgetting the map of degree 1 (a condition called *admissibility* in *loc. cit.*). Moreover, as in the case of $\mathscr{D}(\mathscr{A})$, the inclusion $\mathscr{D}^{\leq n} \subset \mathscr{D}$ (resp. $\mathscr{D}^{\geq n} \subset \mathscr{D}$) admits a right (resp. left) adjoint $\tau_{\leq n}$ (resp. $\tau_{\geq n}$), similar to the canonical truncation functors on $D(\mathscr{A})$; given any X in $\mathscr{D}$, there exists a unique (up to a unique isomorphism) exact triangle $\tau_{\leq 0}X \to X \to \tau_{\geq 1}X \to$ with $\tau_{\leq 0}X \in \mathscr{D}^{\leq 0}$ and $\tau_{\geq 1}X \in \mathscr{D}^{\geq 1}$; finally the functor $H^0 := \tau_{\leq 0}\tau_{\geq 0} (= \tau_{\geq 0}\tau_{\leq 0}) : \mathscr{D} \to \mathscr{C}$ is a cohomological functor: exact triangles $K \to L \to M \to$ give rise to long exact sequences $\cdots \to H^iK \to H^iL \to H^iM \to H^{i+1}K \to \cdots$, where $H^iK := H^0(K[i])$.

The question of reconstructing $\mathscr{D}$ (or rather its full subcategory $\mathscr{D}^b$) from $\mathscr{C}$ is tackled in (*loc. cit.*, 3.1). Let $\mathscr{D}$ be a full triangulated subcategory of $D^+(\mathscr{A})$ for $\mathscr{A}$ an abelian category having enough injectives, $\mathscr{C}$ the heart of a t-structure on $\mathscr{D}$, and $\mathscr{D}^b$ the full subcategory of $\mathscr{D}$ union of the $\mathscr{D}^{[a,b]} := \mathscr{D}^{\leq b} \cup \mathscr{D}^{\geq a}$. Then there is defined a *realization functor* real : $D^b(\mathscr{C}) \to \mathscr{D}^b$, which is an equivalence if an only if an effaceability condition for certain Ext groups is satisfied. See [22] for geometric examples where this is the case. The definition of real is not purely in terms of the t-structure on $\mathscr{D}$, it uses the companion filtered derived categories $\mathscr{D}F$, $\mathscr{D}^bF$. The realization functor also appears in work of Beilinson [21] and M. Saito [222] in mixed Hodge theory.

Let $\mathscr{D}_i$ ($i = 1, 2$) be a triangulated category equipped with a t-structure, with heart $\mathscr{C}_i$, and let $T : \mathscr{D}_1 \to \mathscr{D}_2$ be an exact functor. Then T is said to be *left exact* (resp. *right exact*, resp. *exact*) if $T(\mathscr{D}_1^{\geq 0}) \subset \mathscr{D}_2^{\geq 0}$ (resp. $T(\mathscr{D}_1^{\leq 0}) \subset \mathscr{D}_2^{\leq 0}$, resp. T is both left and right exact). In this case, the functor

$$^pT := H^0T : \mathscr{C}_1 \to \mathscr{C}_2 \tag{135}$$

is left exact (resp. right exact, resp. exact) (*loc. cit.*, 1.3.17).

The Riemann–Hilbert correspondence [139, 140, 191], briefly mentioned at the end of Sect. 3.1 "Higher Dimension: The Riemann–Hilbert Correspondence", provides an example of a non-standard t-structure on a derived category. Let X be a smooth scheme over $\mathbf{C}$, purely of dimension d, let $\mathscr{D}_X$ denote its sheaf of differential operators. The (shifted) de Rham complex functor $DR_X : M \mapsto \Omega^\bullet_{X^{\mathrm{an}}}(M^{\mathrm{an}})[d]$ defines an equivalence of triangulated categories

$$DR_X : D^b_{rh}(\mathscr{D}_X) \xrightarrow{\sim} D^b_c(X^{\mathrm{an}}, \mathbf{C}), \tag{136}$$

where $D^b_{rh}(\mathscr{D}_X) \subset D^b(\mathscr{D}_X)$ (resp. $D^b_c(X^{\mathrm{an}}, \mathbf{C}) \subset D^b(X^{\mathrm{an}}, \mathbf{C})$) is the full subcategory consisting of objects whose cohomology sheaves are regular holonomic (resp. algebraically constructible). The standard t-structure on the left hand side is transformed by DR_X into a t-structure t_X on the right hand side, which is not the standard one if $d > 0$. Its heart consists of the so-called *perverse sheaves* on X. The problem of interpreting it (and more generally the truncation functors) intrinsically, i.e., independently of the equivalence (136), was one of the motivations for the introduction of the above formalism.

For $d = 1$, if Y is a finite closed subset of X, and $U = X - Y$ the open complement, denote by $D^b_l((X,U)^{\mathrm{an}}, \mathbf{C})$ (resp $D^b_l(U^{\mathrm{an}}, \mathbf{C})$) the full subcategory of $D^b_c(X^{\mathrm{an}}, \mathbf{C})$ (resp. $D^b_c(U^{\mathrm{an}}, \mathbf{C})$) consisting of complexes whose cohomology sheaves are lisse on U^{an}. Then the t-structure induced by t_X on $D^b_l((X,U)^{\mathrm{an}}, \mathbf{C})$ appears as glued from the standard t-structure on $D^b_c(Y, \mathbf{C})$ and the shifted t-structure $(D^{\leq -1}, D^{\geq -1})$ on $D^b_l(U^{\mathrm{an}}, \mathbf{C})$. More generally, it is shown in ([D53, 1982], 2.1.4) that, given a suitable stratification $\mathscr{S}$ of X^{an}, the t-structured induced by t_X on the full subcategory $D^b_{\mathscr{S}}(X^{\mathrm{an}}, \mathbf{C})$ consisting of complexes whose cohomology sheaves are lisse along the strata can be obtained by successive gluing from shifted standard t-structures on the corresponding derived categories of the strata.

Here is a model for this gluing process. Let $(X, \mathscr{O})$ be a ringed space, $i : Y \hookrightarrow X$ a closed subspace, $j : U = X - Y \hookrightarrow X$ the open complement. It is proved in (*loc. cit.*, 1.4) that, given t-structures on $D^+(Y, \mathscr{O})$ and $D^+(U, \mathscr{O})$, the pair $(D^{\leq 0}, D^{\geq 0})$ of full subcategories of $D = D^+(X, \mathscr{O})$ defined by $K \in D^{\leq 0}$ if and only if $i^*K \in D^{\leq 0}(Y, \mathscr{O})$ and $j^*K \in D^{\leq 0}(U, \mathscr{O})$, $K \in D^{\geq 0}$ if and only if $i^!K \in D^{\geq 0}(Y, \mathscr{O})$[26] and $j^*K \in D^{\geq 0}(U, \mathscr{O})$, is a t-structure on D, which is said to be obtained by gluing from those on $D^+(Y, \mathscr{O})$ and $D^+(U, \mathscr{O})$. When the t-structures on $D^+(Y, \mathscr{O})$ and $D^+(U, \mathscr{O})$ are the standard ones, the resulting t-structure on D is the standard one. The proof uses only the usual formal properties of the functors i_*, j_* and their adjoints on both sides. It can therefore be transposed into an abstract framework, in which the spaces X, Y, U have disappeared. The result is the following *gluing lemma* (*loc. cit.*, 1.4.10), which is the key technical tool for the constructions in *loc. cit.*:

Theorem 33 *Let $\mathscr{D}, \mathscr{D}_U, \mathscr{D}_Y$ be triangulated categories, $i_* : \mathscr{D}_Y \to \mathscr{D}$, $j_* : \mathscr{D}_U \to \mathscr{D}$ be exact functors having adjoints $(i^*, i^!)$, $(j_!, j^*)$ satisfying the formal properties (loc. cit., (1.4.3.1) up to (1.4.3.5)). Suppose that $\mathscr{D}_Y$ and $\mathscr{D}_U$ are equipped with t-structures. Then*

$$(\mathscr{D}^{\leq 0} := \{K \in \mathscr{D} \mid i^*K \in \mathscr{D}_Y^{\leq 0}, j^*K \in \mathscr{D}_U^{\leq 0}\},$$
$$\mathscr{D}^{\geq 0} := \{K \in \mathscr{D} \mid i^!K \in \mathscr{D}_Y^{\geq 0}, j^*K \in \mathscr{D}_U^{\geq 0}\})$$

is a t-structure on $\mathscr{D}$ (said to be obtained by gluing from those on $\mathscr{D}_Y$ and $\mathscr{D}_U$).

In the next section, we discuss applications given in *loc. cit.* to complexes with constructible cohomology in the étale setting. It should be mentioned, however, that in the past 15 years t-structures on derived categories of complexes of $\mathscr{O}$-modules with bounded, coherent cohomology sheaves on noetherian schemes have played an important role in birational geometry (Bondal, Bridgeland, Orlov, and many others).

[26] Here $i^!$ stands for $Ri^!$.

Perverse Sheaves in the Étale Setting

Theorem 33 is applied to the construction of t-structures on:

(i) derived categories of sheaves of $\mathscr{O}$-modules on ringed spaces $(X, \mathscr{O})$ equipped with a stratification $\mathscr{S}$ and a *perversity function* $p : \mathscr{S} \to \mathbf{Z}$ (*loc. cit.*, 2.1); this covers the case of the t-structure t_X appearing in (136), and in the situations considered by Goresky and MacPherson, where $\mathscr{O}$ is constant of value R, this leads to a description of the p-perverse cohomology groups of X with value in R as cohomology groups of X with values in a certain complex $j_{!*}R$, *intermediate extension* of the constant sheaf R_U on a suitable open subset $j : U \hookrightarrow X$;

(ii) in the étale setting, categories of the form $D^b_c(X, \Lambda)$ for X separated and of finite type over a field k and Λ a ring such as $\mathbf{Z}/\ell^n\mathbf{Z}$, $\mathbf{Z}_\ell$, $\mathbf{Q}_\ell$, E_λ (a finite extension of $\mathbf{Q}_\ell$), or $\overline{\mathbf{Q}}_\ell$, for a prime number ℓ prime to the characteristic exponent of k (*loc. cit.*, 2.2, 4).

Let me briefly describe the constructions in (ii).

A *perversity function* is a function $p : 2\mathbf{Z} \to \mathbf{Z}$, such that both p and the *dual perversity* function p^* defined by $p^*(2n) = -p(2n) - 2n$ are non increasing, i.e., for any integers $n \le m$, one has $0 \le p(2n) - p(2m) \le 2m - 2n$. The *middle perversity* $p_{1/2}$ is the self-dual function $p_{1/2}(2n) = -n$. Let p be a perversity function.

For X/k and Λ as above, define the full subcategories $D^{\le p}_c(X, \Lambda)$ and $D^{\ge p}_c(X, \Lambda)$ of $D^b_c(X, \Lambda)$ by

$$(K \in D^{\le p}_c(X, \Lambda)) \Leftrightarrow (\forall x \in X, H^q i^*_x K = 0 \text{ for } q > p(2\dim(x))) \tag{137}$$

$$(K \in D^{\ge p}_c(X, \Lambda)) \Leftrightarrow (\forall x \in X, H^q i^!_x K = 0 \text{ for } q < p(2\dim(x))).$$

Here $\dim(x)$ means the dimension of the closure $\overline{\{x\}}$ of the point x, $i_x : x = \operatorname{Spec} k(x) \to X$ the canonical map, and $i^!_x := j^* i^!$ for the factorization of i_x into $x \xrightarrow{j} \overline{\{x\}} \xrightarrow{i} X$ (with $i^!$ is short for $Ri^!$). The main result (*loc. cit.*, 2.2.11, 2.2.12) is:

Theorem 34 *The pair $(D^{\le p}_c(X, \Lambda), D^{\ge p}_c(X, \Lambda))$ is a t-structure on $D^b_c(X, \Lambda)$.*

For p the constant function of value 0, this is the standard t-structure. The most interesting one is that relative to $p_{1/2}$.

The *p-perverse* Λ-sheaves on X are by definition the objects of the heart Per(X, Λ) of this t-structure. These sheaves are in fact complexes, but, in a sense, they behave like sheaves, as morphisms and objects can be glued on open coverings of X (*loc. cit.*, 2.2.19).

The proof of Theorem 34 is indirect. One reduces to proving its analogue for $\Lambda = \mathbf{F}_\ell$. Then one writes $D^b_c(X, \Lambda)$ as a filtering union of subcategories $D^b_{\mathscr{S},\mathscr{L}}(X, \Lambda)$ relative to a pair of a suitable stratification $\mathscr{S}$ on X and the choice $\mathscr{L}$ of a finite set of isomorphism classes of lisse irreducible sheaves of Λ-modules on each stratum. On each such category $D^b_{\mathscr{S},\mathscr{L}}(X, \Lambda)$ the desired t-structure is defined inductively by gluing, using Theorem 33. A different, more direct proof (working in other algebro-geometric contexts as well) is given by Gabber in [98].

For an immersion, or more generally, a quasi-finite morphism $f : X \to Y$ between schemes separated and of finite type over k, the *intermediate extension* functor

$$f_{!*} : \mathrm{Per}(X, \Lambda) \to \mathrm{Per}(Y, \Lambda) \tag{138}$$

is defined by

$$f_{!*}K := \mathrm{Im}({}^p f_! K \to {}^p f_* K),$$

with respect to the canonical factorization $f_! K \to {}^p f_! K \to {}^p f_* K \to f_* K$ (with the notation of (135), and the abbreviation of Rf_* into f_*).

From now on, assume $p = p_{1/2}$, $\Lambda = \overline{\mathbf{Q}}_\ell$. We will write Per(−) for Per(−, Λ), and $D^b_c(-)$ for $D^b_c(-, \Lambda)$. We fix an algebraic closure $\bar{k}$ of k. For $a : X \to \operatorname{Spec} k$ separated and of finite type, the dualizing functor $D = R\mathscr{H}om(-, a^! \Lambda)$ exchanges $D_c^{\le p}(X)$ and $D_c^{\ge p}(X)$, and, in particular, induces a self-duality of Per(X).

Artin's affine Lefschetz theorem ([4], XIV) can be reformulated by saying that, for an affine k-morphism $f : X \to Y$ (with X, Y separated of finite type over k), the functor Rf_* is right t-exact. The main result of geometric nature on perverse sheaves is the following theorem, which follows from this re-interpretation:

Theorem 35 *For X/k separated and of finite type, the abelian category* Per(X) *is noetherian and artinian. Every simple object is of the form $j_{!*}\mathscr{L}[\dim(V)]$, for an irreducible subscheme $j : V \hookrightarrow X$ such that $V_{\bar{k},\mathrm{red}}$ is smooth, and an irreducible lisse $\overline{\mathbf{Q}}_\ell$-sheaf $\mathscr{L}$ on V.*

For X equidimensional of dimension d, a remarkable perverse sheaf (simple if X is irreducible) is the *intersection complex*

$$IC_X := j_{!*}(\overline{\mathbf{Q}}_\ell[d]), \tag{139}$$

where $j : V \hookrightarrow X$ is a dense open immersion, with $V_{\bar{k},\mathrm{red}}$ smooth. The cohomology groups $H^i(X_{\bar{k}}, IC_X[-d])$ are the analogues of the intersection cohomology groups constructed by Goresky and MacPherson [102–104].

The Purity and Decomposition Theorems

We keep the conventions at the end of Sect. 5.8 "Perverse Sheaves in the Étale Setting". We assume $k = \mathbf{F}_q$, for q a power of a prime p.

For X_0/k separated and of finite type, the t-structure relative to $p_{1/2}$ on $D^b_c(X_0) = D^b_c(X_0, \overline{\mathbf{Q}})$ induces a t-structure on the full subcategory $D^b_m(X_0)$

consisting of mixed complexes (94), and any subquotient of a mixed perverse sheaf is mixed. A central result is the so-called *purity theorem for the intermediate extension*, first proved by Gabber:

Theorem 36

(i) *If* $\mathscr{F}_0 \in \mathrm{Per}(X_0)$ *is mixed of weights* $\leq w$ *(resp.* $\geq w$*), any subquotient of* $\mathscr{F}_0$ *is mixed of weights* $\leq w$ *(resp.* $\geq w$*).*
(ii) *If* $j : U_0 \hookrightarrow X_0$ *is an affine embedding, and* $\mathscr{F}_0 \in \mathrm{Per}(U_0)$ *is mixed of weights* $\leq w$ *(resp.* $\geq w$*), then* $j_{!*}\mathscr{F}_0$ *is mixed of weights* $\leq w$ *(resp.* $\geq w$*). In particular, if* $\mathscr{F}_0$ *is pure of weight* w*, then so is* $j_{!*}\mathscr{F}_0$*.*

Assertion (ii) follows from (i) by Artin's affine Lefschetz theorem. The proof of (i) given in *loc. cit.* is different from Gabber's original proof. It relies on a criterion (*loc. cit.*, 5.2.1) for a perverse sheaf $\mathscr{F}_0$ to be mixed of weights $\geq w$ involving $H^0(U, \mathscr{F})$ on a variable affine scheme U_0 étale over X_0, namely that, for all such U_0, $H^0(U, \mathscr{F})$ be of weight $\geq w$ (the removing of the index 0 denoting as usual the pull-back over $\overline{k}$).

A corollary of (ii) is that, if X_0 is proper and equidimensional of dimension d, then $\mathrm{IH}^i(X)$ (139) is pure of weight i. Gabber proved later [97] that $\det(1 - Ft, \mathrm{IH}^i(X))$ belongs to $\mathbf{Z}[t]$ and is independent of ℓ, which generalizes (Theorem 22, Corollary 3).

We have seen in (Sect. 5.6 "First Applications", The weight filtration) that, in ([D46, 1980], 3.4.1), as a consequence of Theorem 22, Deligne proved that, if $\mathscr{F}_0$ is a mixed lisse sheaf on X_0, then $\mathscr{F}_0$ admits a unique finite increasing filtration W such that $\mathrm{gr}_i^W \mathscr{F}_0$ is lisse and *punctually* pure of weight i. A better statement holds for perverse sheaves, with "punctually pure" replaced by "pure": Theorem 36 implies:

Theorem 37 *Any perverse mixed sheaf* $\mathscr{F}_0$ *on* X_0 *admits a unique finite increasing filtration* W *(in the category* $\mathrm{Per}(X_0)$*), again called the weight filtration, such that, for all* i*,* $\mathrm{gr}_i^W \mathscr{F}_0$ *is pure (as a complex) of weight* i*. Any morphism* $\mathscr{F}_0 \to \mathscr{G}_0$ *in* $\mathrm{Per}(X_0)$ *is strictly compatible with the weight filtrations.*

In particular, *any simple mixed perverse sheaf is pure*. Moreover, any *pure* perverse sheaf $\mathscr{F}_0$ is *geometrically semisimple*: $\mathscr{F}$ on X is a direct sum of simple perverse sheaves of the form $j_{!*}\mathscr{L}[d]$, for a connected smooth subscheme $j : U \hookrightarrow X$ of dimension d and an irreducible lisse $\overline{\mathbf{Q}}_\ell$-sheaf $\mathscr{L}$ on U.

By definition, an object K_0 of $D_m^b(X_0)$ is of weights $\leq w$ if an only if, for all i, $H^i K_0$ is of weights $\leq w + i$ (cf. (94)). Surprisingly, the same holds with $H^i K_0$ replaced by ${}^p H^i K_0$ $(= {}^p H^0(K_0[i]))$. As $D\,{}^p H^i K_0 = {}^p H^{-i} D K_0$, where D is the dualizing functor on X_0, the assertion with $\leq w$ replaced by $\geq w$ holds, too, while it is not the case with the usual H^i's. In particular, K_0 is pure of weight w if and only if, for all i, ${}^p H^i K_0$ is pure of weight $w + i$. This leads to the other major result in the theory, the so-called *decomposition theorem* (*loc. cit.*, 5.4.5):

Theorem 38 *Let $K_0 \in D^b_m(X_0)$ be a pure complex. Then, in $D^b_c(X)$, $K = K_0|X$ has a decomposition*

$$K \tilde{\to} \oplus_{i \in \mathbf{Z}} {}^pH^i(K)[-i]. \tag{140}$$

This theorem has the following applications. Let $f_0 : X_0 \to Y_0$ be *proper*, then, as recalled at the end of Sect. 5.6 "Mixed Sheaves, Statement of the Main Theorem", Rf_{0*} sends $D^b_{\leq w}$ to $D^b_{\leq w}$ (resp. $D^b_{\geq w}$ to $D^b_{\geq w}$), and in particular, transforms pure complexes into pure complexes. For example, if X_0 is smooth of pure dimension d, then $Rf_{0*}\overline{\mathbf{Q}}_\ell$ is pure of weight 0, hence we have a decomposition

$$Rf_*\overline{\mathbf{Q}}_\ell \tilde{\to} \oplus_{i \in \mathbf{Z}} {}^pR^i f_*\overline{\mathbf{Q}}_\ell[-i], \tag{141}$$

with ${}^pR^i f_*\overline{\mathbf{Q}}_\ell$ pure of weight i. Analyzing the simple components of the (semisimple) pure perverse sheaves ${}^pR^i f_*\overline{\mathbf{Q}}_\ell$ is, in general, a nontrivial task. In the case of a *Hitchin fibration*, the determination of their supports was at the core of Ngo's proof of the fundamental lemma ([201], 7.2.1).

Theorem 38 implies generalizations of the local invariant cycle theorem (Theorem 30) and a global variant, and of the hard Lefschetz theorem (Theorem 29) (see (*loc. cit.*, 5.4.7, 5.4.8, 5.4.10).

Consequences Over C

The results in Sect. 5.8 "Perverse Sheaves in the Étale Setting" produce harmonics in Hodge theory: by the usual *spreading out* arguments, one gets from them analogues of the last theorems (decomposition, hard Lefschetz, etc.) for certain classes of objects of $D^b_c(X, \mathbf{C})$ ($X/\mathbf{C}$ separated and of finite type), called *of geometric origin* (*loc. cit.*, 6). A remarkable application is the following result ([D34, 1975], Théorème 2), ([D53, 1982], 6.2.3)), which shows the *discreteness* of the weight filtration of Hodge theory, as mentioned after Theorem 15:

Theorem 39 *Let $f : X \to S$ be a separated morphism of schemes of finite type over $\mathbf{C}$, and let $n \in \mathbf{N}$. Assume that $R^n f_*\mathbf{Q}$ is locally constant and that, for each closed point $s \in S$, the restriction map $r : (R^n f_*\mathbf{Q})_s \to H^n(X_s, \mathbf{Q})$ is an isomorphism. Then, there exists an increasing filtration of $R^n f_*\mathbf{Q}$ by locally constant subsheaves $W_i R^n f_*\mathbf{Q}$ such that, for each i, r induces an isomorphism $(W_i R^n f_*\mathbf{Q})_s \tilde{\to} W_i H^n(X_s, \mathbf{Q})$ for all s in S.*

The assumption of the theorem is satisfied, for example, if f is the restriction to the complement of a relative normal crossings divisor in a proper and smooth scheme over S.

The formalism of mixed complexes and perverse sheaves over finite fields served as a model for M. Saito's theory of mixed Hodge modules [222] (cf. Sect. 4.2 "Mixed Hodge Theory" and Sect. 4.5 "Link with Mixed Hodge Structures and Regulators"), and T. Mochizuki's theory of mixed twistor $\mathscr{D}$-modules [194].

5.9 Recent Results

In [D114, 2012], [D115, 2013], and [80] Deligne gives applications of Lafforgue's main theorem in [159] to various questions concerning ℓ-adic sheaves.

A Finiteness Theorem

In ([Weil II], 1.2.10) Deligne made a number of conjectures on ℓ-adic sheaves. Namely, given $X_0/\mathbf{F}_q$ of finite type, normal and geometrically irreducible, and a lisse, irreducible $\overline{\mathbf{Q}}_\ell$-sheaf $\mathscr{F}_0$ of rank r on X_0 whose determinant is of finite order, he conjectured the following:

(i) $\mathscr{F}_0$ is pure of weight zero.
(ii) There exists a number field E contained in $\overline{\mathbf{Q}}_\ell$ such that, for all $x \in |X_0|$, the polynomial $\det(1 - F_x t, \mathscr{F}_0)$ has coefficients in E.
(iii) For any finite place λ of E not dividing p, the inverse roots of $\det(1-F_x t, \mathscr{F}_0)$ in $\overline{E}_\lambda$ are λ-adic units.
(iv) For all places λ of E dividing p, and any inverse root α of $\det(1 - F_x t, \mathscr{F}_0)$, the absolute value of the valuation $v(\alpha)$ satisfies the inequality (where $N_x = \sharp k(x)$)

$$|v(\alpha)/v(N_x))| \leq r/2.$$

(v) Up to enlarging E, for any finite place λ of E not dividing p, there exists a lisse E_λ-sheaf $\mathscr{F}'_0$ compatible with $\mathscr{F}_0$, i.e., having the same eigenvalues of Frobenius.
(vi) For λ dividing p, crystalline companions are expected.

In [159], for X_0 a curve, L. Lafforgue proved (i), (ii), (iii), (v), and an asymptotically weaker version (iv′) of (iv), namely $|v(\alpha)/v(N_x))| \leq (r-1)^2/r$. He also showed how to reduce the general case of (i), (iii) and (iv′) to the curve case, but there was a gap in his Bertini argument, which, for (i) and (iii), was filled in by Deligne in [D114, 2012] (see also Drinfeld ([83], Th. 2.15) and Esnault-Kerz's ([89], Prop. 8.1) for alternate arguments and more precise results) . By ([90], B1) one gets (iv′) for $X_0/\mathbf{F}_q$ smooth; the general case follows, using ([264], 2.5). The estimate (iv) on curves was proved by V. Lafforgue ([160], Cor. 2.2). In fact, V. Lafforgue proves a common generalization of (iv) and (iv′), namely (iv″): $|v(\alpha)/v(N_x))| \leq (r-1)/2$, which extends to the general case by the same method.[27] In [D114, 2012], Deligne proves (ii) unconditionally. L. Lafforgue's results, especially (v) (for X_0 a curve), play an essential role in the proof. See [89] for a variant of the exposition. Using (ii), Drinfeld [83] proved (v) for X_0 *smooth*

[27] Drinfeld and Kedlaya [84] recently showed that, moreover, for $X_0/\mathbf{F}_q$ smooth, the slopes of the minimal Newton polygon have gaps ≤ 1, which result also extends to the general case ([265], 2.7).

(the general case is still open). See [265] for generalizations of these results to Artin stacks. For X_0 a curve, conjecture (vi) was proved by T. Abe [10]. For X_0 smooth of higher dimension conjecture (vi) is still open, but variants are discussed in [11] and [151].

Counting Lisse ℓ-Adic Sheaves

In [D115, 2013] and [80] Deligne revisits results of Drinfeld [82]. Given a smooth projective curve $X_0/\mathbf{F}_q$, a reduced closed subset S_0 of X_0, and an integer $n \geq 1$, let $E(\mathscr{R})$ be the set of isomorphism classes of lisse irreducible $\overline{\mathbf{Q}}_\ell$-sheaves $\mathscr{F}$ of rank n on $X - S$ (where as usual, $X = X_0 \otimes_{\mathbf{F}_q} \mathbf{F}$, $S = S_0 \otimes_{\mathbf{F}_q} \mathbf{F}$), with prescribed ramification $\mathscr{R}$ at the points of S (given by a family of isomorphism classes of rank n sheaves on the local fields of the strict localizations of X at the points of S, which is isomorphic to itself by Fr^*, see ([80], 2.1)). The problem at stake is the following ([80], 2.3 (i)). Count the cardinality of the fixed point set $E(\mathscr{R})^V$ of $E(\mathscr{R})$ under the permutation $V = \mathrm{Fr}^*$ given by pull-back by the Frobenius endomorphism $\mathrm{Fr} = F_{X_0} \otimes_{\mathbf{F}_q} \mathbf{F}$ of X. One can view $E(\mathscr{R})^V$ as the set of classes that come from lisse Weil sheaves on $X_0 - S_0$ ([D115, 2013], 1.3, 1.6) (or the set of classes, modulo torsion by a rank one Weil sheaf on $\mathrm{Spec}\, \mathbf{F}_q$, of lisse Weil sheaves $\mathscr{F}_0$ on $X_0 - S_0$ which become irreducible on $X - S$), and have the prescribed ramification $\mathscr{R}$. More generally, one can consider, for any integer $m \geq 1$, the set

$$\mathscr{T} = \mathscr{T}(X_0, S_0, n, m, \mathscr{R})$$

of such isomorphism classes which are fixed under V^m, whose cardinality we denote by $T = T(X_0, S_0, n, m, \mathscr{R})$. The mere finiteness of this set is not obvious (it is a consequence of Lafforgue's main theorem). For $n = 1$ and $S_0 = \emptyset$, class field theory identifies it with the set of characters of the finite group $\mathrm{Pic}^0_{X_0}(\mathbf{F}_{q^m})$. For $n = 2$, and $S_0 = \emptyset$, Drinfeld [82], using an automorphic interpretation of $\mathscr{T}$ and Arthur–Selberg trace formula for GL(2), found that T, as a function of m, is of the form $\sum a_i \beta_i^m$ for suitable q-Weil numbers β_i.

In [D115, 2013] a similar formula is established in arbitrary rank n, under the restriction that S_0 has at least two points, and $\mathscr{R}$ is given by principal unipotent sheaves, i.e., the inertia at each s acts through $\mathbf{Z}_\ell$ with one unipotent Jordan block. The proof relies on the automorphic interpretation of $\mathscr{T}$ given by Lafforgue [159] and a compact case of the trace formula. In [80] Deligne gives an overview of what is known on the general problem mentioned at the beginning of this section, especially the question of finding a geometric interpretation for the formulas obtained for T.

See [90] for an alternate exposition of some of these results and questions.

6 Modular and Automorphic Forms

6.1 Construction of ℓ-Adic Representations

Deligne's proof of the Ramanujan–Petersson conjecture for modular forms relies on his construction of associated ℓ-adic representations (Sect. 5.6 "First Applications"). For a historical sketch of the ideas leading to this construction, involving work of Eichler, Igusa, Kuga, Sato, Shimura, see Serre ([240], *Une interprétation des congruences relatives à la fonction* τ *de Ramanujan*, § 6).

The main result is the following ([D28, 1974], 6.1), where the notation is the same as in Theorem 26:

Theorem 40 *Let* $f = \sum_{n\geq 1} a_n q^n$ *be a modular form of type* $(k+2, \varepsilon)$ *on* $\Gamma_0(N)$, *with* $k \geq 0$, *which is cuspidal and primitive. Then, for any prime number* ℓ, *there exists a lisse irreducible* $\overline{\mathbf{Q}}_\ell$*-sheaf* $\mathscr{F}_{f,\ell}$ *of rank 2 on* $\mathrm{Spec}(\mathbf{Z}[1/N\ell])$, *punctually pure of weight* $k+1$, *such that for any prime* p *not dividing* $N\ell$, *one has*

$$\det(1 - F_p t, \mathscr{F}_{f,\ell}) = 1 - a_p t + \varepsilon(p) p^{k+1} t^2,$$

where F_p *is a geometric Frobenius at* p.

In other words, there exists a (continuous) representation $\rho_{f,\ell} : \mathrm{Gal}(\overline{\mathbf{Q}}/\mathbf{Q}) \to \mathrm{GL}_2(\overline{\mathbf{Q}}_\ell)$ satisfying

$$\mathrm{Tr}(\rho_{f,\ell}(\varphi_p)) = a_p, \quad \det(\rho_{f,\ell}(\varphi_p)) = \varepsilon(p) p^{k+1}$$

for all p not dividing $N\ell$, where $\varphi_p = F_p^{-1}$[28] and $a_p = \alpha_p + \overline{\alpha}_p$, where α_p is a p-Weil number of weight $k+1$. Actually, by the modular interpretation of the space of cusp forms of weight $k+2$ under $\Gamma_1(N)$, there exists a number field K containing all the a_p's and $\varepsilon(p)$'s ([D28, 1974], 2.7), and the $\overline{\mathbf{Q}}_\ell$-sheaf $\mathscr{F}_{f,\ell}$ comes by extension of scalars from an K_λ-sheaf $\mathscr{F}_{f,\lambda}$, where λ is a place above ℓ, i.e., $\rho_{f,\ell}$ comes from $\rho_{f,\lambda} : \mathrm{Gal}(\overline{\mathbf{Q}}/\mathbf{Q}) \to \mathrm{GL}_2(K_\lambda)$.

The desired sheaf $\mathscr{F}_{f,\ell}$ is defined as a suitable direct summand of the lisse sheaf

$$ {}^k_N W_\ell := R^1 a_*(j_* \mathrm{Sym}^k R^1 f_* \overline{\mathbf{Q}}_\ell) \tag{142}$$

on $\mathrm{Spec}(\mathbf{Z}[1/N\ell])$, where

$$a : \mathscr{M}_N \to \mathrm{Spec}(\mathbf{Z}[1/N\ell])$$

[28] Note that for $\sigma \in \mathrm{Gal}(\overline{\mathbf{F}}_p/\mathbf{F}_p)$, the isomorphism $[\sigma] : \mathrm{Spec}\,\overline{\mathbf{F}}_p \to \mathrm{Spec}\,\overline{\mathbf{F}}_p$ deduced by transportation of structure is given by $[\sigma]^* x = \sigma^{-1} x$.

is the restriction to $\mathrm{Spec}(\mathbf{Z}[1/N\ell])$ of the Deligne–Mumford stack $\mathcal{M}_N[1/N]$ classifying generalized elliptic curves over $\mathrm{Spec}(\mathbf{Z}[1/n])$ with a full level N structure (Theorem 6),

$$j : \mathcal{M}_N^0 \hookrightarrow \mathcal{M}_N$$

the inclusion of the open substack consisting of elliptic curves, and $f : E \to \mathcal{M}_N^0$ is the universal elliptic curve. The construction is sketched in [D6, 1969] (and is complete for $N = 1$, $k = 10$, which gives the Ramanujan conjecture). The *Shimura isomorphism* expressing the stalk at $\mathbf{C}$ of ${}^k_N W_\ell^H$ (H a suitable subgroup of $\mathrm{GL}_2(\mathbf{Z}/N\mathbf{Z})$) as a direct sum of a space of cusp forms of weight k and of its conjugate ([D6, 1969], 2.10), and its ℓ-adic counterpart, the *Eichler–Shimura congruence formula*, describing the reduction mod a prime p not dividing $N\ell$ of the Hecke operator T_p on ${}^k_N W_\ell|\mathrm{Spec}\mathbf{F}_p$ as a sum of a Frobenius operator and its twisted dual ([D6, 1969], 4.9) (see also Theorem 7), play a key role.

The definition of ${}^k_N W_\ell$ and its application to the construction of the desired representations was proposed by Serre in a letter to Verdier dated Feb. 11, 1967 ([8], pp. 909–911) (as a first approximation Serre worked with the compactly supported cohomology group $R^1 a_* j_!$ instead of the interior one). A copy of this letter was later sent to Deligne, who solved the problem. However, Deligne hasn't yet written up the details of his construction of $\rho_{f,\ell}$. The reader may consult [230], where Scholl constructs a motive $M(f)$ over $\mathbf{Q}$ with coefficients in K giving rise to the family of $\rho_{f,\lambda}$'s. See also [224].

6.2 The Weil–Deligne Group

Let R be a henselian discrete valuation ring, with fraction field K and finite residue field $\mathbf{F}_q$ of characteristic p. Let $\overline{K}$ be a separable closure of K, $\overline{R}$ the integral closure of R in $\overline{K}$, and $\overline{k}$ the residue field of $\overline{R}$ (which is an algebraic closure of $\mathbf{F}_q$). Let $W(\overline{K}/K) \subset \mathrm{Gal}(\overline{K}/K)$ be the *Weil group*, inverse image of the subgroup $W(\overline{k}/k)$ of $\mathrm{Gal}(\overline{k}/k)$ generated by the geometric Frobenius $F : x \mapsto x^{1/q}$, so that we have an exact sequence

$$0 \to I \to W(\overline{K}/K) \overset{\deg}{\to} W(\overline{k}/k)(= \mathbf{Z}) \to 0,$$

where I is the inertia subgroup, and $W(\overline{k}/k)$ is identified with $\mathbf{Z}$ by $F \mapsto 1$. Let ℓ be a prime number $\neq p$, E_λ a finite extension of $\mathbf{Q}_\ell$, and $\rho : W(\overline{K}/K) \to \mathrm{GL}(V)$ a continuous representation, where V is a finite dimensional E_λ-vector space. As in the discussion following Lemma 1, by *Grothendieck's local monodromy theorem*, there exists an open subgroup I_1 of the inertia group $I \subset W(\overline{\eta}/\eta)$ and a nilpotent morphism $N : V(1) \to V$, such that $\rho(g) = \exp(N.t_\ell(g))$ for all $g \in I_1$, where $t_\ell : I \to \mathbf{Z}_\ell(1)$ is the ℓ-primary component of the tame character. The morphism

$N \in \mathrm{End}(V)(-1)$ is unique, hence Galois equivariant: for $g \in W(\overline{K}/K)$, one has

$$\rho(g)N\rho(g)^{-1} = q^{-\deg(g)}N \tag{143}$$

(as g acts on $\mathbf{Z}_\ell(-1)$ by $a \mapsto q^{\deg(g)}a$). In particular, if $\widetilde{F}$ is a lifting of F in $W(\overline{K}/K)$, one has

$$N\rho(\widetilde{F}) = q\rho(\widetilde{F})N. \tag{144}$$

In ([D25, 1973], 8.3, 8.4), Deligne constructs an algebraic group

$$'W(\overline{K}/K) \tag{145}$$

such that isomorphism classes of *continuous* representations ρ as above correspond bijectively to isomorphism classes of *algebraic* representations of $'W(\overline{K}/K)$ (over E_λ). This group is now called the *Weil–Deligne group* (of K). It is defined as the semidirect product of the (discrete) group $W(\overline{K}/K)$ by $\mathbf{G}_a$ (over $\mathbf{Q}$), $W(\overline{K}/K)$ acting on $\mathbf{G}_a$ by $gxg^{-1} = q^{-\deg(g)}x$, as suggested by (143).[29] If E is a field of characteristic zero, an (algebraic) representation of $'W(\overline{K}/K)$ in a finite dimensional E-vector space W is a pair (ρ', N') consisting of a homomorphism $\rho' : W(\overline{K}/K) \to \mathrm{GL}(W)$ and a nilpotent endomorphism N' of W satisfying the relation $\rho'(g)N'\rho'(g)^{-1} = q^{-\deg(g)}N'$. Starting with a continuous homomorphism $\rho : W(\overline{K}/K) \to \mathrm{GL}(V)$ as above, and choosing an isomorphism $\tau : \mathbf{Q}_\ell(1) \xrightarrow{\sim} \mathbf{Q}_\ell$ (of $\mathbf{Q}_\ell$-vector spaces) and a lift $\widetilde{F}$ of F as above, the pair (ρ', N') associated with ρ is defined by taking $N' = N$ (using τ) and $\rho'(\widetilde{F}^n\sigma) = \rho(\widetilde{F}^n\sigma)\exp(-t_\ell(\sigma)N)$; the isomorphism class of ρ' does not depend on the choices.

The interpretation of ρ in terms of ρ' is purely algebraic, in particular, does not depend of the topology of E_λ. This change of viewpoint had useful consequences.

(a) It enabled to define the notion of F-semisimplification (of a representation ρ) ([D26, 1973), 8.6) (in particular, ρ is F-semisimple if and only if $\rho(\widetilde{F})$ is semisimple (this doesn't depend on the choice of $\widetilde{F}$)), and that of a *strictly compatible system* of λ-adic representations of $W(\overline{K}/K)$ ([D26, 1973), 8.8), and consequently that of a strictly compatible system of λ-adic representations of the Weil group of a global field k. For such a system, and k the function field of a smooth irreducible curve over $\mathbf{F}_q$, Deligne proved a product formula for the global constant of the family (see Sect. 6.3).
(b) The Weil–Deligne group appears in the formulation of the Deligne–Langlands conjecture (see Sect. 10), and, more generally, in the *local Langlands correspondence*. For a local field K as above, and an integer $n \geq 1$, this correspondence, for GL_n, is a bijection $\pi \mapsto \rho_\pi$ between the set of isomorphism classes

[29] A variant of this construction was later considered by Langlands, with $\mathbf{G}_a$ replaced by SL_2, [164].

of smooth irreducible (complex) representations π of $\mathrm{GL}_n(K)$ and the set of isomorphism classes of continuous F-semisimple n-dimensional (complex) representations ρ_π of the Weil–Deligne group $'W(\overline{K}/K)$, characterized by the property of preserving L-functions and ε-factors of pairs and extending the Artin correspondence for $n = 1$. In a seminal letter to Piatetski-Shapiro [65], Deligne sketches how to construct such a correspondence, for $K = \mathbf{Q}_p$, $p \neq 2$, by looking at the action of $\mathrm{GL}_2(\mathbf{Q}_p) \times H^* \times W(\overline{\mathbf{Q}}_p/\mathbf{Q}_\mathbf{p})$ on the group of ℓ-adic vanishing cycles $R^1\Psi(\mathbf{Q}_\ell)$ of the modular curve of p^∞-level (cf. Sect. 2.3 "Reduction mod p") at a supersingular point E of the special fibre over $\overline{\mathbf{F}}_p$ (here $H = \mathrm{End}(E)$). The correspondence was then established in general by Laumon–Rapoport–Stuhler [172] for K of equal characteristic, and by Harris and Taylor [111], and, independently, Henniart [116], in the mixed characteristic case (a simplified proof was recently found by Scholze [232]).

6.3 Local Constants of L-Functions

Construction of Local Constants

Deligne shows in ([D26, 1973], 4.1) that there exists a rule associating with each local field K as in Sect. 6.2, $\overline{K}$ a separable closure of K, dx a Haar measure on K, ψ a nontrivial additive character of K, $\rho : W(\overline{K}/K) \to \mathrm{GL}(V)$ a continuous homomorphism, where V is a finite dimensional complex vector space, a number

$$\varepsilon(V, \psi, dx) \in \mathbf{C}^*, \tag{146}$$

(called a *local constant*) satisfying a number of functoriality and normalization properties which characterize this rule uniquely. The functoriality properties consist of mutiplicativity in short exact sequences, homogeneity in ψ (replacing ψ by $a\psi$ replaces ε by $a^{\dim(V)}\varepsilon$), and compatibility with finite separable induction (see *loc. cit.* for the precise formulation). The normalization is that for $\dim(V) = 1$, i.e., ρ defined by a quasi-character $\chi : W(\overline{K}/K) \to \mathbf{C}^*$, one has

$$\varepsilon(V, \psi, dx) = \varepsilon(\chi, \psi, dx), \tag{147}$$

where the right hand side is the constant appearing in Tate's local functional equation (see (*loc. cit.*, 3.3.1)). In particular, $\varepsilon(\chi, \psi, dx) = 1$ if χ is unramified (and $\int_R dx = 1$, $\psi|R = 1$), and $\varepsilon(\chi, \psi, dx) = \int_{K^*} \chi^{-1}(x)\psi(x)dx$ if χ is ramified (essentially a Gauss sum in the tame case). These constants for V and its dual are related by a simple functional equation.

If now k is a global field, and W a finite dimensional (continuous) complex representation of the Weil group of k, the (Weil) L-function of W satisfies a functional equation where the global constant $\varepsilon(W)$ is equal to a product of local constants $\varepsilon(W_v, \psi_v, dx_v)$ (*loc. cit.* 5.11.3). Thanks to the formal properties of the local constants, this formula is reduced to the case where W is of dimension 1, i.e., to the functional equation for Hecke L-functions.

These results had been obtained earlier by Langlands [162]. His construction of the local constants was achieved by a purely local method and some details remained unpublished.[30] Deligne's proof, which is much simpler, uses a global argument, based on a formula expressing the behavior of the local constant under torsion of W by a very ramified character of K^*.

The Case of Function Fields

Let now k be the function field of a geometrically irreducible smooth projective curve $X_0/\mathbf{F}_q$, $\overline{k}$ a separable closure of k, E_λ a finite extension of $\mathbf{Q}_\ell$, with $\ell \neq p$ ($p = \mathrm{char}(\mathbf{F}_q)$), and $\rho : \mathrm{Gal}(\overline{k}/k) \to \mathrm{GL}(W)$ a finite dimensional (continuous) representation of $\mathrm{Gal}(\overline{k}/k)$ over E_λ, which is almost everywhere unramified. This representation is the stalk $\mathscr{F}_{0\overline{\eta}}$ at the geometric point $\overline{\eta} = \mathrm{Spec}\,\overline{k}$ of a lisse E_λ-sheaf $\mathscr{F}_0$ of rank $r = \dim(W)$ on an open Zariski subset $j_0 : U_0 \hookrightarrow X_0$. The L-function $L(W, t)$ ($t = q^{-s}$) is Grothendieck's L-function (an element of $E_\lambda(t)$)

$$L(W, t) = L(X_0, j_{0*}\mathscr{F}_0, t),$$

defined by the product

$$L(W, t) := \prod_{v \in |X_0|} \det(1 - F_v t^{\deg(v)}, j_*\mathscr{F})^{-1}$$

indexed by the closed points of X_0 (= places of k), where $(j_*\mathscr{F})_{\overline{v}} = W^{I_v}$, I_v denoting an inertia group at $\overline{v}$ over v (cf. (73)). Grothendieck's L-functions $L(G_0, t)$ for G_0 in $D^b(X_0, \overline{\mathbf{Q}}_\ell)$ satisfy a functional equation of the form

$$L(G_0, t) = \det(-Ft, R\Gamma(X, G))^{-1} L(D(G), t^{-1}),$$

whose proof, by reduction to the case of finite coefficients, relies on a trace formula for Frobenius applied to (derived) symmetric powers of $R\Gamma(X, G)$, which itself makes a crucial use of Deligne's symmetric Künneth formula (13). Here, as $D(j_{0*}\mathscr{F}_0) = j_{0*}D\mathscr{F}_0$ (91), as observed by Deligne in ([D26, 1973], 10.8), the functional equation of $L(W, t)$ takes the simple form

$$L(W, t) = \varepsilon(W, t) L(W^\vee(1), t^{-1}) (= \varepsilon(W, t) L(W^\vee, q^{-1}t^{-1})),$$

[30] See Langlands's comments on his website, on Langlands's Notes on Artin L-functions.

where

$$\varepsilon(W,t) = \varepsilon(W)t^{-\chi(X,\mathscr{F})},$$

$$\varepsilon(W) = \det(-F, H^*(X, j_*\mathscr{F}))^{-1}$$

(with the notation of Sect. 5.1 for $(X, \mathscr{F})$, and $\chi(X,\mathscr{F}) = \sum(-1)^i \dim H^i(X,\mathscr{F})$). The case of Weil L-functions discussed above suggested to Deligne a formula for $\varepsilon(W)$ as a product of local constants, namely,

$$\varepsilon(W) = \prod_{v \in |X_0|} \varepsilon(W_v, \psi_v, dx_v), \tag{148}$$

where $dx = \otimes dx_v$ is a decomposition of the Tamagawa measure on $\mathbf{A}_k$ giving mass 1 to $\mathbf{A}_k/k$, with total mass of $\mathscr{O}_v$ (the completed local ring of X_0 at v) equal to 1 for almost all v, and ψ a nontrivial additive character of $\mathbf{A}_k/k$, inducing ψ_v on $k_v = \mathrm{Frac}(\mathscr{O}_v)$. The conjectural formula (148), implicit in [D26, 1973], was stated in [74], where Deligne gave it the equivalent geometric form (*loc. cit*, II 2.3, IV 2.1.3)

$$\varepsilon(W) = q^{(1-g)\dim W} \prod_{v \in |X_0|} \varepsilon(W_v, \psi_v, (dx)_{0v}), \tag{149}$$

where g is the genus of X, and $(dx)_{0v}$ denotes the Haar measure on k_v of total mass 1.

In [D26, 1973] Deligne proved (148) in two cases: (a) $\mathscr{F}_0$ has finite geometric global monodromy, i.e., the restriction of ρ to $\pi_1(U,\overline{\eta}) = \mathrm{Ker}(\pi_1(U_0,\overline{\eta}) \to \mathrm{Gal}(\mathbf{F}/\mathbf{F}_q))$ has finite image, (more generally, for any Weil representation on a global field) (ii) ρ belongs to an *infinite* family $(\rho_\lambda)_{\lambda \in \mathbf{L}}$ of *strictly compatible* representations $\rho_\lambda : \mathrm{Gal}(\overline{k}/k) \to \mathrm{GL}(V_\lambda)$, where $\mathbf{L}$ is an infinite set of finite places $\lambda \neq p$ of a number field E, V_λ a finite dimensional vector space over the completion of E at λ, and strict compatibility means compatibility of the associated Weil–Deligne representations (or, which suffices, of their F-semisimplifications) at each place of X_0, as mentioned in (a) after (145).

In a long letter to Serre [D30, 1974], Deligne described a strategy to prove (148), based on his symmetric Künneth formula (13) and the known relations between symmetric powers of a curve and its Jacobian, and solved the problem in the rank one case. Working at a finite level, with a finite local ring of coefficients Λ, of residue characteristic ℓ, the problem is to determine the *graded invertible* Λ-module $\det(R\Gamma_c(U,\mathscr{F}))$ up to a unique isomorphism. Suppose that $\chi_c(U,\mathscr{F}) = -N$, with $N \geq 1$. Then, using (13) and some general nonsense on the category of graded invertible modules, viewed as a category of *stable* projective modules, one gets the following formula, which is the starting point of the theory:

$$\det R\Gamma_c(U,\mathscr{F}) = (\det R\Gamma_c(\mathrm{Sym}^N(U), \Gamma^N_{\mathrm{ext}}(\mathscr{F})))^{(-1)^{N+1}}. \tag{150}$$

In the case: $U_0 = X_0$, $\mathscr{F}_0$ lisse of rank 1, so that $N = 2g - 2$, where g is the genus of X, the right hand side of (150) can be easily analyzed through the Leray spectral sequence of the canonical morphism

$$\pi : \mathrm{Sym}^{2g-2}(X) \to J^{2g-2} \tag{151}$$

(where J^{2g-2} is the component of degree $2g-2$ of Pic_X), whose fibers are projective spaces, and is smooth outside the canonical class. In fact, by this method, and techniques of geometric class field, Deligne treated more generally the case of any lisse $\mathscr{F}_0$ of *rank one* on U_0, with arbitrary ramification along $X_0 - U_0$ (and obtained (148)).

Elaborating on this, in [74] Deligne proved (148) for $\mathscr{F}_0$ tamely ramified of *any* rank, and in a subsequent seminar, planned to generalize this to the case where $j_*\mathscr{F}$ is *arbitrarily ramified* (using as another ingredient his theory of nearby cycles over general bases Sect. 7.4). However, the details were not written up. One reason is that in [171] Laumon gave a proof of (149) in the general case, by a different method, using Deligne's ℓ-adic Fourier transform, and his ℓ-adic analogue of the stationary phase principle (134). This product formula was later a key ingredient in the proof by Lafforgue of the Langlands correspondence for GL_n over function fields [159]. Note that, in turn, Lafforgue's theorem, combined with the local Langlands correspondence (cf. (b) after (145)), shows that an irreducible lisse $\overline{\mathbf{Q}}_\ell$-sheaf $\mathscr{F}_0$ on U_0 with finite determinant is a member of an infinite strictly compatible system of such sheaves.

Additional Results

Deligne made two additional contributions to the study of the local constants.

(a) In [D37, 1976], he considers the local constant $\varepsilon(V \otimes \omega_s, \psi, dx)$ (where ω_s is the quasi-character $x \mapsto ||x||^s$ of $K^* = W(\overline{K}/K)^{\mathrm{ab}}$), for a *real* virtual representation V of $\mathrm{Gal}(\overline{K}/K)$, of dimension 0 and determinant 1. This local constant depends neither on ψ nor dx, which can therefore be omitted from the notation. He gives a formula for $\varepsilon(V \otimes \omega_{1/2})$, namely

$$\varepsilon(V \otimes \omega_{1/2}) = \exp(2\pi i \,\mathrm{cl}(w_2(V))), \tag{152}$$

where $\mathrm{cl}(w_2(V))$ is the image in $(\mathbf{Q}/\mathbf{Z})_2 = \pm 1$ of the second Whitney class of V, $w_2(V) \in H^2(G, \mathbf{Z}/2\mathbf{Z})$ (G a finite quotient of $\mathrm{Gal}(\overline{K}/K)$ through which V factorizes), by the composite of the natural map from $H^2(G, \mathbf{Z}/2\mathbf{Z})$ to $H^2(\mathrm{Gal}(\overline{K}/K), \overline{K}^*)$, and the map $\mathrm{inv} : H^2(\mathrm{Gal}(\overline{K}/K), \overline{K}^*) \to \mathbf{Q}/\mathbf{Z}$ of local class field theory. By the reciprocity law, this theorem implies earlier results of Fröhlich and Queyrut [95].

(b) In the proof of the existence of a theory of local constants, Deligne used, as a crucial ingredient, a formula expressing the behavior of the local constant under torsion of W by a very ramified character of K^*. In a joint paper with Henniart [D48, 1981], he generalizes this to the torsion of W by a representation of $W(\overline{K}/K)$ of arbitrary dimension.

6.4 Abelian L-Functions and Hilbert–Blumenthal Moduli Spaces

In [63], following a suggestion of Serre, Deligne described how congruences among values at negative integers of abelian L-functions for totally real fields would follow from a theory of p-adic Hilbert modular forms. Such a theory would rely on the construction over $\mathbf{Z}$ of certain Hilbert–Blumenthal moduli schemes, having irreducible geometric fibers in characteristic p. He proposed this construction to Rapoport as a problem for his thesis. Rapoport solved it in [214]. Deligne and Ribet exploited this in [D45, 1980]. The crux of their article is an irreducibility theorem (*loc. cit.*, 4.6) of the above mentioned type, whose proof—in addition to the results of [214]—uses a description of ordinary abelian varieties over finite fields given much earlier by Deligne [D7, 1969]. As a corollary they obtain a q-expansion principle for Hilbert modular forms on $\Gamma_{00}(N)$, giving rise to integrality results for the values at negative integers of the corresponding L-functions, Kummer-type congruences, and the construction of p-adic L-functions.

However, it was later discovered that a key result in [214] used by Deligne and Ribet in [D45, 1980] was wrong: the compactification $\overline{\mathscr{M}}$ constructed by Rapoport for the moduli space $\mathscr{M}$ for g-dimensional abelian varieties with multiplication by the ring of integers of a totally real field of degree g over $\mathbf{Q}$ and level structure $\Gamma(n)$, $n \geq 3$, could not be proper and smooth over $\mathbf{Z}[\mu_n][1/n]$, as asserted, because at the primes p dividing the discriminant Δ of K the corresponding ℓ-adic representation of Gal($\overline{\mathbf{Q}}/\mathbf{Q}$) is ramified ($p$ prime to ℓ and n) (as was already observed for $g = 2$ in [110]). In [D79, 1994] Deligne and Pappas fill in the gap by modifying the moduli problem: the new moduli problem contains the old one as an open subscheme with dense fibers, and can be compactified into a scheme proper over $\mathbf{Z}[\mu_n][1/n]$, whose fibers over the primes p dividing Δ are singular (in fact, they are shown to be local complete intersections, smooth in codimension 1, in particular, normal). With this modification, the desired irreducibility property can be proved, and the main results of [D45, 1980] are restored.

7 Nearby Cycles and Euler Numbers

7.1 The Picard–Lefschetz Formula

In SGA 7 (= [6]) Grothendieck introduced and studied the nearby and vanishing cycles functors $R\Psi$ and $R\Phi$, both in the complex analytic setup for Betti cohomology, and in the algebraic setup, for étale cohomology. He used them to give a proof of Milnor's conjecture on the quasi-unipotency of the monodromy of a Milnor fiber at an isolated critical point of a holomorphic function on a smooth complex analytic space (see Sect. 7.2), and, more generally, of the monodromy theorem,[31] in equal characteristic zero (and conditionally otherwise[32]). Deligne gave a brief account of this in (*loc. cit.*, I), and developed the formalism in more detail in ([7] (= [D19, 1972]), XIII, XIV). In particular, he proved a *comparison theorem* ([7], XIV, 2.8) between Betti and étale nearby cycles, similar to Artin's comparison theorem between étale and Betti higher direct images by a morphism of $\mathbf{C}$-schemes separated and of finite type.

A central result in the theory is the *Picard–Lefschetz formula*, which describes the *variation morphism* for isolated ordinary quadratic singularities. Unable to make sense of the topological arguments of Lefschetz [174], Grothendieck left it to Deligne to write a proof and translate the result into étale cohomology. This was the object of Deligne's exposé ([7], XV). See (*loc. cit.*, 3.2.1, 3.3.5) for a precise statement of the formula in the étale cohomology setup. The datum is a flat, finite type morphism $X \to S$, of relative dimension n, with (S, s, η) a henselian trait, such that the special fibre X_s is smooth except for an ordinary quadratic singularity at a single closed point. For n even, the proof is algebraic. But for n odd, Deligne used a deformation argument to reduce, by means of the comparison theorem mentioned above, to the classical formula over $\mathbf{C}$, of which he gave a transcendental proof in ([7], XIV).[33] Grothendieck used the Picard–Lefschetz formula in relative dimension 1 in the proof of the semistable reduction theorem for abelian varieties ([6], IX 12.5). In arbitrary dimension, the Picard–Lefschetz formula was the basis for the cohomological study of Lefschetz pencils, done by Katz in ([7], XVII, XVIII), which in turn played a critical role in [Weil I], as we have seen (Sect. 5.5).

A p-adic theory of vanishing cycles is still lacking. In the mixed characteristic $(0, p)$ case, analogues of ℓ-adic vanishing or nearby cycles, with $\mathbf{Q}_\ell$ replaced by $\mathbf{Q}_p$, were considered and studied by Bloch–Kato [37], and many others afterwards. They play an important role in p-adic Hodge theory. However, they are far from giving rise to the expected theory, as in the case of good reduction they are already

---

[31] This theorem says that, if (S, s, η) is a henselian trait and X/η is separated and of finite type, $\overline{\eta}$ a geometric point over η, and n an integer invertible on S, an open subgroup I_1 of the inertia group I acts unipotently on $H^*(X_{\overline{\eta}}, \mathbf{Z}/n\mathbf{Z})$ (resp. $H_c^*(X_{\overline{\eta}}, \mathbf{Z}/n\mathbf{Z})$).

[32] Grothendieck gave an unconditional proof for $H_c^*(X_{\overline{\eta}}, \mathbf{Z}/n\mathbf{Z})$ by another argument, of arithmetic nature, see (*loc. cit.*, I).

[33] A purely algebraic proof was found later [126].

highly nontrivial invariants. In [D57, 1984], Deligne proves the following theorem (a generalization of a result of Furstenberg):

Theorem 41 *Let k be a field and* $g = \sum a_{\mathbf{n}} \mathbf{x}^{\mathbf{n}}$ *a formal series in N indeterminates* $\mathbf{x} = (x_1, \cdots, x_N)$ $(\mathbf{n} = (n_1, \cdots, n_N))$. *Consider the series*

$$I_N(g) = \sum_n a_{n,\cdots,n} t^n \in k[[t]].$$

If g is algebraic over $k(x_1, \cdots, x_N)$ *and k is of characteristic* $p > 0$, *then* $I_N(g)$ *is algebraic over* $k(t)$.

The relation between Theorem 41 and the sought for theory of p-adic vanishing cycles comes from the following integral formula for $I_N(g)$ when $k = \mathbf{C}$ and g is convergent:

$$I_N(g) = \int_{Z(t)} g dz_1 \cdots dz_N / dt, \tag{153}$$

where $Z(t)$ is the "vanishing cycle" at 0 for the morphism $\mathbf{A}^N \to \mathbf{A}^1, (x_1, \cdots, x_N) \mapsto x_1 \cdots x_N$, defined by $Z(t) = \{(x_1, \cdots, x_N) | x_1 \cdots x_N = t, |x_1| = r_1, \cdots, |x_N| = r_N\}$ with $\prod r_i = |t|$. The case $N = 2$ is the Picard–Lefschetz situation in relative dimension 1. The proof of Theorem 41 is by induction on N, and, for $N = 2$, uses a form of Grothendieck duality for coherent sheaves on a surface. Deligne expresses the hope that a suitable theory of p-adic vanishing cycles would yield a direct proof in the general case, and, for g with integer coefficients, an estimate in $O(p^M)$ for the degree over $\mathbf{F}_p(t)$ of the reduction mod p of $I_N(g)$.

7.2 *The Milnor Number*

Let $f : (\mathbf{C}^{n+1}, 0) \to (\mathbf{C}, 0)$ be a germ of holomorphic function, with $f(0) = 0$, smooth outside 0. Milnor showed in [192] that if B is a small closed ball around $0 \in \mathbf{C}^{n+1}$, then, for $t \in \mathbf{C}$ sufficiently close to 0, $V_f := B \cap f^{-1}(t)$ is a manifold with boundary having the homotopy type of a bouquet of r n-dimensional spheres, these manifolds V_f (later called *Milnor fibers*) forming a locally trivial fibration over a sufficiently small punctured disc centered at 0. In particular, if $\widetilde{H}^* = \operatorname{Coker} H^*(\mathrm{pt}) \to H^*$, we have $\widetilde{H}^q(V_f, \mathbf{Z}) = 0$ for $q \neq n$, and $\widetilde{H}^n(V_f, \mathbf{Z}) = \mathbf{Z}^r$. Moreover, Milnor gave a differential interpretation of the integer r, namely,

$$r = \dim_{\mathbf{C}} \mathbf{C}[[x_0, \cdots, x_n]] / (\partial f / \partial x_0, \cdots, \partial f / \partial x_n). \tag{154}$$

In Grothendieck's notation, $\widetilde{H}^q(V_f, \mathbf{Z}) = R^q\Phi_f(\mathbf{Z})_{\{0\}}$. If f comes from an algebraic map $\widetilde{f}$ from an étale neighborhood of $\{0\}$ in $\mathbf{A}^{n+1}_{\mathbf{C}}$ to an étale neighborhood of $\{0\}$ in $\mathbf{A}^1_{\mathbf{C}}$, then, by the comparison theorem, $R^q\Phi_f(\mathbf{Z})_{\{0\}}) \otimes \mathbf{Z}/\ell\mathbf{Z} = R^q\Phi_{\widetilde{f}}(\mathbf{Z}/\ell\mathbf{Z})_{\{0\}}$, where the right hand side is taken for the étale topology, so that (154) has a purely algebraic meaning. This suggested to Deligne the following algebraic analogue.

Let (S, s, η) be a strictly local trait, with *algebraically closed* residue field $k = k(s)$, let $\overline{\eta}$ be a geometric point over η, I the inertia group. Let $f : X \to S$ be a flat, finite type morphism, of relative dimension n, with X regular, x a closed point of X_s, and assume that $f|X - \{x\} \to S$ is smooth. Then f is a locally complete intersection morphism, and the cotangent complex $L_{X/S}$ is just $\Omega^1_{X/S}$. The sheaf $\mathscr{D}^1_{X/S} = \mathscr{E}xt^1(\Omega^1_{X/S}, \mathscr{O}_X)$ is a coherent module supported on x. Deligne defined the *Milnor number* of f at x as the length of the $\mathscr{O}_{X,x}$-module which is its stalk at x:

$$\mu(f, x) = \lg(\mathscr{D}^1_{X/S,x}). \tag{155}$$

This generalizes the right hand side of (154). He then conjectured the following formula, generalizing (154):

$$(-1)^n \mathrm{dimtot}(R\Phi_f(\mathbf{Z}/\ell\mathbf{Z})_x) = \mu(f, x). \tag{156}$$

Here

$$\mathrm{dimtot}(R\Phi_f(\mathbf{Z}/\ell\mathbf{Z})_x) := \sum_i (-1)^i \mathrm{dimtot}(R^i\Phi_f(\mathbf{Z}/\ell\mathbf{Z})_x), \tag{157}$$

where, for a finite dimensional $\mathbf{F}_\ell$-representation V of the inertia group I, $\mathrm{dimtot}(V) := \dim(V) + \mathrm{sw}(V)$, $\mathrm{sw}(V)$ denoting the *Swan conductor* of V.[34] In fact, it was later observed that, under the above assumptions, $R^i\Phi_f(\mathbf{Z}/\ell\mathbf{Z})_x = 0$ for $i \neq n$, so that the left hand side of (156) is equal to $\mathrm{dimtot}(R^n\Phi_f(\mathbf{Z}/\ell\mathbf{Z})_x)$. In ([D19, 1972], XVI), Deligne proved (157) for $n = 0$ or S of equal characteristic. The case $n = 1$ was treated by Orgogozo [209]. The general case is still open. Variants in equal characteristic and with coefficients have recently been established by T. Saito, in the case of surfaces in [225], and in arbitrary dimension in [227].

7.3 Wild Ramification and Euler–Poincaré Characteristics

Let X be a projective smooth curve over an algebraically closed field k of characteristic p, and ℓ a prime number $\neq p$. If $\mathscr{F}$ is a constructible $\mathbf{F}_\ell$-sheaf

[34] "dimtot" stands for "dimension totale".

on X, the *Grothendieck–Ogg–Shafarevich formula* ([5], X) expresses the Euler–Poincaré characteristic $\chi(X, \mathscr{F}) = \sum(-1)^i \dim H^i(X, \mathscr{F})$ in terms of the generic rank $r(\mathscr{F})$ of $\mathscr{F}$ and local invariants $a_x(\mathscr{F})$ at the points of non smoothness of $\mathscr{F}$:

$$\chi(X, \mathscr{F}) = \chi(X)r(\mathscr{F}) - \sum_{x \in X(k)} a_x(\mathscr{F}), \tag{158}$$

where $\chi(X) = \chi(X, \mathbf{F}_\ell) = \chi(X, \mathbf{Q}_\ell)$ and $a_x(\mathscr{F}) = r(\mathscr{F}) - \dim(\mathscr{F}_x) + \mathrm{sw}(\mathscr{F}_{\overline{\eta}_x})$ is the *total drop* of rank, $\overline{\eta}_x$ denoting a geometric generic point of the strict localization of X at x.

In a series of letters, to Katz [67] and to me [68–72], Deligne began investigating generalizations of (158) to higher dimensions. Here are the main points. See [129] for more details.

Local Behavior

It follows from (158) that if $\mathscr{F}_1$ and $\mathscr{F}_2$ are constructible $\mathbf{F}_\ell$-sheaves on X having the same local behavior at each point, then they have the same Euler–Poincaré characteristic. In [70], using pencils, Deligne showed that, more generally, if X/k is proper and smooth (of any dimension), and $\mathscr{F}_1$, $\mathscr{F}_2$ are constructible $\mathbf{F}_\ell$ (or $\overline{\mathbf{F}}_\ell$) sheaves on X whose images in the corresponding Grothendieck group are locally equal, then $\chi(X, \mathscr{F}_1) = \chi(X, \mathscr{F}_2)$. A little later, by a different method, inspired by his work with Lusztig [D35, 1976] (see Sect. 8.1 "A Fixed Point Formula"), Deligne proved a strong refinement, namely, that if $\mathscr{F}_1$ and $\mathscr{F}_2$ have the same rank and wild ramification at infinity along the strata of a suitable stratification of X where they are lisse, then $\chi(X, \mathscr{F}_1) = \chi(X, \mathscr{F}_2)$ still holds [123]. In particular, if $\mathscr{F} = j_!\mathscr{G}$, for $j : U \hookrightarrow X$ the complement of a divisor, and $\mathscr{G}$ lisse of rank r on U, and *tamely ramified* along $X - U$, then $\chi_c(U, \mathscr{G}) = r\chi_c(U)$.[35]

It was recently shown by T. Saito and Yatagawa [226] that the equality $\chi(X, \mathscr{F}_1) = \chi(X, \mathscr{F}_2)$ holds more generally for $\mathscr{F}_i$ a constructible $\mathbf{F}_{\ell_i}$-sheaf, with ℓ_1 and ℓ_2 different from p, and possibly unequal, provided that $\mathscr{F}_1$ and $\mathscr{F}_2$ "have the same wild monodromy" (a (weaker) variant of the condition described above[36]). In fact, under this assumption, $\mathscr{F}_1$ and $\mathscr{F}_2$ have the same *characteristic cycle* (in the sense of [227]).

[35]The same holds for χ, as $\chi_c(U, \mathscr{G}) = \chi(U, \mathscr{G})$ by Laumon [168].

[36]For surfaces, this variant is re-interpreted in [141] as an equality of conductors for the restriction to every curve.

Jump of Swan Conductor

A natural way of attacking the problem (of generalizing (158)) was by the usual strategy of fibration into curves, hence a preliminary question was to understand the behavior of the Swan conductor $\mathrm{sw}_{x_t}(\mathscr{F}_t)$ of a sheaf $\mathscr{F}_t$ on a curve X_t, when $(X_t, \mathscr{F}_t, x_t)$ moves in a family over a parameter space S. Deligne proved the following result [69, 167], analogue of the result on irregularities in his letter to Katz in [D107, 2007] (cf. Sect. 3.4):

Theorem 42 *Let $f : X \to S$ be a smooth relative curve, with S excellent noetherian, $j : U \hookrightarrow X$ the complement of a closed subscheme Y, finite and flat over S. Let ℓ be a prime invertible on S, and $\mathscr{F}$ a lisse $\mathbf{F}_\ell$-sheaf on U, of constant rank r. Then the function*

$$\varphi : S \to \mathbf{N}, \quad s \mapsto \varphi(s) = \sum_{x \in Y_{\overline{s}}} (\mathrm{sw}_x(j_!\mathscr{F}|X_{\overline{s}}) + r)$$

(where $\overline{s}$ is a geometric point over s) is constructible and lower semicontinuous. If it is locally constant, f is universally locally acyclic with respect to $j_!\mathscr{F}$.

When S is a strictly local trait, with closed (resp. generic) point s (resp. η), case to which the problem can be reduced by a global to local method, cf. Sect. 1.4 "Finiteness", and Y_s consists of a single point x, Deligne gave a formula for the jump of the Swan conductor:

$$\varphi(s) - \varphi(\eta) = -\dim(R^1\Phi_f(j_!\mathscr{F})_x). \tag{159}$$

This theorem was recently revisited by T. Saito, who gave a simplified proof [227].

Surfaces

Let X/k be a proper and smooth surface (k algebraically closed of characteristic $p \neq \ell$ as above), $D \subset X$ a divisor, $j : U = X - D \hookrightarrow X$, and $\mathscr{F}$ a lisse $\mathbf{F}_\ell$-sheaf of rank r on U. In [70], under some restriction on the ramification of $\mathscr{F}$, Deligne wrote a formula for $\chi(X, j_!\mathscr{F})$ of the form

$$\chi(X, j_!\mathscr{F}) = r\chi(X) - \delta(X, \mathscr{F}), \tag{160}$$

where the error term $\delta(X, \mathscr{F})$ is a sum of terms depending on the generic wild ramification of $\mathscr{F}$ along the components of D, and also at a finite number of exceptional closed points of D. The restrictive hypothesis is that $\mathscr{F}$ has no *fierce* ramification, i.e., at each maximal point of D, the normalization of X in a finite extension trivializing $\mathscr{F}$ does not make appear any purely inseparable extension of the residue field. The proof of (160) uses a method of pencils. The details were written up by Laumon in his thesis (*Compléments à "Caractéristique d'Euler–Poincaré de faisceaux constructibles sur une surface", Orsay, 1983*), see [169] for an overview. In [71] Deligne investigates the fierce case, and studies Artin–Schreier examples in detail.

With these letters Deligne initiated a new line of research to which several mathematicians have brought important contributions (especially K. Kato and T. Saito) and is still active today. In [225] T. Saito treated the general fierce case on a surface. In [79], Deligne sketched a theory of singular support, characteristic cycle, and Euler–Poincaré formulas à la Brylinski–Dubson–Kashiwara, that Beilinson [29] and T. Saito [227] recently developed in full generality. See [129] for a brief report. Further progress in the direction of the ultimate goal, i.e., a Grothendieck–Riemann–Roch type formula for ℓ-adic sheaves, was made by T. Saito in [228].

7.4 Nearby Cycles Over General Bases

In the wake of his proof of the product formula conjecture (148) in the tame case Deligne introduced in [170] a new geometric and cohomological tool, enabling him to study nearby cycles in families. It had been known in the late 1970s that Milnor fibrations didn't generalize well to bases of dimension >1 (see [220]). A fortiori, it looked doubtful that one could construct a reasonable theory of nearby cycles in étale cohomology over bases of arbitrary dimension. This is nevertheless what Deligne did.

Given morphisms of topoi $f : X \to S$, $g : Y \to S$ (the case of interest is when f and g are the morphisms of étale topoi of schemes), Deligne constructs a topos $X \overleftarrow{\times}_S Y$ (the *oriented product* of f and g), together with 1-morphisms $p_1 : X \overleftarrow{\times}_S Y \to X$, $p_2 : X \overleftarrow{\times}_S Y \to Y$, and a 2-morphism $\tau : gp_2 \to fp_1$, which is universal for these properties (see [170], ([131], XI)). In the case $g = \mathrm{Id}_S$ and f comes from a morphism of schemes (still denoted f), $X \overleftarrow{\times}_S S$ is called the *vanishing topos* of f. The pair of projections ($\mathrm{pr}_1 = \mathrm{Id}_X$, $f : X \to S$) defines a morphism $\Psi_f : X \to X \overleftarrow{\times}_S S$ such that $p_1 \Psi_f = \mathrm{Id}_X$, $p_2 \Psi_f = f$, whose derived functor

$$R\Psi : D^+(X, \mathbf{Z}/n\mathbf{Z}) \to D^+(X \overleftarrow{\times}_S S, \mathbf{Z}/n\mathbf{Z}) \tag{161}$$

generalizes the usual functor of nearby cycles when S is a henselian trait. The derived category cokernel $R\Phi_f$ of the canonical morphism $p_1^* \to R\Psi_f$ generalizes the vanishing cycles functor. When f is a morphism of finite type between noetherian schemes, there is a good notion of constructible sheaf of $\mathbf{Z}/n\mathbf{Z}$-modules on $X \overleftarrow{\times}_S S$, and, consequently, of the derived category $D_c^b(X \overleftarrow{\times}_S, \mathbf{Z}/n\mathbf{Z})$, consisting of complexes with bounded, constructible cohomology. Assume now that n is invertible on S. Deligne proved that, for $K \in D_c^b(X, \mathbf{Z}/n\mathbf{Z})$, $R\Psi_f K$ is in $D^b_c(X \overleftarrow{\times}_S S, \mathbf{Z}/n\mathbf{Z})$ and base change compatible provided that the locus of local acyclicity of (K, f) is quasi-finite over S. He conjectured that, with no assumption on the locus of local acyclicity of (K, f), this property would hold after a suitable modification of S (depending on (K, f)). That was proven by Orgogozo [210]. For $f = \mathrm{Id}_S$ and $g : Y \to S$, the topos $S \overleftarrow{\times}_S Y$, called the *co-vanishing topos*, a variant of a topos introduced by Faltings in p-adic Hodge theory, plays a role in the p-adic Simpson correspondence, studied by Abbes–Gros-Tsuji [9].

8 Reductive Groups

8.1 Deligne–Lusztig

Representation theory of finite groups of Lie type has a long history (see, e.g., [56, 177, 238]). Let k be an algebraic closure of the finite field $\mathbf{F}_q$ of characteristic p, $G_0/\mathbf{F}_q$ a (connected) reductive group, G/k deduced from G_0 by base change (with the notational convention of Sect. 5.1). Let F be the Frobenius k-endomorphism of G. The fixed point scheme G^F is the finite group $G_0(\mathbf{F}_q)$. In [D35, 1976] Deligne and Lusztig describe a cohomological procedure to construct irreducible representations of G^F. In fact, for any maximal F-stable torus $T \subset G$ and character θ of T^F, they construct a (virtual) representation R^{θ}_T of G^F, which is irreducible if θ is sufficiently general, and cuspidal if $T/Z(G)$ is anisotropic. In particular, they prove *Macdonald's conjecture*. Further results, pertaining to duality, are discussed in ([D51, 1982], [D54, 1983]). Their method made a breakthrough, and was very influential. An especially rich development is Lusztig's theory of *character sheaves*, already alluded to at the end of Sect. 5.7 "Laumon's Contribution and Applications", which provides a complete solution to the classification of irreducible representations of finite groups of Lie type, in the spirit of the geometric Langlands correspondence.

In what follows, we only give a brief account of some of Deligne–Lusztig's main results, essentially extracted from Serre's Bourbaki report [238].

Deligne–Lusztig Varieties

Let $T = T_0 \otimes k$ be as above, and let W be *the* Weyl group (inverse limit, under conjugation, of $N(T')/T'$ for T' a maximal torus). Let X be the (projective) variety of k-Borel subgroups of G, which, for $B \in X$, is identified with G/B by $gB \mapsto gBg^{-1}$. Let G act on $X \times X$ by diagonal conjugation: $g(B_1, B_2)g^{-1} = (gB_1g^{-1}, gB_2g^{-1})$. By the Bruhat decomposition, for any $(B_1, B_2) \in X \times X$, there exists $g \in G$, $B \in X$, and a *unique* $w \in W$ such that $(B_1, B_2) = g(B, wBw^{-1})g^{-1}$. We then say that B_1, B_2 are in *relative position* w (written $B_1 \overset{w}{-} B_2$). Thus, W

parametrizes the orbits of G on $X \times X$. Let $O(w)$ denote the orbit relative to w, i.e., $O(w) := \{(B_1, B_2) | B_1 \overset{w}{-} B_2\}$. It is a smooth scheme of dimension $\dim(X) + l(w)$, where $l(w)$ is the length of w. The Frobenius endomorphism F of G acts on X by $B \mapsto FB$.[37] For $w \in W$, Deligne–Lusztig define the locally closed subscheme

$$X(w) := \{B \in X | (B, FB) \in X(w)\}, \tag{162}$$

i.e., the (transverse) intersection of $O(w)$ with the graph of Frobenius. The $X(w)$ are smooth subschemes of X of dimension $l(w)$, and they form a stratification of X. For $w = e$, $X(e)$ is the finite set of B's such that $B = FB$, i.e., $(G_0/B_0)(\mathbf{F}_q)$ if $B = B_0 \otimes_{\mathbf{F}_q} k$ is one of them. If $G = GL_n$, X is the variety of complete flags $D = (D_1 \subset \cdots \subset D_{n-1})$ in $\mathbf{A}^n_k$, and for $w \in W = S_n$ the circular permutation $(1, \cdots, n)$, $X(w)$ is identified by $D \mapsto D_1$ with the set of lines in $\mathbf{A}^n_k$ not contained in any $\mathbf{F}_q$-rational hyperplane (D is then the flag $D_1 \subset D_1 + FD_1 \subset D_1 + FD_1 + F^2 D_1 \subset \cdots$).

The representations R^θ_T mentioned above are obtained from certain G^F-equivariant T^F-torsors on the Deligne–Lusztig varieties $X(w)$, whose definition depends on auxiliary choices. Namely, let U be a unipotent subgroup of G such that $B = T.U$ is a Borel subgroup (so that U is the unipotent radical of B). Let w be the element of W such that $B \in X(w)$. Let $\mathscr{L} : G \to G$ be the Lang isogeny, $g \mapsto g^{-1}Fg$, whose kernel is G^F. Define

$$X_U := \mathscr{L}^{-1}(FU) \subset G. \tag{163}$$

This is a G^F-torsor on FU. It also has a compatible action of T^F: $G^F \times T^F$ acts on X_U by $(g, t)x = gxt$. From the projection $G \to X$ one deduces an isomorphism

$$X(w) \xrightarrow{\sim} X_U/T^F.(U \cap FU), \tag{164}$$

which makes $Z_U := X_U/(U \cap FU)$ a G^F-equivariant T^F-torsor on $X(w)$. Let ℓ be a prime $\neq p$, and $\overline{\mathbf{Q}}_\ell$ an algebraic closure of $\mathbf{Q}_\ell$. If Γ is a finite group, we denote by $R(\Gamma, \overline{\mathbf{Q}}_\ell)$ (or $R(\Gamma)$) the Grothendieck group of finite dimensional $\overline{\mathbf{Q}}_\ell$-representations of Γ. A homomorphism $\theta : T^F \to \overline{\mathbf{Q}}^*_\ell$ defines a G^F-equivariant lisse rank 1 $\overline{\mathbf{Q}}_\ell$-sheaf $\mathscr{F}_\theta$ on Z_U/T^F (or $X(w)$ via (164)). The representation R^θ_T is the virtual representation of G^F

$$R^\theta_T := \sum (-1)^i H^i_c(Z_U/T^F, \mathscr{F}_\theta), \tag{165}$$

an element of $R(G^F)$. If $\pi : Z_U \to Z_U/T^F$ is the projection, we have a G^F-equivariant decomposition $\pi_* \overline{\mathbf{Q}}_\ell = \oplus_{\theta : T^F \to \overline{\mathbf{Q}}^*_\ell} \mathscr{F}_\theta$, hence $H^*(Z_U, \overline{\mathbf{Q}}_\ell)$ decomposes

[37] More precisely, $F^{-1}(B^{(q)})$, for $F : G \to G^{(q)}$ the relative Frobenius.

into

$$H^*(Z_U, \overline{\mathbf{Q}}_\ell) = \oplus_{\theta:T^F \to \overline{\mathbf{Q}}_\ell^*} H^*(Z_U/T^F, \mathscr{F}_\theta), \tag{166}$$

and

$$\sum (-)^i H^i(Z_U, \overline{\mathbf{Q}}_\ell) = \sum R_T^\theta. \tag{167}$$

A miracle—whose proof is at the core of [D35, 1976]—is that R_T^θ *does not depend on the choice of* U (which justifies the notation). The map $\theta \mapsto R_T^\theta$ extends to a $\overline{\mathbf{Q}}_\ell$-linear *induction homomorphism*

$$R_T : R(T^F) \to R(G^F), \quad \rho \mapsto R_T^\rho. \tag{168}$$

When one can choose U such that $U = FU$, so that $B = FB$, hence $w = 1$, which is the case, for example, when T_0 is split, then $Z_U = G^F.U/U = G^F/U^F$ is finite (and $Z_U/T^F = G^F/B^F = (G/B)^F$). Then $H^*(Z_U, \overline{\mathbf{Q}}_\ell) = H^0(Z_U, \overline{\mathbf{Q}}_\ell)$, and R_T is the usual procedure, consisting in restricting θ to B^F via the projection $B^F \to T^F$, and inducing it, in the classical sense, to G^F. New representations occur for $w \neq 1$. For $G = \mathrm{SL}_2$, and T_0 a nonsplit torus, then one can show that Z_U is isomorphic to the affine curve of equation $xy^q - x^q y = 1$, with its linear action of G^F. Drinfeld had studied this example, and showed that cuspidal representations of G^F occurred as summands of the form R_T^θ in $\sum(-)^i H^i(Z_U, \overline{\mathbf{Q}}_\ell)$ (167). This was the starting point of Deligne–Lusztig's theory.

A Fixed Point Formula

The main tool in the calculation of the characters of the representations R_T^θ is a Lefschetz fixed point formula for certain finite group actions. The main result is the following ([D35, 1976], 3.2):

Theorem 43 *Let k be an algebraically closed field of characteristic exponent p, ℓ a prime number $\neq p$, and X/k be a scheme separated and of finite type, endowed with an automorphism g of finite order. Write $g = su$, where s (resp. u) is a power of g, and s (resp. u) is of order m (resp. p^r), with $(p, m) = 1$. Then:*

(i) $\mathrm{Tr}(g^*, H_c^*(X, \mathbf{Q}_\ell))$ *is in* $\mathbf{Z}$ *and independent of* ℓ.
(ii) $\mathrm{Tr}(g^*, H_c^*(X, \mathbf{Q}_\ell)) = \mathrm{Tr}(u^*, H_c^*(X^s, \mathbf{Q}_\ell))$.

Here $\mathrm{Tr}(-, H_c^*) := \sum(-1)^i \mathrm{Tr}(-, H_c^i)$.

The key case, to which one is easily reduced, is when k is an algebraic closure of $\mathbf{F}_q$, $X = X_0 \otimes k$, with $X_0/\mathbf{F}_q$ quasi-projective, and g is defined on X_0. The proof of (i) relies on two elementary observations (which since then have been applied to many similar situations): (a) It suffices to show that for $n \geq 1$, $\mathrm{Tr}(F^n g, H_c^*(X, \mathbf{Q}_\ell))$

is in $\mathbf{Z}$ and independent of ℓ (b) For each $n \geq 1$, there exists a (quasi-projective) $Z_n/\mathbf{F}_{q^n}$ and an isomorphism $Z_n \otimes k \xrightarrow{\sim} X$ by which $F_{Z_n} \otimes_{\mathbf{F}_{q^n}} k = F^n g$. Assertion (ii) is reduced to the following theorem, whose proof combines (i) with techniques of perfect complexes and Brauer theory initiated by Grothendieck and Verdier:

Theorem 44 *Let X/k and ℓ be as in Theorem 43, and let G be a finite group acting freely on X. Then $R\Gamma(X, \mathbf{Z}_\ell)$ is a perfect complex of $\mathbf{Z}_\ell[G]$-modules, and for any $g \in G$ whose order is not a power of p,* $\mathrm{Tr}(g, H_c^*(X, \mathbf{Q}_\ell)) = 0$.

This last theorem is at the origin of Deligne's results on $\chi(X, \mathcal{F})$ mentioned in Sect. 7.3 "Local Behavior". The topic has been recently revisited by Serre et al. (see [130]).

Properties of the R_T^θ

They are obtained by a calculation of the corresponding characters, using Theorem 43, which emphasizes the importance of the knowledge of them on the set $\mathcal{U}(G^F)$ of *unipotent* elements $u \in G^F$, i.e., those of order a power of p, (when T_0 is split, and $B = TU$ is an F-stable Borel, then $\mathcal{U}(G^F)$ is just $U^F(\mathbf{F}_q)$). This knowledge is encoded in the so-called *Green function*

$$Q_T : \mathcal{U}(G^F) \to \overline{\mathbf{Q}}_\ell \tag{169}$$

defined by $Q_T(u) = R_T^1(u)$. A simple formula expresses the character of R_T^θ in terms of Q_T and θ ([D35, 1976], th. 4.2), but the determination of Q_T is rather involved. It is given the by *Green polynomials* (in the case of GL_n), and (for sufficiently big p) by the *Springer–Kazhdan formula*. However, we have $R_T^\theta(1) = Q_T(1)$, and $Q_T(1)$ is known in all cases:

$$R_T^\theta(1)(= \dim(R_T^\theta)) = Q_T(1) = \varepsilon_T \varepsilon_G |G^F|_{p'} |T^F|^{-1}, \tag{170}$$

where, for a finite set S, $|S|$ (resp. $|S|_{p'}$) denotes its cardinality (resp. the prime to p factor of its cardinality), ε_G (resp. ε_T) is $(-1)^\rho$, ρ denoting the $\mathbf{F}_q$-rank of G_0 (resp. T_0).

A central result in [D35, 1976] is an orthogonality relation between the characters of the R_T^θ's. In order to formulate it, recall that if Γ is a finite group, and C an algebraically closed field of characteristic zero (here we will take $C = \overline{\mathbf{Q}}_\ell$), there is a pairing on C-valued central functions on Γ given by

$$\langle a, b\rangle_\Gamma = \frac{1}{|\Gamma|} \sum_{g \in \Gamma} a(g) b(g^{-1}),$$

with respect to which characters of irreducible representations form an orthonormal basis. Deligne–Lusztig's orthogonality relation is the following formula:

Theorem 45 *Let T, T' be maximal tori of G defined over $\mathbf{F}_q$, and let θ (resp. θ') be an (irreducible) character of T^F (resp. T'^F). Then*

$$\langle R_T^\theta, R_{T'}^{\theta'} \rangle_{G^F} = N(\theta, \theta'), \tag{171}$$

where $N(\theta, \theta')$ is the number of isomorphisms $T \to T'$ induced by conjugation by an element g of G^F transforming θ into θ', i.e., such that $\theta(gtg^{-1}) = \theta'(t)$ for all $t \in T^F$.

This theorem has several important consequences:

- R_T^θ and $R_{T'}^{\theta'}$ are *orthogonal* if and only if (T, θ) and (T', θ') are not G^F-conjugate.
- If $N(\theta, \theta) = 1$, in which case one says that θ is *general*, then R_T^θ is irreducible.
- Assume that no proper parabolic subgroup of G defined over $\mathbf{F}_q$ contains T (this is the case if T modulo the center of G is *anisotropic*, i.e., of $\mathbf{F}_q$-rank zero). Then R_T^θ is *cuspidal* (*discrete series* in another terminology), i.e., its restriction to the unipotent radical of any proper parabolic subgroup of G does not contain the unit representation.

In addition, Deligne–Lusztig give a criterion for *disjointness* for a pair $(R_T^\theta, R_{T'}^{\theta'})$ in terms of *geometric conjugacy* ([D35,1976], 6.3), and prove:

- Every irreducible representation of G^F is a constituent of at least one representation R_T^θ ([D35,1976], 7.7).

They also calculate the values of R_T^θ on semisimple elements of G^F, thus obtaining the results predicted by Macdonald. In addition, they prove a remarkable relation (*loc. cit. 7.3*) between R_T^θ and the *Steinberg* representation St of G^F, namely:

$$R_T^\theta.\mathrm{St} = \varepsilon_G \varepsilon_T \mathrm{Ind}_{T^F}^{G^F}(\theta), \tag{172}$$

representations being identified with their characters; (172) applied to $g = 1$ gives (170).

Duality

Alvis [12] and Curtis [57] discovered and studied a duality operation on *characters* of groups of type G^F, exchanging the trivial character 1 and that of the Steinberg representation St. In [D51, 1982] and [D54, 1983], Deligne and Lusztig construct an explicit lift of this operation to a map $D_G : R(G^F) \to R(G^F)$. They prove that with respect to D_G, virtual representations of the form R_G^θ are self-dual up to sign, thus answering positively a question of Alvis (*loc. cit.*). More precisely, they prove that

$$D_G(R_T^\theta) = \varepsilon_G \varepsilon_T R_T^\theta, \tag{173}$$

with the notation of (170). The proof uses a generalization (due to Lusztig) of the induction (168), where the pair (T, B) is replaced by a pair (L, P), L a Levi subgroup of a parabolic subgroup P.

8.2 Central Extensions

Local symbols and central extensions have been a recurrent theme in Deligne's work. We have already mentioned his universal coefficients theorem (Sect. 1.4 "Picard Stacks and Geometric Class Field Theory"), and its analytic variants (Sect. 4.5 "Link with the Tame Symbol"), which give rise to the construction of certain central extensions on compact Riemann surfaces with boundary. In 1977–1978 Deligne ran a seminar at the IHÉS in which, given an absolutely simple, simply connected algebraic group G over a field k, such that G, as a scheme over k, is a rational variety, he constructed a canonical extension of $G(k)$ by $K_2(k)$ generalizing that constructed by Matsumoto for G split [189]. The seminar was not written up, but two related papers arose from it.

(a) In [D84, 1996], for G/k as above, but without the additional assumption of rationality, and given an integer n invertible in k, Deligne constructs a central extension

$$0 \to H^2(k, \mathbf{Z}/n\mathbf{Z}(2)) \to \widetilde{G}(k) \to G(k) \to 0, \tag{174}$$

which is defined up to a unique isomorphism, and is functorial in both k and G (where $H^2(k, -)$ means $H^2(\mathrm{Spec}\, k, -)$ for the étale topology, i.e., $H^2(\mathrm{Gal}(\bar{k}/k), -)$ for $\bar{k}$ a separable closure of k). For G split, it is deduced from Matsumoto's extension by pushing out via the Tate symbol

$$K_2(k) \to H^2(k, \mathbf{Z}/n\mathbf{Z}(2)), \tag{175}$$

defined by $\{x, y\} \mapsto dx.dy$, for x, y in k^*, and $d : k^* \to H^1(k, \mathbf{Z}/n\mathbf{Z}(1))$ the boundary of the Kummer sequence. However, Deligne's construction does not require G to be split, and for G split does not use Matsumoto's extension. The key ingredient of his construction is that, for $k = \bar{k}$, $H^4(BG, \mathbf{Z}/n\mathbf{Z}(2)) = \mathbf{Z}/n\mathbf{Z}$, which he deduces from known results for $k = \mathbf{C}$. It follows that, for k arbitrary, the relative (étale) cohomology group $H^4(BG \bmod Be, \mathbf{Z}/n\mathbf{Z}(2))$ is canonically isomorphic to $\mathbf{Z}/n\mathbf{Z}$, with a canonical generator c_Q associated with a certain quadratic form Q on the cocharacter group of a maximal torus of G. This generator has an image in $H^2(BG(k), H^2(k, \mathbf{Z}/n\mathbf{Z}(2))$, which defines the extension (174).

As a by-product of his construction, for k a global field, Deligne defines an extension of *metaplectic* type, namely a central extension of the adelic group

$G(\mathbf{A}_k)$ by the group $\mu(k)$ of roots of unity in k, canonically trivialized along $G(k)$:

$$\begin{array}{ccccccccc} & & & & & & G(k) & & \\ & & & & & \swarrow & \downarrow & & \\ 0 & \longrightarrow & \mu(k) & \longrightarrow & \widetilde{G}(\mathbf{A}_k) & \longrightarrow & G(\mathbf{A}_k) & \longrightarrow & 0 \end{array} \tag{176}$$

(b) Shortly after Deligne completed this work, Brylinski independently found that the restrictive hypothesis of rationality in Deligne's construction in his seminar was superfluous. The upshot was the joint paper [D98, 2001], in which, for any (connected) reductive group G over a field k, the authors classify central extensions of G by the sheaf $\mathbf{K}_2$ on the big Zariski site of $\operatorname{Spec} k$. If G is simple and simply connected, the group of isomorphism classes of such central extensions is $\mathbf{Z}$, and $1 \in \mathbf{Z}$ defines a canonical central extension

$$0 \to K_2(k) \to \widetilde{G}(k) \to G(k) \to 0, \tag{177}$$

which, for G split is the one constructed by Matsumoto. For $G = \mathrm{SL}_n$, and k infinite, $\widetilde{G}$ is the *Steinberg group* St_n, and (177) is the *universal central extension* constructed by Milnor. Deligne's extension (174) associated with c_Q is shown to be obtained from (177) by push-out by the Tate symbol (175). We refer the reader to the introduction of [D98, 2001] for the statement of the classification theorem, whose formulation would require too many preliminaries.

8.3 Braid Groups

Braid groups and the geometry and topology of related buildings or hyperplane arrangements have been a frequent topic in Deligne's work. I will discuss only two contributions.

In [D22, 1972], Deligne proves the following theorem:

Theorem 46 *Let V be a finite dimensional real vector space, $\mathscr{M}$ a finite set of linear hyperplanes of V, and $Y = V_{\mathbf{C}} - \cup_{M\in\mathscr{M}} M_{\mathbf{C}}$, where $V_{\mathbf{C}} = V \otimes_{\mathbf{R}} \mathbf{C}$, $M_{\mathbf{C}} = M \otimes_{\mathbf{R}} \mathbf{C}$. Assume that the connected components of $V - \cup_{M\in\mathscr{M}} M$ are open simplicial cones. Then Y is a $K(\pi, 1)$.*

First examples: (a) $V = \mathbf{R}$, $\mathscr{M}$ consisting of the single element $\{0\}$, $Y = \mathbf{C}^* = K(\mathbf{Z}, 1)$; (b) $V = \mathbf{R}^2 = \mathbf{C}$, $\mathscr{M}$ consisting of the m lines $\mathbf{R}e^{ki\pi/m}$, $m \geq 2$, $0 \leq k < m$, $Y = \mathbf{C}^* \times (\mathbf{P}^1(\mathbf{C}) - \mu_m)$, a $K(\pi, 1)$ with $\pi = \mathbf{Z} \times F_{m-1}$. In example (b), the dihedral group D_m (with $2m$ elements) acts freely on Y, and $\pi_1(Y/D_m)$ is the generalized braid group associated with the Coxeter group D_m. More generally, let W be a finite subgroup of $\mathrm{GL}(V)$ such that $V^W = \{0\}$, and assume that W is generated by orthogonal reflections for a W-invariant euclidian structure on V. Then the set $\mathscr{M}$ of hyperplanes M such that the orthogonal reflection through M belongs to W satisfies the condition of Theorem 46, W acts freely on $Y_W = V_{\mathbf{C}} - \cup_{M \in \mathscr{M}} M_{\mathbf{C}}$, and $X_W = Y_W/W$ is a $K(\pi, 1)$ with π the generalized braid group G_W associated with W (defined by the generators g_i $(1 \leq i \leq n = \dim(V))$ subject to the sole relations $g_i g_j g_i \cdots = g_j g_i g_j \cdots$ where the number of factors on each side is m_{ij}, (m_{ij}) denoting the Coxeter matrix of W). This result had been conjectured by Brieskorn [43], and proved but for a small number of cases. Deligne's proof of Theorem 46, however, is direct, and does not proceed by reduction to the case of a Coxeter complex. It does not involve any braid group, though some of the arguments were inspired by Garside's work on the word problem [99]. It consists in the construction of a certain *building* I associated with the data $(V, \mathscr{M})$, having the homotopy type of a bouquet of spheres, and a contractible covering $\widetilde{Y}$ of Y defined in terms of I.

In [D86, 1997], Deligne used the constructions of [D22, 1972] and its key contractibility result 2.9 to prove the following homotopical uniqueness theorem in positive braid monoids. Let B_n^+ be the monoid (without unit) of strictly positive braids on n strands, $n \geq 2$, presented by generators g_i $(1 \leq i \leq n-1)$, subject to the relations $g_i g_{i+1} g_i = g_{i+1} g_i g_{i+1}$ and $g_i g_j = g_j g_i$ for $j \geq i+2$. The canonical homomorphism from B_n^+ to the symmetric group S_n sends g_i to the transposition $s_i = (i, i+1)$. Let $\tau : S_n - \{e\} \to B_n^+$ be the set-theoretic section characterized by $\tau(s_i) = g_i$, $\tau(st) = \tau(s)\tau(t)$ if $l(st) = l(s) + l(t)$. Then the elements $\tau(w)$ for $w \in S_n - \{e\}$ and the previous relations make another presentation of B_n^+ (*loc. cit.*, 1.4.4). For $b \in B_n^+$, let $E(b)$ the set of ways of writing b as a product of $\tau(w)$'s (see (*loc. cit.*, 1.5) for a formal definition). Then $E(b)$ has a natural order relation, and the main result (*loc. cit.*, Th. 1.7) is:

Theorem 47 *The geometric realization of $E(b)$ is contractible.*

The theorem holds, in fact, more generally, for positive braid monoids associated with finite Coxeter groups. Deligne deduces from it a convenient description, by generators and relations, of an action of such a positive braid monoid on a monoidal category. He applies this to give refinements of Bondal–Kapranov's theory of exceptional systems [42], and Broué–Michel's theory of correspondences on flag manifolds [45].

8.4 Reductive Groups Over Local Fields

Aside from his letter to Piatetski-Shapiro [65] discussed above (Sect. 6.2 (b)), Deligne's work on this topic is essentially contained in the monograph [D58, 1984]. In addition to writing up Bernstein's exposé on the Bernstein center, Deligne made two contributions:

(a) In [D59, 1984] (joint with D. Kazhdan and M.-F. Vigneras), the Jacquet–Langlands correspondence [135] is generalized to higher ranks. Given an integer $m \geq 1$, a non archimedian local field F, and an F-central division algebra D of dimension d^2, the main theorem constructs[38] a bijection between isomorphism classes of square-integrable representations of $\mathrm{GL}(n, F)$ and $\mathrm{GL}(m, D)$, where $n = md$, preserving (up to sign) characters, L-functions, ε-factors. The Jacquet–Langlands case was $n = d = 2$, and the case $m = 1$ had been treated by Rogawski [216]. The proof uses a global argument, and a form of the Selberg trace formula.

(b) In [D60, 1984], Deligne transposed to the Galois side a principle of Kazhdan that the theory of representations of a local field of positive characteristic should be a limit of the corresponding theories for local fields having the same residue characteristic and absolute ramification index tending to infinity. Let F be a complete discrete valuation field, with perfect residue field k of characteristic $p > 0$, ring of integers $\mathcal{O}$, and maximal ideal $\mathbf{m}$. Elaborating on ideas of Krasner, Deligne shows that the category $\mathrm{ext}(F)^e$ of finite separable extensions E of F of ramification $\leq e$ (in the upper numbering notation)[39] depends only on the truncated discrete valuation ring $\mathcal{O}/\mathbf{m}^e$ and the pair D consisting of the invertible $\mathcal{O}/\mathbf{m}^e$-module $\mathbf{m}/\mathbf{m}^{e+1}$ and the canonical map $\mathbf{m}/\mathbf{m}^{e+1} \to \mathcal{O}/\mathbf{m}^e$. Thus, if F' is a second complete discrete valuation field, with residue field k, ring of integers $\mathcal{O}'$, and maximal ideal $\mathbf{m}'$, for any integer $e \geq 1$, the datum of an isomorphism $(\mathcal{O}/\mathbf{m}^e, D) \xrightarrow{\sim} (\mathcal{O}'/(\mathbf{m}')^e, D')$ defines an equivalence from $\mathrm{ext}(F)^e$ to $\mathrm{ext}(F')^e$, and, in particular, an isomorphism

$$\mathrm{Gal}(\overline{F}/F)/\mathrm{Gal}(\overline{F}/F)^e \xrightarrow{\sim} \mathrm{Gal}(\overline{F'}/F')/\mathrm{Gal}(\overline{F'}/F')^e, \tag{178}$$

unique, in fact, up to an inner automorphism, where $\overline{F}$ (resp. $\overline{F'}$) is a separable closure of F (resp. F'). The isomorphism (178) preserves Herbrand's functions, and, when k is finite (resp. algebraically closed), is compatible with the isomorphisms of class field theory (resp. geometric class field theory). As the truncations of the ring of integers of a local field of characteristic zero can be killed by p, one can thus "approximate" a local field of characteristic p by ramified local fields of characteristic zero. For example, if $F = \mathbf{F}_p((x))$, $F' = \mathbf{Q}_p[t]/(t^e - p)$ $(e > 0)$, one defines an isomorphism $\alpha : (\mathcal{O}/\mathbf{m}^e, D) \xrightarrow{\sim} (\mathcal{O}'/(\mathbf{m}')^e, D')$ by sending x to the uniformizing parameter π defined by the class of t (and one can make the correspondence $\mathrm{ext}(F)^e$ and $\mathrm{ext}(F')^e$ explicit, by lifting coefficients of Eisenstein polynomials via α).

At about the same time, and independently, Fontaine and Wintenberger [263], with motivations coming from Sen's theory and the theory of Fontaine's rings, gave

[38] At least, for F of characteristic zero: the case of positive characteristic was treated later by Badulescu [17].

[39] I.e., such that $\mathrm{Gal}(E_1/F)^e = 1$, where E_1 is the Galois closure of E.

another illustration of the same philosophy: given a local field K (of any characteristic) with perfect residue field k of characteristic $p > 0$, and a "big" algebraic separable extension L of K (for example, a totally ramified Galois extension with Galois group a positive dimensional p-adic Lie group), they construct a local field $X_K(L)$ of characteristic p with residue field isomorphic to that of L (called the *field of norms*), in such a way that $M \mapsto X_K(M)$ is an equivalence from the category of algebraic separable extensions of L to that of $X_K(L)$. This construction played an important role in p-adic Hodge theory, and far reaching generalizations have been recently developed by Scholze, in his theory of perfectoid spaces [231].

A few years later, the objects $(\mathscr{O}/\mathbf{m}^e, D)$ inspired to Deligne a theory of generalized Cartier divisors, called *"divisors"*, that he sketched in [77]. A "divisor" D on a scheme X is the datum of an invertible sheaf $\mathscr{L}$ and an $\mathscr{O}_X$-linear map $u : L \to \mathscr{O}_X$ (D corresponds to an effective Cartier divisor when the image by u of any local basis of $\mathscr{L}$ is a nonzero divisor). Together with a similar notion devised by Faltings, and the classical theory of de Rham complexes with log poles, it is at the origin of the theory of logarithmic structures (Fontaine–Illusie, Kato et al.), see [129] for historical remarks on this.

9 Motives and Periods

9.1 *Tensor Categories*

The Tannakian Formalism

In his second talk on motives at the IHÉS in 1967, Grothendieck introduced what he later called a *Tannakian category*: given a field k, a k-linear abelian category $\mathscr{A}$ equipped with a tensor product $\otimes$ together with data of associativity, commutativity and unity satisfying certain constraints, each object X having a dual $X^\vee$ (satisfying certain obvious axioms), and such that there exists a *fiber functor* ω from $\mathscr{A}$ to the category of vector spaces over an extension K of k.[40] When one can take $K = k$, in which case one says that $\mathscr{A}$ is *neutral*, $\mathscr{A}$ turns out to be equivalent to the category of representations of an affine k-group scheme G, the automorphism group of ω (which plays the role of a fundamental group $\pi_1(\mathscr{A}, \omega)$); in general, representations of G have to be replaced by representations of a certain groupoid. Admitting the standard conjectures, Grothendieck applied this formalism to $k = \mathbf{Q}$ and $\mathscr{A}$ the $\mathbf{Q}$-linear abelian category of (pure) motives M over a field F, up to isogeny; here fiber functors come from ℓ-adic realizations, or Betti realizations for F of characteristic

[40] I.e., a k-linear, exact functor, with an isomorphism $\omega(X) \otimes \omega(Y) \xrightarrow{\sim} \omega(X \otimes Y)$ compatible with the associativity and commutativity data; such a functor necessarily has values in the category of finite dimensional K-vector spaces.

zero, and give rise to the so-called *motivic Galois groups*. He left it to Saavedra to write up the details of the theory in his thesis, which appeared in [218, 219].

However, in the course of writing a report on Saavedra's work in ([D52, 1982], II), Deligne and Milne discovered an error in the proof of the main result in the non-neutral case ([219], Introduction, th. 3), and conjectured a corrected statement, with the added hypothesis that the endomorphism ring End(**1**) of the unit object **1** of $\mathscr{A}$ is k (*loc. cit.*, 3.15). This conjecture remained open for a few years, and was finally positively solved by Deligne in [D71, 1990]. The main result is the following (*loc. cit.*, 1.12):

Theorem 48 *Let k be a field, and $\mathscr{A}$ be a Tannakian category over k, such that* $\mathrm{End}(\mathbf{1}) = k$. *Let K be an extension of k and ω a fiber functor of $\mathscr{A}$ over* $S = \mathrm{Spec}\, K$. *Let $\mathscr{A}ut_k^{\otimes}(\omega)$ be the k-groupoid whose object of objects is S and object of arrows G, where, for a k-scheme T, $G(T)$ is the set of triples $(a : T \to S, b : T \to S, u : b^*\omega \xrightarrow{\sim} a^*\omega)$, with the obvious composition law. Then:*

(i) *G is represented by a scheme faithfully flat over $S \times_k S$, and ω gives an equivalence between $\mathscr{A}$ and the category of representations of G, by which ω corresponds to the forgetful functor.*
(ii) *Two fiber functors ω_1, ω_2 over S are locally isomorphic for the fppf topology.*

In fact, (ii) follows from (i), and in *loc. cit.* there is a more general statement where S can be replaced by a nonempty k-scheme,[41] and (i) has a natural converse. It also follows from (i) that, if $\mathscr{A}$ has a $\otimes$-generator, it admits a fiber functor over a finite extension of k.

The interest of Theorem 48 is that, with this correction, the results of Saavedra's thesis—which had many applications—are validated. The proof uses a theorem of Barr–Beck on pairs of adjoint functors (a generalization of fpqc descent), and (for k not perfect) a theorem of representability of quotients by a groupoid action, essentially due to Artin. It also introduces new ingredients and ideas: (i) construction of the *tensor product* of abelian categories satisfying certain finiteness conditions ([D71, 1990], 5) (ii) a *geometric language* in a k-tensor category $\mathscr{A}$[42]: notions of $\mathscr{A}$-affine scheme, S-affine scheme (for S an affine $\mathscr{A}$-scheme), S-affine group scheme, etc., leading, in particular, to the definition of a *fundamental group* of $\mathscr{A}$. This language plays an important role in [D69, 1989], written almost at the same time.

[41] A fiber functor is then defined as an exact k-linear functor ω from $\mathscr{A}$ to the category of quasi-coherent $\mathscr{O}_S$-modules, with a compatible isomorphism $\omega(X) \otimes_{\mathscr{O}_S} \omega(Y) \xrightarrow{\sim} \omega(X \otimes Y)$; it is shown that it has values in the category of locally free $\mathscr{O}_S$-modules.

[42] I.e., a category having the data and satisfying the axioms of a Tannakian k-linear category, with $\mathrm{End}(\mathbf{1}) = k$, but without the requirement of existence of a fiber functor.

Further Results

- **Super representations** At the end of [D71, 1990], Deligne briefly discusses the case of the (tensor) category of finite dimensional *super vector spaces* over k. He comes back to this in [D100, 2002], where, assuming k algebraically closed, he characterizes k-tensor categories of $\otimes$-finite generation which are equivalent to a category of finite dimensional super representations of an affine super group scheme over k by the fact that any object is annihilated by a suitable Schur functor, a condition satisfied for example if $\mathscr{A}$ has only finitely many simple objects (in the non super case, exterior powers suffice). Here the dimension of an object V of $\mathscr{T}$ is that of $\omega(V)$ over K, for a fiber functor ω on $\mathscr{T}$ with values on K-vector spaces, K an extension of k.
- **The symmetric group** S_t The Tannakian formalism can be viewed as a techique of construction of groups (or group-like objects), which is reminiscent of that of the construction of schemes (or algebraic spaces, or algebraic stacks) by representing functors. Mumford–Tate groups, differential Galois groups (discussed at the end of [D71, 1990]), and, most importantly, motivic Galois groups (see, e.g., Sect. 9.2 "Mixed Tate Motives", (b) below) are classical examples. In [D106, 2007], an article with an intriguing title, Deligne gives an exotic illustration of this philosophy. Fix a field k of characteristic zero. For $t \in k$, Deligne defines a category denoted $\mathrm{Rep}(S_t)$, playing the role of the category of k-linear representations of a symmetric group S_t on t letters. The category $\mathrm{Rep}(S_t)$ is k-linear (*Hom*'s have a k-linear structure, and the composition is k-bilinear), pseudo-abelian (= additive and karoubian: idempotents are projections on direct summands), and is endowed with a k-bilinear tensor product satisfying the usual compatibilities with the ACU data (associativity, commutativity, and unity), which is rigid,[43] with $\mathrm{End}(1) = k$. Moreover, if t is *not* in $\mathbf{N}$, $\mathrm{Rep}(S_t)$ is abelian and semisimple (in particular, is a tensor category).

  The category $\mathrm{Rep}(S_t)$ is deduced by linear extension via $\mathbf{Z}[T] \to k$, $T \mapsto t$, from a universal $\mathbf{Z}[T]$-linear category $\mathrm{Rep}(S_T)$, whose definition is reminiscent of Grothendieck's definition of Chow motives: one starts with the category $\mathscr{C}$ having for objects the finite sets U, and morphisms from U to V the free $\mathbf{Z}[T]$-module generated by *gluing data* $U \subset C \supset V$ (and a certain combinatorial formula for the composition of morphisms, see (*loc. cit.*, 2.12); then $\mathrm{Rep}(S_T)$ is defined as the additive, pseudo-abelian envelope of $\mathscr{C}$. If $t \in k$ is an integer $n \geq 0$, and $\mathrm{Rep}(\mathbf{S}_n, k)$ denotes the category of k-linear representations of the symmetric group $\mathbf{S}_n$ on n letters, Deligne constructs a functor

$$\mathrm{Rep}(S_t) \to \mathrm{Rep}(\mathbf{S}_n, k),$$

  sending the finite set U to the permutation representation of $\mathbf{S}_n$ on the set of injections from U to $\{1, \cdots, n\}$. This functor induces an equivalence from the

[43] I.e., dual objects exist and satisfy the same axioms as in a tensor category.

quotient of $\mathrm{Rep}(S_t)$ by the ideal of *negligible* morphisms.[44] For t not an integer, irreducible objects of $\mathrm{Rep}(S_t)$ are classified, and their dimensions calculated. Similar results for orthogonal and general linear groups are discussed (essentially due to Wenzl [262] in the orthogonal case). In the case of general linear groups, the dimension formulas for the irreducible objects of the analogous category $\mathrm{Rep}(\mathrm{GL}(t))$ involve polynomials described in Deligne's work on the exceptional series [D82, 1996] (see Sect. 10).

- **Characteristic** $p > 0$ Quite recently, Deligne studied Tannakian categories over a field k of characteristic $p > 0$ [D116, 2014]. He proved that, given a finite family $(V_i)_{i \in I}$ of semisimple objects in such a category $\mathscr{T}$ such that $\sum(\dim(V_i) - 1) < p$, then $\otimes_{i \in I} V_i$ is semisimple. This generalizes a result of Serre, for $\mathscr{T}$ the category of representations of a smooth affine group scheme over k.

9.2 Periods

Deligne promoted the idea[45] that, rather than sticking to Grothendieck's conjectural construction of an abelian category of motives, one should instead exploit what is sometimes called the *philosophy of motives*, i.e., the rich expected compatibilities between cohomological realizations of algebraic varieties. We have already mentioned aspects of this in his theory of absolute Hodge cycles (Sect. 4.4) and his proof of the Weil conjecture for K3 surfaces (Sect. 5.3). Very roughly, his work consists of two (closely related) main contributions.

Values of L-Functions at Critical Integers

In [D43, 1979], Deligne defines the notion of *critical* integers n for a pure motive M over $\mathbf{Q}$ with coefficients in a number field E, and gives a conjectural formula for the value of the L-function of M at such integers in terms of certain explicit periods, up to multiplication by (unknown) elements of E^*. *Motive* in *loc. cit.* is taken in a loose sense: M appears through its realizations $\mathscr{H}(M)$, $\mathscr{H}$ being an ℓ-adic (H_ℓ), Betti (H_B), or de Rham (H_{dR}) realization. For $E = \mathbf{Q}$, a typical example is furnished by a proper smooth scheme $X/\mathbf{Q}$, and $M = H^i(X)(m)$, for $m \in \mathbf{Z}$, whose realizations are $H_\ell(M) = H^i(X \otimes \overline{\mathbf{Q}}, \mathbf{Q}_\ell(m))$ (with its Galois action), $H_B(M) = H^i(X(\mathbf{C}), (2\pi i)^m \mathbf{Q})$ (with its Hodge structure), $H_{dR}(M) = H^i_{dR}(X/\mathbf{Q})$ (with the shifted Hodge filtration $F(m)$, cf. (47)). Another typical example is the motive $M(f)$ (over $\mathbf{Q}$, with coefficients in K) associated by Scholl to a cusp form f (see the end of Sect. 6.1).

[44] I.e., $f : X \to Y$ such that $\mathrm{Tr}(fu) = 0$ for all $u : Y \to X$.

[45] Developed in [D76, 1994], but used by him and other authors much earlier.

The L-function of a motive M over $\mathbf{Q}$ is the function of $s \in \mathbf{C}$ defined by the Euler product

$$L(M, s) = \prod L_p(M, s), \tag{179}$$

where p runs through all places of $\mathbf{Q}$, including ∞, and

$$L_p(M, s) = \det(1 - F_p t, H_\ell(M)^{I_p})^{-1}_{t=p^{-s}} \tag{180}$$

for p finite, $\ell \neq p$, I_p denoting the inertia subgroup of a decomposition group at p of $\mathrm{Gal}(\overline{\mathbf{Q}}/\mathbf{Q})$ acting on $H_\ell(M)$ and $F_p \in \mathrm{Gal}(\overline{\mathbf{Q}}/\mathbf{Q})$ a geometric Frobenius, and $L_\infty(M, s)$ is a product of Γ-factors depending on the Hodge decomposition of $H_B(M) \otimes \mathbf{C}$, together with its involution F_∞ defined by complex conjugation. One assumes that (180) has coefficients in $\mathbf{Q}$ and is independent of ℓ (this is the case, by the Weil conjecture, if $M = H^i(X)(m)$ as above, if X has good reduction at p, but unknown otherwise). Whatever the notion of motive which is adopted, the Dirichlet series defined by (179) converges for $\mathscr{R}(s)$ sufficiently large, and one assumes that it admits an analytic continuation to $\mathbf{C}$, and[46] satisfies a (conjectural) functional equation of the form $\Lambda(M, s) = \varepsilon(M, s)\Lambda(M^\vee, 1 - s)$, where $M^\vee$ is the dual motive (of realizations the duals of the realizations of M), and the constant $\varepsilon(M, s)$ is (up to a power of $\sqrt{-1}$) a product of local constants of the form (146), i.e., the product of a constant by an exponential function.

Deligne defines $n \in \mathbf{Z}$ to be *critical* for M if neither $L_\infty(M, s)$ nor $L_\infty(M^\vee, 1 - s)$ have a pole at $s = n$. As $L_\infty(M(n), s) = L_\infty(M, s + n)$, n is critical for M if an only if 0 is critical for $M(n)$, which turns out to mean that the Hodge numbers h^{pq} of $M(n)$, for $p \neq q$ are nonzero only for $(p < 0, q \geq 0)$ or $(p \geq 0, q < 0)$, and F_∞ acts on $H^{p,p}$ by 1 if $p < 0$, and -1 if $p \geq 0$. For example, for $M = \mathrm{Spec}\,\mathbf{Q}$, so that $L(M, s) = \zeta(s)$, n is critical if and only if n is even > 0 or odd < 0.

For a motive M, the *periods* of M are the numbers $\langle \omega, c \rangle$, for $\omega \in H_{dR}(M)$ and $c \in H_B(M)^\vee$ (for $M = H^i(X)$ as above, $\omega \in H^i_{dR}(X)$, and c defined by $\gamma \in H_i(X, \mathbf{Q})$, this is $\int_\gamma \omega$). Suppose 0 is critical for M. Deligne then calls M *critical*, and defines a period $c^+(M)$ in the following way. Let $H^+_B(M)$ denote the $\mathbf{Q}$-subspace of $H_B(M)$ fixed by F_∞. As 0 is critical, and F_∞ exchanges $H^{p,q}$ and $H^{q,p}$ for $p \neq q$, its dimension $d^+(M)$ is equal to that of the $\mathbf{Q}$-vector space $H^+_{dR}(M) := (H_{dR}(M)/F^0$. Consider the inverse of the period isomorphism (a special, simple case of (48)):

$$I : H_B(M) \otimes \mathbf{C} \xrightarrow{\sim} H_{dR}(M) \otimes \mathbf{C}. \tag{181}$$

[46] Assuming that the local F-semisimplified representations of the local decomposition groups are compatible, cf. (Sect. 6.2, (a)).

It induces a composite isomorphism

$$I^{+} : H_{B}^{+}(M) \otimes \mathbf{C} \to H_{B}(M) \otimes \mathbf{C} \xrightarrow{\sim} H_{dR}(M) \otimes \mathbf{C} \to H_{dR}^{+}(M) \otimes \mathbf{C}, \tag{182}$$

which is defined over $\mathbf{R}$. Then Deligne defines $c^{+}(M)$ as the determinant

$$c^{+}(M) = \det(I^{+}), \tag{183}$$

calculated in rational bases of $H_{B}^{+}(M)$, $H_{dR}^{+}(M)$. This is an element of $\mathbf{R}^{*}$, well defined up to multiplication by an element of $\mathbf{Q}^{*}$. Deligne conjectures (*loc. cit.*, 1.8):

$$L(M, 0) = c^{+}(M) \tag{184}$$

in $\mathbf{R}^{*}/\mathbf{Q}^{*}$. A similar (more general) conjecture is formulated for motives over a number field k with coefficients in a number field E (*loc. cit.*, 2.8, 2.9). Deligne shows that the conjecture is compatible with the functional equation of the L function and the Birch and Swinnerton-Dyer conjecture. He proves it for Artin motives (as a consequence of results of Siegel), and motives $M(f)$ attached to modular forms (from classical results on Eichler integrals), though he makes no attempt to define $M(f)$ as a Grothendieck motive—which was later tackled by Scholl.

In (*loc. cit.*, 8), Deligne examines avatars of his conjecture for motives M over a number field k with coefficients in a number field E, and of *rank* 1, i.e., such that the rational Betti realization $H_{B}(M)$ is of dimension 1 over E. He first conjectures the general shape of these motives. Namely:

Conjecture 4

(i) For any algebraic Hecke character χ of k with values in E, there is associated with χ a motive $M(\chi)$ over k, with coefficients in E, and of rank 1, such that for any finite place λ of E, the λ-adic realization $H_{\lambda}(M(\chi))$ (an E_{λ}-vector space of dimension 1) has its Galois action given by χ, i.e., for any closed point x of $\operatorname{Spec}(\mathscr{O}_{k})$ where both λ and the conductor of χ are invertible, $\operatorname{Tr}(F_{x}, H_{\lambda}(M(\chi))) = \chi(x)$ (where F_{x} is a geometric Frobenius). The motive $M(\chi)$ is characterized up to isomorphism by this property.
(ii) Every motive over k with coefficients in E and of rank 1 is of this form.

Deligne also gives an explicit (conjectural) formula for the Hodge filtration of the de Rham realization of $M(\chi)$. The primitive part of the middle dimension cohomology of a Fermat hypersurface gives rise to such motives of rank 1. Unraveling the period conjecture (*loc. cit.*, 2.8, 2.9) in this case led him to formulate, with Gross, a formula relating, for certain subHodge structures (of the cohomology of a smooth, projective variety over $\overline{\mathbf{Q}}$) with multiplication by a finite abelian extension of $\mathbf{Q}$, the periods with products of values of the Γ function ([105], p. 205). A weak form of this conjecture was proved by Maillot and Roessler [188]. These results were recently revisited and improved by Fresàn [94].

The conjectural formula (184) (and its generalization just mentioned) raised the questions of formulating analogues for the leading terms of L-functions at not necessarily critical integers, and eventually led Beilinson to his celebrated conjectures [20]. Soulé reported on this at the Bourbaki seminar [241]. Shortly before the oral exposé, Deligne wrote him a letter [76], in which he explained how to rephrase Beilinson's conjectures using extension groups of (conjectural) mixed motives, instead of K-theoretic invariants. Scholl [229] showed how to give a unified formulation in this setup of both Beilinson's conjectures and Deligne's conjecture (184) (and its generalization by Bloch [36] and Beilinson [23]). An alternative approach was to use Bloch's higher Chow groups instead of extensions of mixed motives, as described in [81]. On the other hand, the $\mathbf{Q}^*$ indeterminacy in (184) attracted great attention, leading to the *motivic Tamagawa number conjectures* of Bloch and Kato [38], involving Fontaine's theory of p-adic period rings. As for the Beilinson's conjectures, they were reformulated by Fontaine and Perrin-Riou in the language of mixed motives, see Fontaine's Bourbaki report [93], and Flach's survey [92] for an overview of further conjectures and results in connection with Iwasawa theory and Stark's conjecture.

Mixed Tate Motives

The search for the definition of a suitable category of mixed motives, forming an analogue of the category of mixed Hodge structures, initiated by Deligne in his letter to Soulé mentioned above, generated a huge amount of work during the past 30 years. A definition using absolute Hodge classes as correspondences was studied by Jannsen [136], and, independently, and in a less precise but more flexible form, by Deligne in [D69, 1989], which brought new inputs and ideas.

This long monograph has two main parts. The first one (discussed in (a)) presents the general setup needed to study the motivic fundamental groups considered in the second one (discussed in (b)). Due to the contributions of several people, some of its imprecise definitions have now been made rigorous and related conjectures have been proven. We will give a brief update in (c).

(a) Mixed motives: definitions and conjectures Deligne "defines" and works with mixed motives not just over Spec($\mathbf{Q}$), but over more general bases S, like Spec(k) for k a number field, or a Zariski open subset of the spectrum of its ring of integers, or a smooth scheme over $\mathbf{Z}$, and also takes into account integral structures. Over $\mathbf{Q}$, as in [136], the starting point is the notion of a *system of realizations* $(M_B, M_{dR}, M_\ell, M_{\mathrm{cris},p})$ and comparison isomorphisms between them, satisfying a number of compatibilities, similar to the systems of realizations $(H_B(M), \cdots)$ considered above, except that these realizations are not necessarily pure, but mixed, i.e., equipped with a *weight filtration*, making a mixed Hodge structure on the Betti side, and satisfying the expected conjectural properties with respect to the action of Frobenius on the ℓ-adic side.[47] Deligne shows that systems of realizations form a Tannakian category, and "defines" the *category of (mixed) motives* over $\mathbf{Q}$ as the full subcategory generated by sum, tensor product, dual, and subquotient from the systems of geometric origin (he also "defines", in the same vein, the generalizations mentioned above). The quotation marks come from the fact that no definition is given (or even suggested) for "of geometric origin". Despite (and often, because of) this imprecision, the notion turned out to provide a useful guideline, suggesting conjectures, certain consequences of them being amenable to a proof. In this respect, basic examples of "lisse" motives over $\mathrm{Spec}(\mathbf{Z})$ with "integer coefficients" (or "integral structure") are the *Tate motives* $\mathbf{Z}(n) = \mathbf{Z}(1)^{\otimes n}$ (sometimes denoted $\mathbf{Q}(n)$, the integral structure being omitted from the notation). Deligne studies *torsors* under Tate motives, i.e., extensions of $\mathbf{Z}$ by $\mathbf{Z}(n)$, and, more generally, iterated extensions of Tate motives, examples of which appear in motivic fundamental groups and their Lie algebras. In particular, for each $n \geq 2$, he (unconditionally) constructs a system of (lisse) realizations of a canonical torsor under $\mathbf{Z}(n)$

$$P_{1,n} \in H^1(\mathrm{Spec}(\mathbf{Z}), \mathbf{Z}(n)), \tag{185}$$

which plays a crucial role in his study of the fundamental group of $\mathbf{P}^1 - \{0, 1, \infty\}$. Its Betti realization is $-(n-1)!\zeta(n) + (2\pi i)^n\mathbf{Z}$. He shows that it is of torsion for n even, of order the denominator of $\frac{1}{2}\zeta(1-n)$ (a property related to Kummer's congruences on Bernoulli numbers).

Beilinson's conjectural formalism in ([23], 5) suggested to Deligne the following conjectures, which, despite their imprecise form, turned out to have both striking and verifiable consequences:

Conjecture 5

(i) Let k be a number field, S an open subset of the spectrum of the ring of integers of k. Then, for any $n \geq 1$,

$$\mathrm{Ext}^1(\mathbf{Q}(0), \mathbf{Q}(n)) = K_{2n-1}(S) \otimes \mathbf{Q}, \tag{186}$$

where the left hand side is taken in the abelian category of lisse motives over S.

(ii) In the abelian category of lisse motives over $\mathrm{Spec}(\mathbf{Z})$, we have

$$\dim_{\mathbf{Q}} \mathrm{Ext}^1(\mathbf{Q}(0), \mathbf{Q}(n)) = 1 \tag{187}$$

for n odd ≥ 3, with the extension defined by $P_{1,n}$ (185) as a generator, and

$$\mathrm{Ext}^1(\mathbf{Q}(0), \mathbf{Q}(n)) = 0 \tag{188}$$

otherwise.

---

[47] The crystalline data and axioms in *loc. cit.* are in a rudimentary form, reflecting the status of the p-adic comparison theorems at the time; Deligne made a *caveat* on this.

By the known structure of $K_i(\mathbf{Z})$ these conjectures are compatible for $k = \mathbf{Q}$. They could not really be tackled, because of their formulation, involving Deligne's "definition" of mixed motives. However, Deligne introduced a smaller category than that of all lisse motives over $\mathrm{Spec}(\mathbf{Z})$, namely the full subcategory $\mathscr{T}$ whose objects M are successive extensions of Tate motives $\mathbf{Q}(n)$, i.e., (because of the structure of $\mathbf{Z}(n)_{dR}$) such that, for all $n \in \mathbf{Z}$, $\mathrm{gr}^W_{-2n+1} M = 0$ and $\mathrm{gr}^W_{-2n} M$ is a sum of copies of $\mathbf{Q}(n)$, where W is the weight filtration. This category is a Tannakian category, and although not being more precisely defined than the previous one, it was later given a rigorous construction, see (c). As any Tannakian category, it has a fundamental group $G = \pi(\mathscr{T})$ (defined in [D71, 1990], see the end of Sect. 9.1, using the language of algebraic geometry in Tannakian categories). This is an affine $\mathscr{T}$-group scheme,[48] which is an extension

$$0 \to U \to G \to \mathbf{G}_m \to 0, \tag{189}$$

where U is the pro-unipotent radical. The natural grading of M_{dR} for M in $\mathscr{T}$ makes its de Rham realization G_{dR} a semidirect product $\mathbf{G}_m \cdot U_{dR}$. From (187) Deligne deduces that the Lie algebra of U_{dR} is the completion of a graded Lie algebra $\mathrm{Lie}_{gr}\, U_{dR} = \oplus_k \mathrm{Lie}\, U^k_{dR}$, generated by one element for each odd degree $k \geq 3$, and he conjectures that it is a *free* Lie algebra.

(b) Motivic fundamental group of $\mathbf{P}^1 - \{0, 1, \infty\}$ Given a scheme X separated and of finite type over $\mathbf{Q}$ (or, more generally, a number field k), it is tempting to try to define a motivic fundamental group $\pi_1(X)$, at least in terms of a compatible system of realizations. However, there are obvious obstacles: (i) the classical fundamental group $\pi_1(X(\mathbf{C}), b)$ is in general too noncommutative to be encoded in cohomological data: the seemingly closest approximations of it amenable to such an encoding are its nilpotent quotients (studied from a Hodge theoretic or ℓ-adic viewpoint by Deligne–Morgan–Sullivan, cf. Sect. 5.6 "First Applications", $\mathbf{Q}_\ell$-homotopy type); (ii) how to make sense of a "motivic" choice of base-points and loops.

Let $\overline{X}$ be a proper, smooth, geometrically irreducible scheme over $\mathbf{Q}$, and let $X = \overline{X} - D$ the complement of a divisor with normal crossings. Assuming that $H^1(\overline{X}, \mathscr{O}) = 0$, and that we are given a base-point $x \in X(\mathbf{Q})$, Deligne constructs a pro-system of compatible realizations

$$\pi_1(X, x)_{\mathrm{mot}} = (\pi_1(X, x)^{(N)}_{\mathrm{mot}})_{N \geq 1}. \tag{190}$$

For example, the Betti realization of $\pi_1(X, x)^{(N)}_{\mathrm{mot}}$ is $\pi_N := \pi_1(X(\mathbf{C}), x)^{[N]\mathrm{alg\,un}}$, with the integral structure given by the congruence subgroup image of $\pi_1(X(\mathbf{C}), x)^{[N]}$ in $\pi_N(\mathbf{Q})$. Here, for a nilpotent group Γ, $\Gamma^{\mathrm{alg\,un}}$ denotes its

[48] I.e., the datum, for every fiber functor ω on a scheme S, an affine group scheme G_ω on S, functorial in ω, and compatible with base change $S' \to S$.

unipotent algebraic envelope over $\mathbf{Q}$, where for a group A, $A^{[N]}$ is the largest torsion free quotient of $A^{(N)}$, with $A^{(N)} := A/Z^{n+1}(A)$, $A = Z^1(A) \supset Z^2(A) \supset \cdots$ the descending central series. Representations of the de Rham realization correspond to vector bundles on X with an integrable connection which is regular (cf. Sect. 3.1 "Higher Dimension: The Riemann–Hilbert Correspondence") and nilpotent along D, i.e., has nilpotent residues along the branches of D.

When X is a curve, and D consists of $\mathbf{Q}$-rational points, Deligne explains how to give a motivic meaning to the local monodromy around the points of D. Starting with the observation that a "simple loop" around $y \in D(\mathbf{C})$ is only "well defined" when its base point b is "close" to y, he introduces a notion, which since then has turned out to be quite fruitful and popular, namely that of *tangential base point*: the local monodromy at a point $y \in D(\mathbf{Q})$ is a morphism (in the Tannakian category of realizations) from $\mathbf{Z}(1)$ to $\widehat{\pi} := \text{proj lim}_N\, \pi_1(X, b)^{(N)}_{\text{mot}}$, where b is a tangential point at y (the datum of a non zero tangent vector). For $\overline{X} = \mathbf{P}^1_{\mathbf{Q}}$, Deligne shows that each $\pi_1(X, x)^{(N)}_{\text{mot}}$ is an iterated extension of (systems of realizations of) Tate motives and $\widehat{\pi}$ a pro-unipotent group scheme in this Tannakian category. He finally shows that, for $D = \{0, 1, \infty\}$, the torsors defining these successive extensions are precisely the $P_{1,n}$'s (185).

The construction of the $P_{1,k}$'s ($k \geq 2$) involves the classical *polylogarithm function*

$$Li_k(x) = \sum_{n \geq 1} \frac{z^n}{n^k}. \tag{191}$$

These functions have a long history (see Oesterlé's Bourbaki report [205]). Their relation with higher regulators and K-groups, discovered by Bloch [35], and their relations with mixed Tate motives and values of zeta (or multizeta) functions at integers have since then been the focus of an extremely active line of research, in which many arithmeticians have participated (Beilinson, Bloch, Deligne, Goncharov, Ihara, Levine, Soulé, Zagier, to mention only a few names). An important conjecture in this domain, due to Zagier, was re-interpreted by Beilinson and Deligne [D78, 1994] in motivic terms. A weak form of this conjecture was established in an unpublished joint work by Beilinson and Deligne (*Motivic polylogarithm and Zagier's conjecture*, 1992). A detailed exposition is given by Huber-Wildeshaus [118, 119]. This weak Zagier conjecture was independently proved (by a different method) by de Jeu [59].

(c) Update Thanks to the work of Levine and Voevodsky, the problem of giving a rigorous definition of the Tannakian category $\mathcal{T}$ of (lisse) mixed Tate motives over $\mathbf{Z}$ (or an open subset of the ring of integers of a number field) could be solved. Deligne and Goncharov explain how in [D102, 2005]. First, let k be a field of characteristic zero. Let $\text{DM}(k)$ be the triangulated category of motives constructed by Voevodsky (that Levine has shown to be equivalent to that which he constructed independently), and $\text{DM}(k)_{\mathbf{Q}}$ the category deduced by tensorisation with $\mathbf{Q}$. It contains Tate objects $\mathbf{Z}(n)$ and admits operations of tensor product and taking a dual. Let $\mathrm{DMT}(k)_{\mathbf{Q}}$ be the triangulated subcategory of $\mathrm{DM}(k)_{\mathbf{Q}}$ generated by the $\mathbf{Q}(n)$'s. Suppose now that k is a *number field*. The vanishing conjecture of Beilinson-Soulé, which is known in this case (from Borel's work on the K-theory of number fields), can be reformulated as

$$\mathrm{Hom}^j(\mathbf{Q}, \mathbf{Q}(i)) = 0 \tag{192}$$

for $i > 0$ and $j \leq 0$ (Hom taken in $\mathrm{DM}(k)_{\mathbf{Q}}$). It follows (cf. Sect. 5.8 "t-Structures") that $DMT(k)_{\mathbf{Q}}$ admits a t-structure, whose heart is an abelian category

$$\mathrm{MT}(k) \tag{193}$$

consisting of iterated extensions of $\mathbf{Q}(n)$'s, called the category of *mixed Tate motives* over k (this part is due to Levine [175]). It is shown in [D102, 2005] that it is Tannakian, and it has the realizations[49] and enjoys the properties stated in [D69, 1989]. In particular, the analogue of (186) is true, namely

$$\mathrm{Ext}^1(\mathbf{Q}(0), \mathbf{Q}(n)) = K_{2n-1}(k) \otimes \mathbf{Q}, \tag{194}$$

where the left hand side is taken in the category $\mathrm{MT}(k)$. Moreover,

$$\mathrm{Ext}^2(\mathbf{Q}(0), \mathbf{Q}(n)) = 0. \tag{195}$$

The category $\mathrm{MT}(\mathbf{Z})$ of (lisse) *mixed Tate motives* over $\mathbf{Z}$ (resp. that of mixed motives over an open subset of $\mathrm{Spec}(k)$) was defined in [D102, 2005] as a certain Tannakian subcategory of $\mathrm{MT}(\mathbf{Q})$ (resp. $\mathrm{MT}(k)$). With this definition, Deligne's conjecture of freeness of the graded Lie algebra of the de Rham realization of the corresponding unipotent group U could be proven (*loc. cit.*, 2.3), as a consequence of (195).

The mysterious relations between $\mathrm{MT}(\mathbf{Z})$ and the motivic pro-unipotent fundamental group $\pi_1(\mathbf{P}^1_{\mathbf{Q}} - \{0, 1, \infty\})_{\mathrm{mot}}$ have recently been elucidated by Brown [46] (see also Deligne's Bourbaki report [D113, 2012], and Brown's Seoul ICM talk). He proved the following conjecture of Deligne:

Conjecture 6 For $X = \mathbf{P}^1_{\mathbf{Q}} - \{0, 1, \infty\}$, $\mathrm{MT}(\mathbf{Z})$, as a Tannakian category, is generated by $\pi_1(X)_{\mathrm{mot}}$, i.e., by the affine algebra of a certain motivic $\pi_1(X, 0)_{\mathrm{mot}}$-torsor of paths $\pi_1(X; 1, 0)_{\mathrm{mot}}$ from tangential points (1 at 0) to (-1 at 1), whose algebra is an ind-object of $\mathrm{MT}(\mathbf{Z})$, see *loc. cit.* for a precise statement.

As an application, he proved a conjecture of Deligne and Ihara on the outer action of $\mathrm{Gal}(\overline{\mathbf{Q}}/\mathbf{Q})$ on the pro-ℓ-fundamental group of X, and a conjecture of Hoffman,

[49]Except for the crystalline ones.

to the effect that multizeta values

$$\zeta(n_1, \cdots, n_r) = \sum_{0<k_1<\cdots<k_r} \frac{1}{k_1^{n_1} \cdots k_r^{n_r}}, \tag{196}$$

($n_i \geq 1$, $n_r \geq 2$) are $\mathbf{Q}$-linear combinations of numbers of the form (196) where each n_i is equal to 2 or 3. Generalizations of all this, with $\mathbf{P}^1 - \{0, 1, \infty\}$ replaced by $X = \mathbf{G}_m - \mu_N$ for certain values of N were studied by Deligne in [D111, 2010]. For $N = 2, 3, 4$, or 8, the motivic fundamental groupoid $P(X, \{0, \infty\} \cup \mu_N)$ is mixed Tate over $k = \mathbf{Q}(\mu_N)$, has good reduction outside N, and generates the Tannakian category of mixed Tate motives over k having good reduction outside N. However, this generation statement is no longer true for many other values of N (e.g., N prime and ≥ 5, as shown by Goncharov [101]). Current work of Brown on multiple modular motives [47] could shed a new light on the problem.

For the past 30 years this topic has been at the junction of many different lines of research in analysis, geometry and number theory, and, more recently, turned out to have deep connections with high energy physics (Feynman diagrams and integrals), an interaction which is fast developing today.

10 Deligne's Conjectures

We have already discussed some of them:

- **1-motives** (Conjecture 1)
- **Motives of rank 1 and Gross–Deligne's conjecture** (Conjecture 4)
- **Du Bois complex** (Conjecture 2)
- **Absolute Hodge cycles** (Conjecture 3)
- **Companions conjecture** (Sect. 5.9 "A Finiteness Theorem")
- **Values of L functions at critical points** (184)
- **The Deligne–Milnor conjecture** (156)
- **Mixed motives** (Conjecture 5: (186), (187), (188), (195), Conjecture 6 (Sect. 9.2 "Mixed Tate Motives", (c))

  Here are a few others.

- **Deligne–Grothendieck's conjecture on discrete Riemann–Roch in characteristic zero**
  In SGA 5 [5], Grothendieck defined and studied homology in the context of étale cohomology: for k algebraically closed, X/k separated and of finite type, and ℓ a prime number different from $p = \mathrm{char}(k)$, and $\Lambda = \mathbf{Z}/\ell^n\mathbf{Z}$, the homology groups of X are the groups $H_i(X, \Lambda) := H^{-i}(X, K_X)$, where K_X is the dualizing complex, i.e., $Ra^!\Lambda$ for $a : X \to \mathrm{Spec}(k)$. At the time, the formalism was unconditional only for $p = 0$. For $p > 0$, it depended on a number of conjectures (resolution, purity), which Deligne managed to

get rid of in ([D39,1977], Th. finitude). Grothendieck's definition superseded that of Borel–Moore, and he showed how to use it to define cycle classes on singular varieties, and prove their compatibility with direct images by proper morphisms (see ([D39, 1977], Cycle), and [165]). Later in the seminar, he proved the Grothendieck–Ogg–Shafarevich formula on curves. As recalled by Sullivan [244], in the early 1970s, a conjectural common generalization in characteristic zero (where wild ramification phenomena disappear) was proposed to him by Deligne (who thought of using Hironaka's resolution, as in Hodge theory), and then forwarded to MacPherson, who solved it in [186] (without using resolution). As mentioned in Sect. 7.3 "Surfaces", the generalization to characteristic $p > 0$ is an open problem.

- **Local terms of the trace formula for Frobenius twisted correspondences**
  The computation of the local terms of the Lefschetz–Verdier formula (SGA 5 III) [5] is in general intractable. Inspired partly by his fixed point formula with Lusztig (Sect. 8.1 "A Fixed Point Formula"), and partly by Drinfeld's work on elliptic modules, Deligne conjectured a simple formula for these local terms, over finite fields, provided that the given correspondence is twisted by a sufficiently high power of Frobenius. After partial results by Pink and Shpiz, the conjecture was proven by Fujiwara [96], using a contracting property of Frobenius in the rigid analytic setting. A simpler proof (of a slightly more general result) was obtained later by Varshavsky [251], using again a contracting property, but in the algebraic setting, together with an argument of deformation to the normal cone.
- **Deligne–Langlands's conjecture**
  Let K be a nonarchimedian local field, with residue field $\mathbf{F}_q$, G be a simple split adjoint group over K, and $\mathcal{I}$ an Iwahori subgroup of $G(K)$. Let $\mathcal{U}(G)$ be the set of isomorphism classes of irreducible admissible complex representations of $G(K)$ which admit nonzero vectors invariant under $\mathcal{I}$. Let $G^\vee/\mathbf{C}$ be the simply connected complex group over $\mathbf{C}$ with root system dual to that of G. Deligne and Langlands conjectured that there should be a bijection between $\mathcal{U}(G)$ and the set S of pairs (s, N), with s semisimple in $G^\vee(\mathbf{C})$, $N \in \mathrm{Lie}(G^\vee(\mathbf{C}))$, such that $sNs^{-1} = q^{-1}N$, modulo conjugation by elements of $G^\vee(\mathbf{C})$. The set S is the set of isomorphism classes of F-semisimple representations of the Weil–Deligne group $'W(\overline{K}/K)$ (145) into $G^\vee(\mathbf{C})$.

  This conjecture, a particular case of the local Langlands conjecture for the unramified principal series, was proved by Bernstein and Zelevinski for G of type A_n, and in the general case, in a modified form due to Lusztig ([178], 1.5), by Kazhdan–Lusztig [150].
- **The weight-monodromy conjecture**
  In ([D15, 1971], 9), Deligne considers an (analytic) projective morphism $f : X \to D$, where D is the unit disc in $\mathbf{C}$, such that its restriction to the punctured disc $D^* = D - \{0\}$ is smooth. Let $t \in D^*$, and let $\widetilde{X}$ be the pull-back of X to a universal cover $\widetilde{D}^*$ of D^*. Then, as $\widetilde{X}$ is homotopically equivalent to X_t, $H^n(\widetilde{X}, \mathbf{Z}) = H^n(X_t, \mathbf{Z})$, and, by Grothendieck's local monodromy theorem, a subgroup of finite index of $\pi_1(D^*)$ acts unipotently on $H^n(\widetilde{X}, \mathbf{Z})$. Up to passing to a finite cover of D^*, one can assume that X has semistable reduction, in which

case a generator T of $\pi_1(D^*)$ acts unipotently, hence, for any n, $N := \log T : H^n(\widetilde{X}, \mathbf{Q}) \to H^n(\widetilde{X}, \mathbf{Q})$ is nilpotent. Under this assumption, Deligne asserts in *loc. cit.* that, for each tangent vector u to D at $\{0\}$, one can define a mixed Hodge structure H_u on $H^n(\widetilde{X}, \mathbf{Z})$ (with weight filtration W and Hodge structures on the graded pieces independent of u) such that N induces a morphism of mixed Hodge structures

$$N : H_u \otimes \mathbf{Q} \to H_u \otimes \mathbf{Q}(-1), \tag{197}$$

and, for all i,

$$N^i : \mathrm{gr}_W^{n+i}(H_u \otimes \mathbf{Q}) \xrightarrow{\sim} \mathrm{gr}_W^{n-i}(H_u \otimes \mathbf{Q})(-i). \tag{198}$$

His original proof was not published. In [242] Steenbrink constructed the desired Hodge structure H_u and the morphism (197), but his proof of (198) was flawed. A correct proof was made by Deligne [75], and, independently, by M. Saito ([221], 4.2.2, 4.2.5).

This suggested to Deligne the following algebraic variant, in étale cohomology. Let (S, s, η) be a henselian trait, $\overline{s}$ a geometric point over s, $S_{(\overline{s})}$ the corresponding strict localization, and $\overline{\eta}$ a geometric point over the generic point η_{ur} of $S_{(\overline{s})}$. Let $f : X \to S$ be a proper morphism, with strict semistable reduction: X is regular and flat over S, X_η is smooth, and X_s is a strict normal crossings divisor in X. Let ℓ be a prime number invertible on S. In this situation, the following (a) had been conjectured since the late 1960s:

(a) The inertia group $I \subset \mathrm{Gal}(\overline{\eta}/\eta)$ acts tamely on the sheaves of nearby cycles $R^q\Psi(\mathbf{Z}_\ell)$.
 In view of the calculation of the tame nearby cycles in SGA 7 ([7], I), (a) is equivalent to saying that I acts *trivially*.
 It follows from (a) that I acts unipotently on $R\Psi(\mathbf{Z}_\ell)$, through its tame quotient $t_\ell : I \to \mathbf{Z}_\ell(1)$, so that on $R\Psi(\mathbf{Q}_\ell)$ the action of σ in I is given by $\exp(Nt_\ell(\sigma))$, for a (unique) nilpotent operator $N : R\Psi(\mathbf{Q}_\ell) \to R\Psi(\mathbf{Q}_\ell)(-1)$. In the early 1980s, (a) was still open, but Gabber discovered the commutation of $R\Psi$ with duality, which implies that $R\Psi\mathbf{Q}_\ell[d]$ is perverse (d denoting the relative dimension of X/S). The nilpotent operator N therefore defines a monodromy filtration M_i of $R\Psi\mathbf{Q}_\ell$, which gives rise to a $\mathrm{Gal}(\overline{\eta}/\eta)$-equivariant spectral sequence

$$E_1^{ij} = H^{i+j}(X_{\overline{s}}, \mathrm{gr}_{-i}^M R\Psi\mathbf{Q}_\ell) \Rightarrow H^{i+j}(X_{\overline{\eta}}, \mathbf{Q}_\ell). \tag{199}$$

 It was conjectured by Deligne that:

(b) The spectral sequence (199) degenerates at E_2;
(c) (*Weight monodromy conjecture*) The abutment filtration of (199) on $H^n(X_{\overline{\eta}}, \mathbf{Q}_\ell)$ is the monodromy filtration (of the nilpotent operator N).

For char(k) $= 0$, comparison theorems reduce the problem to $k = \mathbf{C}$ and Betti cohomology, where (b) and (c) follow from the results of Steenbrink and Deligne–M. Saito mentioned above.

Assume now char(k) $= p > 0$. Conjecture (a) was proved by Rapoport–Zink ([215], 2.23). Imitating Steenbrink's method, they re-write the E_1-term of (199) as a sum of cohomology groups of m-fold intersections of components of the special fiber, with d_1 induced by restriction and Gysin morphisms. When k is finite, it then follows from Deligne's main theorem in [Weil II] that (199) degenerates at E_2 and the abutment filtration $\widetilde{M}$ on $H^n(X_{\overline{\eta}}, \mathbf{Q}_\ell)$ is the *weight filtration* W of ([D46, 1980], 1.7.5) (cf. Sect. 5.6 "Ingredients of the Proof", Weight monodromy theorem), up to shift, i.e., $\widetilde{M}_i = W_{n+i}$, hence the name of (c), sometimes rephrased as *purity of the monodromy filtration*. It is in Rapoport–Zink's paper that conjecture (c) is mentioned for the first time (*loc. cit.*, l. 3 above 2.12).

The status of (b) and (c) is as follows.

(b) was proved by reduction to k finite by Nakayama [200], and, independently, Ito [132].

(c) was proved in the following cases:

- for k finite, X/S coming by localization from a proper, flat scheme over a smooth curve over k, with semistable reduction at a closed point, by Deligne's Theorem 25;
- in the general equicharacteristic p case, by Ito [133];
- for k finite and $\dim(X/S) \leq 2$, by Rapoport–Zink (*loc. cit.*, 2.13, 2.23);
- for certain three-folds X_η, and certain p-adically uniformized varieties X_η [132, 134];
- for X_η a set-theoretic complete intersection in a projective space (or in a smooth projective toric variety), by Scholze [231].

- **Operads**

In a letter to Stasheff et al. [78], Deligne expressed the hope that, given an associative algebra A over a commutative ring k, the complex

$$C^*(A, A) = \oplus_{n \geq 0} \mathrm{Hom}_{k-\mathrm{mod}}(A^{\otimes n}, A)$$

calculating (for A projective over k) Hochschild cohomology

$$HH^*(A) = \mathrm{Ext}^*_{A \otimes A^0}(A, A)$$

(a graded algebra equipped with an extra structure (a Lie bracket of degree -1), making it a so-called *Gerstenhaber algebra*) should be an algebra over a suitable operad $\mathscr{S}$, a chain version of the little disks operad. This hope attracted much attention, and was made true by several mathematicians (and different methods). For an extensive report on this, see the featured review by A. Voronov (MR1890736) on the article by McClure and Smith [185], giving a solution to Deligne's conjecture.

- **Exceptional Lie groups**
  In [D82, 1996], Deligne discovered strange uniformity and symmetry phenomena in a list of virtual representations of the automorphism group G of a split, adjoint group G^0 over $\mathbf{Q}$ of one of the types A_1, A_2, G_2, D_4, F_4, E_6, E_7, E_8. These representations are zero, or irreducible up to sign, and they include the trivial representation and the adjoint representation. The symmetry properties are relations in the Grothendieck group of representations of G, which involve exchanging k and $-(1/6)-k$, where $k = \Phi(\alpha,\alpha)$, for Φ the Killing form on the dual of the Lie algebra of a maximal torus, and α the longest root.[50] Moreover, the dimensions of these representations are rational functions in λ whose numerators and denominators are products of linear factors,[51] where $\lambda = 6a$, for $a = k$ or $a = -(1/6) - k$. To explain these phenomena, he conjectured the existence of a semisimple abelian rigid[52] tensor category $\mathscr{C}_t$ over $\mathbf{Q}(t)$, having certain additional data (action of a certain Lie algebra $\underline{g}$ on objects of $\mathscr{C}_t$), such that, in a suitable sense, the category of representations of G would be a specialization of $\mathscr{C}_t$ at $t = a$.
  This conjecture is still open. Computational evidence was given by Cohen and de Man [55]. Further uniformity properties in the behavior of the above exceptional series – with the super group $\mathrm{SOSp}(1,2)$ added – were established by Deligne and de Man [D83, 1996]. In [D99, 2002], Deligne and Gross put these results into a new perspective, by organizing the groups of the exceptional series into a *magic triangle*, whose entry at a pair $H \subset K$ is the centralizer G of H in the automorphism group of $\mathrm{Lie}(K)$, a generalization of Freudenthal–Tits's magic square. According to Deligne,[53] work of Dylan Thurston suggests that the above conjecture is false, as its analogue for some other lines of the magic triangle is.

11 Expository Articles

Work of P. A. Griffiths [D10, 1970]
Non-rational unirational varieties (Artin and Mumford) [D14, 1970]
Modular forms and representations of GL(2) [D23, 1973]
Elliptic curves (after J. Tate) [D32, 1975]
Diffeomorphisms of the circle (Herman) [D38, 1977]
Introduction to étale cohomology [D39, 1977]
Cubic Gauss sums and coverings of SL(2) **(Patterson)** [D41, 1979]

---

50 k is the inverse of the dual Coxeter number $h^\vee$.

51 For the adjoint representation, such formulas had been found by P. Vogel; according to Deligne, that was the beginning of the story.

52 I.e., objects have duals, hence a dimension with value in $\mathrm{End}(1) = \mathbf{Q}(t)$.

53 Private communication, June 2017.

Fundamental group of the complement of a plane nodal curve (Fulton) [D44, 1979]
Faltings's proof of the Mordell conjecture [D56, 1983], [D61, 1985], [D62, 1985]
Drinfeld's modules [D67, 1987]
Grothendieck's main ideas [D87, 1998]
Quantum fields and strings [D88, 1999], [D89, 1999], [D90, 1999], [D91, 1999], [D92, 1999], [D93, 1999], [D94, 1999], [D95, 1999]
The Hodge conjecture [D104, 2006]
Voevodsky's motivic cohomology [D110, 2009]
F. Brown's work on multizeta values [D113, 2012]

Acknowledgements I warmly thank Alexander Beilinson, Francis Brown, Pierre Deligne, Hélène Esnault, Ragni Piene, Takeshi Saito, Jean-Pierre Serre, Claire Voisin, and Weizhe Zheng for useful comments on preliminary versions of this report, and Gérard Laumon for helpful discussions and his assistance in collecting the references.

References

1. *Dix exposés sur la cohomologie des schémas.* Advanced Studies in Pure Mathematics, Vol. 3. North-Holland Publishing Co., Amsterdam; Masson & Cie, Editeur, Paris, 1968.
2. *Théorie des topos et cohomologie étale des schémas. Tome 1. Théorie des topos.* Lecture Notes in Mathematics, Vol. 269. Springer-Verlag, Berlin-New York, 1972. Séminaire de Géométrie Algébrique du Bois-Marie 1963–1964 (SGA 4), Dirigé par M. Artin, A. Grothendieck, et J. L. Verdier. Avec la collaboration de N. Bourbaki, P. Deligne et B. Saint-Donat.
3. *Théorie des topos et cohomologie étale des schémas. Tome 2.* Lecture Notes in Mathematics, Vol. 270. Springer-Verlag, Berlin-New York, 1972. Séminaire de Géométrie Algébrique du Bois-Marie 1963–1964 (SGA 4), Dirigé par M. Artin, A. Grothendieck et J. L. Verdier. Avec la collaboration de N. Bourbaki, P. Deligne et B. Saint-Donat.
4. *Théorie des topos et cohomologie étale des schémas. Tome 3.* Lecture Notes in Mathematics, Vol. 305. Springer-Verlag, Berlin, 1973. Séminaire de Géométrie Algébrique du Bois-Marie 1963–1964 (SGA 4), Dirigé par M. Artin, A. Grothendieck et J. L. Verdier. Avec la collaboration de P. Deligne et B. Saint-Donat.
5. *Cohomologie ℓ-adique et fonctions L.* Lecture Notes in Mathematics, Vol. 589. Springer-Verlag, Berlin-New York, 1977. Séminaire de Géometrie Algébrique du Bois-Marie 1965–1966 (SGA 5), Dirigé par A. Grothendieck. Avec la collaboration de I. Bucur, C. Houzel, L. Illusie, J.-P. Jouanolou et J.-P. Serre. Édité par Luc Illusie.
6. *Groupes de monodromie en géométrie algébrique. I.* Lecture Notes in Mathematics, Vol. 288. Springer-Verlag, Berlin, 1972. Séminaire de Géométrie Algébrique du Bois-Marie 1967–1969 (SGA 7 I), Dirigé par A. Grothendieck. Avec la collaboration de M. Raynaud et D. S. Rim.
7. *Groupes de monodromie en géométrie algébrique. II.* Lecture Notes in Mathematics, Vol. 340. Springer-Verlag, Berlin, 1973. Séminaire de Géométrie Algébrique du Bois-Marie 1967–1969 (SGA 7 II), Dirigé par P. Deligne et N. Katz.
8. *Correspondance Serre-Tate. Vol. II.* Documents Mathématiques (Paris) [Mathematical Documents (Paris)], 14. Société Mathématique de France, Paris, 2015. Edited, and with notes and commentaries by Pierre Colmez and Jean-Pierre Serre.
9. A. Abbes, M. Gros, and T. Tsuji. *The p-adic Simpson Correspondence*, Annals of Math. Studies, vol. 193, Princeton Univ. Press, 2016.

10. T. Abe. Langlands correspondence for isocrystals and existence of crystalline companion for curves. *J. Amer. Math. Soc.*, 31(4):921–1057, 2018.
11. T. Abe and H. Esnault. A Lefschetz theorem for over convergent isocrystals with Frobenius structure. *arXiv* 1607.07112, 2016.
12. D. Alvis. The duality operation in the character ring of a finite Chevalley group. *Bull. Amer. Math. Soc. (N.S.)*, 1(6):907–911, 1979.
13. Y. André. Pour une théorie inconditionnelle des motifs. *Inst. Hautes Études Sci. Publ. Math.*, (83):5–49, 1996.
14. M. Artin. Algebraization of formal moduli. I. In *Global Analysis (Papers in Honor of K. Kodaira)*, pages 21–71. Univ. Tokyo Press, Tokyo, 1969.
15. M. Artin. Versal deformations and algebraic stacks. *Invent. Math.*, 27:165–189, 1974.
16. M. Artin and G. Winters. Degenerate fibres and stable reduction of curves. *Topology*, 10:373–383, 1971.
17. A. I. Badulescu. Correspondance de Jacquet-Langlands pour les corps locaux de caractéristique non nulle. *Ann. Sci. École Norm. Sup. (4)*, 35(5):695–747, 2002.
18. L. Barbieri-Viale, A. Rosenschon, and M. Saito. Deligne's conjecture on 1-motives. *Ann. of Math. (2)*, 158(2):593–633, 2003.
19. L. Barbieri-Viale. On the theory of 1-motives. In *Algebraic cycles and motives. Vol. 1*, volume 343 of *London Math. Soc. Lecture Note Ser.*, pages 55–101. Cambridge Univ. Press, Cambridge, 2007.
20. A. A. Beilinson. Higher regulators and values of L-functions. In *Current problems in mathematics, Vol. 24*, Itogi Nauki i Tekhniki, pages 181–238. Akad. Nauk SSSR, Vsesoyuz. Inst. Nauchn. i Tekhn. Inform., Moscow, 1984.
21. A. A. Beilinson. Notes on absolute Hodge cohomology. In *Applications of algebraic K-theory to algebraic geometry and number theory, Part I, II (Boulder, Colo., 1983)*, volume 55 of *Contemp. Math.*, pages 35–68. Amer. Math. Soc., Providence, RI, 1986.
22. A. A. Beilinson. On the derived category of perverse sheaves. In *K-theory, arithmetic and geometry (Moscow, 1984–1986)*, volume 1289 of *Lecture Notes in Math.*, pages 27–41. Springer, Berlin, 1987.
23. A. A. Beilinson. Height pairing between algebraic cycles. In *K-theory, arithmetic and geometry (Moscow, 1984–1986)*, volume 1289 of *Lecture Notes in Math.*, pages 1–25. Springer, Berlin, 1987.
24. A. Beilinson and J. Bernstein. A proof of Jantzen conjectures. In *I. M. Gelfand Seminar*, volume 16 of *Adv. Soviet Math.*, pages 1–50. Amer. Math. Soc., Providence, RI, 1993.
25. A. Beilinson, S. Bloch, and H. Esnault. ϵ-factors for Gauss-Manin determinants. *Mosc. Math. J.*, 2(3):477–532, 2002. Dedicated to Yuri I. Manin on the occasion of his 65th birthday.
26. A. Beilinson, S. Bloch, P. Deligne, and H. Esnault. Periods for irregular connections on curves. unpublished, http://www.mi.fu-berlin.de/users/esnault/preprints/helene/69-preprint-per051206.pdf.
27. A. Beilinson. $\mathscr{E}$-factors for the period determinants of curves. In *Motives and algebraic cycles*, volume 56 of *Fields Inst. Commun.*, pages 15–82. Amer. Math. Soc., Providence, RI, 2009.
28. A. Beilinson. p-adic periods and derived de Rham cohomology. *J. Amer. Math. Soc.*, 25(3):715–738, 2012.
29. A. Beilinson. Constructible sheaves are holonomic. *Selecta Math. (N.S.)*, 22(4):1797–1819, 2016.
30. P. Berthelot. *Cohomologie cristalline des schémas de caractéristique $p > 0$*. Lecture Notes in Mathematics, Vol. 407. Springer-Verlag, Berlin-New York, 1974.
31. B. Bhatt. Completions and derived de Rham cohomology. *arXiv* 1207.6193v1, 2012.
32. B. Bhatt and P. Scholze. The pro-étale topology for schemes. *Astérisque*, (369):99–201, 2015.
33. A. Blanchard. Sur les variétés analytiques complexes. *Ann. Sci. Ecole Norm. Sup. (3)*, 73:157–202, 1956.
34. S. Bloch. Algebraic K-theory and crystalline cohomology. *Inst. Hautes Études Sci. Publ. Math.*, (47):187–268 (1978), 1977.

35. S. Bloch. Applications of the dilogarithm function in algebraic K-theory and algebraic geometry. In *Proceedings of the International Symposium on Algebraic Geometry (Kyoto Univ., Kyoto, 1977)*, pages 103–114. Kinokuniya Book Store, Tokyo, 1978.
36. S. Bloch. Height pairings for algebraic cycles. In *Proceedings of the Luminy conference on algebraic K-theory (Luminy, 1983)*, volume 34, pages 119–145, 1984.
37. S. Bloch and K. Kato. p-adic étale cohomology. *Inst. Hautes Études Sci. Publ. Math.*, (63):107–152, 1986.
38. S. Bloch and K. Kato. L-functions and Tamagawa numbers of motives. In *The Grothendieck Festschrift, Vol. I*, volume 86 of *Progr. Math.*, pages 333–400. Birkhäuser Boston, Boston, MA, 1990.
39. S. Bloch and H. Esnault. Gauss-Manin determinants for rank 1 irregular connections on curves. *Math. Ann.*, 321(1):15–87, 2001. With an appendix in French by P. Deligne.
40. S. Bloch and H. Esnault. Homology for irregular connections. *J. Théor. Nombres Bordeaux*, 16(2):357–371, 2004.
41. S. Bloch and H. Esnault. Local Fourier transforms and rigidity for $\mathscr{D}$-modules. *Asian J. Math.*, 8(4):587–605, 2004.
42. A. I. Bondal and M. M. Kapranov. Representable functors, Serre functors, and reconstructions. *Izv. Akad. Nauk SSSR Ser. Mat.*, 53(6):1183–1205, 1337, 1989.
43. E. Brieskorn. Sur les groupes de tresses [d'après V. I. Arnold]. In *Séminaire Bourbaki, 24ème année (1971/1972), Exp. No. 401*, pages 21–44. Lecture Notes in Math., Vol. 317. Springer, Berlin, 1973.
44. S. Brochard. Foncteur de Picard d'un champ algébrique. *Math. Ann.*, 343(3):541–602, 2009.
45. M. Broué and J. Michel. Sur certains éléments réguliers des groupes de Weyl et les variétés de Deligne-Lusztig associées. In *Finite reductive groups (Luminy, 1994)*, volume 141 of *Progr. Math.*, pages 73–139. Birkhäuser Boston, Boston, MA, 1997.
46. F. Brown. Mixed Tate motives over $\mathbf{Z}$. *Ann. of Math. (2)*, 175(2):949–976, 2012.
47. F. Brown. Multiple Modular Values for $SL_2(\mathbf{Z})$. Preprint, 2014.
48. J.-L. Brylinski. Transformations canoniques, dualité projective, théorie de Lefschetz, transformations de Fourier et sommes trigonométriques. *Astérisque*, (140–141):3–134, 251, 1986. Géométrie et analyse microlocales.
49. M. A. A. de Cataldo. Hodge-theoretic splitting mechanisms for projective maps. *J. Singul.*, 7:134–156, 2013. With an appendix containing a letter from P. Deligne.
50. E. Cattani and A. Kaplan. Algebraicity of Hodge loci for variations of Hodge structure. In *Hodge theory, complex geometry, and representation theory*, volume 608 of *Contemp. Math.*, pages 59–83. Amer. Math. Soc., Providence, RI, 2014.
51. F. Charles. The Tate conjecture for $K3$ surfaces over finite fields. *Invent. Math.*, 194(1):119–145, 2013.
52. F. Charles. Erratum to: The Tate conjecture for $K3$ surfaces over finite fields. *Invent. Math.*, 202(1):481–485, 2015.
53. F. Charles and C. Schnell. Notes on absolute Hodge classes. In *Hodge theory*, volume 49 of *Math. Notes*, pages 469–530. Princeton Univ. Press, Princeton, NJ, 2014.
54. L. Clozel, M. Harris, and R. Taylor. Automorphy for some ℓ-adic lifts of automorphic mod ℓ Galois representations. *Publ. Math. Inst. Hautes Études Sci.*, (108):1–181, 2008. With Appendix A, summarizing unpublished work of Russ Mann, and Appendix B by Marie-France Vignéras.
55. A. M. Cohen and R. de Man. Computational evidence for Deligne's conjecture regarding exceptional Lie groups. *C. R. Acad. Sci. Paris Sér. I Math.*, 322(5):427–432, 1996.
56. C. W. Curtis. Representations of finite groups of Lie type. *Bull. Amer. Math. Soc. (N.S.)*, 1(5):721–757, 1979.
57. C. W. Curtis. Truncation and duality in the character ring of a finite group of Lie type. *J. Algebra*, 62(2):320–332, 1980.
58. A. D'Agnolo and M. Kashiwara. Riemann-Hilbert correspondence for holonomic D-modules. *Publ. Math. Inst. Hautes Études Sci.*, 123:69–197, 2016.

59. R. De Jeu. Zagier's conjecture and wedge complexes in algebraic K-theory. *Compositio Math.*, 96(2):197–247, 1995.
60. P. Deligne. Lettre à M. F. Atiyah, May 26, 1968.
61. P. Deligne. Le complexe cotangent. Handwritten notes, 1970, unpublished.
62. P. Deligne. Cristaux discontinus Handwritten notes, 1970, unpublished.
63. P. Deligne. Letters to J.-P. Serre, October, 1972, and December, 1973.
64. P. Deligne. Letter to L. Illusie, October 9, 1973.
65. P. Deligne. Letter to I. Piatetski-Shapiro, March 25, 1973.
66. P. Deligne. V-complexes de de Rham, Handwritten notes, 1975, unpublished.
67. P. Deligne. Letter to N. Katz, Oct. 18, 1976.
68. P. Deligne. Letter to L. Illusie, Oct. 28, 1976. Incipit: "Voici une preuve globale ..."
69. P. Deligne. Letter to L. Illusie, Oct. 28, 1976. Incipit: "Voici une semi-continuité ..."
70. P. Deligne. Letter to L. Illusie, Oct. 28, 1976. Incipit: "Troisième lettre ..."
71. P. Deligne. Letter to L. Illusie, Nov. 4, 1976. Incipit: "J'ai réfléchi à ce que donne la méthode des pinceaux ... dans le cas général ..."
72. P. Deligne. Letter to L. Illusie, Nov. 4, 1976. Incipit: "Les pinceaux donnent la conjecture de ma lettre à Katz ..."
73. P. Deligne. Letter to D. Kazhdan, Nov. 29, 1976.
74. P. Deligne. Les constantes des équations fonctionnelles des fonctions L. Séminaire à l'IHES, 1980. Partial notes by L. Illusie.
75. P. Deligne. Positivité, signes: I, Feb. 16, 1984; II, Nov. 6, 1985. Unpublished notes.
76. P. Deligne. Letter to C. Soulé, Jan. 20, 1985.
77. P. Deligne. Letter to L. Illusie, June 1, 1988.
78. P. Deligne. Letter to Stasheff et al. May 17, 1993.
79. P. Deligne. Notes sur Euler-Poincaré: brouillon project. Handwritten notes, Feb. 8, 2011.
80. P. Deligne. Comptage de faisceaux ℓ-adiques. *Astérisque*, (369):285–312, 2015.
81. C. Deninger and A. J. Scholl. The Beilinson conjectures. In *L-functions and arithmetic (Durham, 1989)*, volume 153 of *London Math. Soc. Lecture Note Ser.*, pages 173–209. Cambridge Univ. Press, Cambridge, 1991.
82. V. Drinfeld. The number of two-dimensional irreducible representations of the fundamental group of a curve over a finite field. *Funktsional. Anal. i Prilozhen.*, 15(4):75–76, 1981.
83. V. Drinfeld. On a conjecture of Deligne. *Mosc. Math. J.*, 12(3):515–542, 668, 2012.
84. V. Drinfeld and K. Kedlaya. Slopes of indecomposable F-isocrystals. *arXiv* 1604.00660v11, 2016.
85. P. Du Bois. Complexe de de Rham filtré d'une variété singulière. *Bull. Soc. Math. France*, 109(1):41–81, 1981.
86. B. Dwork. On the rationality of the zeta function of an algebraic variety. *Amer. J. Math.*, 82:631–648, 1960.
87. T. Ekedahl. On the adic formalism. In *The Grothendieck Festschrift, Vol. II*, volume 87 of *Progr. Math.*, pages 197–218. Birkhäuser Boston, Boston, MA, 1990.
88. H. Esnault and E. Viehweg. Logarithmic de Rham complexes and vanishing theorems. *Invent. Math.*, 86(1):161–194, 1986.
89. H. Esnault and M. Kerz. Notes on Deligne's letter to Drinfeld dated March 5, 2007. Notes for the Forschungsseminar in Essen, summer 2011.
90. H. Esnault and M. Kerz. A finiteness theorem for Galois representations of function fields over finite fields (after Deligne). *Acta Math. Vietnam.*, 37(4):531–562, 2012.
91. H. Esnault, C. Sabbah, and J.D. Yu. E_1-degeneration of the irregular Hodge filtration (with an appendix by Morihiko Saito). *J. Reine und Ang. Math.*, 729, 171–227, 2017.
92. M. Flach. The equivariant Tamagawa number conjecture: a survey. In *Stark's conjectures: recent work and new directions*, volume 358 of *Contemp. Math.*, pages 79–125. Amer. Math. Soc., Providence, RI, 2004. With an appendix by C. Greither.
93. J.-M. Fontaine. Valeurs spéciales des fonctions L des motifs. *Astérisque*, (206):Exp. No. 751, 4, 205–249, 1992. Séminaire Bourbaki, Vol. 1991/92.

94. J. Fresán. Periods of Hodge structures and special values of the gamma function. *Invent. Math.*, 208(1):247–282, 2017.
95. A. Fröhlich and J. Queyrut. On the functional equation of the Artin L-function for characters of real representations. *Invent. Math.*, 20:125–138, 1973.
96. K. Fujiwara. Rigid geometry, Lefschetz-Verdier trace formula and Deligne's conjecture. *Invent. Math.*, 127(3):489–533, 1997.
97. K. Fujiwara. Independence of ℓ for intersection cohomology (after Gabber). In *Algebraic geometry 2000, Azumino (Hotaka)*, volume 36 of *Adv. Stud. Pure Math.*, pages 145–151. Math. Soc. Japan, Tokyo, 2002.
98. O. Gabber. Notes on some t-structures. In *Geometric aspects of Dwork theory. Vol. I, II*, pages 711–734. Walter de Gruyter GmbH & Co. KG, Berlin, 2004.
99. F. A. Garside. The braid group and other groups. *Quart. J. Math. Oxford Ser. (2)*, 20:235–254, 1969.
100. J. Giraud. *Cohomologie non abélienne*. Springer-Verlag, Berlin-New York, 1971. Die Grundlehren der mathematischen Wissenschaften, Band 179.
101. A. B. Goncharov. The dihedral Lie algebras and Galois symmetries of $\pi_1^{(l)}(\mathbf{P}^1 - (\{0, \infty\} \cup \mu_N))$. *Duke Math. J.*, 110(3):397–487, 2001.
102. M. Goresky and R. MacPherson. La dualité de Poincaré pour les espaces singuliers. *C. R. Acad. Sci. Paris Sér. A-B*, 284(24):A1549–A1551, 1977.
103. M. Goresky and R. MacPherson. Intersection homology theory. *Topology*, 19(2):135–162, 1980.
104. M. Goresky and R. MacPherson. Intersection homology. II. *Invent. Math.*, 72(1):77–129, 1983.
105. B. H. Gross. On the periods of abelian integrals and a formula of Chowla and Selberg. *Invent. Math.*, 45(2):193–211, 1978. With an appendix by David E. Rohrlich.
106. A. Grothendieck. On the de Rham cohomology of algebraic varieties. *Inst. Hautes Études Sci. Publ. Math.*, (29):95–103, 1966.
107. A. Grothendieck. *Catégories cofibrées additives et complexe cotangent relatif.* Lecture Notes in Mathematics, No. 79. Springer-Verlag, Berlin-New York, 1968.
108. A. Grothendieck. Crystals and the De Rham Cohomology of schemes (notes by J. Coates and O. Jussila), IHÉS, Décembre 1966. Adv. Stud. Pure Math., 3, Dix exposés sur la cohomologie des schémas, 306–358, North-Holland, Amsterdam, 1968.
109. A. Grothendieck. Standard conjectures on algebraic cycles. In *Algebraic Geometry (Internat. Colloq., Tata Inst. Fund. Res., Bombay, 1968)*, pages 193–199. Oxford Univ. Press, London, 1969.
110. G. Harder, R. P. Langlands, and M. Rapoport. Algebraische Zyklen auf Hilbert-Blumenthal-Flächen. *J. Reine Angew. Math.*, 366:53–120, 1986.
111. M. Harris and R. Taylor. *The geometry and cohomology of some simple Shimura varieties*, volume 151 of *Annals of Mathematics Studies*. Princeton University Press, Princeton, NJ, 2001. With an appendix by Vladimir G. Berkovich.
112. M. Harris, N. Shepherd-Barron, and R. Taylor. A family of Calabi-Yau varieties and potential automorphy. *Ann. of Math. (2)*, 171(2):779–813, 2010.
113. R. Hartshorne. *Residues and duality*. Lecture notes of a seminar on the work of A. Grothendieck, given at Harvard 1963/64. With an appendix by P. Deligne. Lecture Notes in Mathematics, No. 20. Springer-Verlag, Berlin-New York, 1966.
114. R. Hartshorne. On the De Rham cohomology of algebraic varieties. *Inst. Hautes Études Sci. Publ. Math.*, (45):5–99, 1975.
115. J. Heinloth, B.-C. Ngô, and Z. Yun. Kloosterman sheaves for reductive groups. *Ann. of Math. (2)*, 177(1):241–310, 2013.
116. G. Henniart. Une preuve simple des conjectures de Langlands pour GL(n) sur un corps p-adique. *Invent. Math.*, 139(2):439–455, 2000.
117. W. V. D. Hodge and M. F. Atiyah. Integrals of the second kind on an algebraic variety. *Ann. of Math. (2)*, 62:56–91, 1955.

118. A. Huber and J. Wildeshaus. Classical motivic polylogarithm according to Beilinson and Deligne. *Doc. Math.*, 3:27–133, 1998.
119. A. Huber and J. Wildeshaus. Corrections to the paper: "Classical motivic polylogarithm according to Beilinson and Deligne". *Doc. Math.*, 3:297–299, 1998.
120. L. Illusie. *Complexe cotangent et déformations. I.* Lecture Notes in Mathematics, Vol. 239. Springer-Verlag, Berlin-New York, 1971.
121. L. Illusie. *Complexe cotangent et déformations. II.* Lecture Notes in Mathematics, Vol. 283. Springer-Verlag, Berlin-New York, 1972.
122. L. Illusie. Complexe de de Rham-Witt et cohomologie cristalline. *Ann. Sci. École Norm. Sup. (4)*, 12(4):501–661, 1979.
123. L. Illusie. Théorie de Brauer et caractéristique d'Euler-Poincaré (d'après P. Deligne). In *The Euler-Poincaré characteristic (French)*, volume 82 of *Astérisque*, pages 161–172. Soc. Math. France, Paris, 1981.
124. L. Illusie. Deligne's ℓ-adic Fourier transform. In *Algebraic geometry, Bowdoin, 1985 (Brunswick, Maine, 1985)*, volume 46 of *Proc. Sympos. Pure Math.*, pages 151–163. Amer. Math. Soc., Providence, RI, 1987.
125. L. Illusie. Autour du théorème de monodromie locale. *Astérisque*, (223):9–57, 1994. Périodes p-adiques (Bures-sur-Yvette, 1988).
126. L. Illusie. Sur la formule de Picard-Lefschetz. In *Algebraic geometry 2000, Azumino (Hotaka)*, volume 36 of *Adv. Stud. Pure Math.*, pages 249–268. Math. Soc. Japan, Tokyo, 2002.
127. L. Illusie. Miscellany on traces in ℓ-adic cohomology: a survey. *Jpn. J. Math.*, 1(1):107–136, 2006.
128. L. Illusie. "La descente galoisienne…". *Mosc. Math. J.*, 9(1):47–55, backmatter, 2009.
129. L. Illusie. From Pierre Deligne's secret garden: looking back at some of his letters. *Jpn. J. Math.*, 10(2):237–248, 2015.
130. L. Illusie and W. Zheng. Odds and ends on finite group actions and traces. *Int. Math. Res. Not. IMRN*, (1):1–62, 2013.
131. L. Illusie, Y. Laszlo, F. Orgogozo (eds.) *Gabber's work on local uniformization and étale cohomology of quasi-excellent schemes. Seminar at École Polytechnique 2006–2008. (Travaux de Gabber sur l'uniformisation locale et la cohomologie étale des schémas quasi-excellents. Séminaire à l'École Polytechnique 2006–2008.)* Astérisque 363–364. Paris: Société Mathématique de France (SMF) (2014).
132. T. Ito. Weight-monodromy conjecture for certain threefolds in mixed characteristic. *Int. Math. Res. Not.*, (2):69–87, 2004.
133. T. Ito. Weight-monodromy conjecture over equal characteristic local fields. *Amer. J. Math.*, 127(3):647–658, 2005.
134. T. Ito. Weight-monodromy conjecture for p-adically uniformized varieties. *Invent. Math.*, 159(3):607–656, 2005.
135. H. Jacquet and R. P. Langlands. *Automorphic forms on* GL(2). Lecture Notes in Mathematics, Vol. 114. Springer-Verlag, Berlin-New York, 1970.
136. U. Jannsen. *Mixed motives and algebraic K-theory*, volume 1400 of *Lecture Notes in Mathematics*. Springer-Verlag, Berlin, 1990. With appendices by S. Bloch and C. Schoen.
137. P. T. Johnstone. *Topos theory*. Academic Press [Harcourt Brace Jovanovich, Publishers], London-New York, 1977. London Mathematical Society Monographs, Vol. 10.
138. A. J. de Jong. Smoothness, semi-stability and alterations. *Inst. Hautes Études Sci. Publ. Math.*, (83):51–93, 1996.
139. M. Kashiwara. Faisceaux constructibles et systèmes holonômes d'équations aux dérivées partielles linéaires à points singuliers réguliers. In *Séminaire Goulaouic-Schwartz, 1979–1980 (French)*, pages Exp. No. 19, 7. École Polytech., Palaiseau, 1980.
140. M. Kashiwara. The Riemann-Hilbert problem for holonomic systems. *Publ. Res. Inst. Math. Sci.*, 20(2):319–365, 1984.
141. H. Kato. Wild ramification and restriction to curves. *Internat. J. Math.*, 29(8):1850052, 20 pp., 2018.

142. N. M. Katz. The regularity theorem in algebraic geometry. In *Actes du Congrès International des Mathématiciens (Nice, 1970), Tome 1*, pages 437–443. Gauthier-Villars, Paris, 1971.
143. N. M. Katz. An overview of Deligne's proof of the Riemann hypothesis for varieties over finite fields. In *Mathematical developments arising from Hilbert problems (Proc. Sympos. Pure Math., Vol. XXVIII, Northern Illinois Univ., De Kalb, Ill., 1974)*, pages 275–305. Amer. Math. Soc., Providence, R.I., 1976.
144. N. M. Katz. *Sommes exponentielles*, volume 79 of *Astérisque*. Société Mathématique de France, Paris, 1980. Course taught at the University of Paris-Sud, Orsay, Fall 1979, With a preface by Luc Illusie, Notes written by Gérard Laumon, With an English summary.
145. N. Katz. Serre-Tate local moduli. In *Algebraic surfaces (Orsay, 1976–78)*, volume 868 of *Lecture Notes in Math.*, pages 138–202. Springer, Berlin-New York, 1981.
146. N. M. Katz. *Gauss sums, Kloosterman sums, and monodromy groups*, volume 116 of *Annals of Mathematics Studies*. Princeton University Press, Princeton, NJ, 1988.
147. N. M. Katz and G. Laumon. Transformation de Fourier et majoration de sommes exponentielles. *Publ. Math. Inst. Hautes Études Sci.*, (62):361–418, 1985.
148. N. M. Katz and B. Mazur. *Arithmetic moduli of elliptic curves*, volume 108 of *Annals of Mathematics Studies*. Princeton University Press, Princeton, NJ, 1985.
149. N. M. Katz and W. Messing. Some consequences of the Riemann hypothesis for varieties over finite fields. *Invent. Math.*, 23:73–77, 1974.
150. D. Kazhdan and G. Lusztig. Proof of the Deligne-Langlands conjecture for Hecke algebras. *Invent. Math.*, 87(1):153–215, 1987.
151. K. Kedlaya. Étale and crystalline companions. Preprint, 2017.
152. S. L. Kleiman. Algebraic cycles and the Weil conjectures. In *Dix exposés sur la cohomologie des schémas*, pages 359–386. North-Holland, Amsterdam; Masson, Paris, 1968.
153. S. L. Kleiman. The standard conjectures. In *Motives (Seattle, WA, 1991)*, volume 55 of *Proc. Sympos. Pure Math.*, pages 3–20. Amer. Math. Soc., Providence, RI, 1994.
154. F. F. Knudsen and D. Mumford. The projectivity of the moduli space of stable curves. I. Preliminaries on "det" and "Div". *Math. Scand.*, 39(1):19–55, 1976.
155. F. F. Knudsen. The projectivity of the moduli space of stable curves. II. The stacks $M_{g,n}$. *Math. Scand.*, 52(2):161–199, 1983.
156. F. F. Knudsen. The projectivity of the moduli space of stable curves. III. The line bundles on $M_{g,n}$, and a proof of the projectivity of $\overline{M}_{g,n}$ in characteristic 0. *Math. Scand.*, 52(2):200–212, 1983.
157. D. Knutson. *Algebraic spaces*. Lecture Notes in Mathematics, Vol. 203. Springer-Verlag, Berlin-New York, 1971.
158. M. Kuga and I. Satake. Abelian varieties attached to polarized K_3-surfaces. *Math. Ann.*, 169:239–242, 1967.
159. L. Lafforgue. Chtoucas de Drinfeld et correspondance de Langlands. *Invent. Math.*, 147(1):1–241, 2002.
160. V. Lafforgue. Estimées pour les valuations p-adiques des valeurs propes des opérateurs de Hecke. *Bull. Soc. Math. France*, 139(4):455–477, 2011.
161. W. E. Lang and N. O. Nygaard. A short proof of the Rudakov-Šafarevič theorem. *Math. Ann.*, 251(2):171–173, 1980.
162. R. P. Langlands. On the functional equation of the Artin L-functions. Mimeographed notes (incomplete), Yale, 1970. a-ps.pdf
163. R. P. Langlands. Problems in the theory of automorphic forms. In *Lectures in modern analysis and applications, III*, pages 18–61. Lecture Notes in Math., Vol. 170. Springer, Berlin, 1970.
164. R. P. Langlands. Letter to P. Deligne, March 31, 1974, rl-deligne-ps.pdf.
165. G. Laumon. Homologie étale. In *Séminaire de géométrie analytique École Norm. Sup., Paris, 1974–75)*, pages 163–188. Astérisque, No. 36–37. Soc. Math. France, Paris, 1976.
166. G. Laumon. Majorations de sommes trigonométriques (d'après P. Deligne et N. Katz). In *The Euler-Poincaré characteristic (French)*, volume 83 of *Astérisque*, pages 221–258. Soc. Math. France, Paris, 1981.

167. G. Laumon. Semi-continuité du conducteur de Swan (d'après P. Deligne). In *The Euler-Poincaré characteristic (French)*, volume 83 of *Astérisque*, pages 173–219. Soc. Math. France, Paris, 1981.
168. G. Laumon. Comparaison de caractéristiques d'Euler-Poincaré en cohomologie ℓ-adique. *C. R. Acad. Sci. Paris Sér. I Math.*, 292(3):209–212, 1981.
169. G. Laumon. Caractéristique d'Euler-Poincaré des faisceaux constructibles sur une surface. In *Analysis and topology on singular spaces, II, III (Luminy, 1981)*, volume 101 of *Astérisque*, pages 193–207. Soc. Math. France, Paris, 1983.
170. G. Laumon. Vanishing cycles over a base of dimension $\geq$ 1. In *Algebraic geometry (Tokyo/Kyoto, 1982)*, volume 1016 of *Lecture Notes in Math.*, pages 143–150. Springer, Berlin, 1983.
171. G. Laumon. Transformation de Fourier, constantes d'équations fonctionnelles et conjecture de Weil. *Inst. Hautes Études Sci. Publ. Math.*, (65):131–210, 1987.
172. G. Laumon, M. Rapoport, and U. Stuhler. D-elliptic sheaves and the Langlands correspondence. *Invent. Math.*, 113(2):217–338, 1993.
173. G. Laumon. Exponential sums and ℓ-adic cohomology: a survey. *Israel J. Math.*, 120(part A):225–257, 2000.
174. S. Lefschetz. *L'analysis situs et la géométrie algébrique*. Gauthier-Villars, Paris, 1950.
175. M. Levine. Mixed motives. In *Handbook of K-theory. Vol. 1, 2*, pages 429–521. Springer, Berlin, 2005.
176. Y. Liu and W. Zheng. Gluing restricted nerves of infinity-categories. arXiv:1211.5294 [math.CT], 2014.
177. G. Lusztig. *Representations of finite Chevalley groups*, volume 39 of *CBMS Regional Conference Series in Mathematics*. American Mathematical Society, Providence, R.I., 1978. Expository lectures from the CBMS Regional Conference held at Madison, Wis., August 8–12, 1977.
178. G. Lusztig. Some examples of square integrable representations of semisimple p-adic groups. *Trans. Amer. Math. Soc.*, 277(2):623–653, 1983.
179. G. Lusztig. *Characters of reductive groups over a finite field*, volume 107 of *Annals of Mathematics Studies*. Princeton University Press, Princeton, NJ, 1984.
180. G. Lusztig. Character sheaves. I. *Adv. in Math.*, 56(3):193–237, 1985.
181. G. Lusztig. Character sheaves. II, III. *Adv. in Math.*, 57(3):226–265, 266–315, 1985.
182. G. Lusztig. Character sheaves. IV. *Adv. in Math.*, 59(1):1–63, 1986.
183. G. Lusztig. Character sheaves. V. *Adv. in Math.*, 61(2):103–155, 1986.
184. G. Lusztig. Erratum: "Character sheaves. V". *Adv. in Math.*, 62(3):313–314, 1986.
185. J. E. McClure and J. H. Smith. A solution of Deligne's Hochschild cohomology conjecture. In *Recent progress in homotopy theory (Baltimore, MD, 2000)*, volume 293 of *Contemp. Math.*, pages 153–193. Amer. Math. Soc., Providence, RI, 2002.
186. R. D. MacPherson. Chern classes for singular algebraic varieties. *Ann. of Math. (2)*, 100:423–432, 1974.
187. K. Madapusi Pera. The Tate conjecture for K3 surfaces in odd characteristic. *Invent. Math.*, 201(2):625–668, 2015.
188. V. Maillot and D. Roessler. On the periods of motives with complex multiplication and a conjecture of Gross-Deligne. *Ann. of Math. (2)*, 160(2):727–754, 2004.
189. H. Matsumoto. Sur les sous-groupes arithmétiques des groupes semi-simples déployés. *Ann. Sci. École Norm. Sup. (4)*, 2:1–62, 1969.
190. D. Maulik. Supersingular K3 surfaces for large primes. *Duke Math. J.*, 163(13):2357–2425, 2014. With an appendix by Andrew Snowden.
191. Z. Mebkhout. Sur le problème de Hilbert-Riemann. In *Complex analysis, microlocal calculus and relativistic quantum theory (Proc. Internat. Colloq., Centre Phys., Les Houches, 1979)*, volume 126 of *Lecture Notes in Phys.*, pages 90–110. Springer, Berlin-New York, 1980.
192. J. Milnor. *Singular points of complex hypersurfaces*. Annals of Mathematics Studies, No. 61. Princeton University Press, Princeton, N.J.; University of Tokyo Press, Tokyo, 1968.

193. T. Mochizuki. Holonomic $\mathscr{D}$-modules with Betti structure. *Mém. Soc. Math. Fr. (N.S.)*, (138–139):viii+205, 2014.
194. T. Mochizuki. *Mixed twistor $\mathscr{D}$-modules*, volume 2125 of *Lecture Notes in Mathematics*. Springer, Cham, 2015.
195. T. Mochizuki. Curve test for enhanced ind-sheaves and holonomic $\mathscr{D}$-modules. arXiv:1610.08572.
196. J. W. Morgan. The rational homotopy theory of smooth, complex projective varieties (following P. Deligne, P. Griffiths, J. Morgan, and D. Sullivan) (Invent. Math. **29** (1975), no. 3, 245–274). In *Séminaire Bourbaki, Vol. 1975/76, 28ème année, Exp. No. 475*, pages 69–80. Lecture Notes in Math., Vol. 567. Springer, Berlin, 1977.
197. G. D. Mostow. Existence of nonarithmetic monodromy groups. *Proc. Nat. Acad. Sci. U.S.A.*, 78(10, Phys. Sci.):5948–5950, 1981.
198. D. Mumford. Picard groups of moduli problems. In *Arithmetical Algebraic Geometry (Proc. Conf. Purdue Univ., 1963)*, pages 33–81. Harper & Row, New York, 1965.
199. D. Mumford. Stability of projective varieties. *Enseignement Math. (2)*, 23(1–2):39–110, 1977.
200. C. Nakayama. Degeneration of ℓ-adic weight spectral sequences. *Amer. J. Math.*, 122(4):721–733, 2000.
201. B. C. Ngô. Le lemme fondamental pour les algèbres de Lie. *Publ. Math. Inst. Hautes Études Sci.*, (111):1–169, 2010.
202. N. O. Nygaard. A p-adic proof of the nonexistence of vector fields on $K3$ surfaces. *Ann. of Math. (2)*, 110(3):515–528, 1979.
203. N. O. Nygaard. The Tate conjecture for ordinary $K3$ surfaces over finite fields. *Invent. Math.*, 74(2):213–237, 1983.
204. N. Nygaard and A. Ogus. Tate's conjecture for $K3$ surfaces of finite height. *Ann. of Math. (2)*, 122(3):461–507, 1985.
205. J. Oesterlé. Polylogarithmes. *Astérisque*, (216):Exp. No. 762, 3, 49–67, 1993. Séminaire Bourbaki, Vol. 1992/93.
206. A. Ogus. F-crystals, Griffiths transversality, and the Hodge decomposition. *Astérisque*, (221):ii+183, 1994.
207. A. Ogus and V. Vologodsky. Nonabelian Hodge theory in characteristic p. *Publ. Math. Inst. Hautes Études Sci.*, (106):1–138, 2007.
208. M. C. Olsson. Deformation theory of representable morphisms of algebraic stacks. *Math. Z.*, 253(1):25–62, 2006.
209. F. Orgogozo. Conjecture de Bloch et nombres de Milnor. *Ann. Inst. Fourier (Grenoble)*, 53(6):1739–1754, 2003.
210. F. Orgogozo. Modifications et cycles proches sur une base générale. *Int. Math. Res. Not.*, pages Art. ID 25315, 38, 2006.
211. D. Quillen. Rational homotopy theory. *Ann. of Math. (2)*, 90:205–295, 1969.
212. N. Ramachandran. One-motives and a conjecture of Deligne. *J. Algebraic Geom.*, 13(1):29–80, 2004.
213. R. A. Rankin. Contributions to the theory of Ramanujan's function $\tau(n)$ and similar arithmetical functions. I. The zeros of the function $\sum_{n=1}^{\infty} \tau(n)/n^s$ on the line $\mathscr{R}s = 13/2$. II. The order of the Fourier coefficients of integral modular forms. *Proc. Cambridge Philos. Soc.*, 35:351–372, 1939.
214. M. Rapoport. Compactifications de l'espace de modules de Hilbert-Blumenthal. *Compositio Math.*, 36(3):255–335, 1978.
215. M. Rapoport and T. Zink. Über die lokale Zetafunktion von Shimuravarietäten. Monodromiefiltration und verschwindende Zyklen in ungleicher Charakteristik. *Invent. Math.*, 68(1):21–101, 1982.
216. J. D. Rogawski. Representations of GL(n) and division algebras over a p-adic field. *Duke Math. J.*, 50(1):161–196, 1983.
217. A. N. Rudakov and I. R. Šafarevič. Inseparable morphisms of algebraic surfaces. *Izv. Akad. Nauk SSSR Ser. Mat.*, 40(6):1269–1307, 1439, 1976.

218. N. Saavedra Rivano. Catégories tannakiennes. *Bull. Soc. Math. France*, 100:417–430, 1972.
219. N. Saavedra Rivano. *Catégories Tannakiennes*. Lecture Notes in Mathematics, Vol. 265. Springer-Verlag, Berlin-New York, 1972.
220. C. Sabbah. Morphismes analytiques stratifiés sans éclatement et cycles évanescents. In *Analysis and topology on singular spaces, II, III (Luminy, 1981)*, volume 101 of *Astérisque*, pages 286–319. Soc. Math. France, Paris, 1983.
221. M. Saito. Modules de Hodge polarisables. *Publ. Res. Inst. Math. Sci.*, 24(6):849–995 (1989), 1988.
222. M. Saito. Mixed Hodge modules. *Publ. Res. Inst. Math. Sci.*, 26(2):221–333, 1990.
223. T. Saito. Vanishing cycles and geometry of curves over a discrete valuation ring. *Amer. J. Math.*, 109(6):1043–1085, 1987.
224. T. Saito. An introduction to Galois representations and modular forms. Autour des motifs—École d'été Franco-Asiatique de Géométrie Algébrique et de Théorie des Nombres/Asian-French Summer School on Algebraic Geometry and Number Theory. Vol. III, 1–27, Panor. Synthèses, 49, Soc. Math. France, Paris, 2016.
225. T. Saito. Characteristic cycle and the Euler number of a constructible sheaf on a surface. *J. Math. Sci. Univ. Tokyo*, 22(1):387–441, 2015.
226. T. Saito and Y. Yatagawa Wild ramification determines the characteristic cycle. *Ann. Sci. Éc. Norm. Supér.*, (4) 50(4):1065–1079, 2017.
227. T. Saito. The characteristic cycle and the singular support of a constructible sheaf. *Invent. Math.*, 207(2):597–695, 2017.
228. T. Saito Characteristic cycles and the conductor of direct image. arXiv:1704.04832.
229. A. J. Scholl. Remarks on special values of L-functions. In *L-functions and arithmetic (Durham, 1989)*, volume 153 of *London Math. Soc. Lecture Note Ser.*, pages 373–392. Cambridge Univ. Press, Cambridge, 1991.
230. A. J. Scholl. Motives for modular forms. *Invent. Math.*, 100(2):419–430, 1990.
231. P. Scholze. Perfectoid spaces. *Publ. Math. Inst. Hautes Études Sci.*, 116:245–313, 2012.
232. P. Scholze. The local Langlands correspondence for GL_n over p-adic fields. *Invent. Math.*, 192(3):663–715, 2013.
233. K. Schwede. A simple characterization of Du Bois singularities. *Compos. Math.*, 143(4):813–828, 2007.
234. M. Sebastiani and R. Thom. Un résultat sur la monodromie. *Invent. Math.*, 13:90–96, 1971.
235. J.-P. Serre. *Groupes algébriques et corps de classes*. Publications de l'institut de mathématique de l'université de Nancago, VII. Hermann, Paris, 1959.
236. J.-P. Serre. Analogues kählériens de certaines conjectures de Weil. *Ann. of Math. (2)*, 71:392–394, 1960.
237. J.-P. Serre and J. Tate. Good reduction of abelian varieties. *Ann. of Math. (2)*, 88:492–517, 1968.
238. J.-P. Serre. Représentations linéaires des groupes finis "algébriques" (d'après Deligne-Lusztig). In *Séminaire Bourbaki, Vol. 1975/76, 28ème année, Exp. No. 487*, pages 256–273. Lecture Notes in Math., Vol. 567. Springer, Berlin, 1977.
239. J.-P. Serre. Propriétés conjecturales des groupes de Galois motiviques et des représentations ℓ-adiques. In *Motives (Seattle, WA, 1991)*, volume 55 of *Proc. Sympos. Pure Math.*, pages 377–400. Amer. Math. Soc., Providence, RI, 1994.
240. J.-P. Serre. *Œuvres/Collected papers. II. 1960–1971*. Springer Collected Works in Mathematics. Springer, Heidelberg, 2013. Reprint of the 2003 edition [of the 1986 original].
241. C. Soulé. Régulateurs. *Astérisque*, (133–134):237–253, 1986. Seminar Bourbaki, Vol. 1984/85.
242. J. Steenbrink. Limits of Hodge structures. *Invent. Math.*, 31(3):229–257, 1975/76.
243. J. Suh. Symmetry and parity in Frobenius action on cohomology. *Compos. Math.*, 148(1):295–303, 2012.
244. D. Sullivan. Combinatorial invariants of analytic spaces. In *Proceedings of Liverpool Singularities—Symposium, I (1969/70)*, pages 165–168. Springer, Berlin, 1971.

245. D. Sullivan. Differential forms and the topology of manifolds. In *Manifolds—Tokyo 1973 (Proc. Internat. Conf., Tokyo, 1973)*, pages 37–49. Univ. Tokyo Press, Tokyo, 1975.
246. A. Suslin and V. Voevodsky. Singular homology of abstract algebraic varieties. *Invent. Math.*, 123(1):61–94, 1996.
247. S. Sun and W. Zheng. Parity and symmetry in intersection and ordinary cohomology. *Algebra Number Theory* 10(2):235–307, 2016.
248. J. T. Tate. Algebraic cycles and poles of zeta functions. In *Arithmetical Algebraic Geometry (Proc. Conf. Purdue Univ., 1963)*, pages 93–110. Harper & Row, New York, 1965.
249. J. Tate. Conjectures on algebraic cycles in l-adic cohomology. In *Motives (Seattle, WA, 1991)*, volume 55 of *Proc. Sympos. Pure Math.*, pages 71–83. Amer. Math. Soc., Providence, RI, 1994.
250. R. Taylor. Automorphy for some ℓ-adic lifts of automorphic mod ℓ Galois representations. II. *Publ. Math. Inst. Hautes Études Sci.*, (108):183–239, 2008.
251. Y. Varshavsky. Lefschetz-Verdier trace formula and a generalization of a theorem of Fujiwara. *Geom. Funct. Anal.*, 17(1):271–319, 2007.
252. J.-L. Verdier. A duality theorem in the étale cohomology of schemes. In *Proc. Conf. Local Fields (Driebergen, 1966)*, pages 184–198. Springer, Berlin, 1967.
253. J.-L. Verdier. Indépendance par rapport à ℓ des polynômes caractéristiques des endomorphismes de Frobenius de la cohomologie ℓ-adique (d'après P. Deligne). In *Séminaire Bourbaki, 25ème année (1972/1973), Exp. No. 423*, pages 98–115. Lecture Notes in Math., Vol. 383. Springer, Berlin, 1974.
254. J.-L. Verdier. Des catégories dérivées des catégories abéliennes. *Astérisque*, (239):xii+253 pp. (1997), 1996. With a preface by Luc Illusie, Edited and with a note by Georges Maltsiniotis.
255. G. Vezzosi. A note on the cotangent complex in derived algebraic geometry. *arXiv* 1008.0601v3, 2010.
256. C. Voisin. *Hodge theory and complex algebraic geometry. I*, volume 76 of *Cambridge Studies in Advanced Mathematics*. Cambridge University Press, Cambridge, 2002. Translated from the French original by Leila Schneps.
257. C. Voisin. Hodge loci. In *Handbook of moduli. Vol. III*, volume 26 of *Adv. Lect. Math. (ALM)*, pages 507–546. Int. Press, Somerville, MA, 2013.
258. A. Weil. *Sur les courbes algébriques et les variétés qui s'en déduisent.* Actualités Sci. Ind., no. 1041 = Publ. Inst. Math. Univ. Strasbourg **7** (1945). Hermann et Cie., Paris, 1948.
259. A. Weil. *Variétés abéliennes et courbes algébriques.* Actualités Sci. Ind., no. 1064 = Publ. Inst. Math. Univ. Strasbourg 8 (1946). Hermann & Cie., Paris, 1948.
260. A. Weil. Numbers of solutions of equations in finite fields. *Bull. Amer. Math. Soc.*, 55:497–508, 1949.
261. A. Weil. Abstract versus classical algebraic geometry. In *Proceedings of the International Congress of Mathematicians, 1954, Amsterdam, vol. III*, pages 550–558. Erven P. Noordhoff N.V., Groningen; North-Holland Publishing Co., Amsterdam, 1956.
262. H. Wenzl. On the structure of Brauer's centralizer algebras. *Ann. of Math. (2)*, 128(1):173–193, 1988.
263. J.-P. Wintenberger. Le corps des normes de certaines extensions infinies de corps locaux; applications. *Ann. Sci. École Norm. Sup. (4)*, 16(1):59–89, 1983.
264. W. Zheng. Sur la cohomologie des faisceaux ℓ-adiques entiers sur les corps locaux. *Bull. Soc. Math. France*, 136(3):465–503, 2008.
265. W. Zheng. Companions on Artin stacks. arXiv:1512.08929v7, 2015.
266. S. Zucker. Variation of mixed Hodge structure. II. *Invent. Math.*, 80(3):543–565, 1985.

List of Publications for Pierre R. Deligne

1966

[1] Congruences sur le nombre de sous-groupes d'ordre p^k dans un groupe fini. *Bull. Soc. Math. Belg.* 18:129–132.

[2] Cohomologie à support propre et construction du foncteur $f^!$. Appendix to: R. Hartshorne. *Residues and duality,* pp. 404–421. Lecture Notes in Math., Vol. 20. Springer, Berlin.

1968

[3] Théorème de Lefschetz et critères de dégénérescence de suites spectrales. *Inst. Hautes Études Sci. Publ. Math.*, 35:259–278.

1969

[4] Cohomologie à supports propres. In *Séminaire de Géométrie Algébrique du Bois-Marie 1963/64, Théorie des topos et cohomologie étale des schémas (SGA 4), dirigé par M. Artin, A. Grothendieck, J.-L. Verdier, Tome 3, Exp. No. XVII*, pp. 250–461. Lecture Notes in Math., Vol. 305. Springer, Berlin.

[5] La formule de dualité globale. In *Séminaire de Géométrie Algébrique du Bois-Marie 1963/64 (SGA 4), dirigé par M. Artin, A. Grothendieck, J.-L. Verdier, Tome 3, Exp. No. XVIII*, pp. 481–587. Lecture Notes in Math., Vol. 305. Springer, Berlin.

[6] Formes modulaires et représentations ℓ-adiques. In *Séminaire Bourbaki, 1968/69, Exp. No. 355*, pp. 139–172. Lecture Notes in Math., Vol. 179. Springer, Berlin.

[7] Variétés abéliennes ordinaires sur un corps fini. *Invent. Math.*, 8:238–243.

[8] (with D. Mumford). The irreducibility of the space of curves of given genus. *Inst. Hautes Études Sci. Publ. Math.* 36:75–109.

H. Holden, R. Piene (eds.), *The Abel Prize 2013–2017*,
https://doi.org/10.1007/978-3-319-99028-6_3

1970

[9] Les constantes des équations fonctionnelles. In *Séminaire Delange–Pisot–Poitou: 1969/70, Théorie des Nombres, Fasc. 2, Exp. 19 bis*, 13 pp. Secrétariat mathématique, Paris.

[10] Travaux de Griffiths. In *Séminaire Bourbaki, 1969/70, Exp. No. 376*, pp. 213–237. Lecture Notes in Math., Vol. 180. Springer, Berlin.

[11] *Équations différentielles à points singuliers réguliers.* Lecture Notes in Math., Vol. 163. Springer, Berlin.

[12] Une congruence entre coefficients multinomiaux. *Mathematica & Paedagogia,* 46:34–37.

1971

[13] Travaux de Shimura. In *Séminaire Bourbaki, 1970/71, Exp. No. 389*, pp. 123–165. Lecture Notes in Math., Vol. 244. Springer, Berlin.

[14] Variétés unirationnelles non rationnelles (d'après M. Artin et D. Mumford). In *Séminaire Bourbaki, 1971/1972, Exp. No. 402*, pp. 45–57. Lecture Notes in Math., Vol. 317. Springer, Berlin.

[15] Théorie de Hodge. I. In *Actes du Congrès International des Mathématiciens (Nice, 1970), Tome 1*, pp. 425–430. Gauthier-Villars, Paris.

[16] Théorie de Hodge. II. *Inst. Hautes Études Sci. Publ. Math.*, 40:5–57. Also in *Matematika: Periodicheskii sbornik perevodov inostrannykh statei.* Tom 17, vyp. 5. Izdat. "Mir", Moscow, 1973 (in Russian).

[17] Appendix to S. Lang: The group of automorphisms of the modular function field. *Invent. Math.*, 14:253–254.

1972

[18] Résumé des premiers exposés de Grothendieck. In *Groupes de monodromie en géométrie algébrique. I.* Séminaire de Géométrie Algébrique du Bois-Marie 1967–1969 (SGA 7 I), dirigé par A. Grothendieck, pp. 1–24. Lecture Notes in Math., Vol. 288. Springer, Berlin.

[19] (with N. Katz, eds.) *Groupes de monodromie en géométrie algébrique. II.* Séminaire de Géométrie Algébrique du Bois-Marie 1967–1969 (SGA 7 II). Lecture Notes in Math., Vol. 340. Springer, Berlin.

[20] La conjecture de Weil pour les surfaces K3. *Invent. Math.*, 15:206–226.

[21] Les intersections complètes de niveau de Hodge un. *Invent. Math.*, 15:237–250.

[22] Les immeubles des groupes de tresses généralisés. *Invent. Math.*, 17:273–302.

1973

[23] Formes modulaires et représentations de GL(2). In *Modular functions of one variable, II (Proc. Internat. Summer School, Univ. Antwerp, Antwerp, 1972)*, pp. 55–105. Lecture Notes in Math., Vol. 349. Springer, Berlin. Correction in *Modular functions of one variable, IV (Proc. Internat. Summer School, Univ.*

Antwerp, Antwerp, 1972), p. 148. Lecture Notes in Math., Vol. 476, Springer, Berlin, 1975.

[24] (with M. Rapoport). Les schémas de modules de courbes elliptiques. In *Modular functions of one variable, II (Proc. Internat. Summer School, Univ. Antwerp, Antwerp, 1972)*, pp. 143–316. Lecture Notes in Math., Vol. 349. Springer, Berlin. Correction in *Modular functions of one variable, IV (Proc. Internat. Summer School, Univ. Antwerp, Antwerp, 1972)*, p. 149. Lecture Notes in Math., Vol. 476, Springer, Berlin, 1975.

[25] (with W. Kuyk, eds.) *Modular functions of one variable, II (Proc. Internat. Summer School, Univ. Antwerp, Antwerp, 1972)*, pp. 501–597. Lecture Notes in Math., Vol. 349. Springer, Berlin.

[26] Les constantes des équations fonctionnelles des fonctions L. In *Modular functions of one variable, II (Proc. Internat. Summer School, Univ. Antwerp, Antwerp, 1972)*, pp. 501–597. Lecture Notes in Math., Vol. 349. Springer, Berlin. Correction in *Modular functions of one variable, IV (Proc. Internat. Summer School, Univ. Antwerp, Antwerp, 1972)*, p. 149. Lecture Notes in Math., Vol. 476, Springer, Berlin, 1975.

1974

[27] La conjecture de Weil. I. *Inst. Hautes Études Sci. Publ. Math.* 43:273–307. Also in *Uspehi. Mat. Nauk*, 30(5 (185)):159–196, 1975 (in Russian).

[28] (with J.-P. Serre). Formes modulaires de poids 1. *Ann. Sci. École Norm. Sup. (4)*, 7:507–530.

[29] Théorie de Hodge. III. *Inst. Hautes Études Sci. Publ. Math.*, 44:5–77.

[30] Lettre à J.-P. Serre. Appendix to S. Bloch, H. Esnault: Gauss–Manin determinants for rank 1 irregular connections on curves. *Math. Ann.*, 321(1):15—87.

1975

[31] (with P. Griffiths, J. Morgan, and D. Sullivan). Real homotopy theory of Kähler manifolds. *Invent. Math.*, 29(3):245–274. Also in *Uspehi Mat. Nauk*, 32(3 (195)):119–152, 1977 (in Russian).

[32] Courbes elliptiques: formulaire d'après J. Tate. In *Modular functions of one variable, IV (Proc. Internat. Summer School, Univ. Antwerp, Antwerp, 1972)*, pp. 53–73. Lecture Notes in Math., Vol. 476. Springer, Berlin.

[33] (with D. Sullivan). Fibrés vectoriels complexes à groupe structural discret. *C. R. Acad. Sci. Paris Sér. A-B*, 281(24):1081–1083.

[34] Poids dans la cohomologie des variétés algébriques. In *Proceedings of the International Congress of Mathematicians (Vancouver, B.C., 1974)*, Vol. 1, pp. 79–85. Canad. Math. Congress, Montreal, Que.

1976

[35] (with G. Lusztig). Representations of reductive groups over finite fields. *Ann. of Math. (2)*, 103(1):103–161.

[36] Le support du caractère d'une représentation supercuspidale. *C. R. Acad. Sci. Paris Sér. A-B*, 283(4):155–157.

[37] Les constantes locales de l'équation fonctionnelle de la fonction *L* d'Artin d'une représentation orthogonale. *Invent. Math.*, 35:299–316.

1977

[38] Les difféomorphismes du cercle (d'après M.R. Herman). In *Séminaire Bourbaki, 1975/76, Exp. No. 477*, pp. 99–121. Lecture Notes in Math., Vol. 567. Springer, Berlin.

[39] *Cohomologie étale*. Lecture Notes in Mathematics, Vol. 569. Springer, Berlin. Séminaire de Géométrie Algébrique du Bois-Marie SGA 4$^{1}/_{2}$, Avec la collaboration de J.F. Boutot, A. Grothendieck, L. Illusie et J.L. Verdier.

1978

[40] Extensions centrales non résiduellement finies de groupes arithmétiques. *C. R. Acad. Sci. Paris Sér. A-B*, 287(4):203–208.

1979

[41] Sommes de Gauss cubiques et revêtements de SL(2) (d'après S.J. Patterson). In *Séminaire Bourbaki, 1978/1979, Exp. No. 539*, pp. 244–277. Lecture Notes in Math., Vol. 770. Springer, Berlin.

[42] Variétés de Shimura: interprétation modulaire, et techniques de construction de modèles canoniques. In *Automorphic forms, representations and L-functions (Proc. Sympos. Pure Math., Oregon State Univ., Corvallis, Ore., 1977), Part 2*, Proc. Sympos. Pure Math., XXXIII, pp. 247–289. Amer. Math. Soc., Providence, RI.

[43] Valeurs de fonctions *L* et périodes d'intégrales. In *Automorphic forms, representations and L-functions (Proc. Sympos. Pure Math., Oregon State Univ., Corvallis, Ore., 1977), Part 2*, Proc. Sympos. Pure Math., XXXIII, pp. 313–346. Amer. Math. Soc., Providence, RI. With an appendix by N. Koblitz and A. Ogus.

[44] Le groupe fondamental du complément d'une courbe plane n'ayant que des points doubles ordinaires est abélien (d'après W. Fulton). In *Séminaire Bourbaki, 1979/80*, pp. 1–10. Lecture Notes in Math., Vol. 842. Springer, Berlin.

1980

[45] (with K.A. Ribet). Values of abelian *L*-functions at negative integers over totally real fields. *Invent. Math.*, 59(3):227–286.

[46] La conjecture de Weil. II. *Inst. Hautes Études Sci. Publ. Math.*, 52:137–252.

1981

[47] Cycles de Hodge absolus et périodes des intégrales des variétés abéliennes. *Mém. Soc. Math. France (N.S.)*, (2):23–33. Abelian functions and transcendental numbers (Colloq., École Polytech., Palaiseau, 1979).

[48] (with G. Henniart). Sur la variation, par torsion, des constantes locales d'équations fonctionnelles de fonctions *L*. *Invent. Math.*, 64(1):89–118.

[49] Relèvement des surfaces $K3$ en caractéristique nulle. In J. Giraud, L. Illusie, M. Raynaud (eds.), *Surfaces algébriques (Séminaire de Géométrie Algébrique d'Orsay, 1976–78)*, pp. 58–79. Lecture Notes in Math., Vol. 868. Springer, Berlin. Rédigé par L. Illusie.

[50] Cristaux ordinaires et coordonnées canoniques. In J. Giraud, L. Illusie, M. Raynaud (eds.), *Surfaces algébriques (Séminaire de Géométrie Algébrique d'Orsay, 1976–78)*, pp. 80–137. Lecture Notes in Math., Vol. 868. Springer, Berlin. With the collaboration of L. Illusie, and with an appendix by N. Katz.

1982

[51] (with G. Lusztig). Duality for representations of a reductive group over a finite field. *J. Algebra*, 74(1):284–291.

[52] (with J.S. Milne, A. Ogus, and K. Shih). *Hodge cycles, motives, and Shimura varieties*, Lecture Notes in Math., Vol. 900. Springer, Berlin.

[53] (with A.A. Beilinson and J. Bernstein). Faisceaux pervers. In *Analysis and topology on singular spaces, I (Luminy, 1981)*, *Astérisque*, 100:5–171.

1983

[54] (with G. Lusztig). Duality for representations of a reductive group over a finite field. II. *J. Algebra*, 81(2):540–545.

[55] (with D. Sullivan). Division algebras and the Hausdorff–Banach–Tarski paradox. *Enseign. Math. (2)*, 29(1–2):145–150.

[56] Preuve des conjectures de Tate et de Shafarevitch (d'après G. Faltings). In *Séminaire Bourbaki, 1983/84, Exp. No. 216. Astérisque* 121–122:25–41 (1985).

1984

[57] Intégration sur un cycle évanescent. *Invent. Math.*, 76(1):129–143.

[58] (ed.) *Representations of reductive groups over a local field*, Travaux en Cours. Hermann, Paris.

[59] (with J.-N. Bernstein, D. Kazhdan, and M.-F. Vignéras). Représentations des algèbres centrales simples p-adiques. In *Representations of reductive groups over a local field*, Travaux en Cours, pp. 33–117. Hermann, Paris.

[60] Les corps locaux de caractéristique p, limites de corps locaux de caractéristique 0. In *Representations of reductive groups over a local field*, Travaux en Cours, pp. 119–157. Hermann, Paris.

1985

[61] Le lemme de Gabber. Séminaire sur les pinceaux arithmétiques: La conjecture de Mordell (Paris, 1983/84), dirigé par L. Szpiro. *Astérisque*, 127:131–150.

[62] Représentations ℓ-adiques. Séminaire sur les pinceaux arithmétiques: La conjecture de Mordell (Paris, 1983/84), dirigé par L. Szpiro. *Astérisque*, 127:249–255.

1986

[63] (with G.D. Mostow). Monodromy of hypergeometric functions and nonlattice integral monodromy. *Inst. Hautes Études Sci. Publ. Math.*, 63:5–89.

[64] (with J. Bernstein and D. Kazhdan). Trace Paley–Wiener theorem for reductive p-adic groups. *J. Analyse Math.*, 47:180–192.

1987

[65] (with L. Illusie). Relèvements modulo p^2 et décomposition du complexe de de Rham. *Invent. Math.*, 89(2):247–270.

[66] Un théorème de finitude pour la monodromie. In *Discrete groups in geometry and analysis (New Haven, Conn., 1984)*, Progr. Math. 67, pp. 1–19. Birkhäuser Boston, Boston, MA.

[67] (with D. Husemoller). Survey of Drinfel'd modules. In *Current trends in arithmetical algebraic geometry (Arcata, Calif., 1985), Contemp. Math.*, 67:25–91. Amer. Math. Soc., Providence, RI.

[68] Le déterminant de la cohomologie. In *Current trends in arithmetical algebraic geometry (Arcata, Calif., 1985), Contemp. Math.*, 67:93–177. Amer. Math. Soc., Providence, RI.

1989

[69] Le groupe fondamental de la droite projective moins trois points. In *Galois groups over $\mathbb{Q}$ (Berkeley, CA, 1987)*, Math. Sci. Res. Inst. Publ., 16, pp. 79–297. Springer, New York.

1990

[70] (with A. Dimca). Filtrations de Hodge et par l'ordre du pôle pour les hypersurfaces singulières. *Ann. Sci. École Norm. Sup. (4)*, 23(4):645–656.

[71] Catégories tannakiennes. In *The Grothendieck Festschrift, Vol. II*, Progr. Math. 87, pp. 111–195. Birkhäuser Boston, Boston, MA.

1991

[72] Groupe de Heisenberg et réalité. *J. Amer. Math. Soc.*, 4(1):197–206.

[73] Le symbole modéré. *Inst. Hautes Études Sci. Publ. Math.*, 73:147–181.

1993

[74] (with G.D. Mostow). *Commensurabilities among lattices in* $\mathrm{PU}(1, n)$, *Annals of Mathematics Studies* 132. Princeton University Press, Princeton, NJ.

1994

[75] Décompositions dans la catégorie dérivée. In *Motives (Seattle, WA, 1991), Proc. Sympos. Pure Math.* 55, pp. 115–128. Amer. Math. Soc., Providence, RI.

[76] À quoi servent les motifs? In *Motives (Seattle, WA, 1991), Proc. Sympos. Pure Math.* 55, pp. 143–161. Amer. Math. Soc., Providence, RI.

[77] Structures de Hodge mixtes réelles. In *Motives (Seattle, WA, 1991), Proc. Sympos. Pure Math.* 55, pp. 509–514. Amer. Math. Soc., Providence, RI.

[78] (with A. Beilinson). Interprétation motivique de la conjecture de Zagier reliant polylogarithmes et régulateurs. In *Motives (Seattle, WA, 1991), Proc. Sympos. Pure Math.* 55, pp. 97–121. Amer. Math. Soc., Providence, RI.

[79] (with G. Pappas). Singularités des espaces de modules de Hilbert, en les caractéristiques divisant le discriminant. *Compositio Math.*, 90(1):59–79.

1995

[80] (with E. Cattani and A. Kaplan). On the locus of Hodge classes. *J. Amer. Math. Soc.*, 8(2):483–506.

[81] Déformations de l'algèbre des fonctions d'une variété symplectique: comparaison entre Fedosov et De Wilde, Lecomte. *Selecta Math. (N.S.)*, 1(4):667–697.

1996

[82] La série exceptionnelle de groupes de Lie. *C. R. Acad. Sci. Paris Sér. I Math.*, 322(4):321–326.

[83] (with R. de Man). La série exceptionnelle de groupes de Lie. II. *C. R. Acad. Sci. Paris Sér. I Math.*, 323(6):577–582.

[84] Extensions centrales de groupes algébriques simplement connexes et cohomologie galoisienne. *Inst. Hautes Études Sci. Publ. Math.*, 84:35–89 (1997).

1997

[85] Local behavior of Hodge structures at infinity. In *Mirror symmetry, II, AMS/IP Stud. Adv. Math.* 1, pp. 683–699. Amer. Math. Soc., Providence, RI.

[86] Action du groupe des tresses sur une catégorie. *Invent. Math.*, 128(1):159–175.

1998

[87] Quelques idées maîtresses de l'œuvre de A. Grothendieck. In *Matériaux pour l'histoire des mathématiques au XXe siècle (Nice, 1996), Sémin. Congr.* 3, pp. 11–19. Soc. Math. France, Paris.

1999

[88] (with P. Etingof, D. S. Freed, L. C. Jeffrey, D. Kazhdan, J. W. Morgan, D. R. Morrison, E. Witten, eds.) *Quantum fields and strings: a course for mathematicians, Vol. 1, 2 (Princeton, NJ, 1996/1997)*. Amer. Math. Soc., Providence, RI.

[89] (with J.W. Morgan). Notes on supersymmetry (following Joseph Bernstein). In *Quantum fields and strings: a course for mathematicians, Vol. 1, 2 (Princeton, NJ, 1996/1997)*, pp. 41–97. Amer. Math. Soc., Providence, RI.

[90] Notes on spinors. In *Quantum fields and strings: a course for mathematicians, Vol. 1, 2 (Princeton, NJ, 1996/1997)*, pp. 99–135. Amer. Math. Soc., Providence, RI.

[91] (with D.S. Freed). Classical field theory. In *Quantum fields and strings: a course for mathematicians, Vol. 1, 2 (Princeton, NJ, 1996/1997)*, pp. 137–225. Amer. Math. Soc., Providence, RI, 1999.

[92] (with D.S. Freed). Supersolutions. In *Quantum fields and strings: a course for mathematicians, Vol. 1, 2 (Princeton, NJ, 1996/1997)*, pp. 227–355. Amer. Math. Soc., Providence, RI.

[93] (with D.S. Freed). Sign manifesto. In *Quantum fields and strings: a course for mathematicians, Vol. 1, 2 (Princeton, NJ, 1996/1997)*, pp. 357–363. Amer. Math. Soc., Providence, RI.

[94] Note on quantization. In *Quantum fields and strings: a course for mathematicians, Vol. 1, 2 (Princeton, NJ, 1996/1997)*, pp. 367–375. Amer. Math. Soc., Providence, RI.

[95] Super space descriptions of super gravity. In *Quantum fields and strings: a course for mathematicians, Vol. 1, 2 (Princeton, NJ, 1996/1997)*, pp. 1013–1016. Amer. Math. Soc., Providence, RI.

2000

[96] (with M. Goresky and R. MacPherson). L'algèbre de cohomologie du complément, dans un espace affine, d'une famille finie de sous-espaces affines. *Michigan Math. J.*, 48:121–136.

[97] (with J. Cogdell, S. Gindikin, P. Sarnak, S. Gelbart, R. Howe, S. Rallis, eds.) *Selected works of Ilya Piatetski-Shapiro*. American Mathematical Society, Providence, RI, 2000. xxvi+821 pp.

2001

[98] (with J.-L. Brylinski). Central extensions of reductive groups by $\mathbf{K}_2$. *Publ. Math. Inst. Hautes Études Sci.*, 94:5–85.

2002

[99] (with B.H. Gross). On the exceptional series, and its descendants. *C. R. Math. Acad. Sci. Paris*, 335(11):877–881.

[100] Catégories tensorielles. *Mosc. Math. J.*, 2(2):227–248.

[101] Appendix to: A. Neeman. A counterexample to a 1961 "theorem" in homological algebra. *Invent. Math.*, 148(2):397–420.

2005

[102] (with A.B. Goncharov). Groupes fondamentaux motiviques de Tate mixte. *Ann. Sci. École Norm. Sup. (4)*, 38(1):1–56.

2006

[103] (with H. Esnault). Appendix to: H. Esnault: Deligne's integrality theorem in unequal characteristic and rational points over finite fields. *Ann. of Math.*, 164(2):726–730.

[104] The Hodge conjecture. In *The Millennium Prize Problems*, pp. 45–53. Clay Math. Inst., Cambridge, MA.

[105] On ζ-functions of several variables. In *Globus. General mathematics seminar.* No. 3 (Russian), pp. 142–151. Mosk. Tsentr Nepreryv. Mat. Obraz., Moscow.

2007

[106] La catégorie des représentations du groupe symétrique S_t, lorsque t n'est pas un entier naturel. In *Algebraic groups and homogeneous spaces*, Tata Inst. Fund. Res. Stud. Math., pp. 209–273. Tata Inst. Fund. Res., Mumbai.

[107] (with B. Malgrange and J.-P. Ramis). *Singularités irrégulières*. Documents Mathématiques, 5. Société Mathématique de France, Paris.

2009

[108] Extended Euler congruence. *Funct. Anal. Other Math.*, 2(2–4):249–250.

[109] (rédigé par L. Clozel). Equivalence numérique, équivalence cohomologique, et théorie de Lefschetz des variétés abéliennes sur les corps finis. *Pure Appl. Math. Q.*, 5(1):495–506.

[110] Voevodsky's lectures on motivic cohomology 2000/2001. In *Algebraic topology, Abel Symp.* 4, pages 355–409. Springer, Berlin.

2010

[111] Le groupe fondamental unipotent motivique de $\mathbf{G}_m - \mu_N$, pour $N = 2, 3, 4, 6$ ou 8. *Publ. Math. Inst. Hautes Études Sci.*, 112:101–141.

[112] Le théorème de plongement de Nagata. *Kyoto J. Math.*, 50(4):661–670.

2012

[113] Multizêtas, d'après Francis Brown, viii. In *Séminaire Bourbaki, Exp. No. 1048*, Vol. 2011/2012. Exposés 1043–1058. *Astérisque*, 352:161–185.

[114] Finitude de l'extension de $\mathbb{Q}$ engendrée par des traces de Frobenius, en caractéristique finie. *Mosc. Math. J.*, 12(3):497–514, 668.

2013

[115] (with Y.Z. Flicker). Counting local systems with principal unipotent local monodromy. *Ann. of Math. (2)*, 178(3):921–982.

2014

[116] Semi-simplicité de produits tensoriels en caractéristique p. *Invent. Math.*, 197(3):587–611.

2015

[117] Comptage de faisceaux l-adiques. *Astérisque*, 369:285–312.

2017

[118] (with V. Balji and A. J. Parameswaran). On complete reducibility in characteristic p. With an appendix by Shiwei Yun. *Épijournal Geom. Algébrique*, 1, Art. 3, 27 pp.

[119] (with W. Yun Chen). Arithmetic monodromy actions on pro-metabelian fundamental groups of once-punctured elliptic curves. *arXiv:1710.05532.*

2018

[120] (with G. I. Lehrer and R. B. Zhang). The first fundamental theorem of invariant theory for the orthosymplectic super group. *Adv. Math.*, 327:4–24.

[121] Extension des scalaires par le morphisme de Frobenius, pour les groupes réductifs. *C. R. Math. Acad. Sci. Paris*, 356:717–719.

Curriculum Vitae for Pierre R. Deligne

Born:	October 3, 1944 at Etterbeek, Belgium
Degrees/education:	Licence en mathématiques, Université Libre de Bruxelles, 1966
	Doctorat en mathématiques, Université Libre de Bruxelles, 1968
	Doctorat d'État ès Sciences Mathématiques, Université de Paris-Sud, 1972
Positions:	Aspirant, Fonds National de la Recherche Scientifique, Bruxelles, 1967–1968
	Guest, Institut des Hautes Études Scientifiques (IHÉS), 1967–1968
	Visiting member, IHÉS, 1968–1970
	Permanent member, IHÉS, 1970–1984
	Professor, Institute for Advanced Study, Princeton, 1984–, emeritus since 2008
Visiting positions:	Member, Institute for Advanced Study, Princeton, 1972–73, 1977
	Visitor, Institute for Advanced Study, Princeton, 1981

H. Holden, R. Piene (eds.), *The Abel Prize 2013–2017*,
https://doi.org/10.1007/978-3-319-99028-6_4

Honors/Memberships:

Académie des Sciences, membre associé étranger, 1978
American Academy of Arts and Sciences, Foreign honorary member, 1978
Académie Royale de Belgique, membre associé, 1994
Moscow Mathematical Society, honorary member, 1995
London Mathematical Society, honorary member, 2003
Accademia Nazionale dei Lincei, foreign member, 2003
Élevé au rang de vicomte par le roi Albert II, 2006
National Academy of Sciences, foreign associate, 2007
Royal Swedish Academy of Sciences, foreign member, 2009
American Philosophical Society, 2009
Norwegian Academy of Science and Letters, 2013
Russian Academy of Sciences, foreign member, 2016

Awards and prizes:

Francois Deruyts Prize, 1974
Henri Poincaré Medal, 1974
Dr. De Leeuw-Damry-Bourlart Prize, 1975
Fields Medal, 1978
Crafoord Prize, 1988
Balzan Prize in Mathematics, 2004
Wolf Prize, 2008
Abel Prize, 2013

Honorary degrees:

Vrije Universiteit Brussel, 1989
École Normale Supérieure, 1995
Université Libre de Bruxelles, 2010

Part II
2014 Yakov G. Sinai

"for his fundamental contributions to dynamical systems, ergodic theory, and mathematical physics"

ABEL
PRISEN

Citation

The Norwegian Academy of Science and Letters has decided to award the Abel Prize for 2014 to **Yakov G. Sinai**, Princeton University and Landau Institute for Theoretical Physics, The Russian Academy of Sciences

> for his fundamental contributions to dynamical systems, ergodic theory, and mathematical physics

Ever since the time of Newton, differential equations have been used by mathematicians, scientists and engineers to explain natural phenomena and to predict how they evolve. Many equations incorporate stochastic terms to model unknown, seemingly random, factors acting upon that evolution. The range of modern applications of deterministic and stochastic evolution equations encompasses such diverse issues as planetary motion, ocean currents, physiological cycles, population dynamics, and electrical networks, to name just a few. Some of these phenomena can be foreseen with great accuracy, while others seem to evolve in a chaotic, unpredictable way. Now it has become clear that order and chaos are intimately connected: we may find chaotic behavior in deterministic systems, and conversely, the statistical analysis of chaotic systems may lead to definite predictions.

Yakov Sinai made fundamental contributions in this broad domain, discovering surprising connections between order and chaos and developing the use of probability and measure theory in the study of dynamical systems. His achievements include seminal works in ergodic theory, which studies the tendency of a system to explore all of its available states according to certain time statistics; and statistical mechanics, which explores the behavior of systems composed of a very large number of particles, such as molecules in a gas.

Sinai's first remarkable contribution, inspired by Kolmogorov, was to develop an invariant of dynamical systems. This invariant has become known as the Kolmogorov–Sinai entropy, and it has become a central notion for studying the complexity of a system through a measure-theoretical description of its trajectories. It has led to very important advances in the classification of dynamical systems.

Sinai has been at the forefront of ergodic theory. He proved the first ergodicity theorems for scattering billiards in the style of Boltzmann, work he continued with Bunimovich and Chernov. He constructed Markov partitions for systems defined by iterations of Anosov diffeomorphisms, which led to a series of outstanding works showing the power of symbolic dynamics to describe various classes of mixing systems.

With Ruelle and Bowen, Sinai discovered the notion of SRB measures: a rather general and distinguished invariant measure for dissipative systems with chaotic behavior. This versatile notion has been very useful in the qualitative study of some archetypal dynamical systems as well as in the attempts to tackle real-life complex chaotic behavior such as turbulence.

Sinai's other pioneering works in mathematical physics include: random walks in a random environment (Sinai's walks), phase transitions (Pirogov–Sinai theory), one-dimensional turbulence (the statistical shock structure of the stochastic Burgers equation, by E–Khanin–Mazel–Sinai), the renormalization group theory (Bleher–Sinai), and the spectrum of discrete Schrödinger operators.

Sinai has trained and influenced a generation of leading specialists in his research fields. Much of his research has become a standard toolbox for mathematical physicists. His works had and continue to have a broad and profound impact on mathematics and physics, as well as on the ever-fruitful interaction of these two fields.

Autobiography

Ya. G. Sinai

I was born on the 21st of September, 1935 in Moscow to a family of scientists. My mother, Nadezka Kagan, was a virologist. She worked on vaccines against encephalitis and died in November of 1938 after she became infected by the vaccine on which she was working. My father was a professor of microbiology in one of Moscow's medical institutions. He participated in World War II working as an epidimeologist from 1941 through 1945. He married again in 1940, and my stepmother, E.N. Levkovich, was also a famous virologist. I lived under her warm care for many years. She sometimes took me to her laboratory, which was staffed completely by women. Therefore, for many years I believed that biology was purely a women's field.

I lived in a big family, headed by my grandfather, V.F. Kagan. He was a mathematician working on the foundations of geometry. He also did a great deal of work for the popularization of Lobachevsky geometry and the Lobachevsky proof of the fifth Euclid postulate.

At the beginning of the Twentieth Century, our family lived in Odessa, a major city in Ukraine. My grandfather worked in a college there, where he gave the first lecture course on Einstein's special relativity theory. The course was very popular among students, some of whom later became leading physicists in the Soviet Union.

V.F. Kagan was also seriously involved in mathematics and physics education. He served as the chief editor of the journal, "Mathesis", which was oriented to younger

Electronic Supplementary Material The online version of this chapter (https://doi.org/10.1007/978-3-319-99028-6_5) contains supplementary material, which is available to authorized users.

Y. G. Sinai (✉)
Department of Mathematics, Princeton University, Princeton, NJ, USA
e-mail: sinai@math.princeton.edu

H. Holden, R. Piene (eds.), *The Abel Prize 2013–2017*,
https://doi.org/10.1007/978-3-319-99028-6_5

students. He also conducted research during that time. (I recently saw a reference to one of his papers from 1916, which was quoted in a current research paper.)

I have one brother, G.I. Barenblatt. He is a well-known expert in fluid dynamics and the theory of fractures and the author of several monographs on scaling methods in fluid dynamics. His first thesis was on turbulence and was written under the supervision of A.N. Kolmogorov.

In the beginning of the 1920s, our family moved to Moscow. My grandfather had become a professor in the Mathematics Department of Moscow State University and also served as the Chair of Differential Geometry. Once, when I was about 15 years old, my grandfather decided to teach me mathematics and gave me a lecture on quaternions. He then asked me to write a composition on them. I assume my results were unsuccessful because his attempts were never repeated.

During my school year I participated in several Oympiads, always without success. (This might be useful for high school students who sometimes exaggerate the role of Olympiads.)

Due to my grandfather's strong support, I entered Moscow State University in 1952. (The gold medal I received after my graduation from high school apparently was not enough!) The first lecture course that I attended made a strong impression on me. It was on classical mechanics and was taught by N.G. Chetaev, a famous expert in this field. My first junior thesis was written under his supervision. Another popular professor there was E.B. Dynkin, who organized a working seminar for first year students. It was attended by many people, including I. Girsanov and L. Seregin, who both later became famous probabilists.

E.B. Dynkin gave me the first serious problem, which I worked on for the next couple of years. Once it was solved, it became my first publication.

In 1957, A.N. Kolmogorov announced that he was giving a lecture course on dynamical systems. In the beginning, he explained von Neumanns theory of systems with pure point spectrum using a purely probabilistic approach. I later found a similar approach in a book written for engineers by Fortet and Blanc-Lapierre. The whole theory looked extremely beautiful. People believed at the time that the theory of dynamical systems with continuous spectrum would be some extension of von Neumann's theory of systems with pure point spectrum. However, Kolmogorov surprised us one day by showing the definition of the entropy of dynamical systems. Using modern language, one can say that he proposed the definition of entropy of Bernoulli shifts and proved that Bernoulli shifts with different value of entropy were metrically non-isomorphic. This was a great breakthrough. In a text that Kolmogorov later submitted for publication, though, he introduced a new class of dynamical systems which he called quasi-regular and provided a definition of entropy for this class. (I shall not discuss his motivations.) Shortly after that, V.A. Rokhlin proposed an example that showed that the entropy proposed by Kolmogorov was not a metric invariant. Now it is easy to construct similar examples. At that time, I was working on the definition of entropy which could be applied to arbitrary dynamical systems. The text by B.M. Gurevich in this book explains many details. Various publications of entropy can be found in the papers by Bunimovich, Szasa, Simanyi, and Pesin, which are also in this book.

I married my wife, Elena Vul, in 1956. She was also a mathematics student in the same year with me at Moscow State University. Her father, B.M. Vul, was a remarkable physicist who worked on semiconductors. We spoke many times of various scientific problems.

Ya. G. Sinai with his friends in Tel-Aviv. (Photo: private)

Over the years I have been very fortunate to have scientific contact with outstanding mathematicians in Moscow like I.M. Gelfand, B.A. Rokhlin, V.I. Arnold, S.P. Novikov, R.L. Dobrushin, R.A. Minols, V.E. Zakharov, F.A. Berezin, D.V. Anosov, and others.

In 1973 I received an invitation to become a member of the Landau Institute of Theoretical Physics. The Director of the Institute was I.M. Khalatnikov, a former student of L.D. Landau. He attracted many physicists, mathematicians, and mathematical physicists. The general atmosphere in the Institute was very friendly. Every paper that was done by people in the Institute would be discussed in the form of a Colloquium talk so that it could be understood by others.

At the current time, the IPPI (or the Institute of Information Transmission) plays a big role in mathematical physics in Moscow. Its Director, A.P. Kuleshov, supports the research of many mathematicians and physicists. It is quite common for Russian mathematicians who are working in the West to give talks at the Institute during their visits to Moscow. My seminar on problems of dynamical systems and statistical physics continues to meet there during my stays there.

Over the past 50 years, I have had many students and I am proud of all of them. They have played an important role in my life.

Prof. Sinai together with his students during the celebration of his 80th birthday party, 2015. (Photo: private)

Today I am a professor in the Princeton University Mathematics Department. I have many colleagues with whom I share warm scientific and personal contacts there as well as at the Institute for Advanced Study.

Prof. Sinai in his office in Fine Hall, 2015. (Photo: private)

I would like to thank L. Bunimovich, B. Gurevich, K. Khanin, D. Li, Ya. Pesin, N. Simanyi, and D. Szasz for their excellent texts that were prepared specifically for this edition.

Prof. Sinai near his house in Princeton, 2005. (Photo: private)

Haakon, The Crown Prince of Norway giving the Abel Prize to Prof. Sinai, 2014. (Scanpix)

Sinai's Dynamical System Perspective on Mathematical Fluid Dynamics

Carlo Boldrighini and Dong Li

Abstract We review some of the most remarkable results obtained by Ya.G. Sinai and collaborators on the difficult problems arising in the theory of the Navier–Stokes equations and related models. The survey is not exhaustive, and it omits important results, such as those related to "Burgers turbulence". Our main focus in on acquainting the reader with the application of the powerful methods of dynamical systems and statistical mechanics to this field, which is the main original feature of Sinai's contribution.

1 Introduction

One of the fundamental unsolved problems in mathematical fluid dynamics is whether smooth solutions to the three-dimensional incompressible Navier–Stokes System (NSS) can develop singularities in finite time. Sinai has a remarkable intuition that the formation of finite time singularities is possible for the 3D Navier–Stokes system: NSS without external forcing can be regarded a reasonable approximation to the dynamics of a dry air in a big desert, and in deserts such phenomena as tornados are possible due to purely kinematic mechanisms. Mathematically speaking, the most notable difficulties of NSS are its non-locality and super-criticality. The system is nonlocal due to the incompressibility constraint and supercritical with respect to the basic energy conservation law. Super-criticality can also be derived through a scaling analysis on the life-span of solutions.

C. Boldrighini (✉)
Istituto Nazionale di Alta Matematica (INdAM), GNFM, Unità locale Università Roma Tre, Rome, Italy
e-mail: boldrigh@mat.uniroma1.it

D. Li
Department of Mathematics, Hong Kong University of Science and Technology, Kowloon, Hong Kong
e-mail: madli@ust.hk

H. Holden, R. Piene (eds.), *The Abel Prize 2013–2017*,
https://doi.org/10.1007/978-3-319-99028-6_6

Over the years, Sinai and his collaborators have developed several original and powerful methods to tackle many difficult wellposedness and regularity questions in hydrodynamics. Unlike the usual practitioners of PDEs, his approach to these problems is highly original, and his incredible technical power and remarkable insight from dynamical systems has led to substantial progress on the understanding of NSS at fine scales, which is the key to the global regularity conjecture.

The list of results surveyed below is certainly not exhaustive and only represents a small fraction of his many important works. For example, we do not discuss Dinaburg–Sinai's Fourier space model of the NSS and Euler systems (see [15, 16] and see also Friedlander–Pavlovic [22] for further developments), and we do not include a detailed survey on Sinai's ground-breaking work on Burgers turbulence, stochastic hydrodynamics and further developments. Nevertheless, we hope that what we report reflects his unique dynamical system perspective on mathematical fluid dynamics. The topics selected here include: a geometric trapping method for wellposedness and regularity of solutions to NSS [35], power series and diagrams [36–38], complex solutions and renormalization group for the three-dimensional NSS [32], bifurcation of solutions for two-dimensional NSS [33, 34] and stochastic dynamics of two-dimensional NSS [18].

2 A Geometric Trapping Method for NSS

Consider the d-dimensional incompressible Navier–Stokes system on the periodic torus $\mathbb{T}^d = \mathbb{R}^d/\mathbb{Z}^d$,

$$\begin{cases} \partial_t u + (u \cdot \nabla)\, u = -\nabla p + \nu \Delta u, & (t,x) \in (0,\infty) \times \mathbb{T}^d, \\ \nabla \cdot u = 0, \\ u|_{t=0} = u_0. \end{cases} \tag{1}$$

Here $u = u(t,x) = (u_1(t,x), \ldots, u_d(t,x))$ represents the velocity of the fluid and $p = p(t,x)$ denotes the pressure. When $\nu = 0$ the system (1) becomes the incompressible Euler equation. The first equation in (1) is just the usual Newton's law: the left-hand side describes the acceleration of the fluid in Eulerian frame, whereas the right-hand side represents the force. The second equation in (1) is the usual incompressibility (divergence-free) condition. It can also be regarded a constraint through which the pressure gradient term emerges as a Lagrange multiplier. To reduce the complexity of the system one can use the vorticity formulation. In two dimensions, define $w = \nabla^{\perp} \cdot u = -\partial_{x_2} u_1 + \partial_{x_1} u_2$. Then the equation governing w takes the form

$$\partial_t w + (u \cdot \nabla)\, w = \nu \Delta w, \tag{2}$$

where, under suitable regularity assumptions, u is connected to w by the Biot–Savart law:

$$u = \Delta^{-1}\nabla^{\perp} w = \left(-\Delta^{-1}\partial_{x_2} w, \Delta^{-1}\partial_{x_1} w\right).$$

It is evident from the vorticity form that for smooth solutions the L^p-norm $\|w\|_p$ is preserved in time for all $1 \le p \le \infty$ in 2D. On the other hand, in three dimensions, one can introduce the vorticity vector $w = \nabla \times u$ for which the vorticity equation takes the form:

$$\partial_t w + (u \cdot \nabla)\, w = (w \cdot \nabla)\, u + \nu \Delta w, \tag{3}$$

with

$$u = -\Delta^{-1}\nabla \times w.$$

Compared with two dimensions, the vorticity stretching term $(w \cdot \nabla u)$ is the main obstruction to global wellposedness in three dimensions. In the whole plane $\mathbb{R}^2$ case, the first existence and uniqueness results for weak solutions of (1) were obtained in Leray's thesis in 1933. For the three-dimensional whole space case Leray [30] proved the existence of weak solutions. Hopf in [23] then obtained the existence of weak solutions in arbitrary open subsets Ω of $\mathbb{R}^n$, $n \ge 2$. Ladyzenskaya [26] in 1962 proved existence and uniqueness of solutions for two-dimensional domains. Since then many other strong methods were developed in [10, 39, 40, 42], providing deep insights into the fine behavior of solutions to (1).

In [35], Mattingly and Sinai developed a novel geometric trapping method for proving existence, uniqueness and regularity of solutions to the Navier–Stokes system. To describe this method, consider the two-dimensional vorticity equation (2). Expand the vorticity w in Fourier series:

$$w(x,t) = \sum_{k \in \mathbb{Z}^2} w_k(t) e^{2\pi i k \cdot x}, \quad x = (x_1, x_2)$$

where w_k denote the Fourier coefficients. Since w is real-valued, we have $w_{-k} = \overline{w_k}$. One can then write a coupled ODE-system for the modes $w_k(t)$ as

$$\frac{d}{dt} w_k + 2\pi i \sum_{l_1 + l_2 = k} w_{l_1} w_{l_2} \frac{k \cdot l_2^{\perp}}{|l_2|^2} = -4\pi^2 \nu |k|^2 w_k, \tag{4}$$

where $|k| = \sqrt{k_1^2 + k_2^2}$, $l^{\perp} = (l^{(1)}, l^{(2)})^{\perp} = (-l^{(2)}, l^{(1)})$.

A more general version of (2) is the case where the Laplacian is replaced by the fractional Laplacian $|\nabla|^{\alpha}$ with $\alpha > 0$. Correspondingly, (4) can be generalized as:

$$\frac{d}{dt} w_k + 2\pi i \sum_{l_1 + l_2 = k} w_{l_1} w_{l_2} \frac{k \cdot l_2^{\perp}}{|l_2|^2} = -4\pi^2 \nu |k|^{\alpha} w_k. \tag{5}$$

Without loss of generality one can assume $w_0 = 0$ since the mean value of w is preserved by the dynamics.

The results obtained in [35] can be formulated as follows.

Theorem 1 ([35]) *Let* $\alpha > 1$ *in* (5). *Suppose for some constant* $0 < D_1 < \infty$, $1 < r < \infty$,

$$|w_k(0)| \le \frac{D_1}{|k|^r}, \quad \forall k \in \mathbb{Z}^2 \setminus \{0\}.$$

Then one can find a finite constant $D_1' > 0$, *depending only on* (D_1, ν), *such that any solution to* (5) *with these initial conditions satisfies*

$$|w_k(t)| \le \frac{D_1'}{|k|^r}, \quad \forall k \in \mathbb{Z}^2 \setminus \{0\}$$

for all $t > 0$.

A few remarks are now in order. First, the main theorems stated in [35] are more general and include the case with external forcing under suitable decay assumptions on the Fourier modes which are uniform in time. By using some refined estimates, Mattingly and Sinai also proved that the solutions become real analytic for $t > 0$ (i.e., $|w_k(t)| \le \text{const} \cdot e^{-\text{const}\cdot|k|}$, for $t > t_0 > 0$). Statements close to these were also proved in [17, 21, 24], but the methods are quite different and more function analytic in nature.

In the three-dimensional setting, one can introduce

$$u(x,t) = \sum_{k \in \mathbb{Z}^3} u_k(t) e^{2\pi i k\cdot x},$$
$$w(x,t) = \sum_{k \in \mathbb{Z}^3} w_k(t) e^{2\pi i k\cdot x}.$$

By using (3), we obtain

$$\begin{aligned} \frac{d}{dt} w_k(t) &= -2\pi i \sum_{l_1+l_2=k} \left[(u_{l_1} \cdot l_2) w_{l_2} - (w_{l_1} \cdot l_2) u_{l_2} \right] - 4\pi^2 \nu |k|^2 w_k \\ &= -2\pi i \sum_{l_1+l_2=k} \left[(u_{l_1} \cdot k) w_{l_2} - (w_{l_1} \cdot k) u_{l_2} \right] - 4\pi^2 \nu |k|^2 w_k, \end{aligned}$$

where the second equality follows from the incompressibility condition. Similar to the two-dimensional case, one can replace the Laplacian with the fractional Laplacian $|\nabla|^\alpha$, and obtain

$$\frac{d}{dt} w_k(t) = -2\pi i \sum_{l_1+l_2=k} \left[(u_{l_1} \cdot k) w_{l_2} - (w_{l_1} \cdot k) u_{l_2} \right] - 4\pi^2 \nu |k|^\alpha w_k. \tag{6}$$

For this nonlocal system, the following theorem was proved in [35].

Theorem 2 ([35]) *Consider* (6) *with* $\alpha > \frac{5}{2}$. *If the initial data* $\{w_k(0)\}$ *are such that for some* $0 < D < \infty$, $r > \frac{3}{2}$,

$$|w_k(0)| \leq \frac{D}{|k|^r}, \qquad \forall k \in \mathbb{Z}^3 \setminus \{0\},$$

then there exists a constant D' *depending only on* (D, r, α), *such that for any* $t \geq 0$,

$$|w_k(t)| \leq \frac{D'}{|k|^r}, \qquad \forall k \in \mathbb{Z}^3 \setminus \{0\}.$$

Remark One should note that $\alpha = 2$ corresponds to the usual Navier–Stokes case. Analogous statements can also be proved for that situation, provided the constant D is sufficiently small, which will become a typical small data global wellposedness result for 3D NSS. For large data global wellposedness, one can lower the constant $\alpha > 2.5$ to $\alpha = 2.5$ or even with some logarithmic damping of the symbol. All of these difficulties are ultimately connected with the lack of globally coercive quantities stronger than energy.

We now focus on the two-dimensional case and describe in more detail the geometric trapping method of Mattingly and Sinai. Roughly speaking, the idea is to consider a finite Galerkin system of coupled ODEs for the Fourier coefficients. One can write a finite approximation of (5) abstractly as

$$\frac{d}{dt} w_k(t) = B_k(w, w) - 4\pi^2 \nu |k|^\alpha w_k.$$

By using the basic enstrophy inequality

$$\sum_k |w_k(t)|^2 \leq \mathscr{E}_0, \qquad \forall t > 0,$$

one can trap the low modes, i.e., for any $K_0 > 0$, there exists $D_1(K_0)$, such that

$$|w_k(t)| \leq \frac{D_1}{|k|^r}, \qquad \forall |k| \leq K_0.$$

One then defines a trapping region for all modes as

$$\Omega = \left\{ (w_k) : |w_k| \leq \frac{D_1}{|k|^r}, \forall 0 \neq k \in \mathbb{Z}^2 \right\}$$

It is evident that the low modes $\{|k| \leq K_0\}$ are already in the trapping region, and the boundary of the trapping region is given by

$$\partial\Omega = \left\{ (w_k) \,:\, |w_k| \leq \frac{D_1}{|k|^r},\ \forall\, 0 \neq k \in \mathbb{Z}^2,\ \text{and equality holds for some } k = k^* \right\}.$$

By choosing D_1 large, Ω contains the initial data in its interior. Then one endeavors to show that the dynamics will always trap the sequence of Fourier modes inside $\overline{\Omega}$. Geometrically speaking, it amounts to showing that the vector field on the boundary $\partial\Omega$ always points into the interior of Ω. More precisely one checks that for K_0 sufficiently large, if there are $|k^*| > K_0$, with $w_{k^*} = \frac{D_1}{|k^*|^r}$ (the case $w_{k^*} = -\frac{D_1}{|k^*|^r}$ is similar), then

$$\left.\frac{d}{dt} w_k(t)\right|_{k=k^*} < 0.$$

By using the enstrophy estimate together with the trapping estimate, one can estimate the nonlinear term as

$$|B_k(w, w)(t)| \leq \text{const} \cdot \sqrt{\mathcal{E}_0} \cdot \frac{D_1}{|k^*|^{r-1}} \cdot \log |k^*|.$$

Thus

$$\left.\frac{d}{dt} w_k(t)\right|_{k=k^*} \leq \text{const} \cdot \sqrt{\mathcal{E}_0} \cdot \frac{D_1}{|k^*|^{r-1}} \cdot \log |k^*| - 4\pi^2 \nu \frac{1}{|k^*|^{r-\alpha}} < 0, \tag{7}$$

if K_0 is chosen sufficiently large.

This concludes the trapping argument. One should note from (7) that the restriction $\alpha > 1$ is purely technical, and due to the fact that only enstrophy conservation and L_t^∞-type breakthrough scenario enter the argument. By using more time integrability, one can obtain analyticity also for $\alpha = 1$ (for global wellposedness we do not need any constraint on α since 2D Euler is globally wellposed by using $\|w\|_{L_x^\infty}$).

One can also rephrase in typical PDE language the trapping argument of Mattingly and Sinai, as a sort of maximum principle in Fourier space. It is a beautiful geometric dynamical system proof, which has since been generalized and developed to many other situations (cf. [2, 4, 11–14] and the references therein).

3 Power Series and Diagrams

In the seminal works [36–38], Sinai developed a power series and diagram representation for the Navier–Stokes system. These works can be viewed as a precursor to the renormalization group approach developed later. Consider the three-dimensional Navier–Stokes system (1), with viscosity $\nu = 1$ and on the whole space $\mathbb{R}^3$. After the Fourier transform

$$v(k,t) = \int_{\mathbb{R}^3} u(x,t)e^{-ik\cdot x}\,dx,$$

it becomes a nonlinear non-local equation:

$$v(k,t) = e^{-|k|^2 t}v(k,0) + i\int_0^t e^{-(t-s)|k|^2}\int_{\mathbb{R}^3}\langle k, v(k-k',s)\rangle\, P_k v(k',s)dk'ds. \tag{8}$$

The incompressibility condition enforces $v(k,t) \perp k$ for any $k \neq 0$. The operator P_k is the orthogonal projection to the subspace orthogonal to k. In this way the pressure does not appear and we consider the space of functions $\{v(k)\colon v(k) \perp k\}$ as the main phase space of the dynamical system defined by (1).

Classical (strong) solutions to (8) on the time interval $[0, t_0]$ are functions $v(k,t)$, $0 \le t \le t_0$, such that the integrals

$$\int_0^t e^{-(t-s)|k|^2}\int_{\mathbb{R}^3}|v(k-k',s)|\cdot|v(k',s)|\,dk'ds,$$

are bounded for any $0 \le t \le t_0$ and the left-hand side is equal to the right-hand side. A more convenient (easily checkable), but stronger condition, is to require the integrals

$$\int_{\mathbb{R}^3}|v(k-k',s)|\cdot|v(k',s)|\,dk'$$

to be uniformly bounded in s. The latter definition was adopted in [38].

Sinai considered (8) in the space of functions which can have singularities near $k = 0$ or $k = \infty$. The following space $\Phi(\alpha, w)$ was introduced in [38].

Definition 4 $\{v(k),\ k \in \mathbb{R}^3\} \in \Phi(\alpha, w)$ if for some constants $0 < C, D < \infty$,

$$|v(k)| \le \begin{cases} \frac{C}{|k|^\alpha}, & \text{if } |k| \le 1,\\ \frac{D}{|k|^w}, & \text{if } |k| > 1. \end{cases}$$

The cut-off "1" for $|k|$ can be replaced by any positive number. The parameters α and w satisfy the inequalities $\alpha \ge 2$, $w < 3$. One can endow the space $\Phi(\alpha, w)$ with a norm by taking the infimum of all possible $C + D$.

In [38], Sinai proved a short-time local existence theorem in the space $\Phi(\alpha, w)$, $\alpha > 2$, $w < 3$. Namely, for any initial data (in the Fourier space) $v(k,0) \in \Phi(\alpha, w)$, there exists $T_0 > 0$ sufficiently small, such that (8) admits a unique solution on $[0, T_0]$ in the space $\Phi(\alpha, w)$. One should note that in this theorem, $v(k,0)$ is allowed to be an arbitrary complex ($\mathbb{C}^3$-valued) vector function. When $v(k,0) = \overline{v(-k,0)}$ for any $k \in \mathbb{Z}^3$, the corresponding velocity $u(x,0)$ is a $\mathbb{R}^3$-valued vector function.

In the space $\Phi(2,2)$ one can prove a small data global wellposedness result. Namely, let $v(k,0) = \frac{C(k,0)}{|k|^2}$, with $\sup_k |C(k,0)| \le C_0$ and C_0 is sufficiently small. Then there exists a unique solution $v(k,t)$ of (8) defined for all $t > 0$.

One can see the references [8, 28, 38] for short proofs of this theorem. Recently, Lei and Lin [29] discovered a remarkable fact, that for Eq. (1) with $\nu > 0$ and on $\mathbb{R}^3$ one can have global wellposedness as long as $\sup_k |C(k,0)| \le C\nu$, where C is an absolute constant.

In [36], Sinai considered the space $\Phi(\alpha,\alpha)$ with $\alpha = 2+\epsilon$ and $\epsilon > 0$ sufficiently small. Denote $v(k,0) = \frac{C(k,0)}{|k|^\alpha}$ where $C(k,0)$ is continuous everywhere outside $k=0$, and $\|C(k,0)\|_{L_k^\infty} = \sup_{k\neq 0} |C(k,0)| = 1$. Introduce a one-parameter family of initial conditions $v_A(k,0) = \frac{AC(k,0)}{|k|^\alpha}$, where A is a complex-valued parameter. For given A, the time of existence for the local solution will depend on A. More precisely, the following theorem was proven in [36].

Theorem 5 ([36]) *There exists a constant $\lambda_0 = \lambda_0(\alpha) > 0$ depending only on α such that if $|\lambda| = |AT^{\frac{\epsilon}{2}}| \le \lambda_0$, then there exists a unique local solution in the space $\Phi(\alpha,\alpha)$ on the time interval* $[0,T]$.

To prove this theorem Sinai used the method of iterations. In terms of the unknown $C_A(k,t) = |k|^\alpha v_A(k,t)$, one can define the iterations $C_A^{(n)}(k,t)$ via the formula

$$\begin{aligned} &C_A^{(n)}(k,t) \\ &\quad = Ae^{-|k|^2 t} C(k,0) \\ &\qquad + i|k|^\alpha \int_0^t e^{-|k|^2(t-s)} \int_{\mathbb{R}^3} \frac{\langle k, C_A^{(n-1)}(k-k',s)\rangle P_k C_A^{(n-1)}(k',s)}{|k-k'|^\alpha |k'|^\alpha} dk' ds, \qquad n \ge 1, \end{aligned}$$

with

$$C_A^{(0)}(k,t) = Ae^{-|k|^2 t} C(k,0).$$

By splitting into low and high frequencies, Sinai showed that if $|\lambda| \le \lambda_0(\alpha) \ll 1$, then $\|C^{(n)}\|_\infty \le 2A$ for all $n \ge 1$, and the sequence of iterations $(C^{(n)})$ is a contraction. From the point of view of dynamical systems, the scalar λ is a ruling parameter in the current situation. In the same paper, Sinai then went on to construct a power series for the solution $C_A(k,t)$, namely:

$$C_A(k,t) = AC(k,0)e^{-t|k|^2} + \sum_{p\ge 1} A^p \int_0^t e^{-(t-s)|k|^2} s^{\frac{p\epsilon}{2}} h_p(k,s)\, ds, \tag{9}$$

where

$$s^{\frac{\epsilon}{2}}h_1(k,s) = i|k|^\alpha \int_{\mathbb{R}^3} \frac{\langle k, C(k-k',0)\rangle P_k C(k',0) e^{-s|k-k'|^2 - s|k'|^2}}{|k-k'|^\alpha \cdot |k'|^\alpha} dk',$$

$$s^\epsilon h_2(k,s)$$
$$= i|k|^\alpha \cdot \Big[\int_0^s s_1^{\frac{\epsilon}{2}} ds_1 \int_{\mathbb{R}^3} \frac{\langle k, h_1(k-k',s_1)\rangle P_k C(k',0) \cdot e^{-(s-s_1)|k-k'|^2 - s|k'|^2} dk'}{|k-k'|^\alpha |k'|^\alpha}$$
$$+ \int_0^s s_2^{\frac{\epsilon}{2}} ds_2 \int_{\mathbb{R}^3} \frac{\langle k, C(k-k',0)\rangle P_k h_1(k',s_2) e^{-s|k-k'|^2 - (s-s_2)|k'|^2} dk'}{|k-k'|^\alpha \cdot |k'|^\alpha}\Big],$$

and

$$s^{\frac{p\epsilon}{2}} h_p(k,s)$$
$$= i|k|^\alpha \cdot \Big[\int_0^s s_1^{\frac{p-1}{2}\epsilon} ds_1 \cdot \int_{\mathbb{R}^3} \frac{\langle k, h_{p-1}(k-k',s_1)\rangle P_k C(k',0) e^{-(s-s_1)|k-k'|^2 - s|k'|^2} dk'}{|k-k'|^\alpha \cdot |k'|^\alpha}$$
$$+ \int_0^s s_2^{\frac{p-1}{2}\epsilon} ds_2 \cdot \int_{\mathbb{R}^3} \frac{\langle k, C(k-k',0)\rangle P_k h_{p-1}(k',s_1) e^{-s|k-k'|^2 - (s-s_2)|k'|^2} dk'}{|k-k'|^\alpha \cdot |k'|^\alpha}$$
$$+ \sum_{\substack{p_1,p_2\ge 1\\ p_1+p_2=p-1}} \int_0^s s_1^{\frac{p_1\epsilon}{2}} ds_1 \int_0^s s_2^{\frac{p_2\epsilon}{2}} ds_2$$
$$\times \int_{\mathbb{R}^3} \frac{\langle k, h_{p_1}(k-k',s_1)\rangle P_k h_{p_2}(k',s_2) e^{-(s-s_1)|k-k'|^2 - (s-s_2)|k'|^2} dk'}{|k-k'|^\alpha \cdot |k'|^\alpha}\Big].$$

Now use the ansatz $h_p(k,s) = s^{\frac{\epsilon}{2}}|k|^\alpha g_p(k\sqrt{s}, s)$ and make the change of variables: $s_1 = s\tilde{s}_1$, $s_2 = s\tilde{s}_2$, $k\sqrt{s} = \tilde{k}$, $k'\sqrt{s} = \tilde{k}'$. Then $h_p(k,s) = s^{\frac{\epsilon}{2}}|k|^\alpha g_p(\tilde{k}, s)$. The system of recurrent relations governing the functions $g_p(\tilde{k}, s)$ then takes the form:

$$g_1(\tilde{k},s) = i\int_{\mathbb{R}^3} \frac{\langle \tilde{k}, C(\frac{\tilde{k}-\tilde{k}'}{\sqrt{s}},0)\rangle P_{\tilde{k}} C(\frac{\tilde{k}'}{\sqrt{s}},0) e^{-|\tilde{k}-\tilde{k}'|^2 - |\tilde{k}'|^2} d\tilde{k}'}{|\tilde{k}-\tilde{k}'|^\alpha \cdot |\tilde{k}'|^\alpha},$$

$$g_2(\tilde{k},s)$$
$$= \int_0^1 \tilde{s}_1^\epsilon d\tilde{s}_1 \int_{\mathbb{R}^3} \frac{\langle \tilde{k}, g_1((\tilde{k}-\tilde{k}')\sqrt{\tilde{s}_1}, s\cdot\tilde{s}_1)\rangle \cdot P_{\tilde{k}} C(\frac{\tilde{k}'}{\sqrt{s}},0) e^{-(1-\tilde{s}_1)|\tilde{k}-\tilde{k}'|^2 - |\tilde{k}'|^2} d\tilde{k}'}{|\tilde{k}'|^\alpha}$$
$$+ \int_0^1 \tilde{s}_2^\epsilon d\tilde{s}_2 \int_{\mathbb{R}^3} \frac{\langle \tilde{k}, C(\frac{\tilde{k}-\tilde{k}'}{\sqrt{s}},0)\rangle P_{\tilde{k}} g_1(\tilde{k}'\sqrt{\tilde{s}_2}, s\tilde{s}_2) e^{-|\tilde{k}-\tilde{k}'|^2 - (1-\tilde{s}_2)|\tilde{k}'|^2} d\tilde{k}'}{|\tilde{k}-\tilde{k}'|^\alpha},$$

and for $p \geq 3$

$$
\begin{aligned}
&g_p(\tilde{k}, s)\\
&= i\Big[\int_0^1 \tilde{s}_1^{\frac{p\epsilon}{2}} d\tilde{s}_1 \int_{\mathbb{R}^3} \frac{\langle \tilde{k}, g_{p-1}((\tilde{k}-\tilde{k}')\sqrt{\tilde{s}_1}, s\tilde{s}_1)\rangle P_{\tilde{k}} C(\frac{\tilde{k}'}{\sqrt{s}}, 0) e^{-(1-\tilde{s}_1)|\tilde{k}-\tilde{k}'|^2 - |\tilde{k}'|^2} d\tilde{k}'}{|\tilde{k}'|^\alpha}\\
&+ \int_0^1 \tilde{s}_2^{\frac{p\epsilon}{2}} d\tilde{s}_2 \int_{\mathbb{R}^3} \frac{\langle \tilde{k}, C(\frac{\tilde{k}-\tilde{k}'}{\sqrt{s}}, 0)\rangle P_{\tilde{k}} g_{p-1}(\tilde{k}'\sqrt{\tilde{s}_2}, s\tilde{s}_2) e^{-|\tilde{k}-\tilde{k}'|^2 - (1-\tilde{s}_2)|\tilde{k}'|^2} d\tilde{k}'}{|\tilde{k}-\tilde{k}'|^\alpha}\\
&+ \sum_{\substack{p_1, p_2 \geq 1\\ p_1+p_2=p-1}} \int_0^1 \tilde{s}_1^{\frac{p_1\epsilon}{2}} d\tilde{s}_1 \int_0^1 \tilde{s}_2^{\frac{p_2\epsilon}{2}} d\tilde{s}_2 \int_{\mathbb{R}^3} \langle \tilde{k}, g_{p_1}((\tilde{k}-\tilde{k}')\sqrt{\tilde{s}_1}, s\cdot\tilde{s}_1)\rangle \cdot\\
&\qquad \cdot P_{\tilde{k}} g_{p_2}(\tilde{k}'\sqrt{\tilde{s}_2}, s\tilde{s}_2) \cdot e^{-(1-\tilde{s}_1)|\tilde{k}-\tilde{k}'|^2 - (1-\tilde{s}_2)|\tilde{k}'|^2} d\tilde{k}'\Big].
\end{aligned} \tag{10}
$$

It follows from these recurrent relations that each $g_p(\tilde{k}, s)$ depends on the initial conditions $C(k, 0)$ via the sum of not more than b^p $4p$-dimensional integrals where b is some constant. The main assumption is that $C(k, 0)$ is compactly supported in $\{|k| \leq R_0\}$, where R_0 is a positive constant.

By using a sophisticated inductive analysis together with some combinatorics, Sinai proved the following theorem.

Theorem 6 ([36]) *The functions $g_p(\tilde{k}, s)$ satisfy the inequality:*

$$|g_p(\tilde{k}, s)| \leq C_p f(|\tilde{k}|) e^{-\frac{|\tilde{k}|^2}{p+1}},$$

where $f(x) = \min\{x, \frac{1}{x}\}$ for $x > 0$, and $C_p \leq b_1 b_2^p$ for some constants $0 < b_1, b_2 < \infty$ depending only on α.

It follows that if $At^{\frac{\epsilon}{2}} < b_2^{-1}$, then the series (9) converges for every $0 \neq k \in \mathbb{R}^3$.

In [37], Sinai analyzed in more detail the recurrent system (10) and introduced diagrams, corresponding to each multi-dimensional integral in the series. Each diagram is determined by a scheme, and any scheme is a sequence of partitions of the set starting from $[1, 2, \ldots, p+1] = \Delta^{(0)}$. By using a deep analogy with statistical mechanics, Sinai then estimated several classes of diagrams and showed that the partition functions of short diagrams decay exponentially. In [37], one can find a systematic approach to study and estimate short diagrams for large p. This approach has a striking resemblance of the renormalization group method in statistical mechanics.

4 Complex Valued Solutions and Renormalization Group

Consider the Navier–Stokes system (1) on $\mathbb{R}^3$ with viscosity $\nu = 1$. By using the Fourier transform

$$\widetilde{v}(k,t) = \int_{\mathbb{R}^3} u(x,t)e^{-ik\cdot x}\,dx,$$

one obtains an equivalent non-local nonlinear system

$$\widetilde{v}(k,t) = e^{-|k|^2 t}\widetilde{v}(k,0) + i\int_0^t e^{-(t-s)|k|^2}\int_{\mathbb{R}^3}\left\langle \widetilde{v}(k-k',s),k\right\rangle P_k\widetilde{v}(k',s)dk'ds, \tag{11}$$

where P_k is the solenoidal projection operator

$$P_k\widetilde{v} = \widetilde{v} - \frac{\langle \widetilde{v},k\rangle}{|k|^2}k,$$

and $\langle\cdot,\cdot\rangle$ denotes the scalar product

$$\langle a,b\rangle = a\cdot b, \quad \text{if } a,\, b\in\mathbb{C}^3.$$

Introduce the change of variable

$$\widetilde{v}(k,t) = -i\,v(k,t).$$

Then in terms of $v(k,t)$, the integral equation (11) now takes the form

$$v(k,t) = e^{-|k|^2 t}v(k,0) + \int_0^t e^{-(t-s)|k|^2}\int_{\mathbb{R}^3}\left\langle v(k-k',s),k\right\rangle P_k v(k',s)dk'ds. \tag{12}$$

This non-local integral equation is the main object of study. In general, $\mathbb{R}^3$-valued solutions to (12) will correspond to complex solutions $u(x,t)$ in (1). If one restricts to the class of $v(k,0)$ such that $v(k,0) = -v(k,0)$ for all $k\in\mathbb{Z}^3$, then $v(k,t)$ will also be odd in k and such solutions correspond to $\mathbb{R}^3$-valued real (and physical) fluid flows.

In [32], a Renormalization Group type method was developed to show that there exists a class of $\mathbb{R}^3$-valued initial data $v(k,0)$ which are compactly supported such that the corresponding solution to (12) blows up in finite time. The velocity field $u(x,0)$ corresponding to $v(k,0)$ is, however, $\mathbb{C}^3$-valued. As such, these solutions do not obey energy conservation and correspond to non-physical flows. Nevertheless the behavior of these solutions in some sense resemble the forward cascade of

Fourier modes and they are a show-case of some important fine structures of the Navier–Stokes system.

We now review in more detail the results of [32].

Consider a one-parameter family of initial data in the form $v_A(k,0) = Av_0(k)$, where $v_0(k)$ will be a fixed profile and A is a positive parameter. The corresponding solution to (12) can then be represented as a power series

$$v_A(k,t) = Ae^{-t|k|^2}v_0(k) + \int_0^t e^{-|k|^2(t-s)}\left[\sum_{p=2}^{\infty} A^p g^{(p)}(k,s)\right] ds. \tag{13}$$

Set $g^{(1)}(k,s) = e^{-s|k|^2}v_0(k)$. Substituting (13) into (12), we then obtain

$$g^{(2)}(k,s) = \int_{\mathbb{R}^3} \langle v_0(k-k'), k\rangle P_k v_0(k')e^{-s|k-k'|^2 - s|k'|^2}\, dk',$$

and for $p > 2$

$$\begin{aligned}
g^{(p)}(k,s) = &\int_0^s ds_2 \int_{\mathbb{R}^3} \langle v_0(k-k', k\rangle P_k g^{(p-1)}(k', s_2)e^{-s|k-k'|^2-(s-s_2)|k'|^2}dk' \\
&+ \int_0^s ds_1 \int_{\mathbb{R}^3} \langle g^{(p-1)}(k-k', s_1), k\rangle P_k v_0(k')e^{-(s-s_1)|k-k'|^2 - s|k'|^2}dk' \\
&+ \sum_{\substack{p_1+p_2=p \\ p_1,p_2>1}} \int_0^s ds_1 \int_0^s ds_2 \langle g^{(p_1)}(k-k', s_1), k\rangle \\
&\qquad \times P_k g^{(p_2)}(k', s_2)e^{-(s-s_1)|k-k'|^2-(s-s_2)|k'|^2}dk'.
\end{aligned} \tag{14}$$

The initial data v_0 will be assumed to have support localized in a sphere around some $K^{(0)} = (0,0,k_0)$, $k_0 \gg 1$. The radius of the sphere is much smaller than k_0. By a deep analogy with probability theory, the support of the functions $g^{(p)}$ is then expected to be localized about the point $pK^{(0)} = (0,0,pk_0)$ with a fattened size $\sqrt{p}$ for large p. From these considerations, one can then introduce the change of variable and ansatz:

$$k = pK^{(0)} + \sqrt{p}Y, \quad h^{(p)}(Y,s) = g^{(p)}(pK^{(0)} + \sqrt{p}Y, s),$$

where the new variable Y typically takes values $O(1)$. In all integrals over s_1, s_2 in (14), make another change of variables $s_j = s(1 - \frac{\theta_j}{p_j^2})$, $j = 1, 2$. Instead of the integration over k', we introduce Y' such that $k' = p_2 k_0 + \sqrt{pk_0} Y'$. Denote $\gamma = \frac{p_1}{p}$. Then we obtain from (14) the recurrent relation

$$h^{(p)}(Y,s) = p^{5/2} \sum_{\substack{p_1+p_2=p \\ p_1,p_2>\sqrt{p}}} \frac{1}{p_1^2 p_2^2} \int_{\mathbb{R}^3} P_{e_3+\frac{Y}{\sqrt{p}}} h^{(p_2)}(\frac{Y'}{\sqrt{1-\gamma}}, s) \cdot \\ \times \langle h^{(p_1)}(\frac{Y-Y'}{\sqrt{\gamma}}, s), e_3 + \frac{Y}{\sqrt{p}} \rangle dY' \cdot (1 + o(1)),$$

where $e_3 = (0, 0, 1)$. In coordinates one can write

$$h^{(p)}(Y,s) = \left(h_1^{(p)}(Y,s), h_2^{(p)}(Y,s), \frac{F^{(p)}(Y,s)}{\sqrt{p}} \right). \tag{15}$$

For large p the incompressibility condition $\langle h^{(p)}(Y,s), k \rangle = 0$ enforces

$$Y_1 h_1^{(p)}(Y,s) + Y_2 h_2^{(p)}(Y,s) + F^{(p)}(Y,s) = O(p^{-1/2}).$$

It follows that $F^{(p)} = O(1)$ and the vector $h^{(p)}(Y,s)$ is almost orthogonal to the k_3-axis for large p.

Make the ansatz

$$h^{(p)}(Y,s) = pA(s)^p \prod_{j=1}^{3} g^{(3)}(Y) \left(H(Y) + \delta^{(p)}(Y,s) \right), \tag{16}$$

where $\Lambda(s)$ is a positive function, $g^{(3)}(Y) = (2\pi)^{-3/2} e^{-|Y|^2/2}$ is the standard Gaussian density, and the remainder term $\delta^{(p)}$ tends to zero as $p \to \infty$. The vector function

$$H(Y) = (H_1(Y_1, Y_2), H_2(Y_1, Y_2), 0)$$

will correspond to the fixed point of the renormalization group. The fact that it is two-dimensional and depends only on (Y_1, Y_2), can be traced back to (15), which is a consequence of the divergence-free condition.

As we take the limit $p \to \infty$, the discrete sum over p_1 in the recurrent relation becomes an integral over $\gamma = \frac{p_1}{p}$. The fixed point equation for the renormalization group then takes the form

$$g_1^{(2)}(Y) H(Y) = \int_0^1 d\gamma \int_{\mathbb{R}^2} g_\gamma^{(2)}(Y - Y') g_{1-\gamma}^{(2)}(Y') \mathscr{L}(H; \gamma, Y, Y') \\ \times H\left(\frac{Y'}{\sqrt{1-\gamma}} \right) dY', \tag{17}$$

where, by abuse of notation, $H(Y) = (H_1(Y_1, Y_2), H_2(Y_1, Y_2))$, $g_0^{(2)}(Y) = \frac{1}{2\pi\sigma} e^{-\frac{Y_1^2+Y_2^2}{2\sigma}}$, and

$$\mathscr{L}(H; \gamma, Y, Y') = -(1-\gamma)^{3/2} \left\langle \frac{Y-Y'}{\sqrt{\gamma}}, H\left(\frac{Y-Y'}{\sqrt{\gamma}}\right) \right\rangle + \gamma^{1/2}(1-\gamma) \left\langle \frac{Y'}{\sqrt{1-\gamma}}, H\left(\frac{Y'}{\sqrt{1-\gamma}}\right) \right\rangle.$$

In Eq. (17), the Y_3-variable was integrated out since it is just the usual convolution. By using the theory of Hermite polynomials, one can classify the solutions to the functional equation (17). Amongst all such solutions, a particular simple one is

$$H^{(0)}(Y_1, Y_2) = C(Y_1, Y_2),$$

where the pre-factor $C > 0$ can be determined from the equation. One can then linearize around this fixed point and study the spectrum of the linearized operator. As it turns out, there are 6 unstable directions and 4 neutral directions. The following theorem was proven in [32].

Theorem 7 ([32]) *For $K^{(0)} = (0, 0, k_0)$ and k_0 large enough, there exists a 10-parameter family of initial data and a time interval $[s_-, s_+]$ such that the ansatz (16) holds for $H = H^{(0)}$ and $s \in [s_-, s_+]$.*

As observed in [5, 6], the recurrent relations and the fixed point equation remain unchanged if $h^{(p)}$ is replaced by $(-1)^p h^{(p)}$. This consideration then leads to two types of solutions, with type I corresponding to the solution described before and type II corresponding to $(-1)^p h^{(p)}$. Note that if the initial data v_0 leads to a type I solution with the fixed point $H^{(0)}$, then $-v_0$ leads to a type II solution with the same fixed point.

In [5], it was shown that the solutions corresponding to type I and type II will have energy and enstrophy diverging as

$$E(t) = \frac{1}{2} \int_{\mathbb{R}^3} |u(x,t)|^2 \, dx = \frac{(2\pi)^3}{2} \int_{\mathbb{R}^3} |v(k,t)|^2 \, dk \sim \frac{C_E^{(\alpha)}}{(\tau - t)^{\beta_\alpha}},$$

$$S(t) = \int_{\mathbb{R}^3} |\nabla u(x,t)|^2 \, dx = (2\pi)^3 \int_{\mathbb{R}^3} |k|^2 |v(k,t)|^2 \, dk \sim \frac{C_S^{(\alpha)}}{(\tau - t)^{\beta_\alpha + 2}},$$

where τ is the blowup time, $\alpha = \text{I, II}$ denotes the type of function, $\beta_{\text{I}} = 1$, $\beta_{\text{II}} = \frac{1}{2}$ and $C_E^{(\alpha)}$, $C_S^{(\alpha)}$ are constants depending on the initial data.

Numerical simulations of the complex-valued singular solutions reveal very interesting features [5, 6] some of which are similar to those of related real-valued energy-preserving solutions.

5 Bifurcations of Solutions to Two-Dimensional Navier–Stokes Systems

The usual bifurcation theory in dynamical systems deals with one-parameter families of smooth maps or vector fields. In that situation fixed points or periodic orbits become functions of this parameter. Bifurcations appear when their linearized spectrum changes its structure. The classical approach is to use versal deformations, i.e., special families such that arbitrary families can be represented as some projections of versal deformations [3]. In such kind of approach the positions of the bifurcating orbits and their dependence on the parameter are known. In [33, 34] a new approach is developed to study deformations produced by solutions of a PDE system and construct bifurcations using properties of the dynamical flow. The construction is nonlinear and does not rely on any knowledge of special fixed points. As a model case, one can study the bifurcation of critical points for a stream function driven by a two-dimensional incompressible viscous flow. Unlike the usual scenario the profile of the function can display quite disparate patterns at different time intervals due to the nonlocal nature of the dynamics.

Consider the Cauchy problem for the two-dimensional Navier–Stokes System written for the stream function $\psi = \psi(t, x, y)$:

$$\begin{cases} \frac{\partial \psi}{\partial t} + \Delta^{-1}\Big(\frac{\partial \psi}{\partial x} \cdot \frac{\partial \Delta\psi}{\partial y} - \frac{\partial \psi}{\partial y} \cdot \frac{\partial \Delta\psi}{\partial x}\Big) = \Delta\psi, \\ \psi(t, x + 2\pi, y) = \psi(t, x, y + 2\pi) = \psi(t, x, y), \qquad \forall\,(x, y) \in \mathbb{T}^2, \end{cases} \tag{18}$$

where $\mathbb{T}^2$ is the two-dimensional periodic torus with period 1 in each directions. The velocity u of the fluid is given by $u = \nabla^{\perp}\psi = (-\partial_y\psi, \partial_x\psi)$. For general initial data the global wellposedness and regularity of solutions to (1) is well-known by using Mattingly–Sinai's geometric trapping method or energy type estimates. The main problem is to study the dynamics of critical points of the stream function ψ. In [33] it was proposed that if the critical points of the stream function (i.e., stagnation points of the velocity field) are points of maxima or minima, then these points are called viscous vortices because near these points the velocity u is tangent to the level sets of ψ which is a closed curve. The nonlocal operator Δ^{-1} in front of the nonlinear term in (18) is of prime importance (i.e., used in an essential way) in the construction of the bifurcation. On the other hand, such construction does not seem to carry over directly to the vorticity formulation. This is deeply connected with the fact that vorticity only obeys a transport equation and during such processes the local maxima or minima of the vorticity function are simply transported.

The following theorem establishes in some sense the splitting (bifurcation) of vortices. It was first proved in [33] under a symmetry assumption and then in [34] for the general case.

Theorem 8 ([33, 34] Existence of bifurcations) *There exists an open set $\mathscr{A}$ in the space of stream functions such that the following holds true: For each stream* *function* $\psi_0 \in \mathscr{A}$, *there is an open neighborhood* U *of the origin, two moments of time* $0 < t_1 < t_2$ *such that the corresponding stream function* $\psi = \psi(t, x, y)$ *solves* (18) *with initial data* ψ_0 *and has critical points which bifurcate from 1 to 2 on* $[0, t_1]$, *and 2 to 3 on* $(t_1, t_2]$ *in the neighborhood* U.

Although the Navier–Stokes equation is not time-reversible, by using a different construction one can reverse the above scenario and also show the merging of vortices (see [33, 34] for more details). The bifurcation method devised in [33, 34] is quite robust and has been generalized to a number of other situations (cf. [31, 43]). In general the behavior of the critical points is not well studied in multi-dimensional situations. For parabolic equations, one can show that the number of critical points decreases as a function of time (see [1]), and estimate the size of critical points (see [9]).

6 Stochastic Hydrodynamics

Stochastic fluid mechanics is an important tool in the study of real fluid flows, and a huge physical literature is devoted to it. The traditional approach deals with space or time averages of some relevant physical quantities. For a deeper insight one needs information on the typical behavior of the solutions, such as can come from the knowledge of the invariant measures and their space-time properties.

A brilliant contribution of Sinai and collaborators in this sense is given by the paper [18], which deals with the two-dimensional Navier–Stokes equations on the 2D torus $\mathbb{T}^2$ with random forcing on a finite set of modes:

$$\begin{cases} \partial_t u + (u \cdot \nabla) u + \nabla p - \nu \Delta u = \frac{\partial}{\partial t} W(x, t), & (t, x) \in (0, \infty) \times \mathbb{T}^2, \\ \nabla \cdot u = 0. \end{cases} \tag{19}$$

$$W(x, t) = \sum_{0 \neq |k| \leq N} \sigma_k w_k(t, \omega) e_k(x), \quad k \in \mathbb{Z}^2, \quad e_k(x) = i \frac{k^{\perp}}{|k|}.$$

Here the $\{w_k\}$'s are standard i.i.d. complex Wiener processes such that $w_{-k}(t) = \overline{w_k(t)}$ and $\sigma_{-k} = \overline{\sigma_k}$, $|\sigma_k| > 0$. Let $u(x) = \sum_k u_k e_k(x)$ with $u_0 = 0$, be the Fourier expansion, and consider the space $\mathbb{L}^2 = \{\sum_{k \in \mathbb{Z}^2} |u_k|^2 < \infty\}$. Projecting on $\mathbb{L}^2$ we get a system of Ito stochastic equations

$$du(x, t) + \nu \Lambda^2 u(x, t) dt = B(u, u) dt + dW(x, t) \tag{20}$$

where, denoting by P the projection on the subspace of the divergence-free functions, we write $\Lambda^2 u = -P \Delta u$, $B(u, v) = -P(u \cdot \nabla) v$. Equation (20) defines a Markovian stochastic semi-flow $\varphi^{\omega}_{s,t}$, $s < t$, on $\mathbb{L}^2$, for all $\omega \in \Omega$, the canonical space generated by $\{dw_k(t)\}$. A measure μ on $\mathbb{L}^2$ is said to be invariant if for any bounded continuous function F on $\mathbb{L}^2$ and $t > 0$ we have

$$\int_{\mathbb{L}^2} F(u)\mu(du) = \int_{\mathbb{L}^2} \mathbb{E}F(\varphi^{\omega}_{0,t}u)\mu(du) \tag{21}$$

were $\mathbb{E}$ denotes expectation with respect to the measure $\mathbb{P}$ on Ω.

The existence of stationary measures was established by compactness in [19, 41]. Uniqueness was proved, under restrictive assumptions, when all modes are forced, as in the papers by Kuksin and Shirikyan [25] and by Bricmont, Kupiainen and Lefevere [7]. The main result of E, Mattingly and Sinai is the following theorem.

Theorem 9 ([18]) *There is an absolute constant $\mathscr{C}$ such that if $N^2 \geq \mathscr{C}\frac{\mathscr{E}_0}{\nu^3}$, where $\mathscr{E}_0 = \sum_{|k|\leq N} |u_k|^2$ then Eq. (20) has a unique stationary measure on $\mathbb{L}^2$.*

Some comment is here in order. Following the seminal work of Ladyzhenskaya [27] we know that the 2-dimensional Navier–Stokes equations in a bounded domain, with no forcing, or with a bounded finite-dimensional force, has a finite-dimensional attractor, of dimension depending on the Reynolds number [20]. There is a finite number of "determining" modes, and for large times the other modes are determined by the past history of the determining ones. The main theorem of [18] states that uniqueness of the stationary measure holds under the condition that all determining modes are forced, and is a natural extension of the above results.

A main step in the proof is a representation of the high modes as functionals of the time-history of the low modes. Let $\mathbb{L}^2_\ell = \text{span}\{e_k : |k| \leq N\}$, $\mathbb{L}^2_h = \text{span}\{e_k : |k| > N\}$ define the subspaces of low and high modes, and denote by P_ℓ, P_h the corresponding projectors in $\mathbb{L}^2$. Setting $\ell(t) = P_\ell u$, $h(t) = P_h u$, Eq. (20) becomes

$$\begin{aligned} d\ell(t) &= \Big[-\nu\Lambda^2\ell + P_\ell B(\ell,\ell)\Big]dt \\ &\quad + [P_\ell B(\ell,h) + P_\ell B(h,\ell) + P_\ell B(h,h)]\,dt + dW(t), \end{aligned} \tag{22}$$

$$\frac{dh(t)}{dt} = \Big[-\nu\Lambda^2 h + P_h B(h,h)\Big] + P_h B(\ell,h) + P_h B(h,\ell) + P_h B(\ell,\ell). \tag{23}$$

If $\ell(t)$ is assigned, Eq. (23) can be solved for h, and let $\Phi_{s,t}(\ell, h_0)$ be the solution of (23) at time t with initial condition h_0 at time s and fixed ℓ.

By stationarity, one can represent the initial data as coming from a distant past. Let $C((-\infty, 0], \mathbb{L}^2)$ be the path space of the past and $\psi^{\omega}_t u \in C((-\infty, t], \mathbb{L}^2)$ the evolution of $u \in C((-\infty, 0], \mathbb{L}^2)$ induced by the semi-group: $(\psi^{\omega}_t u)(s) = u(s)$ for $s \leq 0$ and $(\psi^{\omega}_t u)(s) = \varphi_{0,s}u(0)$ for $s \in [0, t]$.

There is an obvious measure μ_p on $C((-\infty, 0], \mathbb{L}^2)$, induced by the product measure $\mathbb{P}\times\mu$ on $\Omega\times\mathbb{L}^2$. Defining the shift on the trajectories as $(\theta_t v)(s) = v(s+t)$, the operator $\theta_t\psi^{\omega}_t$ maps $C((-\infty, 0], \mathbb{L}^2)$ into itself. If μ is stationary, then μ_p is also stationary in the sense that for any bounded function $F(u)$ on $C((-\infty, 0], \mathbb{L}^2)$ we have

$$\int_{C((-\infty,0],\mathbb{L}^2)} F(u)d\mu_p(u) = \mathbb{E}\int_{C((-\infty,0],\mathbb{L}^2)} F(\theta_t\psi_t^\omega u)d\mu_p(u).$$

Moreover, it is clear that if μ and ν are two stationary measures for the stochastic flow (20), then $\mu_p = \nu_p$ implies $\mu = \nu$.

The proof further shows that there is a subset $U \subset C((-\infty, 0], \mathbb{L}^2)$ of full measure consisting of functions $v\colon (-\infty, 0] \to \mathbb{H}$ where $\mathbb{H} = \{u \in \mathbb{L}^2\colon \sum_k k^2|u_k|^2 < \infty\}$, and moreover the energy has the correct average in time and the fluctuations are typical.

The reconstruction of the high modes as a function of the past stretching to $-\infty$ is given by the following lemma.

Lemma 10 ([18]) *There is some absolute constant* $\mathscr{C}$ *such that if* $N^2 \geq \mathscr{C}\frac{\mathscr{E}_0}{\nu^3}$ *then the following holds*

(i) If there are two solutions $u_1(t) = (\ell(t), h_1(t))$, $u_2(t) = (\ell(t), h_2(t))$ *corresponding to some (maybe different) realization of the forcing and such that* $u_1, u_2 \in U$, *then* $h_1 = h_2$.
(ii) Given a solution $u(t) = (\ell(t), h(t)) \in U$, *any* $h_0 \in \mathbb{L}_h^2$ *and* $t < 0$, *the limit* $\lim_{t_0\to-\infty} \Phi_{t_0,t}(\ell, h_0) = h^*$ *exists and* $h^* = h$.

The lemma implies that there is a map Φ_t giving the high modes at time t in terms of the past trajectory of the low modes $L^t = \{\ell(s)\colon s \in (-\infty, t]\} \in C((-\infty, t], \mathbb{L}^2)$: $h(t) = \Phi_t(L^t)$. Equation (22) then becomes

$$d\ell(t) = \Big[-\nu\Lambda^2\ell + P_\ell B(\ell, \ell) + G(\ell(t), \Phi_t(L^t))\Big]dt + dW(t) \tag{24}$$

where $G(\ell, h) = P_\ell B(\ell, h) + P_\ell B(h, \ell) + P_\ell B(h, h)$. Equation (20) is thus reduced to a dynamics of the low modes: it is a finite-dimensional process with memory extending back to $-\infty$, which is not Markovian, but rather Gibbsian.

In the final part of the proof one shows that the memory is not so strong as to violate ergodicity. A crucial fact is that for a set of full measure of "nice" past histories of the low modes $L \in C((-\infty, t], \mathbb{L}_\ell^2)$ and for any $t > 0$, the conditional distribution of $\ell(t) \in \mathbb{L}_\ell$ has a component equivalent to the Lebesgue measure. This fact is shown to imply that the assumption that the corresponding stationary measures on the path space of the past $\mu_{p,i}, i = 1, 2$ are different, leads to a contradiction.

We remark that Kuksin and Shirikyan [25] who deal with a forcing given by a bounded kicked noise acting on all modes, did also introduce a Gibbs construction in their proof of uniqueness.

References

1. S. Angenent. The zero set of a solution of a parabolic equation. *J. Reine Angew. Math.*, 390:79–96, 1988.
2. M. Arnold, Y. Bakhtin, and E. Dinaburg. Regularity of solutions to vorticity Navier–Stokes system on $\mathbb{R}^2$. *Comm. Math. Phys.*, 258(2):339–348, 2005.
3. V.I. Arnol'd. Lectures on bifurcations and versal families (in Russian). *Uspehi Mat. Nauk*, 27(5(167)):119–184, 1972.
4. C. Boldrighini and P. Buttà. Navier–Stokes equations on a flat cylinder with vorticity production on the boundary. *Nonlinearity*, 24(9):2639–2662, 2011.
5. C. Boldrighini, S. Frigio, and P. Maponi. On the blow-up of some complex solutions of the 3D Navier–Stokes equations: theoretical predictions and computer simulations. *IMA J. Appl. Math.*, 82(4):697–716, 2017.
6. C. Boldrighini, D. Li, and Ya.G. Sinai. Complex singular solutions of the 3-d Navier–Stokes equations and related real solutions. *J. Stat. Phys.*, 167(1):1–13, 2017.
7. J. Bricmont, A. Kupiainen, and R. Lefevere. Probabilistic estimates for the two-dimensional stochastic Navier–Stokes equations. *J. Statist. Phys.*, 100(3–4):743–756, 2000.
8. M. Cannone and F. Planchon. On the regularity of the bilinear term for solutions to the incompressible Navier–Stokes equations. *Rev. Mat. Iberoamericana*, 16(1):1–16, 2000.
9. X.-Y. Chen. A strong unique continuation theorem for parabolic equations. *Math. Ann.*, 311(4):603–630, 1998.
10. P. Constantin and C. Foias. *Navier–Stokes Equations*. Chicago Lectures in Mathematics. University of Chicago Press, Chicago, IL, 1988.
11. J. Cortissoz. Some elementary estimates for the Navier–Stokes system. *Proc. Amer. Math. Soc.*, 137(10):3343–3353, 2009.
12. J.C. Cortissoz. On the blow-up behavior of a nonlinear parabolic equation with periodic boundary conditions. *Arch. Math. (Basel)*, 97(1):69–78, 2011.
13. E. Dinaburg, D. Li, and Ya. G. Sinai. A new boundary problem for the two dimensional Navier–Stokes system. *J. Stat. Phys.*, 135(4):737–750, 2009.
14. E. Dinaburg, D. Li, and Ya. G. Sinai. Navier–Stokes system on the flat cylinder and unit square with slip boundary conditions. *Commun. Contemp. Math.*, 12(2):325–349, 2010.
15. E.I. Dinaburg and Ya. G. Sinai. A quasilinear approximation for the three-dimensional Navier–Stokes system. *Mosc. Math. J.*, 1(3):381–388, 471, 2001.
16. E.I. Dinaburg and Ya. G. Sinai. On some approximation of the 3D Euler system. *Ergodic Theory Dynam. Systems*, 24(5):1443–1450, 2004.
17. C.R. Doering and E.S. Titi. Exponential decay rate of the power spectrum for solutions of the Navier–Stokes equations. *Phys. Fluids*, 7(6):1384–1390, 1995.
18. W. E, J.C. Mattingly, and Ya. Sinai. Gibbsian dynamics and ergodicity for the stochastically forced Navier–Stokes equation. *Comm. Math. Phys.*, 224(1):83–106, 2001.
19. F. Flandoli. Dissipativity and invariant measures for stochastic Navier–Stokes equations. *NoDEA Nonlinear Differential Equations Appl.*, 1(4):403–423, 1994.
20. C. Foias, O. Manley, R. Rosa, and R. Temam. *Navier–Stokes Equations and Turbulence*, volume 83 of *Encyclopedia of Mathematics and its Applications*. Cambridge University Press, Cambridge, 2001.
21. C. Foias and R. Temam. Gevrey class regularity for the solutions of the Navier–Stokes equations. *J. Funct. Anal.*, 87(2):359–369, 1989.
22. S. Friedlander and N. Pavlović. Blowup in a three-dimensional vector model for the Euler equations. *Comm. Pure Appl. Math.*, 57(6):705–725, 2004.
23. E. Hopf. Über die Anfangswertaufgabe für die hydrodynamischen Grundgleichungen. *Math. Nachr.*, 4:213–231, 1951.
24. H.-O. Kreiss. Fourier expansions of the solutions of the Navier–Stokes equations and their exponential decay rate. In *Analyse Mathématique et Applications*, pages 245–262. Gauthier-Villars, Montrouge, 1988.

25. S. Kuksin and A. Shirikyan. Stochastic dissipative PDEs and Gibbs measures. *Comm. Math. Phys.*, 213(2):291–330, 2000.
26. O.A. Ladyzhenskaya. *The Mathematical Theory of Viscous Incompressible Flow*, volume 2 of *Mathematics and its Applications*. Gordon and Breach, Science Publishers, New York-London-Paris, 1969.
27. O.A. Ladyzhenskaya. A dynamical system generated by the Navier–Stokes equations. *J. Sov. Math.*, 3:458–479, 1975.
28. Y. Le Jan and A.S. Sznitman. Stochastic cascades and 3-dimensional Navier–Stokes equations. *Probab. Theory Related Fields*, 109(3):343–366, 1997.
29. Z. Lei and F. Lin. Global mild solutions of Navier–Stokes equations. *Comm. Pure Appl. Math.*, 64(9):1297–1304, 2011.
30. Jean Leray. Sur le mouvement d'un liquide visqueux emplissant l'espace. *Acta Math.*, 63(1):193–248, 1934.
31. D. Li. Bifurcation of critical points for solutions of the 2D Euler and 2D quasi-geostrophic equations. *J. Stat. Phys.*, 149(1):92–107, 2012.
32. D. Li and Ya. G. Sinai. Blow ups of complex solutions of the 3D Navier–Stokes system and renormalization group method. *J. Eur. Math. Soc. (JEMS)*, 10(2):267–313, 2008.
33. D. Li and Ya. G. Sinai. Bifurcations of solutions of the 2-dimensional Navier–Stokes system. In *Essays in mathematics and its applications*, pages 241–269. Springer, Heidelberg, 2012.
34. D. Li and Ya. G. Sinai. Nonsymmetric bifurcations of solutions of the 2D Navier–Stokes system. *Adv. Math.*, 229(3):1976–1999, 2012.
35. J.C. Mattingly and Ya. G. Sinai. An elementary proof of the existence and uniqueness theorem for the Navier–Stokes equations. *Commun. Contemp. Math.*, 1(4):497–516, 1999.
36. Ya. Sinai. Power series for solutions of the 3D-Navier–Stokes system on $\mathbf{R}^3$. *J. Stat. Phys.*, 121(5–6):779–803, 2005.
37. Ya G. Sinai. Diagrammatic approach to the 3D Navier–Stokes system. *Russ. Math. Surv.*, 60(5):849–873, 2005.
38. Ya. G. Sinai. On local and global existence and uniqueness of solutions of the 3D Navier–Stokes system on $\mathbb{R}^3$. In *Perspectives in Analysis*, volume 27 of *Math. Phys. Stud.*, pages 269–281. Springer, Berlin, 2005.
39. R. Temam. *Navier–Stokes Equations: Theory and Numerical Analysis*, volume 2 of *Studies in Mathematics and its Applications*. North-Holland Publishing Co., Amsterdam-New York, revised edition, 1979.
40. R. Temam. *Navier–Stokes Equations and Nonlinear Functional Analysis*, volume 66 of *CBMS-NSF Regional Conference Series in Applied Mathematics*. Society for Industrial and Applied Mathematics (SIAM), Philadelphia, PA, second edition, 1995.
41. M.J. Vishik and A.V. Fursikov. *Mathematical Problems of Statistical Hydromechanics*, volume 9 of *Mathematics and its Applications (Soviet Series)*. Kluwer Academic Publishers Group, Dordrecht, 1988.
42. V.I. Yudovich. *The Linearization Method in Hydrodynamical Stability Theory*, volume 74 of *Translations of Mathematical Monographs*. American Mathematical Society, Providence, RI, 1989.
43. T. Zhang. Bifurcations of solutions for the Boussinesq system in $\mathbb{T}^2$. *Nonlinear Anal. Real World Appl.*, 14(1):314–328, 2013.

Billiards

Leonid Bunimovich

Light is what a newborn baby sees first. It is no wonder that people were always interested in light propagation. Fundamental laws were discovered through the centuries. However, it was Ya. G. Sinai who first laid the foundation for the study of global properties of light propagation in media which contain reflectors (scatterers, mirrors, etc) and obtained fundamental results in this area.

It is a great and rare event when Mathematics discovers new laws of Nature. Indeed, it is commonly assumed that mathematicians just prove results which physicists already knew. The area of chaotic billiards, pioneered by Sinai, studies propagation of the rays of light in domains with reflecting boundaries. It turned out that physicists had at best a vague and sometimes wrong understanding of these processes. One should mention though that the analogy between the unstable dynamics of geodesic flows on surfaces of negative curvature and the unstable dynamics of the Boltzmann gas of elastically colliding hard spheres was noted by soviet physicist N. S. Krylov [11]. More than once while the theory of chaotic billiards was developing, did the physics community find rigorous mathematical results on billiards unbelievable. Only after performing experimental studies, both numerically and in a laboratory, did physicists accept these results and start to use them both in their theoretical and experimental research. Mathematical studies of billiards allowed the advancement of geometric optics, acoustics and classical and quantum mechanics. Impressive breakthroughs were made in Statistical Mechanics where the most classical and fundamental models, such as Boltzmann and Lorentz gases, are billiards with specially shaped reflecting boundaries.

L. Bunimovich (✉)
School of Mathematics, Georgia Institute of Technology, Atlanta, GA, USA
e-mail: bunimovh@math.gatech.edu

H. Holden, R. Piene (eds.), *The Abel Prize 2013–2017*,
https://doi.org/10.1007/978-3-319-99028-6_7

1 Mathematical Billiards

In the theory of dynamical systems billiards form arguably the most visual example and demonstrate a great variety of dynamical and statistical behavior. It is no wonder that nowadays billiards have become one of the favorite training grounds for exploring new advances in chaos theory.

All that was started in 1970 with the absolutely remarkable paper of Sinai [16]. It was a truly groundbreaking piece of work. Virtually all the advances in the theory of chaotic billiards can be traced to this paper. Sinai was far ahead of the community. At that time I was his graduate student and decided to start working on billiards. Some of the more senior mathematicians tried to talk me out of that. Having the best intentions, they were trying to prevent me from making a serious mistake. Their argument was that billiard dynamics is a very murky area where only Yasha (Sinai) understands something.

Indeed, until recently the community of mathematical billiards players was very small, mostly because of how difficult the area is to enter. Newcomers have to read a lot of papers that contain quite a few new ideas and employ rather sophisticated techniques. The situation changed rather recently after publication of the book by Chernov and Markarian [6] which provides an excellent introduction to the theory of chaotic billiards.

Mathematically a billiard is a dynamical system generated by the free (frictionless) motion of a point particle within some domain Q belonging to a finite dimensional Euclidian space $\mathbb{R}^d$ or a torus with Euclidian metric. This domain Q is called a configuration space or just a billiard table. The boundary ∂Q consists of a finite number of smooth submanifolds $\partial Q_i, i = 1, \dots, m$ of codimension one called the regular components of the boundary. The traditional assumption is that regular components of the boundary are at least C^3-smooth. This assumption ensures that there are no trajectories that have an infinite number of reflections off the boundary in a finite time. At each point q of a regular component of the boundary there is a normal unit vector $n(q)$ pointing into the interior of Q. The singular part of the boundary ∂Q is formed by the intersections $\partial Q_i \cap \partial Q_j, i \neq j$.

A regular component ∂Q_i of the boundary is called dispersing, focusing or neutral if $K(q) > 0$, $K(q) < 0$, or $K(q) = 0$, respectively, at all points $q \in \partial Q_i$, where $K(q)$ is the operator of the second fundamental form at the point $q \in \partial Q$.

Billiard dynamics is generated by the uniform motion of a point particle with unit speed within a billiard table. Upon reaching the boundary ∂Q the particle gets reflected according to the law of elastic reflections, i.e., "the angle of incidence equals the angle of reflection". Hence billiard orbits in the configuration space are broken lines. If singular part of the boundary of a billiard table is nonempty, then billiard orbits stop (are not defined) after hitting a singular point of the boundary. A billiard is a Hamiltonian system with an infinite potential on the boundary of the billiard table together with the condition that collisions with the boundary are elastic. Therefore the phase volume ν gets preserved under the dynamics. A billiard flow is defined almost everywhere on the phase space $\mathscr{M}$ which is the unit tangent bundle over the billiard table Q. Points of $\mathcal{M}$ are the unit vectors $x = (q, v)$. Denote by π the natural projection of $\mathcal{M}$ onto Q. Then $\partial\mathcal{M} = \pi^{-1}(\partial Q)$.

In the paper [16] Sinai introduced and analyzed billiards with strictly dispersing and smooth boundaries, i.e., regular components of the boundary of a billiard table do not intersect. Such billiards are called Sinai billiards. A billiard flow is defined at all points in the phase space of such billiards. A standard example of Sinai billiards is the torus with a sphere removed. The unfolding of this system is a classical model of Statistical Mechanics called the periodic Lorentz gas.

Because almost all billiard orbits eventually hit the boundary, these dynamical systems have a natural global Poincaré section. Therefore one can introduce a billiard map T which maps the set $M = \{x : \pi(x) \in \partial Q, (v, n(q)) \geq 0\}$ into itself. Draw the geodesic along the direction v up to its first intersection with the boundary. A vector Tx is equal to the reflection of the tangent vector at the intersection point. The phase space M of a billiard map is a product of the boundary of the billiard table and the unit $(d-1)$-dimensional semi-sphere. For two-dimensional Sinai billiards M is a union of cylinders where each cylinder corresponds to a regular component of the boundary of a billiard table. At each cylinder one considers coordinates (ℓ, φ) where ℓ is a coordinate (normalized length) of a point q on the boundary ∂Q and φ, $-\pi/2 \leq \varphi \leq \pi/2$ is the angle between the velocity vector and the inner normal $n(q)$. Billiard map preserves the measure μ which is the projection of ν onto M.

Consider a smooth curve $\tilde{\gamma} \subset Q$ and a continuous family γ of unit vectors normal to $\tilde{\gamma}$. Then γ is a smooth curve in the phase space M. Clearly two curves γ correspond to every curve $\tilde{\gamma}$ according to choice of a field of normal vectors. By fixing the curve γ we define curvature of the curve $\tilde{\gamma}$. We will refer in what follows to the curvature of γ to avoid ambiguities. In modern terms such curves are called wave fronts. It is always assumed that wave fronts are initially short, i.e., one considers narrow beams of rays.

Denote by $\kappa(x_0)$ the curvature of a wave front γ_0 at the point x_0. Let t be so small that no point of the curve γ_0 could reach the boundary on the time interval $[0, t]$. It is easy to see that the curvature of the curve $\gamma_t = T^t\gamma_0$ at the point $x_t = T^t x_0$ equals $\kappa(x_t) = \kappa(x_0)(1 + t\kappa(x_0))^{-1}$. Clearly $\kappa(x_t) > 0$ if $\kappa(x_0) > 0$, that is, if γ_0 is a convex curve, then γ_t is a convex curve too. Consider now what happens upon reflections off the boundary. Let $\tau(x_0) > 0$ be a moment of the first reflection of the trajectory of a point x_0 from the boundary. Classical mirror formula of the geometric optics reads as

$$\kappa(x_{\tau+0}) = \kappa(x_{\tau-0}) + \frac{2k(q_\tau)}{\cos\varphi(x_\tau)} \tag{1}$$

where $k(q_\tau)$ is the curvature of the boundary at the point of reflection and $\varphi(x_\tau)$ is the angle between the reflected ray and the inner unit normal vector to the boundary at the point of reflection.

2 Continued Fractions

A key tool that Sinai introduced to study dynamics of billiards are continued fractions that correspond to positive and negative infinite semi-trajectories of billiards.

Let $0 < t_1 < t_2 < \cdots < t_n < \cdots$ be the moments of the consecutive reflections of the forward trajectory of the point $x \in M$ off the boundary, $t_n \to \infty$ as $n \to \infty$.

Let $\tau_i = t_i - t_{i-1}$, $t_0 = 0$. Denote by $q_i \in \partial Q$ the point of the boundary where the ith reflection occurs and by $A_i \subset \mathbb{R}^d$ the hyperplane tangent to ∂Q at the point q_i. Let v_i^- and v_i^+ be the velocities directly before and after the ith reflection, and define φ_i by $\cos\varphi_i = -(v_i^+, n(q))$. Let K_i be the operator of the second fundamental form of the boundary ∂Q at the point q, and $A_i^- \subset \mathbb{R}^d$, $A_i^+ \subset \mathbb{R}^d$ be the hyperplanes that contain the point q_i and are orthogonal to v_i^- and v_i^+, respectively. Denote by U_i the isometric operator which maps in the direction parallel to $n(q_i)$ onto the hyperplane $A_i^- \subset \mathbb{R}^d$, and by V_i the one which maps A_i^- parallel to v_i^- onto A_i. Let V_i^* be the operator adjoint to V_i, and I be the identity operator.

Infinite operator-valued continued fraction introduced by Sinai [16, 17] has the form

$$B^s(x) = \cfrac{I}{\tau_1 I + \cfrac{I}{2\cos\varphi_1 V_1^* K_1 V_1 + U_1^{-1} \cfrac{I}{\tau_2 I + \cfrac{I}{2\cos\varphi_2 V_2^* K_2 V_2 + \cdots}} U_1}} \tag{2}$$

In two-dimensional case each odd numbered element of the continued fraction is equal to the length of free path between corresponding consecutive reflections off the boundary and according to the mirror formula (1) each even numbered element equals $2k(q_i)/\cos\varphi_i$ where $k(q_i)$ is the curvature of ∂Q at the point of the ith reflection and φ_i is the corresponding angle of incidence.

3 Hyperbolicity of Sinai Billiards

A key fact proved by Sinai [16, 17] is that the operator $B^s(x)$ defines the plane tangent to the local stable manifold (see Pesin's paper [13] on hyperbolicity) of the point x. In two-dimensional case a local stable manifold (LSM) has the slope $\frac{d\varphi}{d\ell} = -B^s(x)\cos\varphi + k(q)$. One gets a continued fraction for local unstable manifold (LUM) by considering in (2) inverse iterates of the billiard map T. Convergence of these continued fractions for Sinai billiards was proved in [17]. In the two-dimensional case convergence immediately follows from the famous Seidel–Stern theorem. This fact alone allows prove positivity of the Kolmogorov–Sinai (KS) entropy (see Gurevich's paper [9] on Ergodic Theory) for Sinai billiards

[16]. Moreover, KS-entropy for Sinai billiards can be computed via the following formula [6, 16]

$$h(T^t) = \int_M \mathrm{tr} B^u(x) d\mu(x). \tag{3}$$

This formula shows that KS-entropy is equal to the average expansion rate of LUMs.

The existence of local stable and unstable manifolds allows immediately prove ergodicity for smooth hyperbolic systems by making use of the Hopf chain. Namely, for any two points x and y in the phase space there exist (non-unique) finite collections $\gamma_1^{(s)}, \gamma_2^{(u)}, \ldots, \gamma_r^{(s)}$ of LSM and LUM such that $\gamma_1^{(s)} \ni x$, $\gamma_r^{(s)} \ni y$, and if γ_i is a LSM, then γ_{i+1} is a LUM, or vice versa. Then ergodicity follows from the Birkhoff ergodic theorem.

However, Sinai billiards are non-smooth dynamical systems. More precisely, billiard flow in Sinai billiards is non-smooth but continuous, and the billiard map is discontinuous. Discontinuities appear on such orbits that are tangent to the boundary of the billiard table (Fig. 1). Therefore from the existence of LSM and LUM follows only a local fact that corresponding ergodic component has positive measure. In order to prove a global ergodicity (uniqueness of ergodic component) one needs principally new tools.

Let $S_0 = \{x \in \partial M \colon (x, n(q)) = 0\}$. Then $S = S_0 \cup T^{-1} S_0$ is a singular set in the phase space of dispersing billiards. It is the only singular set for Sinai billiards because the boundary is smooth. Denote by $\tau(x)$ a time that it takes for a billiard particle to travel from the point x to the point Tx. One gets a billiard with finite horizon if $\tau(x) < \infty$, otherwise a billiard with infinite horizon. In Sinai billiards with finite horizon, the singular set S consists of a finite number of smooth manifolds, while in case when the horizon is infinite, S consists of infinite (countable) number of smooth components. An infinite number of manifolds where a billiard map is discontinuous, is situated in neighborhoods of orbits which have infinitely many tangent collisions with the boundary (scatterers) and at the points

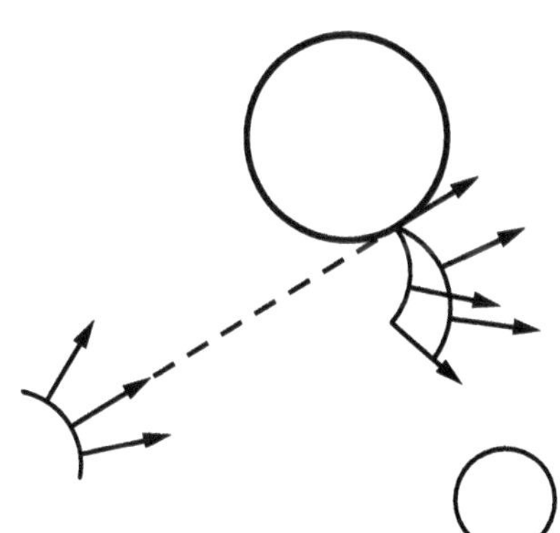

Fig. 1 Formation of discontinuities

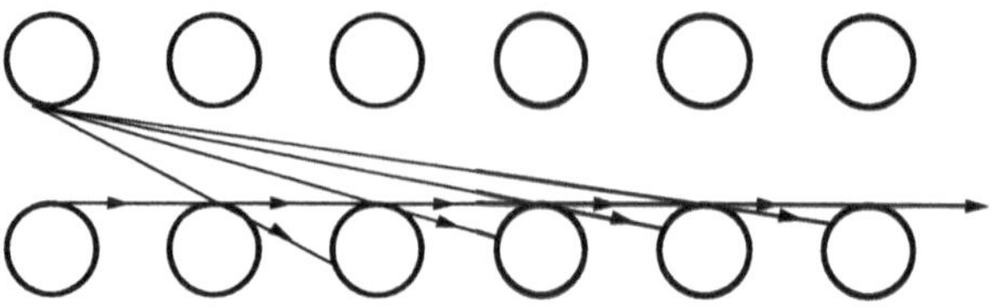

Fig. 2 Sinai billiard with infinite horizon

of tangency the scatterers are situated on one and the same side of the billiard orbit (Fig. 2).

In ergodic billiards the orbit of singularity set is everywhere dense in the phase space. Therefore smooth pieces of LSM and LUM could be arbitrarily small. Therefore a Hopf chain cannot routinely be constructed for Sinai billiards in the same manner as it gets constructed for geodesic flows, Anosov systems, Smale's Axiom A systems, etc. Actually, this is a major principally new problem for proving ergodicity for any non-uniformly hyperbolic system.

The corresponding theorem, proved by Sinai [16] and which allows us to overcome this principal problem, is now called main theorem of the theory of hyperbolic billiards, fundamental theorem of the theory of non-uniformly hyperbolic systems, or just the Main Theorem. The approach to a proof of this theorem is very characteristic of Sinai's scientific style, that is, first to completely clarify the situation and then go to a proof in the most direct manner that is relevant to the intrinsic nature of the problem. Although there are now many technical modifications in the proof of the main theorem, they all follow Sinai's footsteps.

If a smooth peace of LUM is moving under the action of billiard flow then its length grows. Therefore some vectors in the image will be tangent to the boundary of the billiard table (Fig. 1). At such points the image of LUM will have singularities and therefore it will consist of many smooth pieces (Fig. 3). These pieces could be very short, and therefore cannot be used for construction of the Hopf chain. Therefore, the evolution of LUMs is defined by competition of two processes, expansion under dynamics and cutting by the singularities. At the heart of the ergodic theory of non-uniformly hyperbolic systems lies Sinai's principle,

Fig. 3 Image under dynamics of a smooth local unstable manifold

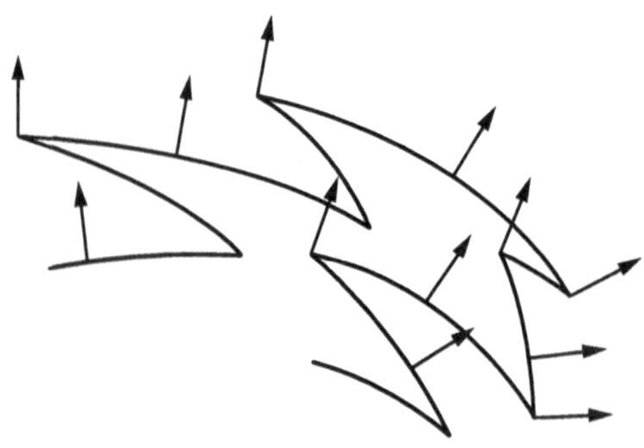

which claims that expansion beats cutting. The Fundamental theorem is an exact mathematical statement of this principle.

The main idea of Sinai's fundamental theorem [16] is to construct an analog of the Hopf chain where each link is not a single (stable or unstable) local manifold but a set of positive measure consisting of sufficiently long local manifolds. Because expansion due to the intrinsic instability of the dynamics is stronger than the process of cutting by singularities, such a "thick" chain must exist.

4 Sinai's Fundamental Theorem of the Theory of Hyperbolic Billiards

A piecewise smooth curve γ in M is called increasing (resp. decreasing) if $\frac{d\varphi}{d\ell} > a_1 > 0$ (resp. $\frac{d\varphi}{d\ell} < a_2 < 0$), where a_1, a_2 are some constants which depend on geometry of the billiard table Q. It easily follows from (2) that LUM (resp. LSM) are increasing (resp. decreasing) curves on the corresponding cylinder. By a quadrilateral we mean a domain with boundary consisting of four piecewise continuously differentiable curves and one non-intersecting pair of curves is increasing while the other pair is decreasing. Denote by $|\gamma|$ the length of the piecewise smooth curve γ.

The Fundamental Theorem Let x_0 be a point of the phase space M such that the positive semi-trajectory $T^i x_0, i = 0, 1, 2, \ldots$ never hits the singular set S. Then for each α $(0 < \alpha < 1)$ and any C $(0 < C < \infty)$ there exists an $\epsilon = \epsilon(x_0, \alpha, C)$ such that the ϵ-neighborhood U_ϵ of x has the following property: for any decreasing curve $\gamma_0 \subset U_\epsilon, |\gamma_0| = \delta_0$ there is a quadrilateral G whose left side is γ_0, and if $G^1 = \{x \in G:$ there is a regular segment of LUM $\gamma^{(u)}(x)$ joining the left and right sides of G and $|\gamma^{(u)}(x)| > C\delta_0\}$, then $\mu(G^1) \geq (1-\alpha)\mu(G)$. Similarly one can construct a quadrilateral $\widetilde{G}$ whose right side is γ_0 and such that the corresponding set $\widetilde{G}^1$ satisfies the inequality $\mu(\widetilde{G}^1) \geq (1-\alpha)\mu(\widetilde{G})$.

Sinai's fundamental theorem for non-uniformly hyperbolic systems allows local ergodicity of the corresponding dynamical system to be established. Local ergodicity means that for almost any point of the phase space there exists an open neighborhood that belongs to one ergodic component (mod 0). To deduce (global) ergodicity from local ergodicity one needs to perform a careful analysis of the singular set of the dynamical system in question. Such an analysis was performed in [16] for Sinai billiards with finite horizon and in [3] for Sinai billiards with infinite horizon, and for general dispersing billiards that may have singularities on the boundary.

Thanks to Sinai's fundamental paper on ergodic theory of hyperbolic dynamical systems [15], we do not need to prove that ergodic hyperbolic systems are mixing and have the K-property. Sinai proved in [15] that these properties follow from the existence of one transversal foliation into unstable leafs. (Usually such foliations appear as pairs of stable and unstable ones.) Thus, thanks to [15], one may just prove ergodicity. Later, after remarkable results by Ornstein on isomorphism of Bernoulli systems, it was proved [7] that the Bernoulli property of hyperbolic systems also follows from ergodicity.

A direct consequence of the proof that Sinai billiards are ergodic, is the ergodicity of the system of two elastically colliding disks on a two-dimensional torus. This breakthrough and very unexpected result was proved in the same paper [16]. Indeed, it was proof of the celebrated Boltzmann hypotheses for two particles while Boltzmann claimed its validity for a gas of extremely many elastically interacting spheres. Since that time the Boltzmann hypothesis has naturally been called the Boltzmann–Sinai hypothesis (see papers by Szasz [18] and Simanyi [14]).

5 Mechanism of Defocusing and Hyperbolic Focusing Billiards

Development of Sinai's ideas and techniques allowed for some absolutely unexpected findings. As a result the theory of billiards took the leading role in a general theory of nonuniformly hyperbolic systems. Perhaps the most unexpected breakthrough was the discovery of the fundamental and quite counterintuitive mechanism of chaos in dynamical systems.

It was well known that billiards in circles, and ellipses demonstrate the most regular possible dynamics, i.e., they are integrable systems. Therefore a common "natural" understanding was that focusing stabilizes dynamics. This general intuition was even more supported when Lazutkin proved [12] that two-dimensional billiards in convex tables with sufficiently smooth boundaries have continuous families of caustics which converge to the boundary ∂Q of a billiard table. (Recall that a curve γ is a caustic if tangency to one link of a billiard orbit implies all other links are also tangent to γ.) Clearly, a billiard is non-ergodic if it has at least one caustic.

It turned out, though, that there are large classes of hyperbolic billiards in the tables having focusing boundary components. Moreover, even billiards with convex tables can be hyperbolic. The reason for such counterintuitive results is the existence of another mechanism causing chaos (hyperbolicity) other than the mechanism of dispersing. The dispersing mechanism generates hyperbolicity (chaos) in geodesic flows in manifolds of negative curvature and in dispersing billiards. It was natural to consider small perturbations of such systems by adding small pieces of manifolds with non-negative curvature for geodesic flows or small focusing pieces of boundary for billiards. Such attempts for geodesic flows were started by Hopf [10] but did not succeed. Only after discovery of a defocusing mechanism [1, 2] were there constructed the ergodic geodesic flows on surfaces with pieces of positive curvature [8]. This is how billiards took a lead in studies of non-uniformly hyperbolic systems.

Suppose that a parallel beam of rays falls onto a dispersing component of the boundary. Then this beam becomes divergent after the reflection off the boundary. In

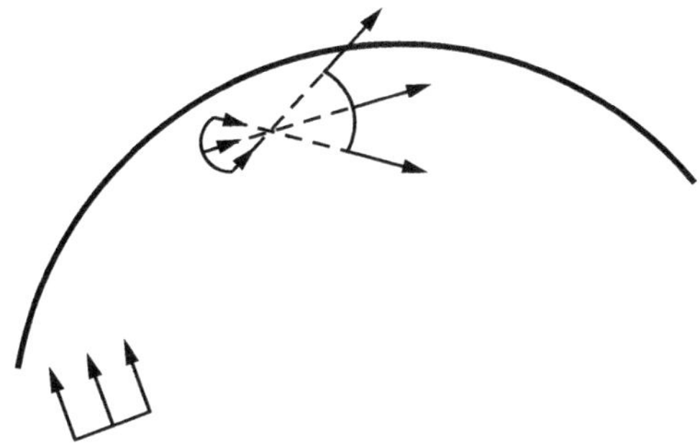

Fig. 4 Divergence of rays by defocusing

dispersing billiards such beams remain divergent for all times and thus nearby orbits permanently diverge in the phase space. This intrinsic instability in the dynamics generates hyperbolicity for dispersing billiards.

Now let a parallel beam of rays fall on a focusing component of the boundary. Then the situation becomes completely different because after reflection such beam becomes convergent. Therefore the distance in phase space between nearby orbits decreases which seems automatically to lead to stability in the dynamics. However, the focused (convergent) beam of rays may become divergent if the time until its next reflection off the boundary is large enough. Moreover, the length of time when this beam was divergent may exceed the length of time when it was convergent. Therefore between two consecutive reflections off the boundary of a billiard table the front of a convergent (focused) beam may increase its size ensuring effective divergence of nearby rays between consecutive reflections (Fig. 4). This is a key to the mechanism of defocusing. Another possibility is to make the focusing stronger but keep the same minimal distance between consecutive points of reflection at different components of the boundary.

It is obvious that the billiard in a circle is nonergodic. Indeed, this fact is equivalent to the fact that all chords of the same length are at the same distance from the center of the circle. Integrability of the billiard in a circle in fact follows from the fact that in this dynamical system convergence (focusing) of orbits and their divergence (defocusing) are completely balanced [2]. Therefore this dynamical system is parabolic. To make it hyperbolic one needs to increase free paths $\tau(x)$ on a set of positive measure.

This has been done in [1, 2] by cutting a circle by a chord and taking the larger part as a billiard table. It was the first example of a chaotic billiard within a convex billiard table. Surprisingly, the most popular focusing chaotic billiard became a degenerate example of the so called stadium which one gets by cutting a circle into two semicircles and connecting them by two parallel straight segments [1]. Such billiards with only focusing or with focusing and neutral components are ergodic, mixing, K- and B-systems [1, 16, 17]. Later, by making use of similar ideas, hyperbolic geodesic flows on spheres [7] were constructed. However, correlations for ergodic hyperbolic billiards which have focusing boundary components decay only power-like while in dispersing billiards correlations decay exponentially [5, 19].

For a long time it was an open question whether or not chaotic focusing billiards exist in dimensions >2. A reason for this question was an optical phenomenon called astigmatism. For the mechanism of defocusing to work, a strong focusing upon reflections off the boundary is needed. However, because of astigmatism the strength of focusing varies over the hyperplanes containing the point of reflection off the boundary of a billiard table. Moreover, the strength of focusing is arbitrarily weak in some hyperplanes. It was proved, though, that the mechanism of defocusing is universal and works in any (finite) dimension $d \geq 2$ [4]. Moreover, hyperbolic and ergodic billiards also exist in higher dimensions. The continued fractions introduced by Sinai are always the main tool to prove hyperbolicity. However, these fractions contain elements with different signs if the billiard table has at least one focusing boundary component. There are no general criteria for convergence of such continued fractions. Therefore this part of the proof, that is trivial for dispersing billiards, often becomes quite involved.

Nowadays billiards are favorite models for physicists. Sinai billiards, stadia, and other chaotic billiards were built as experimental devices in physics labs all over the world. Many questions about chaotic billiards remain unanswered and new findings and surprises continue to come.

References

1. L.A. Bunimovich. On ergodic properties of certain billiards. *Funct. Anal. Appl.*, 8(3):254–255, 1974.
2. L.A. Bunimovich. On billiards close to dispersing. *Math. USSR-Sb.*, 23(1974):45–67, 1975.
3. L.A. Bunimovich and Ya.G. Sinai. On a fundamental theorem in the theory of dispersing billiards. *Math. USSR-Sb.*, 19(1973):407–423, 1974.
4. L.A. Bunimovich and J. Rehacek, How high-dimensional stadia look like. *Comm. Math. Phys.*, 197:277–301, 1998.
5. N. Chernov. Decay of correlations and dispersing billiards. *J. Statist. Phys.*, 94(3–4):513–556, 1999.
6. N. Chernov and R. Markarian. *Chaotic Billiards*, volume 127 of *Mathematical Surveys and Monographs*. American Mathematical Society, Providence, RI, 2006.
7. N.I. Chernov and C. Haskell. Nonuniformly hyperbolic K-systems are Bernoulli. *Ergodic Theory Dynam. Systems*, 16(1):19–44, 1996.
8. V.J. Donnay. Geodesic flow on the two-sphere. I. Positive measure entropy. *Ergodic Theory Dynam. Systems*, 8(4):531–553, 1988.
9. B. Gurevich. Entropy theory of dynamical systems. In H. Holden, R. Piene, editors, *The Abel Prize 2013–2017*, pages 221–242. Springer, 2019.
10. E. Hopf. Statistik der Lösungen geodätischer Probleme vom unstabilen Typus. II. *Math. Ann.*, 117:590–608, 1940.
11. N.S. Krylov. *Works on the Foundations of Statistical Physics*. Princeton University Press, Princeton, N.J., 1979.
12. V.F. Lazutkin. The existence of caustics for the billiard problem in a convex domain. *Math. USSR-Izv.*, 7(1973):185–214, 1974.

13. Y. Pesin. Sinai's work on Markov partitions and SRB measures. In H. Holden, R. Piene, editors, *The Abel Prize 2013–2017*, pages 257–285. Springer, 2019.
14. N. Simányi. Further developments of Sinai's ideas: the Boltzmann–Sinai hypothesis. In H. Holden, R. Piene, editors, *The Abel Prize 2013–2017*, pages 287–298. Springer, 2019.
15. Ya.G. Sinai. Dynamical systems with countably-multiple Lebesgue spectrum. II. In *Ten papers on differential equations and functional analysis*, volume 68 of *American Mathematical Society Translations*, pages 34–88, 1968.
16. Ya.G. Sinai. Dynamical systems with elastic reflections. *Russ. Math. Surv.*, 25(2):137–189, 1970.
17. Ya.G. Sinai. *Development of Krylov's ideas.* Afterword to N.S. Krylov, *Works on the Foundations of Statistical Physic*, pages 239–281. Princeton University Press, Princeton, NJ,1979.
18. D. Szász. Markov approximations and statistical properties of billiards. In H. Holden, R. Piene, editors, *The Abel Prize 2013–2017*, pages 299–319. Springer, 2019.
19. L.-S. Young. Statistical properties of dynamical systems with some hyperbolicity. *Ann. of Math. (2)*, 147(3):585–650, 1998.

Ya.G. Sinai's Work on Number Theory

F. Cellarosi

Abstract In this article we will survey some of the contributions of Ya.G. Sinai to number theory and related fields. The multifacetedness of his work demonstrates Ya.G. Sinai's vision of mathematics as a highly interconnected discipline, rather than a series of compartmented fields.

Ya.G. Sinai has explored some of the deep connections of number theory with other disciplines, such as probability theory, statistical mechanics, and the theory of dynamical systems. In the attempt to illustrate this, we shall focus on [3, 4, 8, 13–15, 22, 23, 59, 60, 62, 63] by Ya.G. Sinai and several coauthors.

The 1990 article by E.I. Dinaburg and Ya.G. Sinai [23] and the more recent papers by Ya.G. Sinai and C. Ulcigrai [62, 63] have a twofold common thread: they study the distribution of certain sequences of number-theoretical interest, and use in their solution a powerful dynamical tool, namely the mixing property of a suitably defined special flow. In [23] E.I. Dinaburg and Ya.G. Sinai study the statistical properties of the solutions to the Diophantine equation $|ax - by| = 1$ for positive integers a and b, with $a < b$. The classical approach is to find all solutions to this equation by first considering the convergents of the continued fraction expansion of the rational number a/b: write

$$\frac{a}{b} = \cfrac{1}{k_1 + \cfrac{1}{k_2 + \cfrac{1}{\ddots + \cfrac{1}{k_{s-1} + \frac{1}{k_s}}}}} = [k_1, k_2, \ldots, k_{s-1}, k_s],$$

F. Cellarosi (✉)
Department of Mathematics and Statistics, Queen's University, Kingston, ON, Canada
e-mail: fcellaro@mast.queensu.ca

H. Holden, R. Piene (eds.), *The Abel Prize 2013–2017*,
https://doi.org/10.1007/978-3-319-99028-6_8

where $k_1, \dots, k_s$ are positive integers and the length of the expansion s depends on a/b. If, for $1 \le j \le s$, we define the j-th convergent of a/b to be $p_j/q_j = [k_1, \dots, k_j]$ with $\gcd(p_j, q_j) = 1$, then $p_s/q_s = a/b$ and the pair $(x_0, y_0) = (q_{s-1}, p_{s-1})$ is a particular solution for the equation $|ax - by| = 1$. All solutions can then be written as $\pm(x_0 + kb, y_0 - ka)$ for $k \in \mathbb{Z}$. The authors study how the particular solution (x_0, y_0) behaves as the coefficients a, b of the Diophantine equation vary. More precisely, the authors study the distribution of the ratio $q_{s-1}/q_s = q_{s-1}/b$ when the coefficients a, b satisfy the inequalities $\alpha_1 N < a < \beta_1 N$ and $\alpha_2 N < b < \beta_2 N$, where $0 < \alpha_1 < \beta_1 < \alpha_2 < \beta_2 < 1$ are fixed and N tends to infinity. We may assume that the fraction a/b is in lowest terms, that is a and b have no common factors. It is known that, as $N \to \infty$, the number S_N of such pairs (a, b) is $\frac{6}{\pi^2}(\beta_1 - \alpha_1)(\beta_2 - \alpha_2)N^2 + O(N \ln N)$. Dinaburg and Sinai define the measure μ_N on the Borel σ-algebra of subsets of $[0, 1]$ as $\mu_N(B) = S_N^{-1}\#\{(a, b) : q_{s-1}/b \in B\}$. This measure is the law of the random variable q_{s-1}/b. They prove that, as $N \to \infty$, this random variable becomes uniformly distributed on the interval $[0, 1]$. More precisely, they show that for every $\varepsilon, \delta > 0$ and any interval $\Delta \subset [0, 1]$ of length $|\Delta| > \delta$ there exists $N_0 = N_0(\varepsilon, \delta) > 0$ such that $|\mu_N(\Delta)/|\Delta| - 1| < \varepsilon$ for all $N > N_0$. In order to achieve this result, the clever idea of the authors is to rephrase the problem in terms of a suitably constructed special flow over the natural extension of the Gauss map, where the roof function is chosen to obtain $\ln q_s$ as an ergodic sum.

It is worth discussing this idea with some details, as it was used successfully used in other papers, such as [63] and [9]. Consider the Gauss map $G : (0, 1] \to (0, 1]$, defined by $G(x) = \{1/x\}$, where $\{x\} = x - \lfloor x \rfloor$. The map G admits an invariant measure μ_G, with density $((1 + x) \ln 2)^{-1}$ with respect to the Lebesgue measure; moreover it acts as a shif in the continued fraction representation: $G([k_1, k_2, k_3, \dots]) = [k_2, k_3, \dots]$. Following Rokhlin [56], Dinaburg and Sinai construct a natural extension $\tilde{G}$ of the Gauss map G. We have $\tilde{G} : (0, 1]^2 \to (0, 1]^2$, acting on pairs of real numbers as $\tilde{G}([k_0, k_{-1}, k_{-2}, \dots], [k_1, k_2, k_3, \dots]) = ([k_1, k_0, k_{-1}, \dots], [k_2, k_3, \dots])$, or $\tilde{G}(x_-, x_+) = ((\lfloor 1/x_+ \rfloor + x_-)^{-1}, \{1/x_+\})$, which preserves the measure $\mu_{\tilde{G}}$ with density $((1 + x_- x_+)^2 \ln 2)^{-1}$ with respect to the two dimensional Lebesgue measure. Let us consider the special flow (or suspension flow) Φ_t over the dynamical system $((0, 1]^2, \tilde{G}, \mu_{\tilde{G}})$, under the roof function $F(x_-, x_+) = -\ln(\pi_-(\tilde{G}(x_-, x_+)))$, where $\pi_- : (0, 1]^2 \to (0, 1]$ is the projection $\pi_-(x_-, x_+) = x_-$. The flow $\{\Phi_t\}_{t\in\mathbb{R}}$ acts on points set of points of the form (x_-, x_+, z) with $(x_-, x_+) \in (0, 1]^2$ and $0 \le z < F(x_-, x_+)$, modulo the identification $(x_-, x_+, F(x_-, x_+)) \sim (F(x_-, x_+), 0)$, by flowing the points in the z-direction with unit speed. The reason for choosing the particular roof function F is that $\pi_-(\tilde{G}^n(x_-, x_+)) = [k_n, k_{n-1}, \dots, k_1, k_0, k_{-1}, \dots]$ yields the continued fraction expansion of (x_-, x_+), read backward from the n-th term. This fact implies that the ergodic sum $\sum_{n=0}^{r-1} F(\tilde{G}^n(x_-, x_+)) = \sum_{n=1}^{r}(-\ln(\pi_-(\tilde{G}^n(x_-, x_+))))$ equals $\ln q_r + \ln\left(1 + x_- \frac{p_r}{q_r}\right)$ and therefore can be used to study the denominators q_r of the continued fraction representation of a/b. Ergodic sums are related to the special flow $\{\Phi_t\}_{t\in\mathbb{R}}$ as follows. By definition, $\Phi_t(x_-, x_+, z) = (\tilde{G}^r(x_-, x_+), z')$, where r and z' depend on (x_-, x_+, z) and t, and are determined by

$$\sum_{n=0}^{r-1} F(\tilde{G}^{n+1}(x_-, x_+)) \le z + t < \sum_{n=0}^{r} F(\tilde{G}^{n+1}(x_-, x_+))$$

$$z' = z + t - \sum_{n=0}^{r-1} F(\tilde{G}^{n+1}(x_-, x_+)),$$

respectively, for $t > 0$.

It is important that the chosen roof function F is integrable over $((0, 1]^2, \mu_{\tilde{G}})$, as this yield an invariant probability measure for the special flow Φ_t. To deduce that the measures μ_N converge to the uniform measure on $[0, 1]$, Dinaburg and Sinai show that the special flow Φ_t is mixing.

The authors actually prove something stronger, namely that the stable and unstable foliations are not integrable. This fact implies that the Pinkser partition is trivial, which in turn implies that the flow has the K-property and, in particular, is mixing.

Dinaburg and Sinai's paper stimulated other researchers to further investigate the distribution of solutions to diophantine equations. For instance, in 1992 A. Fujii [34] used Kloosterman sums to improve Dinaburg and Sinai's result and obtain an effective equidistribution result. This improvement allows Fujii to consider the case when the interval Δ shrinks as $N \to \infty$, provided $|\Delta| \to 0$ not too quickly. Independently, in 1994 D. Dolgopyat [24] was able to study the joint limiting distribution of a/b and x_0/b as $(a, b) \in S_N$ and $N \to \infty$. Although some of the results by Dolgopyat can be derived from those of Fujii [34], the estimate of the number of pairs $(a, b) \in S_N$ such that $(a/b, x_0/b) \in \Delta_1 \times \Delta_2$ and $|\Delta_1| \cdot |\Delta_2| < N^{-1/2}$ does not follow from Fujii's work.

The beautiful method of deriving results of number-theoretical nature from the statistical properties of a flow has become very popular and has shed new light onto several new problems. For instance Ya.G. Sinai and C. Ulcigrai used this method in 2008 in [63] to study the statistical properties of the integer sequence $(q_n)_n$ of denominators of the continued fraction convergents of a typical real number α. Since the integer sequence $q_n = q_n(\alpha)$ is increasing (this follows easily from the fact that $q_{n+1} = k_{n+1}q_n + q_{n-1}$), one can find the first index, say $n_L = n_L(\alpha)$, for which $q_{n_L} > L$. Inspired by renewal theory, it is natural to ask how much larger q_{n_L} is relative to L. It turns out that, for random α, the ratio q_{n_L}/L has a limiting distribution on $(0, \infty)$ as L tends to infinity. Proving the mixing a suitably constructed special flow over the natural extension of the Gauss map (as explained above for [23]), allowed Sinai and Ulcigrai to obtain the existence of the liming distribution. Their method, however, did not provide an explicit expression for the distribution function of the limiting random variable. A beautiful formula was found in 2010 by A.V. Ustinov [68], who showed that for every $0 \leq t_1, t_2 \leq 1$ we have

$$\left|\left\{\alpha \in [0,1]:\ \frac{q_{n_L-1}}{L} \leq t_1,\ \frac{L}{q_{n_L}} \leq t_2\right\}\right| = -\frac{2}{\zeta(2)}\mathrm{Li}_2(-t_1 t_2) + O\left(\frac{\log L}{L}\right),$$

where $|\cdot|$ denotes the Lebesgue measure, and $\mathrm{Li}_2(u) = \sum_{k=1}^{\infty}\frac{u^k}{k^2}$ is Euler's dilogarithm function. The ideas introduced by Dinaburg and Sinai and the technique used by Sinai and Ulcigrai are very flexible, and can be used for other kind of continued fraction expansions, and yield new results of Diophantine type. For instance, a variation the method of [23] and [63] was used by F. Cellarosi in [9] to prove a renewal-type limit theorem for continued fractions expansions with even partial quotients (introduced by F. Schweiger in 1982 [58]). This result was needed in the subsequent paper [10] where the limiting distribution of normalized Weyl sums $N^{-1/2}\sum_{n=1}^{N} e^{2\pi i n^2\alpha}$ was studied. Partial sums of the above sums can be viewed as deterministic walks on the complex plane, depending upon a single real parameter α, as done by several authors, e.g. [6, 21]. In November 2006 Ya.G. Sinai became interested in Weyl sums—and the corresponding geometric patterns called *curlicues*—after attending the a seminar in the Physics department at Princeton. The seminar by F. Klopp, titled "On the multifractal structure of the generalized eigenfunctions of certain sparse Schrödinger operators", discussed a joint paper with A.A. Fedotov [31] and featured several spiral-like curves that triggered Ya.G. Sinai's interest. It became clear that a limit theorem these curves, when suitably rescaled, had to be proven. A geometric way to achieve a weak invariance principle for Weyl sums is outlined in the paper [60] that Ya.G. Sinai dedicated to S. Novikov on his 70th birthday. The idea of [10] and [60] is to use the renormalization group method to establish the existence of finite-dimensional limiting distribution for the partial sums as $N \to \infty$. As it is done in probability theory, limit theorems can be viewed as fixed point theorems. Here the renormalization map is an extension of the continued fraction transformation $\alpha \mapsto -1/\alpha \mod 2$. This renormalization approach traces back to the work of G.H. Hardy and J.E. Littlewood [41], J.R. Wilton [70], L.J. Mordell [52], and more recently of E.A. Coutsias and N.D. Kazarinoff [18, 19]. A key ingredient in [10] and [60] for the proof of the existence of a fixed point (a limiting distribution) is the mixing property of a suitably constructed special flow over the natural extension of the above continued fraction transformation. Recent progress on the distribution of Weyl sums is due to F. Cellarosi and J. Marklof [12]. Among other results, it is shown in [12] that rescaled curlicues have the same Hölder regularity as typical realizations of the Brownian motion, but slightly smaller modulus of continuity.

Another application of the method from [23] can be found in a second paper [62] by Ya.G. Sinai and C. Ulcigrai. A classical result by G.H. Hardy and J.E. Littlewood states that the trigonometric sum $\sum_{n=1}^{N}\sin(n\pi\alpha)^{-1}$ is uniformly bounded if α is a quadratic irrational, [42]. Ya.G. Sinai and C. Ulcigrai studied the distribution of a similar sum, namely $\sum_{n=0}^{N-1}(1-e^{2\pi i(n\alpha+x)})^{-1}$. They were able to show that if (α, x) is a uniformly distributed random point on the unit square, then the normalized sum $N^{-1}\sum_{n=0}^{N-1}(1-e^{2\pi i(n\alpha+x)})^{-1}$ has a limiting distribution on the complex plane as N tends to infinity. This follows from a more general theorem concerning the distribution of normalized ergodic sums of non-integrable observables over rotations of the circle (i.e. maps $R_\alpha(x) = x + \alpha \bmod 1$). The class of observables considered in [62] is characterized by symmetric logarithmic singularities of the form $\frac{1}{x}$ for x in a neighborhood of zero. In 2007 A.V. Kochergin [44] proved that the special flows over rotations R_α under a roof function with symmetric logarithmic singularities are not mixing for any α. This contrasts the case of roof functions with asymmetric logarithmic singularities, considered by K.M. Khanin and Ya.G. Sinai in 1992. They proved [61] that the special flow over a rotation is mixing if one assumes the following Diophantine condition on the rotation angle α: let us write the continued fraction expansion $\alpha = [k_1, k_2, k_3, \ldots]$ and assume that $k_n \le c \cdot n^{1+\gamma}$, where $0 < \gamma < 1$ and the constant c is allowed to depend on α. It is known that this condition holds for a set of angles with full Lebesgue measure. The result by Khanin and Sinai answered affirmatively a question asked by V.I. Arnol'd [1] in 1991 concerning the decomposition of a generic Hamiltonian flow on the torus. A complete understanding of the statistical properties of ergodic sum of observables with asymmetric logarithmic singularities over interval exchange transformations (which generalize rotations) was obtained by C. Ulcigrai [65, 66] and applied in the remarkable paper [67] to the study of area-preserving flows on surfaces.

The results of Sinai and Ulcigrai [63] and Ustinov [68] were used in two more papers, by J. Bourgain and Ya.G. Sinai [8] and V. Shchur, Ya.G. Sinai and A.V. Ustinov [59]. These papers concern the classical Frobenius problem, also known as the "coin change problem": given n relatively prime positive integers $a_1, a_2, \ldots, a_n$, what is the largest integer that is not representable in the form $x_1a_1 + \ldots + x_na_n$, with x_j nonnegative integers? This number is called the Frobenius number $F(a_1, \ldots, a_n)$. In 1884 J.J. Sylvester answered the question when $n = 2$ by giving the formula $F(a_1, a_2) = a_1a_2 - a_1 - a_2$. However, for $n \ge 3$, no formula for $F(a_1, \ldots, a_n)$ is known, see [55]. Ya.G. Sinai and his coauthors approached this problem by randomizing (a_1, a_2, a_3) as follows: if we pick a triple (a_1, a_2, a_3) with $\gcd(a_1, a_2, a_3) = 1$ and $a_i \le N$ uniformly at random, it is natural to ask whether the (suitably normalized) Frobenius number $F(a_1, a_2, a_3)$ has a limiting distribution as N tends to infinity. It turns out that this is true, provided we consider the normalization $N^{-3/2}F(a_1, a_2, a_3)$. This was proven in [8] and [59], along with partial results for $n > 3$. An explicit formula for the limiting distribution in this case $n = 3$ was found by A.V. Ustinov [69]. A formula for the limiting distribution for every $n \ge 3$ was obtained by J. Marklof [47] using the dynamics of a certain flow on the space of lattices $\mathrm{SL}(n-1, \mathbb{Z})\backslash\mathrm{SL}(n-1, \mathbb{R})$ and an equidistribution theorem for a multidimensional Farey sequence on closed horospheres.

In April 2010, P. Sarnak delivered a series of lectures at the Institute for Advanced Study on 'Möbius randomness and dynamics'. After attending Sarnak's lectures, Ya.G. Sinai became interested in this topic and wrote a series of papers with coauthors F. Cellarosi [13–15], and M. Avdeeva and D. Li [3, 4]. The goal for all these papers is to better understand the statistical properties of the famous Möbius function μ (defined as $\mu(n) = (-1)^k$ if n is the product of k distinct primes, and zero otherwise), and shed some light on two conjectures: the first by D.S. Chowla concerning autocorrelations for $\mu(n)$, and the second by P. Sarnak concerning the correlations of $\mu(n)$ with so-called deterministic sequences.

It is known, for instance, that $\sum_{n\leq N}\mu(n) = o(N)$ as $N \to \infty$ is equivalent to the Prime Number Theorem. On the other hand, the more precise statement $\sum_{n\leq N}\mu(n) = O_\varepsilon(N^{1/2+\varepsilon})$ as $N \to \infty$, which provides explicit power savings for the sum, is equivalent to the Riemann hypothesis. S. Chowla [17] conjectured that for every positive integer k, every $h_1, h_2, \ldots, h_k$ distinct integers, and every $\epsilon_1, \epsilon_2, \ldots, \epsilon_k \in \{1, 2\}$ not all even, we have

$$\sum_{n\leq N}\mu^{\epsilon_1}(n+h_1)\mu^{\epsilon_2}(n+h_2)\cdots\mu^{\epsilon_k}(n+h_k) = o(N)$$

as $N \to \infty$, and that the same result should hold for Liouville's function $\lambda(n)$ (which gives the parity of the number of prime factors of n, counted with multiplicity). Even the simplest non-trivial cases when $k = 2$, stating that $\sum_{n\leq N}\mu(n)\mu(n+1) = o(N)$ and $\sum_{n\leq N}\lambda(n)\lambda(n+1) = o(N)$ as $N \to \infty$, are still unproven.

P. Sarnak [57] defines a sequence $(a(n)_{n\geq 1}$ to be deterministic if there exists a topological dynamical system (X, T)—in particular X is compact—with zero topological entropy, a point $x \in X$, and a continuous function $f : X \to \mathbb{C}$ such that $a(n) = f(T^n(x))$ for all $n \geq 1$. Sarnak's conjecture [57] states that for every deterministic sequence $(a(n))_{n\geq 1}$ we have

$$\sum_{n\leq N}\mu(n)a(n) = o(N)$$

as $N \to \infty$. In other words, the sequence $\mu(n)$ is orthogonal to $a(n)$ for every deterministic sequence $a(n)$. Moreover, Chowla's conjecture implies Sarnak's conjecture, [57]. The simplest non-trivial sequences for which Sarnak's conjecture holds are those obtained by rotations (X is the circle $\mathbb{T} = \mathbb{R}/\mathbb{Z}$ and $T(x) = x + \alpha$ modulo 1), orthogonality in this case follows from a classical estimate by Davenport [20]. We will mention several more recent results toward Sarnak's conjecture later.

In [13] and [15] F. Cellarosi and Ya.G. Sinai discuss a probabilistic model for square-free integers, using ideas from statistical mechanics. Let $p_1 < p_2 < \ldots < p_m$ be the first m primes, consider the set Ω_m consisting of all the square-free integers of the form $n = p_1^{\nu_1}p_2^{\nu_2}\cdots p_m^{\nu_m}$ with $\nu_i \in \{0, 1\}$, and equip this set with the discrete probability measure $\mathbb{P}_m(\{n\}) = c_m/n$ for a suitable constant c_m. It is natural to ask how a typical square-free integer in Ω_m looks like as m tends to infinity. It turns out that if we write $n = p_m^{\xi_m(n)}$, then ξ_m has a limiting distribution on $(0, \infty)$ as $m \to \infty$. The limiting random variable is infinitely divisible, has a continuous density and is rather unusual: it is constant on the interval $(0, 1]$ and then it decays to zero faster than exponentially. This density is $e^{-\gamma}\rho(t)$, where γ is Euler's constant and $\rho(t)$ is the so-called Dickman-De Bruijn function. It is determined by $\rho(t) = 0$ for $t \le 0$, $\rho(t) = 1$ for $0 < t \le 1$, and by the integral equation $t\rho(t) = \int_{t-1}^{t} \rho(s)ds$. A consequence of limit theorem in [13] is that for every $s > 0$ we have

$$\lim_{m\to\infty} \mathbb{P}_m\{n \in \Omega_m : n \le p_m^s\} = e^{-\gamma} \int_0^s \rho(t)dt.$$

For $s = 2$ the limit equals $e^{-\gamma}(3 - \log 4) \approx 0.90603$, which means roughly that, although the largest element of Ω_m is $p_1 \cdots p_m = e^{(1+o(1))m \log m}$, approximately 90% of the mass of the probability measure $\mathbb{P}_m$ is concentrated on numbers less than p_m^2, for large m.

The Dickman–De Bruijn function ρ had appeared before in the study of smooth numbers (see the survey by A. Granville [38] and the references therein) and implicitly in work of V.L. Goncharov on random permutations [37]. Another instance is the following: let $(X_j)_{j\ge1}$ be a sequence of random variables such that $\mathbb{P}\{X_j = j\} = \frac{1}{j}$ and $P\{X_j = 0\} = 1 - \frac{1}{j}$. Then $\lim_{n\to\infty} P\{\frac{1}{n}\sum_{j=1}^{n} \le s\} = e^{-\gamma} \int_0^s \rho(t)dt$. It is also worth mentioning the nice form of the characteristic function (inverse Fourier transform) of $e^{-\gamma}\rho(t)$, namely $\phi(\tau) = \exp\Big(\int_0^1 \frac{e^{i\tau v}-1}{v} dv\Big)$.

Another result of [13] is that in the ensemble Ω_m, the number of distinct prime divisors of n (that is $\omega(n) = \sum_{j=1}^{m} \nu_j$) satisfies an Erdős-Kac central limit theorem with expectation and variance $\log\log m$. More precisely, for every $a, b \in \mathbb{R}$, $a \le b$, we have

$$\lim_{m\to\infty} \mathbb{P}_m \left\{ n \in \Omega_m : a \le \frac{\omega(n) - \log\log m}{\sqrt{\log\log m}} \le b \right\} = \frac{1}{\sqrt{2\pi}} \int_a^b e^{-x^2/2} dx.$$

It is natural to ask whether the results of [13] can be generalized to include a larger class of measures other than $\mathbb{P}_m$, such as signed or complex measures. The case of signed measures was considered by M. Avdeeva and D. Li and Ya.G. Sinai in [3]. They slightly modify the definition of the Ω_m to consist of all of all odd square-free integers of the form $n = p_1^{\nu_1} p_2^{\nu_2} \cdots p_m^{\nu_m}$ with $\nu_i \in \{0, 1\}$, where $p_1 < p_2 < \ldots < p_m$ be the first m primes larger than 2. The set Ω_m is then equipped with the measure given by $\mathbb{P}_m(\{n\}) = c_m(-2)^{\omega(n)}/n$, where $\omega(n) = \sum_{j=1}^{m} \nu_j$ and the choice for $c_m = (\log m)^{-2}$ comes from the fact that $\mathbb{P}_m(\Omega_m) = O(1)$. Avdeeva, Li, Sinai show the following local limit theorem for $\omega(n)$:

$$\mathbb{P}_m\{n \in \Omega_m : \omega(n) = k\} = (-1)^k \sqrt{\frac{1}{4\pi \log\log m}} e^{-\frac{(k-2\log\log m)^2}{4\log\log m} + B_m} + \varepsilon_{k,m},$$

where $B_m = \sum_{2<p<m,\ p \text{ prime}} \log(1 + 2/p) - 2\log\log m$ (which is uniformly bounded in m) and $\varepsilon_{k,m} = O(\log\log\log m / \log\log m)$ uniformly in m and k. This shows an almost Gaussian distribution for the measure of the set where $\omega(n) = k$, except for the factor $(-1)^k$. If one considers the cases of even k and odd k separately,

then one gets, for every $a, b \in \mathbb{R}$, $a < b$,

$$\mathbb{P}_m\left\{n \in \Omega_n : \omega(n) \text{ is even and } a \le \frac{\omega(n) - 2\log\log m}{\sqrt{2\log\log m}} \le b\right\} = \frac{C}{\sqrt{2\pi}}\int_a^b e^{-x^2/2}dx,$$

where $C = \lim_{m\to\infty} e^{B_m}$. The analog statement when $\omega(n)$ is odd is also true if we replace C by $-C$. In this setting, it still makes sense to consider $\xi_m(n)$ such that $n = p_m^{\xi_m(n)}$ and ask what is the limiting distribution of $\xi_m \in [0, \infty)$ as $m \to \infty$. Recall that for the probability measure studied in [13] and [15] the limiting distribution was given by the Dickman-de Bruijn probability distribution. In the case of signed measures considered by Avdeeva, Li, and Sinai, the limiting distribution is an explicit tempered distribution involving first and second derivatives of δ functions. The same phenomenon occurs when one studies complex-valued measures P_m, as done by F. Cellarosi [11]. He extended the results in [13] to complex measures on ensembles of k-free integers, and obtained an explicit error term. F. Cellarosi extended the results in [13] to complex measures [11] for k-free integers and, assuming some regularity for the test functions, obtained a general limit theorem for the distribution of ξ_m as $m \to \infty$, with an explicit error term.

The understanding of the autocorrelations for the function $\mu(n)$ (described by Chowla's conjecture) is still far from being complete, although a number of remarkable results have appeared recently. For example, K. Matomäki, M. Radziwiłł, and T. Tao [50] proved in 2015 that for every positive integer k and every $10 \le H \le N$ we have

$$\sum_{h_1,\dots,h_k\le H}\left|\sum_{n\le N}\mu(n+h_1)\cdots\mu(n+h_k)\right| = O\left(k\left(\frac{\log\log H}{\log H} + \frac{1}{\log^{1/3000} N}\right)H^{k-1}N\right).$$

The same year, E.H. El Abdalaoui and X. Ye [25] independently proved that for every $\varepsilon > 0$

$$\sum_{h\le N}\left|\sum_{n\le N}\mu(n)\mu(n+h)\right| = O_\varepsilon\left(\frac{N^2}{(\log N)^\varepsilon}\right).$$

On the other hand, the situation is fully understood for the square of the Möbius function $\mu^2(n)$, i.e. the indicator of square-free integers. In 2013, F. Cellarosi and Ya.G. Sinai [14] considered the r-point correlation functions for the sequence $(\mu^2(n))_{n\ge1}$, namely

$$c_r(h_1, \dots, h_r) = \lim_{N\to\infty}\frac{1}{N}\sum_{n\le N}\mu^2(n+h_1)\mu^2(n+h_2)\cdots\mu^2(n+h_r),$$

which were first studied by L. Mirsky in 1949. For example $c_1(0) = 1/\zeta(2) = 6/\pi^2$, $c_2(0,1) = \prod_{p \text{ prime}}(1 - 2/p^2) \approx 0.3226340989$, and $c_4(0,1,2,3) = 0$.

They use these correlation functions to construct a probability measure Π on the space of binary sequences $X = \{0, 1\}^{\mathbb{Z}}$. For every positive integer r and every integers $k_0 < k_1 < \ldots < k_r$, they set

$$\Pi\{x \in X : x(k_0) = x(k_1) = \ldots = x(k_r) = 1\} := \zeta(2)\, c_r(0, k_1 - k_0, k_2 - k_0, \ldots, k_r - k_0).$$

This defines the measure of arbitrary cylinders (subsets of X in which finitely many coordinates are fixed) and determines uniquely a probability measure on X, which is invariant under the shift $T : X \to X$, $Tx = x'$, $x'(n) = x(n+1)$. In other words, Cellarosi and Sinai construct a dynamical system (X, Π, T) which encodes all the statistics of the number-theoretical sequence $\mu^2(n)$. The main result of [14] is that the dynamical system (X, Π, T) has pure point spectrum, and is isomorphic to the an ergodic translation on the compact abelian group

$$G = \prod_{p \text{ prime}} \mathbb{Z}/p^2\mathbb{Z},$$

equipped with the normalized Haar measure. The ergodic translation on G is given by $\tau : g \mapsto g + (1, 1, 1, \ldots)$, where the first coordinate is considered modulo 4, the second coordinate modulo 9, etc. Dynamical systems of this kind are often referred to as "Kronecker systems". The surprising fact is that the group G can be obtain using only the second correlation function $c_2(h_1, h_2) = c(0, h_2 - h_1)$. In fact, the sequence $(c_2(0, h))_{h \geq 0}$ is positive definite and, by Bochner-Herglotz theorem, $c_2(0, h)$ is the h-th Fourier coefficient of a probability measure ν on the circle $\mathbb{S}^1$. This measure is then shown to be atomic, supported on the "rational" points of $\mathbb{S}^1$ of the form $e^{2\pi i l/d^2}$ where $0 \leq l \leq d^2 - 1$ and d is a square-free integer. The set of such points (i.e. the support of the measure ν) is a discrete group, and its Pontryagin dual is precisely the compact abelian group G.

The results of [14] were independent of an earlier paper by M. Baake, R. V. Moody, and P. A.B. Pleasants [5] in which the spectral measure ν corresponding to the second correlation function for square-free integers had been computed.

The main theorem of [14] strengthen a result by P. Sarnak. He defines a subset A of $\mathbb{Z}$ to be admissible if its reduction modulo p^2 does not cover all of the residue classes modulo p^2 for every prime p. He then considers the set $\mathscr{A}$ of sequences $x \in X = \{0, 1\}^{\mathbb{Z}}$ such that $\{k \in \mathbb{Z} : x(k) \neq 0\}$ is admissible. Sarnak proved [57] that $\mathscr{A}$ coincides with the closure in X of the orbit of the sequence $(\mu^2(n))_n$ under the shift T; furthermore, the topological dynamical system $(\mathscr{A}, T)$—which is a subshift not of finite type—has topological entropy equal to $\frac{6}{\pi^2} \log 2$, is proximal, has no nontrivial Kronecker factors, and has a nontrivial joining (in the sense of Furstenberg, [35]) with (G, τ). Sarnak also defines a shift-invariant probability measure m on $\mathscr{A}$ and proves that the dynamical system $(\mathscr{A}, m, T)$ is a factor of (G, Haar, τ). In particular $(\mathscr{A}, m, T)$ cannot be weak-mixing. It turns out that $(\mathscr{A}, m, T)$ is isomorphic to the dynamical system (X, Π, T) constructed by

Cellarosi and Sinai and, by the main theorem of [14], is isomorphic to the Kroncker system (G, Haar, τ).

More generally, it follows from the main theorem in [14] is that the sequence $(\mu^2(n))_n$ is a typical realization of an ergodic dynamical system with zero Kolmogorov–Sinai entropy, and no mixing properties. In other words, the function $\mu^2(n)$ is almost periodic. This result shows that the randomness in the Möbius function $\mu(n)$ does not come from the locations of zeros and non-zeros (which are described by $\mu^2(n)$), but only from the parity of the number of prime divisors of square-free n's. As pointed out by E.H. El Abdalaoui and M. Disertori [26], the results of [14] and [57] can be rephrased as follows: let (X, m, T) be a uniquely weakly mixing dynamical system, then for every continuous function $f : X \to \mathbb{C}$ with $\int_X f(x)dm(x) = 0$ and every $x \in X$ we have

$$\sum_{n\leq N} f(T^n x)\mu^2(n) = o(N)$$

as $N \to \infty$.

The results in [14] have been generalized to k-free integers in an arbitrary number fields by F. Cellarosi and I. Vinogradov [16]. They showed, for instance, that the two dimensional array obtained by considering the indicator of square-free Gaussian integers in $\mathbb{Z}[i]$ is a typical realization of an ergodic $\mathbb{Z}^2$-action with pure point spectrum on a compact abelian group. In 2013, P.A.B. Pleasants and C. Huck [54] studied the statistics (and entropy) of k-free points in an arbitrary lattice, using a geometric notion of k-freeness that agrees with the one considered in [14] when the lattice is $\mathbb{Z}$.

Another generalization of square-free numbers is given by $\mathscr{B}$-free numbers, that is integers that are not divisible by any of the elements of $\mathscr{B} = \{b_1, b_2, b_3, \ldots\}$, where the b_i's are pairwise relatively prime integers greater than and such that $\sum_{i=1}^{\infty} 1/b_i < \infty$. These integers were introduced in 1966 by P. Erdős [30] and reduce to square-free integers when $\mathscr{B}$ consists of the squares of the primes. The work of Sarnak [57] and Cellarosi and Sinai [14] has been extended in 2015 by E.H. El Abdalaoui, M. Lemańczyk, and T. De la Rue [29] to $\mathscr{B}$-free integers. The paper [29] also discusses a remarkable result, namely that the statistics obtained in the Mirsky-like correlation functions (where one averages over the interval $[1, N]$) can be achieved by averaging over rather short intervals of the form $[N, N + \sqrt{N})$. This partially answers a question by P. Erdős, who conjectured that for every $c > 0$, the interval $[N, N + N^c)$ always contains at least one $\mathscr{B}$-free integer, for large enough N. Moreover, the authors of [29] show that Chowla's conjecture is equivalent to a genericity condition for the sequence $(\mu(n))_{n\geq 1}$ with respect to the so-called "completely random" extension of the $\mathscr{B}$-free analogue of the zero entropy measure m considered by Sarnak in [57] (or the measure Π considered by Cellarosi and Sinai in [14]). Some recent progress on the statistics of $\mathscr{B}$-free integers in short intervals is due to and K. Matomäki [48] and M. Avdeeva [2]. In particular, Avdeeva is able to find find the asymptotic growth of the variance for $\mathscr{B}$-free integers in short intervals of the form $[x, x + N)$, where N is fixed, $1 \leq x \leq X$ and $X \to \infty$. A consequence of her work is a partial improvement of Matomäki's estimate on the number of short intervals containing no $\mathscr{B}$-free integers. In Matomäki's approach, however, N is allowed to grow with X and in this setting her estimates are, to the best of our knowledge, the strongest available. It is worthwhile mentioning that [2] also includes estimates for the variance of k-free integers in arbitrary number fields, previously considered in [16].

As mentioned above, P. Sarnak showed that the topological dynamical system obtained as orbit closure of the sequence $(\mu^2(n))_n$, which is a subshift not of finite type of (X, T), has positive topological entropy. Moreover, it was proven that there exists a unique measure of maximal entropy, as shown by R. Peckner [53] and in a more general setting by J. Kułaga-Przymus, M. Lemańczyk and B. Weiss [45]. Moreover, it is shown in [53] that the Pinsker factor of the measure of maximal entropy is precisely the measure considered by F. Cellarosi and Ya.G. Sinai in [14]. This means that the dynamical system considered in [14] is a fundamental building-block in the study of the thermodynamical formalism of the topological shift associated to $\mu^2(n)$.

Remarkable progress has been made towards Sarnak's conjecture, and many classes of deterministic sequences have been shown to be orthogonal to the Möbius sequence. The works of B. Green and T. Tao [40], J. Bourgain, P. Sarnak and T. Ziegler [7], J. Liu and P. Sarnak [46], B. Green [39], S. Ferenzi, J. Kułaga-Przymus, M. Lemańczyk, and C. Mauduit, C. [32], S. Ferenczi and C. Mauduit [33], C. Mauduit and J. Rivat [51], E.H. El Abdalaoui, M. Lemańczyk, M. and T. de la Rue [28], are just a few of the remarkable achievements in this direction, some of which have been influenced by Ya.G. Sinai's work.

The intimate connection between Sarnak's and Chowla's conjectures has been analyzed from a dynamical point of view by E.H. El Abdalaoui, J. Kułaga-Przymus, M. Lemańczyk, and T. De la Rue [27]. Among many beautiful results, they highlight that their work and the main theorem of F. Cellarosi and Ya.G. Sinai [14] imply that if Chowla's conjecture holds for the Liouville function, then it must also hold for the Möbius function. This may seem quite obvious, given the similarities between the two functions. However, the recent work of K. Matomäki, M. Radziwiłł and T. Tao [49] shows how the sign patterns for the Liouville function may be easier to study than those of the Möbius function.

Amongst the very recent activity toward the relation between Sarnaks and Chowlas conjecture, we refer the reader to the works of T. Tao [64], A. Gomilko, D. Kwietniak, M. Lemanńczyk [36], and E.H. El Abdalaoui [43].

We believe that the interest of Ya.G. Sinai in problems at the intersection between number theory, probability theory, and the theory of dynamical systems has given extraordinary momentum to the research in all these fields. We are confident that the future mathematical endeavours of Ya.G. Sinai will be equally pivotal and inspiring for many generations of mathematicians to come.

References

1. V.I. Arnol'd. Topological and ergodic properties of closed 1-forms with incommensurable periods. *Funct. Anal. Appl.*, 25(2):81–90, 1991.
2. M. Avdeeva. Variance of $\mathscr{B}$-free integers in short intervals. *Preprint. arXiv:1512.00149.*
3. M. Avdeeva, D. Li, and Ya.G. Sinai. New Erdős–Kac type theorems for signed measures on square-free integers. *J. Stat. Phys.*, 154(1–2):327–333, 2014.
4. M. Avdeeva, D. Li, and Ya.G. Sinai. New limiting distributions for the Möbius function. *Mosc. J. Comb. Number Theory*, 4(1):3–51, 2014.
5. M. Baake, R.V. Moody, and P.A.B. Pleasants. Diffraction from visible lattice points and kth power free integers. *Discrete Math.*, 221(1–3):3–42, 2000.
6. M.V. Berry and J. Goldberg. Renormalisation of curlicues. *Nonlinearity*, 1(1):1–26, 1988.
7. J. Bourgain, P. Sarnak, and T. Ziegler. Disjointness of Moebius from horocycle flows. In H.M. Farkas, R.C. Gunning, M.I. Knopp, and B.A. Taylor, editors. *From Fourier Analysis and Number Theory to Radon Transforms and Geometry*, volume 28 of *Dev. Math.*, pages 67–83. Springer, New York, 2013.
8. J. Burgain and Ya.G. Sinai. Limit behavior of large Frobenius numbers. *Russ. Math. Surv.*, 62(4):713–725, 2007.
9. F. Cellarosi. Renewal-type limit theorem for continued fractions with even partial quotients. *Ergodic Theory Dynam. Systems*, 29(5):1451–1478, 2009.
10. F. Cellarosi. Limiting curlicue measures for theta sums. *Ann. Inst. Henri Poincaré Probab. Stat.*, 47(2):466–497, 2011.
11. F. Cellarosi. Smooth sums over smooth k-free integers and statistical mechanics. *Mosc. J. Comb. Number Theory*, 3(2):24–60, 2013.
12. F. Cellarosi and J. Marklof. Quadratic Weyl sums, automorphic functions, and invariance principles. *Preprint. arXiv:1501.07661.*
13. F. Cellarosi and Ya.G. Sinai. The Möbius function and statistical mechanics. *Bull. Math. Sci.*, 1(2):245–275, 2011.
14. F. Cellarosi and Ya.G. Sinai. Ergodic properties of square-free numbers. *J. Eur. Math. Soc. (JEMS)*, 15(4):1343–1374, 2013.
15. F. Cellarosi and Ya.G. Sinai. Non-standard limit theorems in number theory. In *Prokhorov and Contemporary Probability Theory*, volume 33 of *Springer Proc. Math. Stat.*, pages 197–213. Springer, Heidelberg, 2013.
16. F. Cellarosi and I. Vinogradov. Ergodic properties of square-free integers in number fields. *Journal of Modern Dynamics*, 7(3):461–488, 2014.
17. S. Chowla. *The Riemann Hypothesis and Hilbert's Tenth Problem (4th ed.).* Gordon & Breach, New York, 1965.
18. E.A. Coutsias and N.D. Kazarinoff. Disorder, renormalizability, theta functions and Cornu spirals. *Physica D: Nonlinear Phenomena*, 26(1–3):295–310, 1987.
19. E.A. Coutsias and N.D. Kazarinoff. On Cornu spirals. Disorder, self-similarity, and Jacobi's $\theta_3(v,\tau)$. In *Bifurcation: Analysis, Algorithms, Applications (Dortmund, 1986)*, volume 79 of *Internat. Schriftenreihe Numer. Math.*, pages 139–152. Birkhäuser, Basel, 1987.
20. H. Davenport. On some infinite series involving arithmetical functions (II). *Quart. J. Math.*, 8:313–320, 1937.
21. F.M. Dekking and M. Mendès France. Uniform distribution modulo one: a geometrical viewpoint. *J. Reine Angew. Math.*, 329:143–153, 1981.
22. E. Dimitrov and Y. Sinai. On a Problem in Diophantine Approximation. *Preprint. arXiv1305.1200.*
23. E.I. Dinaburg and Ya.G. Sinaĭ. Statistics of the solutions of the integer-valued equation $ax - by = \pm 1$. *Funct. Anal. Appl.*, 24(3):165–171, 1990.
24. D.I. Dolgopyat. On the distribution of the minimal solution to a linear diophantine equation with random coefficients. *Funct. Anal. Appl.*, 28(3):168–177, 1994.

25. E. H. El Abdalaoui and X. Ye. A cubic nonconventional ergodic average with Möbius and Liouville weight. *Preprint. arXiv:1504.00950.*
26. E.H. El Abdalaoui and M. Disertori. Spectral properties of the Möbius function and a random Möbius model. *Stochastics and Dynamics*, 16(01):1650005, 2016.
27. E.H. El Abdalaoui, J. Kułaga-Przymus, M. Lemańczyk, and T. de la Rue. The Chowla and the Sarnak conjectures from ergodic theory point of view. *Preprint. arXiv:1410.1673.*
28. E.H. El Abdalaoui, M. Lemańczyk, and T. de la Rue. Automorphisms with quasi-discrete spectrum, multiplicative functions, and average orthogonality along short intervals. *Preprint. arXiv:1507.04132.*
29. E.H. El Abdalaoui, M. Lemańczyk, and T. de la Rue. A dynamical point of view on the set of $\mathscr{B}$-free integers. *Int. Math. Res. Not.*, 2015(16):7258–7286, 2015.
30. P. Erdős. On the difference of consecutive terms of sequences defined by divisibility properties. *Acta Arith.*, 12:175–182, 1966.
31. A.A. Fedotov and F. Klopp. Renormalization of exponential sums and matrix cocycles. *Séminaire Équations aux dérivées partielles. Centre de mathématiques Laurent Schwartz, École polytechnique*, pages 1–10, 2004–2005.
32. S. Ferenczi, J. Kułaga-Przymus, M. Lemańczyk, and C. Mauduit. Substitutions and Möbius disjointness. In by J. Auslander, A. Johnson, and C.E. Silva, editors. *Ergodic Theory, Dynamical Systems, and the Continuing Influence of John C. Oxtoby*. Contemporary Mathematics, vol. 678, pp. 151–173. American Mathematical Society, 2016.
33. S. Ferenczi and C. Mauduit. On Sarnak's conjecture and Veech's question for interval exchanges. *J. Anal. Math.*, 134(2):543–573, 2018.
34. A. Fujii. On a problem of Dinaburg and Sinaĭ. *Proc. Japan Acad. Ser. A Math. Sci.*, 68(7): 198–203, 1992.
35. H. Furstenberg. Disjointness in ergodic theory, minimal sets, and a problem in Diophantine approximation. *Math. Systems Theory*, 1:1–49, 1967.
36. A. Gomilko, D. Kwietniak, and M. Lemańczyk. Sarnak's conjecture implies the Chowla conjecture along a subsequence *Preprint. arXiv:1710.07049.*
37. V.L. Goncharov. From the realm of combinatorics (in Russian). *Izv. Akad. Nauk SSSR, ser. matem., 8*, 1:3–48, 1944.
38. A. Granville. Smooth numbers: computational number theory and beyond. In J.P. Buhler and P. Stevenhagen, editors. *Algorithmic Number Theory: Lattices, Number Fields, Curves and Cryptography*, volume 44 of *Math. Sci. Res. Inst. Publ.*, pages 267–323. Cambridge Univ. Press, Cambridge, 2008.
39. B. Green. On (not) computing the Möbius function using bounded depth circuits. *Combinatorics, Probability and Computing*, 21:942–951, 11 2012.
40. B. Green and T. Tao. The Möbius function is strongly orthogonal to nilsequences. *Ann. of Math. (2)*, 175(2):541–566, 2012.
41. G.H. Hardy and J.E. Littlewood. Some problems of diophantine approximation. I. The fractional part of $n^k\Theta$. II. The trigonometrical series associated with the elliptic ϑ-functions. *Acta Math.*, 37:155–191, 193–239, 1914.
42. G.H. Hardy and J.E. Littlewood. Some problems of diophantine approximation—a series of cosecants. *Bulletin of Calcutta Mathematical Society*, 20(1):251–267, 1928.
43. E.H. Houcein el Abdalaoui. On Veech's proof of Sarnak's theorem on the Möbius flow. *Preprint. arXiv:1711.06326.*
44. A.V. Kochergin. Nondegenerate saddles and the absence of mixing in flows on surfaces. *Proc. Steklov Inst. Math.*, 256:238–252, 2007.
45. J. Kułaga-Przymus, M. Lemańczyk, and B. Weiss. On invariant measures for $\mathscr{B}$-free systems. *Proc. Lond. Math. Soc. (3)*, 110(6):1435–1474, 2015.
46. J. Liu and P. Sarnak. The Möbius function and distal flows. *Duke Math. J.* 164, no. 7, 1353–1399, 2015.
Preprint. arXiv:1303.4957.
47. J. Marklof. The asymptotic distribution of Frobenius numbers. *Invent. Math.*, 181(1):179–207, 2010.

48. K. Matomäki. On the distribution of -free numbers and non-vanishing Fourier coefficients of cusps forms. *Glasg. Math. J.*, 54:381–397, 5 2012.
49. K. Matomäki, M. Radziwiłł, and T. Tao. Sign patterns of the Liouville and Möbius functions. *Forum Math. Sigma*, 4:e14, 44 pp, 2016.
50. K. Matomäki, M. Radziwiłł, and T. Tao. An averaged form of Chowla's conjecture. *Algebra Number Theory*, 9(9):2167–2196, 2015.
51. C. Mauduit and J. Rivat. Prime numbers along Rudin–Shapiro sequences. *J. Eur. Math. Soc. (JEMS)*, 17(10):2595–2642, 2015.
52. L.J. Mordell. The approximate functional formula for the theta function. *J. London Mat. Soc.*, 1:68–72, 1926.
53. R. Peckner. Uniqueness of the measure of maximal entropy for the squarefree flow. *Israel Journal of Mathematics*, 210(1):335–357, 2015.
54. P.A.B. Pleasants and C. Huck. Entropy and diffraction of the k-free points in n-dimensional lattices. *Discrete Comput. Geom.*, 50(1):39–68, 2013.
55. J.L. Ramírez Alfonsín. *The Diophantine Frobenius Problem*, volume 30 of *Oxford Lecture Series in Mathematics and its Applications*. Oxford University Press, Oxford, 2005.
56. V.A. Rokhlin. Exact endomorphisms of a Lebesgue space (in Russian). *Izv. Akad. Nauk SSSR Ser. Mat.*, 25:499–530, 1961.
57. P. Sarnak. Three lectures on the Möbius function randomness and dynamics (Lecture 1). http://publications.ias.edu/sarnak/paper/506.
58. F. Schweiger. Continued fractions with odd and even partial quotients. *Arbeitsberichte Math. Institut Universität Salzburg*, 4:59–70, 1982.
59. V. Shchur, Ya. Sinai, and A. Ustinov. Limiting distribution of Frobenius numbers for $n = 3$. *J. Number Theory*, 129(11):2778–2789, 2009.
60. Ya.G. Sinai. Limit theorem for trigonometric sums. Theory of curlicues. *Russ. Math. Surv.*, 63(6):1023–1029, 2008.
61. Ya.G. Sinai and K.M. Khanin. Mixing for some classes of special flows over rotations of the circle. *Funct. Anal. Appl.*, 26(3):155–169, 1992.
62. Ya.G. Sinai and C. Ulcigrai. A limit theorem for Birkhoff sums of non-integrable functions over rotations. In *Geometric and Probabilistic Structures in Dynamics*, volume 469 of *Contemp. Math.*, pages 317–340. Amer. Math. Soc., Providence, RI, 2008.
63. Ya.G. Sinai and C. Ulcigrai. Renewal-type limit theorem for the Gauss map and continued fractions. *Ergodic Theory Dynam. Systems*, 28(2):643–655, 2008.
64. T. Tao. Equivalence of the logarithmically averaged Chowla and Sarnak conjectures. In C. Elsholtz and P. Grabner (editors), *Number Theory – Diophantine Problems, Uniform Distribution and Applications*, pages 391–421. Springer, 2017.
65. C. Ulcigrai. *Ergodic properties of some area-preserving flows*. ProQuest LLC, Ann Arbor, MI, 2007. Thesis (Ph.D.)–Princeton University.
66. C. Ulcigrai. Mixing of asymmetric logarithmic suspension flows over interval exchange transformations. *Ergodic Theory Dynam. Systems*, 27(3):991–1035, 2007.
67. C. Ulcigrai. Absence of mixing in area-preserving flows on surfaces. *Ann. of Math. (2)*, 173(3):1743–1778, 2011.
68. A.V. Ustinov. On the statistical properties of elements of continued fractions. *Dokl. Math.*, 79(1):87–89, 2009.
69. A.V. Ustinov. On the distribution of Frobenius numbers with three arguments. *Izv. Math.*, 74(5):1023–1049, 2010.
70. J.R. Wilton. The approximate functional formula for the theta function. *J. London Math. Soc.*, 2(3):177–180, 1927.

Entropy Theory of Dynamical Systems

B. Gurevich

Abstract This section is devoted to Sinai's advances in the entropy theory of dynamical systems and to some developments of his ideas. A history of dynamical entropy is also represented. When describing several events of this history, author's personal recollection is partially used.

Sinai's participation in the creation and development of the entropy theory of dynamical systems was the first direction of his scientific activity that made his name renowned to many mathematicians as early as he was a student. A number of reviews of this field has been published since then. Some of them, especially [18] and [63] were used when writing this section.

This area of ergodic theory the foundations of which were laid by Andrei Nikolaevich Kolmogorov had a dramatic impact on ergodic theory and, more generally, on theory of dynamical systems as a whole. Multiple completely new problems arisen right after introducing the concept of entropy at once attracted to this field several gifted young mathematicians. One of them was Kolmogorov's student Yasha Sinai.

1 Prehistory of Dynamical Entropy (Shannon, Khinchin, Kolmogorov)

In 1948, Claude Shannon published his famous paper "Mathematical theory of communications" [45], where he proposed the quantity

$$H(\mathbf{p}) = -\sum_{i=1}^{n} p_i \log p_i \tag{1}$$

B. Gurevich (✉)
Moscow State University, Moscow, Russia
e-mail: gurevich@mech.math.msu.su

H. Holden, R. Piene (eds.), *The Abel Prize 2013–2017*,
https://doi.org/10.1007/978-3-319-99028-6_9

as a measure of uncertainty associated with a random experiment with n outcomes whose probabilities are $p_1, \ldots, p_n$ (Shannon used the base 2 logarithms, but the base is of no importance). Shannon points out that the same expression for entropy was known in physics—he mentioned Ludwig Boltzmann—but had another meaning. In the mathematical theory of information transmission (or simply information theory) developed by Shannon, the entropy became a very important concept.

Shannon's theory quickly attracted particular interest among specialists in a variety of sciences, in particular, mathematicians. In the Soviet Union, a majority of them grouped around A.N. Kolmogorov and A.Ya. Khinchin. In 1953, Shannon's paper [45] was published in Russian (under another title and with some omissions), and in the same year A.Ya. Khinchin [22] proved rigorously that a few natural properties of entropy determine it uniquely up to a positive factor to be given by the expression (1). Starting from the mid-1950s, Kolmogorov popularized information-theoretic ideas in numerous papers and speeches at scientific meetings, and at about that time he planned for extensive studies in different fields of mathematics with the use of these ideas.

It was repeatedly observed that the decade from 1950 to 1960 was one of the most fruitful periods in Kolmogorov's scientific activity: suffice it to say that Hilbert's 13th problem was solved and the KAM theory was founded just in this period. Kolmogorov worked hard on problems of classical mechanics. According to his own words, he was motivated by John von Neumann's works on spectral theory of dynamical systems and the results by Bogoliubov and Krylov in topological dynamics. In spite of all these facts, it appeared absolutely surprising that Kolmogorov decided to apply the concept of entropy to dynamical systems.

2 Emergence of Dynamical Entropy

The first presentation of Kolmogorov's new concept arose during the course of lectures on dynamical systems he delivered at the Mechanics and Mathematics Department of Moscow State University in the fall 1957. The lectures remained unpublished, and all we know is based on the memories of Kolmogorov's students. Yasha Sinai became a graduate student just at that time, and he witnessed [54, 61] that at one lecture, Kolmogorov unexpectedly introduced a new notion, namely the entropy of a Bernoulli shift, and proved some its properties.

Kolmogorov's approach stated in [23] concerned itself with much more general class of dynamical systems, and it turned out to be quite different. From the very beginning, Kolmogorov uses Rohlin's theory [39] of Lebesgue spaces and their measurable partitions. According to this theory, in a Lebesgue space (a universal example is the interval [0,1] with the Lebesgue measure), there is one-to-one correspondence between the σ-subalgebras of the σ-algebra of measurable sets and measurable partitions. More precisely, one should deal with classes of sets and partitions which coincide up to a set of zero measure (in short mod 0). The above-mentioned approach is as follows (our notation is somewhat different from that in [23]). Let $(M, \mathcal{M}, \mu)$ be a Lebesgue space, where $\mathcal{M}$ is the σ-algebra of measurable sets and μ a probability measure without atoms (below we often omit $\mathcal{M}$ from the notation). Kolmogorov first introduces the joint conditional information of measurable partitions α and γ, given a measurable partition β, and its particular case (when $\gamma = \alpha$), the *conditional entropy of α given β*. It is the random variable $H_x(\alpha|\beta)$ that is defined at a point $x \in M$ via formula (1): if $\alpha = (A_1, A_2, \dots)$ is a countable partition, one puts $p_i := \mu(A_i|C_x)$, where C_x is the atom of β containing x. The integral of $H_x(\alpha|\beta)$ in x (w.r.t. μ) is denoted by $H(\alpha|\beta)$ and called the *mean conditional entropy* of α given β (now one usually omits the word 'mean'). In fact, Kolmogorov defines $H(\alpha|\beta)$ for all measurable partitions α, β or, equivalently, for all σ-subalgebras of the basic σ-algebra $\mathcal{M}$. Then he proves the main properties of the quantities introduced, and turns in the next section to dynamical systems.

He considers a one-parameter group $\{T^t\}$ of measure-preserving transformations of M and introduces a new concept that in the future will play an outstanding part in entropy theory. Namely, he calls $\{T^t\}$ a quasi-regular dynamical system if there exists a σ-algebra $\mathcal{A} \subset \mathcal{M}$ such that (a) $T^t\mathcal{A} \supseteq \mathcal{A}$ for $t \geq 0$; (b) the minimal σ-algebra containing $\cup_t T^t\mathcal{A}$ is $\mathcal{M}$; (c) $\cap_t T^t\mathcal{A}$ is trivial in the sense that every set in this σ-algebra has measure 0 or 1.

Quasi-regularity simulates the regularity property of some stationary processes that gradually forget their past, in which case $\mathcal{A}$ is the collection of events observed until time 0, and T^t is the shift by t in the trajectory space of the process.

The term 'quasi-regularity' was soon abandoned by Kolmogorov's followers in favor of 'K-property', 'K-mixing' or 'Kolmogorov's mixing' and so on, while the systems with this property received the name of 'K-systems'.

Kolmogorov states that for every K-algebra $\mathcal{A}$ (a σ-algebra with properties (a)–(c)) and for every $s \geq 0$, we have $H(T^{t+s}\mathcal{A}|T^t\mathcal{A}) = hs$, where $h \in [0, \infty]$ is independent of $\mathcal{A}$. This h was called the *entropy of the dynamical system* $\{T^t\}$.

However, it was soon discovered by V.A. Rohlin that Kolmogorov's proof of the independence of h from $\mathcal{A}$ was incorrect. Curiously, some other famous mathematicians made later similar mistakes.

Kolmogorov was deeply distressed with his error, presumably the only erroneous statement in his publications by then, and was striving to correct it as soon as possible. In late 1959 he submitted a new paper [24]. Here he deals with the discrete time systems (automorphisms) only, and considers the countable measurable partitions $\alpha = (A_1, A_2, \dots)$ such that

$$H(\alpha) := -\sum_i \mu(A_i) \log \mu(A_i) < \infty$$

and α is generating (or a generator) in the sense that the sets $T^n A_i$, $n \in \mathbf{Z}$, $i = 1, 2, \dots$, generate the σ-algebra $\mathcal{M}$. He proves that for such α the limit

$$\lim_{t\to\infty} n^{-1} H(\alpha_T^{n-1}) =: h(T, \alpha),$$

where

$$H(\alpha_T^{n-1}) := H(\alpha \vee T^1\alpha \vee \cdots \vee T^{n-1}\alpha), \tag{2}$$

is independent of α, and now one refers to this joint value of $h(T,\alpha)$ as $h(T)$, the *entropy of the automorphism* $T := T^1$. By definition the partition α_T^{n-1} is formed by the sets $A_{i_0} \cap T^1 A_{i_1} \cap \ldots T^{n-1} A_{i_{n-1}}$, while the existence of $h(T,\alpha)$ was already mentioned by Shannon in other terms.

Therefore, Kolmogorov [24] defined the entropy for all automorphisms that have a generator of finite entropy. Let us note that from the very beginning, the definitions of the entropy were stated in such a way as to make this quantity invariant under isomorphisms (the dynamical systems $\{T^t\}$ and $\{\tilde{T}^t\}$ acting in the spaces (M,μ) and $(\tilde{M},\tilde{\mu})$, respectively, are isomorphic if there exists a mod 0 isomorphism $\varphi : (M,\mu) \to (\tilde{M},\tilde{\mu})$ that takes T^t to $\tilde{T}^t$ for each t; if t is discrete, it suffices to have this for $t = 1$). The entropy from [24] is clearly invariant, which enabled Kolmogorov to solve a long-stated problem: he showed that *two Bernoulli schemes with distributions* $\mathbf{p}$ *and* $\mathbf{p}'$ *generate the non-isomorphic shifts (Bernoulli shifts) in the corresponding sequence spaces if* $H(\mathbf{p}) \neq H(\mathbf{p}')$. It is necessary to note that this was already stated in [23] and even in Kolmogorov's lectures mentioned above.

However, at that time Kolmogorov already knew that there existed a more general definition of the entropy suggested by Ya. Sinai. Kolmogorov recommended Sinai's paper [46] for publication in 'Doklady Akademii Nauk SSSR' ('Proceedings of the USSR Academy of Sciences') the same day as he submitted [24], and both papers appeared in the same issue of the journal.

Sinai deals with an automorphism T of a Lebesgue space (M,μ) and finite partitions α of M into measurable sets. He defines the entropy of T by

$$h(T) = \sup_\alpha h(T,\alpha), \tag{3}$$

where the supremum is taken over all α as above ($h(T)$ is also said to be the measure-theoretic or metric entropy of T with respect to μ). Then he proves the following important theorem.

Theorem 1 *If* $\alpha = (A_1,\ldots,A_k)$ *and* $\beta = (B_1,\ldots,B_l)$ *are finite partitions such that every* B_j *belongs* mod 0 *to the* σ*-algebra generated by the sets* $T^n A_i$, $i = 1,\ldots,k$, $n \in \mathbf{Z}$, *then* $h(T,\beta) \le h(T,\alpha)$.

This immediately implies that $h(T) = h(T,\alpha)$ if α is a generator, and a similar (and even simpler) argument proves that $h(T) = 0$ if T has a one-sided generator, i.e., a partition $\alpha = (A_1,\ldots,A_k)$ such that the sets $T^n A_i$, $i = 1,\ldots,k$, $n \in \mathbf{Z}_+$ generate mod 0 the basic σ-algebra in M.

The advantages of these general results and such an approach were demonstrated in this paper. Sinai considers a group automorphism T of a 2-dimensional torus, given by a 2×2 matrix $\hat{T}$ with integer entries and $\det \hat{T} = \pm 1$. The Lebesgue measure is T-invariant, and T is ergodic if and only if $\hat{T}$ has a real eigenvalue λ

with $|\lambda| > 1$. It is proved in detail in [46] that $h(T) = \log|\lambda|$. Therefore two ergodic automorphisms of a 2-torus can be isomorphic only if the eigenvalues of the corresponding matrices coincide in absolute value. The multidimensional extension concerns the automorphisms T of an n-torus whose matrices $\hat{T}$ have only real eigenvalues λ_i and reads: $h(T) = \sum_{i:|\lambda_i|>1} \log|\lambda_i|$. This statement is claimed to have a proof similar to that for the 2-torus case.

Sinai's work [46] merited so detailed considerations here, because it not only contained the first definition of entropy applicable to all automorphisms of a Lebesgue space (in fact, of any probability space), but also gave a means for calculating the entropy for specific systems. Group automorphisms of tori were the first nontrivial examples.

Already at this stage one could have suspected that the entropy theory for systems with positive entropy must differ essentially from that for systems with zero entropy (like shifts on compact Abelian groups with Haar's measure). This prediction was confirmed later.

Upon concluding this section, I should mention one more character of the entropy drama. It became a custom in papers and books on entropy theory to mention that D. Arov, the 1957 final-year student of the Odessa State University, suggested in his handwritten thesis to use Shannon's entropy in the study of dynamical systems (this fact was mentioned for the first time in [24]). In fact, Arov introduced a quantity that he named the ε-entropy of a dynamical system with continuous time, but, for an automorphism T, it would read as follows. For every $\varepsilon \in (0, 1/2]$, *the* ε*-entropy of* T *is defined by* $h_\varepsilon(T) = \sup_{\mathfrak{A}(\varepsilon)} h(T, \alpha)$, *where* $\mathfrak{A}(\varepsilon)$ *is the family of measurable partitions with all atoms of measure* $\geq \varepsilon$. This definition remained unpublished for many years (see [6]). Arov had no means for evaluating his entropy, and he considered no examples. Clearly, $\lim_{\varepsilon \to 0} h_\varepsilon(T) = h(T)$, so that $h(T)$ in Sinai's definition is determined by the function $\varepsilon \mapsto h_\varepsilon(T)$. But, for ergodic automorphisms, the opposite is also true: *if* T_1, T_2 *are ergodic and* $h(T_1) = h(T_2)$, *then* $h_\varepsilon(T_1) = h_\varepsilon(T_2)$ *for all* ε. At the same time *for non-ergodic automorphisms, this is not the case.* It seems that these facts cannot be established using only elementary properties of the entropy (see below). The principal value of Arov's achievement is that he guessed that it is useful to take the supremum over partitions. A similar approach was later used in Sinai's definition. The same was earlier done for Bernoulli shifts in Kolmogorov's lecture mentioned above.

3 Early Development of Entropy Theory

Along with Sinai, a key role in this development was played by V.A. Rohlin and his student L.M. Abramov, and a little later by Kolmogorov's student M.S. Pinsker. In the same year, 1959, Rohlin [40] described some useful properties of the entropy. In particular, he introduced the following entropy metric ρ ('Rohlin's metric') in the space of partitions ξ with $H(\xi) < \infty$: $\rho(\xi, \eta) = H(\xi|\eta) + H(\eta|\xi)$, and proved that this space with ρ was a complete separable metric space and that

$|h(T, \xi) - h(T, \eta)| \le \rho(\xi, \eta)$. Also he observed that if $H(\xi) < \infty$, then

$$h(T, \xi) = h(\xi | \xi_T^-), \text{ where } \xi_T^- := \vee_{i=1}^{\infty} T^{-i}\xi. \tag{4}$$

This formula is very useful in evaluating the entropy of specific systems. Moreover, it can be applied to endomorphisms (measurable, but in general not one-to-one maps T such that $\mu(T^{-1}A) = \mu(A)$), and hence makes it possible to deal with the entropy for them as well.

Abramov published two important technical papers, [1] and [2], where he found the entropy of an induced automorphism as well as the entropy of the so-called suspension flow. The next year, Pinsker [38] showed that for every automorphism T of a Lebesgue space there exists a maximal factor-automorphism with zero entropy, i.e., a σ-algebra $\mathscr{A}^0(T)$ such that if α is a countable measurable partition with $H(\alpha) < \infty$, then $h(T, \alpha) = 0$ if and only if all atoms of α belong to $\mathscr{A}^0(T)$. Since then the σ-algebra $\mathscr{A}^0(T)$ is called the Pinsker algebra, and the measurable partition corresponding to it, is called the *Pinsker partition* of T and denoted by $\pi(T)$. If $h(T, \alpha) > 0$ for a given T and every nontrivial α, then T is referred to as an automorphism with *completely positive entropy*; both $\mathscr{A}^0(T)$ and $\pi(T)$ for such T are trivial, i.e., each set in $\mathscr{A}^0(T)$ has measure 0 or 1.

During that period entropy theory was progressing very rapidly. Rohlin and Sinai [44] investigated the σ-algebras (or partitions) that increase under the action of an automorphism T. They proved the following:

Theorem 2 *For every T there exists a measurable partition ξ such that (a) $T\xi \ge \xi$, i.e., each atom of $T\xi$ lies in an atom of ξ; (b) $T^n\xi$ tends in a natural sense to ε, the partition into individual points, as $n \to \infty$; (c) $T^{-n}\xi$ tends to $\pi(T)$; and (d) $H(T\xi|\xi) = h(T)$.*

They referred to such ξ as a *perfect partition*. If $h(T) > 0$, there exists partitions that have only some of the properties (a)–(d). For example, in (d), instead of equality, one can have $H(T\xi|\xi) < h(T)$ (the opposite strict inequality is impossible). If (a) and (b) hold, then $\lim_{n\to\infty} T^{-n} \ge \pi(T)$. By definition, a K-partition ξ satisfies (a), (b), and $\lim_{n\to\infty} T^{-n}\xi$ is the trivial partition. Hence $\pi(T)$ is trivial if T is a K-automorphism (which was earlier stated in [38]). Combined with the existence of a perfect partition, this implies that *the family of K-automorphisms coincides with that of automorphisms with completely positive entropy* (for automorphisms with finite generator this was also stated in [38]). Notice that contrary to what was written in [23], there can be K-partitions ξ with $H(T\xi|\xi) < h(T)$. Moreover, E. Lindenstrauss, Y. Peres and W. Schlag [28] showed much later that, for some dynamical systems, there exist K-partitions ξ with $H(T\xi|\xi) = h$ for every positive $h < h(T)$.

Immediately after emergence of the dynamical entropy, the following two questions came to the center of attention: how are the entropy and spectrum of a dynamical system, especially of a flow, related to each other, and are two Bernoulli shifts with equal entropy isomorphic?

3.1 Entropy and Spectrum

Already Kolmogorov [23] noted that every K-automorphism had the Lebesgue spectrum of infinite multiplicity and conjectured that the same might be true for the K-flows. Rohlin [39] added that if T is an automorphism of (M, μ) with $h(T) > 0$, then the unitary operator U_T defined by $U_T f(x) := f(Tx)$, $f \in L^2(M, \mu)$, $x \in M$, has an invariant subspace $L \subset L^2(M, \mu)$ where it has the Lebesgue spectrum of infinite multiplicity. (T as a whole is said to have such spectrum if L is the orthogonal complement of the one-dimensional subspace of constants.)

But for flows, the question turned out to be much more complicated. Kolmogorov's conjecture was proved by Sinai [50] who used the so called suspension representation, which can be assigned to every aperiodic flow. (This is an abstract form of the Poincaré section and the first return map, or Poincaré map.)

However, not every flow with the Lebesgue spectrum of infinite multiplicity has the K-property or positive entropy. The first example of this kind was the horocycle flow on a compact surface of constant negative curvature. By means of the method used by I. Gelfand and S. Fomin [10] in their study of the spectrum for the geodesic flow, O. Parasyuk [35] showed that the horocycle flow has the Lebesgue spectrum, while by the same method one can show that the spectral multiplicity is infinite. On the other hand, Sinai's conjecture that the entropy of the horocycle flow is zero, was proved in [15].

3.2 Isomorphism Problem: First Results

Shortly thereafter Kolmogorov's student L. Meshalkin [30] showed that two Bernoulli shifts are isomorphic if they have equal entropy and if all the probabilities that determine them, are of the form p^{-k_i}, where p is a positive integer, common for both systems, and k_i are arbitrary positive integers such that $\sum_i p^{-k_i} = 1$. The simplest case of such a situation is provided by the distributions $(1/8, 1/8, 1/8, 1/8, 1/2)$ and $(1/4, 1/4, 1/4, 1/4,)$ with entropy $2 \log 2$ (since then it is called Meshalkin's example). The isomorphism is constructed in the form of coding the realizations of a stationary process into realizations of another one, and care is taken that the shift and measure in the first space are mapped into the shift and measure in the second one, respectively. Some generalizations of Meshalkin's method were made afterward, but no general results emerged.

A completely different example of isomorphism was considered later by R. Adler and B. Weiss [4], who established that two ergodic group automorphisms of the 2-torus are isomorphic if they have the same entropy. For this they first proved that the entropy of such automorphism with respect to the Lebesgue measure is bigger than the entropy with respect to any other invariant measure. The latter fact fits naturally in the thermodynamic formalism as well (see Sect. 4).

But the first general result in the new isomorphism problem was obtained by Sinai [52] (see [55] for details) who suggested the concept of a *weak isomorphism* of dynamical systems. By definition it retains all properties of the isomorphism except that the mappings of Lebesgue spaces onto one another are now not necessarily invertible. Weakly isomorphic systems have the same spectrum and entropy, and, what is much more important, all Bernoulli shifts with equal entropy are weakly isomorphic. This spectacular theorem follows from another remarkable one.

Theorem 3 *If T is an ergodic automorphism of a Lebesgue space and $\mathscr{A}$ is a strictly increasing σ-algebra (such $\mathscr{A}$ exists if and only if $h(T) > 0$), then for any probability distribution $\mathbf{p} = (p_1, \ldots, p_k)$ with $H(\mathbf{p}) \leq H(T\mathscr{A}|\mathscr{A})$, there exists a partition $\alpha = (A_1, \ldots, A_k)$ such that $A_i \in \mathscr{A}$, $\mu(A_i) = p_i$ for all i, and $\{T^n\alpha, n \in \mathbf{Z}\}$ is a sequence of independent partitions, i.e., α generates a Bernoulli factor.*

This fact was of fundamental importance for the whole theory of dynamical systems and its applications. Everybody realized the interplay between dynamical systems and random processes: if T is an automorphism of a Lebesgue space (M, μ) and f a measurable function on M, then $F := \{f(T^n x),\ x \in M,\ n \in \mathbf{Z}\}$ can be treated as a stationary random process. If T itself is of probability origin (to be the shift in the trajectory space of a stationary process), it would be not surprising if T exhibits some stochastic behavior. But if T is a diffeomorphism, i.e., purely deterministic, e.g., a group automorphism of the torus, one could expect that F cannot be 'too random'. Sinai's theorem showed that this is not the case: even the 'most random' of all discrete time stationary processes, a Bernoulli process, can appear as F in this construction. Today this is universally known, but at that time, one had to show great intellectual bravery to imagine something of the kind.

The proof of the theorem on Bernoulli factors consists in successive transitions from one increasing partition to another and taking their limit (intersection). Here the following observation by Rohlin is used: if ξ is an increasing measurable partition, i.e., $T^{-1}\xi < \xi$, and η is a measurable partition such that $T^{-1}\xi < \eta < \xi$, then η is also increasing. Moreover, every partition sequence obtained by this strategy clearly decreases, and one should only make sure that the limit of this sequence be of the form $\vee_{i=0}^{\infty} T^{-i}\alpha$, where α generates a Bernoulli factor and has the prescribed measures of atoms. But this is only a skeleton of the proof. To implement it in detail one should choose a transition between partitions mentioned above at each step. In [55] where the proof was published, Sinai starts with a significant new contribution to his joint work with Rohlin on the theory of increasing partitions. Then he carried out the necessary construction, where a deep insight in the situation helped him to overcome a number of technical difficulties.

3.3 First Examples of Smooth K-Systems

The absolutely first such an example, I think, was an ergodic group automorphism of the 2-torus. Its K-property was discovered by Sinai, but left unpublished (see, however, [53]). Later Rohlin [41] obtained a similar result for automorphisms and even endomorphisms of compact commutative groups (for endomorphisms, the K-property is replaced by that of having completely positive entropy). And again Sinai [48] discovered first smooth K-flows, namely, the geodesic flows on Riemann manifolds of negative constant curvature and finite volume. In [48] he constructed a K-partition for the 2-dimensional case to some extent explicitly and proved the K-property of this partition, making use of the ergodicity of the horocycle flow established by Hedlund. Sinai also evaluated the entropy of the geodesic flow in the compact case. He expressed it in terms of the curvature, the volume of the whole manifold (of dimension n) and the $(n-1)$-dimensional volume of the unit $(n-1)$ sphere.

Almost simultaneously Sinai found another evidence of stochasticity for geodesic flows. Let $\{S_t\}$ be such a flow and f be a real measurable function on its phase space X provided with the corresponding probability measure μ. Then $f_t(x) := f(S_t x)$, $x \in X$, is a stationary random process. In [47] he proved the central limit theorem in the following form: if the flow $\{S_t\}$ acts on a manifold of constant negative curvature and the function f satisfies some regularity conditions, then the random variable $F_\tau(x) := \int_0^\tau f_t(x)dt$, after subtracting its expectation and dividing by the square root of its variance, will converge in distribution (as $\tau \to \infty$) to a standard Gaussian random variable. From a general point of view, this fact is not so surprising: since $\{S_t\}$ is a K-flow, F_τ is a sum of many small weakly dependent random variables and must be asymptotically Gaussian. But to prove this rigorously, one had to overcome considerable difficulties. The only general sufficient condition known by then under which a stationary random process obeys the central limit theorem, was the so-called Rosenblatt strong mixing condition (named after M. Rosenblatt). However, as Sinai writes in [53], he could not check if the above process F_τ fits this condition. That is why he introduced [51] a weaker 'local Rosenblatt' condition and proved that it was also sufficient for the central limit theorem to apply. Moreover, he managed to find out that many processes $\{F_\tau\}$ obey this property, at least the set of functions f generating such processes is dense in $L^2(X, \mu)$.

A little later Sinai [49] extended the results from [48] and [47] to the geodesic flows on some compact surfaces of negative non-constant curvature. Some properties of the horocycles, known by that time and new ones, discovered specially for this purpose, were also used there, but the Hedlund–Hopf approach could not be applied literally.

Looking at Sinai's achievements in entropy theory over the few years since the work [46], everybody would be struck by abundance and depth of his results. So it is hardly surprising that in 1962, he was invited as a speaker to the International Congress of Mathematicians in Stockholm (more surprising is that he was authorized to leave Russia). In his paper [53] he not only discusses his previous results, but also suggests some new directions of investigation. In particular, he introduces the notion of a transversal flow related to a given automorphism T (or flow $\{S_t\}$) acting on a Riemann manifold X. By definition the trajectory partition of the transversal flow $\{Z_s\}$ is invariant in the sense that each trajectory is transferred into another trajectory under the action of T (or every S_t). Another property is that there exists the local contraction coefficient, that is for every $x \in X$ and every interval $\{Z_s x, 0 \leq s \leq u\}$, the time length of the interval $\{T Z_s x,\ 0 \leq s \leq u\}$ divided by u tends to a limit $\lambda(x)$ (for $\{S_t\}$ the definition is similar). The transversal flow relates to bundles of asymptotic trajectories: if two trajectories of T or $\{S_t\}$ start from the same trajectory of $\{Z_s\}$, they approach each other, often exponentially fast. Thus the classes of asymptotic trajectories of the initial system can be identified with the trajectories of the transversal flow. Such a structure in the phase space owes its origin to the instability of the motion: for a point x of the phase space, the majority of trajectories starting from points x' near x move away from the trajectory of x, while the exceptional points x' constitute a 'manifold' of positive codimension.

The simplest example is the ergodic automorphism T of the 2-torus, for which a transversal flow can be taken as the motion with unit velocity along the eigenvector of the corresponding matrix whose eigenvalue λ is less than one. Another example is the geodesic flow on a surface of a constant negative curvature, in which case the part of the transversal flow is played by the horocycle flow. In both examples the entropy is closely related to the contraction coefficients. In particular, the contraction coefficient for an ergodic automorphism of the 2-torus, is the above eigenvalue $\lambda < 1$, while the entropy equals $-\log \lambda$. In the case of geodesic flow the contraction coefficient is $-(-k)^{1/2}$ (where k is the curvature), and the entropy is proportional to it (let us note that by definition the contraction coefficient for a flow is similar not to that for an automorphism, but to its logarithm).

An existence condition for a K-partition can also be expressed in terms of a transversal flow. Moreover, under these conditions one can divide the trajectories of the transversal flow into intervals in such a way as to obtain a K-partition. It is very useful to consider the transversal flows for T and T^{-1} (or $\{S_t\}$ and $\{S_{-t}\}$) together. In some cases it gives a possibility to find the partition π for T (or $\{S_t\}$) almost immediately.

One can observe that the transversal flow as such is not as essential for this approach as the partition into its trajectories. That is why the multidimensional case can also be included in this context, except that the transversal flow should be replaced by a transversal field. The latter can be identified with the partition of the phase space into the orbits of the field, and the above-mentioned properties of the trajectory partition of the transversal flow, its invariance and the existence of the contraction coefficient, should retain their validity.

Forerunners of some further investigations can be found here. For instance, in 1963, D. Anosov introduced a class of dynamical systems that were later named after him. The definition of Anosov's (or uniformly hyperbolic) systems resembles that of the systems with transversal fields, and the theories of these two classes of systems are to some extent close to each other. Moreover, many system studied by Sinai earlier (the ergodic group automorphisms of the 2-torus, geodesic flows on compact manifolds of negative curvature) turned out to be Anosov's systems. This was a motivation for publishing the joint paper [5]. But shortly before, Sinai published another work [57] where he continued the study of flows with Lebesgue's spectrum of infinite multiplicity started in [50] with the method of transversal fields contemplated in [53]. He developed this approach in a measure-theoretic context, much more general than needed for Anosov's flows.

Emergence of Anosov's systems and their generalizations opened up a new field in theory of dynamical systems, and Sinai contributed much to this field (see [37]).

3.4 Generators

Let us recall that Kolmogorov's definition of the entropy stated in [24] made sense only for automorphisms that had generators with finite entropy. A more general definition by Sinai [46] was free of this restriction, but the existence problem for generators remained open for some time. The first result here was obtained by Rohlin [42, 43], who proved that *every aperiodic automorphism* T *with* $h(T) < \infty$ *had a countable generator* ξ *with* $H(\xi) < \infty$. This made Kolmogorov's definition almost as general as Sinai's. But this happened four years after the definitions by Kolmogorov and Sinai appeared.

In 1970, W. Krieger [25] made the next step—he proved that *every ergodic automorphism* T *with* $h(T) < \infty$ *has a finite generator*. Later he refined this statement as follows: *there exists a generator with* $\leq 2^{h(T)} + 1$ *atoms*; note that no generator can have $< 2^{h(T)}$ atoms, so that this estimation can be treated as optimal. It is interesting that Krieger's argument is essentially based on a result of Sinai's student A. Zaslavsky, who also tried to solve the problem of finite generators. A number of alternative proofs appeared containing refinements and generalizations of Krieger's result. The strongest of the statements I know, is due to C. Grillenberger and U. Krengel [13]. The next theorem contains a particular case of their result.

Theorem 4 *Let* T *be an ergodic automorphism with* $h(T) < \infty$, *and* $\mathbf{p} = p_1, \ldots, p_k$ *a probability distribution with* $H(\mathbf{p}) > h(T)$. *Then there exists a generator* $\xi = (C_1, \ldots, C_k)$ *such that* $\mu(C_i) = p_i$ *for* $1 \leq i \leq k$.

This theorem together with Sinai's theorem on Bernoulli factors can be used to prove that $h(T)$ determines Arov's ε-entropy (see Sect. 2) for an ergodic automorphism T.

3.5 Entropy and Periodic Orbits

Periodic orbits of dynamical systems are of traditional interest to various fields of mathematics, especially to geometry. From results by J. Hadamard and M. Morse, it is known that the set of tangent vectors of a compact surface of negative curvature that are tangent to closed geodesics, is everywhere dense. In 1963 Anosov extended this to the multidimensional case. A relation between periodic orbits and entropy was realized rather early. In 1966, Sinai [56] published the following theorem.

Theorem 5 *Let Q be a closed compact Riemann manifold of dimension $d > 1$ and $\nu(t)$, the number of closed geodesics of multiplicity 1 and length $\leq t$, $t > 0$. Assume that the curvature K of Q along every 2-dimensional direction satisfies $-K_2^2 \leq K \leq -K_1^2$. Then*

$$(d-1)K_1 \leq \liminf_{t\to\infty} \frac{\ln \nu(t)}{t} \leq \limsup_{t\to\infty} \frac{\ln \nu(t)}{t} \leq (d-1)K_2. \tag{5}$$

Formally, the proof of the theorem was purely geometric, but the reasoning behind it was closely related to the transversal fields method developed by Sinai in [53] and used by him before in other contexts. He remarks that the constant K_1 in (5) can be replaced by the entropy of the geodesic flow.

This ability to reveal dynamical problems where they are not so evident as above, is typical for Sinai's mathematics. I repeatedly heard him saying: 'this is our question' (or 'ergodic question'), especially as a response to a physics talk at his seminar. The dynamical approach to formally non-dynamical problems was used by many Sinai's followers, the most successful of which is G. Margulis. He strengthened Theorem 5 by replacing lim sup and lim inf for the limit. Later he obtained, for Anosov's flows, an even stronger result: $\lim_{t\to\infty} ht\nu(t)e^{-ht} = 1$, where h is the topological entropy of the flow.

Similar estimates for the exponential grow rate of $P_n(T)$, the number of fixed points of T^n, were proved for hyperbolic diffeomorphisms. For some popular examples, such as topological Markov shifts and torus automorphisms, $P_n(T)$ can be found explicitly. In these cases a counterpart of the above Margulis formula holds. There were attempts to obtain the similar results beyond the hyperbolic systems. Namely, A. Katok proved in [17] a theorem that implies the following: *if T is a $C^{1+\alpha}$ $(\alpha > 0)$ diffeomorphism of a smooth surface, then* $\limsup_{n\to\infty} \frac{P_n(T)}{n} \geq h_{\text{top}}(T)$. Recently this result was extended in a stronger form by Yu. Lima and O. Sarig [27] to flows with positive topological entropy on 3-dimensional smooth manifolds. They used some ideas from thermodynamic formalism for infinite alphabet topological Markov shifts.

4 Topological Entropy and Emergence of Thermodynamic Formalism

The history of topological entropy had in fact started somewhat earlier than when its definition was published in full generality. Sinai did not participate in this development personally, but his students did, and all this was discussed at his seminar. In 1964, W. Parry [36] considered the following question. Let A be an $n \times n$ matrix with entries $a_{i,j} \in \{0, 1\}$, and

$$X_A := \{x = (x_i)_{i \in \mathbf{Z}} \in \{1, \dots, n\}^{\mathbf{Z}} : a_{x_i, x_{i+1}=1}\}. \tag{6}$$

The set X_A (now called a *Markov set* or a *Markov compact*) is clearly T-invariant where T is the one step shift transformation on $\{1, \dots, n\}^{\mathbf{Z}}$. Assuming that A is transitive in the sense that for every pair (i, j), there exists $k \in \mathbf{Z}$ such that the (i, j)th entry of A^k is positive, Parry asks: what is the supremum of $h_\mu(T)$ over the family of T-invariant probability measures μ concentrated on X_A. (We write $h_\mu(T)$ instead of $h(T)$, because now μ is not fixed.) He refers to this supremum as the absolute entropy and shows that it equals log $\lambda(A)$ where $\lambda(A)$ is the maximal in the absolute value eigenvalue (Perron number) of A. Moreover, it is achieved at the unique μ, the Markov measure (x_i, $i \in \mathbf{Z}$, forms a stationary Markov chain with respect to μ) whose transition probabilities are expressed explicitly in terms of A.

It is fair to say that Shannon [45] solved almost the same problem in connection with his definition of capacity for a noiseless channel. His results were improved by Yu. Lyubich [29], but were not noticed in time by specialists in dynamical systems.

In 1965, R. Adler, A. Konheim and M. McAndrew [3] introduced a new topological invariant for continuous maps of a compact topological space M. They named it the *topological entropy* and defined by analogy with $h(T)$, except that countable partitions and their entropy are changed for open covers (of any cardinality) and the logarithm of the cardinality of their minimal subcovers. In [3] and subsequent works the topological entropy for a variety of dynamical systems was evaluated, and it turned out that *Parry's absolute entropy was simply the topological entropy of the Markov shift T on X_A* (X_A is compact in a natural topology, and T is a homeomorphism of X_A).

The challenge immediately arose to discover the relationship between $h(T)$ and $h_{\text{top}}(T)$, the topological entropy of T. Parry's result suggested that

$$h_{\text{top}}(T) = \sup_{\mu \in \mathscr{I}(T)} h_\mu(T), \tag{7}$$

where $\mathscr{I}(T)$ is the family of T-invariant Borel probability measures on M. For several years the variational principle (7) remained a conjecture. In 1969, L. Goodwyn [12] showed that $h_\mu(T) \leq h_{\text{top}}(T)$. A year later, E. Dinaburg [9] proved (7) for homeomorphisms of the spaces whose topological dimension is finite. And at last, T. Goodman [11] established (7) in full generality. It is interesting that another definition of $h_{\text{top}}(T)$, very popular now, arose in [9], but its author is Kolmogorov, who recommended the paper for publication and made a hand-written insert into the text (in [9] Kolmogorov's authorship is indicated). This definition is suitable for a continuous map T of a compact metric space (M, ρ) and is as follows. One can define a sequence of metrics ρ_n on M by $\rho_n(x, y) = \max_{0 \leq i < n} \rho(T^i x, T^i y)$ and, for every $\varepsilon > 0$, denote by $N(T, n, \varepsilon)$ the minimal m such that there is a partition of (M, ρ) into m sets of diameter $\leq 2\varepsilon$.

Then $\lim_{\varepsilon\to 0}\lim_{n\to\infty}\frac{1}{n}\log N(T,n,\varepsilon) = h_{\text{top}}(T)$. Let us note that, by definition, $\log N(T,n,\varepsilon)$ is the Kolmogorov ε-entropy of the compact (M,ρ_n). Since $h_{\text{top}}(T)$ is determined by the topology on M, the equality implies that we can take any metric ρ inducing this topology for evaluating $h_{\text{top}}(T)$ in this way.

At the same time as this development, significant events have occurred in statistical physics. R. Dobrushin, O. Lanford, and D. Ruelle (the two last working together) suggested a new approach to the notion of the limiting Gibbs distribution, including it into the more general concept of a Gibbs random field on $\mathbf{Z}^d$ or $\mathbf{R}^d$ (or a Gibbs measure on the configuration space). This soon resulted in the 'Gibbs explosion' in mathematical methods of statistical physics, comparable with the 'entropy explosion' in ergodic theory. What is more, the two fields turned out to be related to each other due to a variational principle. Although the physicists primarily are interested in the multidimensional case, we consider a 1-dimensional lattice compact spin system. Then we have, as before, a compact metric space M (the space of spin configurations in $\mathbf{Z}$), a continuous map T defined on M (the shift transformation), and a continuous function $f : M \to \mathbf{R}$ (determined by the interaction potential). One defines the functional $\mu \mapsto h_\mu(T)+\int f d\mu$, $\mu \in \mathscr{I}(T)$. Its supremum $P(f)$ is referred to as the *pressure*. The definition of $P(f)$ becomes a variational principle if we take into account that $P(f)$ can be defined independently, by f only. The points of maximum of the functional in question are said to be the *equilibrium measures*, and usually one can prove that these measures are the same as the Gibbs measures mentioned above. All this can be carried out in a general situation, with no mention of physical models; $P(f)$ is often referred to as the pressure (or topological pressure) of f (because T is usually fixed). If $f \equiv 0$, we arrive at the definition of $h_{\text{top}}(T)$. Thus the topological entropy is a special case of the topological pressure. The principal questions now concern the existence and properties of equilibrium measures, in particular, the number of them and the possibility of alternative, more explicit, descriptions for these measures. Gradually, due to these and related problems, a new direction in the theory of dynamical systems arose, which is often referred to as the Thermodynamic Formalism. The term was already used in this meaning by R. Bowen [7], but it was known in statistical physics (where it had another meaning) at least from the first half of the twentieth century.

Sinai's paper [59] became one of the cornerstones of the Thermodynamic Formalism. He suggested an alternative approach to the notion of a Gibbs measure. This measure is obtained as a weak limit of probability measures $\mu_{n,m}$ absolutely continuous with respect to a measure of maximal entropy for T with densities of the form $p_{m,n}(x) = c_{n,m}\exp\sum_{i=-n}^{m} f(T^i x)$. Taking a mixing finite alphabet topological Markov shift for T (such T possesses only one measure of maximal entropy) and a sufficiently 'smooth' f, Sinai proves that the limit $\lim_{n,m\to\infty}\mu_{m,n} =: \mu_f$ exists, and he discovers its properties. Then he considers an Anosov diffeomorphism $\tilde{T}$ defined on a Riemann manifold $\tilde{M}$ and, using his theorem on Markov partitions [58] (improved by Bowen), he transfers μ_f to $\tilde{M}$. If f is determined by the volume expansion coefficient along an unstable manifold, then the $\tilde{T}$-invariant measure $\tilde{\mu}_f$

obtained in this way exhibits some remarkable features, in particular, the measures it induces on unstable manifolds are absolutely continuous with respect to the Riemannian volume on the corresponding manifolds. This $\tilde{\mu}_f$ is called u-Gibbsian ('u' is after 'unstable'). Another f related to stable manifolds yields s-Gibbsian measures. A similar construction is suggested in [59] for Anosov flows. This will be described in more detail and with some generalizations in [37].

5 Ornstein's Theory

Ornstein's solution to the isomorphism problem for Bernoulli shifts was published in 1970. While the result was known already, the paper [31] was not yet available in the Soviet Union. D. Ornstein came to a symposium on information theory in Tallinn. Sinai and a group of his students also came there. I remember that for several consecutive days, instead of attending official meetings, we met in a separate room, and Don tried to explain his proof to us. He was writing on the blackboard nothing but several letters 'a' and 'b' variously ordered, and his arguments appeared to be heuristic and not too clear. Much later, when working on a Russian translation of his book [32], I realized the depth and originality of his approach. It is not necessary to present Ornstein's theory in detail here, because there are good presentations in the literature. I only wish to say a few words (some of his results were obtained together with his collaborates, but for short I mention below only him).

Ornstein deals with an automorphism T of a probability space $(M, \mathscr{M}, \mu)$ (this space does not need to be a Lebesgue space, but $\mathscr{M}$ should be countably generated) and a finite measurable partition ξ of M whose atoms are numbered. He refers to (T, ξ) as a process (in fact, every discrete time stationary process with a finite number of states has such representation). There were several notions of mixing, or regularity, for stationary processes, the weakest of which is the K-mixing (or the K-property of T if ξ is a generator). Ornstein discovered a new mixing condition called by him the *very weak Bernoulli (V.W.B.)* condition. It is intermediate between the K-mixing and the uniformly strong mixing (or φ-mixing) by I. Ibragimov. Similar to the K-mixing, *the V.W.B. holds for all ξ if it does for at least one generating ξ (when T is fixed), and this condition is equivalent to that T is isomorphic to a Bernoulli shift.*

A key part in Ornstein's theory is played by a metric $\overline{d}$ (introduced by him) in the space of processes. Roughly speaking, the $\overline{d}$-distance between processes (T, ξ) and (T_1, ξ_1) such that T, T_1 are ergodic and ξ, ξ_1 have the same number of atoms, is the proportion of those i for which $T^i x$ and $T^i x_1$, when x, x_1 are representative points, belong to differently numbered atoms of ξ and ξ_1 (the Hamming distance). A precise definition is based on the Kantorovich distance between measures on a finite set. Ornstein referred to a process (T, ξ) as a Bernoulli process (B-process) if T is isomorphic to a Bernoulli shift, while ξ is arbitrary. He proved that the set of B-processes is closed in $\overline{d}$.

All familiar stationary processes that could be B-processes because of their mixing property and positive entropy, turned actually out to be B-processes. First of all this holds true for mixing Markov chains.

Ornstein extended his theory to continuous time dynamical systems and defined a flow $\{T^t\}$ to be a Bernoulli flow (B-flow) if T^1 is (i.e., is isomorphic to) a Bernoulli automorphism. And again, many popular examples were found to be B-flows, among which were the geodesic flows on compact manifolds of negative curvature.

Ornstein proved that *every two B-automorphisms or B-flows with the same entropy, finite or infinite, are isomorphic*. On the other hand, together with P. Shields he constructed an uncountable family of K-automorphisms with equal entropy which are pairwise non-isomorphic. The simplest example of a K- but not a B-automorphism was found by S. Kalikow [16] in 1982. This example can also be described as a simple random walk in $\mathbf{Z}$ in a random environment, or as a Markov shift with continuous alphabet and a very simple transition function. These and other results in the field lead to the conclusion that, contrary to initial hope, the classification problem for K-systems is of the same complexity as for all ergodic dynamical systems.

From a technical standpoint, Ornstein's contribution into ergodic theory is that he managed to combine two different lines of thought on a measure preserving map: as on a continuous matter of geometry and as on a collection of symbolic sequences familiar to coding theory. A connecting link, although almost invisible, is dealing with partitions whose atoms are labeled. This made his approach, combinatoric in its nature, much more flexible.

Ornstein wrote in the book [32] that his research was motivated by his wish to gain better insight into the Sinai weak isomorphism theorem. But there is an essential difference between their approaches: Sinai's proof is based on a physically realizable coding method, that is, if a sequence $x = (x_i, i \in \mathbf{Z})$ is coded into a sequence $y = (y_i, i \in \mathbf{Z})$, then y_0 is a function of $x_0, x_{-1}, \ldots$. This is not the case for Ornstein's coding, which requires some anticipation. However, this anticipation, in general inevitable, can be made locally finite. Namely, M. Keane and M. Smorodinsky [19] invented an invertible mod 0 coding of one Bernoulli shift into another one (with the same entropy) with the following property: *there exist two functions, $x \mapsto k^-(x) \in \mathbf{Z}_+$ and $x \mapsto k^+(x) \in \mathbf{Z}_+$ such that y_0, the* 0th *component of the coded sequence y, is a function of the components x_i of the initial sequence x with $-k^-(x) \le i \le k^+(x)$*. Such a coding is said to be *finitary*.

6 Kakutani Equivalence

Ornstein's isomorphism theorem stimulated studying another classification of dynamical systems based on the so-called *Kakutani equivalence* (or *monotone equivalence*). This notion was introduced by S. Kakutani in 1944. For the continuous time case, Kakutani equivalence is a time change in flows. In the discrete time case, one of equivalent definitions is the following: *two ergodic automorphisms, T_1 and T_2, are Kakutani equivalent if there are isomorphic suspension flows constructed over T_1 and T_2, respectively*. This equivalence relation does not seem too interesting for entropy theory, because it does not preserve the entropy $h(T)$, but only the property of having $h(T) = 0$, $h(T) = \infty$, or $0 < h(T) < \infty$. However, the theory of Kakutani equivalence was progressing similarly to Ornstein's isomorphism theory and through the latter it relates to entropy.

In the mid-1970s, A. Katok and J. Feldman (a little later) initiated independently this development (see a detailed exposition in [33]). There is a distance (denoted by $\bar{f}$) in the space of processes that plays the role in the Kakutani equivalence theory that resembles the one played by $\bar{d}$ in the Ornstein theory. Replacing $\bar{d}$ by $\bar{f}$, one obtains the so called *loosely Bernoulli (LB)* systems (introduced by Feldman) instead of the very weak Bernoulli ones (see Sect. 5). This class is invariant with respect to the Kakutani equivalence and closed under going to factors, induced and integral automorphisms, and to suspension flows. But together with all Bernoulli automorphisms of finite entropy, it contains many automorphisms with zero entropy, in particular, the ergodic shifts on compact commutative groups. On the other hand, Feldman constructed a K-automorphism that is not LB and hence is not Bernoulli. Another such an example is the Kalikow automorphism mentioned above. The simplest class of automorphisms equivalent to each other was introduced by Katok under the name of 'standard'. All standard automorphisms have zero entropy.

7 Other Entropy Type Characteristics of Dynamics

In 1967, A. Kushnirenko [26], following an approach suggested by A. Kirillov, defined, for an automorphism T of Lebesgue's space (M, μ) and an arbitrary sequence of integers $A := \{n_1, n_2, \dots\}$, what he called the *A-entropy* of T by

$$h_A(T) = \sup_{\xi} \limsup_{k\to\infty} H(T^{n_1}\xi \vee \cdots \vee T^{n_k}\xi),$$

where sup is taken over all measurable partitions ξ with $h(\xi) < \infty$. It is clear that $h_A(T) = h(T)$ when $A = \{1, 2, \dots\}$ and that $h_A(T)$ is a metric invariant for each A. Kushnirenko proved that $h_A(T)$ has the approximation properties similar to those of $h(T)$, and, as in the case of $h(T)$, these properties enable evaluation of the A-entropy for some important examples. In particular, it turned out that *the horocycle flow and its Cartesian square have different A-entropies for $A = \{2^0, 2^1, \dots\}$ and hence these two flows with zero entropy and Lebesgue spectrum of infinite multiplicity are not isomorphic*.

Another approach to entropy type invariants is developed in recent works of A. Vershik and his group (see [64]).

They note that each finite or countable measurable partition ξ of a Lebesgue space (M, μ) induces a semi-metric ρ_ξ on (M, μ) by $\rho_\xi(x, y) := \delta_{C(x),C(y)}$, where $C(x), C(y)$ are the atoms of ξ that contain x and y, respectively. Then they define a natural class of 'admissible' semi-metrics on (M, μ) (containing ρ_ξ) and suggest one to study not the evolution of measurable partitions under the action of T, but the evolution of admissible metrics and semi-metrics. Some invariants of the latter evolution do not depend on the initial semi-metric and can characterize T itself. Among them is the class of 'scaling' sequences describing the growth rate of the 'ε-entropy' of the space (M, μ) with respect to the average semi-metric $\rho^{\mathrm{av}}(x, y) := n^{-1} \sum_{i=0}^{n-1} \rho(T^i x, T^i y)$, where ρ is an admissible semi-metric. The definition of the ε-entropy is close to that by Kolmogorov, but uses the Kantorovich distance between probability measures on M. In some cases one can define, simultaneously with scaling classes, a numerical invariant, the 'scaling entropy'. Presumably these invariants can distinguish automorphisms with zero entropy. It has been proved that *the automorphisms with discrete spectrum are characterized by bounded scaling sequences*. Similar ideas can be found in earlier works by J. Feldman and S. Ferenczi.

8 Entropy for Actions of General Groups

As early as in the first half of the 1970s, advances in entropy theory for actions of **Z** and **R** resulted in creation of the corresponding theory for more general groups actions.

An action T of a countable group G by automorphisms of Lebesgue's space (M, μ) is a homomorphism of G to the group of automorphisms T_g, $g \in G$, of (M, μ) (for non-countable groups some measurability or continuity conditions should be added).

The first general results were obtained by J.-P. Conze [8] who introduced the entropy and K-property for actions of Abelian finitely generated groups. He also proved that every K-action of such a group has completely positive entropy.

As far as I know, there are only two works by Sinai on entropy for actions of groups more general than **Z** and **R**. In 1985, together with his student N. Chernov, he considered [62] the time evolution of an infinite system of hard spheres in $\mathbf{R}^d$ that elastically collide with each other and move by inertia between collisions. This motion induces an infinite-dimensional dynamical system $\{T_t, t \in \mathbf{R}\}$ that preserves a family of Gibbs measures. Earlier Sinai and his students contributed much in construction of such dynamics. It is clear that the entropy of this system is infinite, but the appropriately normalized entropy of its approximating finite-dimensional system with respect to a natural approximating invariant measure has a finite limit h; this fact was established by Sinai earlier. The authors of [62] observe that the Gibbs measures are invariant not only under the group of time shifts $\{T_t\}$, but also under the action of the space shifts, which commute with $\{T_t\}$. Then they show that h estimates from below the entropy of the system in question under the action of the group $\mathbf{R}^{d+1}$ of space-time shifts.

In 1985, Sinai published the work [60] devoted to the action of the group $\mathbf{Z}^2$ on a sequence space. There he answered a question asked by the lecturer during by J. Milnor's lecture on cellular automata.

Gradually most parts of entropy theory were extended to $\mathbf{Z}^d$, $\mathbf{R}^d$ and then to general amenable groups. A countable amenable group G is characterized by the existence of a sequence of finite sets $F_n \subset G$ (named after Følner) such that $\lim_{n\to\infty} \#(gF_n \Delta F_n)/\#F_n = 0$ for every $g \in G$. (For locally compact amenable groups the cardinality should be replaced by the Haar measure.) Given a Følner sequence $\{F_n\}$, one can define the entropy for an action T of G by

$$h(T) := \sup_{\xi} \lim_{n\to\infty} H(\vee_{g\in F_n} T_g \xi),$$

where sup is taken over all finite measurable partitions or all partitions with finite entropy (the limit does not depend on $\{F_n\}$).

Ornstein's isomorphism theorem for Bernoulli shifts with equal entropy was extended to Bernoulli actions of amenable groups by Ornstein and Weiss [34]. Much earlier, A. Stepin showed that this theorem holds for a group if this is the case for a subgroup. B. Kaminskiy extended the theory of invariant partitions to actions of $\mathbf{Z}^d$ and introduced the notion of K-action. Then he proved that K-action can be characterized by the property of having completely positive entropy.

In parallel with old questions, such as if the entropy can be computed from a generator (Kolmogorov–Sinai theorem), many new ones arose for general groups. For instance, what are the information pasts for a partition with finite entropy? In other words, what should be taken instead of $\{z \in \mathbf{Z} : z < 0\}$ in the definition $\xi_T^- = \vee_{n<0} T^n \xi$ to keep Eq. (4) valid? For actions of amenable groups G information pasts were studied by B. Pitskel in 1975. Some amenable groups have no information pasts at all. On the other hand, if, say, $G = \mathbf{Z}^2$, one should draw a straight line L through the origin and take the set G^- consisting of all $g \in G$ lying on one side of L, and add all $g \in L \cap G$ lying on one side of the origin. These sets G^- form the collection of information pasts for Z^2.

For a number of years it could seem that the applications of entropy theory and ergodic theory as a whole to the setting of group actions, were limited to the case of amenable groups. But recently this class of groups was considerably extended.

In 1999, M. Gromov [14] and, in 2000, B. Weiss [65] (more explicitly) introduced a new class of groups called *sofic groups. A countable group is sofic if there exists a sequence of positive integers* d_n *and a sequence of maps* $\sigma_n : G \to \mathrm{Sym}(d_n)$ *(the symmetric group on* $\Delta_n := \{1, \dots, d_n\}$*) such that (i)* $d_n \to \infty$ *as* $n \to \infty$*; (ii) for each pair of distinct* $g, g' \in G$,

$$\#\{\delta \in \Delta_n : \sigma_n(g)\delta = \sigma_n(g')\delta\} = o(d_n);$$

(iii) for each pair $g, g' \in G$,

$$\#\{\delta \in \Delta_n : \sigma_n(gg')\delta \neq \sigma_n(g)\sigma_n(g')\delta\} = o(d_n).$$

The sequence $\{\sigma_n\}$ is called a *sofic approximation* of G, while properties (ii) and (iii) mean that it is asymptotically free and asymptotically multiplicative, respectively. All amenable and free groups are sofic. Moreover, there are no examples of countable non-sofic groups. A few years later L. Bowen initiated development of entropy theory for actions of these groups. Apart from him, significant contributions were made by D. Kerr and H. Li. As these authors write, they wanted to follow the Kolmogorov and Sinai line of research as far as possible. But, according to Kerr, Bowen replaced "the internal information-theoretic approach of Kolmogorov with the statistical-mechanical idea of counting external finite models". First of all, an alternative to the function $h(T, \xi)$ had to be found. This was done by different authors in different ways (with the help of sofic approximations, but not so simple and natural as for amenable groups with the help of Følner's sequences). A definition of $h(T)$ close to the classical one is due to Kerr [20]. In this definition $h(T) = \sup_\xi h(T, \xi)$, where sup is taken over finite partitions. But the novelty is that $h(T, \xi) = \inf_{\alpha \geq \xi} h'(T, \xi, \alpha)$, where $h'(T, \xi, \alpha)$ is an entropy type quantity depending on ξ and its finite refinement α. With this definition the classical Kolmogorov–Sinai theorem remains true. There are many other equivalent definitions including ones based on operator algebras, topological models and random sofic approximations. Ornstein's isomorphism theorem was extended by L. Bowen to a wide class of non-amenable sofic groups. Kerr and Li [21] defined an extension of the topological entropy and proved a variational principle (see Sect. 4). The theory is progressing rapidly. In particular, there is Bowen's work on actions of sofic groupoids, based on the Rudolf-Weiss invariance theorem for the relative entropy. However, it is unknown if there exists a countable non-sofic group.

References

1. L.M. Abramov. The entropy of a derived automorphism (in Russian). *Dokl. Akad. Nauk SSSR*, 128:647–650, 1959.
2. L.M. Abramov. On the entropy of a flow (in Russian). *Dokl. Akad. Nauk SSSR*, 128:873–875, 1959.
3. R.L. Adler, A.G. Konheim, and M.H. McAndrew. Topological entropy. *Trans. Amer. Math. Soc.*, 114:309–319, 1965.
4. R.L. Adler and B. Weiss. Entropy, a complete metric invariant for automorphisms of the torus. *Proc. Nat. Acad. Sci. U.S.A.*, 57:1573–1576, 1967.
5. D.V. Anosov and Ya.G. Sinai. Some smooth ergodic systems. *Russ. Math. Surv.*, 22(5): 103–167, 1967.
6. D.Z. Arov. The influence of V.P. Potapov and M.G. Krein on my scientific work. In *Matrix and Operator Valued Functions*, volume 72 of *Oper. Theory Adv. Appl.*, pages 1–16. Birkhäuser, Basel, 1994.
7. R. Bowen. Markov partitions for Axiom A diffeomorphisms. *Amer. J. Math.*, 92:725–747, 1970.
8. J.P. Conze. Entropie d'un groupe abélien de transformations. *Z. Wahrscheinlichkeitstheorie und Verw. Gebiete*, 25:11–30, 1972.
9. E.I. Dinaburg. The relation between topological entropy and metric entropy. *Sov. Math., Dokl.*, 11:13–16, 1970.

10. I.M. Gelfand and S.V. Fomin. Geodesic flows on manifolds of constant negative curvature (in Russian). *Uspehi Matem. Nauk (N.S.)*, 7(1(47)):118–137, 1952.
11. T.N.T. Goodman. Relating topological entropy and measure entropy. *Bull. London Math. Soc.*, 3:176–180, 1971.
12. L.W. Goodwyn. Topological entropy bounds measure-theoretic entropy. *Proc. Amer. Math. Soc.*, 23:679–688, 1969.
13. C. Grillenberger and U. Krengel. On marginal distributions and isomorphisms of stationary processes. *Math. Z.*, 149(2):131–154, 1976.
14. M. Gromov. Endomorphisms of symbolic algebraic varieties. *J. Eur. Math. Soc. (JEMS)*, 1(2):109–197, 1999.
15. B.M. Gurevich. The entropy of horocycle flows (in Russian). *Dokl. Akad. Nauk SSSR*, 136:768–770, 1961.
16. S.A. Kalikow. T, T^{-1} transformation is not loosely Bernoulli. *Ann. of Math. (2)*, 115(2): 393–409, 1982.
17. A. Katok. Lyapunov exponents, entropy and periodic orbits for diffeomorphisms. *Inst. Hautes Études Sci. Publ. Math.*, (51):137–173, 1980.
18. A. Katok. Fifty years of entropy in dynamics: 1958–2007. *J. Mod. Dyn.*, 1(4):545–596, 2007.
19. M. Keane and M. Smorodinsky. Bernoulli schemes of the same entropy are finitarily isomorphic. *Ann. of Math. (2)*, 109(2):397–406, 1979.
20. D. Kerr. Sofic measure entropy via finite partitions. *Groups Geom. Dyn.*, 7(3):617–632, 2013.
21. D. Kerr and H. Li. Entropy and the variational principle for actions of sofic groups. *Invent. Math.*, 186(3):501–558, 2011.
22. A.Ya. Khinchin. The concept of entropy in the theory of probability (in Russian). *Uspehi Matem. Nauk (N.S.)*, 8(3(55)):3–20, 1953.
23. A.N. Kolmogorov. A new metric invariant of transient dynamical systems and automorphisms in Lebesgue spaces (in Russian). *Dokl. Akad. Nauk SSSR (N.S.)*, 119:861–864, 1958.
24. A.N. Kolmogorov. Entropy per unit time as a metric invariant of automorphisms (in Russian). *Dokl. Akad. Nauk SSSR*, 124:754–755, 1959.
25. W. Krieger. On entropy and generators of measure-preserving transformations. *Trans. Amer. Math. Soc.*, 149:453–464, 1970.
26. A.G. Kushnirenko. On metric invariants of entropy type. *Russ. Math. Surv.*, 22(5):53–61, 1967.
27. Y. Lima and O. Sarig. Symbolic dynamics for three dimensional flows with positive topological entropy. *J. Eur. Math. Soc. (JEMS)* (to appear).
28. E. Lindenstrauss, Y. Peres, and W. Schlag. Bernoulli convolutions and an intermediate value theorem for entropies of K-partitions. *J. Anal. Math.*, 87:337–367, 2002.
29. Yu.I. Lyubich. A remark on the conductiong capacity of a discrete noiseless channel (in Russian). *Uspehi Mat. Nauk*, 17(1 (103)):191–198, 1962.
30. L.D. Meshalkin. A case of isomorphism of Bernoulli schemes (in Russian). *Dokl. Akad. Nauk SSSR*, 128:41–44, 1959.
31. D. Ornstein. Bernoulli shifts with the same entropy are isomorphic. *Advances in Math.*, 4: 337–352, 1970.
32. D.S. Ornstein. *Ergodic Theory, Randomness, and Dynamical Systems*, volume 5 of *Yale Mathematical Monographs*. Yale University Press, New Haven, Conn.-London, 1974.
33. D.S. Ornstein, D.J. Rudolph, and B. Weiss. Equivalence of measure preserving transformations. *Mem. Amer. Math. Soc.*, 37(262):xii+116, 1982.
34. D.S. Ornstein and B. Weiss. Entropy and isomorphism theorems for actions of amenable groups. *J. Analyse Math.*, 48:1–141, 1987.
35. O.S. Parasyuk. Flows of horocycles on surfaces of constant negative curvature (in Russian). *Uspehi Matem. Nauk (N.S.)*, 8(3(55)):125–126, 1953.
36. W. Parry. Intrinsic Markov chains. *Trans. Amer. Math. Soc.*, 112:55–66, 1964.
37. Y. Pesin. Sinai's work on Markov partitions and SRB measures. In H. Holden, R. Piene, editors, *The Abel Prize 2013–2017*, pages 257–285. Springer, 2019.
38. M.S. Pinsker. Dynamical systems with completely positive or zero entropy. *Soviet Math. Dokl.*, 1:937–938, 1960.

39. V.A. Rohlin. On the fundamental ideas of measure theory (in Russian). *Mat. Sbornik N.S.*, 25(67):107–150, 1949.
40. V.A. Rohlin. Entropy of metric automorphism (in Russian). *Dokl. Akad. Nauk SSSR*, 124: 980–983, 1959.
41. V.A. Rohlin. Exact endomorphisms of a Lebesgue space (in Russian). *Izv. Akad. Nauk SSSR Ser. Mat.*, 25:499–530, 1961.
42. V.A. Rohlin. Generators in ergodic theory (in Russian). *Vestnik Leningrad. Univ.*, 18(1):26–32, 1963.
43. V.A. Rohlin. Generators in ergodic theory. II (in Russian). *Vestnik Leningrad. Univ.*, 20(13): 68–72, 1965.
44. V.A. Rohlin and Ya.G. Sinai. The structure and properties of invariant measurable partitions. *Soviet Math. Dokl.*, 2(4–6):1611–1614, 1961.
45. C.E. Shannon. A mathematical theory of communication. *Bell System Tech. J.*, 27:379–423, 623–656, 1948.
46. Ya. Sinai. On the concept of entropy for a dynamic system (in Russian). *Dokl. Akad. Nauk SSSR*, 124:768–771, 1959.
47. Ya.G. Sinai. The central limit theorem for geodesic flows on manifolds of constant negative curvature. *Soviet Math. Dokl.*, 1:983–987, 1960.
48. Ya.G. Sinai. Geodesic flows on manifolds of negative constant curvature. *Soviet Math. Dokl.*, 1:335–339, 1960.
49. Ya.G. Sinai. Geodesic flows on compact surfaces of negative curvature. *Soviet Math. Dokl.*, 2:106–109, 1961.
50. Ya.G. Sinai. Dynamical systems with countable Lebesgue spectrum. I (in Russian). *Izv. Akad. Nauk SSSR Ser. Mat.*, 25:899–924, 1961.
51. Ya.G. Sinai. Dynamical systems and stationary Markov processes. *Theor. Probab. Appl.*, 5:309–312, 1962.
52. Ya.G. Sinai. Weak isomorphism of transformations with an invariant measure. *Sov. Math., Dokl.*, 3:1725–1729, 1962.
53. Ya.G. Sinai. Probabilistic ideas in ergodic theory. *Amer. Math. Soc. transl. (2)*, 31:62–80, 1963.
54. Ya.G. Sinai. Letter to the editor of Uspekhi Matematicheskikh Nauk (in Russian). *Uspekhi Mat. Nauk*, 20(4(124)):232, 1965.
55. Ya.G. Sinai. On a weak isomorphism of transformations with invariant measures. *Amer. Math. Soc. transl. (2)*, 57:123–149, 1966.
56. Ya.G. Sinai. Asymptotic behavior of closed geodesics on compact manifolds with negative curvature (in Russian). *Izv. Akad. Nauk SSSR Ser. Mat.*, 30:1275–1296, 1966.
57. Ya.G. Sinai. Classical dynamic systems with countably-multiple Lebesgue spectrum. *Amer. Math. Soc. transl. (2)*, 68:34–88, 1968.
58. Ya.G. Sinai. Markov partitions and C-diffeomorphisms. *Funct. Anal. Appl.*, 2:61–82, 1968.
59. Ya.G. Sinai. Gibbs measures in ergodic theory. *Russ. Math. Surv.*, 27(4):21–69, 1972.
60. Ya.G. Sinai. An answer to a question by J. Milnor. *Comment. Math. Helv.*, 60(2):173–178, 1985.
61. Ya.G. Sinai. Author's comments to the paper "On the notion of entropy of a dynamical system". In *Selecta. Volume I. Ergodic theory and dynamical systems*, pages 9–10. Springer, New York, 2010.
62. Ya.G. Sinai and N.I. Chernov. Entropy of a gas of hard spheres with respect to the group of space-time translations. *J. Math. Sci.*, 32(4):389–406, 1986.
63. A.M. Vershik. Information, Entropy, Dynamics. In *Mathematics of the XXth Century: A View from St. Petersburg*, pages 47–76. MCCME, Moscow, 2010.
64. A.M. Vershik, P.B. Zatitskiy, and F.V. Petrov. Geometry and dynamics of admissible metrics in measure spaces. *Cent. Eur. J. Math.*, 11(3):379–400, 2013.
65. B. Weiss. Sofic groups and dynamical systems. *Sankhyā Ser. A*, 62(3):350–359, 2000. Ergodic theory and harmonic analysis (Mumbai, 1999).

Mathematical Physics

Konstantin Khanin

1 Introduction

Sinai was never really far away from Mathematical Physics. Already his first papers on Kolmogorov–Sinai entropy and on the stability of Kolmogorov's flow in 2D hydrodynamics (joint with L. Meshalkin, [19]) were very much in the areas which are closely related to Mathematical Physics. However, in his first research period, roughly in the late 1950s and the early 1960s, Sinai's work was mostly concentrated around Ergodic Theory and Dynamical Systems. It is fair to say that his deep and lasting interest in Mathematical Physics started with his work on Statistical Mechanics in the late 1960s. This period culminated in the celebrated Pirogov–Sinai theory of phase transitions for ferromagnetic systems. After that Mathematical Physics was always one of the main themes of Sinai's research. In general it was a period of very active interaction between mathematicians and physicists in the USSR. It was especially true for Sinai. At the beginning of 1970s Sinai moved to the Landau Institute for Theoretical Physics where he was surrounded by a stellar group of physicists. During the Landau Institute period, Sinai made fundamental contributions to the spectral theory of Schrödinger operators with quasi-periodic potentials, renormalization theory for Dyson's hierarchical models, random walks in random environment, renormalization theory of dynamical systems. Later his interests moved in the direction of the random Burgers equation and Navier–Stokes

K. Khanin (✉)
Department of Mathematics, University of Toronto, Toronto, ON, Canada
e-mail: khanin@math.toronto.edu

H. Holden, R. Piene (eds.), *The Abel Prize 2013–2017*,
https://doi.org/10.1007/978-3-319-99028-6_10

equations. Below we discuss Sinai's contributions in all of the above directions apart from the Navier–Stokes equations which will be discussed in [5]. We decided to include papers on Burgers equation in the Mathematical Physics Section since this area is closer related to Mathematical Physics and Statistical Mechanics then to the Turbulence Theory.

And a final disclaimer. The body of Sinai's work is huge. Below we are discussing only a selection of his work. It goes without saying that all the responsibility for the selection lies entirely with the author of this text.

2 Statistical Mechanics

From the end of the 1960s Statistical Mechanics became one of the main direction of Sinai's research. The famous Seminar on Statistical Mechanics at the Moscow State University was one of the world leading centers in the area of mathematical Statistical Mechanics. Sinai was one of the organisers and leaders of the seminar. Mathematical Statistical Mechanics was still a very "young" area at that time, and it is fair to say that many new directions and ideas were discovered by the participants of the seminar. One of the special issues of the European Physical Journal H is dedicated to the history of the seminar [22].

One of the first contributions of Sinai to the equilibrium statistical mechanics was a series of papers, joint with R. Minlos, on the phenomenon of the separation of phases [20, 21]. The main result, which is often cited as the "droplet theorem", provides a description of the structure of typical configuration of a spin system in $+$ phase which conditioned to have a large component of negative spins. It was proved that at low temperatures the system will have two domains, with a prevalence respectively of positive and negative spins, moreover these two domains are separated by one large contour of approximately square shape. Note that in the late 1980s and the early 1990s, R. Dobrushin, R. Kotecky, and S. Shlosman obtained more precise information on the shape of a droplet [7].

A sophisticated technique developed in [20, 21] is based on analysis of the statistics of contours. It was much further developed in the so-called Pirogov–Sinai theory of phase transitions. This theory, developed by Sinai and his student S. Pirogov, is an outstanding achievement, providing, in a certain sense, a final solution to the problem of the phase transitions at low temperatures. The main result can be formulated in the following way. Consider a statistical mechanics system on the lattice $\mathbb{Z}^d$, $d \geq 2$ with a translation-invariant Hamiltonian H_0, and spin variable taking a finite number of values. Assume that the system has a finite number k of periodic ground states, that is spatially periodic spin configurations minimising the Hamiltonian H_0. We also assume that this k ground states satisfy certain non-degeneracy requirement, called Peierls stability condition. This condition is always easy to check in concrete models. The classical example is provided by the Ising model where one has two ground states where all the spin variables take the same value +1 or −1 respectively. Consider now a $(k-1)$-parameter family of periodic Hamiltonians

$$H_\mu = H_0 + \mu_1 H_1 + \cdots + \mu_{k-1} H_{k-1},$$

which resolves the degeneracy of the ground states. Namely, any subset of the set of k ground states can be realised as a set of ground states for some $\mu = (\mu_1, \ldots, \mu_{k-1})$ from a ball $|\mu| \leq \epsilon$ for ϵ small enough. Then for small enough temperatures locally in a parameter μ one has a full stratification of the set of pure phases, that is ergodic Gibbs states. In other words, there exists a parameter value $\mu(\beta)$ for which the system has exactly k pure phases, corresponding to all k ground states. Then there are k curves γ_i, $1 \leq i \leq k$ originated from the point $\mu(\beta)$ such that on each curve one has $k-1$ pure phases. The numeration corresponds to the ground states. Namely, the curve γ_i has pure phases corresponding to all ground states except the ith one. Every two curves γ_i and γ_j are connected by a smooth two-dimensional surfaces $\gamma_{i,j}$ such that for $\mu \in \gamma_{i,j}$ the number of pure phases is $k-2$, and they correspond to all ground states except the ith and the jth. And so on, this stratification continues further until all $k-2$ dimensional surfaces with exactly two pure phases are constructed. Everywhere else outside of the constructed manifolds a pure phase is unique. This stratification depends on the parameter β which is called inverse temperature which must be large enough. One can say that a μ-dependent linear structure of ground states at zero temperature ($\beta = +\infty$) survives for small positive temperatures and translates into the structure of pure phases. However, the dependence on the parameter μ is not linear anymore. The main technical tool in Pirogov–Sinai theory is based on the method of contour expansions developed by the authors. Counter expansions can be viewed as a far reaching extension of the Peierls approach to the problem of phase transition in Ising model. Note that later, in the 1980s, the Pirogov–Sinai theory was extended by R. Dobrushin and M. Zahradnik to systems with continuous spin variables.

Another important series of papers of Sinai is dedicated to the renormalization group theory. The ideas of scaling invariance which originated in quantum field theory started to play exceptionally important role in statistical physics starting from the 1960s. One should mention here M. Fisher, L. Kadanof, K. Wilson, A. Patashinski, V. Pokrovsky, A. Polyakov, A. Migdal and many others. Renormalization group method became one of the main tools in the studies of critical phenomena. The success of renormalization theory culminated with the 1982 Nobel prize for K. Wilson for his contributions and development of the ϵ-expansion method. At the same time the rigorous mathematical explanations of the scaling invariance and conformal invariance is still an extremely important but very difficult and challenging problem. Sinai jointly with P. Bleher have developed a complete mathematical theory of the renormalization behaviour for the so-called Dyson hierarchical model [3, 4]. Although statistical mechanics systems provided by the hierarchical models are rather simplistic, their renormalization behaviour is highly nontrivial, and a development of the mathematically rigorous theory was a great achievement. I should add that the 2010 Fields Medal was awarded to S. Smirnov for the proof of conformal invariance for several 2D statistical mechanics systems at the critical point.

3 Spectral Theory of Schrödinger Operators

Another important area of Sinai's research starting from the 1970s was connected with the study of spectral properties of one-dimensional Schrödinger operators with quasi-periodic potentials. Two papers of Sinai in this direction were very influential and played an important, perhaps crucial, role in the development of the subject. In the first paper [6], joint with E. Dinaburg, the authors managed to prove existence of a postive measure component of continuous spectrum in the case of small coupling constants, that is, for small quasi-periodic potentials. In the second paper [24], written somewhat 10 years later, Sinai considered the opposite case of large coupling constants and proved the existence of the pure point spectrum with exponentially localized eigenfunctions. This result demonstrates that for large quasi-periodic potentials one has Anderson localization and the spectral behaviour is similar to the case of random potentials. Although the authors in [6] consider the Schrödinger operators $-\frac{d^2}{dx^2}\psi + V(x)\psi$ in continuous setting, below we present results in an equivalent discrete case:

$$(H_\theta^{\alpha,\lambda}\psi)_n = -\psi_{n+1} - \psi_{n-1} + \lambda V(\theta + n\alpha)\psi_n.$$

The operator $H_\theta^{\alpha,\lambda}$ acts as a self-adjoint operator in the Hilbert space $l^2(\mathbb{Z})$. It is assumed that the potential $V(x)$ is a smooth (analytic) function on the unit circle S^1. The main (and most studied) example is provided by the potential $V(x) = \cos 2\pi x$. In the case of rational $\alpha = p/q$, the operator is periodic. It is well known that in this case the spectrum is absolutely continuous and has a so-called zone structure. The results here go back to the classical papers by F. Bloch and G. Floquet. Dinaburg and Sinai proved that in the case of irrational Diophantine α, there exists a set of positive Lebesgue measure of energies E corresponding to Bloch eigenfunctions of the form $\psi_{\alpha,E}(n) = a_{\alpha,E}(n\alpha)e^{ik(\alpha,E)n}$, provided the coupling constant λ is small. These eigenfunctions satisfy the relation $H_\theta^{\alpha,\lambda}\psi_{\alpha,E} = E\psi_{\alpha,E}$. It follows that in the case of small λ the spectrum of the operator $H_\theta^{\alpha,\lambda}$ contains an absolutely continuous component. The most important feature of the paper is the application of the KAM techniques. They are used to deal with that main difficulty of the problem related to the so-called small divisors. To illustrate how KAM theory appears in the spectral problem for the Schrödinger operator, consider an eigenfunction equation

$$(H_\theta^{\alpha,\lambda}\psi)_n = -\psi_{n+1} - \psi_{n-1} + \lambda V(\theta + n\alpha)\psi_n = E\psi_n.$$

Then we have

$$\begin{pmatrix} \psi_{n+1} \\ \psi_n \end{pmatrix} = \begin{pmatrix} \lambda V(\theta + n\alpha) - E & -1 \\ 1 & 0 \end{pmatrix} \begin{pmatrix} \psi_n \\ \psi_{n-1} \end{pmatrix}.$$

Using a notation

$$S_E(\theta) = \begin{pmatrix} \lambda V(\theta) - E & -1 \\ 1 & 0 \end{pmatrix} \in SL(2, \mathbb{R})$$

we obtain

$$\begin{pmatrix} \psi_n \\ \psi_{n-1} \end{pmatrix} = S_E(\theta + (n-1)\alpha) \dots S_E(\theta + \alpha) S_E(\theta) \begin{pmatrix} \psi_0 \\ \psi_{-1} \end{pmatrix}.$$

The resulting object is called the Schrödinger cocycle. The main question here is to study asymptotic properties of the products of the values of a smooth matrix-valued function $S_E(\theta)$ along a trajectory of the rotation of the unit circle by an angle α. It is convenient to think that there exists a bundle over the unit circle with each fiber being the two-dimensional plane $\mathbb{R}^2$. Then a matrix $S_E(\theta)$ can be viewed as a linear operator acting from $\mathbb{R}^2$ over a point θ into $\mathbb{R}^2$ over a point $\theta + \alpha \pmod 1$. If we make a coordinate change in every fibre given by a smooth matrix-valued function $B(\theta)$, then the original cocycle in new coordinates will have the following form:

$$\bar{S}_E(\theta) = B^{-1}(\theta + \alpha) S_E(\theta) B(\theta).$$

The situation will be especially simple if the resulting cocycle is a constant one, i.e., it does not depend on θ. Cocycles which can be reduced to a constant cocycle, are called reducible. The main statement of the Dinaburg–Sinai theorem is that in the case of small coupling constants the Schrödinger cocycle is reducible for a set of energies E of a positive Lebesgue measure. Later H. Eliasson showed that in the case of small coupling constants, Schrödinger cocycles are linearizable for Lebesgue almost all E which implies that in this case the spectrum is pure absolutely continuous.

In the opposite case of large coupling constants, the spectrum is pure point. To prove this statement, one has to construct a basis in $l^2(\mathbb{Z})$ which consists of eigenfunctions for the operator $H_\theta^{\alpha,\lambda}$. This was done by Sinai [24] and independently by J. Fröhlich, T. Spencer, and P. Wittwer [10] for a slightly more special class of potentials. Sinai proved that for Diophantine α and typical θ, the spectrum is pure point provided the coupling constant λ is large. Moreover the eigenfunctions decay exponentially fast, i.e., the phenomenon of Anderson localization holds. In terms of a Schrödinger cocycle the case of large λ corresponds to the hyperbolic behaviour with one positive and one negative Lyapunov exponents. It follows that there exists a unique unit stable vector

$$\begin{pmatrix} \psi_0 \\ \psi_{-1} \end{pmatrix}$$

such that

$$S_E^n(\theta) \begin{pmatrix} \psi_0 \\ \psi_{-1} \end{pmatrix} \to 0$$

exponentially fast as $n \to \infty$. Here $S_E^n(\theta) = S_E(\theta + (n-1)\alpha) \dots S_E(\theta + \alpha) S_E(\theta)$. Also there exists a unique unit unstable vector which contracts backward in time:

$$\left(S_E^n(\theta - n\alpha)\right)^{-1} \begin{pmatrix} \bar{\psi}_0 \\ \bar{\psi}_{-1} \end{pmatrix} \to 0$$

exponentially fast as $n \to \infty$. The eigenvalues are such values of E for which both vectors coincide:

$$\begin{pmatrix} \psi_0 \\ \psi_{-1} \end{pmatrix} = \begin{pmatrix} \bar{\psi}_0 \\ \bar{\psi}_{-1} \end{pmatrix}$$

The proof of the existence of such values of E and of the fact that the corresponding eigenfunctions form a basis in $l^2(\mathbb{Z})$, is the main achievement of the Sinai paper. It should be mentioned that Sinai does not use the cocycle representation in his paper. The proof, in fact, is based on a difficult analysis of the resonances appearing in the problem. As Sinai put it in one of his comments: "Localisation is due to interplay between resonances". The two papers by Sinai which we discussed above laid the foundations for a huge research area which was very actively studied in the last 30 years. Important contributions were made by A. Avila, J. Avron, J. Bellissard, J. Bourgain, D. Damanik, H. Eliasson, J. Fröhlich, M. Goldstein, M. Herman, S. Jitomirskaya, R. Krikorian, Y. Last, J. Puig, W. Schlag, B. Simon, and T. Spencer.

Note that the results of Sinai have a perturbative character. In other words, they are related to asymptotic regimes corresponding to either small or large values of the coupling constant λ. In the last years a large progress was made in studying spectral properties for all values of the coupling constant. In particular, now one has a full description of the transition from the absolute continuous spectrum to the pure point one for the almost Mathieu operator with potential $V(x) = \cos 2\pi x$. It turns out that in this case, the spectrum is absolutely continuous for all $\lambda \in [0, 2)$ for all values of α, θ. For $\lambda = 2$ the spectrum is almost surely continuous and singular. "Almost surely" means that for any irrational α, one cannot exclude the existence of an $l^2(\mathbb{Z})$ eigenfunction for an at most countable exceptional set of θ. Conjecturally such exceptional valued do not exist, but at present it is an open problem. For $\lambda > 2$, the spectrum is almost surely pure point. It is known that the exceptional values of α, θ

exist but they form a set of zero measure. A big contribution to the non-perturbative results formulated above is due to former Sinai's student S. Jitomirskaya. Finally, I should mention the results of A. Avila who developed a global non-perturbative theory for general analytic potentials. These results formed a very significant part of the body of work for which Avila was awarded the Fields Medal in 2014.

4 Random Walks in Random Environment

In 1982, Sinai published two papers on random walks in random environment. In the first of these papers [23] he considered a one-dimensional random walk on a lattice $\mathbb{Z}^1$. In a simplest case when a particle can only jump to one of its neighbors, the random environment is completely determined by a sequence of random variables $0 \le p(x) \le 1$ which represents the probability of a particle at point $x \in \mathbb{Z}^1$ to jump right to a neighboring point $x + 1$. Then the probability to jump left to $x - 1$ is given by $1 - p(x)$. It is assumed that $\{p(x),\ x \in \mathbb{Z}^1\}$ form an independent identically distributed (*iid*) sequence of random variables. It is also assumed that the situation is non-degenerate (elliptic), that is, the probability distribution for $p(x)$ is bounded away from 0 and 1. Such random walks in random environment were first considered by M. Kozlov and F. Solomon in the early 1970s. Later H. Kesten, M. Kozlov, and F. Spitzer [15] proved that the random walk is recurrent almost surely with respect to the environment if and only if $E \log p(x)/q(x) = 0$. However, it was not known how the random walk really behaves in this case. In 1982 Sinai solved this problem and discovered a new important phenomenon of anomalous diffusion, which is in our days called Sinai's random walk. To describe Sinai's result in more details let us assume that $p(x) = 1/2 + \epsilon\xi(x)$, $q(x) = 1/2 - \epsilon\xi(x)$, where $\{\xi(x),\ x \in \mathbb{Z}^1\}$ is an *iid* sequence of random variables with compact support and zero mean value. Then the non-degeneracy condition is satisfied for ϵ small enough. Assume also that the distribution for ξ is even, which guarantees that the recurrence condition holds. For $\epsilon = 0$ the environment is non-random, and we have the usual simple random walk with diffusive behavior. Namely, $x(n) \sim \sqrt{n}$, where $x(n)$ is a position of random walk at time n, and the probability distribution for $x(n)/\sqrt{n}$ converges to the normal (Gaussian) distribution $N(0, 1)$ asymptotically as $n \to \infty$. It turns out that for any $\epsilon > 0$ the behavior of the walk is completely different. Remarkably, Sinai proved that for an arbitrary small ϵ, one has $x(n) \sim \log^2 n$. Moreover, for large values of n, the rescaled position of a random walk $x(n)/\log^2 n$ is located in a small neighborhood of some random point m_n which depends on the realization of the environment $\{\xi(x),\ x \in \mathbb{Z}^1\}$. Sinai also proved that the probability distribution for m_n as a function of the random environment, has a limit as $n \to \infty$. Such seemingly strange behavior of a random walk can be explained by the fluctuation mechanism. Namely, due to the fluctuations of the environment, there are special places on the lattice $\mathbb{Z}^1$ which trap a random walk for a long time. These traps can be viewed as certain potential wells. They are characterized by their depth, or, in other words, by the time required to escape from the trap. The random walk sits in the trap for a long time and waits for a fluctuation which will allow it to escape. After that the random walk relatively quickly reaches another, even deeper, trap which requires even longer escape time. One can say even more. In fact, a random walk is localised inside the trap. As shown by Sinai's student A. Golosov [12], a random walk asymptotically has a limiting distribution inside the trap. We should also mention several generalizations of Sinai's random walks. One can consider a situation when the random walk can jump, not only to the neighboring positions, but also further away. Interesting results in this direction were obtained in the last 15 years by E. Bolthausen, D. Dolgopyat, and I. Goldsheid. One can also consider random walks in random environment in higher dimensions. The trapping mechanism discovered by Sinai is essentially one-dimensional. In dimension 3 and above for small ϵ the random walk will have diffusive behavior. A possibility of anomalous diffusion for large values of ϵ is an open problem. In dimension 2 it is expected that the diffusion will slow down, but only by a logarithmic factor. It is also expected that the probability distribution for a properly normalized random walk converges to the Gaussian law, however at present there are no rigorous results in this case.

Another important paper by Sinai on random walks in random environment, joint with V. Anshelevich and K. Khanin, deals with the case of the so-called symmetric random walks. In this case, the probability to jump along a certain edge of the lattice $\mathbb{Z}^d$ is random, but it depends only on the edge but not on the direction of a jump [1]. It is more convenient to consider random walks with continuous time. Then the environment is given by a collection of positive *iid* random variables $\eta(e)$ labeled by the edges e of the lattice $\mathbb{Z}^d$. These random variables are viewed as rates of jumps along a particular edge. Again, one should also assume that the probability distribution for this random variables $\eta(e)$ is bounded away from 0. In this case, due to the symmetry condition, the traps are not possible, and the behavior of random walk is diffusive. Moreover there exists an effective non-random covariance matrix. Contrary to the previous case, the result about diffusive behavior holds in any dimension.

Closely related problems were actively studied starting from the late 1970s in the context of the averaging theory for parabolic operators with random coefficients—so-called homogenisation problem. The main results in this direction were obtained by V. Zhikov, S. Kozlov, O. Oleinik, and G. Papanicolaou, S. Varadhan. The main achievement of the approach developed in Sinai's paper compare to other results on homogenisation is connected with a possibility to control the effective diffusion. The paper [1] not only proves self-averaging and the existence of diffusive behavior, but also provides a convergent power series for the effective covariance matrix.

5 Renormalization Methods in Dynamical Systems

We have already discussed Sinai's work on renormalization for Dyson hierarchical model. The renormalization ideology became a really important tool in Sinai's approaches to different problems. He applied it even to the problem of singularities for the 3D Navier–Stokes equations (see [5]). Here we discuss Sinai's results related to renormalization in dynamics. Starting from the late 1970s, the renormalization ideas made their way into the theory of dynamical systems. It all started with a work of M. Feigenbaum on the universal mechanism of transition to chaos through the infinite sequences of the period-doubling bifurcations. Very soon renormalization became one of the most important and powerful tools in asymptotic analysis of dynamical systems. In 1984 Sinai in a joint paper with E. Vul and K. Khanin [27] developed the thermodynamic formalism describing universal metrical properties of the Feigenbaum attractor. This was an important paper for the development of the mathematical theory. Note that the Feigenbaum attractor in a modern terminology is a multifractal object. In this context [27] can be considered as the first example of the so-called multifractal formalism.

Another series of papers where renormalization ideas played a very important role is related to the problem of linearization of nonlinear circle diffeomorphisms. In 1961, V. Arnold [2] in a framework of the KAM theory, proved a local theorem on the analytic linearization of analytic circle diffeomorphisms close to the linear ones under condition that their rotation numbers are typical in the Diophantine sense. M. Herman developed a global theory and proved that for typical rotation numbers the conjugacy is smooth for C^3-smooth diffeomorphisms [13]. In 1987, Sinai and K. Khanin [18] suggested a new approach to the Herman theory based on the renormalization ideas. Using the new approach they proved smoothness of linearization for $C^{2+\epsilon}$-smooth diffeomorphisms. It is important to mention that this result is essentially sharp. Indeed for C^2-diffeomorphisms with typical rotation numbers the linearization is singular in general. In a couple of years in another joint paper [26] Sinai and K. Khanin proved much stronger result. Assume that T is a $C^{2+\epsilon}$-smooth diffeomorphism for $\epsilon > 0$. Also assume that T has an irrational rotation number ρ which belongs to the Diophantine class D_δ. Namely, there exists a constant $c(\rho) > 0$ such that for all $p, q \in \mathbb{Z}$, $q \neq 0$ the following inequality holds: $|\rho q - p| \geq c(\rho) q^{-1-\delta}$. Then the conjugacy with the linear rotation by the angle ρ is $C^{1+\epsilon-\delta}$ smooth, provided $\epsilon > \delta$. A simpler proof of this result was given recently by K. Khanin and A. Teplinsky [16]. Note that also at the end of 1980s another approach to the Herman theory was developed independently by Y. Katnelson and D. Ornstein [14].

Concluding our brief discussion of Sinai's work in the area of dynamical renormalization, let us mention two more papers by Sinai. In the first one, joint with K. Khanin, the renormalization was applied to the construction of the KAM invariant curves for area-preserving cylinder maps similar to the Standard map [17].

In the second paper (joint with A. Golberg and K. Khanin) a new phenomenon of complex universality was discovered numerically [11]. The complex universality is a generalization of the Feigenbaum universality. It concerns with universal asymptotic properties of sequences of bifurcations for families of holomorphic maps. It is interesting that the development of a mathematical theory for complex universality is still an open problem.

6 Random Burgers Equation

Sinai always followed with great interest the developments in mathematical hydrodynamics. Of course, that is not surprising for a student of A. Kolmogorov. We have already mentioned his early work (joint with L. Meshalikin) on the linear stability of the so-called Kolmogorov flow [19]. Then it was a long break until the end of the 1980s and the beginning of 1990s, when Sinai started to work on problems related to the random Burgers equation. As we already explained above, we decided to include the work in this direction in the Mathematical Physics chapter since the random Burgers equation and closely related Kardar–Parisi–Zhang (KPZ) equation in our days are more popular in the Mathematical Physics community.

The Burgers equation was initially suggested by J. Burgers as a model nonlinear equation of the hydrodynamics type:

$$u_t + (u \cdot \nabla)u = \upsilon \Delta u + f(x, t),$$

where $\nu > 0$ is the viscosity, and $f(t, x)$ is an external force applied to the system. The main difference with the Navier–Stokes equation is the absence of the pressure term which is responsible for the incompressibility condition. Hence the Burgers equation correspond to compressible flows. Sometimes people studying the Burgers equation even speak about turbulence without pressure. Another name which is used increasingly often is Burgulence. Despite the fact that the dynamics described by the Burgers equation has very little in common with "real" hydrodynamics, there is huge interest in the Burgers equation, and its importance is related to its numerous applications in non-equilibrium statistical mechanics and mathematical physics. The examples of such applications are provided by cosmological models of large scale structures in the universe which goes back to the original approach by Ya. Zeldovich, dynamics of interfaces which is described by the KPZ equation, and many others. The disordered situation when a random element is present in the system, is the most interesting case in this area. Usually the two cases of disordered input are considered. The first one corresponds to the random initial conditions (so-called decaying turbulence). The second setting is provided by systems with the random external force. In both settings the most interesting case is the inviscid Burgers equation when the shock waves (or, simply, shocks) are formed. The shocks are evolving in time and merge with each other which create a physical mechanism for the dissipation of energy even in the case of zero viscosity.

Initially Sinai's interest in the Burgers equation was inspired by his interaction with U. Frisch and with V. Yahot. He wrote several important papers on the random Burgers equation in both the setting of the random initial conditions and the random forcing setting. In the first paper dedicated to the one-dimensional inviscid Burgers equation, Sinai studied the case when the initial condition is given by a realization of the Wiener process [25]. The main result of the paper is a very precise description of the structure of shocks. In one-dimensional case, shocks are located at isolated points. The set of these points has a very complicated structure. Sinai proved that for an arbitrary small t, the shocks form a set of the Hausdorff dimension $1/2$. This important result generated a lot of activity in studying models with different statistical assumptions on the initial conditions.

The second direction of Sinai's work is related to the Burgers equation with a random external force $f^{\omega}(x, t)$. Since the random force is pumping energy into the system, which compensates the dissipation due to the merging of shocks, one can expect a stationary statistical behavior in this case. In a very influential physical paper A. Polyakov suggested to apply the methods of Quantum Field Theory to the problem of Burgulence. The theory developed by Polyakov predicted certain scaling behavior in the stationary regime. In particular, he studied the probability distribution for a random variable $\xi = u_x(x, t)$ which represent the gradient of the velocity field. In the stationary regime, the probability law for ξ is given by a density $p(\xi)$ which is obviously not universal and depends on the statistical properties of the forcing $f^{\omega}(x, t)$. At the same time, it is natural to expect that the asymptotic behavior of $p(\xi)$, namely the tails as $\xi \to \pm\infty$, are universal. It is easy to see that the tails are asymmetric. Indeed, large positive values of ξ have an extremely small probability. It is possible to show that $-\log p(\xi) \sim \xi^3$ as $\xi \to +\infty$. On the other hand, the negative tail of $p(\xi)$ as $\xi \to -\infty$ should behave as $|\xi|^{\alpha}$ for some α negative. Polyakov's theory predicted the value of $\alpha = -5/2$. This prediction was disproved in Sinai's papers, joint with Weinan E, K. Khanin, and A. Mazel [8, 9]. The main result of [9] is the development of the theory of stationary solutions in the one-dimensional case. The random Burgers equation is closely related to the theory of random Lagrangian systems. It turns out that for such Lagrangian systems and for any value of the average drift almost surely, there exists a unique global minimizer. Moreover, this minimizer is a hyperbolic orbit of the random Lagrangian flow with one stable and one unstable direction. One can show that for any given time t the unique global solution $u^{\omega}(x, t)$ corresponds to the unstable manifold of the global minimizer. It follows that the stationary solution is piecewise smooth, and, hence, at any given time, the number of shocks is finite. Using this conclusion one can show that the main contribution to the probability of large negative values of ξ comes from the pre-shock points, i.e., such space-time locations where new shocks are created. The contribution of such preshock points is easy to estimate which gives the right value of $\alpha = -7/2$. The papers [8, 9] were important for the development of this research area. Later the results were extended to the multi-dimensional setting in the papers by R. Iturriaga, K. Khanin and by K. Khanin, K. Zhang. Very interesting and important is the problem of stationary solutions to the random Burgers equation in the non-compact (non-periodic) case. The first results in this direction were obtained recently by Yu. Bakhtin, E. Cator, and K. Khanin. Note that this problem is closely related to the problem of KPZ universality which was extremely actively studied in the last 10 years.

References

1. V.V. Anshelevich, K.M. Khanin, and Ya.G. Sinai. Symmetric random walks in random environments. *Comm. Math. Phys.*, 85(3):449–470, 1982.
2. V.I. Arnold. Small denominators. I. Mapping the circle onto itself (in Russian). *Izv. Akad. Nauk SSSR Ser. Mat.*, 25:21–86, 1961.
3. P.M. Bleher and Ya.G. Sinai. Investigation of the critical point in models of the type of Dyson's hierarchical models. *Comm. Math. Phys.*, 33(1):23–42, 1973.
4. P.M. Bleher and Ya.G. Sinai. Critical indices for Dyson's asymptotically-hierarchical models. *Comm. Math. Phys.*, 45(3):247–278, 1975.
5. C. Boldrighini and D. Li. Sinai's dynamical system perspective on mathematical fluid dynamics. In H. Holden, R. Piene, editors, *The Abel Prize 2013–2017*, pages 175–194. Springer, 2019.
6. E.I. Dinaburg and Ya.G. Sinai. The one-dimensional Schrödinger equation with a quasiperiodic potential. *Funct. Anal. Appl.*, 9:279–289, 1976.
7. R. Dobrushin, R. Kotecký, and S. Shlosman. *Wulff Construction*, volume 104 of *Translations of Mathematical Monographs*. American Mathematical Society, Providence, RI, 1992.
8. W. E, K. Khanin, A. Mazel, and Ya. Sinai. Probability distribution functions for the random forced Burgers equation. *Phys. Rev. Lett.*, 78:1904–1907, March 1997.
9. W. E, K. Khanin, A. Mazel, and Ya. Sinai. Invariant measures for Burgers equation with stochastic forcing. *Ann. of Math. (2)*, 151(3):877–960, 2000.
10. J. Fröhlich, T. Spencer, and P. Wittwer. Localization for a class of one-dimensional quasi-periodic Schrödinger operators. *Comm. Math. Phys.*, 132(1):5–25, 1990.
11. A.I. Golberg, Ya.G. Sinai, and K.M. Khanin. Universal properties for sequences of bifurcations of period three. *Russ. Math. Surv.*, 38(1):187–188, 1983.
12. A.O. Golosov. Localization of random walks in one-dimensional random environments. *Comm. Math. Phys.*, 92(4):491–506, 1984.
13. M.-R. Herman. Sur la conjugaison différentiable des difféomorphismes du cercle à des rotations. *Inst. Hautes Études Sci. Publ. Math.*, (49):5–233, 1979.
14. Y. Katznelson and D. Ornstein. The differentiability of the conjugation of certain diffeomorphisms of the circle. *Ergodic Theory Dynam. Systems*, 9(4):643–680, 1989.
15. H. Kesten, M.V. Kozlov, and F. Spitzer. A limit law for random walk in a random environment. *Compositio Math.*, 30:145–168, 1975.
16. K. Khanin and A. Teplinsky. Herman's theory revisited. *Invent. Math.*, 178(2):333–344, 2009.
17. K.M. Khanin and Ya.G. Sinai. The renormalization group method and the KAM theory. In R.Z. Sagdeev, editor, *Nonlinear Phenomena in Plasma Physics and Hydrodynamics*, pages 93–118. Mir, Moscow, 1986.
18. K.M. Khanin and Ya.G. Sinai. A new proof of M. Herman's theorem. *Comm. Math. Phys.*, 112(1):89–101, 1987.
19. L.D. Meshalkin and Ya.G. Sinai. Investigation of the stability of a stationary solution of a system of equations for the plane movement of an incompressible viscous liquid. *J. Appl. Math. Mech.*, 25:1700–1705, 1961.
20. R.A. Minlos and Ya.G. Sinai. The phenomenon of "separation of phases" at low temperatures in certain lattice models of a gas. I. *Math. USSR-Sb.*, 2(3):335–395, 1967.
21. R.A. Minlos and Ya.G. Sinai. The phenomenon of "separation of phases" at low temperatures in certain lattice models of a gas. II. *Trans. Moscow Math. Soc.*, 19:121–196, 1968.
22. E. Pechersky (guest editor), The European Physical Journal H. *Seminar on Mathematical Statistical Physics in Moscow State University, 1962–1994*, volume 37(4). Springer, 2012.
23. Ya.G. Sinai. The limiting behavior of a one-dimensional random walk in a random medium. *Theory Probab. Appl.*, 27:256–268, 1982.
24. Ya.G. Sinai. Anderson localization for one-dimensional difference Schrödinger operator with quasiperiodic potential. *J. Statist. Phys.*, 46(5–6):861–909, 1987.

25. Ya.G. Sinai. Statistics of shocks in solutions of inviscid Burgers equation. *Comm. Math. Phys.*, 148(3):601–621, 1992.
26. Ya.G. Sinai and K.M. Khanin. Smoothness of conjugacies of diffeomorphisms of the circle with rotations. *Russ. Math. Surv.*, 44(1):69–99, 1989.
27. E.B. Vul, Ya.G. Sinai, and K.M. Khanin. Feigenbaum universality and the thermodynamic formalism. *Russ. Math. Surv.*, 39(3):1–40, 1984.

Sinai's Work on Markov Partitions and SRB Measures

Yakov Pesin

Abstract Some principal contributions of Ya. Sinai to hyperbolic theory of dynamical systems, focusing mainly on constructions of Markov partitions and of Sinai–Ruelle–Bowen measures, are discussed. Some further developments in these directions stemming from Sinai's work, are described.

1 Introduction

In this article I discuss some of the many principal contributions of Ya. Sinai to the hyperbolic theory of smooth dynamical systems. I focus on two related topics: (1) Markov partitions and (2) Sinai–Ruelle–Bowen (SRB) measures. Dynamical systems that admit Markov partitions with finite or countable number of partition elements allow symbolic representations by topological Markov shifts with finite or respectively countable alphabet. As a result these systems exhibit high level of chaotic behavior of trajectories. SRB-measures serve as *natural* invariant measures with rich collection of ergodic properties. Various constructions of Markov partitions as well as of SRB-measures represent an important and still quite active area of research in dynamics that utilizes Sinai's original ideas and develops them further to cover many other classes of dynamical systems. Therefore, along with describing results by Ya. Sinai, I briefly survey some of the latest developments in this area.

I stress that hyperbolic theory of dynamical systems provides a rigorous mathematical foundation for studying models in science that exhibit chaotic motions. For reader's convenience, I begin with an informal discussion of the role that the hyperbolic theory plays in studying various chaotic phenomena.

The author was partially supported by NSF grant DMS–1400027.

Y. Pesin (✉)
Department of Mathematics, Pennsylvania State University, University Park, PA, USA
e-mail: pesin@math.psu.edu

H. Holden, R. Piene (eds.), *The Abel Prize 2013–2017*,
https://doi.org/10.1007/978-3-319-99028-6_11

1.1 From Scientific Determinism to Deterministic Chaos

In the nineteenth century the prevailing view in dynamics was *causal* or *scientific determinism* best expressed by Laplace as follows:

> We may regard the present state of the universe as the effect of its past and the cause of its future. An intellect which at a certain moment would know all forces that set nature in motion, and all positions of all items of which nature is composed, if this intellect were also vast enough to submit these data to analysis, it would embrace in a single formula the movements of the greatest bodies of the universe and those of the tiniest atom; for such an intellect nothing would be uncertain and the future just like the past would be present before its eyes.

It took about a century to shake up this view with the discovery by Poincaré—in his work on the three-body problem—of the existence of *homoclinic tangles* formed by intersections of stable and unstable separatrices of a hyperbolic fixed point.

Poincaré wrote:

> When we try to represent the figure formed by these two curves and their infinitely many intersections... one must be struck by the complexity of this shape, which I do not even attempt to illustrate. Nothing can give us a better idea of the complication of the three-body problem, and in general of all problems of dynamics for which there is no uniform integral.

In 1963, in his talk at the International Conference on Nonlinear Oscillations (Kiev, Ukraine), Smale [66] made the crucial observation that the homoclinic tangle contains a *horseshoe*, i.e., a fractal set that is locally the product of two Cantor sets. One obtains this set by taking the closure of the set of intersections of stable and unstable separatrices near the fixed point. The horseshoe provided the first example of a differential map with infinitely many hyperbolic periodic points.

Smale's discovery was an important step in shaping up a new area of research in dynamical systems—the hyperbolicity theory—that studies relations between chaotic motions, instability of trajectories and fractal structure of invariant sets. The foundation of this new area was built in the 1960s–1970s in seminal works of Anosov, Sinai, and Smale, see [5, 6, 59–62, 66, 67]. I would like also to emphasize an important role for the development of the theory of dynamical systems that was played during this time by two Moscow seminars, one run by Alekseev and Sinai[1] and another one by Anosov and Katok (see [19, 37, 38]) as well as by Smale's school at Berkeley.

The current view on dynamics draws a much richer picture allowing a variety of motions ranging from *regular* to *intermittently chaotic* to *all-time chaotic*. Moreover, a dynamical system, which is typical in a sense, should possess an *invariant fractal set* of complicated self-similar geometric structure, and the trajectories that start on or in a vicinity of this set are unstable (hyperbolic). The combination of fractality of the set and instability of trajectories causes these trajectories to behave unpredictably; such a chaotic behavior can persist all the time or can be intermittent.

[1] After Alekseev's untimely death in 1980, the seminar was run by Sinai only.

Furthermore, one should typically expect to have infinitely many such fractal sets, which are mixed together in one *invariant multi-fractal set*. These fractal sets can occupy either the whole phase space, or a part of it, in which case the dynamics on its complement can be quite regular—the highly non-trivial phenomenon known as the *essential coexistence*, see [25, 26, 35] for a detailed description of the phenomenon and recent examples of systems with discrete and continuous time that exhibit it.

To describe the phenomenon of the appearance of "chaotic" motions in purely deterministic dynamical systems, one uses the controversial but expressive term *deterministic chaos*.[2] Its crucial feature is that the chaotic behavior is not caused by an external random force such as white noise, but by the system itself. The source of the deterministic chaotic behavior is instability along typical trajectories of the system, which drives orbits apart. On the other hand, compactness of the phase space forces them back together; the consequent unending dispersal and return of nearby trajectories is one of the hallmarks of chaos.

After Poincaré, the fact that instability can cause some complicated chaotic behavior was further observed and advanced in works of Birkhoff, Hadamard, Hopf, and Morse. Many years later some systems with chaotic behavior were found and studied numerically by Lorenz, Chirikov, Ford, Zaslavsky, etc. I refer the reader to Sinai's articles [63–65] for a more detailed discussion of the chaos theory, its earlier development and relations between chaotic behavior and instability of trajectories as well as for relevant references. In these papers Sinai also demonstrates how ideas and methods of statistical physics can be used to explain various chaotic phenomena in dynamics.

1.2 Markov Partitions and Symbolic Representations of Chaotic Dynamics

To explain what it means for deterministic trajectories to exhibit chaotic behavior, consider a map f acting on a phase space M and a point $x \in M$. Let us divide the phase space into two parts A and B. Given an orbit $\{f^n(x)\}$, we write 0 if $f^n(x)$ lies in A and 1 otherwise. This way we obtain a *coding* of every trajectory by a two-sided infinite sequence of symbols 0 and 1 that is

$$x \to \omega = \{\dots, \omega_{-2}, \omega_{-1}, \omega_0, \omega_1, \omega_2, \dots\}, \text{ where } \omega_i = 0 \text{ or } 1.$$

The principal question is:

Given a symbolic sequence of 0 *and* 1, *can we find a point* x *whose trajectory is coded by this sequence?*

[2]This term was first used in works of Chirikov, Ford and Yorke.

If so, starting with a random symbolic sequence that is obtained, for example, by flipping a dime, one gets a **random** orbit of the system whose location in either A or B can only be predicted with a certain probability.

Another way to look at this is to say that the system under consideration is modeled by (or equivalent to) the classical Bernoulli process in probability theory.

Smale's horseshoe is a classical example which allows the above coding and hence, a symbolic representation by the full shift on 2 symbols. In many "practical" situations however, one may need more sophisticated partitions of the phase space called *Markov partitions* (the term coined by Sinai). In general, elements of Markov partitions may have very complicated *fractal* structure. These partitions allow one to model the systems by more general Markov (not necessarily Bernoulli) processes with finite or even countable set of states. From the probability theory point of view such processes are chaotic in the strongest possible sense.

The first construction of Markov partitions was obtained by Adler and Weiss [1] in the particular case of hyperbolic automorphisms of the 2-torus (see also Berg, [12] whose work is independent of [1]). As a crucial corollary they observed that the map allowed a symbolic representation by a subshift of finite type and that this can be used to study its ergodic properties.

Sinai's groundbreaking contribution was to realize that existence of Markov partitions is a rather general phenomenon and in [59] he designed a method of successive approximations to construct Markov partitions for general Anosov diffeomorphisms (see Sect. 3 below for more details). Furthermore, in [62] Sinai showed how Markov partitions can be used to study ergodic properties of hyperbolic dynamical systems and he was also the first to observe the analogy between the symbolic models of Anosov diffeomorphisms and lattice gas models in physics—the starting point in developing the thermodynamic formalism.

Using a different approach, Bowen constructed Markov partitions with finitely many elements for Axiom A diffeomorphisms, see [15]. The construction for hyperbolic flows was carried out independently by Bowen [16] and Ratner [54] (see also [15, 17]). Recently, Sarig [57] constructed Markov partitions with a countable number of elements for surface diffeomorphisms with positive topological entropy. Symbolic dynamics associated with hyperbolic systems was also studied by Alekseev [2].

Aside from smooth dynamical systems, Markov partitions with countable number of partition elements were constructed for a particular class of hyperbolic billiards by Bunimovich and Sinai [20] and by Bunimovich, Sinai, and Chernov [21] (see also the article by Szasz [68]).

1.3 Entropy

Introduced by Kolmogorov and Sinai, the metric entropy is one of the most important invariants of dynamics, and this manifests itself in the famous *isomorphism problem*. Given a transformation $T : X \to X$ preserving a measure μ, we say that

(T, μ) is a Bernoulli automorphism if it is metrically isomorphic to the Bernoulli shift (σ, κ) associated to some Lebesgue space (Y, ν), so that ν is metrically isomorphic to Lebesgue measure on an interval together with at most countably many atoms and κ is given as the direct product of $\mathbb{Z}$ copies of ν on $Y^{\mathbb{Z}}$. Bernoulli systems exhibit the highest level of chaotic behavior and entropy is a *complete* invariant that distinguishes one Bernoulli map from another. This statement is known as the isomorphism problem for Bernoulli systems. I refer the reader to the article by Gurevich [32] for a more detailed discussion of this problem, its history, and relevant references, but I would like to emphasize the important role of Sinai's work on weak isomorphism [58] that laid the ground for the famous Ornstein solution of the isomorphism problem for Bernoulli systems, [46, 47].

Since in this paper we are mostly interested in smooth hyperbolic dynamical systems, we will present a formula for the entropy of these systems with respect to smooth or SRB measures. This formula connects the entropy with the Lyapunov exponents (see Theorem 3 below); the latter are asymptotic characteristics of instability of trajectories of the system. We will also discuss the Bernoulli property; establishing it for smooth hyperbolic systems is based on verifying Ornstein's criterium for Bernoullicity.

1.4 Hyperbolicity

Intuitively, hyperbolicity means that the behavior of orbits that start in a small neighborhood of a given one resembles that of the orbits in a small neighborhood of a hyperbolic fixed point. In other words, the tangent space along the orbit $\{f^n(x)\}$ should admit an invariant splitting

$$T_{f^n(x)}M = E^s(f^n(x)) \oplus E^u(f^n(x)) \tag{1}$$

into the *stable subspace* E^s along which the differential of the system contracts and the *unstable subspace* E^u along which the differential of the system expands.

One should distinguish between two types of hyperbolicity: *uniform* and *nonuniform*. In the former case **every** trajectory is hyperbolic and the contraction and expansion rates are **uniform** in x. More generally, one can consider a compact invariant subset $\Lambda \subset M$ and require that f acts uniformly hyperbolic on Λ. Such a set Λ is called *uniformly hyperbolic*. In the case of nonuniform hyperbolicity the set of hyperbolic trajectories has **positive** (in particular, **full**) measure with respect to an invariant measure and the contraction and expansion rates **depend** on x. Thus, nonuniform hyperbolicity is a property of the system as well as of its invariant measure (called *hyperbolic*).

One can extend the notion of hyperbolicity by replacing the splitting (1) along the orbit $\{f^n(x)\}$ with the splitting

$$T_{f^n(x)}M = E^s(f^n(x)) \oplus E^c(f^n(x)) \oplus E^u(f^n(x)) \tag{2}$$

into the *stable* E^s, *unstable* E^u and *central* E^c subspaces with the rates of contraction and/or expansion along the central subspace being slower than the corresponding rates along the stable and unstable subspaces. This is the case of *partial hyperbolicity*.

2 An Overview of Hyperbolicity Theory

In this section I formally introduce three major types of hyperbolicity and briefly discuss some of their basic properties.

2.1 *Uniform Hyperbolicity*

It originated in the work of Anosov and Sinai [5, 6]; see also the book [39] for the state of the art exposition of the uniform hyperbolicity theory.

A diffeomorphism f of a compact Riemannian manifold M is called *uniformly hyperbolic* or *Anosov* if for each $x \in M$ there is a continuous df-invariant decomposition of the tangent space $T_xM = E^s(x) \oplus E^u(x)$ and constants $c > 0$, $\lambda \in (0, 1)$ such that for each $x \in M$:

1. $\|d_x f^n v\| \le c\lambda^n \|v\|$ for $v \in E^s(x)$ and $n \ge 0$;
2. $\|d_x f^{-n} v\| \le c\lambda^n \|v\|$ for $v \in E^u(x)$ and $n \ge 0$.

The distributions E^s and E^u are called *stable* and *unstable*, respectively. One can show that they depend Hölder continuously in x. Clearly, the angle between stable and unstable subspaces is bounded away from zero in x.

Using the classical Hadamard–Perron theorem, for each $x \in M$ one can construct a *local stable manifold* $V^s(x)$ and a *local unstable manifold* $V^u(x)$ such that

(L1) $x \in V^{s,u}(x)$ and $T_x V^{s,u}(x) = E^{s,u}(x)$;
(L2) $f(V^s(x)) \subset V^s(f(x))$ and $f^{-1}(V^u(x)) \subset V^u(f^{-1}(x))$.

Furthermore, define the *global stable manifold* $W^s(x)$ and the *global unstable manifold* $W^u(x)$ by

$$W^s(x) = \bigcup_{n\ge 0} f^{-n}(V^s(f^n(x))), \quad W^u(x) = \bigcup_{n\ge 0} f^n(V^u(f^{-n}(x))).$$

These sets have the following properties:

(G1) they are smooth submanifolds;
(G2) they are invariant under f, that is, $f(W^{s,u}(x)) = W^{s,u}(f(x))$;

(G3) they are characterized as follows:

$$W^s(x) = \{y \in M : d(f^n(y), f^n(x)) \to 0,\ n \to \infty\},$$
$$W^u(x) = \{y \in M : d(f^n(y), f^n(x)) \to 0,\ n \to -\infty\};$$

(G4) they integrate the stable and unstable distributions, that is, $E^{u,s}(x) = T_x W^{u,s}(x)$.

It follows that $W^s(x)$ and $W^u(x)$ form two uniformly transverse f-invariant continuous *stable* and *unstable foliations* W^s and W^u with smooth leaves. In general, the leaves of these foliations depend only continuously on x.[3]

Any sufficiently small perturbation in the C^1 topology of an Anosov diffeomorphism is again an Anosov diffeomorphism. Hence, Anosov diffeomorphisms form an open set in the space of C^1 diffeomorphisms of M.

There are very few particular examples of Anosov diffeomorphisms, namely

1. A linear hyperbolic automorphism of the n-torus given by an $n \times n$-matrix $A = (a_{ij})$ whose entries a_{ij} are integers, $\det A = 1$ or -1, and all eigenvalues $|\lambda| \neq 1$;
2. The Smale automorphism of a compact factor of some nilpotent Lie group (see [67] and also [39]).

A topologically transitive C^2 Anosov diffeomorphism f preserving a smooth measure μ is ergodic, and if f is topologically mixing, then it is a Bernoulli diffeomorphisms with respect to μ. The Bernoulli property was established by Bowen [15], and a much more general result is given by Statement 4 of Theorem 2.

A compact invariant subset $\Lambda \subset M$ is called *hyperbolic* if for every $x \in \Lambda$ the tangent space at x admits an invariant splitting as described above. For each $x \in \Lambda$ one can construct *local* stable $V^s(x)$ and unstable $V^u(x)$ manifolds which have Properties (L1) and (L2).

A hyperbolic set Λ is called *locally maximal* if there exists a neighborhood U of Λ with the property that given a compact invariant set $\Lambda' \subset U$, we have that $\Lambda' \subset \Lambda$. In this case

$$\Lambda = \bigcap_{n \in \mathbb{Z}} f^n(U).$$

Locally maximal hyperbolic sets can be characterized as having *local direct product structure*, that is, given two points $x, y \in \Lambda$, which are sufficiently close to each other, the intersection $[x, y] = V^s(x) \cap V^u(y)$ lies in Λ. If g is a small perturbation in the C^1 topology of a diffeomorphism f with a locally maximal hyperbolic set Λ_f, then g possesses a locally maximal hyperbolic set Λ_g that lies in a small neighborhood of Λ_f.

[3]In fact, the dependence in x is Hölder continuous.

A diffeomorphism f is called an *Axiom A diffeomorphism* if its non-wandering set $\Omega(f)$ is a locally maximal hyperbolic set.

The Spectral Decomposition Theorem claims (see [39]) that the set $\Omega(f)$ of an Axiom A diffeomorphism f can be decomposed into finitely many disjoint closed f-invariant locally maximal hyperbolic sets, $\Omega(f) = \Lambda_1 \cup \cdots \cup \Lambda_m$ such that $f|\Lambda_i$ is topologically transitive. Moreover, for each i there exists a number n_i and a set $A_i \subset \Lambda_i$ such that the sets $f^k(A_i)$ are disjoint for $0 \le k < n_i$, their union is the set Λ_i, $f^{n_i}(A_i) = A_i$, and the map $f^{n_i}|A_i$ is topologically mixing.

2.2 Nonuniform Hyperbolicity

It originated in the work of Pesin [49–51]; see also the books [7, 8] for a sufficiently complete description of the modern state of the theory.

A diffeomorphism f of a compact Riemannian manifold M is *non-uniformly hyperbolic* if there are a measurable df-invariant decomposition of the tangent space $T_xM = E^s(x) \oplus E^u(x)$ and measurable positive functions $\varepsilon(x)$, $c(x)$, $k(x)$ and $\lambda(x) < 1$ such that for almost every $x \in M$ (here $\angle(S_1, S_2)$ denotes the angle between subspaces S_j):

1. $\|df^n v\| \le c(x)\lambda(x)^n \|v\|$ for $v \in E^s(x)$, $n \ge 0$;
2. $\|df^{-n} v\| \le c(x)\lambda(x)^n \|v\|$ for $v \in E^u(x)$, $n \ge 0$;
3. $\angle(E^s(x), E^u(x)) \ge k(x)$;
4. $c(f^m(x)) \le e^{\varepsilon(x)|m|}c(x)$, $k(f^m(x)) \ge e^{-\varepsilon(x)|m|}k(x)$, $\lambda(f^m(x)) = \lambda(x)$, $m \in \mathbb{Z}$.

The last property means that the rates of contraction and expansion (given by $\lambda(x)$) are constant along the trajectory and the estimates in (1) and (2) can deteriorate with a rate which, while exponential, has a sufficiently small exponent.

Non-uniform hyperbolicity can also be expressed in more "practical" terms using the *Lyapunov exponent* of μ:

$$\chi(x, v) = \limsup_{n\to\infty} \frac{1}{n} \log \|df_x^n v\|, \; x \in M, \; v \in T_xM.$$

This means that for all sufficiently large n and a sufficiently small ε,

$$\|df_x^n v\| \sim \exp(\chi(x, v) \pm \varepsilon)n.$$

If $\chi(x, v) > 0$, the differential asymptotically expands v with some exponential rate, and if $\chi(x, v) < 0$, the differential asymptotically contracts v with some exponential rate.

Therefore, f is non-uniformly hyperbolic if for almost every trajectory with respect to μ, the Lyapunov exponent $\chi(x, v)$ is not equal to zero for *every* vector v;

in this case the measure μ is called *hyperbolic*. In other words,

$$E^s(x) = E^s_f(x) = \{v \in T_xM : \chi(x, v) < 0\}, \quad E^u(x) = E^u_f(x) = E^s_{f^{-1}}(x).$$

Nonuniform hyperbolicity is equivalent to the fact that Lyapunov exponents of f are nonzero *almost* everywhere in M (i.e., the smooth invariant measure for f is hyperbolic)—the phenomenon known as the *Anosov rigidity*. One can show that f is an Anosov diffeomorphism if:

1. the Lyapunov exponents for f are nonzero at *every* point $x \in M$, see [33, 45];
2. the Lyapunov exponents for f are nonzero on a set of total measure one, i.e., on a set that has full measure with respect to any invariant measure, see [23, 24].

If μ is a hyperbolic measure, then for almost every $x \in M$ one can construct local stable and unstable manifolds $V^s(x)$ and $V^u(x)$. They depend measurably on x, in particular, their sizes can be arbitrarily small.

An example of a diffeomorphism with nonzero Lyapunov exponents was constructed by Katok [36]. Starting with a hyperbolic automorphism of the 2-torus, he used the *slow-down* procedure in a neighborhood of a hyperbolic fixed point p to turn p into an indifferent fixed point. In particular, the Lyapunov exponents at p are all zero. Katok used this example as a starting point in his construction of area preserving C^∞ diffeomorphisms with nonzero Lyapunov exponents on compact surfaces. This result was extended by Dolgopyat and Pesin [31] who showed that any compact manifold of dimension ≥ 2 admits a volume preserving C^∞ diffeomorphism with nonzero Lyapunov exponents.

2.3 Partial Hyperbolicity

It originated in the work of Brin and Pesin [18] and of Pugh and Shub [34]; see also the book [52] for a sufficiently complete exposition of the core of the theory.

A diffeomorphism f of a compact Riemannian manifold M is called *partially hyperbolic* if for each $x \in M$ there is a continuous df-invariant decomposition of the tangent space $T_xM = E^s(x) \oplus E^c(x) \oplus E^u(x)$ and constants $c_1, c_2, c_3, c_4 > 0$, $\lambda_1 < \mu_1 \leq \mu_2 < \lambda_2$, $\mu_1 \leq 1$ such that for each $x \in M$ and $n \geq 0$:

1. $\|d_x f^n v\| \leq c_1 \lambda_1^n \|v\|$ for $v \in E^s(x)$;
2. $\|d_x f^n v\| \geq c_2 \lambda_2^n \|v\|$ for $v \in E^u(x)$;
3. $c_4 \mu_1^n \|v\| \leq \|d_x f^n v\| \leq c_3 \mu_2^n \|v\|$ for $v \in E^c(x)$.

The distributions E^s, E^u, and E^c are called *stable*, *unstable* and *central*, respectively. They depend continuously in x.[4] Clearly, the angle between any two subspaces $E^s(x)$, $E^u(x)$ and $E^c(x)$ is bounded away from zero uniformly in x.

[4] One can show that the dependence in x is Hölder continuous.

Any sufficiently small perturbation in the C^1 topology of a partially hyperbolic diffeomorphism is again a partially hyperbolic diffeomorphism. Hence, partially hyperbolic diffeomorphisms form an open set in the space of C^1 diffeomorphisms of M.

The stable and unstable distributions E^s and E^u can be integrated to continuous foliations W^s and W^u, respectively, with smooth leaves. The central distribution may or may not be integrable.

Some well-known examples of partially hyperbolic diffeomorphisms are (1) a direct product of an Anosov diffeomorphism with the identity map of a manifold; (2) a group extension over an Anosov diffeomorphism; (3) the time-1 map of an Anosov flow.

A compact invariant subset $\Lambda \subset M$ is called *partially hyperbolic* if the restriction $f|\Lambda$ is partially hyperbolic in the above sense. For each $x \in \Lambda$, one can construct *local* stable $V^s(x)$ and unstable $V^u(x)$ manifolds.

3 Markov Partitions

3.1 Definition of Markov Partitions

Let Λ be a locally maximal hyperbolic set for a diffeomorphism f of a compact smooth Riemannian manifold M. From now on we assume that $f|\Lambda$ is topologically mixing. The general case can be easily reduced to this one by using the Spectral Decomposition Theorem.

A non-empty closed set $R \subset \Lambda$ is called a *rectangle* if

- $\operatorname{diam} R \le \delta$ (where $\delta > 0$ is sufficiently small);
- $R = \overline{\operatorname{int} R}$ where $\operatorname{int} R$ is defined in the relative topology in R;
- $[x, y] \in R$ whenever $x, y \in R$.[5]

A rectangle R has *direct product structure* that is given $x \in R$, there exists a homeomorphism

$$\theta\colon R \to R \cap V^{(s)}(x) \times R \cap V^{(u)}(x).^6$$

One can show that both θ and θ^{-1} are Hölder continuous. A finite cover $\mathscr{R} = \{R_1, \dots, R_p\}$ of Λ by rectangles R_i, $i = 1, \dots, p$ is called a *Markov partition* for f if

[5]We use here the fact that the set Λ is locally maximal.

[6]In other words, θ identifies the rectangle R with the product $R \cap V^{(s)}(x) \times R \cap V^{(u)}(x)$.

1. $\operatorname{int} R_i \cap \operatorname{int} R_j = \emptyset$ unless $i = j$;
2. for each $x \in \operatorname{int} R_i \cap f^{-1}(\operatorname{int} R_j)$ we have

$$f(V^s(x) \cap R_i) \subset V^s(f(x)) \cap R_j, \quad f(V^u(x) \cap R_i) \supset V^u(f(x)) \cap R_j.$$

These relations are called the *Markov property* of the Markov partition. We stress that despite the name a Markov partition $\mathscr{R}$ is a **cover** of Λ which is **almost** a partition: any two elements of the cover can intersect only along their boundaries.

3.2 Symbolic Models

A Markov partition $\mathscr{R} = \{R_1, \dots, R_p\}$ generates a symbolic model of $f|\Lambda$ by a finite Markov shift or a subshift of finite type (Σ_A, σ), where Σ_A is the set of two-sided infinite sequences of numbers $\{1, \dots, p\}$, which are admissible with respect to the transfer matrix of the Markov partition $A = (a_{ij})$ (i.e., $a_{ij} = 1$ if $\operatorname{int} R_i \cap f^{-1}(\operatorname{int} R_j) \neq \emptyset$, and $a_{ij} = 0$ otherwise). Namely, define

$$R^{(u)}_{i_0 \dots i_n} = \bigcap_{j=0}^{n} f^{-j}(R_{i_j}), \quad R^{(s)}_{i_{-n} \dots i_{-1}} = \bigcap_{j=-1}^{-n} f^{-j}(R_{i_j}),$$

$$R_{i_{-n} \dots i_n} = R^{(s)}_{i_{-n} \dots i_{-1}} \cap R^{(u)}_{i_0 \dots i_n}.$$

Now we define the *coding map* $\chi : \Sigma_A \to \Lambda$ by

$$\chi(\dots, i_{-n}, \dots, i_0, \dots i_n, \dots) = \bigcap_{n \geq 0} R_{i_{-n} \dots i_n}.$$

Note that the maps f and σ are conjugate via the coding map χ, i.e., $f \circ \chi = \chi \circ \sigma$. The map χ is Hölder continuous and injective on the set of points whose trajectories never hit the boundary of any element of the Markov partition.

For any points $\omega = (\dots, i_{-1}, i_0, i_1, \dots) \in \Sigma_A$ and $\omega' = (\dots, i'_{-1}, i'_0, i'_1, \dots) \in \Sigma_A$ with the same past (i.e., $i'_j = i_j$ for any $j \leq 0$) we have that $\chi(\omega') \in V^{(u)}(x) \cap R(x)$, where $x = \chi(\omega)$ and $R(x)$ is the element of a Markov partition containing x. Similarly, for any point $\omega'' = (\dots, i''_{-1}, i''_0, i''_1, \dots) \in \Sigma_A$ with the same future as ω (i.e., $i''_j = i_j$ for any $j \geq 0$) we have that $\chi(\omega'') \in V^{(s)}(x) \cap R(x)$. Thus, the set $V^{(u)}(x) \cap R(x)$ can be identified via the coding map χ with the cylinder $C^+_{i_0}$ in the space Σ^+_A of "positive" one-sided infinite sequences of numbers $\{1, \dots, p\}$ and the set $V^{(s)}(x) \cap R(x)$ can be identified via the coding map χ with the cylinder $C^-_{i_0}$ in the space Σ^-_A of "negative" one-sided infinite sequences of numbers $\{1, \dots, p\}$.

3.3 Sinai's Construction of Markov Partitions

I outline here a construction of Markov partition from [59] (see also [27]). For simplicity, I only consider Anosov diffeomorphisms of the two-dimensional torus, in which case the geometry of the construction is rather simple and it produces a partition whose elements are connected subsets with non-empty interior. In the multi-dimensional case the requirement that partition elements are connected cannot be ensured unless one allows partitions with countable number of elements. For the construction of Markov partitions for general Axiom A maps I refer the reader to the works of Bowen [14, 15].[7]

Let f be an Anosov diffeomorphism of the two dimensional torus $\mathbb{T}^2$. Fix $\varepsilon > 0$. We shall construct a Markov partition with diameter of elements $\leq \varepsilon$. It suffices to do so for some power n of f. Indeed, if $\mathscr{R}$ is a Markov partition for f^n, then $\bigcap_{k=-n}^{n} f^k \mathscr{R}$ is a Markov partition for f.

For points in the torus, local stable and unstable manifolds are smooth curves which are called *stable* and *unstable curves*. In the course of our construction every rectangle R is a closed **connected** subset of the torus. Its boundary ∂R is the union of four curves, two of which are stable and the other two are unstable. The union of stable curves forms the *stable boundary* $\partial^s R$ of R while the union of unstable curves forms the *unstable boundary* $\partial^u R$ of R. For every $x \in R$ we denote by $\gamma_R^s(x)$ (respectively, $\gamma_R^u(x)$) the *full length* stable (respectively, unstable) curve through x, i.e., the segment of stable (unstable) curve whose endpoints lie on the unstable (stable) boundary of R.

Let us now fix $\delta > 0$, $n > 0$, and let $\lambda \in (0, 1)$ be the constant in the definition of Anosov diffeomorphisms. A collection of rectangles $\tilde{\mathscr{R}} = \{\tilde{R}_1, \dots, \tilde{R}_p\}$ is called a *sufficient* (n, δ)-collection if

1. $\bigcup_{j=1}^{p} \tilde{R}_j = \mathbb{T}^2$;
2. $\operatorname{diam} \tilde{R}_j \leq \delta$, $j = 1, \dots, p$;
3. given a rectangle $\tilde{R}_j$, one can find two subcollections of rectangles $\{\tilde{R}_{i_1}, \dots, \tilde{R}_{i_k}\}$ and $\{\tilde{R}_{s_1}, \dots, \tilde{R}_{s_t}\}$ such that
   (a) $f^n(\tilde{R}_j) \subset \bigcup_{\ell=1}^{k} \tilde{R}_{i_\ell}$ and $f^{-n}(\tilde{R}_j) \subset \bigcup_{\ell=1}^{t} \tilde{R}_{s_\ell}$;
   (b) for every $x \in \tilde{R}_j$, if $f^n(x)$ lies in some rectangle $\tilde{R}_{i_\ell}$ from the first subcollection, then $f^n(\gamma_{\tilde{R}_j}^s(x)) \subset \gamma_{\tilde{R}_{i_\ell}}^s(f^n(x))$;
   (c) for every $x \in \tilde{R}_j$, if $f^{-n}(x)$ lies in some rectangle $\tilde{R}_{s_\ell}$ from the second subcollection, then $f^{-n}(\gamma_{\tilde{R}_j}^u(x)) \subset \gamma_{\tilde{R}_{s_\ell}}^u(f^{-n}(x))$.

It is not difficult to show that given $\delta > 0$ and a large enough $n > 0$, there is a sufficient (n, δ)-collection $\tilde{\mathscr{R}}$.

[7] In [14], Bowen used a method similar to the original Sinai method known as the method of successive approximations. In [15] he used a different approach based on pseudo-orbits.

Our goal is to slightly "extend" each rectangle of a given sufficient (n, δ)-collection $\tilde{\mathscr{R}}$ in both the stable and unstable directions to ensure the Markov property in these directions. This will produce a cover of the torus by rectangles, which is a *Markov cover*.

To this end fix a rectangle $\tilde{R}_j \in \tilde{\mathscr{R}}$ and consider two subcollections $\{\tilde{R}_{i_1}, \dots, \tilde{R}_{i_k}\}$ and $\{\tilde{R}_{s_1}, \dots, \tilde{R}_{s_t}\}$, which have the properties with respect to $\tilde{R}_j$ mentioned above. We refer to the union of the (un)stable boundaries of all rectangles in the subcollection as the (un)stable boundary of the subcollection.

Consider the set $f^n(\partial^u \tilde{R}_j)$. It consists of two unstable curves $\gamma_1^u = f^n(\tilde{\gamma}_1^u)$ and $\gamma_2^u = f^n(\tilde{\gamma}_2^u)$ where the curves $\tilde{\gamma}_1^u$ and $\tilde{\gamma}_2^u$ form the unstable boundary of $\tilde{R}_j$. Denote by A_1, B_1 and A_2, B_2 the endpoints of these curves. We refer to A_1 and A_2 as the *left* endpoints of γ_1^u and γ_2^u, respectively, and to B_1 and B_2 as the *right* endpoints of γ_1^u and γ_2^u, respectively. If $\tilde{R}_m$ is a rectangle from the subcollection that contains A_1, then it also contains A_2.

Consider now the full length unstable curve $\gamma^u_{\tilde{R}_m}(A_1)$. It intersects the stable boundary of $\tilde{R}_m$ at two points, C_1 and D_1. One of them, say C_1, lies on the "left" of A_1 and does not belong to the curve γ_1^u (while the other one does). We now extend the curve $\tilde{\gamma}_1^u$ to the left by adding the segment $f^{-n}(A_1C_1)$ to its left point $f^{-n}(A_1)$. It is easy to see that the length of this segment does not exceed $\delta\lambda^{-n}$. Similarly, the full length unstable curve $\gamma^u_{\tilde{R}_m}(A_2)$ intersects the stable boundary of $\tilde{R}_m$ at two points, C_2 and D_2 of which C_2 lies on the "left" of A_2 and does not belong to the curve γ_2^u. We again extend the curve $\tilde{\gamma}_2^u$ to the left by adding the segment $f^{-n}(A_2C_2)$ to its left point $f^{-n}(A_2)$. The length of this segment does not exceed $\delta\lambda^{-n}$. As a result we obtain a new rectangle $\tilde{R}_j^l$, which is a left extension of the rectangle $\tilde{R}_j$. The left stable boundary of this new rectangle is the stable curve $f^{-n}(C_1C_2)$.

In a similar manner we can extend the rectangle $\tilde{R}_j$ to the right and obtain a new rectangle which has the Markov property in the unstable direction with respect to the subcollection associated to $\tilde{R}_j$. Continuing in this way, we obtain a new cover $\tilde{\mathscr{R}}^{(1)} = \{\tilde{R}_1^{(1)}, \dots, \tilde{R}_p^{(1)}\}$ which has the Markov property in the unstable direction with respect to the cover $\tilde{R}$. Note that the diameter of each rectangle in the new cover in the unstable direction does not exceed $\delta + \delta\lambda^n$, while the diameter in the stable direction does not exceed δ. Proceeding by induction we obtain a sequence of covers $\tilde{\mathscr{R}}^{(q)} = \{\tilde{R}_1^{(q)}, \dots, \tilde{R}_p^{(q)}\}$ such that

1. rectangles in the cover $\tilde{\mathscr{R}}^{(q)}$ have the Markov property in the unstable direction with respect to the rectangles in the cover $\tilde{\mathscr{R}}^{(q-1)}$;
2. the diameter of each rectangle in the cover $\tilde{\mathscr{R}}^{(q)}$ in the unstable direction does not exceed

$$\frac{1 - \lambda^{-(q+1)n}}{1 - \lambda^{-n}} 2\delta,$$

while the diameter in the stable direction does not exceed δ;

3. for each $q > 0$ and $j = 1, \ldots, p$ we have that $\tilde{R}_j^{(q)} \subset \tilde{R}_j^{(q-1)}$ and the rectangle $\tilde{R}_j^{(q)}$ is a connected subset.

One can show that the sets $R_j^+ = \bigcup_{q>0} \tilde{R}_j^{(q)}$ forms a cover $\mathscr{R}^+$, which has the Markov property in the unstable direction and whose diameter in the unstable direction does not exceed $\frac{1}{1-\lambda^{-n}} 2\delta$, while the diameter in the stable direction does not exceed δ. Furthermore each rectangle in this cover is a connected subset. Replacing f^n with f^{-n} and repeating the above argument, we can slightly extend each element of the cover $\mathscr{R}^+$ in the stable direction to obtain a new cover $\mathscr{R} = (\mathscr{R}^+)^-$ which has the Markov property in both the unstable and stable directions and whose diameter does not exceed $\frac{1}{1-\lambda^{-n}} 2\delta$. Moreover, each rectangle in this cover is a connected subset. One can now subdivide the rectangles of the cover to obtain the desired Markov partition.

4 SRB Measures I: Hyperbolic Attractors

4.1 Topological Attractors

Let f be a diffeomorphism of a compact smooth Riemannian manifold M. A compact invariant subset $\Lambda \subset M$ is called a *topological attractor* for f if there is an open neighborhood U of Λ such that $\overline{f(U)} \subset U$ and

$$\Lambda = \bigcap_{n \geq 0} f^n(U).$$

The set U is said to be a *trapping region* or a *basin of attraction* for Λ. The maximal open set with this property is called the *topological basin of attraction* for Λ. It follows immediately from the definition of the attractor that Λ is locally maximal, i.e., is the largest invariant set in U.

4.2 Natural and Physical Measures

Starting with the volume m in U, consider its evolution under the dynamics, i.e., the sequence of measures

$$m_n = \frac{1}{n} \sum_{k=0}^{n-1} f_*^k m. \tag{3}$$

This sequence is compact in the week* topology, and hence has a convergent subsequence m_{n_k}. Clearly, the limit of m_{n_k} is supported on Λ and by the Bogolubov–Krylov theorem, it is an f-invariant measure called a *natural* measure for f. In general, the measure μ may be quite trivial—just consider the point mass at an attracting fixed point.

Given a measure μ on an attractor Λ, define its *basin of attraction* $B(\mu)$ as the set of μ-generic points $x \in U$, i.e., points such that for every continuous function φ on Λ,

$$\lim_{n\to\infty} \frac{1}{n}\sum_{k=0}^{n-1} \varphi(f^k(x)) = \int_\Lambda \varphi\, d\mu. \tag{4}$$

A natural measure μ on the attractor Λ is a *physical measure* if its basin of attraction $B(\mu)$ has positive volume. An attractor with a physical measure is called a *Milnor attractor*.

4.3 SRB Measures

Let μ be a hyperbolic invariant measure supported on Λ. Using results of nonuniform hyperbolicity theory one can construct for almost every $x \in \Lambda$ a local stable $V^s(x)$ manifold and a local unstable $V^u(x)$ manifold. It is easy to show that for such points x we have $V^u(x) \subset \Lambda$ and consequently, $W^u(x) \subset \Lambda$ (recall that $W^u(x)$ is the global unstable manifold through x). On the other hand, the intersection of Λ with stable manifolds of its points is a Cantor set.

There is a collection $\{\Lambda_\ell\}_{\ell\geq 1}$ of nested subsets of Λ that exhaust Λ (mod 0) such that local stable $V^s(x)$ and unstable $V^u(x)$ manifolds depend continuously on $x \in \Lambda_\ell$. In particular, their "sizes" are bounded uniformly from below. Given $x \in \Lambda_\ell$, set

$$Q_\ell(x) = \bigcup_{y\in B(x,r_\ell)\cap\Lambda_\ell} V^u(y),$$

where $r_\ell > 0$ is sufficiently small, and let ξ_ℓ be the partition of $Q_\ell(x)$ by local unstable leaves $V^u(y)$, $y \in B(x, r_\ell)\cap\Lambda_\ell$. Denote by $\mu^u(y)$ the conditional measure on $V^u(y)$ generated by μ with respect to the partition ξ_ℓ and by $m^u(y)$ the leaf-volume on $V^u(y)$ generated by the Riemannian metric.[8]

A hyperbolic invariant measure μ on Λ is called an *SRB measure* (after Sinai, Ruelle and Bowen) if for every ℓ with $\mu(\Lambda_\ell) > 0$, almost every $x \in \Lambda_\ell$ and almost every $y \in B(x, r_\ell) \cap \Lambda_\ell$ the measures $\mu^u(y)$ and $m^u(y)$ are equivalent. The idea of

[8]Both $\mu^u(x)$ and $m^u(x)$ are probability measures.

describing an invariant measure by its conditional probabilities on the elements of a continuous partition goes back to the classical work of Kolmogorov and later work of Dobrushin on random fields, [30] (see also [62]).

The following result describes the density $d^u(x, y)$ of the conditional measure $\mu^u(x)$ with respect to the leaf-volume $m^u(x)$.

Theorem 1 (Sinai [62], Pesin and Sinai [53], Ledrappier [41]) *For almost every x the density $d^u(x, y)$ is given by $d^u(x, y) = \rho^u(x)^{-1}\rho^u(x, y)$ where for $y \in V^u(x)$*

$$\rho^u(x, y) = \prod_{k=0}^{\infty} \frac{Jac(df|E^u(f^{-k}(y)))}{Jac(df|E^u(f^{-k}(x)))} \tag{5}$$

and

$$\rho^u(x) = \int_{V^u(x)} \rho^u(x, y)\, dm^u(x)(y) \tag{6}$$

is the normalizing factor.

The Eq. (5) can be viewed as an analog of the famous Dobrushin–Lanford–Ruelle equation in statistical physics, see [40] and [62].

Using results of nonuniform hyperbolicity theory one can obtain a sufficiently complete description of ergodic properties of SRB measures.

Theorem 2 *Let f be a $C^{1+\epsilon}$ diffeomorphism of a compact smooth manifold M with an attractor Λ and let μ be an SRB measure on Λ. Then*

1. $\Lambda = \bigcup_{i\geq 0} \Lambda_i, \quad \Lambda_i \cap \Lambda_j = \emptyset;$
2. $\mu(\Lambda_0) = 0$ *and* $\mu(\Lambda_i) > 0$ *for* $i > 0$;
3. $f|\Lambda_i$ *is ergodic for* $i > 0$;
4. *for each* $i > 0$ *there is* $n_i > 0$ *such that* $\Lambda_i = \bigcup_{j=1}^{n_i} \Lambda_{ij}$ *where* $f(\Lambda_{ij}) = \Lambda_{i\,j+1}$, $f(\Lambda_{n_i 1}) = \Lambda_{i1}$ *and* $f^{n_i}|\Lambda_{i1}$ *is Bernoulli.*

For smooth measures this theorem was proved by Pesin in [50] and its extension to SRB measures was obtained by Ledrappier in [41] (see also [7, 8]). We note that the proof of the Bernoulli property in Statement 4 of the theorem is based on the work of Ornstein and Weiss who established the Bernoulli property for geodesic flows on compact manifolds of negative curvature, [48].

SRB measures admit the following characterization.

Theorem 3 *Let μ be a measure on Λ of positive entropy. Then μ is an SRB measure if and only if its entropy is given by the* entropy formula*:*

$$h_\mu(f) = \int_\Lambda \sum_{\chi_i(x)>0} \chi_i(x)\, d\mu(x).$$

For smooth measures (which are particular cases of SRB measures) the entropy formula was proved by Pesin [50] (see also [7]) and its extension to SRB measures was obtained by Ledrappier and Strelcyn [42]. The fact that a hyperbolic measure satisfying the entropy formula is an SRB measure was shown by Ledrappier [41].[9]

It follows from Theorem 3 that any ergodic SRB measure is a physical measure (any ergodic component of an SRB measure is an ergodic SRB measure). In particular, if an attractor supports an ergodic SRB measure then it is a Milnor attractor.

The limit measures for the sequence of measures (3) are natural candidates for SRB measures. The classical *eight figure map*[10] is an example of a diffeomorphism f with an attractor Λ such that the sequence of measures (3) converges to a hyperbolic measure μ whose basin of attraction has full volume, however μ is **not** an SRB measure for f.

4.4 Uniformly Hyperbolic Attractors

An attractor Λ is *hyperbolic* if it is a uniformly hyperbolic set for f.[11] The unstable subspace E^u is integrable: given $x \in \Lambda$, the global unstable manifold $W^u(x)$ lies in Λ, and hence the attractor is the union of the global unstable manifolds of its points, which form a lamination of Λ. On the other hand the intersection of Λ with stable manifolds of its points may be a Cantor set.

Theorem 4 *Assume that the map $f|\Lambda$ is topologically transitive. Then the sequence of measures* (3) *converges to a unique SRB measure on Λ and so does the sequence of measures* (7) *(independently of the starting point x).*

This theorem was proved by Sinai, [60] for the case of Anosov diffeomorphisms, Bowen [15], and Ruelle [56] extended this result to hyperbolic attractors, and Bowen and Ruelle [17] constructed SRB measures for Anosov flows.

Well-known examples of hyperbolic attractors are the DA (derived from Anosov) attractor and the Smale–Williams solenoid (see [39] for definitions and details).

In the following two subsections I will outline two different approaches to prove Theorem 4. The first approach was developed by Sinai in [60] and uses Markov partitions, while the second one deals with the sequence of measures (3) in a straightforward way and hence, is more general. In particular, it can be used to

[9] In this paper we use the definition of SRB measure that requires that it is hyperbolic. One can weaken the hyperbolicity requirement by assuming that some (but not necessarily all) Lyapunov exponents are non-zero (with at least one positive). It was proved by Ledrappier and Young [43, 44] that within the class of such measures, SRB measures are the only ones that satisfy the entropy formula.

[10] This is a two dimensional smooth map with a hyperbolic fixed point whose stable and unstable separatrices form the eight figure. Inside each of the two loops there is a repelling fixed point.

[11] Clearly, the set Λ is locally maximal.

construct some special measures for partially hyperbolic attractors which do not allow Markov partitions; these are so called u-measures which are natural analog of SRB measures in this case, see [53] and Sect. 6. For simplicity of the exposition I only consider the case of Anosov diffeomorphisms, extension to hyperbolic attractors is not difficult.

4.5 First Proof of Theorem 4 (Sinai [60])

Let $\mathscr{R}$ be a Markov partition of sufficiently small diameter and let $\mathscr{R}^- = \bigvee_{n=0}^{\infty} f^{-n}\mathscr{R}$. One can show that the partition $\mathscr{R}^-$ has the following properties:

1. $f\mathscr{R}^- \geq \mathscr{R}^-$;
2. $\bigvee_{k=0}^{\infty} f^k\mathscr{R}^-$ is the trivial partition;
3. there is an $r > 0$ such that every element of the partition $\mathscr{R}^-$ is contained in a local stable manifold and contains a ball in this manifold of radius r.

Given $x \in \Lambda$, denote by $C_{\mathscr{R}^-}(x)$ the element of the partition $\mathscr{R}^-$ containing x. For every $n > 0$ we have that $f^n(C_{\mathscr{R}^-}(x)) = C_{f^n(\mathscr{R}^-)}(f^n(x))$ and that f^n is a bijection between $C_{\mathscr{R}^-}(x)$ and $C_{f^n(\mathscr{R}^-)}(f^n(x))$. Therefore, f^{-n} transfers the normalized leaf-volume on $C_{f^n(\mathscr{R}^-)}(f^n(x))$ to a measure on $C_{\mathscr{R}^-}(x)$ which we denote by μ_n. This measure is equivalent to the leaf-volume on $C_{\mathscr{R}^-}(x)$ and we denote by $\rho_n(y)$ the corresponding density function, which is continuous. One can show that the sequence of functions ρ_n converges uniformly to a continuous function $\tilde{\rho}(y) = \tilde{\rho}_{C_{\mathscr{R}^-}(x)}(y)$, which can be viewed as the density function for a normalized measure $\tilde{\mu}_{C_{\mathscr{R}^-}(x)}$ on $C_{\mathscr{R}^-}(x)$. These measures have the following properties:

1. $\tilde{\mu}_{C_{\mathscr{R}^-}(x)}$ is equivalent to the leaf-volume on $C_{\mathscr{R}^-}(x)$;
2. for every measurable set $A \subset C'_{\mathscr{R}^-} \subset C_{f^{-1}(\mathscr{R}^-)}$ the following Chapman–Kolmogorov relation holds:

$$\tilde{\mu}(A|C_{f^{-1}(\mathscr{R}^-)}) = \tilde{\mu}(A|C'_{\mathscr{R}^-})\tilde{\mu}(C'_{\mathscr{R}^-}|C_{f^{-1}(\mathscr{R}^-)});$$

3. the measures $\tilde{\mu}$ are determined by Properties 1 and 2 uniquely.

One can now show that for any $x \in M$ and any measurable subset $A \subset M$ there is a limit

$$\mu(A) = \lim_{n\to\infty} \tilde{\mu}(A|C_{f^{-n}(\mathscr{R}^-)}(x)),$$

which does not depend on x. The number $\mu(A)$ determines an invariant measure for f which is the desired SRB measure.

4.6 Second Proof of Theorem 4 (Pesin and Sinai, [53])

The way of constructing SRB measures on Λ based on the sequence of measures (3) can be viewed as being "from outside of the attractor". There is another way to construct SRB measures "from within the attractor". Fix $x \in \Lambda$ and consider a local unstable leaf $V = V^u(x)$ at x. One can view the leaf-volume $m^u(x)$ on $V^u(x)$ as a measure on the whole of Λ. Consider the sequence of measures on Λ

$$\nu_n(x) = \frac{1}{n}\sum_{k=0}^{n-1} f_*^k m^u(x). \tag{7}$$

We shall show that every limit measure for the sequence of measures (7) is an SRB-measure. In fact, every SRB-measure μ can be constructed in this way, i.e., it can be obtained as the limit measure for a subsequence of measures ν_n. Furthermore, if $f|\Lambda$ is topologically transitive, then the sequence of measures (7) converges to μ and so does the sequence of measures (3).

We stress that in the definition of the sequence of measures (7) one can replace the local unstable manifold $V^u(x)$ with any *admissible* manifold, i.e., a local manifold passing through x and sufficiently close to $V^u(x)$ in the C^1 topology.

Let μ be a limit measure of the sequence of measures (7) and let z be such that $\mu(B(z,r)) > 0$ for every $r > 0$. Consider a rectangle R of size $r > 0$ containing z and its partition ξ into unstable local manifolds $V^u(y)$, $y \in R$. We identify the factor space R/ξ with $W = V^s(z) \cap \Lambda$ and we denote by $U_n = f^n(V)$. Set

$$\begin{aligned} A_n &= \{y \in W\colon V^u(y) \cap U_n \neq \emptyset\}, \\ B_n &= \{y \in W\colon V^u(y) \cap \partial U_n \neq \emptyset\}, \\ C_n &= A_n \setminus B_n. \end{aligned}$$

Note that C_n is a finite set, and we denote by δ_n the measure on W, which is the uniformly distributed point mass on C_n. If h is a continuous function on Λ with support in R, then

$$\begin{aligned} \int_\Lambda h\,d\nu_n &= \int_R h\,d\nu_n \\ &= \sum_{y\in A_n} \int_{V^u(z)\cap U_n} h\,d\nu_n \\ &= \sum_{y\in C_n} \int_{V^u(z)\cap U_n} h\,d\nu_n + \sum_{y\in B_n} \int_{V^u(z)\cap U_n} h\,d\nu_n \\ &= I_n^{(1)} + I_n^{(2)}. \end{aligned}$$

One can show that $I_n^{(2)} \leq \frac{C}{n}$ where $C > 0$ is a constant. One can further show that

$$
\begin{aligned}
I_n^{(1)} &= c_n \sum_{y \in C_n} \rho^u(f^n(x), y) \int_{V^u(y)} h(w)\rho^u(y, w)\, d\mu^u(w) \\
&= \int_W c_n \rho^u(f^n(x), y)\rho^u(y)\, d\delta_n(y) \int_{V^u(y)} h(w) \frac{\rho^u(y, w)}{\rho^u(y)}\, d\mu^u(w),
\end{aligned}
$$

where

$$
c_n = \Big[\prod_{k=0}^{n-1} \mathrm{Jac}(df | E^u(f^k(x)))\Big]^{-1}
$$

and $\rho^u(y)$ is given by (6).

It follows that for any subsequence $n_\ell \to \infty$ for which the sequence of measures $\nu_{n_\ell}(x)$ converges to a measure μ on Λ, one has that μ is an SRB measure.

The above argument implies that $\mu(\mathrm{int}\, R) > 0$ and hence, the set

$$
E = \bigcup_{n \in \mathbb{Z}} f^n(\mathrm{int}\, R)
$$

is open and is an ergodic component of μ of positive measure (i.e., $f|E$ is ergodic). In fact, every ergodic component of μ can be obtained in this way and hence, is open (mod 0). One can derive from here that there are at most finitely many SRB measures and if $f|\Lambda$ is topological transitive, then there is only one SRB measure.

5 SRB Measures II: Chaotic Attractors

5.1 Chaotic Attractors: The Concept

An attractor Λ for a diffeomorphism f is *chaotic* if there is a natural measure that is *hyperbolic*, i.e., a measure with nonzero Lyapunov exponents (with some being positive and some being negative). In this case using results of nonuniform hyperbolicity theory one can show that for almost every $x \in \Lambda$ there are a local stable $V^s(x)$ and unstable $V^u(x)$ manifolds. It is easy to see that for such points x we have $V^u(x) \subset \Lambda$, so that the attractor contains all the unstable manifolds of its points.

5.2 *Chaotic Attractors: Some Open Problems*

1. Construct an example of a diffeomorphism f with an attractor Λ such that the volume m is a non-invariant hyperbolic measure for f (i.e., for almost every $x \in U$ with respect to m and for every nonzero vector $v \in T_xM$ the Lyapunov exponent $\chi(x, v) \neq 0$) but the sequence of measures (3) converges to a measure μ on Λ for which the Lyapunov exponent $\chi(x, v) = 0$ for almost every $x \in U$ with respect to μ and for every nonzero vector $v \in T_xM$;
2. Construct an example of a diffeomorphism f with an attractor Λ such that for almost every $x \in U$ with respect to volume m and for every nonzero vector $v \in T_xM$ the Lyapunov exponent $\chi(x, v) = 0$ but the sequence of measures (3) converges to a hyperbolic measure μ on Λ.

5.3 *The Hénon Attractor*

Consider the Hénon family of maps given by

$$H_{a,b}(x, y) = (1 - ax^2 + by, x). \tag{8}$$

For $a \in (0, 2)$ and sufficiently small b there is a rectangle in the plane, which is mapped by $H_{a,b}$ into itself. It follows that $H_{a,b}$ has an attractor—*the Hénon attractor*.

Benedicks and Carleson [9] developed a highly sophisticated techniques to describe the dynamics near the attractor. Building on this analysis, Benedicks and Young [10] established existence of SRB measures for the Hénon attractors.

Theorem 5 *There exist $\varepsilon > 0$ and $b_0 > 0$ such that for every $0 < b \leq b_0$ one can find a set $A_b \in (2 - \varepsilon, 2)$ of positive Lebesgue measure with the property that for each $a \in A_b$, the map $H_{a,b}$ admits a unique SRB measure $\mu_{a,b}$.*

Wang and Young [71] introduced and studied some more general 2-parameter families of maps with one unstable direction to which the above result extends.

The underlying mechanism of constructing SRB measures in these systems is the work of Young [72] where she introduced a class of non-unformly hyperbolic diffeomorphisms f admitting a symbolic representation via a tower whose base A is a hyperbolic set with direct product structure and the induced map on the base admits a Markov extension. Assuming that the return time R to the base is integrable, one can show that there is an SRB measure.

As in the case of uniformly hyperbolic systems, the real power of a symbolic representation is not just to help prove existence of SRB-measures but to show the exponential decay of correlations, the Central Limit Theorem, etc. For the Hénon attractor, Benedicks and Young [11] showed that if for all $T > 0$ we have $\int R\, dm \leq C\lambda^T$ where $C > 0$, $0 < \lambda < 1$ and the integral is taken over the set of points $x \in A$

with $R(x) > T$, then f has the exponential decay of correlations for the class of Hölder continuous functions.

5.4 Chaotic Attractors: An Example

Let f be a diffeomorphism with an attractor Λ to be the Smale–Williams solenoid. f has an SRB measure on Λ. In a small ball $B(p, r)$ around a fixed point p, the map f is the time-1 map of the linear system $\dot{x} = Ax$ of ODEs.[12] We wish to perturb f locally by *slowing down* trajectories near p. Define a map g to be the time-1 map for the following nonlinear system of ODEs inside $B(p, r)$

$$\dot{x} = \psi(x)Ax$$

and set $g = f$ outside of $B(p, r)$. Here $\psi(x) = \|x\|^\alpha$ for $\|x\| < 2$ and $\psi(x) = 1$ for $\|x\| \geqslant 2$.

Theorem 6 (Climenhaga, Dolgopyat, Pesin, [28]) *The map g has an SRB-measure.*

5.5 Chaotic Attractors: Constructing SRB-Measures

Consider the set $S \subset U$ (U is a neighborhood of the attractor Λ) of points such that

- $f(S) \subset S$, i.e., S is forward invariant;
- there are two measurable cone families $K^s(x) = K^s(x, E_1(x), \theta(x))$ and $K^u(x) = K^u(x, E_2(x), \theta(x))$,[13] which are *invariant*,[14] i.e.,

$$\overline{Df(K^u(x))} \subset K^u(f(x)), \quad \overline{Df^{-1}(K^s(f(x)))} \subset K^s(x)$$

  and *transverse*, i.e., $T_xX = E_1(x) \oplus E_2(x)$.

Define

- $\lambda^s(x) = \sup\{\log \|Df(v)\| : v \in K^s(x), \|v\| = 1\}$—*coefficient of contraction*;
- $\lambda^u(x) = \inf\{\log \|Df(v)\| : v \in K^u(x), \|v\| = 1\}$—*coefficient of expansion*;
- $d(x) = \max(0, (\lambda^s(x) - \lambda^u(x))$—*defect of hyperbolicity*;
- $\lambda(x) = \lambda^u(x) - d(x)$—*coefficient of effective hyperbolicity*;

---

[12]The matrix A is assumed to be hyperbolic having one positive and two negative eigenvalues.

[13]Recall that given $x \in M$, a subspace $E(x) \subset T_xM$, and $\theta(x) > 0$, the cone at x around $E(x)$ with angle $\theta(x)$ is defined by $K(x, E(x), \theta(x)) = \{v \in T_xM : \angle(v, E(x)) < \theta(x)\}$.

[14]We stress that the subspaces $E_1(x)$ and $E_2(x)$ do note have to be invariant under df.

- $\alpha(x) = \angle(K^s(x), K^u(x)) > 0$—*angle* between the cones;
- $\rho_{\hat\alpha}(x) = \lim_{n\to\infty} \frac{1}{n}\#\{0 \le k < n \colon \alpha(f^k(x)) < \hat\alpha\}$—*average time* the angle between the cones is below a given threshold $\hat\alpha > 0$.

We further assume that for every $x \in S$,

(S1) $\underline{\lim}_{n\to\infty}\frac{1}{n}\sum_{k=0}^{n-1}\lambda(f^k(x)) > 0$;
(S2) $\lim_{\hat\alpha\to 0}\rho_{\hat\alpha}(x) = 0$;
(S3) $\overline{\lim}_{n\to\infty}\frac{1}{n}\sum_{k=0}^{n-1}\lambda^s(f^k(x)) < 0$.

Theorem 7 (Climenhaga, Dolgopyat, Pesin, [28]) *Assume that the set S has positive volume. Then f possesses an SRB-measure.*

In [70], Viana conjectured that *if the set of all points with non-zero Lyapunov exponents for a $C^{1+\alpha}$ diffeomorphism f has positive (in particular, full) volume (which is not necessarily invariant), then f admits an SRB measure*. The above theorem provides some stronger conditions under which the conclusion of Viana'a conjecture holds. An affirmative solution of this conjecture for surface diffeomorphisms, under some general additional assumptions, is obtained in a recent work by Climenhaga, Luzzatto and Pesin [29]. It is conjectured that *if Requirement (S1) is replaced with a stronger requirement that*

$$\underline{\lim}_{n\to\infty}\frac{1}{n}\sum_{k=0}^{n-1}\lambda(f^k(x)) \ge \lambda$$

for some $\lambda > 0$, then f possesses at most finitely many SRB-measures. In [55], F. Rodriguez Hertz, J. Rodriguez Hertz, Tahzibi and Ures showed that any topologically transitive surface diffeomorphism possesses at most one SRB measure.

6 SRB-Measures III: Partially Hyperbolic Attractors

6.1 Partially Hyperbolic Attractors

An attractor Λ is *partially hyperbolic* if $f|\Lambda$ is uniformly partially hyperbolic, i.e., if the tangent space $T\Lambda$ admits an invariant splitting

$$T\Lambda = E^s \oplus E^c \oplus E^u$$

into *strongly stable, central* and *strongly unstable subspaces*, respectively, which satisfy conditions (1)–(3) in Sect. 2.3. The subspace E^u is integrable: given $x \in \Lambda$, a local unstable leaf $V^u(x)$ lies in Λ, and hence so does the *global strongly unstable manifolds W^u*. It follows that the attractor is the union of the global strongly unstable manifolds of its points, which form a lamination of Λ.

One can obtain an example of a partially hyperbolic attractor by considering the product map $F = f_1 \times f_2$ where $f_1: M \to M$ is a map possessing a uniformly hyperbolic attractor and $f_2: S^1 \to S^1$ is an isometry.

If f is a diffeomorphism possessing a uniformly (partially) hyperbolic attractor $\Lambda = \Lambda_f$ then any sufficiently small perturbation g of f in the C^1 topology possesses a uniformly (partially) hyperbolic attractor Λ_g that lies in a small neighborhood of Λ_f. This provides an open set of uniformly (partially) hyperbolic attractors in the spaces of C^1 diffeomorphisms.

6.2 SRB-Measures on Partially Hyperbolic Attractors

Let Λ be a partially hyperbolic attractor for a diffeomorphisms f. A measure μ on Λ is a *u-measure* if for almost every $x \in \Lambda$, the conditional measure $\mu^u(x)$ generated by μ on the global strongly unstable leaf $W^u(x)$ is absolutely continuous with respect to the leaf-volume $m^u(x)$. One can show that the Jacobian of the u-measure in the unstable direction is given by the formula (5). The following result shows that every partially hyperbolic attractor carries a u-measure. Its proof can be obtained by adjusting the argument in the second proof of Theorem 4 to the partial hyperbolicity setting.

Theorem 8 (Pesin, Sinai, [53]) *The following statements hold:*

1. *Any limit measure μ of the sequence of measures* (3) *is a u-measure on Λ;*
2. *Any limit measure μ of the sequence of measures* (7) *is a u-measure on Λ.*

Unlike the case of hyperbolic attractors, the topological transitivity of $f|\Lambda$ (or even topological mixing) does not guarantee uniqueness of u-measures.[15]

Every SRB-measure on a partially hyperbolic attractor Λ is a u-measure but not every u-measure is an SRB-measure. We say that a u-measure ν has *negative (positive) central exponents* if there is an invariant subset $A \subset \Lambda$ with $\nu(A) > 0$ such that the Lyapunov exponents $\chi(x, v) < 0$ (respectively, $\chi(x, v) > 0$) for every $x \in A$ and every nonzero vector $v \in E^c(x)$. A u-measure with negative (positive) central exponents is an SRB-measure.

Below is a result that guarantees existence and uniqueness of SRB-measures for partially hyperbolic attractors with negative central exponents. It requires existence of at least one u-measure with negative central exponents and a strong **transitive** condition. A detailed discussion of these requirements can be found in [22].

[15]Indeed, consider $F = f_1 \times f_2$, where f_1 is a topologically transitive Anosov diffeomorphism and f_2 a diffeomorphism close to the identity. Then any measure $\mu = \mu_1 \times \mu_2$, where μ_1 is the unique SRB-measure for f_1 and μ_2 any f_2-invariant measure, is a u-measure for F. Thus, F has a unique u-measure if and only if f_2 is uniquely ergodic. On the other hand, F is topologically mixing if and only if f_2 is topologically mixing.

Theorem 9 (Bonatti, Viana, [13]; Burns, Dolgopyat, Pollicott, Pesin, [22]) *Let f be a $C^{1+\epsilon}$ diffeomorphism of a compact smooth manifold M with a partially hyperbolic attractor Λ. Assume that:*

1. *there exists a u-measure ν with negative central exponents;*
2. *for every $x \in \Lambda$ the global strongly unstable manifold $W^u(x)$ is dense in Λ.*

Then ν is the unique u-measure for f and is the unique hyperbolic SRB-measure for f whose basin of attraction $B(\nu)$ has full volume in the topological basin of attraction of Λ.

The case of positive central exponents is more difficult and existence of SRB-measures can be established under the stronger requirements that (see [69]):

1. there is a **unique** u-measure ν with positive central exponents on a subset $A \subset \Lambda$ of **full** measure;
2. for every $x \in \Lambda$ the global strongly unstable manifold $W^u(x)$ is dense in Λ.

6.3 Dominated Splitting and SRB-Measures

The key tool in constructing SRB-measures in the uniform hyperbolic setting is presence of a *dominated splitting*, i.e., a decomposition of the tangent bundle $T_xM = E_1(x) \oplus E_2(x)$ for every $x \in \Lambda$ such that

1. $E_1(x)$ and $E_2(x)$ depend continuously on x;
2. $\measuredangle(E_1(x), E_2(x))$ is bounded away from 0;
3. there is $0 < \lambda < 1$ such that

$$\|Df|E_1(x)\| < \lambda, \quad \|Df|E_1(x)\| \cdot \|Df^{-1}|E_2(f(x))\| < \lambda.$$

Construction of SRB measures for systems with dominated splitting was effected in various situations. Here is an (incomplete) list:

- (Alves, Bonatti, Viana, [3]) there is a subset $S \subset U$ of positive volume and $\varepsilon > 0$ such that for every $x \in S$,

$$\limsup_{n\to\infty} \frac{1}{n} \sum_{j=1}^{n} \log \|df^{-1}|E_2(f^j(x))\| < -\varepsilon.$$

  In this case, in addition one can have no more than finitely many distinct SRB measures.
- (Alves, Dias, Luzzatto, Pinheiro, [4]) there is a subset $S \subset U$ of positive volume and $\varepsilon > 0$ such that for every $x \in S$,

$$\liminf_{n\to\infty} \frac{1}{n} \sum_{j=1}^{n} \log \|df^{-1}|E_2(f^j(x))\| < -\varepsilon.$$

In this case, in addition one can have no more than finitely many distinct SRB measures. In fact, if f is topologically transitive and $m(S) = 1$, then the SRB measure is unique.

We stress that for non-uniformly hyperbolic f, the splitting of the tangent space does not have to be dominated.

Acknowledgements I would like to thank B. Gurevich and O. Sarig who read the paper and gave me many valuable comments. I also would like to thank B. Weiss and A. Katok for their remarks on the paper. Part of the work was done when I visited ICERM (Brown University, Providence) and the Erwin Schrödinger Institute (Vienna). I would like to thank them for their hospitality.

References

1. R.L. Adler and B. Weiss. Entropy, a complete metric invariant for automorphisms of the torus. *Proc. Nat. Acad. Sci. U.S.A.*, 57:1573–1576, 1967.
2. V.M. Alekseev. Symbolic dynamics (in Russian). In *Eleventh Mathematical School (Summer School, Kolomyya, 1973)*, pages 5–210. Izdanie Inst. Mat. Akad. Nauk Ukrain. SSR, Kiev, 1976.
3. J.F. Alves, C. Bonatti, and M. Viana. SRB measures for partially hyperbolic systems whose central direction is mostly expanding. *Invent. Math.*, 140(2):351–398, 2000.
4. J.F. Alves, C.L. Dias, S. Luzzatto, and V. Pinheiro. SRB measures for partially hyperbolic systems whose central direction is weakly expanding. *J. Eur. Math. Soc. (JEMS)*, 19(10):2911–2946, 2017.
5. D.V. Anosov. *Geodesic Flows on Closed Riemann Manifolds with Negative Curvature*. Number 90 in Proc. Steklov Inst. Math. American Mathematical Society, Providence, RI, 1967.
6. D.V. Anosov and Ya.G. Sinai. Some smooth ergodic systems. *Russ. Math. Surv.*, 22(5):103–167, 1967.
7. L. Barreira and Ya. Pesin. *Nonuniform Hyperbolicity*, volume 115 of *Encyclopedia of Mathematics and its Applications*. Cambridge University Press, Cambridge, 2007. Dynamics of systems with nonzero Lyapunov exponents.
8. L. Barreira and Ya. Pesin. *Introduction to Smooth Ergodic Theory*, volume 148 of *Graduate Studies in Mathematics*. American Mathematical Society, Providence, RI, 2013.
9. M. Benedicks and L. Carleson. The dynamics of the Hénon map. *Ann. of Math. (2)*, 133(1):73–169, 1991.
10. M. Benedicks and L.-S. Young. Sinai–Bowen–Ruelle measures for certain Hénon maps. *Invent. Math.*, 112(3):541–576, 1993.
11. M. Benedicks and L.-S. Young. Markov extensions and decay of correlations for certain Hénon maps. *Astérisque*, (261):xi, 13–56, 2000.
12. K.R. Berg. *On the conjugacy problem for K-systems*. PhD thesis, University of Minnesota, 1967. Also available as *Entropy of Torus Automorphisms* in *Topological Dynamics (Symposium, Colorado State Univ., Ft. Collins, Colo.)*, pages 67–79, Benjamin, New York, 1968.
13. C. Bonatti and M. Viana. SRB measures for partially hyperbolic systems whose central direction is mostly contracting. *Israel J. Math.*, 115:157–193, 2000.
14. R. Bowen. Markov partitions for Axiom A diffeomorphisms. *Amer. J. Math.*, 92:725–747, 1970.
15. R. Bowen. *Equilibrium States and the Ergodic Theory of Anosov diffeomorphisms*. Lecture Notes in Mathematics, Vol. 470, pages i+108. Springer-Verlag, Berlin-New York, 1975.
16. R. Bowen. Some systems with unique equilibrium states. *Math. Syst. Theory*, 8:193–202, 1975.

17. R. Bowen and D. Ruelle. The ergodic theory of Axiom A flows. *Invent. Math.*, 29(3):181–202, 1975.
18. M.I. Brin and Ya.B. Pesin. Partially hyperbolic dynamical systems (in Russian). *Izv. Akad. Nauk SSSR Ser. Mat.*, 38:170–212, 1974.
19. L.A. Bunimovich, B.M. Gurevich, and Ya. Pesin, editors. *Sinai's Moscow Seminar on Dynamical Systems*, volume 171 of *American Mathematical Society Translations, Series 2*. American Mathematical Society, Providence, RI, 1996.
20. L.A. Bunimovich and Ya.G. Sinai. Markov partitions for dispersed billiards. *Commun. Math. Phys.*, 78:247–280, 1980. Erratum *ibid*, 107(2):357–358, 1986.
21. L.A. Bunimovich, Ya.G. Sinai, and N.I. Chernov. Markov partitions for two-dimensional hyperbolic billiards. *Russ. Math. Surv.*, 45(3):105–152, 1990.
22. K. Burns, D. Dolgopyat, Ya. Pesin, and M. Pollicott. Stable ergodicity for partially hyperbolic attractors with negative central exponents. *J. Mod. Dyn.*, 2(1):63–81, 2008.
23. Y. Cao. Non-zero Lyapunov exponents and uniform hyperbolicity. *Nonlinearity*, 16(4):1473–1479, 2003.
24. Y. Cao, S. Luzzatto, and I. Rios. A minimum principle for Lyapunov exponents and a higher-dimensional version of a theorem of Mañé. *Qual. Theory Dyn. Syst.*, 5(2):261–273, 2004.
25. J. Chen, H. Hu, and Ya. Pesin. The essential coexistence phenomenon in dynamics. *Dyn. Syst.*, 28(3):453–472, 2013.
26. J. Chen, H. Hu, and Ya. Pesin. A volume preserving flow with essential coexistence of zero and non-zero Lyapunov exponents. *Ergodic Theory Dynam. Systems*, 33(6):1748–1785, 2013.
27. N. Chernov. Invariant measures for hyperbolic dynamical systems. In *Handbook of dynamical systems, Vol. 1A*, pages 321–407. North-Holland, Amsterdam, 2002.
28. V. Climenhaga, D. Dolgopyat, and Y. Pesin. Non-stationary non-uniform hyperbolicity: SRB measures for dissipative maps. *Comm. Math. Phys.*, 346(2):553–602, 2016.
29. V. Climenhaga, S. Luzzatto, and Ya. Pesin. Young towers for nonuniformly hyperbolic surface diffeomorphisms. Preprint, PSU, 2018.
30. R.L. Dobrushin. Description of a random field by means of conditional probabilities and conditions for its regularity. *Theor. Probab. Appl.*, 13:197–224, 1968.
31. D. Dolgopyat and Ya. Pesin. Every compact manifold carries a completely hyperbolic diffeomorphism. *Ergodic Theory Dynam. Systems*, 22(2):409–435, 2002.
32. B. Gurevich. Entropy theory of dynamical systems. In H. Holden, R. Piene, editors, *The Abel Prize 2013–2017*, pages 221–242. Springer, 2019.
33. B. Hasselblatt, Ya. Pesin, and J. Schmeling. Pointwise hyperbolicity implies uniform hyperbolicity. *Discrete Contin. Dyn. Syst.*, 34(7):2819–2827, 2014.
34. M.W. Hirsch, C.C. Pugh, and M. Shub. *Invariant Manifolds*, volume 583 of *Lecture Notes in Mathematics*. Springer-Verlag, Berlin-New York, 1977.
35. H. Hu, Ya. Pesin, and A. Talitskaya. A volume preserving diffeomorphism with essential coexistence of zero and nonzero Lyapunov exponents. *Comm. Math. Phys.*, 319(2):331–378, 2013.
36. A. Katok. Bernoulli diffeomorphisms on surfaces. *Ann. of Math. (2)*, 110(3):529–547, 1979.
37. A. Katok. Moscow dynamic seminars of the nineteen seventies and the early career of Yasha Pesin. *Discrete Contin. Dyn. Syst.*, 22(1–2):1–22, 2008.
38. A. Katok. Dmitry Viktorovich Anosov: his life and mathematics. In *Modern theory of dynamical systems*, volume 692 of *Contemp. Math.*, pages 1–21. Amer. Math. Soc., Providence, RI, 2017.
39. A. Katok and B. Hasselblatt. *Introduction to the Modern Theory of Dynamical Systems*, volume 54 of *Encyclopedia of Mathematics and its Applications*. Cambridge University Press, Cambridge, 1995.
40. O.E. Lanford, III and D. Ruelle. Observables at infinity and states with short range correlations in statistical mechanics. *Comm. Math. Phys.*, 13:194–215, 1969.
41. F. Ledrappier. Propriétés ergodiques des mesures de Sinai. *Inst. Hautes Études Sci. Publ. Math.*, (59):163–188, 1984.

42. F. Ledrappier and J.-M. Strelcyn. A proof of the estimation from below in Pesin's entropy formula. *Ergodic Theory Dynam. Systems*, 2(2):203–219 (1983), 1982.
43. F. Ledrappier and L.-S. Young. The metric entropy of diffeomorphisms. I. Characterization of measures satisfying Pesin's entropy formula. *Ann. of Math. (2)*, 122(3):509–539, 1985.
44. F. Ledrappier and L.-S. Young. The metric entropy of diffeomorphisms. II. Relations between entropy, exponents and dimension. *Ann. of Math. (2)*, 122(3):540–574, 1985.
45. R. Mañé. Quasi-Anosov diffeomorphisms and hyperbolic manifolds. *Trans. Amer. Math. Soc.*, 229:351–370, 1977.
46. D. Ornstein. Bernoulli shifts with the same entropy are isomorphic. *Advances in Math.*, 4:337–352, 1970.
47. D.S. Ornstein. *Ergodic Theory, Randomness, and Dynamical Systems*, volume 5 of *Yale Mathematical Monographs*. Yale University Press, New Haven, Conn.-London, 1974.
48. D.S. Ornstein and B. Weiss. Geodesic flows are Bernoullian. *Israel J. Math.*, 14:184–198, 1973.
49. Ya.B. Pesin. Families of invariant manifolds that correspond to nonzero characteristic exponents. *Math. USSR-Izv.*, 40(6):1261–1305, 1976.
50. Ya.B. Pesin. Characteristic Lyapunov exponents and smooth ergodic theory. *Russ. Math. Surv.*, 32(4):55–114, 1977.
51. Ya.B. Pesin. Description of the π-partition of a diffeomorphism with invariant measure. *Math. Notes*, 22(1):506–515, 1978.
52. Ya.B. Pesin. *Lectures on Partial Hyperbolicity and Stable Ergodicity*. Zurich Lectures in Advanced Mathematics. European Mathematical Society (EMS), Zürich, 2004.
53. Ya.B. Pesin and Ya.G. Sinai. Gibbs measures for partially hyperbolic attractors. *Ergodic Theory Dynam. Systems*, 2(3–4):417–438 (1983), 1982.
54. M. Ratner. Markov partitions for Anosov flows on n-dimensional manifolds. *Israel J. Math.*, 15:92–114, 1973.
55. F. Rodriguez Hertz, M.A. Rodriguez Hertz, A. Tahzibi, and R. Ures. Uniqueness of SRB measures for transitive diffeomorphisms on surfaces. *Comm. Math. Phys.*, 306(1):35–49, 2011.
56. D. Ruelle. A measure associated with axiom-A attractors. *Amer. J. Math.*, 98(3):619–654, 1976.
57. O.M. Sarig. Symbolic dynamics for surface diffeomorphisms with positive entropy. *J. Amer. Math. Soc.*, 26(2):341–426, 2013.
58. Ya.G. Sinai. Weak isomorphism of transformations with an invariant measure. *Sov. Math., Dokl.*, 3:1725–1729, 1962.
59. Ya.G. Sinai. Construction of Markov partitions. *Funct. Anal. Appl.*, 2:245–253, 1968.
60. Ya.G. Sinai. Markov partitions and C-diffeomorphisms. *Funct. Anal. Appl.*, 2:61–82, 1968.
61. Ya.G. Sinai. Mesures invariantes des Y-systèmes. Actes Congr. Int. Math. 1970, 2, 929–940, 1971.
62. Ya.G. Sinai. Gibbs measures in ergodic theory. *Russ. Math. Surv.*, 27(4):21–69, 1972.
63. Ya.G. Sinai. Finite-dimensional randomness. *Russ. Math. Surv.*, 46(3):177–190, 1991.
64. Ya.G. Sinai. Hyperbolicity and chaos. In *Boltzmann's Legacy 150 Years after His Birth (Rome, 1994)*, volume 131 of *Atti Convegni Lincei*, pages 107–110. Accad. Naz. Lincei, Rome, 1997.
65. Ya.G. Sinai. How mathematicians study chaos (in Russian). *Mat. Pros.*, 3(5):32–46, 2001.
66. S. Smale. A structurally stable differentiable homeomorphism with an infinite number of periodic points. In *Qualitative Methods in the Theory of Non-Linear Vibrations (Proc. Internat. Sympos. Non-linear Vibrations, Vol. II, 1961)*, pages 365–366. Izdat. Akad. Nauk Ukrain. SSR, Kiev, 1963.
67. S. Smale. Differentiable dynamical systems. *Bull. Amer. Math. Soc.*, 73:747–817, 1967.
68. D. Szász. Markov approximations and statistical properties of billiards. In H. Holden, R. Piene, editors *The Abel Prize 2013–2017*, pages 299–319. Springer, 2019.
69. C.H. Vásquez. Stable ergodicity for partially hyperbolic attractors with positive central Lyapunov exponents. *J. Mod. Dyn.*, 3(2):233–251, 2009.

70. M. Viana. Dynamics: a probabilistic and geometric perspective. In *Proceedings of the International Congress of Mathematicians, Vol. I (Berlin, 1998)*, number Extra Vol. I, pages 557–578, 1998.
71. Q. Wang and L.-S. Young. Strange attractors with one direction of instability. *Comm. Math. Phys.*, 218(1):1–97, 2001.
72. L.-S. Young. Statistical properties of dynamical systems with some hyperbolicity. *Ann. of Math. (2)*, 147(3):585–650, 1998.

Further Developments of Sinai's Ideas: The Boltzmann–Sinai Hypothesis

Nándor Simányi

Abstract In this chapter we present a brief survey of the rich and manifold developments of Sinai's ideas, dating back to 1963, concerning his exact mathematical formulation of Boltzmann's original ergodic hypothesis. These developments eventually lead to the 2013 proof of the so called "Boltzmann-Sinai Ergodic Hypothesis".

1 Preface

In 1963, Ya. G. Sinai [23] formulated a modern version of Boltzmann's ergodic hypothesis, what we now call the "Boltzmann–Sinai Ergodic Hypothesis": The billiard system of N ($N \geq 2$) hard balls of unit mass moving on the flat torus $\mathbb{T}^\nu = \mathbb{R}^\nu/\mathbb{Z}^\nu$ ($\nu \geq 2$) is ergodic after we make the standard reductions by fixing the values of trivial invariant quantities. It took 50 years and the efforts of several people, including Sinai himself, until this conjecture was finally proved. In this short survey we provide a quick review of the closing part of this process, by showing how Sinai's original ideas developed further between 2000 and 2013, eventually leading to the proof of the conjecture.

2 Posing the Problem: The Investigated Models

Non-uniformly hyperbolic systems (possibly, with singularities) play a pivotal role in the ergodic theory of dynamical systems. Their systematic study started several decades ago, and it is not our goal here to provide the reader with a comprehensive review of the history of these investigations but, instead, we opt for presenting in a nutshell a cross section of a few selected results.

N. Simányi (✉)
Department of Mathematics, The University of Alabama at Birmingham, Birmingham, AL, USA
e-mail: simanyi@uab.edu

H. Holden, R. Piene (eds.), *The Abel Prize 2013–2017*,
https://doi.org/10.1007/978-3-319-99028-6_12

In 1939, G.A. Hedlund and E. Hopf [3, 4], proved the hyperbolic ergodicity (i.e., full hyperbolicity and ergodicity) of geodesic flows on closed, compact surfaces with constant negative curvature by inventing the famous method of "Hopf chains" constituted by local stable and unstable invariant manifolds.

In 1963, Ya. G. Sinai [23] formulated a modern version of Boltzmann's ergodic hypothesis, what we call now the "Boltzmann–Sinai Hypothesis": the billiard system of N (≥ 2) hard balls of unit mass moving on the flat torus $\mathbb{T}^\nu = \mathbb{R}^\nu/\mathbb{Z}^\nu$ ($\nu \geq 2$) is ergodic after we make the standard reductions by fixing the values of the trivial invariant quantities. It took seven years until he proved this conjecture for the case $N = 2$, $\nu = 2$ in [24]. Another 17 years later, N. I. Chernov and Ya. G. Sinai [26] proved the hypothesis for the case $N = 2$, $\nu \geq 2$ by also proving a powerful and very useful theorem on local ergodicity.

In the meantime, in 1977, Ya. Pesin [12] laid down the foundations of his theory on the ergodic properties of smooth, hyperbolic dynamical systems. Later on, this theory (nowadays called Pesin theory) was significantly extended by A. Katok and J-M. Strelcyn [5] to hyperbolic systems with singularities. That theory is already applicable for billiard systems, too.

Until the end of the 1970s the phenomenon of hyperbolicity (exponential instability of trajectories) was almost exclusively attributed to some direct geometric scattering effect, like negative curvature of space, or strict convexity of the scatterers. This explains the profound shock that was caused by the discovery of L. A. Bunimovich [1]: Certain focusing billiard tables (like the celebrated stadium) can also produce complete hyperbolicity and, in that way, ergodicity. It was partly this result that led to Wojtkowski's theory of invariant cone fields, [29, 30].

The big difference between the system of two balls in $\mathbb{T}^\nu$ ($\nu \geq 2$, [26]) and the system of N (≥ 3) balls in $\mathbb{T}^\nu$ is that the latter one is merely a so called semi-dispersive billiard system (the scatterers are convex but not strictly convex sets, namely cylinders), while the former one is strictly dispersive (the scatterers are strictly convex sets). This fact makes the proof of ergodicity (mixing properties) much more complicated. In our series of papers jointly written with A. Krámli and D. Szász [7, 8], and [9], we managed to prove the (hyperbolic) ergodicity of three and four billiard balls on the toroidal container $\mathbb{T}^\nu$. By inventing a new topological methods and the Connecting Path Formula (CPF), in the two-part paper [13, 14], I proved the (hyperbolic) ergodicity of N hard balls in $\mathbb{T}^\nu$, provided that $N \leq \nu$.

The common feature of hard ball systems is—as D. Szász pointed this out first in [27] and [28]—that all of them belong to the family of so called cylindric billiards, the definition of which can be found later in this survey. However, the first appearance of a special, 3-D cylindric billiard system took place in [6], where we proved the ergodicity of a 3-D billiard flow with two orthogonal cylindric scatterers. Later D. Szász [28] presented a complete picture (as far as ergodicity is concerned) of cylindric billiards with cylinders whose generator subspaces are spanned by mutually orthogonal coordinate axes. The task of proving ergodicity for the first non-trivial, non-orthogonal cylindric billiard system was taken up in [20].

Finally, in our joint venture with D. Szász [21] we managed to prove the complete hyperbolicity of *typical* hard ball systems on flat tori.

2.1 Cylindric Billiards

Consider the d-dimensional ($d \geq 2$) flat torus $\mathbb{T}^d = \mathbb{R}^d/\mathscr{L}$ supplied with the usual Riemannian inner product $\langle\,.\,,\,.\,\rangle$ inherited from the standard inner product of the universal covering space $\mathbb{R}^d$. Here $\mathscr{L} \subset \mathbb{R}^d$ is supposed to be a lattice, i.e., a discrete subgroup of the additive group $\mathbb{R}^d$ with $\text{rank}(\mathscr{L}) = d$. The reason why we want to allow general lattices other than just the integer lattice $\mathbb{Z}^d$, is that otherwise the hard ball systems would not be covered. The geometry of the structure lattice $\mathscr{L}$ in the case of a hard ball system is significantly different from the geometry of the standard orthogonal lattice $\mathbb{Z}^d$ in the Euclidean space $\mathbb{R}^d$.

The configuration space of a cylindric billiard is $\mathbf{Q} = \mathbb{T}^d \setminus (C_1 \cup \cdots \cup C_k)$, where the cylindric scatterers C_i ($i = 1, \ldots, k$) are defined as follows:

Let $A_i \subset \mathbb{R}^d$ be a so called lattice subspace of $\mathbb{R}^d$, which means that $\text{rank}(A_i \cap \mathscr{L}) = \text{dim}A_i$. In this case the factor $A_i/(A_i \cap \mathscr{L})$ is a subtorus in $\mathbb{T}^d = \mathbb{R}^d/\mathscr{L}$, which will be taken as the generator of the cylinder $C_i \subset \mathbb{T}^d$, $i = 1, \ldots, k$. Denote by $L_i = A_i^{\perp}$ the orthocomplement of A_i in $\mathbb{R}^d$. Throughout this survey we will always assume that $\dim L_i \geq 2$. Let, furthermore, the numbers $r_i > 0$ (the radii of the spherical cylinders C_i) and some translation vectors $t_i \in \mathbb{T}^d = \mathbb{R}^d/\mathscr{L}$ be given. The translation vectors t_i play a crucial role in positioning the cylinders C_i in the ambient torus $\mathbb{T}^d$. Set

$$C_i = \left\{x \in \mathbb{T}^d \,:\, \text{dist}\,(x - t_i,\, A_i/(A_i \cap \mathscr{L})) < r_i\right\}.$$

In order to avoid further unnecessary complications, we always assume that the interior of the configuration space $\mathbf{Q} = \mathbb{T}^d \setminus (C_1 \cup \cdots \cup C_k)$ is connected. The phase space $\mathbf{M}$ of our cylindric billiard flow will be the unit tangent bundle of $\mathbf{Q}$ (modulo some natural gluings at its boundary), i.e., $\mathbf{M} = \mathbf{Q} \times \mathbb{S}^{d-1}$. (Here $\mathbb{S}^{d-1}$ denotes the unit sphere of $\mathbb{R}^d$.)

The dynamical system $(\mathbf{M}, \{S^t\}, \mu)$ that we investigate is called a cylindric billiard flow. Here S^t ($t \in \mathbb{R}$) is the dynamics defined by uniform motion inside the domain $\mathbf{Q}$ and specular reflections at its boundary (at the scatterers), and μ is the Liouville measure.

2.2 Transitive Cylindric Billiards

The main conjecture concerning the (hyperbolic) ergodicity of cylindric billiards is the "Erdőtarcsa conjecture" (named after the picturesque village in rural Hungary where it was initially formulated) that appeared as Conjecture 1 in Section 3 of [22]:

Conjecture 1 (The Erdőtarcsa Conjecture) A cylindric billiard flow is ergodic if and only if it is transitive, i.e., the Lie group generated by all rotations across the constituent spaces of the cylinders acts transitively on the sphere of compound velocities, see Section 3 of [22]. In the case of transitivity the cylindric billiard system is actually a completely hyperbolic Bernoulli flow, see [2] and [11].

The theorem of [15] proves a slightly relaxed version of this conjecture (only full hyperbolicity without ergodicity) for a wide class of cylindric billiard systems, namely the so called "transverse systems", which include every hard ball system.

2.3 Transitivity

Let $L_1, \ldots, L_k \subset \mathbb{R}^d$ be subspaces, $A_i = L_i^{\perp}$, $\dim L_i \geq 2$, $i = 1, \ldots, k$. Set

$$\mathscr{G}_i = \left\{ U \in \mathrm{SO}(d) : U \big| A_i = \mathrm{Id}_{A_i} \right\},$$

and let $\mathscr{G} = \langle \mathscr{G}_1, \ldots, \mathscr{G}_k \rangle \subset \mathrm{SO}(d)$ be the algebraic generate of the compact, connected Lie subgroups $\mathscr{G}_i$ in $\mathrm{SO}(d)$. The following notions appeared in Section 3 of [22].

Definition 2 We say that the system of base spaces $\{L_1, \ldots, L_k\}$ (or, equivalently, the cylindric billiard system defined by them) is *transitive* if and only if the group $\mathscr{G}$ acts transitively on the unit sphere $\mathbb{S}^{d-1}$ of $\mathbb{R}^d$.

Definition 3 We say that the system of subspaces $\{L_1, \ldots, L_k\}$ has the Orthogonal Non-Splitting Property (ONSP) if there is no non-trivial orthogonal splitting $\mathbb{R}^d = B_1 \oplus B_2$ of $\mathbb{R}^d$ with the property that for every index i $(1 \leq i \leq k)$ $L_i \subset B_1$ or $L_i \subset B_2$.

The next result can be found in Section 3 of [22] (see 3.1–3.6 thereof):

Proposition 4 *For the system of subspaces $\{L_1, \ldots, L_k\}$ the following three properties are equivalent:*

1. *$\{L_1, \ldots, L_k\}$ is transitive;*
2. *$\{L_1, \ldots, L_k\}$ has the ONSP;*
3. *the natural representation of $\mathscr{G}$ in $\mathbb{R}^d$ is irreducible.*

2.4 Transverseness

Definition 5 We say that the system of subspaces $\{L_1, \ldots, L_k\}$ of $\mathbb{R}^d$ is *transverse* if the following property holds: For every *non-transitive* subsystem $\{L_i : i \in I\}$ $(I \subset \{1, \ldots, k\})$ there exists an index $j_0 \in \{1, \ldots, k\}$ such that $P_{E^+}(A_{j_0}) = E^+$, where $A_{j_0} = L_{j_0}^{\perp}$, and $E^+ = \mathrm{span}\{L_i : i \in I\}$. We note that in this case, necessarily, $j_0 \notin I$, otherwise $P_{E^+}(A_{j_0})$ would be orthogonal to the subspace $L_{j_0} \subset E^+$. Therefore, every transverse system is automatically transitive.

We note that every hard ball system is transverse, see [15]. The main result of that paper is the following theorem.

Theorem 6 *Assume that the cylindric billiard system is transverse. Then this billiard flow is completely hyperbolic, i.e., all relevant Lyapunov exponents are nonzero almost everywhere. Consequently, in such a dynamical system every ergodic component has positive measure, and the restriction of the flow to each ergodic component has the Bernoulli property, see* [2] *and* [11].

An immediate consequence of this result is the following result.

Corollary 7 *Every hard ball system is completely hyperbolic.*

Thus Theorem 6 above generalizes the main result of [21], where the complete hyperbolicity of *almost every* hard ball system was proven.

3 Toward Ergodicity

In the series of articles [6, 8, 9, 13], and [14] the authors developed a powerful, three-step strategy for proving the (hyperbolic) ergodicity of hard ball systems. First of all, all these proofs are inductions on the number N of balls involved in the problem. Secondly, the induction step itself consists of the following three major steps:

3.1 Step I

To prove that every non-singular (i.e., smooth) trajectory segment $S^{[a,b]}x_0$ with a "combinatorially rich" symbolic collision sequence is automatically sufficient (or, in other words, "geometrically hyperbolic"), provided that the phase point x_0 does not belong to a countable union J of smooth sub-manifolds with codimension at least two. (Containing the exceptional phase points.)

Here combinatorial richness means that the symbolic collision sequence of the orbit segment contains a large enough number of consecutive, connected collision graphs, see also the introductory section of [21].

The exceptional set J featuring this result is negligible in our dynamical considerations—it is a so called slim set, i.e., a subset of the phase space $\mathbf{M}$ that can be covered by a countable union $\bigcup_{n=1}^{\infty} F_n$ of closed, zero-measured subsets F_n of $\mathbf{M}$ that have topological co-dimension at least 2.

3.2 Step II

Assume the induction hypothesis, i.e., that all hard ball systems with N' balls ($2 \leq N' < N$) are (hyperbolic and) ergodic. Prove that then there exists a slim set $S \subset \mathbf{M}$ with the following property: For every phase point $x_0 \in \mathbf{M} \setminus S$ the whole trajectory

$S^{(-\infty,\infty)}x_0$ contains at most one singularity and its symbolic collision sequence is combinatorially rich, just as required by the result of Step I.

3.3 Step III

By using again the induction hypothesis, prove that almost every singular trajectory is sufficient in the time interval (t_0, ∞), where t_0 is the moment of the singular reflection, i.e., a tangential reflection or multiple reflections occurring at the same time. (Here the phrase "almost every" refers to the volume defined by the induced Riemannian metric on the singularity manifolds.)

We note here that the almost sure sufficiency of the singular trajectories (featuring Step III) is an essential condition for the proof of the celebrated Theorem on Local Ergodicity for algebraic semi-dispersive billiards proved by Chernov and Sinai in [26]. Under this assumption the theorem of [26] states that in any algebraic semi-dispersive billiard system (i.e., in a system such that the smooth components of the boundary $\partial\mathbf{Q}$ are algebraic hypersurfaces) a suitable, open neighborhood U_0 of any hyperbolic phase point $x_0 \in \mathbf{M}$ (with at most one singularity on its trajectory) belongs to a single ergodic component of the billiard flow.

In an inductive proof of ergodicity, steps I and II together ensure that there exists an arc-wise connected set $C \subset \mathbf{M}$ with full measure, such that every phase point $x_0 \in C$ is hyperbolic with at most one singularity on its trajectory. Then the cited Theorem on Local Ergodicity (now taking advantage of the result of Step III) states that for every phase point $x_0 \in C$ an open neighborhood U_0 of x_0 belongs to one ergodic component of the flow. Finally, the connectedness of the set C and $\mu(C) = 1$ easily imply that the billiard flow with N balls is indeed ergodic, and actually fully hyperbolic, as well.

In the papers [16, 21], and [17] we investigated systems of hard balls with masses $m_1, m_2, \dots, m_N$ $(m_i > 0)$ moving on the flat torus $\mathbb{T}^\nu_L = \mathbb{R}^\nu / L \cdot \mathbb{Z}^\nu$, $L > 0$.

The main results of the papers [16] and [17] are summarized as follows:

Theorem 8 *For almost every selection $(m_1, \dots, m_N;\ L)$ of the external geometric parameters from the region $m_i > 0$, $L > L_0(r, \nu)$, where the interior of the phase space is connected, it is true that the billiard flow $\left(\mathbf{M}_{\mathbf{m},L}, \{S^t\}, \mu_{\mathbf{m},L}\right)$ of the N-ball system is ergodic and completely hyperbolic. Then, following from the results of* [2] *and* [11]*, such a semi-dispersive billiard system actually enjoys the Bernoulli mixing property, as well.*

Remark 9 We note that the results of the papers [16] and [17] nicely complement each other. They precisely assert the same, almost sure ergodicity of hard ball systems in the cases $\nu = 2$ and $\nu \geq 3$, respectively. It should be noted, however, that the proof of [16] is primarily dynamical-geometric (except the verification of the Chernov–Sinai Ansatz), whereas the novel parts of [17] are fundamentally algebraic. We note that the Chernov–Sinai Ansatz claims that almost every singular trajectory is eventually hyperbolic.

Remark 10 The above inequality $L > L_0(r, \nu)$ corresponds to physically relevant situations. Indeed, in the case $L < L_0(r, \nu)$ the particles would not have enough room to even freely exchange positions.

4 The Conditional Proof

In the paper [18] we again considered the system of N (≥ 2) elastically colliding hard spheres with masses $m_1, \ldots, m_N$ and radius r on the flat unit torus $\mathbb{T}^\nu$, $\nu \geq 2$. We proved the Boltzmann–Sinai Ergodic Hypothesis, i.e., the full hyperbolicity and ergodicity of such systems for every selection $(m_1, \ldots, m_N; r)$ of the external parameters, provided that almost every singular orbit is geometrically hyperbolic (sufficient), i.e., that the so called Chernov–Sinai Ansatz is true. The proof does not use the formerly developed, rather involved algebraic techniques, instead it extensively employs dynamical methods and tools from geometric analysis.

To upgrade the full hyperbolicity to ergodicity, one needs to refine the analysis of the degeneracies, i.e., the set of non-hyperbolic phase points. For hyperbolicity, it was enough that the degeneracies made a subset of codimension ≥ 1 in the phase space. For ergodicity, one has to show that its codimension is ≥ 2, or to find some other ways to prove that the (possibly) arising one-codimensional, smooth submanifolds of non-sufficiency are incapable of separating distinct, open ergodic components from each other. The latter approach was successfully pursued in [18]. In the paper [16], I took the first step in the direction of proving that the codimension of exceptional manifolds is at least two: It was proved there that the systems of $N \geq 2$ disks on a 2D torus (i.e., $\nu = 2$) are ergodic for typical (generic) $(N+1)$-tuples of external parameters $(m_1, \ldots, m_N, r)$. The proof involved some algebro-geometric techniques, thus the result is restricted to generic parameters $(m_1, \ldots, m_N; r)$. But there was a good reason to believe that systems in $\nu \geq 3$ dimensions would be somewhat easier to handle, at least that was indeed the case in early studies.

In the paper [17], I was able to improve further the algebro-geometric methods of [21], and proved that for any $N \geq 2$, $\nu \geq 2$, and for almost every selection $(m_1, \ldots, m_N; r)$ of the external geometric parameters the corresponding system of N hard balls on $\mathbb{T}^\nu$ is (fully hyperbolic and) ergodic.

In the paper [18] the following result was obtained.

Theorem 11 *For any integer values $N \geq 2$, $\nu \geq 2$, and for every $(N+1)$-tuple $(m_1, \ldots, m_N, r)$ of the external geometric parameters the standard hard ball system $\left(\mathbf{M}_{\mathbf{m},r}, \left\{S^t_{\mathbf{m},r}\right\}, \mu_{\mathbf{m},r}\right)$ is (fully hyperbolic and) ergodic, provided that the Chernov–Sinai Ansatz holds true for all such systems.*

Remark 12 The novelty of the theorem (as compared to the result in [17]) is that it applies to every $(N+1)$-tuple of external parameters (provided that the interior of the phase space is connected), without an exceptional zero-measure set. Somehow, the most annoying shortcoming of several earlier results was exactly the fact that those results were only valid for hard sphere systems apart from an undescribed, countable collection of smooth, proper submanifolds of the parameter space $\mathbb{R}^{N+1} \ni (m_1, m_2, \ldots, m_N; r)$. Furthermore, those proofs do not provide any effective means to check if a given $(m_1, \ldots, m_N; r)$-system is ergodic or not, most notably for the case of equal masses in Sinai's classical formulation of the problem.

Remark 13 The present result speaks about exactly the same models as the result of [15], but the statement of this new theorem is obviously stronger than that of the theorem in [15]: It has been known for a long time that, for the family of semi-dispersive billiards, ergodicity cannot be obtained without also proving full hyperbolicity.

Remark 14 As it follows from the results of [2] and [11], all standard hard ball systems (the models covered by the theorems of this survey), once they are proved to be mixing, they also enjoy the much stronger Bernoulli mixing property. However, even the K-mixing property of semi-dispersive billiard systems follows from their ergodicity, as the classical results of Sinai in [24] and [25] show. Here it is worth noting that in his publication [25] Sinai pays tribute to the late Russian physicist N. S. Krylov [10], who appeared to be the first physicist pointing out the potential importance of studying hard sphere systems in order to better understand Boltzmann's ergodic hypothesis.

In the subsequent part of this survey we review the necessary technical prerequisites of the proof, along with some of the needed references to the literature. The fundamental objects of the paper [18] are the so called "exceptional manifolds" or "separating manifolds" J: they are codimension-one submanifolds of the phase space that are separating distinct, open ergodic components of the billiard flow.

In §3 of [18] we proved Main Lemma 3.5, which states, roughly speaking, the following: Every separating manifold $J \subset \mathbf{M}$ contains at least one sufficient (or geometrically hyperbolic) phase point. The existence of such a sufficient phase point $x \in J$, however, contradicts the Theorem on Local Ergodicity of Chernov and Sinai (Theorem 5 in [26]), since an open neighborhood U of x would then belong to a single ergodic component, thus violating the assumption that J is a separating manifold. In §4 this result was exploited to carry out an inductive proof of the (hyperbolic) ergodicity of every hard ball system, provided that the Chernov–Sinai Ansatz holds true for all hard ball systems.

In what follows, we make an attempt to briefly outline the key ideas of the proof of Main Lemma 3.5 of [18]. Of course, this outline will lack the majority of the nitty-gritty details, technicalities, that constitute an integral part of the proof. The proof is a proof by contradiction.

We consider the one-sided, tubular neighborhoods U_δ of J with radius $\delta > 0$. Throughout the whole proof of the main lemma the asymptotics of the measures $\mu(X_\delta)$ of certain (dynamically defined) sets $X_\delta \subset U_\delta$ are studied, as $\delta \to 0$. We fix a large constant $c_3 \gg 1$, and for typical points $y \in U_\delta \setminus U_{\delta/2}$ (having non-singular forward orbits and returning to the layer $U_\delta \setminus U_{\delta/2}$ infinitely many times in the future) we define the arc-length parametrized curves $\rho_{y,t}(s)$ $(0 \le s \le h(y,t))$

in the following way: $\rho_{y,t}$ emanates from y and it is the curve inside the manifold $\Sigma_0^t(y)$ with the steepest descent towards the separating manifold J. Here $\Sigma_0^t(y)$ is the inverse image $S^{-t}\left(\Sigma_t^t(y)\right)$ of the flat, local orthogonal manifold passing through $y_t = S^t(y)$. The terminal point $\Pi(y) = \rho_{y,t}(h(y,t))$ of the smooth curve $\rho_{y,t}$ is either

(a) on the separating manifold J, or
(b) on a singularity of order $k_1 = k_1(y)$.

The case (b) is further split in two sub-cases, as follows:

(b/1) $k_1(y) < c_3$;
(b/2) $c_3 \le k_1(y) < \infty$.

About the set $\overline{U}_\delta(\infty)$ of (typical) points $y \in U_\delta \setminus U_{\delta/2}$ with property (a), it is shown that, actually, $\overline{U}_\delta(\infty) = \emptyset$. Roughly speaking, the reason for this is the following: For a point $y \in \overline{U}_\delta(\infty)$ the iterates S^t of the flow exhibit arbitrarily large contractions on the curves $\rho_{y,t}$, thus the infinitely many returns of $S^t(y)$ to the layer $U_\delta \setminus U_{\delta/2}$ would "pull up" the other endpoints $S^t(\Pi(y))$ to the region $U_\delta \setminus J$, consisting entirely of sufficient points, and showing that the point $\Pi(y) \in J$ itself is sufficient, thus violating the assumed hypothesis of the proof by contradiction.

The set $\overline{U}_\delta \setminus \overline{U}_\delta(c_3)$ of all phase points $y \in U_\delta \setminus U_{\delta/2}$ with the property $k_1(y) < c_3$ are dealt with by a lemma, where it is shown that

$$\mu\left(\overline{U}_\delta \setminus \overline{U}_\delta(c_3)\right) = o(\delta),$$

as $\delta \to 0$. The reason, in rough terms, is that such phase points must lie at the distance $\le \delta$ from the compact singularity set

$$\bigcup_{0 \le t \le 2c_3} S^{-t}\left(\mathcal{SR}^-\right),$$

and this compact singularity set is transversal to J, thus ensuring the measure estimate $\mu\left(\overline{U}_\delta \setminus \overline{U}_\delta(c_3)\right) = o(\delta)$.

Finally, the set $F_\delta(c_3)$ of (typical) phase points $y \in U_\delta \setminus U_{\delta/2}$ with $c_3 \le k_1(y) < \infty$ is dealt with by Lemmas 3.36, 3.37, and Corollary 3.38 of [18], where it is shown that $\mu\left(F_\delta(c_3)\right) \le C\delta$, with constants C that can be chosen arbitrarily small by selecting the constant $c_3 \gg 1$ big enough. The ultimate reason of this measure estimate is the following fact: For every point $y \in F_\delta(c_3)$ the projection

$$\tilde{\Pi}(y) = S^{t_{\overline{k}_1(y)}} \in \partial\mathbf{M}$$

(where $t_{\overline{k}_1(y)}$ is the time of the $\overline{k}_1(y)$-th collision on the forward orbit of y) will have a tubular distance $z_{tub}\left(\tilde{\Pi}(y)\right) \le C_1\delta$ from the singularity set $\mathcal{SR}^- \cup \mathcal{SR}^+$, where the constant C_1 can be made arbitrarily small by choosing the contraction coefficients of the iterates $S^{t_{\overline{k}_1(y)}}$ on the curves $\rho_{y,t_{\overline{k}_1(y)}}$ arbitrarily small with the help of the fine expansion and contraction estimates published in Appendix B of [18]. The upper measure estimate (inside the set $\partial\mathbf{M}$) of the set of such points $\tilde{\Pi}(y) \in \partial\mathbf{M}$ (Lemma 2 in [26]) finally yields the required upper bound $\mu(F_\delta(c_3)) \leq C\delta$ with arbitrarily small positive constants C (if $c_3 \gg 1$ is big enough).

The listed measure estimates and the obvious fact

$$\mu\left(U_\delta \setminus U_{\delta/2}\right) \approx C_2\delta$$

(with some constant $C_2 > 0$, depending only on J) show that there must exist a point $y \in U_\delta \setminus U_{\delta/2}$ with the property (a) above, thus ensuring the sufficiency of the point $\Pi(y) \in J$.

In the closing section of [18] we completed the inductive proof of ergodicity (with respect to the number of balls N) by utilizing Main Lemma 3.5 and earlier results from the literature. Actually, a consequence of the Main Lemma will be that exceptional J-manifolds do not exist, and this will imply the fact that no distinct, open ergodic components can coexist.

5 Proof of Ansatz

Finally, in the paper [19] we proved the Boltzmann–Sinai Hypothesis for hard ball systems on the ν-torus $\mathbb{R}^\nu/\mathbb{Z}^\nu$ ($\nu \geq 2$) without any assumed hypothesis or exceptional model.

As said before, in [18] the Boltzmann–Sinai Hypothesis was proved in full generality (i.e., without exceptional models), by assuming the Chernov–Sinai Ansatz.

The only missing piece of the whole puzzle is to prove that no open piece of a singularity manifold can precisely coincide with a codimension-one manifold describing the trajectories with a non-sufficient forward orbit segment corresponding to a fixed symbolic collision sequence. This is exactly what we claim in our Theorem below.

5.1 Formulation of Theorem

Let $U_0 \subset \mathbf{M} \setminus \partial\mathbf{M}$ be an open ball, $T > 0$, and assume that

(a) $S^T(U_0) \cap \partial\mathbf{M} = \emptyset$,
(b) S^T is smooth on U_0.

Next we assume that there is a *codimension-one*, smooth submanifold $J \subset U_0$ with the property that for every $x \in U_0$ the trajectory segment $S^{[0,T]}x$ is geometrically hyperbolic (sufficient) if and only if $x \notin J$. (J is a so called non-hyperbolicity or degeneracy manifold.) Denote the common symbolic collision sequence of the orbits $S^{[0,T]}x$ ($x \in U_0$) by $\Sigma = (e_1, e_2, \dots, e_n)$, listed in the increasing time order. Let $t_i = t(e_i)$ be the time of the i-th collision, $0 < t_1 < t_2 < \dots < t_n < T$.

Finally we assume that for every phase point $x \in U_0$ the first reflection $S^{\tau(x)}x$ *in the past* on the orbit of x is a singular reflection (i.e., $S^{\tau(x)}x \in \mathscr{S}\mathscr{R}_0^+$) if and only if x belongs to a codimension-one, smooth submanifold K of U_0. For the definition of the manifold of singular reflections $\mathscr{S}\mathscr{R}_0^+$ see, for instance, the end of §1 in [18].

Theorem 15 *Using all the assumptions and notations above, the submanifolds J and K of U_0 do not coincide.*

References

1. L.A. Bunimovich. On the ergodic properties of nowhere dispersing billiards. *Comm. Math. Phys.*, 65(3):295–312, 1979.
2. N.I. Chernov and C. Haskell. Nonuniformly hyperbolic K-systems are Bernoulli. *Ergodic Theory Dynam. Systems*, 16(1):19–44, 1996.
3. G.A. Hedlund. The dynamics of geodesic flows. *Bull. Amer. Math. Soc.*, 45(4):241–260, 1939.
4. E. Hopf. Statistik der geodätischen Linien in Mannigfaltigkeiten negativer Krümmung. *Ber. Verh. Sächs. Akad. Wiss. Leipzig*, 91:261–304, 1939.
5. A. Katok, J.-M. Strelcyn, F. Ledrappier, and F. Przytycki. *Invariant Manifolds, Entropy and Billiards; Smooth Maps with Singularities*, volume 1222 of *Lecture Notes in Mathematics*. Springer-Verlag, Berlin, 1986.
6. A. Krámli, N. Simányi, and D. Szász. Ergodic properties of semi-dispersing billiards. I. Two cylindric scatterers in the 3D torus. *Nonlinearity*, 2(2):311–326, 1989.
7. A. Krámli, N. Simányi, and D. Szász. A "transversal" fundamental theorem for semi-dispersing billiards. *Comm. Math. Phys.*, 129(3):535–560, 1990.
8. A. Krámli, N. Simányi, and D. Szász. The K-property of three billiard balls. *Ann. of Math. (2)*, 133(1):37–72, 1991.
9. A. Krámli, N. Simányi, and D. Szász. The K-property of four billiard balls. *Comm. Math. Phys.*, 144(1):107–148, 1992.
10. N.S. Krylov. *Works on the Foundations of Statistical Physics*. Princeton University Press, Princeton, NJ, 1979.
11. D. Ornstein and B. Weiss. On the Bernoulli nature of systems with some hyperbolic structure. *Ergodic Theory Dynam. Systems*, 18(2):441–456, 1998.
12. Ya.B. Pesin. Characteristic Lyapunov exponents and smooth ergodic theory. *Russ. Math. Surv.*, 32(4):55–114, 1977.
13. N. Simányi. The K-property of N billiard balls. I. *Invent. Math.*, 108(3):521–548, 1992.
14. N. Simányi. The K-property of N billiard balls. II. Computation of neutral linear spaces. *Invent. Math.*, 110(1):151–172, 1992.
15. N. Simányi. The complete hyperbolicity of cylindric billiards. *Ergodic Theory Dynam. Systems*, 22(1):281–302, 2002.
16. N. Simányi. Proof of the Boltzmann–Sinai ergodic hypothesis for typical hard disk systems. *Invent. Math.*, 154(1):123–178, 2003.
17. N. Simányi. Proof of the ergodic hypothesis for typical hard ball systems. *Ann. Henri Poincaré*, 5(2):203–233, 2004.
18. N. Simányi. Conditional proof of the Boltzmann–Sinai ergodic hypothesis. *Invent. Math.*, 177(2):381–413, 2009.

19. N. Simányi. Singularities and non-hyperbolic manifolds do not coincide. *Nonlinearity*, 26(6):1703–1717, 2013.
20. N. Simányi and D. Szász. The *K*-property of 4D billiards with nonorthogonal cylindric scatterers. *J. Statist. Phys.*, 76(1–2):587–604, 1994.
21. N. Simányi and D. Szász. Hard ball systems are completely hyperbolic. *Ann. of Math. (2)*, 149(1):35–96, 1999.
22. N. Simányi and D. Szász. Non-integrability of cylindric billiards and transitive Lie group actions. *Ergodic Theory Dynam. Systems*, 20(2):593–610, 2000.
23. Ya.G. Sinai. On the foundations of the ergodic hypothesis for a dynamical system of statistical mechanics. *Soviet Math. Dokl.*, 4:1818–1822, 1963.
24. Ya.G. Sinai. Dynamical systems with elastic reflections. *Russ. Math. Surv.*, 25(2):137–189, 1970.
25. Ya.G. Sinai. *Development of Krylov's ideas.* Afterword to N.S. Krylov, *Works on the Foundations of Statistical Physic*, pages 239–281. Princeton University Press, Princeton, NJ,1979.
26. Ya.G. Sinai and N.I. Chernov. Ergodic properties of certain systems of two-dimensional discs and three-dimensional balls. *Russ. Math. Surv.*, 42(3):181–207, 1987.
27. D. Szász. Ergodicity of classical billiard balls. *Phys. A*, 194(1–4):86–92, 1993.
28. D. Szász. The *K*-property of "orthogonal" cylindric billiards. *Comm. Math. Phys.*, 160(3):581–597, 1994.
29. M. Wojtkowski. Invariant families of cones and Lyapunov exponents. *Ergodic Theory Dynam. Systems*, 5(1):145–161, 1985.
30. M. Wojtkowski. Principles for the design of billiards with nonvanishing Lyapunov exponents. *Comm. Math. Phys.*, 105(3):391–414, 1986.

Markov Approximations and Statistical Properties of Billiards

Domokos Szász

Abstract Markov partitions designed by Sinai (Funct Anal Appl 2:245–253, 1968) and Bowen (Am J Math 92:725–747, 1970) proved to be an efficient tool for describing statistical properties of uniformly hyperbolic systems. For hyperbolic systems with singularities, in particular for hyperbolic billiards the construction of a Markov partition by Bunimovich and Sinai (Commun Math Phys 78:247–280, 1980) was a delicate and hard task. Therefore later more and more flexible and simple variants of Markov partitions appeared: Markov sieves (Bunimovich–Chernov–Sinai, Russ Math Surv 45(3):105–152, 1990), Markov towers (Young, Ann Math (2) 147(3):585–650, 1998), standard pairs (Dolgopyat). This remarkable evolution of Sinai's original idea is surveyed in this paper.

1 Introduction

Mathematical billiards appeared as early as in the early 1910s in the works of the Ehrenfest couple, [43] (the wind tree model) and of D. König and A. Szűcs, [47] (billiards in a cube) and a bit later in 1927 in the work of G. Birkhoff, [10] (those in an oval).[1] Ergodic theory itself owes its birth to the desire to provide mathematical foundations to Boltzmann's celebrated *ergodic hypothesis* (cf. [11, 61, 65]). I briefly went over its history in my article [62] written earlier on the occasion of Sinai's Abel Prize. Therefore for historic details I recommend the interested reader to consult that freely available article. Here I only mention some most relevant facts from it. In particular, I present here two circumstances:

[1]Actually the Lorentz gas suggested by H. A. Lorentz in [50] is also a billiard in a space with infinite invariant measure, cf. Sect. 3 below.

D. Szász (✉)
Budapest University of Technology and Economics, Budapest, Hungary

H. Holden, R. Piene (eds.), *The Abel Prize 2013–2017*,
https://doi.org/10.1007/978-3-319-99028-6_13

1. The two most significant problems from physics motivating the initial study of mathematical billiards were
   (a) the ergodic hypothesis and
   (b) the goal to understand Brownian motion from microscopic principles.

   (In the last decades quantum billiards have also challenged both mathematicians and physicists and, moreover, in the very last years billiard models of heat transport have also become attackable.)
2. Sinai himself was aware of and highly appreciated the works of the N. S. Krylov, the great Russian statistical physicist who—first of all in Russia—brought hard ball systems, themselves hyperbolic billiards, to the attention of the community of mathematicians as a hopeful candidate for hyperbolic behavior, and possibly for ergodic one as well (cf. [48]).

From the side of mathematics the 1960s saw the birth and rapid development of the theory of smooth hyperbolic dynamical systems with Sinai being one of the leading creators of this theory. For mathematics Sinai's 1970 paper [59] introduced a new object to study: hyperbolic billiards as hyperbolic dynamical systems with singularities. Later it turned out that this theory also covers basic models of *chaos theory*, like the Lorenz system, the Hénon map, logistic maps, etc.

The rich world of hyperbolic billiards and Sinai's emblematic influence on it is demonstrated by the fact that no less than three articles of this volume are devoted to Sinai's achievements in their theory. Thus I will not address here Sinai's main accomplishments in the 1970s and 1980s and some of the most important later expansions, see paper by Leonid Bunimovich [15], neither will I write about the progress related to the Boltzmann–Sinai ergodic hypothesis, see paper by Nándor Simányi [56]. The subject of my article will be restricted to developments related to establishing *statistical properties of hyperbolic billiards*. These results grew out of

- the appearance of the highly efficient method of Markov partitions making possible to create Markov approximations to obtain statistical properties of dynamical systems;
- Sinai's ambition to create a mathematical theory for Brownian motion, a theory also called the dynamical theory of Brownian motion (cf. [51]). Its final goal is to derive Brownian motion from microscopic assumptions, in particular from Newtonian dynamics.

With strong simplifications our topic is the treatment of statistical properties of hyperbolic billiards via Markov approximations. The main steps in the development of this theory are, roughly speaking, the following ones:

1. Markov partitions and Markov approximations for Anosov systems (and Axiom A systems) (cf. [13, 57, 58, 60]);
2. Markov partitions and Markov approximations for 2D Sinai billiards (cf. [16, 17]);
3. Markov sieves and Markov approximations for 2D Sinai billiards (cf. [19, 20]);

4. Young's towers for hyperbolic systems with singularities, in particular for 2D Sinai billiards (cf. [66]);
5. Chernov and Dolgopyat's method of standard pairs (cf. [22]).

Sinai played a founding and instrumental role in the first three steps. Chernov and Young wrote an excellent survey [27] on the fourth step also providing a pithy historical overview about the place of Markov partitions in the theory of dynamical systems. Referring to it permits me to focus here on the mathematical content of the exposition. At this point I note that Pesin's article [53] discusses Markov partitions and their role in the theory of smooth hyperbolic systems in detail. My major goal in this paper will be double folded:

- to put Sinai's most original achievements into perspective;
- provide an idea about the vast and astonishing influence of them.

As the title of this paper suggests we restrict our attention to stochastic properties of billiards obtained via Markov approximations. Consequently we do not discuss results obtained through directly functional analytical approach whose one of most spectacular achievements is the recent work of Baladi–Demers–Liverani, [2] on the exponential decay of correlations for the 2D finite-horizon Sinai billiard *flow*.

2 Markov Partitions for Anosov Maps (and for Axiom A Maps)

Let f be an Anosov diffeomorphism of a compact differentiable manifold M or an Axiom A diffeo on Λ, one of its basic sets. Markov partitions were first constructed by Adler and Weiss [1] (and also by Berg [9]) for ergodic algebraic automorphisms of 2D tori. The goal of [1] was to provide an important and quite spectacular positive example related to the famous isomorphism problem. Sinai's general construction for Anosov maps [57] and its wide-ranging conclusions [58] revealed the sweeping perspectives of the concept. Then Bowen [13] extended the notion to Axiom A maps and also gave a completely different construction. In this section we treat both approaches simultaneously. We also remark that the content of this section finds a broader exposition in [53].

As to fundamental notions on hyperbolic dynamical systems we refer to [35, 45, 46, 53] while here we are satisfied with a brief summary. If a diffeomorphism $f : M \to M$ has a *hyperbolic structure*, i.e., a decomposition into expanding vs. contracting subspaces on its unit tangent bundle $T_1 M$, then it is called an *Anosov map*. Then there exist two *foliations* into (global) stable vs. unstable invariant manifolds $\{W^u\}$ and $\{W^s\}$. Connected, bounded pieces W^u_{loc} (or W^s_{loc}) of a W^u (or of a W^s) are called *local stable (resp. unstable) invariant manifolds*. In particular, for any small $\varepsilon > 0$ denote by $W^u_{(\varepsilon)}(x)$ and $W^s_{(\varepsilon)}(x)$ the ball-like local manifolds of diameter ε around an $x \in W^u$. For sufficiently small ε the foliations possess a *local product structure*: the map $[.,.]: M \times M \to M$ is uniquely defined via

$\{[x, y]\} = W^u_{(\varepsilon)}(x) \cap W^s_{(\varepsilon)}(y)$. (We note that all these notions make also sense on a basic set of an Axiom A diffeo, cf. [53].)

Definition 1 A subset R of small diameter ε is called a *parallelogram* if it is closed for the operation $[.,.]$ and $R = \mathrm{Cl}(\mathrm{Int}\, R)$. (Further notations: $W^u_R(x) = W^u(x) \cap R$ and $W^s_R(x) = W^s(x) \cap R$.)

Definition 2 (Sinai [57]; Bowen [13]) A cover $\mathscr{P} = \{R_1, \ldots, R_k\}$ of M with a finite number of parallelograms with pairwise disjoint interiors is a Markov partition if for $\forall i, j$ and $\forall \in \mathrm{Int}\, R_i \cap f^{-1}\mathrm{Int}\, R_j$ one has

1. $W^s_{R_i}(x) \subset f^{-1} W^s_{R_j}(fx)$
2. $W^u_{R_j}(fx) \subset f W^u_{R_i}(x)$.

The inclusions in the definition imply that, whenever $x \in \mathrm{Int}\, R_i \cap f^{-1}\mathrm{Int}\, R_j$, then $f^{-1}W^s_{R_j}(fx)$ intersects R_i completely and similarly $f W^u_{R_i}(x)$ intersects R_j completely (one can also say that these ways of intersections are Markovian). The simplest example of a Markov partition is the one for the hyperbolic automorphism T of the 2-torus: $\mathbb{T}^2 = \mathbb{R}^2/\mathbb{Z}^2$ defined as follows: $Tx = \begin{pmatrix} 2 & 1 \\ 1 & 1 \end{pmatrix} x \pmod{\mathbb{Z}^2}$ (see Fig. 1, cf. [1]).

Note that a Markov partition determines a symbolic dynamics τ_A. Indeed, let $A = (a_{i,j})_{1 \le i,j \le k}$ be defined as follows: $a_{i,j} = 1$ iff $\mathrm{Int}\, R_i \cap f^{-1}\mathrm{Int}\, R_j \neq \emptyset$ and $=$ 0 otherwise. Let Σ_A be the subset of those $\sigma \in \{1, \ldots, k\}^{\mathbb{Z}} = \Sigma$ such that $\sigma \in \Sigma_A$ iff $\forall n \in \mathbb{Z} \quad a_{x_n x_{n+1}} = 1$. Then the so-called left-shift $\tau_A : \Sigma_A \to \Sigma_A$ is defined for $\sigma \in \Sigma_A$ by $(\tau_A \sigma)_i = \sigma_{i+1}$. Σ_A is a closed subset of the compact metric space Σ, a product of discrete spaces and then σ_A, called a subshift of finite type, is a homeomorphism. For a $\sigma \in \Sigma_A$ the intersection $\cap_{i \in \mathbb{Z}} f^{-i} R_{\sigma_i}$ consists of a single point x which we denote by $\pi(\sigma)$ (Figs. 2 and 3).

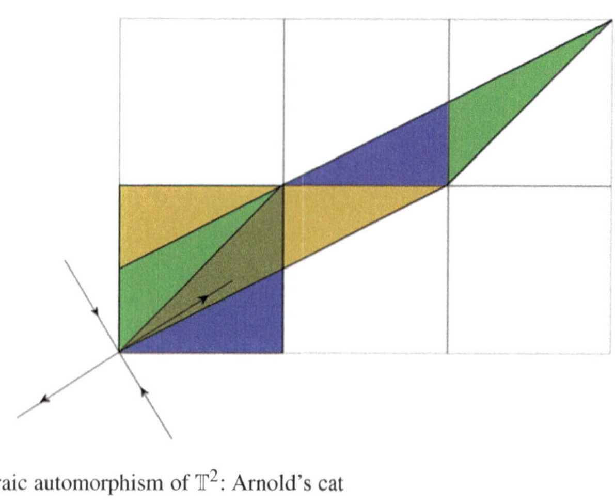

Fig. 1 Algebraic automorphism of $\mathbb{T}^2$: Arnold's cat

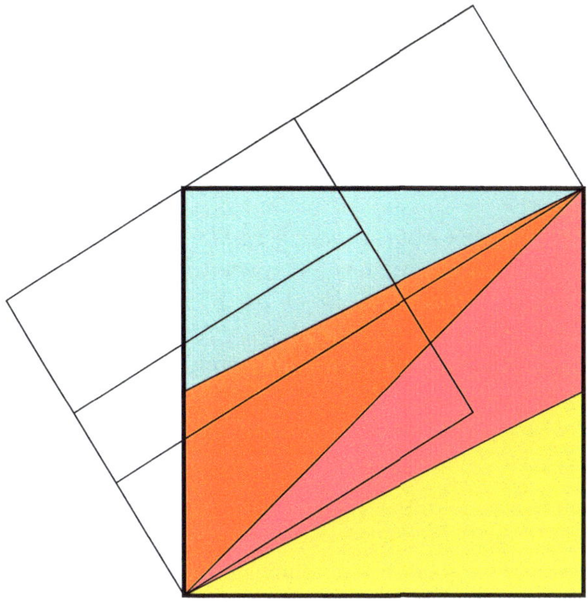

Fig. 2 The four skew parallelograms form the Markov partition of $\mathbb{T}^2$. (They are blank!)

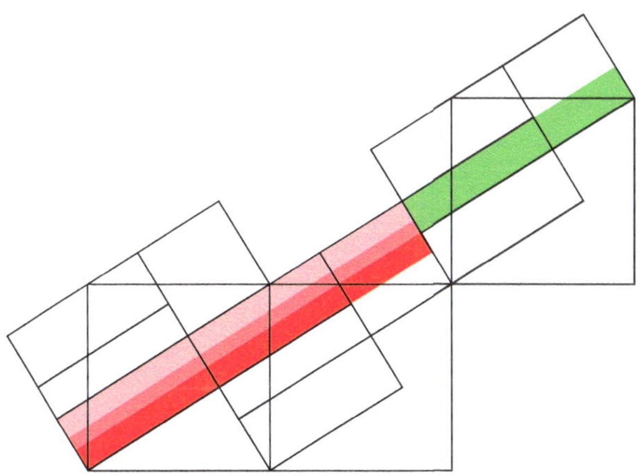

Fig. 3 The four elongated—differently tinted—skew parallelograms are the images of the elements of the Markov partition of Fig. 2

Theorem 3 (Sinai [57]; Bowen [13]) *$\pi : \Sigma_A \to M$ (or Λ) is a continuous surjective map and $f \circ \pi = \pi \circ \tau$.*

Theorem 4 (Sinai [58])

1. *For any transitive Anosov diffeomorphism f there exists a measure μ^s, positive on open subsets, such that it is invariant with respect to f and f is a Kolmogorov-automorphism.*
2. *Let $\xi^s = \{W^s\}$ be the stable foliation of M. Then the conditional measure $\mu^s(. | W^s)$ induced on almost every W^s is equivalent to the Riemannian volume on W^s. (Analogous statement is valid for the unstable foliation, too.)*
3. *If f is an algebraic Anosov automorphism of $M = \mathbb{T}^D$, $D \geq 2$ (its invariant measure is Lebesgue), then f is metrically conjugate (i.e., isomorphic) to a finite Markov chain.*
4. *The previous Markov chain has maximal entropy among all Markov chains on Σ possessing the same possible transitions.*

Claim 1 asserts a very strong ergodic property: Kolmogorov mixing. Nevertheless, it is only a qualitative attribute, similarly to the Bernoulli property, the strongest possible ergodic one. In the topologically mixing case an Anosov map is also Bernoulli (cf. [14]). In typical applications to problems of physics one also needs qualitative control of mixing, for instance when one has to prove a central limit theorem (CLT). In that respect Claim 3 opened principally fruitful perspectives. Indeed, for algebraic automorphisms of $\mathbb{T}^D$, once they are topologically mixing, the finite Markov chain arising via the Markov partition is exponentially mixing. In such cases, if one takes a Hölder observable on $\mathbb{T}^D$, then this smoothness combined with the strong mixing also provides strong stochastic properties, specifically a CLT. In general, for the study of statistical properties of dynamical systems this approach makes it possible to set in the arsenal of probability theory. Later we will see the far-reaching consequences of this development. We note that in the Axiom A case, Bowen [14] established exponential correlation decay for Hölder functions and thus implying the CLT for such functions. Claim 4 was the predecessor of Sinai's great work [60], where by introducing symbolic dynamics in the presence of a potential function he connected the theory of dynamical systems with spin models of equilibrium statistical physics. Later this work led to thermodynamic formalism for hyperbolic systems, cf. for instance [14].

3 Sinai Billiard and Lorentz Process

A billiard is a dynamical system describing the motion of a point particle in a connected, compact domain $Q \subset \mathrm{T}^D = \mathbb{R}^D/\mathbb{Z}^D$. In general, the boundary ∂Q of the domain is assumed to be piecewise C^3-smooth; denote its smooth pieces by

$\{\partial Q_\alpha \mid 1 \le \alpha \le J < \infty\}$. Inside Q the motion is uniform while the reflection at the boundary ∂Q is elastic (by the classical rule "the angle of incidence is equal to the angle of reflection"). This dynamics is called the *billiard flow*. (In what follows we will mainly restrict our review to the discrete time billiard map.) Since the absolute value of the velocity is a first integral of motion, the phase space of the billiard flow is fixed as $M = Q \times S_{D-1}$—in other words, every phase point x is of the form $x = (q, v)$ with $q \in Q$ and $v \in \mathbb{R}^d$, $|v| = 1$. The Liouville probability measure μ on M is essentially the product of Lebesgue measures, i.e., $d\mu = \text{const.}\, dqdv$ (here the constant is $1/(\text{vol}\, Q\ \text{vol}\, S_{D-1})$).

Let $n(q)$ denote the unit normal vector of a smooth component of the boundary ∂Q at the point q, directed inwards Q. Throughout the sequel we restrict our attention to *dispersing billiards*: we require that for every $q \in \partial Q$ the second fundamental form $K(q)$ of the boundary component be positive (in fact, uniformly bounded away from 0).

The boundary ∂Q defines a natural cross-section for the billiard flow. Consider namely

$$\partial M = \{(q, v) \mid q \in \partial Q,\ \langle v, n(q)\rangle \ge 0\}.$$

The *billiard map* T is defined as the first return map on ∂M. The invariant measure for the map is denoted by μ^∂, and we have $d\mu^\partial = \text{const.}\, |\langle v, n(q)\rangle|\, dqdv$ (with const. $= 2/(\text{vol}\partial Q\ \text{vol} S_{D-1})$). Throughout the sequel we work with this discrete time dynamical system.

The *Lorentz process* (cf. [50]) is the natural $\mathbb{Z}^D$ cover of the above-described toric billiard. More precisely: consider $\Pi\colon \mathbb{R}^D \to \mathbb{T}^D$ the factorisation by $\mathbb{Z}^D$. Its fundamental domain D is a cube (semi-open, semi-closed) in $\mathbb{R}^D$, so $\mathbb{R}^D = \cup_{z\in\mathbb{Z}^D}(D+z)$, where $D+z$ is the translated fundamental domain. We also lift the scatterers to $\mathbb{R}^D$ and define the phase space of the Lorentz flow as $\tilde{M} = \tilde{Q} \times S_{D-1}$, where $\tilde{Q} = \cup_{z\in\mathbb{Z}^D}(Q+z)$. In the non-compact space $\tilde{M}$ the dynamics is denoted by $\tilde{S}^t$ and the billiard map on $\partial\tilde{M}$ by $\tilde{T}$; their natural projections to the configuration space $\tilde{Q}$ are denoted by $L(t) = L(t; x), t \in \mathbb{R}_+$ and $L_n^\partial \in \mathbb{Z}_+$ and called (periodic) Lorentz flows or processes with natural invariant measures $\tilde{\mu}$ and $\tilde{\mu}^\partial$, respectively.

The *free flight vector* $\tilde{\psi}: \tilde{M} \to \mathbb{R}^D$ is defined as follows: $\tilde{\psi}(\tilde{x}) = \tilde{q}(\tilde{T}\tilde{x}) - \tilde{q}(\tilde{x})$.

Definition 5 The Sinai-billiard (or the Lorentz process) is said to have *finite horizon* if the free flight vector is bounded. Otherwise the system is said to have *infinite horizon*.

A Sinai billiard with finite horizon (in a—torus-like—cell of the hexagonal lattice) is shown on Fig. 4 and one with infinite horizon in $\mathbb{T}^2$ on Fig. 5.

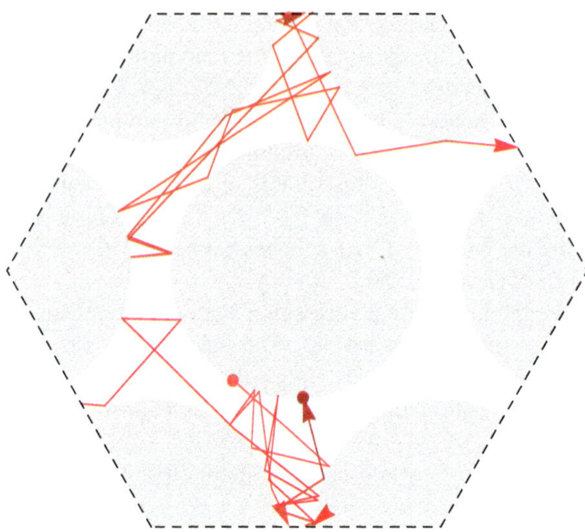

Fig. 4 Sinai billiard with finite horizon (in a—torus-like—cell of the hexagonal lattice)

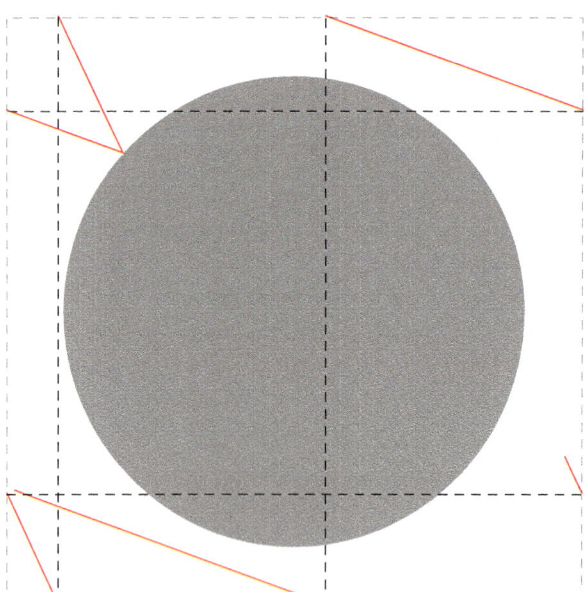

Fig. 5 Sinai billiard with infinite horizon in $\mathbb{T}^2$

3.1 Singularities

Tangential Singularities

Consider the set of tangential reflections, i.e.,

$$\mathscr{S} := \{(q, v) \in \partial M \mid \langle v, n(q)\rangle = 0\}.$$

It is easy to see that the map T is not continuous at the set $T^{-1}\mathscr{S}$. As a consequence, the (tangential) part of the singularity set for iterates T^n, $n \geq 1$ is

$$\mathscr{S}^{(n)} = \bigcup_{i=1}^{n} \mathscr{S}^{-i},$$

where in general $\mathscr{S}^k = T^k \mathscr{S}$.

Multiple Collisions

After the billiard trajectory hits $\partial Q_{\alpha_1} \cap \partial Q_{\alpha_2}$ (for some $\alpha_1 \neq \alpha_2$), the orbit stops to be uniquely defined and there arise two—or more—trajectory branches. Denote

$$\mathscr{R} := \{(q, v) \in \partial M \mid q \in \partial Q_{\alpha_1} \cap \partial Q_{\alpha_2} \text{ for some } \alpha_1 \neq \alpha_2\}.$$

Standing assumption We always assume that if $q \in \partial Q_{\alpha_1} \cap \partial Q_{\alpha_2}$ for some $\alpha_1 \neq \alpha_2$, then these two smooth pieces meet in q in general position (in the planar case this implies a non-zero angle between the pieces).

It is easy to see that the map T is not continuous at the set $T^{-1}\mathscr{R}$. As a consequence, for iterates T^n, $n \geq 1$ the part of the singularity set, caused by multiple collisions, is

$$\mathscr{R}^{(n)} = \bigcup_{i=1}^{n} \mathscr{R}^{-i},$$

where in general $\mathscr{R}^k = T^k \mathscr{R}$.

Handling the Singularities

Here we only give a very rough idea. When hitting any type of singularities the map is not continuous (the flow is still continuous at tangential collisions, but it stops being smooth). Consequently, W^u_{loc} are those connected pieces of W^u which never hit

$$\Sigma_{n=1}^{\infty}(\mathscr{S}^{(n)} \cup \mathscr{R}^{(n)})$$

in the future. (Reversing time one obtains W^s_{loc}). The basic observation in Sinai's approach to billiards was that these smooth pieces are $D-1$-dimensional local manifolds almost everywhere.

An additional difficulty connected to tangential singularities is that the expansion rate in the direction orthogonal to the singularities is infinite, a phenomenon breaking the necessary technical quantitative bounds. The way out was found in [19] where the authors introduced additional—so called secondary—singularities. These will further cut $W^{u,s}_{\text{loc}}$ and in what follows $W^{u,s}_{\text{loc}}$ will denote these smaller pieces, themselves $D-1$-dimensional local manifolds almost everywhere. (Detailed exposition of these can be found in [25] in the planar case, and in [4] in the multidimensional case.)

4 Statistical Properties of 2D Periodic Lorentz Processes

Given the successes of Markov partitions for smooth hyperbolic systems and of Sinai's theory of ergodicity for hyperbolic billiards, being a prototype of hyperbolic systems with singularities, it is a natural idea to extend the method of Markov partitions to Sinai billiards. Yet, when doing so there arise substantial difficulties. The most serious one is that basic tools of hyperbolic theory: properties of the holonomy map (also called canonical isomorphism), distorsion bounds, etc. are only valid for smooth pieces of the invariant manifolds (maximal such components are called local invariant manifolds and denoted by $W^{u,s}_{\text{loc}}$, cf. 3.1.3). These can, however, be arbitrarily short implying that a Markov partition can only be infinite, not finite. Technically this means the construction of a countable set of parallelograms (products of Cantor sets in this case) with an appropriate Markov interlocking; this is a property which is really hard to control.

Assume we are given a Sinai billiard. In the definition of parallelogram we make two changes. First, in the operation $[.,.]$ we only permit $W^{u,s}_{\text{loc}}$ and, second, we do not require $R = \text{Cl}(\text{Int}\, R)$ any more. Now we will denote $W^{u,s}_R(x) = W^{u,s}_{\text{loc}}(x) \cap R$.

Definition 6 (Bunimovich–Sinai [16]; Bunimovich–Chernov–Sinai [20]) A cover $\mathscr{P} = \{R_1, R_2, \ldots,\}$ of M with a countable number of parallelograms, satisfying $\mu_1(R_i \cap R_j) = 0$ $(\forall 1 \le i < j)$, is a Markov partition if one has that if $x \in \text{Int}\, R_i \cap f^{-1}\text{Int}\, R_j$, then

1. $W^s_{R_i}(x) \subset f^{-1} W^s_{R_j}(fx)$
2. $W^u_{R_j}(fx) \subset f W^u_{R_i}(x)$.

From now on we assume that $D = 2$ and that, unless otherwise stated, the *horizon is finite*.

4.1 Bunimovich–Sinai, 1980

Theorem 7 (Bunimovich–Sinai [16]) *Assume that for the billiard in* $Q = \mathbb{T}^2 \setminus \Sigma_{j=1}^{J} \mathcal{O}_j$ *the strictly convex obstacles* $\mathcal{O}_j$ *are closed, disjoint with* C^3*-smooth boundaries. Then for the billiard map* T *there exists a countable Markov partition of arbitrarily small diameter.*

It is worth mentioning that the statement of Theorem 3 still keeps holding for the constructed Markov partition. (We also note that a correction and a simplification of the construction was given in [18, 49].)

In the companion paper [17] the authors elaborate on further important properties of the constructed Markov partition and prove groundbreaking consequences for the Lorentz process (Fig. 6). For $x = (q, v) \in \partial M$ denote $T^n x = (q(n), v(n))$ and for $x \in \tilde{M}$ the—diffusively—rescaled version of the Lorentz process by

$$L_A(x) = \frac{1}{\sqrt{A}} L(At; x) \qquad (t \in \mathbb{R}_+).$$

Theorem 8 (Bunimovich–Sinai [17]) *There exists a constant* $\gamma \in (0, 1)$ *such that for all sufficiently large* n

$$\left|\mathbb{E}_{\mu^\partial}(v(0)v(n))\right| \leq \exp(-n^\gamma).$$

The proof uses Markov approximation. One of its essential elements is that a rank function is introduced on elements of the partition: roughly speaking the smaller the element is, the larger is its rank. Though the tail distribution for the rank is exponentially decaying nevertheless the well-known Doeblin condition of

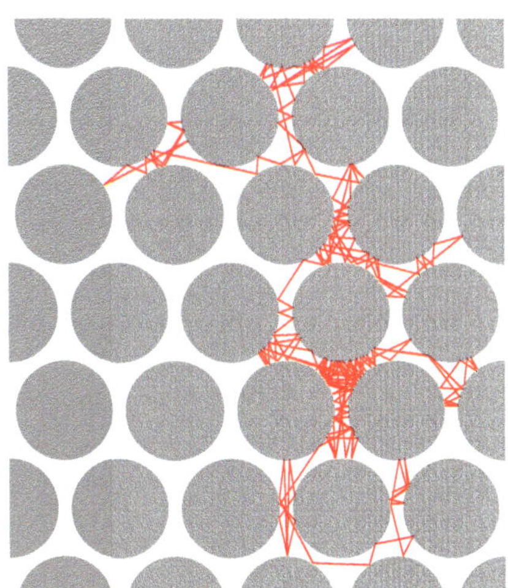

Fig. 6 The orbit of a Lorentz process with finite horizon

probability theory, ensuring exponential relaxation to equilibrium, does not hold for one step transition probabilities. Fortunately it does hold for higher step ones, still with the step size depending on the rank of the element of the partition. This weaker form of Doeblin property implies that γ is necessarily smaller than 1. Yet this is a sufficiently strong decay of correlations to imply convergence to Brownian motion. Assume ν is a probability measure on $\tilde{M}$ supported on a bounded domain and absolutely continuous with respect to $\tilde{\mu}$.

Theorem 9 (Bunimovich–Sinai [17]) *With respect to the initial measure ν, as $A \to \infty$*

$$L_A(t) \Rightarrow B_\Sigma(t)$$

where $B_\Sigma(t)$ is the planar Wiener process with zero shift and covariance matrix Σ and the convergence is weak convergence of measures in $C[0, 1]$ (or in $C[0, \infty]$).

Moreover, if the Lorentz process is not localised and the scatterer configuration is symmetric with respect to the line $q_x = q_y$, then Σ is not singular.

4.2 Bunimovich–Chernov–Sinai, 1990–1991

Ten years after the first construction Bunimovich, Chernov, and Sinai revisited the topic in two companion papers. The authors not only simplified the original constructions and proofs of [16, 17], but also clarified and significantly weakened the conditions imposed. Below we summarize the most important attainments.

Wider class of billiards Consider a planar billiard in $Q \subset \mathrm{T}^2$ with piecewise C^3-smooth boundary. Impose the following conditions:

1. If $q \in \partial Q_{\alpha_1} \cap \partial Q_{\alpha_2}$ for some $\alpha_1 \neq \alpha_2$, then the angle between ∂Q_{α_1} and ∂Q_{α_2} is not zero;
2. There exists a constant $K_0 = K_0(Q)$ such that the multiplicity of the number of curves of $\bigcup_{i=-n}^{n}(\mathscr{S}^{-i} \cup \mathscr{R}^{-i})$ meeting at any point of ∂M is at most $K_0 n$;

Theorem 10 (Bunimovich–Chernov–Sinai [20]) *Assume that a planar billiard with finite horizon satisfies the two conditions above. Then for the billiard map T there exists a countable Markov partition of arbitrarily small diameter.*

Extending Theorems 8 and 9 and simplifying the proofs, the three authors could get the following results. Denote by $\mathscr{H}_\beta$, $\beta > 0$ the class of β-Hölder functions $h \colon \partial M \to \mathbb{R}$ (i.e., $\exists C(h)$ such that $\forall \alpha \leq J$ and $\forall x, y \in \partial Q_\alpha$ we have $|h(x) - h(y)| \leq C(h)|x - y|^\beta$). We note that the sequence $X_n = h(T^n x)$ $(n \in \mathbb{Z})$ is stationary with respect to the invariant measure μ^∂ on ∂M.

Theorem 11 (Bunimovich–Chernov–Sinai [19]) *Assume the billiard satisfies the previous two conditions and take a function* $h \in \mathscr{H}_\beta$ *with* $\mathbb{E}_{\mu^\partial} h = 0$. *Then* $\forall n \in \mathbb{Z}$ *one has*

$$|\mathbb{E}_{\mu^\partial} X_0 X_n| \leq C(h) e^{-a\sqrt{n}}$$

where $a = a(Q) > 0$ *only depends on the billiard table.*

Theorem 12 (Bunimovich–Chernov–Sinai [19]) *Assume the billiard satisfies the previous two conditions. With respect to the initial measure* ν, *as* $A \to \infty$

$$L_A(t) \text{ “} \Rightarrow \text{” } B_\Sigma(t)$$

where $B_\Sigma(t)$ *is the planar Wiener process with zero shift and covariance matrix* Σ *and the convergence “*$\Rightarrow$*” is weak convergence of finite dimensional distributions. Moreover, if the Lorentz process is not localised, then* Σ *is not singular.*

It is important to add that these two papers also discuss semi-dispersing billiards and, in general, provide a lot of important information about the delicate geometry of various examples. The proofs of Theorems 11 and 12 do not use the Markov partition of Theorem 10 directly but build up a Markov approximation scheme by using so-called approximate finite Markov-sieves. An immediate additional success of the method of Markov sieves was a spectacular physical application. In [31, 32] the authors could study a billiard-like models under the simultaneous action a Gaussian thermostat and a small external field. The interesting feature of the model is that the system is not Hamiltonian and has an attractor. Among other beautiful results they derive a formula for the rate of entropy production and verify Einstein's formula for the diffusion coefficient.

5 Further Progress of Markov Methods

5.1 *Markov Towers*

Sinai's ideas on connecting dynamical systems with statistical physics and probability theory have been justified by the works mentioned in Sect. 4. It became evident that billiard models are highly appropriate for understanding classical questions of statistical physics. Having worked out and having simplified the meticulous details of Markov approximations, the way opened for further progress.

Young, who had also had experience with other hyperbolic systems with singularities, like logistic maps and the Hénon map, was able to extract the common roots of the models. She introduced a fruitful and successful system of axioms under which one can construct Markov towers, themselves possessing a Markov partition. A major advantage of her approach was the following: the papers discussed in

Sect. 4 had showed that, though it was indeed possible, but at the same time rather hard to construct Markov partitions for billiards directly. An important idea of [66] is that one can rather use renewal properties of the systems and build Markov towers instead. A remarkable accomplishment of the tower method was that Young could improve the stretched exponential bound on correlation decay of Theorem 11 to the optimal, exponential one. Since this was just one—though much important—from the applications of her method, in [66] she restricted her discussion to the case of planar finite-horizon Sinai billiards with C^3-smooth scatterers.

Theorem 13 (Young [66]) *Assume the conditions of Theorem 7. Then for any* $\beta > 0$ *and for any* $g, h \in \mathscr{H}_\beta$ *there exist* $a > 0$ *and* $C = C(g, h)$ *such that*

1. $\forall n \in \mathbb{Z}$ *one has*

$$|\mathbb{E}_{\mu^\partial}(g \circ f^n)h - E_{\mu^\partial}(g)E_{\mu^\partial}(h)| \le Ce^{-an};$$

2.

$$\frac{1}{\sqrt{n}}\left(\sum_0^{n-1} g \circ f^n - nE_{\mu^\partial}(g)\right) \stackrel{\text{distr}}{\longrightarrow} \mathscr{N}(0, \sigma) \qquad (n \to \infty)$$

where $N(0, \sigma)$ *denotes the normal distribution with* 0 *mean and variance* $\sigma^2 \ge 0$.

Remark 14 Please pay attention to the differences in the assertions of Theorems 8, 12, and 13. The authors of the last two works have not claimed and checked tightness which was, indeed, settled in [17].

Beside the original paper one can also read [27] describing the ideas in a very clear way. It is also worth noting that [67] extended the tower method to systems where the renewal time has a tail decreasing slower than exponential. Young's tower method can be considered as a fulfillment of Sinai's program. Her axioms for hyperbolic systems with singularities serve as an autonomous—and most popular—subject and make it possible to discuss wide-ranging delicate stochastic properties of the systems covered, interesting either from probabilistic or dynamical or physical point of view. Two examples from the numerous applications are [54] proving large deviation theorems for systems satisfying Young's axioms and [52] describing a recurrence type result in the planar Lorentz process setup.

Unfortunately, without further assumptions the method works so far for the planar case only. However, in their paper [8] providing an important achievement, Bálint and Tóth formulated a version of the tower method for multidimensional billiards under the additional 'complexity' hypothesis, whose verification for multidimensional models is a central outstanding question of the theory.

5.2 Standard Pairs

Another astonishing development of Markovian tools was the 'standard pair' method of Chernov and Dolgopyat [22]. This method has already had remarkable applications, but so far it is not easy to see where its limits are. As to a recent utilization we can, for instance, mention that standard pairs have also been applied to the construction of SRB measures for smooth hyperbolic maps in any dimension; section 3 of [33] provides a brief introduction to the tool, too. Since—for systems with singularities—the method of standard pairs does not have until now a clear survey exposition as [27] is for the tower method, we present very briefly a theorem showing how it handles Markovity.

Let $(\partial M, T, \mu^\partial)$ be the billiard ball map—for simplicity for a planar billiard. A standard pair is $\ell = (W, \rho)$ where W is an unstable curve, ρ is a nice probability density on W (an unstable curve is a smooth curve in ∂M whose derivatives at every point lie in the unstable cone). Decompose ∂M into a family of nice standard pairs. Select a standard pair $\ell = (W, \rho)$ from this family. Fix a nice function $A : M \to \mathbb{R}$. Then according to the well-known law of total probability

$$\mathbb{E}_\ell(A \circ T^n) = \sum_\alpha c_{\alpha n} \mathbb{E}_{\ell_{\alpha n}}(A) \tag{1}$$

where $c_{\alpha n} > 0$, $\sum_\alpha c_{\alpha n} = 1$. The T^n-image of W is cut to a finite or countable number of pieces $W_{\alpha n}$. Thus $\ell_{\alpha n} = (W_{\alpha n}, \rho_{\alpha n})$ are disjoint standard pairs with $T^n W = \cup_\alpha W_{\alpha,n}$ where $\rho_{\alpha n}$ is the pushforward of ρ up to a multiplicative factor.

Theorem 15 (Chernov–Dolgopyat [22]; Growth lemma ∼ Markov property) *If* $n \geq \beta_3 |\log \text{length}(\ell)|$, *then*

$$\sum_{\text{length}(\ell_{\alpha n}) < \varepsilon} c_{\alpha n} \leq \beta_4 \varepsilon.$$

Equation 1 expressed how an unstable curve is partitioned after n iterations. Among the arising pieces there are, of course, longer and shorter pieces. The theorem provides a quantitative estimate for the total weight of pieces shorter than ε. This theorem—a quantitative formulation of Sinai's traditional billiard philosophy that 'expansion prevails partitioning'—was not new, in various forms it had appeared in earlier works, too. Its consequent application, however, together with modern formulations of averaging theory and a perturbative study of dynamical systems, was absolutely innovative and most successful.

6 Further Successes of Markov Methods

Because of the abundance of related results my summary will be very much selective. My main guiding principle will be that I try to focus on those developments that are either directly related to Sinai's interests or even to the problems he raised or alternatively show a variety of questions from physics.

6.1 Applications of the Tower Method

1. Already in 1999, Chernov [21] could extend the exponential correlation bound of [66] to planar billiards with infinite horizon: it holds for Hölder observables. (The work also contains precious analysis of the growth lemma, of homogeneity layers, etc.).
2. As mentioned before, the works [31, 32] treated billiards with small external forces. Chernov, in a series of articles (that started with [29] and ended with [28]) worked out a comprehensive theory of these models.
3. The methods initiated by Sinai also made it possible to study Sinai billiards with small holes, a model suggested by physicists. Early answers to the questions were treated by Markov partitions (cf. [24, 26] in case of Anosov maps) whereas later, results for billiards were found by applying Young towers (e.g., in [36, 37]).
4. After the CLT of [17] for the Lorentz process, Sinai formulated the question: is Pólya recurrence true for finite horizon planar billiards? Positive answers were obtained by [34, 55] and [63]. The latter work accomplished that by proving a local version of the CLT for the planar finite horizon Sinai billiard (cf. next point, too).
5. In the planar infinite horizon case the free flight function determining the Lorentz process is not Hölder (not even bounded), so the correlation bound of [21] is not applicable to it. In fact, in this case, as forecasted in [12], the scaling of the Lorentz process, in a limit law like that of Theorems 9 and 12, is different, and it is $\sqrt{n \log n}$ rather than $\sqrt{n}$. This was shown in [64], where, by extending the method of [63], Pólya recurrence was also obtained via an appropriate local limit theorem. The analogous results for the Lorentz flow with many other interesting theorems—also in the presence of external field—were obtained in [30] (Fig. 7 on next page).
6. As this was observed in [5], in limit laws for stadium billiards there may arise limit theorems with both classical and non-classical scaling (cf. previous point).

Applications of the Method of Standard Pairs

Stating the additional step in rough terms, this method, in the context of Markov approximation tools, can be characterized by two main intertwining advantages:

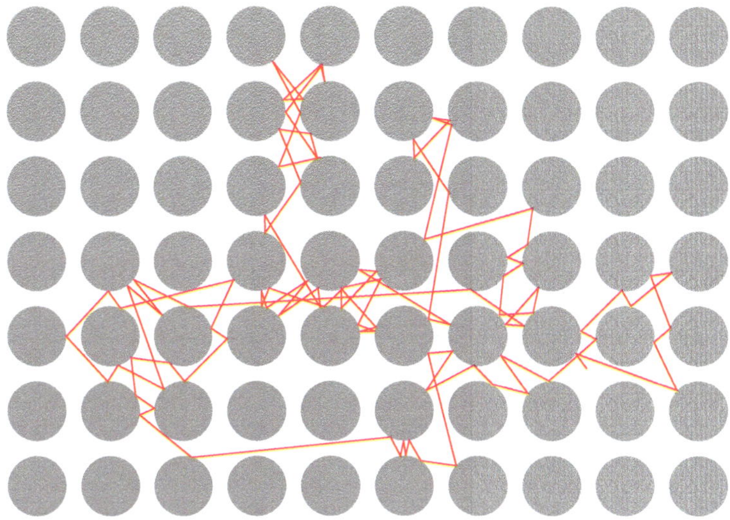

Fig. 7 The orbit of a Lorentz process with infinite horizon

first it makes possible to treat systems with two (or several) time scales and second it is appropriate for a perturbative description of dynamical systems, in particular, of billiards. The basic reference is [22] though the authors started lecturing about it as early as in 2005.[2]

1. Chernov and Dolgopyat [22] provides an important step in the dynamical theory of Brownian motion: two particles move on a planar Sinai billiard table, with one of them being an elastic disk much heavier than the other one which a point particle. Since the motion of the heavy disk is slow, for the point particle—in short time intervals—statistical properties hold (among the scatterers of the original Sinai billiard plus the—temporarily fixed—heavy disk particle). However, when the heavy particle gets close to any of the original scatterers, then additional phenomena appear and so far this is the limit of the applicability of the method. Bálint et al. [3] is the first step toward extending the time interval where the theory is hoped to be applicable.
2. Multiple times scales are treated by standard pairs in [39]. Even though their model is not a billiard one—actually the dynamics is smooth—the work is very successful in deriving a mesoscopic, stochastic process from Newtonian, microscopic laws of motion. This task, also important in a rigorous study of a heat transport model of physicists (cf. [44]), is the subject of [6, 7] for a billiard model.

[2]See, for instance, D. Dolgopyat: Introduction to averaging. Lecture notes, Institut Henri Poincaré, http://www2.math.umd.edu/~dolgop/IANotes.pdf (2005).

3. Another spectacular development was obtained in [23]. The authors were considering a point particle moving in $\mathbb{R}^2$ in the presence of a constant force among periodically situated strictly convex scatterers (the horizon is assumed to be finite). They could derive non-classical limit laws both for the velocity and the position of the particle and moreover, they could also prove the recurrence of the particle.
4. Sinai raised the following problem in 1981: consider a finite horizon, planar Sinai billiard and displace one scatterer a bit. Prove for it an analogue of Theorem 12. This problem was answered by the method of standard pairs in the companion works [41, 42].
5. Returning to heat conduction: in [40] the authors could derive the heat equation for a Lorentz process in a quasi one-dimensional tube being long finite and asymptotically infinite. The boundaries, on the one hand, absorb particles reaching them and, on the other hand, particles are also injected with energies corresponding to different temperatures.

Closing this section, I note that despite the striking successes of Markov approximation techniques, their applicability so far is essentially restricted to two dimensional models, except perhaps for [33]. Sinai's original works addressed explicitly planar models, only. Though the tower method is extended to the multidimensional case, the 'complexity condition' arising in it, has not been checked hitherto for any multidimensional system, yet it is strongly believed that it does hold at least in typical cases.

As mentioned earlier many successes of the theory of hyperbolic billiards were motivated by problems of physics. The recent survey [38], however, shows that, as usual, there are more open problems than those solved.

Acknowledgements The author is highly indebted to Yasha Pesin for his careful reading of the manuscript and many useful remarks and, moreover, to Péter Bálint and Thomas Gilbert for kindly helping him out with the figures and to Ferenc Wettl for technical help with them. This research was supported by Hungarian National Foundation for Scientific Research grant No. K 104745 and OMAA-92öu6 project.

References

1. R.L. Adler and B. Weiss. Entropy, a complete metric invariant for automorphisms of the torus. *Proc. Nat. Acad. Sci. U.S.A.*, 57:1573–1576, 1967.
2. V. Baladi, M.F. Demers, and C. Liverani. Exponential decay of correlations for finite horizon Sinai billiard flows. *Invent. Math.*, 211(1):39–177, 2018.
3. P. Bálint, N. Chernov, and D. Dolgopyat. Limit theorems for dispersing billiards with cusps. *Comm. Math. Phys.*, 308(2):479–510, 2011.
4. P. Bálint, N. Chernov, D. Szász, and I.P. Tóth. Geometry of multi-dimensional dispersing billiards. *Astérisque*, (286):119–150, 2003.
5. P. Bálint and S. Gouëzel. Limit theorems in the stadium billiard. *Comm. Math. Phys.*, 263(2):461–512, 2006.

6. P. Bálint, T. Gilbert, P. Nándori, D. Szász, Domokos, and I.P. Tóth. On the limiting Markov process of energy exchanges in a rarely interacting ball-piston gas. *J. Stat. Phys.* **166**(3–4), 903–925 (2017).
7. P. Bálint, P. Nándori, D. Szász, and I.P. Tóth. Equidistribution for standard pairs in planar dispersing billiard flows. *Ann. Henri Poincaré* **19**(4), 979–1042 (2018).
8. P. Bálint and I.P. Tóth. Exponential decay of correlations in multi-dimensional dispersing billiards. *Ann. Henri Poincaré*, 9(7):1309–1369, 2008.
9. K.R. Berg. *On the conjugacy problem for K-systems.* PhD thesis, University of Minnesota, 1967.
10. G.D. Birkhoff. On the periodic motions of dynamical systems. *Acta Math.*, 50(1):359–379, 1927.
11. G.D. Birkhoff. Proof of the ergodic theorem. *Proc. Natl. Acad. Sci. USA*, 17:656–660, 1931.
12. P. M. Bleher. Statistical properties of two-dimensional periodic Lorentz gas with infinite horizon. *J. Statist. Phys.* **66**(1–2), 315–373 (1992).
13. R. Bowen. Markov partitions for Axiom A diffeomorphisms. *Amer. J. Math.*, 92:725–747, 1970.
14. R. Bowen. *Equilibrium States and the Ergodic Theory of Anosov Diffeomorphisms*, volume 470 of *Lecture Notes in Mathematics*. Springer-Verlag, Berlin, 2008.
15. L. Bunimovich. Billiards. In H. Holden, R. Piene, editors, *The Abel Prize 2013–2017*, pages 195–205. Springer, 2019.
16. L.A. Bunimovich and Ya.G. Sinai. Markov partitions for dispersed billiards. *Commun. Math. Phys.*, 78:247–280, 1980.
17. L.A. Bunimovich and Ya.G. Sinai. Statistical properties of Lorentz gas with periodic configuration of scatterers. *Commun. Math. Phys.*, 78:479–497, 1981.
18. L.A. Bunimovich and Ya.G. Sinai. Erratum [16]. *Comm. Math. Phys.*, 107(2):357–358, 1986.
19. L.A. Bunimovich, Ya.G. Sinai, and N.I. Chernov. Statistical properties of two-dimensional hyperbolic billiards. *Russ. Math. Surv.*, 46(4), 1991.
20. L.A. Bunimovich, Ya.G. Sinai, and N.I. Chernov. Markov partitions for two-dimensional hyperbolic billiards. *Russ. Math. Surv.*, 45(3):105–152, 1990.
21. N. Chernov. Decay of correlations and dispersing billiards. *J. Statist. Phys.*, 94(3–4):513–556, 1999.
22. N. Chernov and D. Dolgopyat. Brownian Brownian motion. I. *Mem. Amer. Math. Soc.*, 198(927):viii+193, 2009.
23. N. Chernov and D. Dolgopyat. The Galton board: limit theorems and recurrence. *J. Amer. Math. Soc.*, 22(3):821–858, 2009.
24. N. Chernov and R. Markarian. Anosov maps with rectangular holes. Nonergodic cases. *Bol. Soc. Brasil. Mat. (N.S.)*, 28(2):315–342, 1997.
25. N. Chernov and R. Markarian. *Chaotic Billiards*, volume 127 of *Mathematical Surveys and Monographs*. American Mathematical Society, Providence, RI, 2006.
26. N. Chernov, R. Markarian, and S. Troubetzkoy. Invariant measures for Anosov maps with small holes. *Ergodic Theory Dynam. Systems*, 20(4):1007–1044, 2000.
27. N. Chernov and L.S. Young. Decay of correlations for Lorentz gases and hard balls. In *Hard Ball Systems and the Lorentz Gas*, volume 101 of *Encyclopaedia Math. Sci.*, pages 89–120. Springer, Berlin, 2000.
28. N. Chernov, H.-K. Zhang, and P. Zhang. Electrical current in Sinai billiards under general small forces. *J. Stat. Phys.*, 153(6):1065–1083, 2013.
29. N.I. Chernov. Sinai billiards under small external forces. *Ann. Henri Poincaré*, 2(2):197–236, 2001.
30. N.I. Chernov and D.I. Dolgopyat. Anomalous current in periodic Lorentz gases with infinite horizon. *Russ. Math. Surv.*, 64(4):651–699, 2009.
31. N.I. Chernov, G.L. Eyink, J.L. Lebowitz, and Ya.G. Sinai. Derivation of ohm's law in a deterministic mechanical model. *Phys. Rev. Lett.*, 70:2209–2212, Apr 1993.
32. N.I. Chernov, G.L. Eyink, J.L. Lebowitz, and Ya.G. Sinai. Steady-state electrical conduction in the periodic Lorentz gas. *Comm. Math. Phys.*, 154(3):569–601, 1993.

33. V. Climenhaga, D. Dolgopyat, and Y. Pesin. Non-stationary non-uniform hyperbolicity: SRB measures for dissipative maps. *Comm. Math. Phys.*, 346(2):553–602, 2016.
34. J.-P. Conze. Sur un critère de récurrence en dimension 2 pour les marches stationnaires, applications. *Ergodic Theory Dynam. Systems*, 19(5):1233–1245, 1999.
35. I.P. Cornfeld, S.V. Fomin, and Ya.G. Sinai. *Ergodic Theory*, volume 245 of *Grundlehren der Mathematischen Wissenschaften*. Springer-Verlag, New York, 1982.
36. M. Demers, P. Wright, and L.-S. Young. Escape rates and physically relevant measures for billiards with small holes. *Comm. Math. Phys.*, 294(2):353–388, 2010.
37. M.F. Demers, P. Wright, and L.-S. Young. Entropy, Lyapunov exponents and escape rates in open systems. *Ergodic Theory Dynam. Systems*, 32(4):1270–1301, 2012.
38. C.P. Dettmann. Diffusion in the Lorentz gas. *Commun. Theor. Phys.*, 62(4):521–540, 2014.
39. D. Dolgopyat and C. Liverani. Energy transfer in a fast-slow Hamiltonian system. *Comm. Math. Phys.*, 308(1):201–225, 2011.
40. D. Dolgopyat and P. Nándori. Nonequilibrium density profiles in Lorentz tubes with thermostated boundaries. *Comm. Pure Appl. Math.*, 69(4):649–692, 2016.
41. D. Dolgopyat, D. Szász, and T. Varjú. Recurrence properties of planar Lorentz process. *Duke Math. J.*, 142(2):241–281, 2008.
42. D. Dolgopyat, D. Szász, and T. Varjú. Limit theorems for locally perturbed planar Lorentz processes. *Duke Math. J.*, 148(3):459–499, 2009.
43. P. Ehrenfest and T. Ehrenfest. Begriffliche Grundlagen der statistischen Auffassung in der Mechanik. Encykl. d. Math. Wissensch. IV 2 II, Heft 6, 90 S (1912)., 1912.
44. P. Gaspard and T. Gilbert. Heat conduction and fourier's law by consecutive local mixing and thermalization. *Phys. Rev. Lett.*, 101:020601, Jul 2008.
45. B. Hasselblatt and A. Katok, editors. *Handbook of Dynamical Systems. Volume 1A*. Amsterdam: North-Holland, 2002.
46. A. Katok and B. Hasselblatt. *Introduction to the Modern Theory of Dynamical Systems*, volume 54 of *Encyclopedia of Mathematics and its Applications*. Cambridge University Press, Cambridge, 1995.
47. D. König and A. Szücs. Mouvement d'un point abandonné à l'intérieur d'un cube. *Rend. Circ. Mat. Palermo*, 36:79–90, 1913.
48. N.S. Krylov. *Works on the Foundations of Statistical Physics*. Princeton University Press, Princeton, NJ, 1979.
49. Y.-E. Lévy. A note on Sinai and Bunimovich's Markov partition for billiards. *J. Statist. Phys.*, 45(1–2):63–68, 1986.
50. H.A. Lorentz. Le mouvement des électrons dans les métaux. *Arch. Néerl.* **10** 336–371 (1905). In *Collected Paper. Vol. III*, pages 180ff.
51. E. Nelson. *Dynamical Theories of Brownian Motion*. Princeton University Press, Princeton, N.J., 1967.
52. F. Pène and B. Saussol. Back to balls in billiards. *Comm. Math. Phys.*, 293(3):837–866, 2010.
53. Y. Pesin. Sinai's work on Markov partitions and SRB measures. In H. Holden, R. Piene, editors, *The Abel Prize 2013–2017*, pages 257–285. Springer, 2019.
54. L. Rey-Bellet and L.-S. Young. Large deviations in non-uniformly hyperbolic dynamical systems. *Ergodic Theory Dynam. Systems*, 28(2):587–612, 2008.
55. K. Schmidt. On joint recurrence. *C. R. Acad. Sci. Paris Sér. I Math.*, 327(9):837–842, 1998.
56. N. Simányi. Further developments of Sinai's ideas: the Boltzmann–Sinai hypothesis. In H. Holden, R. Piene, editors, *The Abel Prize 2013–2017*, pages 287–298. Springer, 2019.
57. Ya.G. Sinai. Construction of Markov partitions. *Funct. Anal. Appl.*, 2:245–253, 1968.
58. Ya.G. Sinai. Markov partitions and C-diffeomorphisms. *Funct. Anal. Appl.*, 2:61–82, 1968.
59. Ya.G. Sinai. Dynamical systems with elastic reflections. *Russ. Math. Surv.*, 25(2):137–189, 1970.
60. Ya.G. Sinai. Gibbs measures in ergodic theory. *Russ. Math. Surv.*, 27(4):21–69, 1972.
61. D. Szász. John von Neumann, the mathematician. *Math. Intelligencer*, 33(2):42–51, 2011.
62. D. Szász. Mathematical billiards and chaos. *Eur. Math. Soc. Newsl.*, (93):22–29, 2014.

63. D. Szász and T. Varjú. Local limit theorem for the Lorentz process and its recurrence in the plane. *Ergodic Theory Dynam. Systems*, 24(1):257–278, 2004.
64. D. Szász and T. Varjú. Limit laws and recurrence for the planar Lorentz process with infinite horizon. *J. Stat. Phys.*, 129(1):59–80, 2007.
65. J. von Neumann. Proof of the quasi-ergodic hypothesis. *Proc. Natl. Acad. Sci. USA*, 18:70–82, 1932.
66. L.-S. Young. Statistical properties of dynamical systems with some hyperbolicity. *Ann. of Math. (2)*, 147(3):585–650, 1998.
67. L.-S. Young. Recurrence times and rates of mixing. *Israel J. Math.*, 110:153–188, 1999.

List of Publications for Yakov G. Sinai

1957

[1] On the distribution of the first positive sum for the sequence of independent random variables. *Theory Probab. Appl.*, 2(1):122–129. Also available in *Teor. Verojatnost. i Primenen.*, 2:126–135 (in Russian).

1959

[2] On the notion of entropy for a dynamic system. *Dokl. Akad. Nauk SSSR*, 124(4):768–771 (in Russian).

[3] Flows with finite entropy. *Dokl. Akad. Nauk SSSR*, 125(6):1200–1202 (in Russian).

1960

[4] Geodesic flows on manifolds of negative constant curvature. *Soviet Math. Dokl.*, 1(2):335–339. Also available in *Dokl. Akad. Nauk SSSR*, 131(4):752–755 (in Russian).

[5] The central limit theorem for geodesic flows on manifolds of constant negative curvature. *Soviet Math. Dokl.*, 1(4):983–987. Also available in *Dokl. Akad. Nauk SSSR*, 133(6):1303–1306 (in Russian).

[6] Dynamical systems and stationary Markov processes. *Theory Probab. Appl.*, 5(3):305–308. Also available in *Teor. Verojatnost. i Primenen.*, 5:335–338 (in Russian).

1961

[7] (with L.D. Mešalkin). Investigation of the stability of a stationary solution of a system of equations for the plane movement of an incompressible viscous liquid. *J. Appl. Math. Mech.*, 25:1700–1705. Also available in *Prikl. Mat. Meh.*, 25:1140–1143 (in Russian).

H. Holden, R. Piene (eds.), *The Abel Prize 2013–2017*,
https://doi.org/10.1007/978-3-319-99028-6_14

[8] Geodesic flows on compact surfaces of negative curvature. *Soviet Math. Dokl.*, 2(1):106–109. Also available in *Dokl. Akad. Nauk SSSR*, 136(3):549–552 (in Russian).

[9] (with V.A. Rohlin). The structure and properties of invariant measurable partitions. *Soviet Math. Dokl.*, 2(4–6):1611–1614. Also available in *Dokl. Akad. Nauk SSSR*, 141(5):1038–1041 (in Russian).

1962

[10] (with V.I. Arnol'd). Small perturbations of the automorphisms of a torus. *Soviet Math. Dokl.*, 3(3):783–786. Also available in *Dokl. Akad. Nauk SSSR*, 144(4):695–698 (in Russian).

[11] A weak isomorphism of transformations with invariant measure. *Soviet Math. Dokl.*, 3(6):1725–1729. Also available in *Dokl. Akad. Nauk SSSR*, 147(4):797–800 (in Russian).

[12] On limit theorems for stationary processes. *Theory Probab. Appl.*, 7(2):205–210. Also available in *Teor. Verojatnost. i Primenen.*, 7:213–219 (in Russian).

1963

[13] Probabilistic ideas in ergodic theory. *Amer. Soc. transl. (2)*, 31:62–80. Also available in *Proc. Internat. Congr. Mathematicians (Stockholm, 1962)*, pages 540–559. Inst. Mittag-Lefler, Djursholm, 1962 (in Russian).

[14] On the foundations of the ergodic hypothesis for a dynamical system of statistical mechanics. *Soviet Math. Dokl.*, 4(6):1818–1822. Also available in *Dokl. Akad. Nauk SSSR*, 153(6):1261–1264 (in Russian).

[15] Some remarks on the spectral properties of ergodic dynamical systems. *Russ. Math. Surv.*, 18(5):37–50. Also available in *Uspehi Mat. Nauk*, 18(5(113)):41–54 (in Russian).

[16] On a "physical" system with positive "entropy." *Vestnik Moskov. Univ. Ser. I Mat. Meh.*, 1963(5):6–12 (in Russian).

1964

[17] Dynamical systems with countable Lebesgue spectrum. I. *Amer. Soc. transl. (2)*, 39:83–110. Also available in *Izv. Akad. Nauk SSSR Ser. Mat.*, 25:899–924, 1961.

[18] On higher order spectral measures of ergodic stationary processes. *Theory Probab. Appl.*, 8(4):429–436. Also available in *Teor. Verojatnost. i Primenen*, 8:463–470 (in Russian).

1965

[19] Letter to the editor of Uspekhi Matematicheskikh Nauk. *Uspekhi Mat. Nauk*, 20(4(124)):232 (in Russian).

1966

[20] On a weak isomorphism of transformations with invariant measure. *Amer. Math. Soc. transl. (2)*, 57:123–149. Also available in *Mat. Sb. (N.S.)*, 63(105):23–42, 1964 (in Russian).

1967

[21] (with R.A. Minlos). The phenomenon of "separation of phases" at low temperatures in certain lattice models of a gas. I. *Math. USSR-Sb.*, 2(3):335–395. Also available in *Mat. Sb. (N.S.)*, 73(115):375–448 (in Russian).

[22] A lemma from measure theory. *Math. Notes*, 2(4):715–718. Also available in *Mat. Zametki*, 2:373–378 (in Russian).

[23] (with D.V. Anosov). Certain smooth ergodic systems. *Russ. Math. Surv.*, 22(5):103–167. Also available in *Uspehi Mat. Nauk*, 22(5):107–172 (in Russian).

1968

[24] Classical dynamic systems with countably-multiple Lebesgue spectrum. II. *Amer. Math. Soc. transl. (2)*, 68:34–88. Also available in *Izv. Akad. Nauk SSSR Ser. Mat.*, 30:15–68, 1966 (in Russian).

[25] The asymptotic behavior of the number of closed geodesics on a compact manifold of negative curvature. *Amer. Math. Soc. transl. (2)*, 73:227–250. Also available in *Izv. Akad. Nauk SSSR Ser. Mat.*, 30:1275–1296, 1966 (in Russian).

[26] Markov partitions and U-diffeomorphisms. *Funct. Anal. Appl.*, 2(1):61–82. Also available in *Funkcional. Anal. i Priložen*, 2(1):64–89 (in Russian).

[27] Construction of Markov partitionings. *Funct. Anal. Appl.*, 2(3):245–253. Also available in *Funkcional. Anal. i Priložen.*, 2(3):70–80 (in Russian).

[28] (with R.A. Minlos). The phenomenon of "separation of phases" at low temperatures in certain lattice models of a gas. II. *Trans. Moscow Math. Soc.*, 19:121–196. Also available in *Trudy Moskov. Mat. Obš.*, 19:113–178 (in Russian).

1970

[29] *Theory of dynamical systems. Part I: Ergodic theory*, volume 1969 of *Lectures held in Warsaw, Spring, 1967*. Lecture Notes Series, No. 23, Matematisk Institut, Aarhus Universitet, Aarhus.

[30] Dynamical systems with elastic reflections. Ergodic properties of dispersing billiards. *Russ. Math. Surv.*, 25(2):137–189. Also available in *Uspehi Mat. Nauk*, 25(2(152)):141–192 (in Russian).

[31] (with R.A. Minlos). Spectra of stochastic arising operators in lattice models of a gas. *Theoret. Math. Phys.*, 2(2):167–176. Also available in *Teoret. Mat. Fiz.*, 2(2):230–243 (in Russian).

1971

[32] Mesures invariantes des Y-systèmes. In *Actes du Congrès International des Mathématiciens (Nice, 1970), Tome 2*, pages 929–940. Gauthier-Villars, Paris.

[33] (with K.L. Volkovysskiĭ). Ergodic properties of an ideal gas with an infinite number of degrees of freedom. *Funct. Anal. Appl.*, 5(3):185–187. Also available in *Funkcional. Anal. i Priložen.*, 5(3):19–21 (in Russian).

1972

[34] (with A.N. Livšic). Invariant measures that are compatible with smoothness for transitive U-systems. *Soviet Math. Dokl.*, 13(6):1656–1659. Also available in *Dokl. Akad. Nauk SSSR*, 207:1039–1041 (in Russian).

[35] Ergodic properties of a gas of one-dimensional hard rods with an infinite number of degrees of freedom. *Funct. Anal. Appl.*, 6(1):35–43. Also available in *Funkcional. Anal. i Priložen.*, 6(1):41–50 (in Russian).

[36] (with A.Ya. Khelemski). A description of differentiations in algebras the type of algebras of local observables of spin systems. *Funct. Anal. Appl.*, 6(4):343–344. Also available in *Funkcional. Anal. i Priložen.*, 6(4):99–100 (in Russian).

[37] Construction of the dynamics for one-dimensional systems of statistical mechanics. *Theoret. Math. Phys.*, 11(2):487–494. Also available in *Teoret. Mat. Fiz.*, 11(2):248–258 (in Russian).

[38] Gibbs measures in ergodic theory. *Russ. Math. Surv.*, 27(4):21–69. Also available in *Uspehi Mat. Nauk*, 27(4(166)):21–64 (in Russian).

1973

[39] (with P.M. Bleher). Investigations of the critical points in models of type Dyson's hierarchical models. *Comm. Math. Phys.*, 33(1):23–42.

[40] (with L.A. Bunimovich). The fundamental theorem of the theory of scattering billiards. *Math. USSR-Sb.*, 19(3):407–423. Also available in *Mat. Sb. (N.S.)*, 90(132):415–431 (in Russian).

[41] (with B.M. Gurevich and Yu.M. Suhov). Invariant measures of dynamical systems of one-dimensional statistical mechanics. *Russ. Math. Surv.*, 28:49–86. Also available in *Uspehi Mat. Nauk*, 28(5(173)):45–82 (in Russian).

1974

[42] (with S.A. Pirogov). Phase transitions of the first kind for small perturbations of the Ising model. *Funct. Anal. Appl.*, 8(1):21–25. Also available in *Funkcional. Anal. i Priložen.*, 8(1):25–30 (in Russian).

[43] (with Yu.M. Sukhov). On an existence theorem for the solutions of Bogoljubov's chain of equations. *Theoret. Math. Phys.*, 19(3):560–573. Also available in *Teoret. Mat. Fiz.*, 19:344–363 (in Russian).

[44] Construction of a cluster dynamic for the dynamical systems of statistical mechanics. *Moscow Univ. Math. Bull.*, 29(1):124–129. Also available in *Vestnik Moskov. Univ. Ser. I Mat. Meh.*, 29(1):152–158 (in Russian).

1975

[45] (with P.M. Bleher). Critical indices for Dyson's asymptotically-hierarchal models. *Comm. Math. Phys.*, 45(3):247–278.

[46] (with E.I. Dinaburg). The one-dimensional Schrödinger equation with quasiperiodic potential. *Funct. Anal. Appl.*, 9(4):279–289. Also available in *Funkcional. Anal. i Priložen.*, 9(4):8–21 (in Russian).

[47] (with S.A. Pirogov). Phase diagrams of classical lattice systems. *Theoret. Math. Phys.*, 25(3):1185–1192. Also available in *Teoret. Mat. Fiz.*, 25(3):358–369 (in Russian).

1976

[48] *Introduction to Ergodic Theory*, volume 18 of *Mathematical Notes*. Princeton University Press, Princeton, NJ.

[49] Self-similar probability distributions. *Theory Probab. Appl.*, 21(1):64–80. Also available in *Teor. Verojatnost. i Primenen.*, 21(1):63–80 (in Russian).

[50] (with S.A. Pirogov). Phase diagrams of classical lattice systems (continuation). *Theoret. Math. Phys.*, 26(1):39–49. Also available in *Teoret. Mat. Fiz.*, 26(1):61–76 (in Russian).

[51] (with V.V. Anshelevich). Some questions on noncommutative ergodic theory. *Russ. Math. Surv.*, 31(4):157–174. Also available in *Uspehi Mat. Nauk*, 31(4(190)):151–167 (in Russian).

[52] (with P.S. Aleksandrov, I.M. Gel'fand, A.N. Kolmogorov, E.V. Maikov, V.P. Maslov, O.A. Oleinik, O.G. Smolyanov, and V.M. Tikhomirov). In memory of Sergei Vasilevich Fomin. *Russ. Math. Surv.*, 31(4):205–220. Also available in *Uspehi Mat. Nauk*, 31(4(190)):199–212 (in Russian).

[53] Some rigorous results in the theory of phase transitions. In *Statistical Physics (Proc. IUPAP Internat. Conf. Statist. Phys., Budapest, 1975)*, pages 139–149. North-Holland, Amsterdam.

1977

[54] (with A.B. Katok and A.M. Stepin). Theory of dynamical systems and general transformation groups with invariant measure. *J. Sov. Math.*, 7(6):974–1065. Also available in *Itogi Nauki i Tekhn. Ser. Mat. Anal.*, 13:129–262, 1975 (in Russian).

[55] (with S.A. Pirogov). Ground states in two-dimensional boson quantum field-theory. *Ann. Physics*, 109(2):393–400.

1978

[56] Billiard trajectories in a polyhedral angle. *Russ. Math. Surv.*, 33(1):219–220. Also available in *Uspehi Mat. Nauk*, 33(1(199)):229–230 (in Russian).

[57] Mathematical foundations of the renormalization group method in statistical physics. In *Mathematical Problems in Theoretical Physics (Proc. Internat. Conf., Univ. Rome, Rome, 1977)*, volume 80 of *Lecture Notes in Phys.*, pages 303–311. Springer, Berlin-New York.

1979

[58] Ergodic properties of a Lorentz gas. *Funct. Anal. Appl.*, 13(3):192–202. Also available in *Funktsional. Anal. i Prilozhen.*, 13(3):46–59 (in Russian).

[59] (with K.M. Khanin). Existence of free energy for models with long-range random Hamiltonians. *J. Stat. Phys.*, 20(6):573–584.

[60] *Development of Krylov's ideas.* Afterword to N.S. Krylov, *Works on the Foundations of Statistical Physic*, pages 239–281. Princeton University Press, Princeton, NJ.

1980

[61] Ergodic and kinetic properties of the Lorenz gas. In *Nonlinear dynamics*, volume 357 of *Annals of the New York Academy of Sciences*, pages 143–149. New York Academy of Sciences, New York.

[62] (with E.B. Vul). Discovery of closed orbits of dynamical systems with the use of computers. *J. Stat. Phys.*, 23(1):27–47.

[63] (with E.B. Vul). A structurally stable mechanism of appearance of invariant hyperbolic sets. In *Multicomponent random systems*, volume 6 of *Adv. Probab. Related Topics*, pages 595–606. Dekker, New York.

[64] (with R.L. Dobrushin). Mathematical problems in statistical mechanics. *Soviet Sci. Rev. Sect. C*, 1:55–106.

1981

[65] (with L. A. Bunimovich). Markov partitions for dispersed billiards. *Comm. Math. Phys.*, 78(2):247–280. Erratum in *Comm. Math. Phys.*, 107(2):357–358, 1986.

[66] (with L.A. Bunimovich). Statistical properties of Lorentz gas with a periodic configuration of scatterers. *Comm. Math. Phys.*, 78(4):479–497.

[67] (with Y.B. Pesin). Hyperbolicity and stochasticity of dynamical systems. *Soviet Sci. Rev. Sect. C*, 2:53–115.

[68] (with E.B. Vul). Hyperbolicity conditions for the Lorenz model. *Phys. D*, 2(1):3–7. Nonlinear and turbulent processes in physics (Proc. Internat. Workshop, Kiev, 1979).

[69] The stochasticity of dynamical systems. *Selecta Math. Soviet.*, 1(1):100–119.

[70] (with D.V. Anosov, V.I. Arnol′d, M.I. Zelikin, A.N. Kolmogorov, O.V. Lokutsievskii, Y.S. Osipov, V.M. Tikhomirov, and M.V. Yakobson). Vladimir Mikhailovich Alekseev. Obituary. *Russ. Math. Surv.*, 36(4):201–206. Also available in *Uspekhi Mat. Nauk*, 36(4(220)):177–182 (in Russian).

[71] (with N.N. Bogolyubov, I.M. Gel'fand, R.L. Dobrushin, A.A. Kirillov, M.G. Krein, D.A. Leites, R.A. Minlos, and M.A. Shubin). Feliks Aleksandrovich Berezin. Obituary. *Russ. Math. Surv.*, 36(4):209–216. Also available in *Uspekhi Mat. Nauk*, 36(4(220)):185–190 (in Russian).

[72] Random walks and some problems concerning Lorentz gas. In *Random systems and dynamical systems (Proc. Symp., Kyoto 1981)*, volume 439 of *RIMS Kokyuroku*, pages 6–17.

1982

[73] *Theory of Phase Transitions. Rigorous Results*, volume 108 of *International Series in Natural Philosophy*, Pergamon Press, Oxford. Also available in Russian.

[74] (with I.P. Cornfel'd and S.V. Fomin). *Ergodic Theory*, volume 245 of *Grundlehren der Mathematischen Wissenschaften*, Springer-Verlag, New York. Also available in Polish and Russian.

[75] Lorentz gas and random walks. In *Mathematical problems in theoretical physics*, volume 153 of *Lecture Notes in Physics*, pages 12–14. Springer-Verlag, Berlin.

[76] (with V.V. Anshelevich and K.M. Khanin). Symmetric random walks in random environments. *Comm. Math. Phys.*, 85(3):449–470.

[77] Commensurate-incommensurate phase transitions in one-dimensional chains. *J. Stat. Phys.*, 29(3):401–425.

[78] The limit behavior of a one-dimensional random walk in a random environment. *Theory Probab. Appl.*, 27(2):256–268. Also available in *Teor. Veroyatnost. i Primenen.*, 27(2):247–258 (in Russian).

[79] (with Ya.B. Pesin). Gibbs measures for partially hyperbolic attractors. *Ergodic Theory Dynam. Systems*, 2(3–4):417–438.

[80] The "commensurability-noncommensurability" transition in one-dimensional chains. *Soviet Phys. JETP*, 83(3):1223–1231.

1983

[81] (with A.I. Gol'berg and K.M. Khanin). Universal properties of sequences of period-tripling bifurcations. *Russ. Math. Surv.*, 38(1):187–188. Also available in *Uspekhi Mat. Nauk*, 38(1(229)):159–160 (in Russian).

[82] (with C. Boldrighini and L.A. Bunimovich). On the Boltzmann equation for the Lorentz gas. *J. Stat. Phys.*, 32(3):477–501.

[83] Ground states of the Dyson antiferromagnetic model. *Theoret. Math. Phys.*, 57(1):1014–1019. Also available in *Teoret. Mat. Fiz.*, 57(1):97–104 (in Russian).

[84] (with S.E. Burkov). Phase diagrams of one-dimensional lattice models with long-range antiferromagnetic interaction. *Russ. Math. Surv.*, 38(4):235–257. Also available in *Uspekhi Mat. Nauk*, 38(4(232)):205–225 (in Russian).

1984

[85] (with S.E. Burkov). Rotation numbers that do not relate to KAM theory. *Funct. Anal. Appl.*, 18(4):327–328. Also available in *Funktsional. Anal. i Prilozhen.*, 18(4):73–74 (in Russian).

[86] (with N.G. Khimchenko). On the description of classical reflectionless potentials. *Rep. Math. Phys.*, 20(1):53–63.

[87] (with E.B. Vul and K.M. Khanin). Feigenbaum universality and thermodynamic formalism. *Russ. Math. Surv.*, 39(3):1–40. Also available in *Uspekhi Mat. Nauk*, 39(3(237)):3–37 (in Russian) and *Alkalmaz. Mat. Lapok*, 11(1–2):201–237, 1985 (in Hungarian).

[88] (with I.M. Khalatnikov, E.M. Lifshitz, K.M. Khanin, and L.N. Shchur). On the stochastic properties of relativistic cosmological models near the singularity. In *General relativity and gravitation (Padova, 1983)*, Fund. Theories Phys., pages 343–349. Reidel, Dordrecht.

1985

[89] (with E.I. Dinaburg). An analysis of ANNNI model by Peierls' contour method. *Comm. Math. Phys.*, 98(1):119–144.

[90] (with I.M. Khalatnikov, E.M. Lifshitz, K.M. Khanin, and L.N. Shchur). On the stochasticity in relativistic cosmology. *J. Stat. Phys.*, 38(1–2):97–114.

[91] An answer to a question by J. Milnor. *Comment. Math. Helv.*, 60(2):173–178.

[92] *Lectures in Probability Theory. Part 1*. Moskov. Gos. Univ., Mekh.-Mat. Fak., Moscow.

[93] (with E. Presutti and M.R. Soloviechik). Hyperbolicity and Möller-morphism for a model of classical statistical mechanics. In *Statistical physics and dynamical systems (Köszeg, 1984)*, volume 10 of *Progr. Phys.*, pages 253–284. Birkhäuser Boston, Boston, MA.

[94] (with C. Boldrighini, A. Pellegrinotti, E. Presutti, and M.R. Soloveichik). Ergodic properties of a semi-infinite one-dimensional system of statistical mechanics. *Comm. Math. Phys.*, 101(3):363–382.

[95] Structure of the spectrum of a Schrödinger difference operator with almost periodic potential near the left boundary. *Funct. Anal. Appl.*, 19(1):34–39. Also available in *Funktsional. Anal. i Prilozhen.*, 19(1):42–48, 96 (in Russian).

1986

[96] (with N.I. Chernov). Entropy of a gas of hard spheres with respect to the group of space-time shifts. *J. Math. Sci.*, 32(4):389–406. Also available in *Trudy Sem. Petrovsk.*, (8):218–238, 1982 (in Russian).

[97] *Lectures in Probability Theory. Part 2*. Moskov. Gos. Univ., Mekh.-Mat. Fak., Moscow.

[98] (with K.M. Khanin and L.N. Shchur). A new approach to the construction of fixed points of the renormalization group in dynamical systems. *Radiophys. Quantum Electron.*, 29(9):1061–1066. Also available in *Izv. Vyssh. Uchebn. Zaved. Radiofiz.*, 29(9):1061–1066 (in Russian).

[99] (with M.R. Solovechik). One-dimensional classical massive particle in the ideal gas. *Comm. Math. Phys.*, 104(3):423–443.

[100] (with V.I. Arnol'd, A.M. Vershik, O.Y. Viro, A.N. Kolmogorov, S.P. Novikov, and D.B. Fuks). Vladimir Abramovich Rokhlin. *Russ. Math. Surv.*, 41(3):189–195. Also available in *Uspekhi Mat. Nauk*, 41(3(249)):159–163 (in Russian).

[101] (with D. Ruelle). From dynamical systems to statistical mechanics and back. *Phys. A*, 140(1–2):1–8.

1987

[102] (with K.M. Efimov). Hydrodynamic modes for a Lorentz gas with a periodic configuration of scatterers. *Amer. Math. Soc. transl. (2)*, 137:101–112. Also available in *Some questions in modern analysis*, pages 102–119. Moskov. Gos. Univ., Mekh.-Mat. Fak., Moscow, 1984.

[103] (with E.I. Dinaburg and A.E. Mazel'). The ANNNI model and contour models with interaction. *Soviet Sci. Rev. Sect. C*, 6:113–168.

[104] (with K.M. Khanin). A new proof of M. Herman's theorem. *Comm. Math. Phys.*, 112(1):89–101.

[105] Anderson localization for one-dimensional difference Schrödinger operator with quasiperiodic potential. *J. Stat. Phys.*, 46(5):861–909.

[106] (with N.I. Chernov). Ergodic properties of some systems of two-dimensional disks and three-dimensional balls. *Russ. Math. Surv.*, 42(3):181–207. Also available in *Uspekhi Mat. Nauk*, 42(3(255)):153–174 (in Russian).

[107] (with K.M. Khanin). The renormalization group method in the theory of dynamical systems. In *Conference "Renormalization Group—86" (Dubna, 1986)*, pages 203–222, 397–398 and 407. Joint Inst. Nuclear Res., Dubna.

[108] Anderson localization for one-dimensional difference Schrödinger operators with quasiperiodic potentials. In *VIIIth International Congress on Mathematical Physics (Marseille, 1986)*, pages 283–291. World Sci. Publishing, Singapore.

1988

[109] Dynamics of local equilibrium Gibbs distributions and Euler equations. The one-dimensional case. *Selecta Math. Soviet.*, 7(3):279–289.

[110] (with E.I. Dinaburg). Contour models with interaction and their applications. *Selecta Math. Soviet.*, 7(3):291–315.

[111] (with Y.B. Pesin). On stable manifolds for class of two-dimensional diffeomorphisms. In *Topology and geometry—Rohlin Seminar*, volume 1346 of *Lecture Notes in Math.*, pages 113–125. Springer, Berlin.

[112] About A.N. Kolmogorov's work on the entropy of dynamical systems. *Ergodic Theory Dynam. Systems*, 8(4):501–502.

[113] The absence of the Poisson distribution for spacings between quasi-energies in the quantum kicked-rotator model. *Phys. D*, 33(1–3):314–316.

[114] (with K.M. Khanin). Renormalization group method in the theory of dynamical systems. *Internat. J. Modern Phys. B*, 2(2):147–165.

[115] (with L.A. Bunimovich). Spacetime chaos in coupled map lattices. *Nonlinearity*, 1(4):491–516.

[116] (with Y.B. Pesin). Space-time chaos in the system of weakly interacting hyperbolic systems. *J. Geom. Phys.*, 5(3):483–492.

1989

[117] (with V.A. Chulaevsky). Anderson localization for the 1-D discrete Schrödinger operator with two-frequency potential. *Comm. Math. Phys.*, 125(1):91–112.

[118] (with K.M. Khanin). Smoothness of conjugacies of diffeomorphisms of the circle with rotations. *Russ. Math. Surv.*, 44(1):66–99. Also available in *Uspekhi Mat. Nauk*, 44(1(265)):57–82, 247 (in Russian).

[119] Kolmogorov's work on ergodic theory. *Ann. Probab.*, 17(3):833–839.

[120] (with A.N. Shiryaev). The Fiftieth Anniversary of the Creation of the Department of Probability Theory of the Faculty of Mechanics and Mathematics at Moscow State University by A.N. Kolmogorov. *Theory Probab. Appl.*, 34(1):164–165. Also available in *Teor. Veroyatnost. i Primenen.*, 34(1):190–191 (in Russian).

[121] (with V.A. Malyshev). Some works on ergodic theory and mathematical problems of statistical mechanics at the Department of Probability Theory of Moscow State University. *Theory Probab. Appl.*, 34(1):186–193. Also available in *Teor. Veroyatnost. i Primenen.*, 34(1):215–222 (in Russian).

[122] (with A. Jaffe). Editorial. *Comm. Math. Phys.*, 125(1):1.

[123] (with L.A. Bunimovich, I.P. Cornfeld, R.L. Dobrushin, M.V. Jakobson, N.B. Maslova, Y.B. Pesin, Y.M. Sukhov, and A.M. Vershik). *Dynamical Systems. II*, volume 2 of *Encyclopaedia of Mathematical Sciences*. Springer-Verlag, Berlin. Also available in Russian.

[124] (with I.P. Kornfel'd, and A.M. Vershik). General ergodic theory of groups of measure preserving transformations. In *Dynamical Systems. II*, volume 2 of *Encyclopaedia of Mathematical Sciences*, pages 1–98, Springer-Verlag, Berlin. Also available in *Dynamical systems. II*, Itogi Nauki Tekh., Ser. Sovrem. Probl. Mat., Fundam. Napravleniya 2, pages 5–111. Moscow, 1985 (in Russian).

[125] (with Ya.B. Pesin, L.A. Bunimovich, and M.V. Yakobson). Ergodic theory of smooth dynamical systems. In *Dynamical Systems. II*, volume 2 of *Encyclopaedia of Mathematical Sciences*, pages 99–206, Springer-Verlag, Berlin. Also available in *Dynamical systems. II*, Itogi Nauki Tekh., Ser. Sovrem. Probl. Mat., Fundam. Napravleniya 2, pages 113–231. Moscow, 1985 (in Russian).

[126] (with R.L. Dobrushin, Yu.M. Sukhov, and N.B. Maslova). Dynamical systems of statistical mechanics and kinetic equations. In *Dynamical Systems. II*, volume 2 of *Encyclopaedia of Mathematical Sciences*, pages 207–278, Springer-Verlag, Berlin. Also available in *Dynamical systems. II*, Itogi Nauki Tekh., Ser. Sovrem. Probl. Mat., Fundam. Napravleniya 2, pages 233–307. Moscow, 1985 (in Russian).

1990

[127] Mathematical problems in the theory of quantum chaos. In *Chaos (Woods Hole, MA, 1989)*, pages 395–414. Amer. Inst. Phys., New York. Also

available in *Geometric aspects of functional analysis (1989–90)*, volume 1469 of *Lecture Notes in Math.*, pages 41–59. Springer, Berlin, 1991.

[128] (with C.M. Series). Ising models on the Lobachevsky plane. *Comm. Math. Phys.*, 128(1):63–76.

[129] (with L.A. Bunimovich and N.I. Chernov). Markov partitions for two-dimensional hyperbolic billiards. *Russ. Math. Surv.*, 45(3):105–152. Also available in *Uspekhi Mat. Nauk*, 45(3(273)):97–134, 221 (in Russian).

[130] (with V.A. Chulaevsky). Anderson localization and KAM-theory. In *Analysis, et cetera*, pages 237–249. Academic Press, Boston, MA.

[131] (with E.I. Dinaburg). The statistics of the solutions of the integer equation $ax - by = \pm 1$. *Funct. Anal. Appl.*, 24(3):165–171. Also available in *Funktsional. Anal. i Prilozhen.*, 24(3):1–8, 96 (in Russian).

[132] Some mathematical problems in the theory of quantum chaos. *Phys. A*, 163(1):197–204. Erratum in *Phys. A*, 173(1–2):302, 1991.

1991

[133] (with S.K. Nechaev). Scaling behavior of random walks with topological constraints. In *New Trends in Probability and Statistics, Vol. 1 (Bakuriani, 1990)*, pages 683–693. VSP, Utrecht.

[134] Hyperbolic billiards. In *Proceedings of the International Congress of Mathematicians, Vol. I, II (Kyoto, 1990)*, pages 249–260. Math. Soc. Japan, Tokyo.

[135] Two results concerning asymptotic behavior of solutions of the Burgers equation with force. *J. Stat. Phys.*, 64(1–2):1–12.

[136] Finite-dimensional randomness. *Russ. Math. Surv.*, 46(3):177–190. Also available in*Uspekhi Mat. Nauk*, 46(3(279)):147–159, 240 (in Russian).

[137] (with L.A. Bunimovich and N.I. Chernov). Statistical properties of two-dimensional hyperbolic billiards. *Russ. Math. Surv.*, 46(4):47–106. Also available in *Uspekhi Mat. Nauk*, 46(4(280)):43–92, 192 (in Russian).

[138] (with Y.B. Pesin). Space-time chaos in chains of weakly interacting hyperbolic mappings. In *Dynamical Systems and Statistical Mechanics (Moscow, 1991)*, volume 3 of *Adv. Soviet Math.*, pages 165–198. Amer. Math. Soc., Providence, RI.

[139] Poisson distribution in a geometric problem. In *Dynamical systems and statistical mechanics (Moscow, 1991)*, volume 3 of *Adv. Soviet Math.*, pages 199–214. Amer. Math. Soc., Providence, RI.

[140] (with V.A. Chulaevski). The exponential localization and structure of the spectrum for 1D quasi-periodic discrete Schrödinger operators. *Rev. Math. Phys.*, 3(3):241–284.

[141] (with L.B. Koralov and S.K. Nechaev). Limiting probability distribution for a random walk with topological constraints. *Chaos*, 1(2):131–133.

[142] (with S.K. Nechaev). Limiting-type theorem for conditional distributions of products of independent unimodular 2×2 matrices. *Bol. Soc. Brasil. Mat. (N.S.)*, 21(2):121–132.

1992

[143] Limit theorems for multiple trigonometric Weyl sums. *Proc. Steklov Inst. Math.*, 191(2):131–143. Also available in *Trudy Mat. Inst. Steklov.*, 191:118–129, 1989 (in Russian).

[144] Distribution of some functionals of the integral of a random walk. *Theoret. Math. Phys.*, 90(3):219–241. Also available in *Teoret. Mat. Fiz.*, 90(3):323–353 (in Russian).

[145] *Probability Theory. An Introductory Course*. Springer Textbook. Springer-Verlag, Berlin.

[146] Statistics of shocks in solutions of inviscid Burgers equation. *Comm. Math. Phys.*, 148(3):601–621.

[147] (with Y.M. Berezanskii, V.P. Maslov, Y.A. Mitropol'skiĭ, and L.A. Pastur). Vladimir Aleksandrovich Marchenko (on the occasion of his seventieth birthday). *Russ. Math. Surv.*, 47(4):239–242. Also available in *Uspekhi Mat. Nauk*, 47(4(286)):219–221 (in Russian).

[148] (with R.L. Dobrushin, A.A. Kirillov, V.A. Malyshev, and A.Y. Khelemskii). Robert Adol'fovich Minlos (on the occasion of his sixtieth birthday). *Russ. Math. Surv.*, 47(1):275–277. Also available in *Uspekhi Mat. Nauk*, 47(1(283)):225–226 (in Russian).

[149] (with A.E. Mazel'). A limiting distribution connected with fractional parts of linear forms. In *Ideas and methods in mathematical analysis, stochastics, and applications (Oslo, 1988)*, pages 220–229. Cambridge Univ. Press, Cambridge.

[150] (with K.M. Khanin). Mixing of some classes of special flows over rotations of the circle. *Funct. Anal. Appl.*, 26(3):155–169. Also available in *Funktsional. Anal. i Prilozhen.*, 26(3):1–21 (in Russian).

[151] (with D.V. Kosygin). On the number of lattice points inside a random domain. In *From Phase Transitions to Chaos*, pages 547–566. World Sci. Publ., River Edge, NJ.

1993

[152] (with D.V. Kosygin and A.A. Minasov). Statistical properties of the spectra of Laplace–Beltrami operators on Liouville surfaces. *Russ. Math. Surv.*, 48(4):1–142. Also available in *Uspekhi Mat. Nauk*, 48(4(292)):3–130 (in Russian).

[153] (with N.I. Chernov, G.L. Eyink, and J.L. Lebowitz). Steady-state electrical conduction in the periodic Lorentz gas. *Comm. Math. Phys.*, 154(3):569–601.

[154] (with L.B. Koralov and S.K. Nechaev). Limit behavior of a two-dimensional random walk with topological constraints. *Theory Probab. Appl.*, 38(2):296–306. Also available in *Teor. Veroyatnost. i Primenen.*, 38(2):331–344 (in Russian).

[155] A random walk with a random potential. *Theory Probab. Appl.*, 38(2):382–385. Also available in *Teor. Veroyatnost. i Primenen.*, 38(2):457–460 (in Russian).

[156] (with N.S. Bakhvalov, A.A. Borovkov, R.L. Dobrushin, A.V. Zabrodin, V.M. Zolotarev, I.A. Ibragimov, Y.V. Prokhorov, B.A. Sevast'yanov, A.V. Skorokhod, V.A. Statulyavichus, R.Z. Khas'minskii, A.S. Kholevo, D.M. Chibisov, and A.N. Shiryaev). Nikolai Nikolaevich Chentsov (February 19, 1930–July 5, 1992). *Theory Probab. Appl.*, 38(3):506–515. Also available in *Teor. Veroyatnost. i Primenen.*, 38(3):613–623 (in Russian).

1994

[157] (with K.M. Khanin, A.E. Mazel', and S.B. Shlosman). Loop condensation effects in the behavior of random walks. In *The Dynkin Festschrift*, volume 34 of *Progr. Probab.*, pages 167–184. Birkhäuser Boston, Boston, MA.

[158] *Topics in Ergodic Theory*, volume 44 of *Princeton Mathematical Series*. Princeton University Press, Princeton, NJ.

[159] A probabilistic approach to the analysis of the statistics of convex polygonal lines. *Funct. Anal. Appl.*, 28(2):108–113. Also available in *Funktsional. Anal. i Prilozhen.*, 28(2):41–48, 96 (in Russian).

1995

[160] (with P.M. Bleher and D.V. Kosygin). Distribution of energy levels of quantum free particle on the Liouville surface and trace formulae. *Comm. Math. Phys.*, 170(2):375–403.

[161] Geodesic flows on manifolds of negative curvature. In *Algorithms, Fractals, and Dynamics (Okayama/Kyoto, 1992)*, pages 201–215. Plenum, New York.

[162] A remark concerning random walks with random potentials. *Fund. Math.*, 147(2):173–180.

[163] (with W. E and Y.G. Rykov). The Lax–Oleinik variational principle for some one-dimensional systems of quasi-linear equations. *Russ. Math. Surv.*, 50(1):220–222. Also available in *Uspekhi Mat. Nauk*, 50(1(301)):193–194 (in Russian).

[164] (with K.M. Khanin, D.L. Lebovits, and A.E. Mazel'). Self-avoiding random walks in five or more dimensions: an approach using polymer expansions. *Russ. Math. Surv.*, 50(2):403–434. Also available in *Uspekhi Mat. Nauk*, 50(2(302)):175–206 (in Russian).

1996

[165] (with W. E and Y.G. Rykov). Generalized variational principles, global weak solutions and behavior with random initial data for systems of conservation laws arising in adhesion particle dynamics. *Comm. Math. Phys.*, 177(2):349–380.

[166] (with B.R. Hunt, K.M. Khanin, and J.A. Yorke). Fractal properties of critical invariant curves. *J. Stat. Phys.*, 85(1–2):261–276.

[167] (with H. Spohn). Remarks on the delocalization transition for heteropolymers. In *Topics in Statistical and Theoretical Physics*, volume 177 of *Amer. Math. Soc. Transl. Ser. 2*, pages 219–223. Amer. Math. Soc., Providence, RI.

[168] Burgers system driven by a periodic stochastic flow. In *Itô's stochastic calculus and probability theory*, pages 347–353. Springer, Tokyo.

[169] A remark concerning the thermodynamical limit of the Lyapunov spectrum. *Internat. J. Bifur. Chaos Appl. Sci. Engrg.*, 6(6):1137–1142.

[170] (with R. Minlos and S. Shlosman). Obituary: Roland L. Dobrushin (1929–1995). *Ergodic Theory Dynam. Systems*, 16(5):863–869.

[171] (with L.A. Bassalygo, S.I. Gel'fand, G.K. Golubev, R.L. Dobrushin, V.V. Prelov, R.Z. Khas'minskii, and A.M. Yaglom). Review of the scientific activity of M.S. Pinsker. *Probl. Inf. Transm.*, 32(1):3–14. Also available in *Probl. Peredachi Inf.*, 32(1):5–19 (in Russian).

1997

[172] (with B.M. Gurevich, I.A. Ibragimov, R.A. Minlos, I.A. Ovseevich, V.I. Oseledets, M.S. Pinsker, V.V. Prelov, Y.V. Prokhorov, A.S. Kholevo, and A.N. Shiryaev). Roland L'vovich Dobrushin (July 20, 1929–December 12, 1995). *Theory Probab. Appl.*, 41(1):132–136. Also available in *Teor. Veroyatnost. i Primenen.*, 41(1):164–169, 1996 (in Russian).

[173] (with E.I. Dinaburg and A.B. Soshnikov). Splitting of the low Landau levels into a set of positive Lebesgue measure under small periodic perturbations. *Comm. Math. Phys.*, 189(2):559–575.

[174] (with W. E, K.M. Khanin and A. Mazel). Probability distribution functions for the randomly forced Burgers equation. *Phys. Rev. Lett.*, 78(10):1904–1907.

[175] On the distribution of the maximum of fractional Brownian motion. *Russ. Math. Surv.*, 52(2):359–378. Also available in *Uspekhi Mat. Nauk*, 52(2(314)):119–138 (in Russian).

[176] (with K.M. Khanin). Hyperbolicity of minimizing trajectories for two-dimensional Hamiltonian systems with random forcing. *Proc. Steklov Inst. Math.*, 1(216):169–173. Also available in *Tr. Mat. Inst. Steklova*, 216(Din. Sist. i Smezhnye Vopr.):176–180 (in Russian).

[177] (with A. Knauf). *Classical Nonintegrability, Quantum Chaos*, volume 27 of *DMV Seminar*. Birkhäuser Verlag, Basel.

[178] (with A. Jaffe and J. Lebowitz). Roland L. Dobrushin. *Comm. Math. Phys.*, 189(2):259–261.

[179] (with V.I. Arnol'd, A.A. Bolibrukh, R.V. Gamkrelidze, V.P. Maslov, E.F. Mishchenko, S.P. Novikov, Y.S. Osipov, A.M. Stepin, and L.L. Faddeev). Dmitrii Viktorovich Anosov (on the occasion of his sixtieth birthday). *Russ. Math. Surv.*, 52(2):437–445. Also available in *Uspekhi Mat. Nauk*, 52(2(314)):193–200 (in Russian).

[180] Hyperbolicity and chaos. In *Boltzmann's Legacy 150 Years after His Birth (Rome, 1994)*, volume 131 of *Atti Convegni Lincei*, pages 107–110. Accad. Naz. Lincei, Rome.

1998

[181] (with S. Goldstein and J.L. Lebowitz). Remark on the (non)convergence of ensemble densities in dynamical systems. *Chaos*, 8(2):393–395. Chaos and irreversibility (Budapest, 1997).

[182] Parabolic perturbations of Hamilton–Jacobi equations. *Fund. Math.*, 157(2–3):299–303.

[183] (with A.B. Soshnikov). A refinement of Wigner's semicircle law in a neighborhood of the spectrum edge for random symmetric matrices. *Funct. Anal. Appl.*, 32(2):114–131. Also available in *Funktsional. Anal. i Prilozhen.*, 32(2):56–79, 96 (in Russian).

[184] (with A.B. Soshnikov). Central limit theorem for traces of large random symmetric matrices with independent matrix elements. *Bol. Soc. Brasil. Mat. (N.S.)*, 29(1):1–24.

[185] Asymptotic behavior of solutions of 1D-Burgers equation with quasi-periodic forcing. *Topol. Methods Nonlinear Anal.*, 11(2):219–226.

[186] Convex hulls of random processes. In *Geometry of differential equations*, volume 186 of *Amer. Math. Soc. Transl. Ser. 2*, pages 153–159. Amer. Math. Soc., Providence, RI.

[187] (with V.P. Gurarii, V.E. Zakharov, V.A. Marchenko, V.P. Maslov, A.V. Pogorelov and E.Ya. Khruslov). Leonid Andreevich Pastur (on his 60th birthday). *Russ. Math. Surv.*, 53(2):427–429. Also available in *Uspekhi Mat. Nauk*, 53(2(320)):180–181 (in Russian).

1999

[188] Simple random walks on tori. *J. Stat. Phys.*, 94(3–4):695–708.

[189] Mathematical problems of turbulence. *Physica A*, 263(1–4):565–566.

[190] (with J.C. Mattingly). An elementary proof of the existence and uniqueness theorem for the Navier–Stokes equations. *Commun. Contemp. Math.*, 1(4):497–516.

[191] Navier–Stokes systems with periodic boundary conditions. *Regul. Chaotic Dyn.*, 4(2):3–15.

[192] Anomalous transport in almost periodic media. *Russ. Math. Surv.*, 54(1):181–208. Also available in *Uspekhi Mat. Nauk*, 54(1(325)):181–208 (in Russian).

[193] Dynamics of a massive particle surrounded by a finite number of light particles. *Theoret. Math. Phys.*, 121(1):1351–1357. Also available in *Teoret. Mat. Fiz.*, 121(1):110–116 (in Russian).

[194] A mechanism of ergodicity in standard-like maps. In *Hamiltonian systems with three or more degrees of freedom (S'Agaró, 1995)*, volume 533 of *NATO Adv. Sci. Inst. Ser. C Math. Phys. Sci.*, pages 242–243. Kluwer Acad. Publ., Dordrecht.

2000

[195] (with W. E, K. Khanin, and A. Mazel). Invariant measures for Burgers equation with stochastic forcing. *Ann. of Math. (2)*, 151(3):877–960.

[196] (with J. Piasecki). A model of non-equilibrium statistical mechanics. In *Dynamics: Models and Kinetic Methods for Non-Equilibrium Many Body Systems (Leiden, 1998)*, volume 371 of *NATO Sci. Ser. E Appl. Sci.*, pages 191–199. Kluwer Acad. Publ., Dordrecht.

[197] (with J. Lebowitz and J. Piasecki). Dynamics of an adiabatic piston. *Dokl. Math.*, 62(3):398–401. Also available in *Dokl. Akad. Nauk*, 375(6):734–736 (in Russian).

[198] (with W. E). New results in mathematical and statistical hydrodynamics. *Russ. Math. Surveys*, 55(4):25–58. Also available in *Uspekhi Mat. Nauk*, 55(4(334)):25–58 (in Russian).

[199] (with V.Yu. Kaloshin). Nonsymmetric simple random walks along orbits of ergodic automorphisms. In *On Dobrushin's Way. From Probability Theory to Statistical Physics*, volume 198 of *Amer. Math. Soc. Transl. Ser. 2*, pages 109–115. Amer. Math. Soc., Providence, RI.

[200] (with V.Yu. Kaloshin). Simple random walks along orbits of Anosov diffeomorphisms. *Proc. Steklov Inst. Math.*, 1(228):224–233. Also available in *Tr. Mat. Inst. Steklova*, 228(Probl. Sovrem. Mat. Fiz.):236–245 (in Russian).

[201] (with J.L. Lebowitz and J. Piasecki). Scaling dynamics of a massive piston in an ideal gas. In *Hard Ball Systems and the Lorentz Gas*, volume 101 of *Encyclopaedia Math. Sci.*, pages 217–227. Springer, Berlin.

[202] Remembrances of A.N. Kolmogorov. In *Kolmogorov in Perspective*, volume 20 of *Hist. Math.*, pages 117–120. Amer. Math. Soc., Providence, RI.

[203] (with L.A. Bunimovich, S.G. Dani, R.L. Dobrushin, M.V. Jakobson, I.P. Kornfeld, N.B. Maslova, Y.B. Pesin, J. Smillie, Y.M. Sukhov, and A.M. Vershik). *Dynamical Systems, Ergodic Theory and Applications*, volume 100 of *Encyclopaedia of Mathematical Sciences*. Springer-Verlag, Berlin, revised edition. Also available in Russian.

[204] Some problems in the theory of dynamical systems and mathematical physics. *Geom. Funct. Anal.*, (Special Volume, Part I):425–433. GAFA 2000 (Tel Aviv, 1999).

2001

[205] (with Y. Ilyashenko, V. Malyshev, S. Shlosman, and M. Tsfasman). Robert A. Minlos. *Mosc. Math. J.*, 1(3):305–306.

[206] (with E.I. Dinaburg). A quasilinear approximation for the three-dimensional Navier–Stokes system. *Mosc. Math. J.*, 1(3):381–388, 471.

[207] (with W. E and J.C. Mattingly). Gibbsian dynamics and ergodicity for the stochastically forced Navier–Stokes equation. *Comm. Math. Phys.*, 224(1):83–106.

[208] (with D.V. Anosov, A.A. Bolibrukh, A.T. Fomenko, A.A. Gonchar, R.I. Grigorchuk, V.V. Kozlov, V.P. Maslov, E.F. Mishchenko, V.M. Tikhomirov, P.L. Ul'yanov and A.M. Vershik). Anatolii Mikhailovich Stepin (on his 60th birthday). *Russ. Math. Surv.*, 56(1):201–205. Also available in *Uspekhi Mat. Nauk*, 56(1):201–204 (in Russian).

[209] (with S.A. Aivazyan, L.G. Afanas'eva, Yu.N. Blagoveshchenskij, V.M. Bukhshtaber, B.M. Gurevich, Yu.V. Prokhorov, V.M. Tikhomirov and A.N. Shiryaev). Lev Dmitrievich Meshalkin. *Russ. Math. Surv.*, 56(3):563–568. Also available in *Uspekhi Mat. Nauk*, 56(3(339)):146–150 (in Russian).

[210] (with R.Z. Khas'minskii, N.A. Kuznetsov, V.A. Malyshev, I.A. Ovseevich, S.A. Pirogov, A.N. Rybko, S.B. Shlosman, A.M. Vershik, N.D. Vvedenskaya and E.A. Zhizhina). Robert Adol'fovich Minlos (on his 70th birthday). *Russ. Math. Surv.*, 56(4):789–792. Also available in *Uspekhi Mat. Nauk*, 56(4(340)):173–176 (in Russian).

[211] How mathematicians study chaos. *Mat. Pros.*, 3(5):32–46 (in Russian).

2002

[212] What, in my opinion, David Ruelle should do in the coming years? *J. Stat. Phys.*, 108(5–6):729–732.

[213] (with N. Chernov and J.L. Lebowitz). Scaling dynamics of a massive piston in a cube filled with ideal gas: exact results. *J. Stat. Phys.*, 109(3–4):529–548.

[214] (with A.V. Kontorovich). Structure theorem for (d, g, h)-maps. *Bull. Braz. Math. Soc. (N.S.)*, 33(2):213–224.

[215] (with D.V. Anosov, V.L. Ginzburg, A.B. Zhizhchenko, M.I. Monastyrskii, S.P. Novikov and M.A. Solov'ev). Naum Natanovich Meiman. *Russ. Math. Surv.*, 57(2):399–405. Also available in *Uspekhi Mat. Nauk*, 57(2):179–184 (in Russian).

[216] (with N. Chernov and J. Lebowitz). Dynamics of a massive piston immersed in an ideal gas. *Russ. Math. Surv.*, 57(6):1045–1125. Also available in *Uspekhi Mat. Nauk*, 57(6(348)):3–86 (in Russian).

2003

[217] Statistical $(3x + 1)$ problem. *Comm. Pure Appl. Math.*, 56(7):1016–1028.

[218] (with E.I. Dinaburg and V.S. Posvyanskii). On some approximations of the quasi-geostrophic equation. *Astérisque*, No. 287, xvii, 19–32. Geometric methods in dynamics. II.

[219] Uniform distribution in the $(3x + 1)$-problem. *Mosc. Math. J.*, 3(4):1429–1440.

[220] (with E.I. Dinaburg). Existence and uniqueness of solutions of a quasilinear approximation of the three-dimensional Navier–Stokes system. *Probl. Inf. Transm.*, 39(1):53–57. Also available in *Problemy Peredachi Informatsii*, 39(1):53–57 (in Russian).

2004

[221] (with A.I. Neishtadt). Adiabatic piston as a dynamical system. *J. Stat. Phys.*, 116(1–4):815–820.

[222] What is ... a billiard? *Notices Amer. Math. Soc.*, 51(4):412–413.

[223] A theorem about uniform distribution. *Comm. Math. Phys.*, 252(1–3):581–588.

[224] (with E.I. Dinaburg). On some approximation of the 3D Euler system. *Ergodic Theory Dynam. Systems*, 24(5):1443–1450.

[225] Mathematical hydrodynamics. *Russ. J. Math. Phys.*, 11(3):355–358.

[226] (with Y.Y. Bakhtin and E.I. Dinaburg). On solutions with infinite energy and enstrophy of the Navier–Stokes system. *Russ. Math. Surv.*, 59(6):1061–1078. Also available in *Uspekhi Mat. Nauk*, 59(6(360)):55–72 (in Russian).

[227] (with A.A. Borovkov, I.A. Ibragimov, V.P. Maslov, Y.V. Prokhorov, and B.A. Sevast'yanov). A.N. Shiryaev (on the occasion of his seventieth birthday). *Teor. Veroyatn. Primen.*, 49(4):775–778 (in Russian). Also available in *Theor. Prob. Appl.*, 49(4):691–694, 2005 (in English).

2005

[228] On local and global existence and uniqueness of solutions of the 3D Navier–Stokes system on $\mathbb{R}^3$. In *Perspectives in analysis*, volume 27 of *Math. Phys. Stud.*, pages 269–281. Springer, Berlin.

[229] Power series for solutions of the 3D-Navier-Stokes system on $\mathbf{R}^3$. *J. Stat. Phys.*, 121(5–6):779–803.

[230] A diagrammatic approach to the 3D Navier–Stokes system. *Russ. Math. Surv.*, 60(5):849–873. Also available in *Uspekhi Mat. Nauk*, 60(5(365)):47–70 (in Russian).

[231] (with C. Ulcigrai). Weak mixing in interval exchange transformations of periodic type. *Lett. Math. Phys.*, 74(2):111–133.

[232] Absence of the local existence theorem in the critical space for the 3D-Navier–Stokes system. *Internat. J. Bifur. Chaos Appl. Sci. Engrg.*, 15(11):3635–3637.

2006

[233] (with A. Bufetov and C. Ulcigrai). A condition for continuous spectrum of an interval exchange transformation. In *Representation Theory, Dynamical Systems, and Asymptotic Combinatorics*, volume 217 of *Amer. Math. Soc. Transl. Ser. 2*, pages 23–35. Amer. Math. Soc., Providence, RI.

[234] How mathematicians and physicists found each other in the theory of dynamical systems and in statistical mechanics. In *Mathematical Events of the Twentieth Century*, pages 399–407. Springer, Berlin.

[235] Mathematicians and physicists = cats and dogs? *Bull. Amer. Math. Soc. (N.S.)*, 43(4):563–565.

2007

[236] On a separating solution of a recurrent equation. *Regul. Chaotic Dyn.*, 12(5):490–501.

[237] (with D.V. Anosov, V.I. Arnol'd, V.A. Vladimirov, and V.V. Kozlov). Viktor Iosifovich Yudovich. *Russ. Math. Surv.*, 62(2):375–378. Also available in *Uspekhi Mat. Nauk*, 62(2(374)):165–168 (in Russian).

[238] (with J. Bourgain). Limit behavior of large Frobenius numbers. *Russ. Math. Surv.*, 62(4):713–725. Also available in *Uspekhi Mat. Nauk*, 62(4(376)):77–90 (in Russian).

[239] (with D.V. Anosov, S.I. Bezuglyĭ, A.M. Vershik, S.L. Gefter, A.I. Danilenko, R.S. Ismagilov, V.A. Marchenko, N.I. Nessonov, S.P. Novikov, S.D. Sinel' shchikov and A.M. Stepin). Valentin Yakovlevich Golodets (on his 70th birthday). *Russian Math. Surv.*, 62(6):1231–1232. Also available in *Uspekhi Mat. Nauk*, 62(6(378)):191–192 (in Russian).

[240] (with L.B. Koralov). *Theory of Probability and Random Processes*. Universitext. Springer, Berlin, second edition.

[241] Congratulations to Professor K. Itô. *Japan J. Math.*, 2(1):129–131.

[242] A new approach to the study of the $3\mathscr{D}$-Navier–Stokes system. In *Prospects in Mathematical Physics*, volume 437 of *Contemp. Math.*, pages 223–229. Amer. Math. Soc., Providence, RI.

[243] (with D.V. Kosygin). From Kolmogorov's work on entropy of dynamical systems to non-uniformly hyperbolic dynamics. In *Kolmogorov's Heritage in Mathematics*, pages 239–252. Springer, Berlin.

2008

[244] (with E. Dinaburg). Existence theorems for the 3D-Navier–Stokes system having as initial conditions sums of plane waves. In *Instability in Models Connected with Fluid Flows. I*, volume 6 of *Int. Math. Ser. (N. Y.)*, pages 289–300. Springer, New York.

[245] Two limit theorems. *Nonlinearity*, 21(12):T253–T254.

[246] Mathematical results related to the Navier–Stokes system. In *SPDE in Hydrodynamic: Recent Progress and Prospects*, volume 1942 of *Lecture Notes in Math.*, pages 151–164. Springer, Berlin.

[247] (with D. Li). Blow ups of complex solutions of the 3D Navier–Stokes system and renormalization group method. *J. Eur. Math. Soc. (JEMS)*, 10(2):267–313.

[248] (with C. Ulcigrai). Renewal-type limit theorem for the Gauss map and continued fractions. *Ergodic Theory Dynam. Systems*, 28(2):643–655.

[249] (with C. Ulcigrai). A limit theorem for Birkhoff sums of non-integrable functions over rotations. In *Geometric and Probabilistic Structures in Dynamics*, volume 469 of *Contemp. Math.*, pages 317–340. Amer. Math. Soc., Providence, RI.

[250] (with S. Albeverio and F. Flandoli). *SPDE in Hydrodynamic: Recent Progress and Prospects*, volume 1942 of *Lecture Notes in Mathematics*. Springer-Verlag, Berlin; Fondazione C.I.M.E., Florence.

[251] (with M.D. Arnold). Global existence and uniqueness theorem for 3D-Navier–Stokes system on $\mathbb{T}^3$ for small initial conditions in the spaces $\Phi(\alpha)$. *Pure Appl. Math. Q.*, 4(1, Special Issue: In honor of Grigory Margulis. Part 2):71–79.

[252] (with D. Li). Complex singularities of solutions of some 1D hydrodynamic models. *Phys. D*, 237(14–17):1945–1950.

[253] (with D. Li). Complex singularities of the Burgers system and renormalization group method. In *Current Developments in Mathematics, 2006*, pages 181–210. Int. Press, Somerville, MA.

[254] (with Yu.M. Berezanskii, V.P. Kotlyarov, V.A. Marchenko, K.V. Maslov, S.P. Novikov, F.S. Rofe-Beketov, A.M. Samoilenko, E.Ya. Khruslov and M.V. Shcherbina). Leonid Andreevich Pastur (on his 70th brithday). *Russ. Math. Surv.*, 63(1):197–199. Also available in *Uspekhi Mat. Nauk* 63(1(379)):190–191 (in Russian).

[255] (with Yu M. Berezanskii, V.A. Marchenko, K.V. Maslov, S.P. Novikov, L.A. Pastur, F.S. Rofe-Beketov, L.D. Faddeev and E.Ya. Khruslov). Aleksandr Yakovlevich Povzner. *Russian Math. Surv.*, 63(4):771–772. Also available in *Uspekhi Mat. Nauk* 63(4(382)):175–176 (in Russian).

[256] Limit theorem for trigonometric sums. Theory of curlicues. *Russ. Math. Surv.*, 63(6):1023–1029. Also available in *Uspekhi Mat. Nauk*, 63(6(384)):31–38 (in Russian).

[257] (with A. Gabrielov, V. Keilis-Borok, and I. Zaliapin). Statistical properties of the cluster dynamics of the systems of statistical mechanics. In *Boltzmann's Legacy*, ESI Lect. Math. Phys., pages 203–215. Eur. Math. Soc., Zürich.

2009

[258] (with D. Li). Asymptotic behavior of generalized convolutions. *Regul. Chaotic Dyn.*, 14(2):248–262.

[259] (with I. Vinogradov). Separating solution of a quadratic recurrent equation. *J. Stat. Phys.*, 136(4):603–613.

[260] (with E. Dinaburg and D. Li). A new boundary problem for the two dimensional Navier–Stokes system. *J. Stat. Phys.*, 135(4):737–750.

[261] (with V. Shchur and A. Ustinov). Limiting distribution of Frobenius numbers for $n = 3$. *J. Number Theory*, 129(11):2778–2789.

[262] (with C. Ulcigrai). Estimates from above of certain double trigonometric sums. *J. Fixed Point Theory Appl.*, 6(1):93–113.

[263] (with A.V. Gurevich, E.A. Kuznetsov, G.A. Mesyats, A.V. Mikhailov, S.L. Musher, S.P. Novikov, A.M. Rubenchik, R.Z. Sagdeev, G.I. Smirnov, A.N. Skrinskii and V.E. Fortov). Vladimir Evgenevich Zakharov (on his 70th birthday). *Phys.-Usp.* 2(8):863–864. Also available in *Uspekhi fiz. nauk*, 179(8):917–918

2010

[264] *Selecta. Volume I. Ergodic Theory and Dynamical Systems*. Springer, New York. [Contains also comments to the papers.]

[265] *Selecta. Volume II. Probability Theory, Statistical Mechanics, Mathematical Physics and Mathematical Fluid Dynamics*. Springer, New York. [Contains also comments to the papers.]

[266] (with E. Dinaburg and D. Li). Navier-Stokes system on the unit square with no slip boundary condition. *J. Stat. Phys.*, 141(2):342–358.

[267] (with D. Li). Blowups of complex-valued solutions for some hydrodynamic models. *Regul. Chaotic Dyn.*, 15(4–5):521–531.

[268] (with E. Dinaburg and D. Li). Navier–Stokes system on the flat cylinder and unit square with slip boundary conditions. *Commun. Contemp. Math.*, 12(2):325–349.

[269] Chaos theory yesterday, today and tomorrow. *J. Stat. Phys.*, 138(1–3):2–7.

[270] (with D. Li). Singularities of complex-valued solutions of the two-dimensional Burgers system. *J. Math. Phys.*, 51(1):015205, 16.

[271] (with D.V. Anosov, M.Ya. Antonovskii, V.M. Buchstaber, V.Z. Grines, A.Yu. Zhirov, E.V. Zhuzhoma, N.E. Klinshpont, A.A. Mal'tsev, E.A. Sataev, A.M. Stepin and L.P. Shil'nikov). Anatolii Gordeevich Kostyuchenko. *Russ. Math. Surv.*, 65(4):767–780. Also available in *Uspekhi Mat. Nauk*, 65(4):179–190 (in Russian).

2011

[272] (with F. Cellarosi). The Möbius function and statistical mechanics. *Bull. Math. Sci.*, 1(2):245–275.

[273] (with D.V. Anosov, M.Ya. Antonovskii, V.M. Buchstaber, V.Z. Grines, A.Yu. Zhirov, E.V. Zhuzhoma, N.E. Klinshpont, A.A. Mal'tsev, E.A. Sataev, A.M. Stepin and L.P. Shil'nikov). Romen Vasilevich Plykin. *Russ. Math. Surv.*, 66(3):627–631. Also available in *Upsekhi Mat. Nauk*, 66(3(399)):199–202 (in Russian).

2012

[274] (with D. Li). Bifurcations of solutions of the 2-dimensional Navier–Stokes system. In *Essays in mathematics and its applications*, pages 241–269. Springer, Heidelberg.

[275] Statistics of gaps in the sequence $\{\sqrt{n}\}$. In *Dynamical systems and group actions*, volume 567 of *Contemp. Math.*, pages 185–189. Amer. Math. Soc., Providence, RI.

[276] (with D. Li). Nonsymmetric bifurcations of solutions of the 2D Navier-Stokes system. *Adv. Math.*, 229(3):1976–1999.

[277] From the seminar on Mathematical Statistical Physics in Moscow State University, 1962–1994. How everything started. *Eur. Phys. J. H*, 37(4):567–569.

2013

[278] Statistical properties of the Möbius function. *Autom. Remote Control*, 74(10):1607–1613. Also available in *Avtom. Telemekh.*, 10:6–14 (in Russian).

[279] (with F. Cellarosi). Ergodic properties of square-free numbers. *J. Eur. Math. Soc. (JEMS)*, 15(4):1343–1374.

[280] (with F. Cellarosi). Non-standard limit theorems in number theory. In *Prokhorov and contemporary probability theory*, volume 33 of *Springer Proc. Math. Stat.*, pages 197–213. Springer, Heidelberg.

[281] (with E. Dimitrov). On a problem in Diophantine approximations. *arXiv:1305.1200*.

2014

[282] (with V.M. Buchstaber, M.I. Gordin, I.A. Ibragimov, V.A. Kaimanovich, A.A. Kirillov, A.A. Lodkin, S.P. Novikov, A.Yu. Okounkov, G.I. Olshanskii, F.V. Petrov, L.D. Faddeev, S.V. Fomin, N.V. Tsilevich and Yu.V. Yakubovich). Anatolii Moiseevich Vershik (on his 80th birthday). *Russ. Math. Surv.*, 69(1):165–179. Also available in *Uspekhi Mat. Nauk*, 69(1(415)):173–186 (in Russian).
[283] (with M. Avdeeva and D. Li). New Erdős–Kac Type Theorems for Signed Measures on Square-Free Integers. *J. Stat. Phys.*, 154(1–2):327–333.
[284] (with D. Li). An application of the renormalization group method to stable limit laws. *J. Stat. Phys.*, 157(4–5):915–930.
[285] (with M. Avdeeva and D. Li). New limiting distributions for the Möbius function. *Mosc. J. Comb. Number Theory*, 4(1):3–51.

2015

[286] (with S.M. Aseev, V.M. Buchstaber, R.I. Grigorchuk, V.Z. Grines, B.M. Gurevich, A.A. Davydov, A.Yu Zhirov, E.V. Zhuzhoma, M.I. Zelikin, A.B. Katok, A.V. Klimenko, V.V. Kozlov, V.P. Leksin, M.I. Monastyrskii, A.I. Neishtadt, S.P. Novikov, E.A. Sataev, and A.M. Stepin). Obituary: Dmitrii Viktorovich Anosov. *Russ. Math. Surv.*, 70(2):369–381. Also available in *Uspekhi Mat. Nauk* 70(2):181–191 (in Russian).

2017

[287] (with M. Avdeeva and F. Cellarosi). Ergodic and statistical properties of $\mathcal{B}$-free numbers. *Theory Probab. Appl.*, 61(4):569–589. Also available in *Teor. Veroyatn. Primen.*, 61(4):805–829, 2016 (in Russian).
[288] (with A.I. Aptekarev, A.L. Afendikov, F.I. Ataullakhanov, N.K. Balabaev, V.N. Biktashev, I.V. Biktasheva, R.M. Borisyuk, N.D. Vvedenskaya, R.D. Dagkesamanskii, Yu.G. Zarkhin, Yu.S. Ilyashenko, V.D. Lakhno, V.Yu. Lunin, N.L. Lunina, E.V. Nikolaev, V.S. Posvyanskii, M.A. Roitberg, V.S. Ryaben'kii, L.B. Ryashko, V.M. Tikhomirov, A.A. Tokarev, A.G. Urzhumtsev, and A.I. Khibnik). In memoriam: Èmmanuil Èl'evich Shnol' (1928–2014). *Russ. Math. Surv.*, 72(1):185–198. Also available in *Uspekhi Mat. Nauk* 72(1):197–208 (in Russian).
[289] (with C. Boldrighini and D. Li). Complex singular solutions of the 3-d Navier–Stokes equations and related real solutions. *J. Stat. Phys.*, 167(1): 1–13.
[290] (with I. Vinogradov). Eigenfunctions of Laplacians in some two-dimensional domains. In *Dynamical Systems, Ergodic Theory, and Probability: In Memory of Kolya Chernov*, volume 698 of *Contemp. Math.*, pages 195–199. Amer. Math. Soc., Providence, RI.

2018

[291] (with C. Boldrighini, V. Malyshev, A. Pellegrinotti, S. Poghosyan, V. Zagrebnov, E. Zhizhina). Robert Adol'fovich Minlos (1931–2018) – his work and legacy. *Eur. Math. Soc. Newsl.*, 108:22–27.

Curriculum Vitae for Yakov Grigorevich Sinai

Born:	September 21, 1935 in Moscow, USSR
Degrees/education:	B.S., Moscow State University, 1957
	PhD, Moscow State University, 1960
	Doctor Degree, Moscow State University, 1963
Positions:	Scientific Researcher, Laboratory of Probabilistic & Statistical Methods, Moscow State University, 1960–1971
	Senior Researcher, Landau Institute of Theoretical Physics, Academy of Sciences of Russia, 1971–
	Professor, Moscow State University, 1971–1993
	Professor, Princeton University, 1993–
Visiting positions:	Thomas Jones Professor of Princeton University, 1997–1998
	Galileo Chair, Piza, Italy, June 2006

H. Holden, R. Piene (eds.), *The Abel Prize 2013–2017*,
https://doi.org/10.1007/978-3-319-99028-6_15

Memberships:	American Academy of Arts and Science, foreign honorary member, 1983
	Russian Academy of Sciences, 1991
	London Mathematical Society, honorary member, 1992
	Hungarian Academy of Sciences, foreign member, 1993
	National Academy of Sciences, foreign associate, 1999
	Brazilian Academy of Sciences, foreign member, 2000
	Academia Europaea, 2008
	Polish Academy of Sciences, foreign member, 2009
	Royal Society of London, foreign member, 2009
	American Mathematical Society, Fellow, 2013
	Norwegian Academy of Science and Letters, 2014
Awards and prizes:	Boltzmann Gold Medal, 1986
	Heineman Prize, 1989
	Markov Prize, 1990
	Dirac Medal, 1992
	Wolf Prize, 1997
	Brazilian Award of Merits in Sciences, 2000
	Moser Prize, 2001
	Nemmers Prize in Mathematics, 2002
	Lagrange Prize, 2008
	Henri Poincaré Prize, 2009
	Dobrushin International Prize, 2009
	Leroy P. Steele Prize for Lifetime Achievement, 2012
	Abel Prize, 2014
Honorary degrees:	Warsaw University, 1993
	Budapest University of Science and Technology, 2002
	Hebrew University, Jerusalem, 2005
	Warwick University, 2010

Part III
2015 John F. Nash, Jr. and Louis Nirenberg

"for striking and seminal contributions to the theory of nonlinear partial differential equations and its applications to geometric analysis"

ABEL
PRISEN

Citation

The Norwegian Academy of Science and Letters has decided to award the Abel Prize for 2015 to **John F. Nash, Jr.**, Princeton University, and **Louis Nirenberg**, Courant Institute, New York University

> for striking and seminal contributions to the theory of nonlinear partial differential equations and its applications to geometric analysis

Partial differential equations are used to describe the basic laws of phenomena in physics, chemistry, biology, and other sciences. They are also useful in the analysis of geometric objects, as demonstrated by numerous successes in the past decades. John Nash and Louis Nirenberg have played a leading role in the development of this theory, by the solution of fundamental problems and the introduction of deep ideas. Their breakthroughs have developed into versatile and robust techniques, which have become essential tools for the study of nonlinear partial differential equations. Their impact can be felt in all branches of the theory, from fundamental existence results to the qualitative study of solutions, both in smooth and non-smooth settings. Their results are also of interest for the numerical analysis of partial differential equations. Isometric embedding theorems, showing the possibility of realizing an intrinsic geometry as a submanifold of Euclidean space, have motivated some of these developments. Nash's embedding theorems stand among the most original results in geometric analysis of the twentieth century. By proving that any Riemannian geometry can be smoothly realized as a submanifold of Euclidean space, Nash's smooth (C^∞) theorem establishes the equivalence of Riemann's intrinsic point of view with the older extrinsic approach. Nash's non-smooth (C^1) embedding theorem, improved by Kuiper, shows the possibility of realizing embeddings that at first seem to be forbidden by geometric invariants such as Gauss curvature; this theorem is at the core of Gromov's whole theory of convex integration, and has also inspired recent spectacular advances in the understanding of the regularity of incompressible fluid flow. Nirenberg, with his fundamental embedding theorems for the sphere S^2 in $\mathbb{R}^3$, having prescribed Gauss curvature or Riemannian metric, solved the classical problems of Minkowski and Weyl (the latter being also treated, simultaneously, by Pogorelov). These solutions were important, both because the problems were representative of a developing area, and because the methods created were the right ones for further applications. Nash's work on realizing manifolds as real algebraic varieties and the Newlander–Nirenberg theorem on complex structures further illustrate the influence of both laureates in geometry. Regularity issues are a daily concern in the study of partial differential equations, sometimes for the sake of rigorous proofs and sometimes for the precious qualitative insights that they provide about the solutions. It was a breakthrough in the field when Nash proved, in parallel with De Giorgi, the first Hölder estimates for solutions of linear elliptic equations in general dimensions without any regularity assumption on the coefficients; among other consequences, this provided a solution to Hilbert's 19th problem about the analyticity of minimizers of analytic elliptic integral functionals. A few years after Nash's proof, Nirenberg, together with

Agmon and Douglis, established several innovative regularity estimates for solutions of linear elliptic equations with L^p data, which extend the classical Schauder theory and are extremely useful in applications where such integrability conditions on the data are available. These works founded the modern theory of regularity, which has since grown immensely, with applications in analysis, geometry and probability, even in very rough, non-smooth situations. Symmetry properties also provide essential information about solutions of nonlinear differential equations, both for their qualitative study and for the simplification of numerical computations. One of the most spectacular results in this area was achieved by Nirenberg in collaboration with Gidas and Ni: they showed that each positive solution to a large class of nonlinear elliptic equations will exhibit the same symmetries as those that are present in the equation itself. Far from being confined to the solutions of the problems for which they were devised, the results proved by Nash and Nirenberg have become very useful tools and have found tremendous applications in further contexts. Among the most popular of these tools are the interpolation inequalities due to Nirenberg, including the Gagliardo–Nirenberg inequalities and the John–Nirenberg inequality. The latter governs how far a function of bounded mean oscillation may deviate from its average, and expresses the unexpected duality of the BMO space with the Hardy space H^1. The Nash–De Giorgi–Moser regularity theory and the Nash inequality (first proven by Stein) have become key tools in the study of probabilistic semigroups in all kinds of settings, from Euclidean spaces to smooth manifolds and metric spaces. The Nash–Moser inverse function theorem is a powerful method for solving perturbative nonlinear partial differential equations of all kinds. Though the widespread impact of both Nash and Nirenberg on the modern toolbox of nonlinear partial differential equations cannot be fully covered here, the Kohn–Nirenberg theory of pseudo-differential operators must also be mentioned. Besides being towering figures, as individuals, in the analysis of partial differential equations, Nash and Nirenberg influenced each other through their contributions and interactions. The consequences of their fruitful dialogue, which they initiated in the 1950s at the Courant Institute of Mathematical Sciences, are felt more strongly today than ever before.

John F. Nash, Jr. and Louis Nirenberg at the Abel Monument, 2015. (Photo: Harald Hanche-Olsen)

Autobiography

John Forbes Nash

My beginning as a legally recognized individual occurred on June 13, 1928 in Bluefield, West Virginia, in the Bluefield Sanitarium, a hospital that no longer exists. Of course I can't consciously remember anything from the first 2 or 3 years of my life after birth. (And, also, one suspects, psychologically, that the earliest memories have become "memories of memories" and are comparable to traditional folk tales passed on by tellers and listeners from generation to generation.) But facts are available when direct memory fails for many circumstances.

My father, for whom I was named, was an electrical engineer and had come to Bluefield to work for the electrical utility company there which was and is the Appalachian Electric Power Company. He was a veteran of WW1 and had served in France as a lieutenant in the supply services and consequently had not been in actual front lines combat in the war. He was originally from Texas and had obtained his B.S. degree in electrical engineering from Texas Agricultural and Mechanical (Texas A. and M.).

My mother, originally Margaret Virginia Martin, but called Virginia, was herself also born in Bluefield. She had studied at West Virginia University and was a school teacher before her marriage, teaching English and sometimes Latin. But my mother's later life was considerably affected by a partial loss of hearing resulting from a scarlet fever infection that came at the time when she was a student at WVU.

From *Les Prix Nobel. The Nobel Prizes 1994*, Editor Tore Frängsmyr, [Nobel Foundation], Stockholm, 1995.

Electronic Supplementary Material The online version of this chapter (https://doi.org/10.1007/978-3-319-99028-6_16) contains supplementary material, which is available to authorized users.

John Forbes Nash was deceased at the time of publication.

H. Holden, R. Piene (eds.), *The Abel Prize 2013–2017*,
https://doi.org/10.1007/978-3-319-99028-6_16

Her parents had come as a couple to Bluefield from their original homes in western North Carolina. Her father, Dr. James Everett Martin, had prepared as a physician at the University of Maryland in Baltimore and came to Bluefield, which was then expanding rapidly in population, to start up his practice. But in his later years Dr. Martin became more of a real estate investor and left actual medical practice. I never saw my grandfather because he had died before I was born but I have good memories of my grandmother and of how she could play the piano at the old house which was located rather centrally in Bluefield.

A sister, Martha, was born about two and a half years later than me on November 16, 1930.

I went to the standard schools in Bluefield but also to a kindergarten before starting in the elementary school level. And my parents provided an encyclopedia, Compton's Pictured Encyclopedia, that I learned a lot from by reading it as a child. And also there were other books available from either our house or the house of the grandparents that were of educational value.

Bluefield, a small city in a comparatively remote geographical location in the Appalachians, was not a community of scholars or of high technology. It was a center of businessmen, lawyers, etc. that owed its existence to the railroad and the rich nearby coal fields of West Virginia and western Virginia. So, from the intellectual viewpoint, it offered the sort of challenge that one had to learn from the world's knowledge rather than from the knowledge of the immediate community.

By the time I was a student in high school I was reading the classic "Men of Mathematics" by E.T. Bell and I remember succeeding in proving the classic Fermat theorem about an integer multiplied by itself p times where p is a prime.

I also did electrical and chemistry experiments at that time. At first, when asked in school to prepare an essay about my career, I prepared one about a career as an electrical engineer like my father. Later, when I actually entered Carnegie Tech. in Pittsburgh I entered as a student with the major of chemical engineering.

Regarding the circumstances of my studies at Carnegie (now Carnegie Mellon U.), I was lucky to be there on a full scholarship, called the George Westinghouse Scholarship. But after one semester as a chem. eng. student I reacted negatively to the regimentation of courses such as mechanical drawing and shifted to chemistry instead. But again, after continuing in chemistry for a while I encountered difficulties with quantitative analysis where it was not a matter of how well one could think and understand or learn facts but of how well one could handle a pipette and perform a titration in the laboratory. Also the mathematics faculty were encouraging me to shift into mathematics as my major and explaining to me that it was not almost impossible to make a good career in America as a mathematician. So I shifted again and became officially a student of mathematics. And in the end I had learned and progressed so much in mathematics that they gave me an M.S. in addition to my B.S. when I graduated.

I should mention that during my last year in the Bluefield schools that my parents had arranged for me to take supplementary math. courses at Bluefield College, which was then a 2-year institution operated by Southern Baptists. I didn't get official advanced standing at Carnegie because of my extra studies but I had advanced knowledge and ability and didn't need to learn much from the first math. courses at Carnegie.

When I graduated I remember that I had been offered fellowships to enter as a graduate student at either Harvard or Princeton. But the Princeton fellowship was somewhat more generous since I had not actually won the Putnam competition and also Princeton seemed more interested in getting me to come there. Prof. A.W. Tucker wrote a letter to me encouraging me to come to Princeton and from the family point of view it seemed attractive that geographically Princeton was much nearer to Bluefield. Thus Princeton became the choice for my graduate study location.

But while I was still at Carnegie I took one elective course in "International Economics" and as a result of that exposure to economic ideas and problems, arrived at the idea that led to the paper "The Bargaining Problem" which was later published in Econometrica. And it was this idea which in turn, when I was a graduate student at Princeton, led to my interest in the game theory studies there which had been stimulated by the work of von Neumann and Morgenstern.

As a graduate student I studied mathematics fairly broadly and I was fortunate enough, besides developing the idea which led to "Non-Cooperative Games", also to make a nice discovery relating to manifolds and real algebraic varieties. So I was prepared actually for the possibility that the game theory work would not be regarded as acceptable as a thesis in the mathematics department and then that I could realize the objective of a Ph.D. thesis with the other results.

But in the event the game theory ideas, which deviated somewhat from the "line" (as if of "political party lines") of von Neumann and Morgenstern's book, were accepted as a thesis for a mathematics Ph.D. and it was later, while I was an instructor at M.I.T., that I wrote up Real Algebraic Manifolds and sent it in for publication.

I went to M.I.T. in the summer of 1951 as a "C.L.E. Moore Instructor". I had been an instructor at Princeton for 1 year after obtaining my degree in 1950. It seemed desirable more for personal and social reasons than academic ones to accept the higher-paying instructorship at M.I.T.

I was on the mathematics faculty at M.I.T. from 1951 through until I resigned in the spring of 1959. During academic 1956–1957 I had an Alfred P. Sloan grant and chose to spend the year as a (temporary) member of the Institute for Advanced Study in Princeton.

During this period of time I managed to solve a classical unsolved problem relating to differential geometry which was also of some interest in relation to the geometric questions arising in general relativity. This was the problem to prove the isometric embeddability of abstract Riemannian manifolds in flat (or "Euclidean") spaces. But this problem, although classical, was not much talked about as an outstanding problem. It was not like, for example, the 4-color conjecture.

So as it happened, as soon as I heard in conversation at M.I.T. about the question of the embeddability being open I began to study it. The first break led to a curious result about the embeddability being realizable in surprisingly low-dimensional ambient spaces provided that one would accept that the embedding would have only limited smoothness. And later, with "heavy analysis", the problem was solved in terms of embeddings with a more proper degree of smoothness.

While I was on my "Sloan sabbatical" at the IAS in Princeton I studied another problem involving partial differential equations which I had learned of as a problem that was unsolved beyond the case of 2 dimensions. Here, although I did succeed in solving the problem, I ran into some bad luck since, without my being sufficiently informed on what other people were doing in the area, it happened that I was working in parallel with Ennio de Giorgi of Pisa, Italy. And de Giorgi was first actually to achieve the ascent of the summit (of the figuratively described problem) at least for the particularly interesting case of "elliptic equations".

It seems conceivable that if either de Giorgi or Nash had failed in the attack on this problem (of a priori estimates of Holder continuity) then that the lone climber reaching the peak would have been recognized with mathematics' Fields medal (which has traditionally been restricted to persons less than 40 years old).

Now I must arrive at the time of my change from scientific rationality of thinking into the delusional thinking characteristic of persons who are psychiatrically diagnosed as "schizophrenic" or "paranoid schizophrenic". But I will not really attempt to describe this long period of time but rather avoid embarrassment by simply omitting to give the details of truly personal type.

While I was on the academic sabbatical of 1956–1957 I also entered into marriage. Alicia had graduated as a physics major from M.I.T. where we had met and she had a job in the New York City area in 1956–1957. She had been born in El Salvador but came at an early age to the U.S. and she and her parents had long been U.S. citizens, her father being an M.D. and ultimately employed at a hospital operated by the federal government in Maryland.

The mental disturbances originated in the early months of 1959 at a time when Alicia happened to be pregnant. And as a consequence I resigned my position as a faculty member at M.I.T. and, ultimately, after spending 50 days under "observation" at the McLean Hospital, travelled to Europe and attempted to gain status there as a refugee.

I later spent times of the order of 5–8 months in hospitals in New Jersey, always on an involuntary basis and always attempting a legal argument for release.

And it did happen that when I had been long enough hospitalized that I would finally renounce my delusional hypotheses and revert to thinking of myself as a human of more conventional circumstances and return to mathematical research. In these interludes of, as it were, enforced rationality, I did succeed in doing some respectable mathematical research. Thus there came about the research for "Le Problème de Cauchy pour les Équations Différentielles d'un Fluide Général"; the idea that Prof. Hironaka called "the Nash blowing-up transformation"; and those of "Arc Structure of Singularities" and "Analyticity of Solutions of Implicit Function Problems with Analytic Data".

But after my return to the dream-like delusional hypotheses in the later 1960s I became a person of delusionally influenced thinking but of relatively moderate behavior and thus tended to avoid hospitalization and the direct attention of psychiatrists.

Thus further time passed. Then gradually I began to intellectually reject some of the delusionally influenced lines of thinking which had been characteristic of my orientation. This began, most recognizably, with the rejection of politically-oriented thinking as essentially a hopeless waste of intellectual effort.

So at the present time I seem to be thinking rationally again in the style that is characteristic of scientists. However this is not entirely a matter of joy as if someone returned from physical disability to good physical health. One aspect of this is that rationality of thought imposes a limit on a person's concept of his relation to the cosmos. For example, a non-Zoroastrian could think of Zarathustra as simply a madman who led millions of naive followers to adopt a cult of ritual fire worship. But without his "madness" Zarathustra would necessarily have been only another of the millions or billions of human individuals who have lived and then been forgotten.

Statistically, it would seem improbable that any mathematician or scientist, at the age of 66, would be able through continued research efforts, to add much to his or her previous achievements. However I am still making the effort and it is conceivable that with the gap period of about 25 years of partially deluded thinking providing a sort of vacation my situation may be atypical. Thus I have hopes of being able to achieve something of value through my current studies or with any new ideas that come in the future.

John Nash, His Life

Sylvia Nasar

Abstract The life and work of John Forbes Nash, Jr.

A few years ago another journalist and I went to St. Petersburg to track down the Russian mathematician who had solved the Poincare Conjecture. Described in the media as a hermit with wild hair and long nails, Grigori Perelman had dropped out of the mathematics community, and given every indication of intending to turn down a Fields medal. His extraordinary decision to refuse the ne plus ultra of honors for a young mathematician—and a Chinese-American rival's attempt to claim credit for solving the 200-year-old problem—was a terrific story... but only if we could find Perelman and convince him to talk to us.

After four frustrating days of searching St. Petersburg we had found no one who had seen Perelman in years or had any clue to his whereabouts. The notes we left outside what we thought might be his apartment remained untouched. A neighbor told us that she had never seen the flat's occupant. But then, by chance, after we had given up, we stumbled onto his mother's apartment... A moment or two later, I was introducing myself to the alleged "hermit," a scholarly looking, youngish man neatly dressed in a sports jacket and Italian loafers. We had apparently interrupted him while he was watching a soccer match on big TV.

I started to say that we were doing a piece for the *New Yorker* magazine when Perelman interrupted: "You're a writer?" he asked in flawless English. "I didn't read the book, but I saw the movie with Russell Crowe."

> I shall not look upon his like again.
> Hamlet, Act 1, Scene 2

A father once asked me after a talk if John Nash's life was more important than that of his son who also suffered from schizophrenia. Of course not, I answered. But

S. Nasar (✉)
Graduate School of Journalism, Columbia University, New York, NY, USA
e-mail: szn1@columbia.edu

H. Holden, R. Piene (eds.), *The Abel Prize 2013–2017*,
https://doi.org/10.1007/978-3-319-99028-6_17

some lives resonate more, touch more of us. John Nash's life was one of these partly because it was so many things: a drama about the mystery of the human mind, an epic of a creative genius, a tale of triumph over incredible adversity, and, not least, a love story.

At one point in the movie, when it looked as if things were all over for Nash, his wife Alicia took his hand, placed it over her heart, and said, "I have to believe that something extraordinary is possible."

Something extraordinary was possible.

Those of you who are mathematicians have probably studied or used one of Nash's stunning contributions to mathematics. I'm going to tell you about the man. Not, almost certainly, what he would have said about himself had he lived to write an autobiographical essay, but some of the things I learned, first, as a New York Times reporter, then, his un-authorized biographer, and, later, simply as a friend.

Before I studied economics, I majored in literature. Starting with the myths of Icarus and Faust, there are many, many stories about the meteoric rise and equally meteoric fall of a remarkable individual. There are very few stories—much less true ones—with a genuine third act. But Nash's life had such a third act.

That third act drew me to his story in the first place. In the early 1990s at the Times, I heard a rumor that a mad mathematician at Princeton University was probably on a short list for a Nobel prize in economics. Nash was hardly a household name, but everyone who had studied economics, as I did in graduate school, was familiar with game theory and the so-called "Nash equilibrium."

Two or three phone calls later, I had learned that by the time he was 30 years old, Nash was a celebrity in the rarified world of mathematics. As a brilliant student at Princeton in the late 1940s, and a rising star on the MIT faculty in the 1950s, before he had succumbed to the most devastating of mental illnesses, he made major contributions not only in game theory for which he would one day win a Nobel, but to several branches of pure mathematics.

Over the next three decades, the ideas Nash had when he was in his twenties had become influential in disciplines as disparate as economics and biology, algebraic geometry and partial differential equations. But, Nash, the man, was all but forgotten.

Generations of students at Princeton University knew him only as the Phantom of Fine Hall, a silent, ghost-like figure who left mysterious messages on the blackboards of Fine Hall. A lot of people like me who knew of Nash's work simply assumed that he had died long ago.

I was naturally intrigued to learn that Nash was alive, apparently recovered from a disease widely considered incurable, and possibly soon to be the recipient of the ultimate intellectual honor. That someone who had been lost for so long could be found again—that someone who had fallen so far could come back—struck me as incredible, something plucked from a fairy tale, a Greek myth, or a Shakespeare tragedy.

> He was a man. Take him as all in all

John with sister Martha circa 1939. (Courtesy of Martha Nash Legg and John D. Stier)

Act One of Nash's life is the story of creative genius. John Forbes Nash Jr. was born in Bluefield in West Virginia coal country on the eve of the Great Depression. He was a peculiar, solitary, precocious child. Other children called him Bug Brains. He amused himself in un-childlike ways. At 10, he was doing sophisticated chemical experiments and tricking other children with electrical shocks. At 15, he was building pipe bombs... and simultaneously re-proving classical theorems by great mathematicians of the past such as Fermat and Gauss.

The summer that World War II ended, the 16 year old Nash went off to Carnegie Tech in Pittsburgh, Pennsylvania to become an engineer like his father. Within months, his professors spotted him as "a young Gauss"—a mathematical prodigy of extraordinary promise.

Three years later they sent him off to Princeton with what was likely the shortest letter of recommendation in the university's history. It consisted of a single line: "This man is a genius."

By the late 1940s, Princeton had become home to the popes of Twentieth-century science: Albert Einstein, Kurt Goedel, Robert Oppenheimer, John von Neumann. A classmate of Nash's, the mathematician John Milnor, recalled, "The notion was that the human mind could accomplish anything with mathematical ideas."

Nash attracted attention as soon as he landed at the center of the mathematical universe. "Genius" was not then the overused term that it has since become. The old Webster's Dictionary defined genius as "transcendent mental superiority," but added that such superiority had to be of a "peculiar, distinctive or identifying character."

At 19, Nash was conspicuous for his movie star looks and his Olympian manner. Over 6 ft tall and heavily muscled, he spoke in a soft southern drawl.

His manners and dress were also southern, slightly formal. But his classmates considered him "weird" "haughty" "spooky." He wore his fingernails unusually long. His conversation had a stilted, ornamental quality. He avoided classes as a matter of principle. He rarely opened a book, telling classmates that he did not wish to endanger his originality. On the few occasions when he was spotted in the Fine Hall library, he would be lying on one of the tables, his arms folded behind his head, staring up at the ceiling.

Like the Cambridge mathematician GH Hardy, Nash thought of mathematics as a ferociously competitive sport. "I imagine that by now you are indeed used to miscalculation," sneers the Russell Crowe's character to a rival. "What if you never come up with your original idea? What if you lose?," says the other man as he beats Nash at Go. For Nash, who craved recognition, mathematics was about winning. He wasn't alone either. "Competitiveness, It was sort of like breathing," another graduate student told me. "We thrived on it." Nash may have skipped lectures, but he never missed afternoon tea. That's where the graduate students and professors played Kriegspiel and Go and traded put downs and mathematical gossip. "Trivial" was Nash's pet putdown. "Hacker" was another. Ranking students and professors—with himself in the Number One spot—was a favorite pastime. He was by no means a brilliant chess player, only an unusually aggressive one. "He managed not just to overwhelm me but to destroy me by pretending to have made a mistake," recalled a man who had made the mistake of challenging Nash to a game.

Outside of the common room, Nash was always pacing. Always whistling Bach. Or riding a bicycle peremptorily commandeered from one of the racks outside the graduate students' residence in tight, concentric circles. Always, it seemed, he was working inside his own head. Lloyd Shapley, a game theoriest and friendly rival of Nash's at Princeton who won a Nobel in 2012, admitted, "He was obnoxious, immature, a brat. What redeemed him was a keen, logical, beautiful mind."

His ambition was awesome. Milnor, a freshman the year that Nash entered the Ph.D. program, 'It was as if he wanted to rediscover, for himself, 300 years of mathematics.' Always on the lookout for a straight line to fame, Nash would corner visiting lecturers, clipboard and writing pad in hand. "He was very much aware of unsolved problems," said Milnor. "He really cross-examined people."

But he was also bursting with his own ideas. Norman Steenrod, Nash's faculty adviser, recalled:

"During his first year of graduate work, he presented me with a characterization of a simple closed curve in the plane. This was essentially the same one given by Wilder in 1932. Some time later he devised a system of axioms for topology based on the primitive concept of connectedness. I was able to refer him to papers by Wallace. During his second year, he showed me a definition of a new kind of homology group which proved to be the same as the Reidemeister group based on homotopy chains."

One afternoon during Nash's first term at Princeton, John von Neumann, the great, the Hungarian polymath best known as a father of the atomic bomb and the digital computer, was in the common room when he noticed two students hunched over a rhombus covered in hexagons and black and white go stones . "What they were playing, he asked a colleague?" "Nash," came the answer, "Nash." Parker Bros. later called Nash's nifty game, which was invented independently by the Danish mathematician and poet Piet Hein, "Hex."

Nash proved a beautiful and surprising theorem showing that the player who makes the first move can always win. But his own story proves that in real life—as opposed to the game—outcomes aren't necessarily determined by the first move, or the second, or even the 50th.

Rebecca West, the English novelist and lover of H. G. Wells, once described genius as "the abnormal justifying itself." Excluded and isolated the genius tries to win acceptance, she speculated, by "some magnificent act of creation." For John Nash several such magnificent acts were to follow before the curtain fell.

Nash's playful foray into mathematical games foreshadowed a far more serious involvement in a novel branch of mathematics. Today, the language of game theory permeates the social sciences. In 1948, game theory was brand-new and very much in the air at Princeton's Fine Hall.

The notion that games could be used to analyze strategic thinking has a long history. Such games as Kriegspiel, a form of blind chess, were used to train Prussian officers. And renowned mathematicians like Emile Borel, Ernst Zermelo, and Hugo Steinhaus studied parlor games to derive novel mathematical insights. The first formal attempt to create a theory of games was von Neumann's 1928 article, "Zur Theorie der Gesellschaftsspiele," in which he developed the concept of strategic interdependence.

But game theory as a basic paradigm for studying decision making in situations where one actor's best options depend on what others do did not come into its own until World War II when the British navy used it to improve its hit rate in the campaign against German submarines. Social scientists discovered it in 1944 when von Neumann and the Princeton economist Oskar Morgenstern published their masterpiece, *Theory of Games and Economic Behavior*, the first attempt to derive logical and mathematical rules about social dynamics, strategies involving conflict and cooperation. The authors predicted that game theory would eventually do for the study of market what calculus had done for physics in Newton's day. Von Neumann's interest in the field lent it irresistible cache for Nash and his fellow graduate students in mathematics.

Nash wrote his first major paper—his now-classic article on bargaining—while attending Albert Tucker's weekly game theory seminar during his first year at Princeton. That is also where he met von Neumann and Morgenstern for the first time. But he had come up with the basic idea as an undergraduate at Carnegie Tech in the only economics course—international trade—he ever took.

Bargaining had long posed a conundrum for economists. Despite the rise of the marketplace with millions of buyers and sellers who never interact directly, one-on-one deals—between individuals, corporations, governments, or unions—have always been a ubiquitous feature economic life. Yet, before Nash, economists assumed that the outcome of a two-way bargaining was determined by psychology and was therefore outside the realm of economics. (Think of Donald Trump's *The*

Art of the Deal.) They had no formal framework for thinking about how parties to a bargain would interact or how they would split the pie.

Obviously, each participant in a negotiation expects to benefit more by cooperating than by acting alone. Equally obviously, the terms of the deal depend on the bargaining power of each. Beyond this, economists had little to add. No one had discovered principles by which to winnow unique predictions from a large number of potential outcomes. Little if any progress had been made since Edgeworth conceded, in 1881, "The general answer is ...contract without competition is indeterminate."

In their game theory opus, von Neumann and Morgenstern suggested that "a real understanding" of bargaining lay in defining bilateral exchange as a "game of strategy." But they, too, came up empty. It is easy to see why: real-life negotiators have an overwhelming number of potential strategies to choose from—what offers to make, when to make them, what information, threats, or promises to communicate, and so on.

Nash took a novel tack: he simply finessed the process. He visualized a deal as the outcome of either a process of negotiation or else independent strategizing by individuals each pursuing his own interest. Instead of defining a solution directly, he asked what reasonable conditions any division of gains from a bargain would have to satisfy. He then posited four conditions and, using an ingenious mathematical argument, showed that, if the axioms held, a unique solution existed that maximized the product of the participants' utilities.

Essentially, he reasoned, how gains are divided reflects how much the deal is worth to each party and what other alternatives each has. By formulating the bargaining problem simply and precisely, Nash showed that a unique solution exists for a large class of such problems. His approach has become the standard way of modeling the outcomes of negotiations in a huge theoretical literature spanning many fields, including labor-management negotiations and international trade agreements.

Nash was naturally irreverent and iconoclastic. When Princeton asked him, on his graduate school application, for his religion, he wrote "Shinto." When he cast about for a thesis topic, he zeroed in on a problem that he knew had eluded the great von Neumann.

A mere 14 months after he enrolled at Princeton, Nash discovered the original idea that got him a Princeton doctorate in 1950 a few days short of his 21st birthday and would ultimately lead to a Nobel. Ironically, it failed to impress Princeton's pure mathematicians. Most considered game theory slightly déclassé because it was actually... useful.

Since 1950, the Nash equilibrium has become "the analytical structure for studying all situations of conflict and cooperation." Nash made his breakthrough at the beginning of his second year at Princeton. As soon as he described his idea David Gale, a fellow graduate student, the latter insisted Nash "plant a flag" by submitting the result as a note to the Proceedings of the National Academy of Sciences. In the note, "Equilibrium Points in n-Person Games," Nash gives the general definition of equilibrium for a large class of games and provides a proof using the Kakutani fixed

Graduation from Princeton 1950. (Courtesy of Martha Nash Legg and John D. Stier)

point theorem to establish that equilibria in randomized strategies must exist for any finite normal form game.

After wrangling for months with Al Tucker, his thesis adviser, Nash provided an elegantly concise doctoral dissertation which contained a second, alternative proof, using the Brouwer fixed point theorem. In his thesis, titled "Non-Cooperative Games," Nash drew the all-important distinction between games where players act on their own "without collaboration or communication with any of the others," and ones where players have opportunities to share information, make deals, and join coalitions. Nash's theory of games—especially his notion of equilibrium for such games—significantly extended the boundaries of economics as a discipline.

All social, political, and economic theory is about interaction among individuals, each of whom pursues his own objectives (whether altruistic or selfish). Before Nash, economics had only one way of formally describing how economic agents interact, namely, the impersonal market. Classical economists like Adam Smith assumed that each participant regarded the market price beyond his control and simply decided how much to buy or sell. By some means—i.e., Smith's famous Invisible Hand—a price emerged that brought overall supply and demand into balance.

Even in economics, the market paradigm sheds little light on less impersonal forms of interaction between individuals with greater ability to influence outcomes. For example, even in markets with vast numbers of buyers and sellers, individuals have information that others do not, and decide how much to reveal or conceal and how to interpret information revealed by others. And in sociology, anthropology, and political science, the market as explanatory mechanism was even more undeveloped. A new paradigm was needed to analyze a wide array of strategic interactions and to predict their results.

Nash's solution concept for games with many players provided that alternative. Economists usually assume that each individual will act to maximize his or her own objective. The concept of the Nash equilibrium, as Roger Myerson has pointed out, is essentially the most general formulation of that assumption. Nash formally defined equilibrium of a non-cooperative game to be "a configuration of strategies, such that no player acting on his own can change his strategy to achieve a better outcome for himself." The outcome of such a game must be a Nash equilibrium if it is to conform to the assumption of rational individual behavior. That is, if the predicted behavior doesn't satisfy the condition for Nash equilibrium, then there must be at least one individual who could achieve a better outcome if she were simply made aware of her own best interests.

In one sense, Nash made game theory relevant to economics by freeing it from the constraints of von Neumann and Morgenstern's two-person, zero-sum theory. By the time he was writing his thesis, even the strategists at RAND had come to doubt that nuclear warfare, much less post-war reconstruction, could usefully be modeled as a game in which the enemy's loss was a pure gain for the other side.

Nash had the critical insight that most social interactions involve neither pure competition nor pure cooperation but rather a mix of both. From a perspective of half a century later, Nash did much more than that. After Nash, the calculus of rational choice could be applied to situations beyond the market itself to analyze the system of incentives created by any social institution. Myerson's eloquent assessment of Nash's influence on economics is worth quoting at length:

Before Nash, price theory was the one general methodology available to economics. The power of price theory enabled economists to serve as highly valued guides in practical policy making to a degree that was not approached by scholars in any other social science. But even within the traditional scope of economics, price theory has serious limits. Bargaining situations where individuals have different information ... the internal organization of a firm ... the defects of a command economy ... crime and corruption that undermine property rights. ... and so on.

The broader analytical perspective of non-cooperative game theory has liberated practical economic analysis from these methodological restrictions. Methodological limitations no longer deter us from considering market and non-market systems on an equal footing, and from recognizing the essential interconnections between economic, social, and political institutions in economic development. By accepting non-cooperative game theory as a core analytical methodology alongside price theory, economic analysis has returned to the breadth of vision that characterized the ancient Greek social philosophers who gave economics its name.

Von Neumann, the dominant figure in mathematics at the time, didn't think much of the Nash equilibrium. When Nash met with him, the Hungarian polymath dismissed the younger man's result as "trivial." The second edition of *The Theory of Games and Economic Behavior* included only a perfunctory mention of "non-cooperative games" in the Preface. Nash didn't care: "If you're going to develop exceptional ideas, it requires a type of thinking that is not simply practical thinking."

His doctorate in his pocket, Nash headed off to RAND, the ultra-secret cold war think tank, in the summer of 1950. He would be part of "the Air Force's big-brain-buying venture"—whose stars would eventually serve as models for Dr. Strangelove—for the next 4 years, spending every other summer in Santa Monica. With the Cold War and the nuclear arms race in full swing, game theory was considered RAND's secret weapon in a war of wits against the Soviet Union. "We hope [the theory of games] will work, just as we hoped in 1942 that the atomic bomb would work," a Pentagon official told Fortune magazine.

At Rand, Nash got an excited reception. Researchers like Kenneth Arrow, who later won a Nobel for his social choice theory, were already chafing at RAND's "preoccupation with the two-person zero-sum game." As weapons became ever more destructive, all-out war could not be seen as a situation of pure conflict in which opponents shared no common interests. Nash's model thus seemed more promising than von Neumann's.

Probably the single most important work Nash did at RAND involved an experiment. Designed with a team that included Milnor and published as "Some Experimental *n*-Person Games," it anticipated by several decades the now-thriving field of experimental economics. At the time the experiment was regarded as a failure, Alvin Roth has pointed out, casting doubt on the predictive power of game theory. But it later became a model because it drew attention to two aspects of interaction.

First, it highlighted the importance of information possessed by participants. Second, it revealed that players' decisions were, more often than not, motivated by concerns about fairness. Despite the experiment's simplicity, it showed that watching how people actually play a game drew researchers' attention to elements of interaction—such as signaling and implied threats—that weren't part of the original model. Nash, whose own interests were rapidly shifting away from game theory to pure mathematics, became fascinated with computers at RAND. Of the dozen or so working papers he wrote during his summers in Santa Monica, none is more visionary than one, written in his last summer at the think tank, called "Parallel Control."

Yet the image that stuck with one of his Rand colleagues for decades afterwards was of Nash running down a street trying to kick some pigeons.

Nash left California determined to prove his prowess as a pure mathematician. Even before completing his doctoral thesis, he turned his attention to the trendy topic of geometric objects called manifolds. Manifolds play a role in many physical problems, including cosmology. Right off the bat, he made what he called "a nice discovery relating to manifolds and real algebraic varieties." Hoping for an appointment at Princeton, he returned there for a post-doctoral year and devoted himself to working out the details of the difficult proof.

Many breakthroughs in mathematics come from seeing unsuspected connections between objects that appear intractable and ones that are already well understood. Dismissing conventional wisdom, Nash argued that manifolds were closely related to a simpler class of objects called algebraic varieties. Loosely speaking, Nash asserted that for any manifold it was possible to find an algebraic variety one of whose parts corresponded in some essential way to the original object. To do this, he showed, one has to go to higher dimensions.

Nash's theorem was initially greeted with skepticism. Experts found the notion that every manifold could be described by a system of polynomial equations simply implausible. "I didn't think he would get anywhere," said his Princeton adviser.

Nash completed "Real Algebraic Manifolds," his favorite paper and the only one he later considered nearly perfect, in the fall of 1951. Its significance was instantly recognized. "Just to conceive the theorem was remarkable," said Michael Artin, an algebraic geometer at MIT. Artin and Barry Mazur, who was a protégé of Nash's as an undergraduate at MIT and later proved the generalized Schoenflies conjecture used Nash's result to resolve a basic problem in dynamics, the estimation of periodic points. Artin and Mazur proved that any smooth map from a compact manifold to itself could be approximated by a smooth map such that the number of periodic points of period p grows at most exponentially with p. The proof relied on Nash's work by translating the dynamic problem into an algebraic one of counting solutions to polynomial equations.

Nash's hoped-for appointment at Princeton did not materialize. Instead, he was forced to accept an offer at MIT, America's leading engineering school but far from the great research university that it was to become. Once there someone dared him to solve a deep problem that had baffled mathematicians since the nineteenth century. So he did.

In 1955, he told a disbelieving audience at the University of Chicago where he had been invited to give a talk, "I did this because of a bet." Two years earlier, a skeptical rival challenged him. "If you're so good, why don't you solve the embedding problem?"

He did. In this instance, he simplified a complex problem that seemed to defy solution by pursuing a strategy that the 'experts' pronounced impossible, if not outlandish. A colleague recalled: 'Everyone else would climb a peak by looking for a path somewhere on the mountain, Nash would climb another mountain altogether and from a distant peak would shine a searchlight back on the first peak.'

When Nash announced that "he had solved it, modulo details," the consensus around Cambridge, Massachusetts was that "he is getting nowhere." The precise question that Nash was posing—"Is it possible to embed any Riemannian manifold in a Euclidian space?"—was a challenge that had frustrated the efforts of eminent mathematicians for three-quarters of a century.

By the early 1950s, interest was shifting to geometric objects in higher dimensions, partly because of the large role played by distorted time and space relationships in Einstein's theory of relativity. Embedding means presenting a given geometric object as a subset of a space of possibly higher dimension, while preserving its essential topological properties. Take, for instance, the surface of a balloon, which is two-dimensional. You cannot put it on a blackboard, which is two-dimensional, but you can make it a subset of a space of three or more dimensions. John Conway, the Princeton mathematician who invented the cellular automaton, the *Game of Life*, called Nash's result "one of the most important pieces of mathematical analysis in this century."

Nash's theorem stated that any surface that embodied a special notion of smoothness could actually be embedded in a Euclidean space. He showed, essentially, that you could fold a manifold like a handkerchief without distorting it. Nobody would have expected Nash's theorem to be true. In fact, most people who heard the result for the first time couldn't believe it. "It took enormous courage to attack these problems," said Paul Cohen, famous for his work on the continuum hypothesis, who knew Nash at MIT.

After the publication of "The Imbedding Problem for Riemannian Manifolds" in the Annals of Mathematics, the earlier perspective on partial differential equations was completely altered. "Many of us have the power to develop existing ideas," said Mikhail Gromov, a geometer and Abel laureate whose work was influenced by Nash. "We follow paths prepared by others. But most of us could never produce anything comparable to what Nash produced. It's like lightening striking . . . there has been some tendency in recent decades to move from harmony to chaos. Nash said that chaos was just around the corner."

A few years after he published his embedding paper, Nash once again stunned the mathematics profession by solving an equally difficult, contemporary problem.

Nominally attached to the Institute for Advanced Study in Princeton during a leave from MIT in the academic year 1956–1957, Nash gravitated to the grittier Courant Institute at New York University, "the national capital of applied mathematical analysis." At Courant, then housed in a former hat factory off Washington Square in Greenwich Village, a group of young mathematicians, including Louis Nirenberg who later shared the 2015 Abel prize with Nash, was responsible for the rapid progress stimulated by World War II in the field of partial differential equations. Such equations were useful in modeling a wide variety of physical phenomena, from air passing under the wings of a jet to heat passing through metal.

By the mid-1950s, mathematicians knew simple routines for solving ordinary differential equations using computers. But straightforward methods for solving most nonlinear partial differential equations—the kind potentially useful for describing large or abrupt changes—did not exist. Stanislaw Ulam, inventor of the Monte Carlo method and, with Edward Teller, the first hydrogen bomb design, complained that such systems of equations were "baffling analytically," noting that they defied "even qualitative insights by present methods." Nash proved basic local existence, uniqueness, and continuity theorems (and also speculated about relations with statistical mechanics, singularities, and turbulence.) He used novel methods of his own invention.

Nash was convinced that deep problems would never yield to a frontal attacks. Taking an ingeniously roundabout approach, he first transformed the non-linear equations into linear ones and then attacked them with non-linear means. Today rocket scientists on Wall Street use Nash inspired methods for solving a particular class of parabolic partial differential equations that arise in finance problems. When he returned to MIT the following fall, there were still gaps in the proof. "It was as if he was a composer and could hear the music, but he didn't know how to write it down," a colleague recalled. Instead of struggling on alone, Nash organized a team of mathematicians to help him get the paper ready for publication. "It was like building the atom bomb … a kind of factory," said one of them later. The complete proof was published in 1958 in "Continuity of Solutions of Parabolic and Elliptic Equations."

To his peers, Nash's was a "bad boy, but a great one." As his 30th birthday approached, he was about to become a full professor. He was singled out by Fortune magazine as the most brilliant of the younger generation of American mathematicians. He seemed poised to make more groundbreaking contributions. He told colleagues of "an idea of an idea" about a possible solution to the Riemann hypothesis, the deepest puzzle in all of mathematics. He set out "to revise quantum theory," along lines he had once, as a first-year graduate student, described to Einstein. Writing to Robert Oppenheimer, the physicist who directed the Manhattan Project and subsequently ran the Institute for Advanced Study, in 1957, Nash had proclaimed, "To me one of the best things about the Heisenberg paper is its restriction to observable quantities … I want to find a different and more satisfying under-picture of a non-observable reality."

To most observers, Nash's private life seemed as enviable as his professional accomplishments. He had succeeded in getting a stunningly beautiful, intelligent glamorous woman to fall madly in love with him. "An El Salvadoran princess with a sense of noblesse oblige," Alicia Larde was one of just 16 women in a class of 800 at MIT. She was a physics major and, a trifle incongruously, a cheerleader. They married in 1958 and within a few months they were expecting a baby. Despite her delicate build, high heels and Elizabeth-Taylor-Butterfield-8 looks, Alicia possessed "a certain steely resolve." She would need all of the metal she had.

Beneath the shiny facade of John Nash's successes lurked chaos and confusion. A neglected illegitimate son. A secret former lover. Ambivalence toward his new marriage and his wife's pregnancy. An undercurrent of anxiety about his abilities as a mathematician.

The first signs of Nash's slide from eccentricity to psychosis were so ambiguous that most of his colleagues assumed he was making one of his weird private jokes. On New Year's Eve, 1958, Nash showed up at a costume party wearing a diaper and spent the night sitting in Alicia's lap, alternately sucking on a pacifier and taking swigs from a baby's bottle filled with bourbon and milk. One morning, he walked into the math common room carrying a copy of the New York Times and announced that a story on the front page contained encrypted messages from inhabitants of another galaxy that only he could decipher. Another time, he pulled one of his doctoral students aside to hand him an intergalactic driver's license and offer him a seat on Nash's newly organized world government…

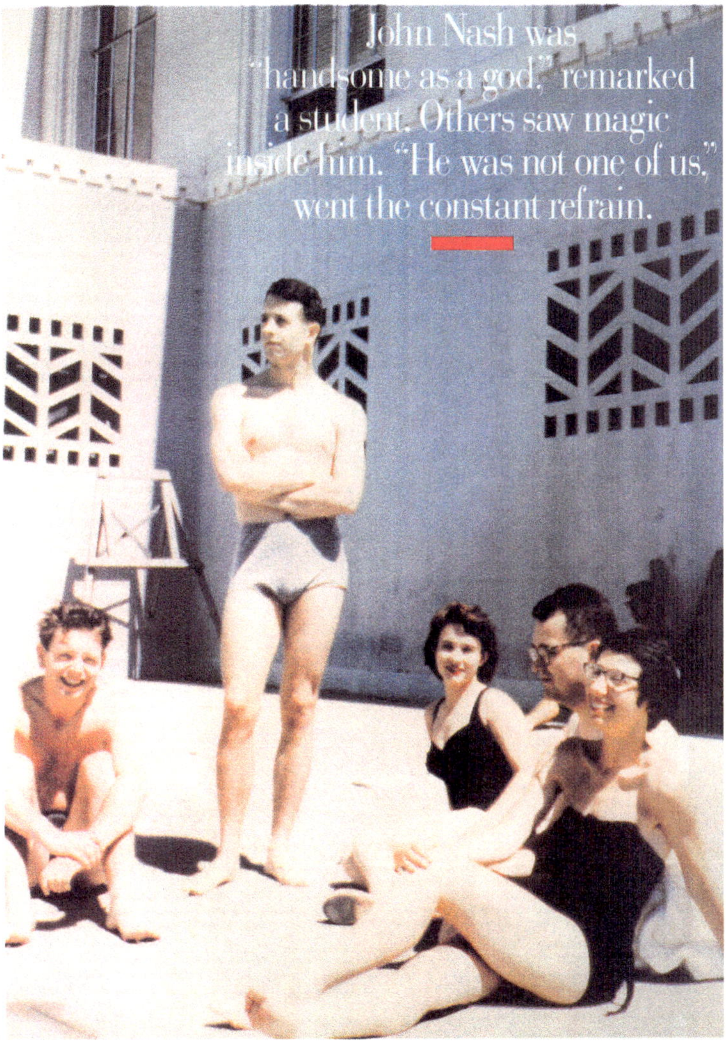

Left to right: Unidentified person, John, Alicia, Felix and Eva Browder. (From Vanity Fair. Courtesy of John D. Stier)

Initially Alicia tried to cover up or explain away her husband's increasingly bizarre behavior. But soon things spun out of control. In February Nash gave a highly anticipated lecture at Columbia University, claiming that he'd solved the Riemann Hypothesis, the third of the trio of "greatest" then-unsolved mathematics problems. The lecture began normally enough, but soon degenerated into a disjointed series of non-sequiturs.

Something was clearly horribly wrong. Alicia had little choice but to turn to psychiatrists at MIT who urged her to commit her husband to a hospital for

John and Alicia. (Courtesy of John D. Stier)

John with John David. (Courtesy of John D. Stier)

observation... against his will if necessary. Nash insisted that he was persecuted not ill. It was a tough call.

In May, 1959, a few weeks before his 31st birthday, two Cambridge police officers took Nash to McLean Hospital, the asylum outside Boston that became the setting for *Girl, Interrupted*. The doctors there diagnosed him with the most devastating and intractable of mental illnesses, paranoid schizophrenia.

A Harvard mathematician who visited Nash at Maclean asked him, "How could you, a mathematician committed to rationality, how could you believe that aliens from outer space were recruiting you to save the world?" Nash replied, "These ideas came to me the same way my mathematical ideas did, so I took them seriously."

The inability to distinguish between delusion and reality, between voices and ones own thoughts, is the tragedy of schizophrenia. We now know that it is a brain disorder, rooted in biology like diabetes or cancer. But when Nash got sick psychiatry was relatively primitive and so were the available treatment. Psychoanalysis, which has since been discredited as an effective treatment for schizophrenia, was in vogue. Psychotic illnesses were supposed to be the fault of bad mothers.

Many of Nash's colleagues and students were appalled by Alicia's decision to have Nash hospitalized. They feared the effects of treatment and confinement on the beautiful mind. Others, however, were shocked by his condition. One recalled his last visit:

"Robert Lowell, the poet, walked in, manic as hell. There's Mrs. Nash, sitting there, pregnant as hell. [Lowell] looks at her and starts quoting the begat sequences in the Bible... And there was John, very quiet and almost not moving. He wasn't even listening. He was totally withdrawn. I focused mostly on his wife and the coming child. I've had that picture in my mind for years. "It's all over for him," I thought."

For a very, very long time, it looked as if it was all over for Nash.

O, what a noble mind is here o'erthrown!

Act Two of Nash's life is the all too common story of a life wrecked by a chronic disease for which there is no adequate treatment, much less cure.

At times Nash believed he was the Prince of Peace, at others a Palestinian refugee. He heard voices and sensed divine revelation. He abandoned mathematics for numerology and prophecy. He wrote letters compulsively to government officials, newspapers and former colleagues. He scribbled mysterious messages on blackboards. He was obsessed with complicated calculations such as converting Nelson Rockefeller's name into base 26 and factoring the result.

He was repeatedly hospitalized, always involuntarily. He was subjected to extreme and futile treatments like insulin shock therapy. He resigned from MIT in order to pursue a quest to give up his US citizenship to become a citizen of the world.

Yet for several years, during temporary remissions, he continued to do mathematics.. "Le problème de Cauchy pour les équations différentielles d'une fluide générale," which appeared in 1962, is described as "basic and noteworthy" by *The Encyclopedic Dictionary of Mathematics* and inspired a good deal of subsequent work by others. He continued to tackle new subjects. Heisuke Hironaka, an algebraic geometer at Harvard and Fields medalist, eventually wrote up a 1964 conjecture as "Nash Blowing Up." In 1966, Nash published "Analyticity of Solutions of Implicit Function Problems with Analytic Data," which pursued his ideas about partial differential equations to their natural conclusion. And in 1967 he completed a much-cited draft, "Arc Structure of Singularities," that was eventually published in a 1995 special issue of the *Duke Journal of Mathematics*.

By the time Nash turned 40, an age at which most mathematicians are at their most productive, almost everything that had once made his life worthwhile was lost. He couldn't work. He had virtually no income. His health suffered. Before long, his front teeth were rotted down nearly to the gums. Old acquaintances avoided him on the street. He was shooed out of stores and coffee shops. Outside Princeton, scholars who built on his work didn't realize he was still alive.

But as Nash sank deeper into obscurity, his ideas were becoming more and more influential. While he was lost in his dreams, his name surfaced more and more often in journals and textbooks in fields as far-flung as economics and biology, mathematics and political science: "Nash equilibrium," "Nash bargaining solution," "Nash program," "De Georgi–Nash," "Nash embedding," "Nash–Moser theorem," "Nash blowing up."

Nash's contributions to pure mathematics—embedding of Riemannian manifolds, existence of solutions of parabolic and elliptic partial differential equations—paved the way for important new developments. By the 1980s, his early work in game theory had permeated economics and helped create new fields within the discipline, including experimental economics. Philosophers, biologists, and political scientists adopted his insights. The growing impact of his ideas was not limited to academe. Advised by game theorists, governments around the world began to auction "public" goods from oil drilling rights to radio spectra, reorganize markets for electricity, and devise systems for matching doctors and hospitals. In business schools, game theory was becoming a staple of management training.

During Nash's "lost years," the brilliant ideas Nash had in his twenties about conflict and cooperation had been widely adopted in the world of economics... Nash published only four game theory papers, but had a bigger impact on economics than any other game theorist. Before Nash, economists could analyze only two kinds of market environments, neither representative: monopolies or markets with so many buyers and sellers that no single individual or firm can affect the behavior of competitors. Most modern markets—cars, oil, airlines, utilities, pharma, housing, healthcare, social media—fall somewhere in between these extremes. Because players must take each others' strategies into account, predicting how they will behave is more complicated. The Nash equilibrium made it possible to cut through the infinite I think therefore he thinks that I think that he thinks... hence the game theory revolution of the 1970s. The impact wasn't confined to economics either but extended to political science, psychology, sociology, and biology.

The contrast between the influential ideas and the bleak reality of Nash's existence was extreme. The usual honors passed him by. He wasn't affiliated with a university. He had virtually no income. He haunted the Princeton campus, in the thrall of a delusion that he was "a religious figure of great, but secret importance."

> I shall not look upon his like again.

Then, after three decades, something extraordinary happened. Act Three began. Freeman Dyson told me later, "It was beautiful. Slowly, he just somehow woke up."

People ask how Nash could recover from an illness almost universally regarded as a life sentence. Was it with the help of "the modern drugs," as Russell Crowe says in the movie? It was not. Like one in ten individuals who suffer from chronic schizophrenia, typically for decades, Nash recovered thanks to the natural chemistry of aging. He also attributed his remission to his own struggle against his delusions and hallucinations which he referred to as "going on a diet of the mind," and the support of a few people who refused to give up on him.

In 1994, Nash's extraordinary story was about to become public with the announcement of the Nobel Prize in economics.

Incidentally, Nash was almost denied the Nobel. One hour before the prize was scheduled to be announced, it was nearly voted down in an unprecedented refusal of many members of the Swedish Academy of Sciences to affirm the prize committee's choice. They feared that giving the prize to a "madman" would sully the Nobel "brand" and spoil the televised prize ceremony hosted by the King and Queen of Sweden in December. Ultimately, those who insisted that a mental illness ought not be a greater bar to the prize than, say, cancer or heart disease, prevailed, but only narrowly.

A small band of contemporaries had always recognized the importance of Nash's work. By the late 1980s, their ranks were swelled by younger scholars who launched a fight to get Nash long-overdue recognition. The prize, that Nash shared with game theorists and experimental economists Reinhard Selten of the University of Bonn and John Harsanyi of the University of California at Berkeley was more than an intellectual triumph. A Nobel rarely changes winners' lives profoundly. Nash was an exception. "We helped lift him into daylight," said Assar Lindbeck, chairman of the Nobel prize committee. "We resurrected him in a way."

When Nash met Russell Crowe for the first time, he told the actor, "You're going to have to go through all these transformations." But the transformation in Nash's own life was as remarkable as any the actor portrayed on the screen. He could not, of course, recover the lost years. He could however repair broken ties with his sister Martha, and his older son John David, travel to conferences, have dinner with friends, see his first Broadway play. He could enjoy the thrill of having a passport, and a drivers license again, of getting a credit card. Then there were the little things like being able to afford a $2 latte at Starbucks. "Lots of academics do that," he told me. "If I was really poor, I couldn't."

To get your life back is a marvelous thing, he told an audience at the world psychiatry conference, but he could never recover the lost years of creativity. Still, he was able to get a grant from the National Science Foundation to develop a new "evolutionary" solution concept for cooperative games. He worked with some

John, Russell Crowe and Ron Howard. (Photo: C. J. Mozzochi)

graduate students. He published papers on ideal money and coalition formation in experimental games.

Most Nobel laureates, while celebrated within their disciplines, remain invisible to the public at large. Recognition not only redeemed the man—bringing him back to society and mathematics—but turned Nash into something of a cultural hero. Since winning the Nobel, the mathematician who spent his life "thinking, always thinking" has been mobbed by reporters and fans from Boston to Mumbai to Beijing.

His story particularly appealed to young people. One of my favorite letters was this one:

> Dear Mr. Nash,
> Hi! I am 9 years old. My name is Ellie Stilson. I am a girl. I really admire you. You are my roll (sic) model for a lot of things. I think you are the smartest person who ever lived. I really wish to be like you. I would love to study math. The only problem with that is that I am not very good at math. I can do it. I like it. I am just not good at it. Was that what it was like for you when you were a kid? Please write back. Love, Ellie P.S. I LOVE your name.

The most unforgettable, though, was addressed to me, arrived in a dirty envelope with no return address and it was scrawled on neon orange paper. It was signed "Berkeley Baby." It would never have made it past the New York Times mailroom after the anthrax scare.

The sender turned out to be the former night rewrite editor on the metro desk, a rising young star at the New York Times in the mid-1970s before he, too, was diagnosed with paranoid schizophrenia. Since then, he had adopted the name Berkeley Baby and lived on the streets of Berkeley, California near the university, a forlorn figure not unlike the Phantom of Fine Hall. He wrote, "John Nash's story give me hope that one day the world will come back to me too." Reading that line always made me cry.

Extraordinary things happen when individuals make extraordinary choices. That is why I dedicated the biography to Alicia Nash. To me, she is very much the hero of Nash's life.

She set out to marry a golden boy who she was convinced was a genius who would be famous one day. Only a few months after the wedding, however, Alicia's girlish notions of romance were shattered by her husband's illness. She acted courageously—and with great compassion. But half a dozen years after Nash got sick, when the husband she was trying to help began to regard her, because of his paranoia, as his worst enemy—she determined to raise their son on her own and got a divorce.

But she never let him go. Five years after they separated, when Nash had no one on earth left to whom he could turn, he wrote to Alicia from a state hospital in Virginia. I beg you "to save me from future hospitalizations and from homelessness." Thirty five and still lovely with most of her life still ahead, she took him in.

What made Alicia do it? It wasn't, I think, masochism, as some suggested. It was love. Not the romantic kind of love, but down to earth, grown up love. She couldn't bear to turn him away. It was "a pretty lean life," her sister-in-law Martha told me. For years, Alicia got up at 4:30 in the morning and commuted 2 h into Manhattan. She did it to support John and their son Johnny, who, at age 15, was diagnosed with the same illness that afflicted his father. She did it to keep her small family together.

Alicia understood—years before research confirmed her intuition—that Nash's only hope lay in living at home in a community where at least a few people knew who he'd been. Nash may have all but disappeared from the world, but Alicia never lost sight of who he was. She saw past the mismatched clothes and expressionless demeanor. For her, Nash was always "a very fine man," someone who had made great contributions, someone for whom "something extraordinary" was always possible.

Recognition is a cure for many ills, but love gave Nash something to come back to: a home, family, a reason to live after his grandiose delusions faded. Alicia was the rock on which he rebuilt his life. Together they experienced the extremes of human existence: genius and madness, sickness and health, obscurity and fame. Together they cared for their disabled son, renewed family ties and friendships, savored what Joan Didion, in her New York Review of Books piece on Nash, called "life's bright pennies."

In 2001, after a nearly 40 year gap in their marriage, John and Alicia said "I do" a second time. "The divorce shouldn't have happened," Nash said. Alicia added,"We saw this as a kind of retraction of that. After all we've been together most of our lives." When the mayor of Princeton Junction pronounced them man and wife, I asked Nash to kiss his bride again for the camera. He looked up, grinning: "A second take? Just like the movies!"

It was Alicia who wanted Nash's story to be told. He was more ambivalent. A friend once asked him about Alicia's whereabouts. "Having dinner with Sylvia," he answered. After a pause he added without much conviction, "I hope they aren't talking about me." Well, 20 years later, people are still talking about him and no doubt will be for a very long time to come.

In 2015 Nash received an honor that meant even more to him than the economics Nobel, the Niels Henrik Abel's Prize in Mathematics. He shared it, as I mentioned, with an old friend from the Courant Institute, Louis Nirenberg. After the ceremony in Oslo, that Nash's older son, John David, was able to attend, Louis, John and Alicia traveled back to the U.S. together. Their flight was cancelled and they were booked on a later one. When they arrived at Newark airport, the Nashes discovered that the driver who usually picked them up had already left. After bidding Louis goodbye, they took one of the cabs lined up outside of the arrivals terminal. Princeton Junction is less than an hour from Newark, but they never made it home. On the New Jersey turnpike, their taxi crashed into the guard rail at high speed, hitting another car. Nash and his wife were both pronounced dead at the scene. He was 86 years old. Alicia was 82.

John Nash's life was tragic, sublime and, now, suddenly, over. The third act shouldn't have ended the way it did. Nonetheless that act, like the whole drama, was truly grand. We will not see the like of him again, but his story belongs to the ages.

References

This essay draws on *A Beautiful Mind*, *The Essential John Nash*, and various articles and lectures ... For the descriptions of Nash's contributions to mathematics, I borrowed heavily from pieces by Avinash Dixit, Barry Nalebuff, Harold Kuhn, Eric Maskin, John Milnor, Roger Myerson, Al Roth and Ariel Rubinstein. All errors, of course, are mine.

1. Dixit, A., and B. Nalebuff. *Thinking Strategically*. New York: W.W. Norton, 1991.
2. Kuhn, H.W., "Introduction," to A Celebration of John F. Nash, Jr., ed. H.W. Kuhn, L. Nirenberg, and P. Sarnak, pp. i–v. In *Duke Mathematical Journal*, 81, nos. 1 and 2 (1995).
3. Kuhn, H.W., "Foreword." In *Classics in Game Theory*, ed. H. W. Kuhn, pp. ix–x. Princeton: Princeton University Press, 1997.
4. Maskin, E., "John Nash's Contributions to Economics and Game Theory," talk at Nash Memorial Celebration, Princeton University, October 24, 2015.
5. Milnor, J. "A Nobel Prize for John Nash." *The Mathematical Intelligencer* 17, no. 3 (1995): 11–17.
6. Myerson, R. B. "Nash Equilibrium and the History of Economic Theory." *Journal of Economic Literature* 37 (1999): 1067–82.
7. Nasar, S., "John Nash, the Life," talk at Nash Memorial Celebration, Princeton University, October 24, 2015.
8. Nasar, S. "Introduction," Harold Kuhn and Sylvia Nasar, eds., *The Essential John Nash*, Princeton University Press, 2001.

9. Nasar, S., "The Lost Years of the Nobel Laureate," *New York Times*, November 13, 1994, sec. F, pp. 1, 8.
10. Nasar, S., *A Beautiful Mind*. New York: Simon & Schuster, 1998.
11. Roth, A. "Game Theory as a Tool for Market Design" (1999). Available at http://web.stanford.edu/~alroth/design.pdf.
12. Rubinstein, A. "John Nash: The Master of Economic Modeling." *The Scandinavian Journal of Economics* 97, no. 1 (1995): 9–13.

Autobiography

Louis Nirenberg

I was born in Hamilton, Ontario, Canada, in 1925. My parents emigrated there from Ukraine, where my father was a Hebrew teacher. When my parents married they immediately crossed the border into Romania, illegally and were promptly arrested. Relatives managed to get them out of jail, and they slowly made their way across Europe, to Antwerp, Belgium, where they hoped to get immigration visas to the US. After a long time, during which my mother worked as a seamstress they decided to go to Canada, and my father continued teaching there.

I can't say when my interest in mathematics began. My father tried to teach me Hebrew but I foolishly resisted, and he finally hired a friend to give me lessons. The friend happened to love mathematical puzzles, and half of each lesson was devoted to them. Perhaps that was the beginning of mathematics for me. To my shame, I never learned Hebrew. When I was 5, or so, my family moved to St. Catherine's, Ontario. There, my older sister taught me what she was learning at school, so I knew how to read when I entered school. Because of the economic depression we could not manage in St. Catherine's, and in 1933 we moved to Montreal.

There, my father had a difficult time finding a position, and he supported the family by giving private lessons. My parents tried selling things, and I have a recollection of going from house to house trying to sell light bulbs. Eventually my parents opened a small gift shop, where they sold English China, crystal etc.

During the depression, to be a high school teacher was considered a very good position, and the excellent high school I went to, in 1937, Baron Byng, had very good teachers. Also, my fellow students were extremely bright. My favorite subjects were Euclidean geometry and physics. The physics teacher even had a PhD. I decided I would like to study physics. I had no idea that mathematics was a living

---

L. Nirenberg (✉)
Department of Mathematics, Courant Institute of Mathematical Sciences, New York University, New York, NY, USA
e-mail: nirenberg@cims.nyu.edu

H. Holden, R. Piene (eds.), *The Abel Prize 2013–2017*,
https://doi.org/10.1007/978-3-319-99028-6_18

subject. Some years ago a Montreal newspaper wrote about the school (long closed) and about some of the graduates who later had distinguished careers. (I was not mentioned.) During my last year in high school I applied for a scholarship to McGill University, but did not succeed. The high school at that time ended with 11th grade. But the school offered 12th grade—equivalent to a first year at college—and I attended that, and this time I received a scholarship to go to McGill in the second year. There, I entered in the Honors course in Mathematics and Physics, in 1942. At the time, young refugees from Europe who had been kept in internment camps in Canada were allowed to leave if they were accepted at some university. Several entered the honors course at the same time as myself. One was Jim Lambek; he knew more mathematics than I did. Eventually he became a mathematician, and spent most of his academic career at McGill.

Louis around 1942 (3rd row, 2nd from right). (Photo: private)

The program at McGill was quite good, though there were no research physicists or mathematicians there at the time, with an exception, Professor Gordon Pall. He worked in number theory, and was very kind and encouraging. I graduated in 1945, just when the war in Europe ended. I was determined to do graduate work in physics, but somehow no one suggested that I apply to any university.

That summer I got a job in a National Research Council Lab in Montreal, where they were doing research on atomic energy. A son of Richard Courant, Ernst, was working at the lab. Courant was a famous mathematician. He had been the head of the Mathematics Institute in Göttingen, Germany, until he was dismissed by the Nazis. Ernst's wife, Sarah, also worked at the lab. She was from Montreal, and I knew her. One day she said that they were going to New York to visit Courant, and I asked her if she would ask him to suggest somewhere where I might apply to study theoretical physics. On her return she said that Courant had suggested that I might come to New York University, where he had set up a graduate mathematics department 10 years earlier, to get a master's degree in mathematics, and then, perhaps, go on to study physics. I went for an interview by Courant and Professor Kurt Friedrichs, and was offered an assistantship. (Later I was told that when Courant saw my record at McGill he commented "This guy has only As. There must be something wrong with him.") So, in September I arrived as a graduate student in mathematics at New York University. I never left. My entire professional career was there.

Incredible luck. I have always been grateful to Sarah Courant. I must say, I feel that I have had a very lucky life.

Friedrichs had been a student of Courant in Göttingen, and came to America a few years after him. When I arrived at NYU he was the principal research mathematician there. At the time I studied there there was a small, but very strong, group of graduate students: Eugene Isaacson, Anneli Kahn, who later married Peter Lax, Joseph Keller, Martin Kruskal, Cathleen Morawetz, Peter Lax, Harold Grad, and Avron Douglis. Many interesting courses were offered.

In my early years at NYU, Courant felt that younger faculty members should teach less than the older ones, so they would have more time for research. He also encouraged people to teach a variety of subjects. In addition to various courses in mathematical analysis, I sometimes taught algebraic topology, differential geometry and differential topology. Courant once asked me to teach a course on elliptic functions. Why, I don't know. I looked at various books, and liked most the one by Francesco Tricomi, in Italian.

Courant arranged that some old friends, then at Princeton, give some courses at NYU. I heard beautiful lectures by Carl Ludwig Siegel and Emil Artin. Artin's wife, Natascha, became the scientific editor of the mathematics journal that Courant and Friedrichs set up, *Communications on Pure and Applied Mathematics*. She was a beloved member of the department and was editor for many years.

After I received a master's degree I asked Friedrichs for a possible doctor's thesis topic. He suggested a general subject connected with ordinary differential equations, but I had no ideas. In fact, around that time I read G.H. Hardy's book, *A mathematician's apology*. It discouraged me. He wrote that a real mathematician thought of his, or her, own research problems and did not rely on others. I couldn't think of any original problem. Only after several years was I able to come up with problems.

One of the courses I took was Differential Geometry, given by Jim Stoker. He later suggested that I look at a problem on which Herman Weyl had worked: Consider a 2-dimensional sphere with a given Riemannian metric on it, having positive Gauss curvature. Can it be embedded isometrically, i.e., preserving lengths, as a closed convex surface in 3-dimensional space? In 1916 Weyl formulated the problem, using partial differential equations, PDEs, and derived some basic estimates for the conjectured solution. But further estimates were needed in order to settle the problem. I started then to learn about nonlinear partial differential equations, and was able to supply the missing estimates. Conversations with Friedrichs played a crucial role in this, and I have always considered him the chief influence in my mathematical development, though Stoker was my official adviser. In particular, Friedrichs instilled in me a love for inequalities.

Stoker and Friedrichs had very different personalities. Stoker was particularly encouraging. He and his wife gave lovely parties, which always ended in dancing; he loved to dance. Friedrichs was more reserved; one had to make an appointment to see him. Once he was in a hospital for some days, and made appointments with students to visit him there.

Louis around 1949. (Photo: private)

I received the PhD in 1949. Around the same time that I settled the Weyl problem, a Soviet mathematician, Aleksei V. Pogorelov, also provided a solution. At that time I had a block about writing mathematics, and my first published papers did not appear until 1953.

Peter Lax also played an important role in my mathematical development. After his army service he came to NYU, in 1946, as an undergraduate, but immediately took graduate courses. He always knew more mathematics than I did—true to this day—and was happy to explain things to me. But he once said "I am willing to explain things more than once but not more than 10 times." Over the years I also learned many jokes from him.

Some months after I arrived in New York my father wrote me that an old friend of his, S.L. Blank, from his town in Ukraine, was living in Philadelphia, and that I might go there for a visit. It was arranged. He had three daughters, and the middle one, Susan, met me at the station. Two and a half years later we got married. . .we might have met in Ukraine.

Louis with his wife Susan. (Photo: private)

My thesis involved PDE; ever since they have played a role in all my work. Many mathematical formulations of problems from other fields, physics, but also engineering, biology, economics, etc. are often expressed in terms of such equations. In my first year at NYU I had taken an excellent course in PDE by Max Shiffman. At the end he assigned term paper topics. Mine was the famous thesis of Tricomi, a long paper In Italian. I told Shiffman that I did not know Italian. He said "So get a dictionary." I've been grateful to him ever since; I love Italy, go there often, and have very close friends there.

In 1953 I published four papers. Two were devoted to my thesis, one was on a strong maximum principle for parabolic equations—I had discovered this while still a graduate student—and one was on a maximum principle for a particular kind of hyperbolic equation, jointly with Shmuel Agmon and Murray Protter. Various forms of the maximum principle play a central role in the study of second order elliptic equations, especially in obtaining estimates for solutions. I have sometimes remarked "I have made a living from the maximum principle."

After I received the PhD I was appointed Instructor at NYU. In general, at American universities, after receiving a doctorate, a student normally went to work at a different university. Courant however tended to keep the best students after graduation. In 1951 I received a fellowship to go to Europe for a year, and Courant arranged that I go to ETH in Zürich, to be with Heinz Hopf, and to Göttingen for a short period. I spent most of the year writing up my thesis for publication, and I attended lectures by Hopf, van der Waerden, and others. Hopf was a wonderful person, extremely warm and kind; he often invited me and my wife to his home. He was also a wonderful speaker. In fact he was my favorite lecturer for many years. We had little social life in Zürich, but Hopf introduced us to a young colleague, Ernst Specker, a logician, and we became, and remained, good friends. During that year my wife and I travelled a lot in Europe and fell in love with Italy.

In 1952, when I returned to NYU I found a new faculty member, Lipman Bers. He was full of energy and enthusiasm, and brought a spark to the department. We quickly became friends; we wrote two papers together, on quasi-conformal mappings. Encouraged by Courant and other colleagues who lived there, he moved to New Rochelle, a suburb. I preferred to live in Manhattan. After my two children were born Courant sometimes said to me "How can you raise children there? It would be better for them to live in a suburb." I finally got him to stop, when I said "Our motto is: give your children happy parents."

In 1954 Courant and Gaetano Fichera, in Trieste, organized a meeting there. When I heard about it I asked Courant to include me among the participants. That was my first contact with Italian colleagues. I met Guido Stampacchia, Enrico Magenes, Carlo Miranda, Carlo Pucci, and Tricomi, among others, and made many friends. It was a wonderful experience.

One of the joys of being a mathematician is that you get to meet many interesting people. Some time ago, at a party I was talking with the wife of a mathematician. She spoke about their 29 year old daughter and asked me if I could introduce her to some young mathematician. I asked what she did—a lawyer. I said "Surely she must meet some young lawyers." She replied "Yes, but I would like her to marry a mathematician. They are such nice people." Indeed, I think that by and large, they are extremely nice.

Guido Stampacchia was full of life, and I came to consider him a brother. I have two other "brothers", Peter Lax and Shmuel Agmon. Guido was born in Naples, and we often said we would visit there together, and he would show me the real Naples. It never happened. Sadly he died in his fifties.

There are essentially two kinds of mathematicians. The first, develop new theories, and the second, are primarily problem solvers. I belong to the second. I also belong to the category of those who when they come to a fork in the road they take it. "Hey, this looks interesting! Let's explore it." A graduate student once asked me how one thinks of problems. I had no simple answer but I said that sometimes I saw a proof I didn't like and I began to look for another. This may lead to new mathematics. He remarked "I never saw a proof I didn't like." I thought: "he's hopeless."

In an early paper I solved a problem posed by Courant, concerning regularity at the boundary of solutions of general elliptic PDEs, including systems, of any order. Regularity in the interior was known, but not at the boundary, under suitable boundary conditions. I remember that I proved this while visiting the University of Chicago in January 1955. There I met Antoni Zygmund, André Weil, Saunders Mac Lane, and others. Zygmund was particularly friendly; we had lunch together often, and he told me about his work with Alberto Calderón—work that had much influence.

One day Weil visited NYU and pointed out a basic problem in analysis of several complex variables. He said "Why are PDE people not working on this?" Shortly afterward S.S. Chern also called my attention to the problem.I thought "Why not give it a try?" To a very talented graduate student August Newlander, I suggested we look at the problem in 2 complex, 4 real, dimensions, and the very simplest case. He saw how to solve it, using the method of characteristics, but in the complex domain. His idea then worked in the general case in 4 dimensions. But, to our surprise, it didn't work in higher dimensions. We then came up with a completely different proof—for any dimension. Sadly, shortly afterwards, Newlander gave up mathematics.

In the fifties I visited a number of universities. In particular, I spent several summers in Berkeley, California, where I made many friends. These were fruitful periods. Charles Morrey was there, a leading figure in elliptic PDEs and in the Calculus of Variations. We wrote a rather technical paper together, proving analyticity of solutions of linear elliptic systems having analytic coefficients.

During the spring semester in 1958 I visited the Institute for Advanced Study in Princeton. One day Kunihiko Kodaira and Donald Spencer asked me for some possible help in connection with their work on deformation of complex structure on manifolds. Spencer described the problem in full generality at the blackboard. I couldn't understand, and asked if he could describe a simple example. Finally, Kodaira, without saying a word stepped up to the blackboard and wrote one down. Using elliptic theory I was able to furnish the needed estimates, and we published a paper together. Later I heard Spencer lecture on the work. I understood nothing. Whenever he lectured I understood very little but his love and enthusiasm were contagious. One always left buoyed up. He was one of the nicest people I ever met; equally courteous to janitors and to university presidents.

My colleague, Fritz John, who had also been a student of Courant in Göttingen, was an extremely original and deep mathematician. In connection with his work on elasticity he introduced a space of functions that he called functions of bounded mean oscillation, BMO, and he asked me if I could prove that they were in L^p for every finite p. I succeeded; we then improved the result and published a joint paper. Peter Lax and I discussed mathematics all the time, but we only wrote one paper together. It concerns difference schemes in numerical analysis. Peter did much fundamental work in both pure and applied problems. I did very little in applied mathematics. Courant was a great believer in the equality of the two, and their nonseparation.

Following work of Hans Lewy, François Trèves and I wrote several papers on local solvability of linear PDEs. Our results were later improved by Charles Fefferman. A conjecture of ours was proved many years later, in 2006, by Nils Dencker. Several of my papers grew out of work by Lewy. He had also been a Göttingen student of Courant, and one paper was with Sidney Webster and Paul Yang, on regularity at the boundary of holomorphic maps.

I spent the academic year 1958–1959, on sabbatical, in Rome. I did not have much mathematical contact there, and because of shortage of space I had no office. I worked in the mathematics library—which closed for a long lunch period.

With Stampacchia, we organized a one-weekend-a-month seminar in Pisa, where various topics were discussed. This was a great pleasure. I complained that after the weekend, on Mondays, the museum was closed. Finally Sandro Faedo, the rector of the University of Pisa at the time, said "Oh, I'll have it opened for you." And he did. Another impressive incident was when, one day, I went to make a coffee in the mathematics department—they had a cafe-sized espresso machine. There were no beans in the container, and the portiere finally appeared with some. He then changed the degree to which the beans were ground. When I asked him why, he said, rubbing two fingers together, "It's a bit humid today."

The year in Rome began first with a conference in Pisa. In one of my lectures there I presented some interpolation inequalities. Afterwards I was introduced to a young mathematician, Emilio Gagliardo, and was told that he had obtained similar results. These are often referred to as Gagliardo—Nirenberg inequalities. At the time I also met Ennio de Giorgi. He was a remarkable mathematician; over the years his work had enormous influence.

Avron Douglis and I undertook to extend the Schauder theory of regularity for second order elliptic equations to higher order ones and also to systems of equations. We obtained interior estimates and then started to work on corresponding estimates up to the boundary. We learned that Agmon was also working on that problem and we joined forces. The resulting two papers have been used by many people. Recently John Ball kindly wrote that I had contributed much to the toolbox of people working in, and applying, PDEs.

Almost all my work has been jointly with others. This has been a great pleasure for me and I warmly recommend to young people to collaborate with others.

In the 1960s Joseph Kohn gave his famous proof of regularity at the boundary for the so-called d-bar Neumann problem. Here the boundary conditions at the boundary do not satisfy the criteria in the theory of Agmon, Douglis and myself. So, though there is regularity, it involves some loss of smoothness. Kohn and I decided to extend his results to other systems. In doing so we needed some extension of the Calderon–Zygmund theory of singular integral operators. This led us to construct an algebra of similar operators, which, as suggested by Friedrichs, we called "pseudo differential operators"—a contribution to the toolbox. Lars Hörmander, a wonderful mathematician, then generalized the theory considerably, inventing also Fourier Integral Operators. He visited NYU after he received his PhD; we became very close friends.

In August 1963, I attended a joint American–Soviet conference in PDE, in Novosibirsk, organized by Courant and M.M. Lavrentiev. It took place in Akademgorodok, the Academic City. There were about 20 Americans attending and about 100 Soviets, from all over the Soviet Union. For almost all of them it was their first contact with western mathematicians. It was like being on a ship for 2 weeks. We made friends immediately, and some of the friends from that time have remained friends to this day. We stayed in different hotels, but ate together. In the evening some of the Russians would walk us back to our hotel then, some of us accompanied them back to theirs. Then they walked us back to ours. This sometimes went on to 2 in the morning. It was a most exhilarating experience. Afterwards I visited Moscow and Leningrad. In Moscow I attended Gelfand's famous seminar, where a friend translated the proceedings for me. I was invited to Gelfand's home for his 50th birthday party. I was also invited to some other homes, despite the instructions they had received not to invite foreigners.

After that, I visited Russia a number of times. Each time Gelfand asked me what I considered the major new developments in mathematics, and where mathematics was heading. Each time I was embarrassed, since I could never come up with good answers. That period was a golden one for mathematics in Moscow.

In the published proceedings of the Novosibirsk conference there is the striking 3-page paper by Calderón which describes how to derive the L^p estimates of Agmon, Douglis and myself, by reducing the problem to one on the boundary.

In 1970 I became director of the Courant Institute—it acquired that title in 1965—succeeding Jürgen Moser. Earlier that year some students, angry about the Vietnam war, stormed the institute and attempted to destroy the rather large computer by setting a primitive bomb. After they left, some colleagues managed to defuse it. Actually, no work connected with the war was done at the computing centre.

Being director was very stressful for me. At the time the university sold its campus in the Bronx, and the mathematicians there were to join the department in Washington Square. With some financial inducement, the university tried to get some faculty to take early retirement, and I was asked to speak to those people. Happily, none accepted the offer. After one year as director, I said that I would continue for just one more year. Afterwards, Peter Lax took over .

In 1972 I published a paper on an abstract form of the Cauchy–Kowalewski theorem. Later, Takaaki Nishida improved it by showing that one of my conditions could be dropped. The result has been applied in the study of water waves.

David Kinderlehrer and I wrote several papers on regularity in free boundary problems, some with Joel Spruck. These are boundaries separating two physical states, like ice and water, and they are not known, a priori. With Basilis Gidas and Wei Ming Ni I wrote two papers on symmetry of solutions of second order elliptic equations in bounded regions and in all of space. We extended the "method of moving planes," originally introduced by A.D. Alexandroff to treat a geometric problem. Since then the method has been developed, and applied, in ways that have sometimes proved surprising to me. In 1991, Henri Berestycki and I extended the results to domains with nonsmooth boundary, such as a cube, and also developed the "sliding method" to prove monotonicity of solutions in some direction. To do this we made use of a result in a paper we had written with Raghu Varadhan. It is about the first eigenvalue of second order linear elliptic operators in general domains. The paper contained various new estimates. The maximum principle plays a central role.

Though the reader, if any remains, may be tired of it I will still write a bit more about my work. Incidentally, there is a common belief that mathematicians do their best work before they are 30. I know many, and I don't think so.

Haim Brezis and I wrote a number of papers together on nonlinear PDE. One involved a semilinear, second order, operator with a nonlinear term having a critical power exponent. Some unexpected things turned up, depending on whether the dimension is 3 or more. We also wrote two papers on topological degree for maps which need not be continuous. This grew out of work by Louis Boutet de Monvel and O. Gabber. The maps we considered belong to VMO, vanishing mean oscillation—related to BMO.

Louis receiving the Crafoord prize (1982) from King Carl Gustaf of Sweden and Mrs Crafoord. (Photo: private)

One day in 1982, Luis Caffarelli, Bob Kohn and I were walking to Chinatown for lunch, and I suggested we work on something together. Shortly before, Vladimir Scheffer had published an interesting paper on the possible dimension of singularities of solutions of the initial value problem for the Navier–Stokes equations, for fluid flow in 3 dimensions. I suggested that we try to extend what he had done. We succeeded; we showed that possible singularities of "suitable solutions"—uniqueness of solutions is not known—have zero one dimensional measure, so cannot be a curve. So far, this result has not been improved. The long standing open problem is whether singularities develop at all. This is one of the problems for which the Clay Foundation has offered one million dollars for a solution.

Caffarelli, Spruck and I wrote eight papers on fully nonlinear elliptic equations, starting with the Monge–Ampere equation: the determinant of the matrix of second derivatives of the unknown function equals a given function. In 1974, at a congress in Vancouver, I spoke on work with Eugene Calabi, in which we had solved the Monge–Ampere equation, or so we thought. Later, Charles Fefferman asked me to explain how we estimated the derivatives at the boundary. When I showed some estimates for third order derivatives and he said "These are upper bounds. How do you get lower bounds?" I then realized that we had neglected to do this and I spent several years trying. Finally, with Caffarelli and Spruck, we derived, instead, estimates of the Holder continuity of the second order derivatives, and these sufficed. Together with Joe Kohn we also treated complex Monge–Ampere equations.

Berestycki and I wrote several papers on travelling fronts. With Caffarelli we wrote a series of papers on semilinear second order elliptic equations in a half space and in other unbounded regions. Yan Yan Li and I wrote a number of papers on different subjects: Finsler space and Hamilton–Jacobi equations; estimates for composite materials; extension of a result of Alexandroff on embedded hypersurfaces with constant mean curvature...

Over the years I enjoyed teaching but by the time I retired, in 1989, I was a bit tired of it. A great pleasure was directing the 45 PhD students I had. My wife, Susan, died in 1998. She suffered from Alzheimer's disease. Later I met my partner, Nanette. Another proof of my good fortune.

Left: Louis with Nanette in 2008 (Photo: Teresa Ludlow). Right: In 2015. (Photo: Harald Hanche-Olsen)

The Masterpieces of John Forbes Nash Jr.

Camillo De Lellis

Abstract In this set of notes I follow Nash's four groundbreaking works on real algebraic manifolds, on isometric embeddings of Riemannian manifolds and on the continuity of solutions to parabolic equations. My aim has been to stay as close as possible to Nash's original arguments, but at the same time present them with a more modern language and notation. Occasionally I have also provided detailed proofs of the points that Nash leaves to the reader.

1 Introduction

John Nash has written very few papers: if for each mathematician in the twentieth century we were to divide the depth, originality, and impact of the corresponding production by the number of works, he would most likely be on top of the list, and even more so if we were to divide by the number of pages. In fact all his fundamental contributions can be stated in very few lines: certainly another measure of his genius, but making any survey of his theorems utterly useless. Discussing the impact of Nash's work is certainly redundant, since all his fundamental contributions have already generated a large literature and an impressive number of surveys and lecture notes. "Reworking" his proofs in my own way, or giving my personal perspective, would be of little interest: much better mathematicians have already developed deep and well-known theories from his seminal papers.

When I was asked to write this contribution to the Abel Volumes I felt enormously honored, but precisely for the reasons listed above it took me very little to realize how difficult it would have been to write something even modestly useful. This note is therefore slightly unusual: I have just tried to rewrite the original papers in a more modern language while adhering as much as possible to the original

C. De Lellis (✉)
School of Mathematics, Institute for Advanced Study, Princeton University, Princeton, NJ, USA

Universität Zürich, Zürich, Switzerland
e-mail: camillo.delellis@math.uzh.ch

H. Holden, R. Piene (eds.), *The Abel Prize 2013–2017*,
https://doi.org/10.1007/978-3-319-99028-6_19

arguments. In fact Nash used often a rather personal notation and wrote in a very informal way, here and there a few repetitions can be avoided and the discussions of some, nowadays standard, facts can be removed. In a sense my role has been simply that of a translator: I just hope to have been a decent one, namely that I have not introduced (too many) errors and wrong interpretations. In particular I hope that these notes might save some time to those scholars who want to work out the details of Nash's original papers, although I strongly encourage anybody to read the source: any translation of any masterpiece always loses something compared to the original and the works of Nash are true masterpieces of the mathematics of the twentieth century!

These notes leave aside Nash's celebrated PhD thesis on game theory and focus on the remaining four fundamental papers that have started an equal number of revolutions in their respective topics, namely the 1952 note on real algebraic varieties, the 1954 paper on C^1 isometric embeddings, the 1956 subsequent work on smooth isometric embeddings and finally the 1958 Hölder continuity theorem for solutions to linear (uniformly) parabolic partial differential equations with bounded nonconstant coefficients. Even the casual reader will realize that everything can be understood up to the smallest detail with a very limited amount of knowledge: I dare say that any good graduate student in mathematics will be able to go through the most relevant arguments with little effort.

I have decided to leave aside the remaining works of Nash in "pure mathematics" either because their impact has not been as striking as that of the four mentioned above (as it is the case for the works [74, 77, 78]) or because, as it is the case for [79], although its impact has been major, this is mainly due to the questions raised by Nash rather than to the actual theorems proved by him. However, for completeness I have included a last section with a brief discussion of these remaining four (short!) notes in pure mathematics and of the "Nash blowup".

2 Real Algebraic Manifolds

2.1 *Introduction*

After his famous PhD thesis in game theory (and a few companion notes on the topic) Nash directed his attention to geometry and specifically to the classical problem of embedding smooth manifolds in the Euclidean space.[1] Consider a smooth closed manifold Σ of dimension n (where with *closed* we mean, as

[1] In a short autobiographical note, cf. [80, Ch. 2], Nash states that he made his important discovery while completing his PhD at Princeton. In his own words "...I was fortunate enough, besides developing the idea which led to "NonCooperative Games", also to make a nice discovery relating manifolds and real algebraic varieties. So, I was prepared actually for the possibility that the game theory work would not be regarded as acceptable as a thesis in the mathematics department and then that I could realize the objective of a Ph.D. thesis with the other results."

usual, that Σ is compact and has no boundary). A famous theorem of Whitney (cf. [105, 106]) shows that Σ can be embedded smoothly in $\mathbb{R}^{2n}$, namely that there exists a smooth map $w : \Sigma \to \mathbb{R}^{2n}$ whose differential has full rank at every point (i.e., w is an *immersion*) and which is injective (implying therefore that w is an homeomorphism of Σ with $w(\Sigma)$).

Clearly $w(\Sigma)$ is a smooth submanifold of $\mathbb{R}^{2n}$ diffeomorphic to Σ. Whitney showed also that w can be perturbed smoothly to a second embedding v so that $v(\Sigma)$ is a *real analytic* submanifold, namely for every $p \in v(\Sigma)$ there is a neighborhood U of p and a real analytic map $u : U \to \mathbb{R}^n$ such that $\{u = 0\} = U \cap v(\Sigma)$ and Du has full rank. Whitney's theorem implies, in particular, that any closed smooth manifold Σ can be given a real analytic structure, namely an atlas $\mathscr{A}$ of charts where the changes of coordinates between pairs of charts are real analytic mappings.

In his only note on the subject, the famous groundbreaking paper [72] published in 1952, Nash gave a fundamental contribution to real algebraic geometry, showing that indeed it is possible to realize any smooth closed manifold of dimension n as an *algebraic* submanifold of $\mathbb{R}^{2n+1}$. We recall that, classically, any subset of $\mathbb{R}^N$ consisting of the common zeros of a collection of polynomial equations is called an *algebraic subvariety*. We can assign a dimension to any algebraic subvariety using a purely algebraic concept (see below) and the resulting number coincides with the usual metric definitions of dimension for a subset of the Euclidean space (for instance with the Hausdorff dimension, see [31, Ch. 2] for the relevant definition). The main theorem of Nash's note is then the following.

Theorem 1 (Existence of real algebraic structures) *For any closed connected smooth n-dimensional manifold Σ there is a smooth embedding $v : \Sigma \to \mathbb{R}^{2n+1}$ such that $v(\Sigma)$ is a connected component of an n-dimensional algebraic subvariety of $\mathbb{R}^{2n+1}$.*

It turns out that for any point $p \in v(\Sigma)$ there is a neighborhood U such that $U \cap v(\Sigma)$ is the zero set of $n + 1$ polynomials with linearly independent gradients. In his note Nash proved also the following approximation statement, see Theorem 9: any smooth embedding $w : \Sigma \to \mathbb{R}^m$ can be smoothly approximated by an embedding $\bar{v}$ so that $\bar{v}(\Sigma)$ is a portion of an n-dimensional algebraic subvariety of $\mathbb{R}^m$. However, in order to achieve the stronger property in Theorem 1, namely that $\bar{v}(\Sigma)$ is a *connected component* of the subvariety, Nash's argument needs to increase the target. He conjectured that this is not necessary, cf. [72, p. 420], a fact which was proved much later by Akbulut and King, see [1]. He also conjectured the existence of a smooth embedding z (in *some* Euclidean space $\mathbb{R}^N$) such that $z(\Sigma)$ is the whole algebraic subvariety, not merely a connected component, and this was proved by Tognoli in [99]. Both [99] and [1] build upon a previous work of Wallace, [101].

As it happens for the real analytic theorem of Whitney, it follows from Theorem 1 that any smooth closed manifold can be given a real algebraic structure, see below for the precise definition. In his note Nash proved also that such structure is indeed unique, cf. Theorem 10.

As already mentioned in the previous paragraph, Nash left a few conjectures and open questions in his paper, which were subsequently resolved through the works of Wallace, Tognoli, and Akbulut and King: we refer the reader to King's paragraph in Nash's memorial article [26] for further details. The ideas of his paper have generated a large body of literature in real algebraic geometry and terms like Nash manifolds, Nash functions, and Nash rings are commonly used to describe some of the objects arising from his argument for Theorem 1, see for instance [9, 91].

2.2 *Real Algebraic Structures and Main Statements*

Following Nash we introduce a suitable algebraic structure on closed real analytic manifolds Σ. In [72] such structures are called *real algebraic manifolds*. Since however nowadays the latter expression is used for a different object, in order to avoid confusion and to be consistent with the current terminology, we will actually use the term "Nash manifolds" for the objects introduced by Nash.

Note that, by the classical Whitney's theorem recalled in the previous section, there is no loss of generality in assuming the existence of a real analytic atlas for any differentiable manifold Σ. The notion of Nash manifold allows Nash to recast Theorem 1 in an equivalent form. The latter will be given in this section, together with several other interesting conclusions, whose proofs will all be postponed to the next sections.

Definition 2 (Basic sets) Any finite collection $\{f_1, \ldots, f_N\}$ of smooth real valued functions over Σ is called a *basic set* if the map $f = (f_1, \ldots, f_N)$ is an embedding of Σ into $\mathbb{R}^N$.

Definition 3 (Nash manifolds) A *Nash manifold* is given by a pair $(\Sigma, \mathscr{R})$ where Σ is a real analytic manifold of dimension n and $\mathscr{R}$ a ring of real valued functions over Σ satisfying the following requirements:

(a) Any $f \in \mathscr{R}$ is real analytic;
(b) $\mathscr{R}$ contains a basic set;
(c) The transcendence degree of $\mathscr{R}$ must be n, more precisely for any collection of $n+1$ distinct elements $f_1, \ldots, f_{n+1} \in \mathscr{R}$ there is a nontrivial polynomial P in $n+1$ variables such that $P(f_1, \ldots, f_{n+1}) = 0$;
(d) $\mathscr{R}$ is maximal in the class of rings satisfying (a), (b), (c).

An important (and not difficult) fact following from the definitions is that the algebraic structure of the ring determines in a suitable sense the manifold Σ and hence that the structure as Nash manifold is essentially unique for every Σ.

Proposition 4 (Algebraic description of Nash manifolds) *On any Nash manifold $(\Sigma, \mathscr{R})$ there is a one-to-one correspondence between maximal ideals of $\mathscr{R}$ and points of Σ, more precisely:*

(I) $\mathscr{I} \subset \mathscr{R}$ *is a maximal ideal if and only if* $\mathscr{I} = \{f \in \mathscr{R} : f(p) = 0\}$ *for some* $p \in \Sigma$.

Moreover, if $(\Sigma_1, \mathscr{R}_1)$ *and* $(\Sigma_2, \mathscr{R}_2)$ *are two Nash manifolds, then a map* $\phi : \mathscr{R}_1 \to \mathscr{R}_2$ *is a ring isomorphism if and only if there is a real analytic diffeomorphism* $\varphi : \Sigma_1 \to \Sigma_2$ *such that* $\phi(f) = f \circ \varphi^{-1}$ *for any* $f \in \mathscr{R}_1$.

Consider now a Nash manifold $(\Sigma, \mathscr{R})$ and recall that by Definition 3(b) we prescribe the existence of a basic set $\mathscr{B} = \{f_1, \ldots, f_N\} \subset \mathscr{R}$: it follows that $f = (f_1, \ldots, f_N)$ is an analytic embedding of Σ into $\mathbb{R}^N$. On the other hand by Definition 3(c) there is a set of nontrivial polynomial relations between the f_i's (because $N > n$) and so it appears naturally that $f(\Sigma)$ is in fact a subset of a real algebraic variety. Following Nash we will call $f(\Sigma)$ a *representation* of the corresponding Nash manifold.

Definition 5 (Representations) If $(\Sigma, \mathscr{R})$ is a Nash manifold, $\mathscr{B} = \{f_1, \ldots, f_N\} \subset \mathscr{R}$ a basic set and $f = (f_1, \ldots, f_N) : \Sigma \to \mathbb{R}^N$, then $f(\Sigma)$ is called an *algebraic representation* of $(\Sigma, \mathscr{R})$.

In order to relate representations with algebraic subvarieties of the Euclidean space we need to introduce the concept of *sheets* of an algebraic subvariety.

Definition 6 (Sheets) A sheet of a real algebraic subvariety $V \subset \mathbb{R}^N$ is a subset $S \subset V$ satisfying the following requirements:

(a) For any $p, q \in S$ there is a real analytic arc $\gamma : [0, 1] \to S$ with $\gamma(0) = q$ and $\gamma(1) = p$.
(b) S is a maximal subset of V with property (a).
(c) There is at least one point $p \in S$ with a neighborhood U such that $U \cap V \subset S$.

Clearly, if $V \subset \mathbb{R}^N$ is an algebraic subvariety and $S \subset V$ a connected component which happens to be a submanifold of $\mathbb{R}^N$, then S is in fact a sheet of A. However:

(i) there might be "smooth" sheets which go across singularities, for instance, if we take $V = \{(x, y) : xy = 0\} \subset \mathbb{R}^2$ and $S = \{(x, y) : x = 0\}$, cf. Fig. 1;
(ii) there might be sheets which are connected components of V but are singular, for instance Bernoulli's lemniscate $\{(x, y) : (x^2 + y^2)^2 = 2y^2 - 2y^2\}$ is a connected subvariety of the plane consisting of one single sheet, singular at the origin (Fig. 2).

An important observation by Nash is that, by simple considerations, any representation of a Nash manifold is in fact a sheet of an irreducible algebraic subvariety with dimension equal to that of the manifold. Recall that an algebraic subvariety V is called *irreducible* if it cannot be written as the union of two proper subsets which are also subvarieties. More precisely we have.

Proposition 7 (Characterization of representations) *A representation of a connected Nash manifold* $(\Sigma, \mathscr{R})$ *is always a sheet of an irreducible subvariety* V *whose dimension is the same as that of* Σ. *Conversely, if* $v : \Sigma \to \mathbb{R}^N$ *is a real analytic embedding of a closed real analytic manifold* Σ *whose image* $v(\Sigma)$ *is a*

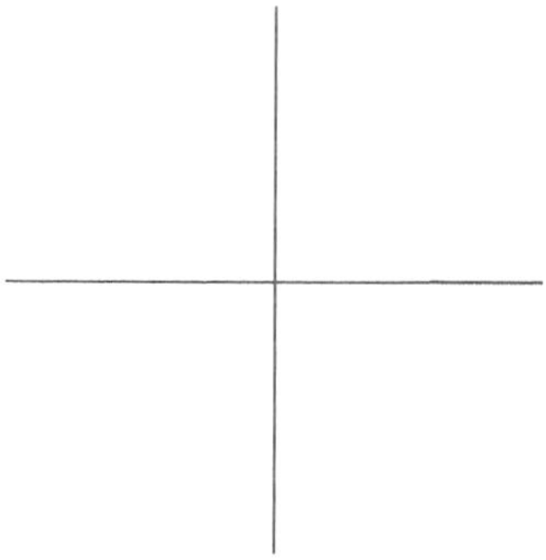

Fig. 1 The set $S = \{(x, y) \in \mathbb{R}^2 : x = 0\}$ is a sheet of the algebraic subvariety $V = \{(x, y) : xy = 0\}$. Note that, although the origin is a singular point of V, it is not a singular point of S. Moreover S is *not* a connected component of V

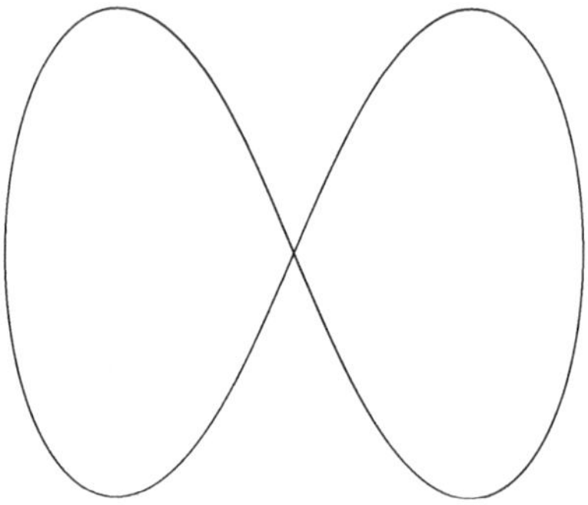

Fig. 2 Bernoulli's lemniscate is an algebraic subvariety of $\mathbb{R}^2$ which consists of a single sheet. Note that it is singular at the origin

sheet of an algebraic subvariety, then there is a structure of Nash manifold $(\Sigma, \mathscr{R})$ *for which the components* $\{v_1, \ldots, v_N\}$ *of* v *form a basic subset of* $\mathscr{R}$.

The outcome of the discussion above is that Theorem 1 can now be equivalently stated in terms of Nash manifolds. However note that Theorem 1 requires the representation to be more than just a sheet of an algebraic subvariety: it really has to be a connected component. For this reason Nash introduces a special term: a representation $v(\Sigma)$ will be called *proper* if it is a *connected component* of the corresponding algebraic subvariety in Proposition 7. Hence we can now rephrase Theorem 1 in the following way.

Theorem 8 (Existence of proper representations) *For any connected smooth closed differentiable manifold* Σ *of dimension* n *there is a structure of Nash manifold* $(\Sigma, \mathscr{R})$ *with a basic set* $\{v_1, \ldots, v_{2n+1}\} = \mathscr{B} \subset \mathscr{R}$ *such that* $v(\Sigma)$ *is a proper representation in* $\mathbb{R}^{2n+1}$.

Giving up the stronger requirement of "properness" of the representation, Nash is able to provide an approximation with algebraic representations of any smooth embedding, without increasing the dimension of the ambient space. As a matter of fact Theorem 8 will be proved as a corollary of such an approximation theorem, whose statement goes as follows.

Theorem 9 (Approximation theorem) *Let* Σ *be a connected closed differentiable manifold and* $w : \Sigma \to \mathbb{R}^m$ *a smooth embedding. Then for any* $\varepsilon > 0$ *and any*

$k \in \mathbb{N}$ there is a structure of Nash manifold $(\Sigma, \mathscr{R})$ with a basic set $\{v_1, \ldots, v_m\}$ such that $\|w - v\|_{C^k} < \varepsilon$.

In the theorem above $\| \cdot \|_{C^k}$ denotes a suitably defined norm measuring the uniform distance between derivatives of w and v up to order k. The norm will be defined after fixing a finite smooth atlas on Σ, we refer to the corresponding section for the details.

As a final corollary of his considerations, Nash also reaches the conclusion that the structure of Nash manifold is in fact determined uniquely by the differentiable one. More precisely we have the following result.

Theorem 10 (Uniqueness of the Nash ring) *If two connected Nash manifolds $(\Sigma_1, \mathscr{R}_1)$ and $(\Sigma_2, \mathscr{R}_2)$ are diffeomorphic as differentiable manifolds, then they are also isomorphic as Nash manifolds, namely there is a real analytic $\varphi : \Sigma_1 \to \Sigma_2$ for which the map $\phi(f) := f \circ \varphi^{-1}$ is a ring isomorphism of $\mathscr{R}_1$ with $\mathscr{R}_2$.*

2.3 Technical Preliminaries

In this section we collect some algebraic and analytical technical preliminaries, standard facts which will be used in the proofs of the statements contained in the previous sections. We begin with a series of basic algebraic properties.

Definition 11 Given an algebraic subvariety $V \subset \mathbb{R}^N$ and a subfield $\mathbb{F} \subset \mathbb{R}$ we say that $\mathbb{F}$ is a *field of definition* of V if there is a set S of polynomials with coefficients in $\mathbb{F}$ such that $V = \{x \in \mathbb{R}^N : P(x) = 0, \ \forall P \in S\}$.

Proposition 12 (Cf. [102, Cor. 3, p. 73, and Prop. 5, p. 76]) *For any algebraic subvariety $V \subset \mathbb{R}^N$ there is a unique minimal field $\mathbb{F} \subset \mathbb{R}$ of definition, namely a field of definition of V which does not contain any smaller field of definition. $\mathbb{F}$ is, moreover, finitely generated over $\mathbb{Q}$.*

Definition 13 We will say that a certain collection of coordinates $\{x_{i_1}, \ldots, x_{i_m}\}$ is algebraically independent over a field $\mathbb{F}$ at a point $p = (p_1, \ldots, p_N)$ if there is no nontrivial polynomial P with coefficients in $\mathbb{F}$ such that $P(p_{i_1}, \ldots, p_{i_m}) = 0$.

Given a point p in an algebraic subvariety $V \subset \mathbb{R}^n$ with minimal field of definition $\mathbb{F}$ we define the algebraic dimension $\dim_V(p)$ of p with respect to V as the maximal number of coordinates which are algebraically independent over $\mathbb{F}$ at p. The algebraic dimension of V is $\dim(V) = \max\{\dim_V(p) : p \in V\}$ and a point $p \in V$ is called a *general point* of V if $\dim_V(p) = \dim(V)$.

Proposition 14 *Let $V \subset \mathbb{R}^N$ be an algebraic subvariety of algebraic dimension n with minimal field of definition $\mathbb{F}$. Then the following holds.*

(a) *Any collection of $n + 1$ coordinates satisfy a nontrivial polynomial relation (as real functions with domain V);*

(b) *For any general point p of V there is a neighborhood U where V is an n-dimensional (real analytic) submanifold and where any collection of coordinates which are algebraically independent at p over $\mathbb{F}$ gives a (real analytic) parametrization of V.*
(c) *If* $\dim_V(p) = m$*, then there is an algebraic subvariety $W \subset V$ of algebraic dimension m which contains p and whose minimal field of definition is contained in $\mathbb{F}$.*

The proofs of the statements (a), (b), and (c) can be found in [102, Ch. II and Ch. IV], more precisely see the discussion at [102, p. 72, Th. 3].

We state here a simple corollary of the above proposition, for which we give the elementary proof.

Corollary 15 *The algebraic dimension of a subvariety V coincides with its Hausdorff dimension as a subset of $\mathbb{R}^N$. In fact, for any $j \leq$* $\dim(V)$*, the subset* $V_j := \{v \in V : \dim_V(p) = j\}$ *is a set of Hausdorff dimension j.*

Proof The second part of the statement obviously implies the first. We focus therefore on the second, which we prove by induction over $\dim(V)$. The 0-dimensional case is obvious: if V is a 0-dimensional subvariety of $\mathbb{R}^N$, then $V_0 = V$ must be contained in $\mathbb{F}^N$, which is necessarily a countable set ($\mathbb{F}$ denotes the minimal field of definition of V and recall that it is finitely generated over $\mathbb{Q}$).

Assume therefore that the statement holds when the dimension of the variety is no larger than $n-1$: we now want to show that the claim holds when $\dim(V) = n$. By Proposition 14, the subset V_n of points $p \in V$ with maximal algebraic dimension is covered by countably many real analytic n-dimensional manifolds. Hence V_n has Hausdorff dimension at most n (cf. [31, Sec. 3.3]). On the other hand by Proposition 14(b) the Hausdorff dimension must be at least n. Next, let $j < n$. By Proposition 14(c) any point $p \in V_j$ is contained in an algebraic subvariety W of algebraic dimension j with minimal field of definition contained in $\mathbb{F}$. Each such W has Hausdorff dimension j, by inductive assumption. On the other hand, since any such W is defined through a finite set of polynomials with coefficient in $\mathbb{F}$, the set of such W is countable. We have therefore shown that V_j has Hausdorff dimension at most j.

Now consider a point $q \in V$ with $\dim_V(q) = j$ and an algebraic subvariety $W \subset V$ as above. Let $\mathbb{F}'$ be its minimal field of definition and consider any $p = (p_1, \ldots, p_N) \in W$. The algebraic dimension $\dim_W(p)$ is at most j, which means that for any collection of $j+1$ distinct coordinates $p_{i_1}, \ldots, p_{i_{j+1}}$ there is a nontrivial polynomial P with coefficients in $\mathbb{F}'$ such that $P(p_{i_1}, \ldots, p_{i_{j+1}}) = 0$. Since $\mathbb{F}' \subset \mathbb{F}$, we must necessarily have $\dim_V(p) \leq j$. Thus, $W \subset V_0 \cup V_1 \cup \ldots \cup V_j$. On the other hand, we know by inductive assumption that W has Hausdorff dimension j and we have shown that the dimension of each V_i is at most i. We then conclude that the Hausdorff dimension of j must be j.

We are now ready to state the two technical facts in analysis needed in the rest of the section. The first is a standard consequence of the implicit function theorem for real analytic mappings, see for instance [59, Th. 1.8.3]. As usual, the tubular neighborhood of size δ of a subset $S \subset \mathbb{R}^N$ is the open set consisting of those points whose distance from S is smaller than δ. In this section we will denote it by $U_\delta(S)$.

Proposition 16 *If $\Sigma \subset \mathbb{R}^N$ is a closed real analytic submanifold, then there is a $\delta > 0$ with the following two properties:*

(a) *For any $x \in U_\delta(\Sigma)$ there is a unique point $u(x) \in \Sigma$ of least distance to x.*
(b) *The map $x \mapsto u(x)$ is real analytic.*

The first statement needs in fact only the C^2 regularity of Σ, cf. [49]. Moreover the proof therein uses the implicit function theorem to give that u is smooth when Σ is smooth: the real analyticity of u follows then directly from [59, Th. 1.8.3].

The following is a classical Weierstrass type result. As usual, given a smooth function g defined in a neighborhood of a compact set $K \subset \mathbb{R}^m$ we denote by $\|g\|_{C^0(K)}$ the number $\max\{|u(x)| : x \in K\}$ and we let

$$\|g\|_{C^j(K)} := \sum_{|I|\leq j} \|\partial^I g\|_{C^0(K)},$$

where, given a multiindex $I = (i_1, \ldots, i_m) \in \mathbb{N}^m$, we let $|I| = i_1 + \cdots + i_m$ and

$$\partial^I f = \frac{\partial^{|I|} f}{\partial x_1^{i_1} \partial x_2^{i_2} \cdots \partial x_m^{i_m}}.$$

Proposition 17 *Let $U \subset \mathbb{R}^N$ be an open set, $K \subset U$ a compact set and $f : U \to \mathbb{R}$ a smooth function. Given any $j \in \mathbb{N}$ and any $\varepsilon > 0$, there is a polynomial P such that $\|f - P\|_{C^j(K)} \leq \varepsilon$.*

Proof Using a partition of unity subordinate to a finite cover of K we can assume, without loss of generality, that $f \in C_c^\infty(U)$. The classical Weierstrass theorem corresponds to the case $j = 0$, see for instance [87]: however the proof given in the latter reference, which regularizes f by convolution with suitable polynomials, gives easily the statement above for general j. Nash in [72] provides instead the following elegant argument. Consider first a box $[-M/2, M/2]^N \subset \mathbb{R}^N$ containing the support of f and let $\tilde{f}$ be the M-periodic function which coincides with f on the box. If we expand $\tilde{f}$ in Fourier series as

$$\tilde{f}(x) = \sum_{\lambda \in \mathbb{Z}^N} a_\lambda e^{2\pi M \lambda \cdot x}$$

and consider the partial sums

$$S_m(x) := \sum_{|\lambda| \leq m} a_\lambda e^{2\pi M \lambda \cdot x},$$

then clearly $\|S_m - f\|_{C^j(K)} = \|S_m - \tilde{f}\|_{C^j(K)} \leq \varepsilon/2$ for m large enough. On the other hand S_m is an entire analytic function and thus for a sufficiently large degree d the Taylor polynomial T_m^d of S_m at 0 satisfies $\|S_m - T_m^d\|_{C^j(K)} \leq \varepsilon/2$.

2.4 The Algebraic Description of Nash Manifolds and the Characterization of Representations as Sheets

Proof (Proof of Proposition 4*)* First of all, for any (proper) ideal $\mathscr{I}$ the set $Z = Z(\mathscr{I})$ of points of Σ at which all elements of $\mathscr{I}$ vanish must be nonempty. Otherwise, for any point $p \in \Sigma$ there would be an element $f_p \in \mathscr{I}$ such that $f_p(p) \neq 0$. Choose then an open neighborhood U_p such that $f_p \neq 0$ on U_p and cover Σ with finitely many U_{p_i}. The function $f := \sum_i f_{p_i}^2$ would belong to the ideal $\mathscr{I}$ and would be everywhere nonzero. But then $\frac{1}{f}$ would belong to $\mathscr{R}$, $f \cdot \frac{1}{f} = 1$ would belong to the ideal $\mathscr{I}$ and the latter would coincide with $\mathscr{R}$, contradicting the assumption that $\mathscr{I}$ s a proper ideal.

Given a point p and a basic set $\mathscr{B} = \{v_1, \ldots, v_N\} \subset \mathscr{R}$, the function

$$g(y) := \sum_i (v_i(y) - v_i(p))^2$$

vanishes only at p and belongs to $\mathscr{R}$. Thus we have:

(i) The set $\mathscr{I}(p)$ of all elements which vanish at p must be nonempty. Moreover, it cannot be the whole ring $\mathscr{R}$ because it does not contain the constant function 1. It is thus a proper ideal and it must be maximal: any larger ideal $\mathscr{J}$ would necessarily have $Z(\mathscr{J}) = \emptyset$.

(ii) If $\mathscr{I}$ is a maximal ideal, then there must be an element $p \in Z(\mathscr{I})$ and, since $\mathscr{I} \subset \mathscr{I}(p)$, we must necessarily have $\mathscr{I} = \mathscr{I}(p)$.

This shows the first part of the proposition. Next, let $(\Sigma_1, \mathscr{R}_1)$ and $(\Sigma_2, \mathscr{R}_2)$ be two Nash manifolds. Clearly, if $\varphi : \Sigma_1 \to \Sigma_2$ is a real analytic diffeomorphism such that $\phi(f) := f \circ \varphi^{-1}$ maps $\mathscr{R}_1$ onto $\mathscr{R}_2$, then ϕ is a ring isomorphism. Vice versa, let $\phi : \mathscr{R}_1 \to \mathscr{R}_2$ be a ring isomorphism. Using the correspondence above, given a point $p \in \Sigma_1$ we have a corresponding maximal ideal $\mathscr{I}(p) \subset \mathscr{R}$, which is mapped by ϕ into a maximal ideal of $\mathscr{R}_2$: there is then a point $\varphi(p) \in \Sigma_2$ such that $\phi(\mathscr{I}(p)) = \mathscr{I}(\varphi(p))$. We now wish to show that

$$\phi(f)(\varphi(p)) = f(p) . \tag{1}$$

First observe that:

$$\text{if } f \text{ vanishes at } p, \text{ then } \phi(f) \text{ must vanish at } \varphi(p). \tag{2}$$

This follows from the property $\phi(f) \in \phi(\mathscr{I}(p)) = \mathscr{I}(\varphi(p))$.

Next we follow the convention that, given a number $q \in \mathbb{R}$, we let q denote both the function constantly equal to q on Σ_1 and that equal to q on Σ_2. Since 1 is the multiplicative unit of $\mathscr{R}_1$ and $\mathscr{R}_2$, then $\phi(1) = 1$. Hence, using the ring axioms, it follows easily that $\phi(q) = q$ for any $q \in \mathbb{Q}$. Observe next that if $f \in \mathscr{R}_1$ is a positive function on Σ_1, then $g := \sqrt{f}$ is a real analytic function and it must belong to $\mathscr{R}_1$, otherwise the latter ring would not satisfy the maximality condition of Definition 3(d). Hence, if $f > 0$, then $\phi(f) = (\phi(\sqrt{f}))^2 \geq 0$. Thus $f > g$ implies $\phi(f) \geq \phi(g)$. Fix therefore a constant real α and two rational numbers $q > \alpha > q'$. We conclude $q = \phi(q) \geq \phi(\alpha) \geq \phi(q') = q'$. Since q and q' might be chosen arbitrarily close to α, this implies that $\phi(\alpha) = \alpha$.

Having established the latter identity, we can combine it with (2) to conclude (1). Indeed, assume $f(p) = \alpha$. Then $g = f - \alpha$ vanishes at p and thus, by (2), $\phi(g) = \phi(f) - \alpha$ vanishes at $\varphi(p)$: thus $\phi(f)(\varphi(p)) = f(p)$.

Next, φ^{-1} is the map induced by the inverse of the isomorphism ϕ, from which we clearly conclude $\phi(f) = f \circ \varphi^{-1}$. It remains to show that φ is real analytic: the same argument will give the real analyticity of φ^{-1} as well, thus completing the proof. Let $\mathscr{B}_1 = \{f_1, \ldots, f_N\}$ be a basic set for $(\Sigma_1, \mathscr{R}_1)$ and $\mathscr{B}_2 = \{g_{N+1}, \ldots, g_{N+M}\}$ be a basic set for $(\Sigma_2, \mathscr{R}_2)$. Set $g_i := f_i \circ \varphi^{-1} = \phi(f_i)$ for $i \leq N$ and $f_j := g_j \circ \varphi = \phi^{-1}(g_j)$ for $j \geq N + 1$. Then $\{f_1, \ldots, f_{N+M}\}$ and $\{g_1, \ldots, g_{N+M}\}$ are basic sets for Σ_1 and Σ_2 respectively. The map $f = (f_1, \ldots, f_{N+M}) : \Sigma_1 \to \mathbb{R}^{N+M}$ is a real analytic embedding of Σ_1 and $g = (g_1, \ldots, g_{N+M})$ a real analytic embedding of Σ_2 with the same image S. We therefore conclude that $\varphi = (g|_S)^{-1} \circ f$ is real analytic, which completes the proof.

Proof (Proof of Proposition 7) **Representation $\Longrightarrow$ Sheet.** We consider first a Nash manifold $(\Sigma, \mathscr{R})$ of dimension n and a representation $\mathscr{B} = \{f_1, \ldots, f_N\} \subset \mathscr{R}$. Our goal is thus to show that, if we set $f = (f_1, \ldots, f_N)$, then $S := f(\Sigma)$ is a sheet of an n-dimensional algebraic subvariety $V \subset \mathbb{R}^N$. First, recalling that $\mathscr{B}$ is a basic set, we know that for each choice of $1 \leq i_1 < i_2 < \ldots < i_{n+1} \leq N$ there is a (nontrivial) polynomial $P = P_{i_1 \ldots i_{n+1}}$ such that $P(f_{i_1}, \ldots, f_{i_{n+1}}) = 0$. Let then V_0 be the corresponding algebraic subvariety, namely the set of common zeros of the polynomials $P_{i_1 \ldots i_{n+1}}$. Clearly, by Proposition 14(b) the dimension of V_0 can be at most n. Otherwise there would be a point $q \in V_0$ of maximal dimension $d \geq n + 1$ and there would be a neighborhood U of q such that $V_0 \cap U$ is a real analytic d-dimensional submanifold of $\mathbb{R}^N$. This would mean that, up to a relabeling of the coordinates and to a possible restriction of the neighborhood, $U \cap V_0$ is the graph of a real analytic function of the first d variables $x_1, \ldots, x_d$. But then this would contradict the existence of a nontrivial polynomial of the first $n + 1 \leq d$ variables which vanishes on V_0.

Note moreover that, since f is a smooth embedding of Σ, by Corollary 15 the dimension must also be at least n. Hence we have concluded that the dimension of V is precisely n.

Next, if V_0 is reducible, then there are two nontrivial subvarieties V and W of V_0 such that $V_0 = V \cup W$. One of them, say V, must intersect S on a set A of positive n-dimensional volume. If P is any polynomial among the ones defining V, we then must have $P(f_1, \dots, f_N) = 0$ on A: however, since $P(f_1, \dots, f_N)$ is real analytic, A has positive measure and S is a connected submanifold of $\mathbb{R}^N$, we necessarily have $P(f_1, \dots, f_N) = 0$ on the whole S. We thus conclude that $S \subset V =: V_1$. If V_1 were reducible, we can go on with the above procedure and create a sequence $V_0 \supset V_1 \supset \dots$ of algebraic varieties containing S: however, by the well-known descending chain condition in the Zariski topology (cf. [45]), this procedure must stop after a finite number of steps. Thus, we have achieved the existence of an n-dimensional irreducible subvariety V such that $S \subset V$.

We claim that S is a sheet of V. First of all, by Corollary 15, S must contain a general point p of V because its dimension is n. Moreover, by Proposition 14 we know that there is a neighborhood U of p such that $U \cap V$ is an n-dimensional submanifold. By further restricting U we can assume that both $U \cap V$ and $U \cap S$ are connected n-dimensional submanifolds. Since $S \subset V$, we must obviously have $S \cap U = V \cap U$. Hence p is a point which satisfies condition (c) in Definition 6. Next, fix a second point $q \in S$ and let $\bar{p} = f^{-1}(p)$ and $\bar{q} = f^{-1}(q)$. Since Σ is a connected real analytic manifold, we clearly know that there is $\bar{\gamma} : [0, 1] \to \Sigma$ real analytic[2] such that $\bar{\gamma}(0) = \bar{q}$ and $\bar{\gamma}(1) = \bar{q}$. Thus $\gamma := f \circ \bar{\gamma}$ is a map as in Definition 6(a). It remains to show that S is maximal among the subsets of V satisfying Definition 6(a).

So, let $\tilde{S}$ be the maximal one containing S and fix $p \in \tilde{S}$: we claim that indeed $p \in S$. By assumption we know that there is a real analytic curve $\gamma : [0, 1] \to \tilde{S}$ such that $\gamma(0) \in S$ is a general point and $\gamma(1) = p$. First of all, since $\gamma(0)$ is a general point of V, there is a neighborhood U of p where S and V coincide. Hence there is $\delta > 0$ such that $\gamma([0, \delta[) \subset S$. Set next

$$s := \sup\{s \in [0, 1] : \gamma([0, s[) \subset S\}\,.$$

Clearly, s is a maximum. Moreover, by compactness of S, $q := \gamma(s) \in S$: we need then to show that $s = 1$. Assume, instead, that $s < 1$. Let U be some coordinate chart in the real analytic manifold Σ containing $f^{-1}(q)$ and $y : U \to \mathbb{R}^n$ corresponding real analytic coordinates. There is $\delta > 0$ such that $f^{-1}(\gamma([s - \delta, s])) \subset U$. The map $\tilde{\gamma} := y \circ f^{-1} \circ \gamma : [s - \delta, s] \to \mathbb{R}^n$ is then real analytic. Hence there is $\eta > 0$ so that $\tilde{\gamma}(t)$ can be expanded in power series of $(t - s)$ on the interval $]s - \eta, s]$. Such power series converges then on

[2]Here we are using the nontrivial fact that in a connected real analytic manifold any pair of points can be joined by a real analytic arc. One simple argument goes as follows: use first Whitney's theorem to assume, without loss of generality, that Σ is a real analytic submanifold of $\mathbb{R}^N$. Fix two points p and q and use the existence of a real analytic projection in a neighborhood of Σ to reduce our claim to the existence of a real analytic arc connecting any two points inside a connected open subset of the Euclidean space. Finally use the Weierstrass polynomial approximation theorem to show the latter claim.

$]s-\eta, s+\eta[$ and extends $\tilde{\gamma}$ to a real analytic map on $]s-\eta, s+\eta[$. Now, $\gamma|_{]s-\eta,s+\eta[}$ and $\bar{\gamma} := f \circ y^{-1} \circ \tilde{\gamma}$ are two maps which coincide on the interval $]s-\eta, s]$: since they are both real analytic, they must then coincide on the whole $]s-\eta, s+\eta[$. Hence $\gamma([0, s+\eta[) \subset S$, contradicting the maximality of s.

Sheet $\Longrightarrow$ Representation. Let $v : \Sigma \to \mathbb{R}^N$ be a real analytic embedding of an n-dimensional real analytic manifold such that $S = v(\Sigma)$ is a sheet of an algebraic subvariety V with minimal field of definition $\mathbb{F}$. Pick now a point $q \in S$ for which there is neighborhood U with $U \cap S = V \cap U$. By Corollary 15 there must necessarily be a point $p \in V \cap U = S \cap U$ with $m := \dim(V) = \dim_V(p) \geq n$. By Proposition 14(c) there is an algebraic subvariety $W \subset V$ with algebraic dimension m containing p and with field of definition $\mathbb{F}' \subset \mathbb{F}$. Note that by the latter property we must necessarily have $\dim_W(p) \geq m$ and thus p is a general point of W. Therefore, by Proposition 14(b) applied to W, there is a neighborhood $U' \subset U$ of p such that $U' \cap W$ is an m-dimensional connected submanifold: since $U' \cap S = U' \cap V \supset U' \cap W$ and S is an n-dimensional submanifold, $m = n$ and there is a neighborhood of p where W and S coincide.

We claim now that $v(\Sigma) = S \subset W$. Fix $p' \in S$: we know that there is an analytic function $\gamma : [0, 1] \to S$ such that $\gamma(0) = p$ and $\gamma(1) = p'$. If P is a polynomial of N variables which vanishes on W, then $P \circ \gamma$ vanishes on a neighborhood of 0. Since $P \circ \gamma$ is real analytic, it must thus vanish on the whole interval $[0, 1]$ and thus $P(p') = 0$. This shows that p' is a zero of any polynomial which vanishes on W, which implies that $p' \in W$. From the very definition of sheet, it follows that S is not only a sheet of the subvariety V, but also a sheet of the subvariety W.

Having established that $v(\Sigma)$ is a sheet of an n-dimensional subvariety of $\mathbb{R}^N$, it follows that any collection of $n+1$ functions chosen among the coordinates $v_1, \ldots, v_N$ must satisfy a nontrivial polynomial relation. Thus $\mathscr{B} := \{v_1, \ldots, v_N\}$ is a basic set. Now consider the ring $\mathscr{R}'$ of real analytic functions generated by $\mathscr{B}$: such ring obviously satisfies the requirements (a) and (c) of Definition 3. Choosing a maximal one (among those satisfying these two requirements and containing $\mathscr{B}$) we achieve the desired structure $(\Sigma, \mathscr{R})$ of which v is a representation.

2.5 *Proof of the Existence of Representations and of the Approximation Theorem*

The proofs of the two theorems follow indeed the same path and will be given at the same time. Before coming to them we need however the following very important lemma.

Lemma 18 *Let Q and R be two monic polynomials in one variable of degrees d_1 and d_2 with real coefficients and no common factors. Let $P = QR$ be their product. Then any monic polynomial $\tilde{P}$ of degree $d = d_1 + d_2$ with real coefficients in a suitable neighborhood U of P can be factorized in two monic polynomials $\tilde{Q}$ and $\tilde{R}$ of degrees d_1 and d_2, with real coefficients and which lie near Q and R respectively.*

Such decomposition is unique and the coefficients of the polynomials of each factor depend analytically upon those of $\tilde{P}$.

Proof First of all we show that the decomposition is unique. Note that two polynomials have no common factors if and only if they have no (complex) root in common. Let $z_1, \ldots, z_{d_1}$ be the roots of Q and $w_1, \ldots, w_{d_2}$ those of R (with repetitions, accounting for multiplicities). If $\tilde{P}$ is close to $P = QR$, then its roots will be close to $z_1, \ldots z_{d_1}, w_1, \ldots, w_{d_2}$ and thus they can be divided in unique way in two groups: d_1 roots close to the roots of Q and d_2 roots close to those of R. Clearly the zeros of the factor $\tilde{Q}$ must be close to those of Q and thus $\tilde{Q}$ is uniquely determined, which in turn determines also the other factor $\tilde{R}$. Note moreover that the coefficients of both $\tilde{Q}$ and $\tilde{R}$ must be real: it suffices to show that if a (nonreal) root ζ of $\tilde{P}$ is a root of $\tilde{Q}$, then its complex conjugate $\bar{\zeta}$ is also a root of $\tilde{Q}$. Indeed, either ζ is close to a real root of Q, in which case $\bar{\zeta}$ is close to the same root, or ζ is close to a nonreal root of z_i of Q, in which case $\bar{\zeta}$ is close to $\bar{z}_i$, which must be a root of Q because Q has real coefficients.

In order to show the existence and the real analytic dependence, set

$$Q(x) = x^{d_1} + \sum_{i=1}^{d_1} a_i x^{d_1 - i},$$

$$R(x) = x^{d_2} + \sum_{i=1}^{d_2} b_i x^{d_2 - i},$$

$$P(x) = x^{d} + \sum_{i=1}^{d} c_i x^{d - i}.$$

We then desire to find a neighborhood U of $c = (c_1, \ldots, c_d) \in \mathbb{R}^d$ and a real analytic map $(\alpha, \beta) : U \to \mathbb{R}^{d_1} \times \mathbb{R}^{d_2}$ with the properties that

(a) $x^d + \sum_i \tilde{c}_i x^{d-i} = (x^{d_1} + \sum_j \alpha_j(\tilde{c}) x^{d_1 - j})(x^{d_2} + \sum_k \beta_k(\tilde{c}) x^{d_2 - k})$ for any $\tilde{c} \in U$;
(b) $\alpha(c) = a$ and $\beta(c) = b$.

Given α, β vectors in some neighborhoods U_1 and U_2 of a and b, let $Q_\alpha := x^{d_1} + \sum_j \alpha_j x^{d_1 - j}$, $R_\beta := x^{d_2} + \sum_k \beta_k x^{d_2 - k}$ and $Q_\alpha R_\beta = x^d + \sum_i \gamma_i x^{d-i}$. This defines a real analytic (in fact polynomial!) map $U_1 \times U_2 \ni (\alpha, \beta) \mapsto \gamma(\alpha, \beta) \in \mathbb{R}^d$ with the property that $\gamma(a, b) = c$. Our claim will then follow from the inverse function theorem if we can show that the determinant of the matrix of partial derivatives of γ at the point (a, b) is nonzero. The latter matrix is however the Sylvester matrix of the two polynomials Q and R: the determinant of the Sylvester matrix of two polynomials (called the resultant), vanishes if and only if the two polynomials have a common zero, see [2].

We are now ready to prove the two main theorems, namely Theorems 9 and 8.

Proof (Proof of Theorem 9) We start with Theorem 9 and consider therefore a smooth embedding $w : \Sigma \to \mathbb{R}^m$ of a smooth closed connected manifold Σ of dimension n. By Whitney's theorem we can assume, without loss of generality, that w is real analytic. Consider now a tubular neighborhood $U := U_{4\delta}(\Sigma)$ so that the nearest point projection $x \mapsto \pi(x) \in \Sigma$ is real analytic on U and let $v : U \to \mathbb{R}^m$ be the function $v(x) := \pi(x) - x$. For each x let also $T_{\pi(x)}\Sigma$ be the n-dimensional tangent space to Σ at $\pi(x)$ (considered as a linear subspace of $\mathbb{R}^m$) and let $\xi \mapsto \mathbf{K}(x)\xi$ be the orthogonal projection from $\mathbb{R}^m$ onto $T^{\perp}_{\pi(x)}\Sigma$, namely the orthogonal complement of the tangent $T_{\pi(x)}\Sigma$. We therefore consider $\mathbf{K}(x)$ to be a symmetric $m \times m$ matrix with coefficients which depend analytically upon x. Let next u and $\mathbf{L}$ be two maps with polynomial dependence on x which on $U_{3\delta}(\Sigma)$ approximate well the maps v and $\mathbf{K}$. More precisely

(i) $\mathbf{L}(x)$ is an $m \times m$ symmetric matrix for every x, with entries which are polynomial functions of the variable x; similarly the components of $u(x)$ are polynomial functions of x;
(ii) $\|u - v\|_{C^N(U_{3\delta}(\Sigma))} + \|\mathbf{L} - \mathbf{K}\|_{C^N(U_{3\delta}(\Sigma))} \leq \eta$, where N is a large natural number and η a small real number, whose choices will be specified later.

The characteristic polynomial of $\mathbf{K}$ is $P(\lambda) = (\lambda - 1)^{m-n}\lambda^n$. We can then apply Lemma 18 and, assuming η is sufficiently small, the characteristic polynomial $P_x(\lambda)$ of $\mathbf{L}(x)$ can be factorized as $Q_x(\lambda)R_x(\lambda)$ where

(iii) $R_x(\lambda)$ is close to λ^n;
(iv) $Q_x(\lambda)$ is close to $(\lambda - 1)^{m-n}$;
(v) The coefficients of R_x and Q_x depend analytically upon x.

It turns out that both Q_x and R_x have all real roots (since $\mathbf{L}(x)$ is symmetric, its eigenvalues are all real). Moreover, the eigenvectors with eigenvalues which are roots of R_x span an n-dimensional vector subspace $\tau(x)$ of $\mathbb{R}^m$ which is close to $T_{\pi(x)}(\Sigma)$. On the other hand the eigenvectors with eigenvalues which are roots of Q_x span the orthogonal of $\tau(x)$, which we will denote by $\tau(x)^{\perp}$ (recall that $\mathbf{L}(x)$ is a symmetric matrix!). Consider next the symmetric matrix $\mathbf{P}(x) = R_x(\mathbf{L}(x))$. Then the kernel of the linear map $\xi \mapsto \mathbf{P}(x)\xi$ is $\tau(x)$. Moreover $|\mathbf{P}(x)\xi - \xi| \leq C\eta|\xi|$ for every $\xi \in \tau(x)^{\perp}$, where C is only a dimensional constant: this happens because $\mathbf{P}(x)$ is close to $(\mathbf{K}(x))^n$, whose linear action on $T^{\perp}_{\pi(x)}\Sigma$ is the identity.

Consider now the map

$$z(x) := x + v(x) - \underbrace{\mathbf{K}(x)\mathbf{P}(x)u(x)}_{=:\psi(x)} .$$

The map $x \mapsto z(x)$ is clearly real analytic on $U_{2\delta}$. Moreover, as $\eta \downarrow 0$, the map $v - \psi$ converges to $x \mapsto v(x) - \mathbf{K}(x)\mathbf{K}(x)v(x) = 0$, because $\mathbf{K}(x)v(x) = v(x)$. The latter convergence is in C^N. Since N is larger than 1, for η sufficiently close to 0 this will imply the local invertibility of the function z. In fact, by the inverse function theorem and compactness of $\overline{U_{3\delta}(\Sigma)}$ we conclude the existence of a $\sigma > 0$ and an

η_0 such that, if $\eta < \eta_0$, then z is injective in $B_\sigma(y)$ for every $y \in U_{2\delta}(\Sigma)$. Then, choosing $\eta < \min\{\eta_0, \sigma/(3C)\}$ for a suitable dimensional constant C we conclude the global injectivity of z on $U_{2\delta}(\Sigma)$: if we have $z(x) = z(x')$ and $x \neq x'$, then necessarily $|x - x'| \geq \sigma$. On the other hand the C^0 norm of the difference between z and the identity map is given by $C\eta$ and thus we can estimate

$$|z(x) - z(x')| \geq |x - x'| - |z(x) - x| - |z(x') - x'| \geq \sigma - \frac{2\sigma}{3}.$$

Finally, by possibly choosing η even smaller, we can assume that $U_\delta(\Sigma)$ is contained in $z(U_{2\delta}(\Sigma))$.

Let now z^{-1} be the inverse of z on $U_\delta(\Sigma)$, which is analytic by the inverse function theorem. We claim that the real analytic subvariety $\Gamma = z^{-1}(\Sigma)$ is a sheet of an algebraic subvariety: this would complete the proof of Theorem 9, provided N is large enough and η small enough.

Note now that, for any choice of x, $x + v(x) = \pi(x)$ belongs to Σ and $\psi(x)$ is orthogonal to $T_{\pi(x)}\Sigma$, by definition of $\mathbf{K}(x)$. Hence $z(x)$ belongs to Σ if and only if $\psi(x) = 0$. We conclude therefore that Γ is indeed the set where ψ vanishes. Recall moreover that, choosing η sufficiently small, $\mathbf{P}(x)u(x)$ belongs to the plane $\tau(x)^\perp$ which is close to $T^\perp_{\pi(x)}\Sigma$: hence $\mathbf{K}(x)\mathbf{P}(x)u(x) = 0$ is equivalent to the condition $\mathbf{P}(x)u(x) = 0$. Γ is therefore the zero set of

$$R_x(\mathbf{L}(x))u(x) = 0.$$

Note however that the coefficients of the polynomial $R_x(\lambda)$ are just analytic functions of x and *not* polynomial functions of x: it is therefore not obvious that Γ is a sheet of an algebraic subvariety. From now on we let $\phi(x)$ be the map $R_x(\mathbf{L}(x))u(x)$.

Consider now $\mathbb{R}^{m+n}$ as a product of $\mathbb{R}^m$ with the linear space of polynomials of degree n and real coefficients in the unknown λ. In other words, to every point $(x, a) \in \mathbb{R}^{m+n}$ we associate the pair $x \in \mathbb{R}^n$ and $p_a(\lambda) = \lambda^n + a_1\lambda^{n-1} + \ldots + a_n$. For any (x, a) consider the polynomial $q_{x,a}(\lambda)$ which is the remainder of the division of $P_x(\lambda)$, the characteristic polynomial of $\mathbf{L}(x)$, by the polynomial $p_a(\lambda)$. In particular, let $\eta_j(x, a)$ be the coefficients of $q_{x,a}$, namely

$$q_{x,a}(\lambda) = \eta_1(x, a)\lambda^{n-1} + \eta_2(x, a)\lambda^{n-2} + \ldots + \eta_n(x, a).$$

The corresponding map $(x, a) \mapsto \eta(x, a) = (\eta_1(x, a), \ldots, \eta_n(x, a))$ is a polynomial map, because the coefficients of $P_x(\lambda)$ depend polynomially on x! For any element (x, a), define $\varphi(x, a) := p_a(\mathbf{L}(x))u(x)$ and consider thus the system of polynomial equations

$$\begin{cases} \eta(x, a) = 0 \\ \varphi(x, a) = 0 \end{cases} \tag{3}$$

Such system defines a real algebraic subvariety V of $\mathbb{R}^{m+n}$. Now, consider the analytic map $x \mapsto \Psi(x) = (x, R_x) \in \mathbb{R}^{m+n}$. Since the remainder of the division of P_x by R_x is 0, we clearly have $\eta(\Psi(x)) = 0$. Moreover, since $\varphi(x, R_x) = \phi(x)$, we conclude that $\Psi(\Gamma)$ is a subset of the set of solutions of (3), namely a subset of V. Moreover $\Psi(\Gamma)$ is a real analytic embedding of Γ and hence also a real analytic embedding of Σ. We next claim that $\Psi(\Gamma)$ is in fact an isolated sheet of V. The only thing we need to show is that in a neighborhood of $\Psi(\Gamma)$ the only solutions of (3) must be images of Γ through Ψ. If (x', a) is a zero of (3) near an element of $(x, R_x) \in \Psi(\Gamma)$, it then follows that the polynomial p_a must be close to the polynomial R_x and must be a factor of $P_{x'}$. Recall however that $R_x(\lambda)$ is close to the polynomial λ^n and, by Lemma 18, nearby λ^n there is a unique factor of $P_{x'}$ which is a monic polynomial of degree n close to λ^n: such factor is $R_{x'}$! This implies that $p_a = R_{x'}$ and hence that $\varphi(x', a) = \phi(x')$. But then $\phi(x') = 0$ implies that $x' \in \Gamma$, which completes the proof that $\Psi(\Gamma)$ is an isolated sheet of the real algebraic subvariety V of $\mathbb{R}^{m+n}$.

In particular, $\Psi(\Gamma)$ is a proper representation, by Proposition 7. But Γ is a projection of such representation, which is still an analytic submanifold and thus it is easy to see that Γ is also a representation of Σ: namely the components of $z^{-1} : \Sigma \to \mathbb{R}^n$ give a basic set $\mathscr{B}$ of Σ and, using the same procedure of the proof of Proposition 7 we can find a Nash ring $\mathscr{R}$ containing $\mathscr{B}$, concluding the proof of Theorem 9.

Proof (Proof of Theorem 8*)* Fix a connected smooth closed differentiable manifold of dimension n and, following the previous proof, consider the isolated sheet $\Psi(\Gamma)$ of the algebraic subvariety V of $\mathbb{R}^{m+n}$ constructed above. We next use the classical projection argument of Whitney, cf. [105], to show that, if π is the orthogonal projection of $\mathbb{R}^{m+n}$ onto a generic (in the sense of Baire category) $2n+1$-dimensional subspace of $\mathbb{R}^{m+n}$, $\pi(\Psi(\Gamma))$ is still a submanifold, it is a connected component of $\pi(V)$ and that $\pi(V)$ is a an algebraic subvariety[3] of $\mathbb{R}^{2n+1}$. The latter claim would then give a proper representation in $\mathbb{R}^{2n+1}$ and would thus show Theorem 8.

In order to accomplish this last task, we first observe that it suffices to show the existence of a projection onto an hyperplane, provided $m + n > 2n + 1$: we can then keep reducing the dimension of the ambient Euclidean space until we reach $2n + 1$. Next, for each hyperplane $\tau \subset \mathbb{R}^{m+n}$ we denote by P_τ the orthogonal projection onto it. The classical argument of Whitney implies that:

(a) For a dense open subset of τ's in the Grassmanian G of hyperplanes of $\mathbb{R}^{m+n}$ the map P_τ restricted on $\Psi(\Gamma)$ is an immersion (i.e. its differential has full rank at every $p \in \Psi(\Gamma)$).
(b) For a generic subset of τ's, P_τ is injective on $\Psi(\Gamma)$.

[3]The projection of an algebraic subvariety is not always an algebraic subvariety: here as well we are taking advantage of the genericity of the projection.

Thus for a dense open subset of τ's, $P_\tau \circ \Psi$ is an embedding of Γ. However, note that point (b) cannot be obviously extended to give injectivity of P_τ on the whole subvariety W, because $W \setminus \Psi(\Gamma)$ is not necessarily a submanifold. We claim that, nonetheless,

$$P_\tau(\Psi(\Gamma)) \cap P_\tau(W \setminus \Psi(\Gamma)) = \emptyset \qquad \text{for } \tau \text{ in a dense open subset of } G. \tag{4}$$

Indeed, by Proposition 14, we know that $W \setminus \Psi(\Gamma)$ can be covered by countably many submanifolds W_i, of dimension $d_i \le n$. Without loss of generality we can assume that each W_i is compact, has smooth boundary and does not intersect $\Psi(\Gamma)$. Consider the map $\Psi(\Gamma) \times W_i \ni (x, y) \mapsto z(x, y) := \frac{x-y}{|x-y|}$. Since z is smooth, $z(\Psi(\Gamma) \times W_i)$ is a (closed) set of Hausdorff dimension at most $n + d_i \le 2n < m+n-1$ and thus it is meager. In particular we conclude that the set $K := z(\Psi(\Gamma) \times (W \setminus \Psi(\Gamma)))$ is a countable union of meager sets and thus a set of first category. Hence the set $U \subset \mathbb{S}^{m+n-1}$ of points p for which neither p nor $-p$ belongs to K is a generic subset of $\mathbb{S}^{m+n-1}$. Clearly, the set of hyperplanes τ orthogonal to $\{p, -p\} \subset U$ is a generic subset of hyperplanes for which $P_\tau(\Psi(\Gamma)) \cap P_\tau(W \setminus \Psi(\Gamma)) = \emptyset$.

Finally, it is a classical fact in real algebraic geometry that, for a generic subset of τ, $P_\tau(W)$ is a real algebraic subvariety. Nash refers to the "classical algebraic geometrical method of generic linear projection", cf. [73, p. 415]. However, it is possible to conclude the existence of a good projection directly with an algebraic variant of Whitney's argument.[4] For completeness we report this alternative possibility in the next two paragraphs.

Consider the complexification $W_{\mathbb{C}} \subset \mathbb{C}^{m+n}$ of W (i.e., $W_{\mathbb{C}}$ is the smallest complex algebraic subvariety of $\mathbb{C}^{m+n}$ containing W). We have that $W_{\mathbb{C}}$ has (real) dimension $2n$, $W = W_{\mathbb{C}} \cap \mathbb{R}^{m+n}$ and $\Psi(\Gamma)$ is contained in the set $W^*_{\mathbb{C}}$ of nonsingular points of $W_{\mathbb{C}}$: for any point $p \in \Psi(\Gamma)$ there is a neighborhood U of p in $\mathbb{C}^{m+n}$ such that $U \cap W_{\mathbb{C}}$ is the zero set of m polynomials with linearly independent gradients. We identify $\mathbb{P}^{m+n-1}(\mathbb{C})$ with the hyperplane at infinity of $\mathbb{C}^{m+n}$. Thus, we can consider $\mathbb{P}^{m+n}(\mathbb{C})$ as the union $\mathbb{C}^{m+n} \cup \mathbb{P}^{m+n-1}(\mathbb{C})$. For each nonzero vector τ of $\mathbb{C}^{m+n}$ we indicate by $[\tau]$ the corresponding point of $\mathbb{P}^{m+n-1}(\mathbb{C})$. Let S denote the set of all $[\tau]$ of the form $\tau = x - y$ with $x, y \in W_{\mathbb{C}}$ and $x \neq y$. Note that S has Hausdorff dimension at most $4n$ and the same is true for its closure[5] T in $\mathbb{P}^{m+n-1}(\mathbb{C})$. The set T contains all points at infinity of $W_{\mathbb{C}}$ (i.e. T contains the intersection between $\mathbb{P}^{m+n-1}(\mathbb{C})$ and the closure of $W_{\mathbb{C}}$ in $\mathbb{P}^{m+n}(\mathbb{C})$). It is immediate to verify that T contains also all the points $[\tau]$ such that τ is a nonzero vector of $\mathbb{C}^{m+n}$ tangent to the complex manifold $W^*_{\mathbb{C}}$ at some of its points. Since $2(m + n - 1) > 4n$, T turns out to be a proper (i.e. $T \subsetneq \mathbb{P}^{m+n-1}(\mathbb{C})$) complex algebraic subvariety of $\mathbb{P}^{m+n-1}(\mathbb{C})$. Thus, the subset $\mathbb{P}^{m+n-1}(\mathbb{R})$ of

[4]Many thanks to Riccardo Ghiloni for suggesting this argument, which follows closely the proof of [55, Lem. 3.2].

[5]Observe that in this context the closure in the Euclidean topology coincides with the Zariski closure.

$\mathbb{PP}^{m+n-1}(\mathbb{C})$ cannot be completely contained in T. Choose $[\nu] \in \mathbb{PP}^{m+n-1}(\mathbb{R}) \setminus T$. Denote by H the hyperplane of $\mathbb{R}^{m+n}$ orthogonal to τ and by $H_{\mathbb{C}} \subset \mathbb{C}^{m+n}$ its complexification.

Observe that the orthogonal projection $\rho : \mathbb{R}^{m+n} \to H$ extends to the projection $\rho_{\mathbb{C}} : \mathbb{C}^{m+n} \to H_{\mathbb{C}}$ which maps each point x into the unique point of the intersection between $H_{\mathbb{C}}$ and the projective line joining $[\nu]$ and x. Since $[\nu] \notin T$, the restriction $\rho'_{\mathbb{C}}$ of $\rho_{\mathbb{C}}$ to $W_{\mathbb{C}}$ is proper and injective, and it is an immersion on $W^*_{\mathbb{C}}$. In particular, $\rho'_{\mathbb{C}}(W_{\mathbb{C}})$ is a complex algebraic subvariety of $H_{\mathbb{C}}$ and $\rho'_{\mathbb{C}}(x)$ is a nonsingular point of $\rho'_{\mathbb{C}}(W_{\mathbb{C}})$ for each $x \in \Psi(\Gamma)$. It follows immediately that the restriction ρ' of ρ to W is an homeomorphism onto its image and it is a real analytic embedding on $\Psi(\Gamma)$. It remains to prove that $\rho'(W)$ is a real algebraic subvariety of H. It suffices to show that $\rho'(W) = \rho'_{\mathbb{C}}(W_{\mathbb{C}}) \cap H$ or, equivalently, that $\rho'_{\mathbb{C}}(W_{\mathbb{C}}) \cap H \subset \rho'(W)$. Let $x \in \rho'_{\mathbb{C}}(W_{\mathbb{C}}) \cap H$ and let $y \in W_{\mathbb{C}}$ with $\rho'_{\mathbb{C}}(y) = x$. We must prove that $y \in \mathbb{R}^{m+n}$. Note that the conjugate point $\overline{y}$ of y belongs to $W_{\mathbb{C}}$, because $W_{\mathbb{C}}$ can be described by real polynomial equations. In this way, since $[\nu]$ is real (i.e. $[\nu] \in \mathbb{PP}^{m+n-1}(\mathbb{R})$), $\rho'_{\mathbb{C}}(\overline{y}) = \overline{x} = x = \rho'_{\mathbb{C}}(y)$. On the other hand, $\rho'_{\mathbb{C}}$ is injective and hence $y \in \mathbb{R}^{m+n}$ as desired.

2.6 *Proof of the Uniqueness of the Nash Ring*

We finally turn to Theorem 10. Let $(\Sigma, \mathscr{R}_1)$ and $(\Gamma, \mathscr{R}_2)$ be two structures of Nash manifolds on two diffeomorphic manifolds and consider two corresponding proper representations $v_1 : \Sigma \to \mathbb{R}^{n_1}$ and $v_2 : \Gamma \to \mathbb{R}^{n_2}$. Let $\alpha : \Gamma \to \Sigma$ be a diffeomorphism and, using Whitney's theorem, assume without loss of generality that α is real analytic and define $a := v_1 \circ \alpha \circ v_2^{-1}$ on $v_2(\Gamma)$. Consider a neighborhood $U_\delta(v_2(\Gamma))$ where the nearest point projection π_2 on $v_2(\Gamma)$ is real analytic and let $w := a \circ \pi_2$: w is a real analytic mapping from $U_\delta(v_2(\Gamma))$ onto $v_1(\Sigma)$. We can then approximate w in C^1 with a map z whose coordinate functions are polynomials. If the approximation is good enough, we can assume that w takes values in a neighborhood U_η of $v_1(\Sigma)$ where the nearest point projection π_1 is real analytic and well defined. Now the nearest point projection $\pi_1(y)$ of a point y onto $v_1(\Sigma)$ is in fact characterized by the orthogonality of $y - \pi_1(y)$ to the tangent space to $v_1(\Sigma)$ at $\pi_1(y)$. It is easy to see that this is a set of polynomial conditions when $v_1(\Sigma)$ is, as in this case, a smooth real algebraic submanifold. Thus π_1 is an algebraic function. Hence $\zeta := \pi_1 \circ z$ is also an algebraic function. If z is close enough to w in the C^k norm, then the restriction of z to $v_2(\Gamma)$ will be close enough to a in the C^1 norm: in particular when this norm is sufficiently small the restriction of z to $v_2(\Sigma)$ must be a diffeomorphism of $v_2(\Gamma)$ with $v_1(\Sigma)$. By the implicit function theorem, the inverse will also be real analytic. Since, however, z is algebraic, its inverse will also be algebraic. Thus z gives the desired isomorphism between the two algebraic structures.

3 C^1 Isometric Embeddings

3.1 *Introduction*

Consider a smooth n-dimensional manifold Σ with a smooth Riemannian tensor g on it. If $U \subset \Sigma$ is a coordinate patch, we write g as customary in local coordinates:

$$g = g_{ij} dx_i \otimes dx_j ,$$

where we follow the Einstein's summation convention. The smoothness of g means that, for any chart of the smooth atlas, the coefficients g_{ij} are C^∞ functions.

An isometric immersion (resp. embedding) $u : \Sigma \to \mathbb{R}^n$ is an immersion (resp. embedding) which preserves the length of curves, namely such that

$$\ell_g(\gamma) = \ell_e(u \circ \gamma) \qquad \text{for any } C^1 \text{ curve } \gamma : I \to \Sigma.$$

Here $\ell_e(\eta)$ denotes the usual Euclidean length of a curve η, namely

$$\ell_e(\eta) = \int |\dot{\eta}(t)| \, dt ,$$

whereas $\ell_g(\gamma)$ denotes the length of γ in the Riemannian manifold (Σ, g): if γ takes values in a coordinate patch $U \subset \Sigma$ the explicit formula is

$$\ell_g(\gamma) = \int \sqrt{g_{ij}(\gamma(t))\dot{\gamma}_i(t)\dot{\gamma}_j(t)} \, dt . \tag{5}$$

The existence of isometric immersions (resp. embeddings) is a classical problem, whose formulation is attributed to the Swiss mathematician Schläfli, see [88]. At the time of Nash's works [73, 75] comparatively little was known about the existence of such maps. Janet [54], Cartan [15] and Burstin [14] had proved the existence of local isometric embeddings in the case of analytic metrics. For the very particular case of 2-dimensional spheres endowed with metrics of positive Gauss curvature, Weyl in [103] had raised the question of the existence of isometric embeddings in $\mathbb{R}^3$. The Weyl's problem was solved by Lewy in [62] for analytic metrics and, only shortly before Nash's work, another brilliant young mathematician, Louis Nirenberg, had settled the case of smooth metrics (in fact C^4, see Nirenberg's PhD thesis [81] and the note [82]); the same problem was solved independently by Pogorolev [85], building upon the work of Alexandrov [3] (see also [86]).

In his two papers on the topic written in the 1950s (he wrote a third contribution in the 1960s, cf. [78]), Nash completely revolutionized the subject. He first proved a very counterintuitive fact which shocked the geometers of his time, namely the existence of C^1 isometric embeddings in codimension 2 in the absence of topological obstructions. He then showed the existence of smooth embeddings in sufficiently high codimension, introducing his celebrated approach to "hard implicit function theorems". In this section we report the main statements and the arguments of the first paper [73].

We start by establishing the following useful notation. First of all we will use the Einstein summation convention on repeated indices. We then will denote by e the standard Euclidean metric on $\mathbb{R}^N$, which in the usual coordinates is expressed by the tensor

$$\delta_{ij} dx_i \otimes dx_j \, .$$

If $v : \Sigma \to \mathbb{R}^N$ is an immersion, we denote by $v^\sharp e$ the pull-back metric on Σ. When $U \subset \Sigma$ is a coordinate patch, the pull-back metric in the local coordinates is then given by

$$v^\sharp e = (\partial_i v \cdot \partial_j v) dx_i \otimes dx_j \, ,$$

where $\partial_i v$ is the i-th partial derivative of the map v and $\cdot$ denotes the usual Euclidean scalar product. The obvious necessary and sufficient condition in order for a C^1 map u to be an isometry is then given by $u^\sharp e = g$, which amounts to the identities

$$g_{ij} = \partial_i u \cdot \partial_j u \, . \tag{6}$$

Note that this is a system of $\frac{n(n+1)}{2}$ partial differential equations in N unknowns (if the target of u is $\mathbb{R}^N$).

In order to state the main theorems of Nash's 1954 note, we need to introduce the concept of "short immersion".

Definition 19 (Short maps) Let (Σ, g) be a Riemannian manifold. An immersion $v : \Sigma \to \mathbb{R}^N$ is short if we have the inequality $v^\sharp e \leq g$ in the sense of quadratic forms: more precisely $h \leq g$ means that

$$h_{ij} w^i w^j \leq g_{ij} w^i w^j \qquad \text{for any tangent vector } w. \tag{7}$$

Analogously we write $h < g$ when (7) holds with a *strict* inequality for any nonzero tangent vector. Hence, if the immersion $v : \Sigma \to \mathbb{R}^N$ satisfies the inequality $v^\sharp e < g$, we say that it is *strictly* short.

Using (5) we see immediately that a short map shrinks the length of curves, namely $\ell_e(v(\gamma)) \leq \ell_g(\gamma)$ for every smooth curve γ. The first main theorem of Nash's paper is then the following result.

Theorem 20 (Nash's C^1 isometric embedding theorem) *Let (Σ, g) be a smooth closed n-dimensional Riemannian manifold and $v : \Sigma \to \mathbb{R}^N$ a C^∞ short immersion with $N \geq n+2$. Then, for any $\varepsilon > 0$ there is a C^1 isometric immersion $u : \Sigma \to \mathbb{R}^N$ such that $\|u - v\|_{C^0} < \varepsilon$. If v is, in addition, an embedding, then u can be assumed to be an embedding as well.*

The closedness assumption can be removed, but the corresponding statement is slightly more involved and in particular we need the notion of "limit set".

Definition 21 (Limit set) Let Σ be a smooth manifold and $v : \Sigma \to \mathbb{R}^N$. Fix an exhaustion of compact sets $\Gamma_k \subset \Sigma$, namely $\Gamma_k \subset \Gamma_{k+1}$ and $\cup_k \Gamma_k = \Sigma$. The limit set of v is the collection of points q which are limits of any sequence $\{v(p_k)\}$ such that $p_k \in \Sigma \setminus \Gamma_k$.

Theorem 22 (C^1 isometric embedding, nonclosed case) *Let (Σ, g) be a smooth n-dimensional Riemannian manifold. The same conclusions of Theorem 20 can be drawn if the map v is short and its limit set does not intersect its image. Moreover, we can impose that the nearby isometry u has the same limit set as v if v is strictly short.*

Combined with the classical theorem of Whitney on the existence of smooth immersions and embeddings, the above theorems have the following corollary.

Corollary 23 *Any smooth n-dimensional Riemannian manifold has a C^1 isometric immersion in $\mathbb{R}^{2n}$ and a C^1 isometric embedding in $\mathbb{R}^{2n+1}$. If in addition the manifold is closed, then there is a C^1 isometric embedding*[6] *in $\mathbb{R}^{2n}$.*

Remark 24 In Nash's original paper the C^0 estimate of Theorem 20 is not mentioned, but it is an obvious outcome of the proof. Moreover, Nash states explicitly that it is possible to relax the condition $N \geq n + 2$ to the (optimal) $N \geq n + 1$ using more involved computations, but he does not give any detail. Indeed, such a statement was proved shortly after by Kuiper in [60], with a suitable adaptation of Nash's argument. The final result is then often called the Nash–Kuiper Theorem.

The Nash–Kuiper C^1 isometric embedding theorem is often cited as one of the very first instances of Gromov's h-principle, cf. [29, 39]. Note that it implies that any closed 2-dimensional oriented Riemannian manifold can be embedded in an arbitrarily small ball of the Euclidean 3-dimensional space with a C^1 isometry. This statement is rather striking and counterintuitive, especially if we compare it to the classical rigidity for the Weyl's problem (see the classical works of Cohn-Vossen and Herglotz [18, 47]): if Σ is a 2-dimensional sphere and g a C^2 metric with positive Gauss curvature, the image of every C^2 isometric embedding $u : \Sigma \to \mathbb{R}^3$ is the boundary of a convex body, uniquely determined up to rigid motions of $\mathbb{R}^3$. Nash's proof of Theorem 20 (and Kuiper's subsequent modification) generates indeed a C^1 isometry which has no further regularity. It is interesting to notice that a sufficiently strong Hölder continuity assumption on the first derivative is still enough for the validity of the rigidity statement in the Weyl's problem (see [10, 19]), whereas for a sufficiently low Hölder exponent α the Nash–Kuiper Theorem still holds in $C^{1,\alpha}$ (see [11, 19, 25]). The existence of a threshold exponent distinguishing between the two different behaviors in low codimension is a widely open problem, cf. [39, p. 219] and [107, Problem 27], which bears several relations with a well-

[6] Closed manifolds can be C^1 isometrically *immersed* in lower dimension: already at the time of Nash's paper this could be shown in $\mathbb{R}^{2n-1}$ (for $n > 1$!) using Whitney's immersion theorem. Nowadays we can use Cohen's solution of the immersion conjecture to lower the dimension to $n - a(n)$, where $a(n)$ is the number of 1's in the binary expansion of n, cf. [17].

known conjecture in the theory of turbulence, solved very recently with methods inspired by Nash's approach to Theorem 20, cf. [12, 13, 27, 51, 97].

3.2 Main Iteration

We start by noticing that Theorem 20 is a "strict subset" of Theorem 22: if Σ is closed, then the limit set of any map is empty. Moreover, the following simple topological fact will be used several times:

Lemma 25 *Let Σ be a differentiable n-dimensional manifold and $\{V_\lambda\}$ an open cover of Σ. Then there is an open cover $\{U_\ell\}$ with the properties that:*

(a) *each U_ℓ is contained in some V_λ;*
(b) *the closure of each U_ℓ is diffeomorphic to an n-dimensional ball;*
(c) *each U_ℓ intersects at most finitely many other elements of the cover;*
(d) *each point $p \in \Sigma$ has a neighborhood intersecting at most $n+1$ elements of the cover;*
(e) *$\{U_\ell\}$ can be subdivided into $n+1$ classes $\mathscr{C}_i$ consisting of pairwise disjoint U_ℓ's.*

Proof By a classical theorem Σ can be triangulated (see [104]) and by locally refining the triangulation we can assume that each simplex is contained in some V_λ. Denote by S such triangulation and enumerate its vertices as $\{S_i^0\}$, its 1-dimensional edges as $\{S_i^1\}$ and so on. Then take the barycentric subdivision of S and call it T (cf. Fig. 3). We notice the following facts:

(i) For each vertex S_i^0 consider the interior U_i^0 of the star of S_i^0 in the triangulation T, see Fig. 4 (recall that the star of S_i^0 is usually defined as the union of all simplices of the triangulation which contain S_i^0, cf. for instance [46, p. 178]). Observe that the U_i^0 are pairwise disjoint.

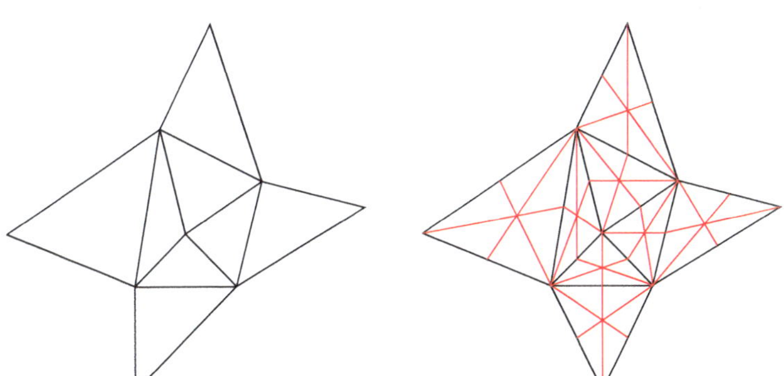

Fig. 3 A planar triangulation S and its barycentric subdivision T

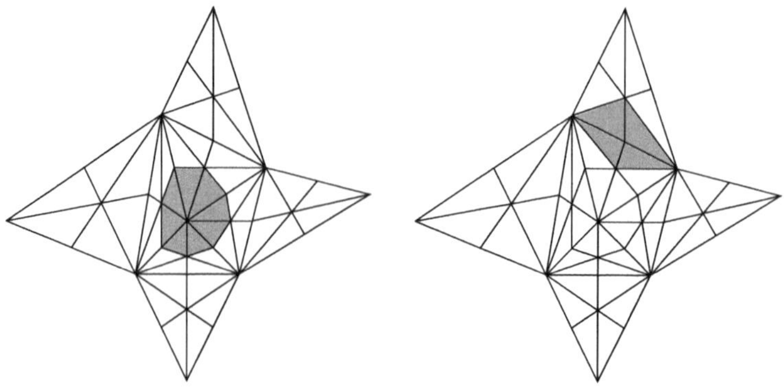

Fig. 4 The shaded area on the left depicts one of the sets U_i^0, whereas the shaded area on the right depicts one of the sets U_j^1

(ii) For each edge S_i^1 consider the interior U_i^1 of the star of S_i^1 in the triangulation T, see Fig. 4. The U_i^1 are pairwise disjoint. Moreover, observe that if $U_i^1 \cap U_j^0 \neq \emptyset$, then $S_j^0 \subset S_i^1$.

(iii) Proceed likewise up to $n-1$. Complete the collection $\{U_i^t : 0 \leq t \leq n-1\}$ with the interiors U_i^n of the n-dimensional simplices S_i^n of S and denote such final collection by $\mathscr{C}$.

The family $\mathscr{C}$ is obviously an open cover of the manifolds which satisfies (a) and (e) by construction. If two distinct elements U_i^s and U_j^t have nonempty intersection and $s \geq t$, then $s > t$ and S_j^t is a face of S_i^s: this implies that $\mathscr{C}$ satisfies (c). Each U_s^j is diffeomorphic to the open Euclidean n-dimensional ball, but its closure is only homeomorphic to the closed ball: however, it suffices to choose an appropriate smaller open set for each U_s^j to achieve an open cover which satisfies (b), while keeping (a), (c) and (e). In fact such open cover can be chosen so that a stronger version of (e) holds, namely that the closures of the elements of $\mathscr{C}_i$ are pairwise disjoint. Statement (d) is then obvious consequence of the latter property: for each class i we either have that p is contained in the closure of an element of $\mathscr{C}_i$, and hence there is a neighborhood of p which intersects only that element, or it is not contained in any of such closures, and hence it has a neighborhood that does not intersect any element of $\mathscr{C}_i$.

From now on we fix therefore a smooth manifold Σ as in Theorem 22 and a corresponding smooth atlas $\mathscr{A} = \{U_\ell\}$ (which is either finite or countably infinite) where the U_ℓ's have compact closure and satisfy the properties (b), (c), and (d) of Lemma 25.

Given any symmetric $(0,2)$ tensor h on Σ we write $h = h_{ij} dx_i \otimes dx_j$ and denote by $\|h\|_{0,U_\ell}$ the supremum of the Hilbert–Schmidt norm of the matrices $h_{ij}(p)$ for $p \in U_\ell$. Similarly, if $v : \Sigma \to \mathbb{R}^N$ is a C^1 map, we write $\|Dv\|_{0,U_\ell}$ for the supremum of the Hilbert–Schmidt norms of the matrices

$Dv(p) = (\partial_1 v(p), \ldots, \partial_n v(p))$, where $p \in U_\ell$. Finally, we set

$$\|h\|_0 := \sup_\ell \|h\|_{0,U_\ell}\,,$$
$$\|Dv\|_0 := \sup_\ell \|Dv\|_{0,U_\ell}\,.$$

We are now ready to state the main inductive statement[7] whose iteration will prove Theorem 22.

Proposition 26 (Iteration stage) *Let (Σ, g) be as in Theorem 22 and $w : \Sigma \to \mathbb{R}^N$ a smooth strictly short immersion. For any choice of positive numbers $\eta_\ell > 0$ and any $\delta > 0$ there is a smooth short immersion $z : \Sigma \to \mathbb{R}^N$ such that*

$$\|z - w\|_{0,U_\ell} < \eta_\ell \qquad \forall \ell\,, \tag{8}$$
$$\|g - z^\sharp e\|_0 < \delta\,, \tag{9}$$
$$\|Dw - Dz\|_0 < C\sqrt{\|g - w^\sharp e\|_0}\,, \tag{10}$$

for some dimensional constant C. If w is injective, then we can choose z injective.

Note that the right-hand side of (10) might be ∞ (because Σ is not necessarily compact), in which case the condition (10) is an empty requirement. We show first how to conclude Theorem 22 from the proposition above. Subsequently we close this section by proving Corollary 23. The rest of the section will then be dedicated to prove Proposition 26.

Proof (Proof of Theorem 22) Let $v_0 := v$ and ε be as in the statement and assume for the moment that v is an immersion. Moreover, without loss of generality we can assume that v is strictly short: it suffices to multiply v by a constant smaller than (but sufficiently close to) 1. Note that such operation will change the limit set of v, which explains why in the last claim of the theorem we assume directly that v is strictly short.

We will produce a sequence of maps v_q by applying iteratively Proposition 26. Since the limit set of v is closed and $v(\overline{U}_\ell)$ compact, there is a positive number β_ℓ such that any point of $v(\overline{U}_\ell)$ is at distance at least β_ℓ from the limit set of v. We then define the numbers

$$\bar{\eta}_{q,\ell} := 2^{-q-1}\min\{\varepsilon, \beta_\ell, 2^{-\ell}\}\,,$$
$$\delta_q := 4^{-q}\,.$$

[7] This is what Nash calls "a stage", cf. [73, p. 391].

At each $q \geq 1$ we apply Proposition 26 with $w = v_{q-1}$, $\eta_\ell = \bar{\eta}_{q,\ell}$, and $\delta = \delta_q$ to produce $z =: v_q$. We then conclude immediately that:

(a) $\|v_q - v_{q-1}\|_0 \leq 2^{-q-1}\varepsilon$ and thus v_q converges uniformly to some u with $\|u - v\|_0 \leq \varepsilon \sum_{q\geq 1} 2^{-q-1} = \frac{\varepsilon}{2}$;
(b) similarly $\|v - u\|_{0,U_\ell} \leq \beta_\ell \sum_{q\geq 1} 2^{-q-1} = \frac{\beta_\ell}{2}$;
(c) again by a similar computation $\|u - v\|_{0,U_\ell} \leq 2^{-\ell}$ and thus the limit set of u coincides with the limit set of v; combined with the estimate above, this implies that the limit set of u does not intersect the image of u;
(d) $\|Dv_q - Dv_{q-1}\|_0 \leq C2^{-q+1}$ for every $q \geq 2$ and thus u is a C^1 map (observe that we claim no bound on $\|Dv_1 - Dv_0\|_0$; on the other hand we do not need it!);
(e) since v_q converges to u in C^1, we have $g - u^\sharp e = \lim_q (g - v_q^\sharp e) = 0$ and thus u is an isometry, from which we also conclude that the differential of u has everywhere full rank and hence u is an immersion.

It remains to show that, if v is injective, then the iteration above can be arranged so to guarantee that u is also injective. To this aim, notice first that all the conclusions above certainly hold in case we implement the same iteration applying Proposition 26 with parameters $\eta_{q,\ell}$ smaller than $\bar{\eta}_{q,\ell}$. Moreover the proposition guarantees the injectivity at each step: we just need to show that the limit map is also injective. For each q consider the compact set $V_q := \cup_{\ell\leq q}\overline{U}_\ell$ and the positive numbers

$$2\gamma_i := \min\{|v_i(x) - v_i(y)| : d(x,y) \geq 2^{-i}, x, y \in V_i\} \qquad \text{for } i < q,$$

where d is the geodesic distance induced by the Riemannian metric g. We then set $\eta_{q,\ell} := \min\{\bar{\eta}_{q,\ell}, 2^{-q-1}\gamma_1, 2^{-q-1}\gamma_2, \ldots, 2^{-q-1}\gamma_{q-1}\}$ and apply the iteration as above with $\eta_{q,\ell}$ in place of $\bar{\eta}_{q,\ell}$. We want to check that the resulting u is injective. Fix $x \neq y$ in Σ and choose $q \geq 1$ such that $2^{-q} \leq d(x,y)$ and $x, y \in V_q$. We can then estimate

$$|u(x)-u(y)| \geq |v_q(x)-v_q(y)| - \sum_{k\geq q} \|v_{k+1} - v_k\|_{0,V_q} \geq 2\gamma_q - \sum_{k\geq q} 2^{-k-1}\gamma_q \geq \gamma_q > 0\,.$$

Hence $u(x) \neq u(y)$. The arbitrariness of x and y shows that u is injective and completes the proof.

Proof (Proof of Corollary 23) Recall that, according to Whitney's embedding theorem in its strong form (see [106]), any smooth differentiable manifold Σ of dimension n can be embedded in $\mathbb{R}^{2n}$. If the manifold in addition is closed, then it suffices to multiply the corresponding map by a sufficiently small positive constant to make it short and the existence of a nearby C^1 isometry with the desired property follows from Theorem 20.

The general case requires somewhat more care. Fix a smooth Riemannian manifold (Σ, g) of dimension n, *not closed*. Below we will produce a suitable smooth embedding $z : \Sigma \to \mathbb{R}^N$ for $N = (n+1)(n+2)$, with the additional properties that

(i) z is a short map;
(ii) the limit set of z is $\{0\}$ and does not intersect the image of z.

We then can follow the standard procedure of the proof of the Whitney's embedding theorem in its weak form (cf. [105]): if we consider the Grassmannian of $2n+1$ dimensional planes π of $\mathbb{R}^N$, we know that, for a subset of full measure, the projection P_π onto π is injective and has injective differential on $z(\Sigma)$. A similar argument shows that, for a set of planes π of full measure, $P_\pi(z(\Sigma))$ does not contain the origin. Since clearly $P_\pi \circ z$ is also short, the map $v := P_\pi \circ z$ satisfies the assumptions of Theorem 22. If we drop the injectivity assumption on π (namely we restrict to immersions), we can project on a suitable $2n$-dimensional plane.

Coming to the existence of z, we use the atlas $\{U_\ell\}$ of Σ given by Lemma 25 and we let $\Phi_\ell : U_\ell \to \mathbb{R}^n$ be the corresponding charts. Observe that, since Σ is not closed, the atlas is necessarily (countably) infinite. After further multiplying each Φ_ℓ by a positive scalar we can assume, without loss of generality, that $|\Phi_\ell| \leq 1$. Recall the $n+1$ classes $\mathscr{C}_i$ of Lemma 25(e). Consider then a family of smooth functions φ_ℓ, each supported in U_ℓ, with $0 \leq \varphi_\ell \leq 1$ and such that for any point $p \in \Sigma$ there is at least one φ_ℓ which is equal to 1 in some neighborhood of p. Finally, after numbering the elements of the atlas, we fix a vanishing sequence ε_ℓ of strictly monotone positive numbers, whose choice will be specified in a moment.

We are now ready to define our map z, which will be done specifying each component z_j. Fix $p \in \Sigma$ and $i \in \{1, \ldots, n+1\}$. If p does not belong to any element of $\mathscr{C}_i$, then we set $z_{(i-1)(n+2)+1}(p) = \ldots = z_{i(n+2)}(p) = 0$. Otherwise, there is a unique $U_\ell \in \mathscr{C}_i$ with $p \in U_\ell$ and we set:

$$z_{(i-1)(n+2)+j}(p) = \varepsilon_\ell^2 \varphi_\ell(p)(\Phi_\ell(p))_j \qquad \text{for } j \in \{1, \ldots, n\}, \tag{11}$$

$$z_{(i-1)(n+2)+n+1}(p) = \varepsilon_\ell^2 \varphi_\ell(p)\,, \tag{12}$$

$$z_{(i-1)(n+2)+n+2}(p) = \varepsilon_\ell \varphi_\ell(p)\,. \tag{13}$$

Now, for any point p there is at least one ℓ for which φ_ℓ is identically equal to 1 in a neighborhood of p: this will have two effects, namely that the differential of z at p is injective and that $z(p) \neq 0$. Since the limit set of z is obviously 0, condition (ii) above is satisfied. To prove that z is an embedding we need to show that z is injective. Fix two points p and q and fix a $U_\ell \in \mathscr{C}_i$ for which $\varphi_\ell(p) = 1$. If $q \in U_\ell$, then either $\varphi_\ell(q) \neq 1$, in which case $z_{(i-1)(n+2)+n+1}(p) \neq z_{(i-1)(n+2)+n+1}(q)$, or $\varphi_\ell(q) = 1$. In the latter case we then conclude $z(q) \neq z(p)$ because $\Phi_\ell(p) \neq \Phi_\ell(q)$. If $q \notin U_\ell$ and $\varphi_{\ell'}(q) = 0$ for any other $U_{\ell'} \in \mathscr{C}_i$, then $z_{(i-1)(n+2)+n+1}(q) = 0 \neq z_{(i-1)(n+2)+n+1}(p)$. Otherwise there is a $U_{\ell'} \in \mathscr{C}_i$ distinct from U_ℓ such that

$\varphi_{\ell'}(q) \neq 0$. In this case we have

$$\frac{z_{(i-1)(n+2)+n+1}(p)}{z_{i(n+2)}(p)} = \varepsilon_\ell \neq \varepsilon_{\ell'} = \frac{z_{(i-1)(n+1)+n+1}(q)}{z_{i(n+2)}(q)}.$$

Thus z is injective.

Finally, by choosing the ε_ℓ inductively appropriately small, it is easy to show that we can ensure the shortness of z.

3.3 *Decomposition in Primitive Metrics*

We will call "primitive metric"[8] any $(0, 2)$ tensor having the structure $a^2 d\psi \otimes d\psi$ for some pair of smooth functions a and ψ. Note that such two tensor is only positive semidefinite and thus it is certainly not a Riemannian metric. The next fundamental lemma shows that any Riemannian metric can be written as a (locally finite) sum of primitive metrics satisfying some additional technical requirements.

Proposition 27 *Let Σ be a smooth n-dimensional manifold, h a smooth positive definite $(0, 2)$ tensor on it and $\{U_\ell\}$ a cover of Σ. Then there is a countable collection h_j of primitive metrics such that $h = \sum_j h_j$ and*

(a) *Each h_j is supported in some U_ℓ.*
(b) *For any $p \in \Sigma$ there are at most*[9] *$K(n) = \frac{n(n+1)^2}{2}$ h_j's whose support contains p.*
(c) *The support of each h_j intersects the supports of at most finitely many other h_k's.*

Proof First of all, for each point $p \in \Sigma$ we find a neighborhood $V_p \subset U_\ell$ (for some ℓ) and $J(n) = \frac{n(n+1)}{2}$ primitive metrics $h_{p1}, \ldots, h_{pJ}$ on V_p such that $h = h_{p1} + \ldots + h_p$. In order to do this fix coordinates on $U_\ell \ni p$ and write h as $h = h_{ij} dx_i \otimes dx_j$. Consider the space $\mathrm{Sym}_{n\times n}$ of symmetric $n \times n$ matrices and let M be the matrix with entries $h_{ij}(p)$. Now, since the set of all matrices of the form $v \otimes v$ is a linear generator of $\mathrm{Sym}_{n\times n}$, there are J such matrices $A'_i = w_i \otimes w_i$ which are linearly independent. Consider $M' := \sum_i A'_i$. By standard linear algebra we can find a linear isomorphism L of $\mathbb{R}^n$ such that $L^T M' L = M$: indeed, since both M and M' are symmetric we can find O and O_1 orthogonal such that $D = O^T M O$ and $D_1 = O_1^T M' O_1$ are diagonal matrices. Since M and M' are both positive definite, the entries of D and D_1 are all positive. Let therefore $D^{-1/2}$ and $D_1^{-1/2}$ be the diagonal matrices whose entries are the reciprocal of the square roots of

---

[8] Although the term is nowadays rather common, it was not introduced by Nash, neither in [73] nor in the subsequent paper [75].

[9] In his paper Nash claims indeed a much larger $K(n)$, cf. [73, bottom of p. 386].

the entries of D and D_1, respectively. If we set $U := OD^{-1/2}$ and $U_1 := O_1 D_1^{-1/2}$, then clearly $U^T M U = U_1^T M' U_1$ is the identity matrix. Thus $L := U_1 U^{-1}$ is the linear isomorphism we were looking for. Having found L, if we set $A_i = L^T A_i' L = (Lw_i) \otimes (Lw_i) = v_i \otimes v_i$, we conclude that $M = \sum_i A_i$.

Next, there are unique linear maps $\mathscr{L}_i : \mathrm{Sym}_{n\times n} \to \mathbb{R}$ such that $A = \sum_i \mathscr{L}_i(A) v_i \otimes v_i$ for every A. Thus, if we consider the maps $\psi_i(x) = v_i \cdot x$ in local coordinates, we find smooth functions $\alpha_i : U_\ell \to \mathbb{R}$ such that

$$h = \sum_{i=1}^{J} \alpha_i d\psi_i \otimes d\psi_i .$$

Note that $\alpha_i(p) = \mathscr{L}_i(M) = 1$ for every $i \in \{1, \ldots, J\}$ and thus in a neighborhood V_p of p each α_i is the square of an appropriate smooth function a_i. The tensors $h_{pi} := a_i^2 d\psi_i \otimes d\psi_i$ are the required primitive metrics.

Finally we apply Lemma 25 and refine the covering V_p to a new covering W_ℓ with the properties listed in the lemma. For each W_ℓ we consider a $V_p \supset W_\ell$ and define the corresponding primitive metrics $h_{(\ell 1)} = h_{p1}, \ldots, h_{(\ell J)} = h_{pJ}$ (we use the subscript (ℓj) in order to avoid confusions with the explicit expression of the initial tensor h in a given coordinate system!). We then consider compactly supported functions $\beta_\ell \in C_c^\infty(W_\ell)$ with the property that for any point p there is at least a β_ℓ which does not vanish at p and we set

$$\varphi_\ell := \frac{\beta_\ell}{\sqrt{\sum_j \beta_j^2}} .$$

The tensors $\varphi_\ell^2 h_{(\ell j)}$ satisfy all the requirements of the proposition.[10]

3.4 *Proof of the Main Iterative Statement*

To complete the proof of the Proposition 26 we still need one technical ingredient.

Lemma 28 *Let B be a closed subset of $\mathbb{R}^n$ diffeomorphic to the n-dimensional closed ball and $\omega : B \to \mathbb{R}^N$ a smooth immersion with $N \geq n+2$. Then there are two smooth maps $\nu, b : B \to \mathbb{R}^N$ such that*

(a) $|\nu(q)| = |b(q)| = 1$ *and* $\nu(q) \perp b(q)$ *for every* $q \in B$;
(b) $\nu(q)$ *and* $b(q)$ *are both orthogonal to* $T_{\omega(q)}(\omega(B))$ *for every* $q \in B$.

[10] The argument of Nash is slightly different, since it covers the space of positive definite matrices with appropriate simplices.

Proof For any point p there exists a neighborhood of it and a pair of maps as above defined on the neighborhood: first select two orthonormal vectors $\nu(p)$ and $b(p)$ which are normal to $T_{\omega(p)}(\omega(B))$ and, by smoothness of ω, observe that they are almost orthogonal to $T_{\omega(q)}(\omega(B))$ for every q in a neighborhood of p. By first projecting on the normal bundle and then using the standard Gram–Schmidt orthogonalization procedure we then produce the desired pair. The problem of passing from the local statement to the global one can be translated into the existence of a suitable section of a fiber bundle: since B is topologically trivial, this is a classical conclusion.[11]

However, one can also use the following elementary argument.[12] We first observe that it suffices to produce ν and b continuous: we can then smooth them by convolution, project on the normal bundle, and use again a Gram–Schmidt procedure to produce a pair with the desired properties. We just have to ensure that the projection on the normal bundle still keeps the two vectors linearly independent at each point. Since ν and b are orthonormal and orthogonal to $\omega(B)$, this is certainly the case if the smoothings are ε-close to them in the uniform topology, where $\varepsilon > 0$ is a fixed geometric constant. Next, in order to show the existence of a continuous pair with properties (a) and (b), assume without loss of generality that $B = \overline{B}_1(0) \subset \mathbb{R}^n$ and consider the set R of all radii r for which there is at least one such pair on $\overline{B}_r(0)$. As observed above R is not empty. Let ρ be the supremum of R: we claim that $\rho \in R$. Indeed choose $\rho_k \in R$ with $\rho_k \uparrow \rho$ and let ν_k, b_k be two corresponding continuous maps on $\overline{B}_{\rho_k}(0)$ satisfying (a) and (b). We define $\tilde{\nu}_k$ and $\tilde{b}_k$ on B_1 by setting them equal to ν_k and b_k on $B_{\rho_k}(0)$ and extending them further by

$$\tilde{\nu}_k(x) = \nu_k\left(\rho_k \frac{x}{|x|}\right) \qquad \text{and} \qquad \tilde{b}_k(x) = b_k\left(\rho_k \frac{x}{|x|}\right) \qquad \text{for } |x| \geq \rho_k .$$

Note that the two maps satisfy (a). As for (b), by the smoothness of ω, for any $\eta > 0$ there is $\delta > 0$ such that, if $|x| \leq \rho_k + \delta$, then the angle between $\tilde{\nu}_k(x)$ (resp. $\tilde{b}_k(x)$) and the tangent space $T_{\omega(x)}\omega(B)$ is at least $\frac{\pi}{2} - \eta$. On the other hand, once η is smaller than a geometric constant, we can project $\tilde{\nu}_k$ and $\tilde{b}_k$ on the normal bundle and apply Gram–Schmidt to produce a continuous pair which satisfies the desired requirements on $\overline{B}_\sigma(0)$ for $\sigma_k = \min\{1, \rho_k + \delta\}$. Thus σ_k belongs to R. By definition $\rho \geq \sigma_k$ for every k: letting $k \uparrow \infty$ and using that $\rho_k \uparrow \rho$, we conclude $\rho \geq \min\{1, \rho + \delta\}$, namely $\rho = 1$. Thus $\sigma_k = 1$ for k large enough, which implies $1 \in R$ and concludes the proof.

Proof (Proof of Proposition 26*)* Fix a partition of unity φ_ℓ subordinate to U_ℓ. Now, each fixed U_ℓ intersects a finite number of other U_j's: denote the set of relevant indices by $I(\ell)$. We can therefore choose $\delta_\ell > 0$ in such a way that $(1 - \delta_\ell)g - w^\sharp e$

[11] Nash cites Steenrod's classical book, [94].

[12] Nash writes *Also they could be obtained by orthogonal propagation*, cf. [73, top of p. 387].

is positive definite and

$$\|\delta_\ell g\|_{0,U_j} < \frac{\delta}{2} \qquad \text{for every } j \in I(\ell). \tag{14}$$

Construct now the function $\varphi := \sum_\ell \delta_\ell \varphi_\ell$ and set $h := (1-\varphi)g - w^\sharp e$. Clearly

$$\|g - (h + w^\sharp e)\|_0 < \frac{\delta}{2} \tag{15}$$

and

$$g - (h + w^\sharp e) > 0\,. \tag{16}$$

In particular, if we choose δ'_ℓ appropriately and we impose that the final map z satisfies

$$\|z^\sharp e - (w^\sharp e + h)\|_{0,U_\ell} < \delta'_\ell \qquad \text{for every } \ell, \tag{17}$$

we certainly conclude that z is short and satisfies (9). Moreover, we will impose the stronger condition

$$\|Dw - Dz\|^2_{0,U_\ell} < 2K(n)^2 \|g - w^\sharp e\|_{0,U_\ell} \tag{18}$$

in place of (10), where $K(n)$ is the constant in Proposition 27. Hence from now on we focus on producing a map z satisfying the local conditions (8), (17), and (18).

Next, we apply Proposition 27 to write $h = \sum_j h_j$, where each h_j is a primitive metric and is supported in some U_ℓ. We assume the index j starts with 1 and follows the progression of natural numbers (note that the h_j's are either finite or countably infinite). Recall, moreover, that at any point of Σ at most $K(n)$ of the h_j's are nonzero and that, for any fixed j, only finitely many U_ℓ intersect the support of h_j, since the latter is a compact set: the corresponding set of indices will be denoted by $L(j)$. We next order the h_j's and we inductively add to the map w a smooth "perturbation" map w^p_j, whose support coincides with that of h_j. If we let $w_j := w + w^p_1 + \ldots + w^p_j$ be the "resulting map" after j steps, we then claim the following estimates:

$$\|w^p_j\|_{0,U_\ell} < \frac{\eta_\ell}{K(n)} \qquad \text{for all } \ell \in L(j), \tag{19}$$

$$\|Dw^p_j\|^2_{0,U_\ell} < 2\|h\|_{0,U_\ell} \qquad \text{for all } \ell \in L(j), \tag{20}$$

$$\|w^\sharp_j e - (w^\sharp_{j-1} e + h_j)\|_{0,U_\ell} < \frac{\delta'_\ell}{K(n)} \qquad \text{for all } \ell \in L(j). \tag{21}$$

We will prove below the existence of w_j^p, whereas we first show how to conclude. We set $z = w + \sum_j w_j^p$. Fix any U_ℓ and any point $q \in U_\ell$. Observe that, since $\overline{U}_\ell$ is compact, only finitely many perturbations w_j^p are nonzero in U_ℓ and thus z is smooth in U_ℓ. Next, note that at most $K(n)$ h_j's (and hence at most $K(n)$ w_j^p's) are nonzero at q. Thus we can sum up all the estimates in (19) and (20) to conclude

$$|w(q) - z(q)| \leq \sum_j \|w_j^p\|_{0,U_\ell} < \eta_\ell\,, \tag{22}$$

$$|Dw(q) - Dz(q)| \leq \sum_j \|Dw_j^p\|_{0,U_\ell} < \sqrt{2}K(n)\|h\|_{0,U_\ell} < \sqrt{2}K(n)\|g - w^\sharp e\|_{0,U_\ell}\,, \tag{23}$$

where in the last inequality we can use (16). Finally, we write

$$z^\sharp e - (w^\sharp e + h) = z^\sharp e - w^\sharp e - \sum_j h_j = \sum_{j\geq 1} (w_j^\sharp e - (w_{j-1}^\sharp e + h_j)) \tag{24}$$

(where $w_0 := w$) and thus we can use (21) to conclude, at the point q and using the coordinate pach U_ℓ,

$$|(z^\sharp e - (g + h))(q)| < \delta'_\ell\,.$$

This completes the proof of (8), (17) and (18).

In order to define w_j^p, select a U_ℓ and apply Lemma 28 on U_ℓ with $\omega = w_{j-1}$ to find two orthonormal smooth vector fields $\nu, b : U_\ell \to \mathbb{R}^N$ with the property that ν and b are normal to $w_{j-1}(U_\ell)$. Recall that $h_j = a_j^2 d\psi_j \otimes d\psi_j$ and set

$$w_j^p(x) = a_j(x)\frac{\nu(x)}{\lambda}\cos\lambda\psi_j(x) + a_j(x)\frac{b(x)}{\lambda}\sin\lambda\psi_j(x)\,,$$

where λ is a positive parameter, which will be chosen very large.

Note first that (19) is obvious provided λ is large enough. Next compute, in the coordinate patch U_ℓ,

$$Dw_j^p(x) = \underbrace{-a_j(x)\sin\lambda\psi_j(x)\,\nu(x)\otimes d\psi_j(x)}_{A(x)} + \underbrace{a_j(x)\cos\lambda\psi_j(x)\,b(x)\otimes d\psi_j(x)}_{B(x)} + E(x)\,,$$

where $|E(x)| \leq C_{j-1}\lambda^{-1}$, for a constant C_{j-1} which depends on the smooth functions a_j, ψ_j, b and ν, but not on λ (note that in the line above we understand all summands as $N \times n$ matrices). We then obviously have

$$|Dw_j^p(x)|^2 \leq a_j(x)^2|d\psi_j(x)|^2 + C_{j-1}\lambda^{-1} \leq \|h_j\|_{0,U_\ell} + C_{j-1}\lambda^{-1} \leq \|h\|_{0,U_\ell} + C_{j-1}\lambda^{-1}$$

(here and in what follows, C_{j-1} denotes constants which might change from line to line but are independent of the parameter λ). Since $\|h\|_{0,U_\ell}$ is positive, it suffices to choose λ large enough to achieve (20).

Next write the tensor $\bar{h} := w_j^\sharp e - w_{j-1}^\sharp e$ in coordinates as $\bar{h} = \bar{h}_{ik} dx_i \otimes dx_k$ and observe that the $\bar{h}_{ik}$ are simply the entries of the symmetric matrix

$$Dw_j^T Dw_j - Dw_{j-1}^T Dw_{j-1}\,.$$

Recall that $Dw_j = Dw_{j-1} + A + B + E$. By the conditions on ν and b we have

$$0 = A^T B = B^T A = A^T Dw_{j-1} = Dw_{j-1}^T A = B^T Dw_{j-1} = Dw_{j-1}^T B\,.$$

We thus conclude that

$$|Dw_j^T Dw_{j-1} - Dw_{j-1}^T Dw_j - (A^T A + B^T B)| \leq C_{j-1}\lambda^{-1}\,.$$

On the other hand,

$$A^T A + B^T B = a_j^2(\cos^2\lambda\psi_j + \sin^2\lambda\psi_j) d\psi_j \otimes d\psi_j = a_j^2 d\psi_j \otimes d\psi_j = h_j\,.$$

Hence (21) follows at once for λ large.

It must be noticed that so far we have shown (19), (20), and (21) only for the chosen coordinate patch which contains the support of h_j, whereas the estimates are claimed in all coordinate patches which intersect the support of h_j. On the other hand, on these other coordinate patches the same computations yield the same estimates, and since there are only finitely many such patches to take into account, our claims readily follow for an appropriate choice of λ.

It remains to show that, if w is injective, then z too can be chosen to be injective. Fix $p, q \in \Sigma$. For j sufficiently large we have $z(p) = w_j(p)$ and $z(q) = w_j(q)$. Thus it suffices to show the injectivity of w_j. We will show, inductively on j, that this can be achieved by choosing λ sufficiently large. Thus assume that w_{j-1} is injective. If p, q are not contained in the support of h_j, then $w_{j-1}(q) = w_j(q)$ and $w_{j-1}(p) = w_j(p)$ and thus we are done. Since the support of h_j is a compact subset of U_ℓ, there is a constant β such that $|w_{j-1}(p) - w_{j-1}(q)| \geq 2\beta$ for every q in the support of h_j and $p \notin U_\ell$. For such pairs of points $w_j(p) \neq w_j(q)$ as soon as $\|w_j - w_{j-1}\|_0 \leq \beta$, which can be achieved by choosing λ sufficiently large. It remains to check $w_j(p) \neq w_j(q)$ when one point belongs to the support of h_j and the other to U_ℓ (and they are distinct!). Consider that $\overline{U}_\ell$ is a compact set and, since w_{j-1} is injective, its restriction to $\overline{U}_\ell$ is a smooth embedding. It then follows that, for a sufficiently small $\eta > 0$, there is a well-defined orthogonal projection π from the normal tubular neighborhood T of thickness η of $w_{j-1}(U_\ell)$ onto $w_{j-1}(U_\ell)$. Of course if λ is sufficiently large $w_j(U_\ell)$ takes values in T and thus, by definition of $w_j - w_{j-1}$, $\pi(w_j(q)) = w_{j-1}(q) \neq w_{j-1}(p) = \pi(w_j(p))$. Obviously this implies $w_j(p) \neq w_j(q)$ and completes the proof.

4 Smooth Isometric Embeddings

4.1 Introduction

Two years after his counterintuitive C^1 theorem (see Theorem 20), Nash addressed and solved the general problem of the existence of smooth isometric embeddings in his other celebrated work [75]. As in the previous section we consider Riemannian manifolds (Σ, g), but this time of class C^k with $k \in \mathbb{N} \cup \{\infty\} \setminus \{0\}$: this means that there is a C^∞ atlas for Σ and that, in any chart the coefficients g_{ij} of the metric tensor in the local coordinates are C^k functions. Nash's celebrated theorem in [75] is then the following result.

Theorem 29 (Nash's smooth isometric embedding theorem) *Let $k \geq 3$, $n \geq 1$ and $N = \frac{n(3n+11)}{2}$. If (Σ, g) is a closed C^k Riemannian manifold of dimension n, then there is a C^k isometric embedding $u : \Sigma \to \mathbb{R}^N$.*

In [75], Nash covered also the case of nonclosed manifolds as a simple corollary of Theorem 29, but with a much weaker bound on the codimension. More precisely he claimed the existence of isometric embeddings for $N' = (n+1)N$. His proof contains however a minor error (Nash really proves the existence of an isometric *immersion*) which, as pointed out by Robert Solovay (cf. Nash's comment in [80, p. 209]), can be easily fixed using the same ideas, but at the price of increasing slightly the dimension N'.

Corollary 30 (C^∞ isometric embedding, nonclosed case) *Let $k \geq 3$, $n \geq 1$,*

$$N' = (n+1)N = (n+1)\frac{n(3n+11)}{2} \quad \text{and} \quad N'' = N' + 2n + 2\,.$$

If (Σ, g) is a C^k Riemannian manifold of dimension n, then there is a C^k isometric embedding $u : \Sigma \to \mathbb{R}^{N''}$ and a C^k isometric immersion $z : \Sigma \to \mathbb{R}^{N'}$.

The dimension of the ambient space in the theorems above has been lowered by subsequent works of Gromov and Günther. Moreover, starting from Gromov's work, Nash's argument has been improved to show statements similar to Theorem 20. More precisely, Gromov and Rokhlin first proved in [40] that any short map on a smooth compact Riemannian manifold can be approximated by isometric embeddings of class C^∞ if the dimension of the ambient Euclidean space is at least $\frac{n(n+1)}{2} + 4n + 5$. The latter threshold was subsequently lowered by Gromov in [39] to $\frac{n(n+1)}{2} + 2n + 3$ and by Günther in [41] to $\frac{n(n+1)}{2} + \max\{2n, 5\}$ (see also [42]). If g is real analytic and $m \geq \frac{n(n+1)}{2} + 2n + 3$, then any short embedding in $\mathbb{R}^m$ can be uniformly approximated by analytic isometric embeddings: in [78] Nash extended Theorem 29, whereas the approximation statement was shown first by Gromov for $m \geq \frac{n(n+1)}{2} + 3n + 5$ in [38] and lowered to the threshold above in [39]. Corresponding theorems can also be proved for noncompact manifolds M, but they are more subtle; for instance the noncompact case with real analytic metrics was left in [78] as an open problem; we refer the reader to [38, 39] for more details.

On the regularity side, Jacobowitz in [53] extended Nash's theorem to $C^{k,\beta}$ metrics (achieving the existence of $C^{k,\beta}$ embeddings) for $k + \beta > 2$. However, the case of C^2 metrics is still an open problem (it is also interesting to notice that Källen in [57] used a suitable improvement of Nash's methods for Theorem 20 in order to show the existence of $C^{1,\alpha}$ isometric embeddings with $\alpha < \frac{k+\beta}{2}$ when $k + \beta \leq 2$: the existence of a C^2 isometric embedding for C^2 metrics is thus an endpoint result for two different "scales").

The starting point of Nash in proving Theorem 29 is first to solve the linearization of the corresponding system of PDEs (6): in particular he realized that a suitable "orthogonality Ansatz" reduces the linearization to a system of linear equations which *does not involve derivatives of the linearization of the unknown*, cf. (29) and (30). The latter system can then be solved via linear algebra when the dimension of the target space is sufficiently high.

Having at hand a (simple) solution formula for the linearized system, one would like to recover some implicit (or inverse) function theorem to be able to assert the existence of a solution to the original nonlinear system (6). There are of course several iterative methods in analysis to prove implicit function theorems, but in Nash's case there is a central analytic difficulty: his solution of the linearized system experiences a phenomenon which in the literature is usually called *loss of derivative*. This problem, which was very well known and occurs in several other situations, looked insurmountable. Mathematics needed the genius of Nash in order to realize that one can deal with it by introducing a suitable regularization mechanism, see in particular the discussion of Sect. 4.5.

This key idea has numerous applications in a wide range of problems in partial differential equations where a purely functional–analytic implicit function theorem fails. The first author to put Nash's ideas in the framework of an abstract implicit function theorem was J. Schwartz, cf. [89]. However, the method became known as the Nash–Moser iteration shortly after Moser succeeded in developing a general framework going beyond an implicit function theorem, which he applied to a variety of problems in his fundamental papers [68, 70, 71], in particular to the celebrated KAM theory. Subsequently several authors generalized these ideas and a thorough mathematical theory has been developed by Hamilton in [43], who defined the categories of "tame Fréchet spaces" and "tame nonlinear maps". Such ideas are usually presented in the framework of a Newton iteration scheme. However, although Nash's original argument is in some sense close in spirit, in practice Nash truly constructs a smooth "curve" of approximate solutions solving a suitable infinite dimensional ordinary differential equation: the curve starts with a map which is close to be a solution and brings it to a final one which is a solution. This "smooth flow" idea seems to have been lost in the subsequent literature.

It is rather interesting to notice that, in order to solve the isometric embedding problem, Nash did not really need to resort to the very idea which made his work so famous in the literature of partial differential equations: Günther has shown in

[41] that the linearization of the isometric embedding system can be solved via a suitable elliptic operator. Hence, one can ultimately appeal to standard contraction arguments in Banach spaces via Schauder estimates, at least if we replace the C^k assumption of Nash's Theorem 29 with a $C^{k,\alpha}$ assumption for some α contained in the open interval (0, 1).

4.2 The Perturbation Theorem

As in the previous sections, we use Einstein's convention on repeated indices. From now on, given a closed n-dimensional manifold Σ, we fix an atlas $\{U_\ell\}$ as in Lemma 25. Given a function f on Σ, we define then $\|D^k f\|_0$ and $\|f\|_k$ as in Sect. 3.2. Given an (i, j) tensor T, consider its expression in coordinates in the patch U_ℓ, namely

$$T^{\alpha_1\ldots\alpha_i}_{a_1\ldots a_j}(u)\frac{\partial}{\partial u_{\alpha_1}}\otimes\cdots\otimes\frac{\partial}{\partial u_{\alpha_i}}\otimes du_{a_1}\otimes\cdots\otimes du_{a_j}\,.$$

We then define

$$\|D^k T\|_{0,U_\ell} := \sum_{\alpha_r,a_s}\|D^k T^{\alpha_1\ldots\alpha_i}_{a_1\ldots a_j}\|_{0,U_\ell}\,,\quad \|D^k T\|_0 := \sup_\ell \|D^k T\|_{0,U_\ell} \text{ and } \|T\|_k := \sum_{i\leq k}\|D^i T\|_0\,.$$

It is easy to see that these norms satisfy the Leibnitz-type inequality

$$\|D^k(T\otimes S)\|_0 \leq \sum_{i\leq k}\|D^i T\|_0\|D^{k-i}\|_0 \tag{25}$$

and, when contracting a given tensor, namely for $\bar{T}^{\alpha_2\ldots\alpha_i}_{a_2\ldots a_j} = \sum_k T^{k\alpha_2\ldots\alpha_i}_{ka_2\ldots a_j}$, we have the corresponding inequality

$$\|\bar{T}\|_0 \leq n\|T\|_0\,. \tag{26}$$

Nash's strategy to attack Theorem 29 is to prove first a suitable perturbation result. Let us therefore start with a smooth embedding $w_0 = (w_1, \ldots, w_N) : \Sigma \to \mathbb{R}^N$ and set $h := g - w_0^\sharp e$. Assuming h small we look for a (nearby) map $u : \Sigma \to \mathbb{R}^N$ such that $u^\sharp e = g$, namely $u^\sharp e - w_0^\sharp e = h$. In fact, we would like to build u as right endpoint of a path of maps starting at w_0. More precisely, consider a smooth curve $[t_0, \infty) \ni t \mapsto h(t)$ in the space of smooth (0, 2) tensors joining $0 = h(t_0)$ and $h = h(\infty)$; we would like to find a corresponding smooth deformation $w(t)$ of $w(t_0) = w_0$ to $w(\infty) = u$ so that

$$w(t)^\sharp e = w_0^\sharp e + h(t) \qquad \text{for all } t. \tag{27}$$

Following Nash's convention, we denote with an upper dot the differentiation with respect to the parameter t.

If we fix local coordinates $x_1, \ldots, x_n$ in a patch U and differentiate (27), we then find the following linear system of partial differential equations for the velocity $\dot{w}(t)$:

$$\frac{\partial w_\alpha}{\partial x_i}\frac{\partial \dot{w}_\alpha}{\partial x_j} + \frac{\partial \dot{w}_\alpha}{\partial x_i}\frac{\partial w_\alpha}{\partial x_j} = \dot{h}_{ij}\,. \tag{28}$$

In fact, since the expression in the right-hand side of (28) will appear often, we introduce the shorthand notation $2dw \odot d\dot{w}$ for it, more precisely:

Definition 31 If $u, v \in C^1(\Sigma, \mathbb{R}^N)$, we let $du \odot dv$ be the $(0, 2)$ tensor $\frac{1}{2}((u + v)^\sharp e - v^\sharp e - u^\sharp e)$, which in local coordinates is given by

$$\frac{1}{2}\left(\frac{\partial v_\alpha}{\partial x_i}\frac{\partial u_\alpha}{\partial x_j} + \frac{\partial u_\alpha}{\partial x_i}\frac{\partial v_\alpha}{\partial x_j}\right).$$

A second important idea of Nash is to assume that $\dot{w}$ is orthogonal to $w(\Sigma)$, namely

$$\frac{\partial w_\alpha}{\partial x_j}\dot{w}_\alpha = 0 \qquad \forall j \in \{1, \ldots, n\}\,. \tag{29}$$

Under this condition we have

$$0 = \frac{\partial}{\partial x_i}\left(\frac{\partial w_\alpha}{\partial x_j}\dot{w}_\alpha\right) = \frac{\partial \dot{w}_\alpha}{\partial x_i}\frac{\partial w_\alpha}{\partial x_j} + \dot{w}_\alpha\frac{\partial^2 w_\alpha}{\partial x_i \partial x_j}\,,$$

and we can rewrite (28) as

$$-2\frac{\partial^2 w_\alpha}{\partial x_j \partial x_i}\dot{w}_\alpha = \dot{h}_{ij}\,. \tag{30}$$

Clearly, in order to solve (29) and (30), it would be convenient if the resulting system of linear equations were linearly independent, which motivates the following definition.

Definition 32 A C^2 map $w : \Sigma \to \mathbb{R}^{\bar{N}}$ is called *free*[13] if, on every system of local coordinates $x_1, \ldots, x_n$, the following $n + \frac{n(n+1)}{2}$ vectors are linearly independent at

[13]The term free was not coined by Nash, but introduced later in the literature by Gromov.

every $p \in \Sigma$:

$$\frac{\partial w}{\partial x_j}(p), \frac{\partial^2 w}{\partial x_i \partial x_j}(p), \qquad \forall i \leq j \in \{1, \ldots, n\}. \tag{31}$$

Although the condition (31) is stated in local coordinates, the definition is independent of their choice. Observe moreover that a free map is necessarily an immersion and that we must have $\bar{N} \geq \frac{n(n+3)}{2}$. If a free map is injective, then we will call it a free embedding. The main "perturbation theorem" of Nash's paper (and in fact the most spectacular part of his celebrated work) is then the following statement. In order to prove it, Nash introduced his famous regularization procedure to overcome the most formidable obstruction posed by (28).

Theorem 33 (Perturbation theorem) *Assume $w_0 : \Sigma \to \mathbb{R}^N$ is a C^∞ free embedding. Then there is a positive constant ε_0, depending upon w_0, such that, if h is a C^k $(0, 2)$ tensor with $\|h\|_3 \leq \varepsilon_0$ and $k \geq 3$ (with possibly $k = \infty$), then there is a C^k embedding $\bar{u} : \Sigma \to \mathbb{R}^N$ such that $\bar{u}^\sharp e = w_0^\sharp e + h$.*

Solving the embedding problem using Theorem 33 certainly requires to produce maps which are "close" to be an isometric embedding. However note that there is a rather subtle issue: since the threshold ε_0 depends upon w_0, producing a "good starting" w_0 is not at all obvious. We will tackle this issue immediately in the next sections and then come to the proof of Theorem 33 afterwards.

4.3 *Proof of the Smooth Isometric Embedding Theorem*

In order to exploit Theorem 33, Nash constructs an embedding u_0 of Σ which is the cartesian product of two smooth maps w and $\bar{w}$, which he calls, respectively, the Z-embedding and the Y-embedding. One crucial elementary ingredient is the following remark.

Remark 34 If $f_1 : \Gamma \to \mathbb{R}^n$ and $f_2 : \Gamma \to \mathbb{R}^m$ are two C^1 maps, then $(f_1 \times f_2)^\sharp e = f_1^\sharp e + f_2^\sharp e$, where we just understand $f_1^\sharp e$, $f_2^\sharp e$ and $(f_1 \times f_2)^\sharp e$ as $(0, 2)$ tensors (note that they are positive semidefinite, but not necessarily positive *definite*).

The strategy of Nash can be summarized as follows:

(i) fix first a free C^∞ smooth embedding w_0 (the Z-embedding) which is (strictly) short with respect to g (cf. Definition 19), and consider the threshold ε_0 needed to apply Theorem 33;
(ii) then use a construction somewhat reminiscent of the proof of Theorem 20 to build a smooth $\bar{w}$ such that $h := g - w_0^\sharp - \bar{w}^\sharp$ satisfies $\|h\|_3 \leq \varepsilon_0$;
(iii) if $\bar{u}$ is finally the map produced by Theorem 33 applied to w_0 and h, we then set $u := \bar{u} \times \bar{w}$ and conclude Theorem 29.

It is indeed not difficult to produce the Z- and Y-embeddings if we allow very large dimensions. In order to achieve the dimension N claimed in Theorem 29, Nash follows a much subtler argument which requires the metric difference $g - w_0^\sharp e$ to satisfy a certain nontrivial property: an important ingredient is the following proposition, whose proof is postponed to the end of the section.

Proposition 35 *There are $N_0 := \frac{n(n+3)}{2}$ smooth functions ψ^r on Σ such that, for each p in Σ, $\{d\psi^r(p) \otimes d\psi^r(p) : r \in \{1, \dots N_0\}\}$ spans the space $S_p = \mathrm{Sym}\,(T_p^*\Sigma \otimes T_p^*\Sigma)$.*

In fact, if we had the more modest goal of proving the above statement with a much larger N_0, we could use the same arguments of Proposition 27. In the proof of Theorem 29 we still need two technical lemmas, whose proofs will also be postponed. The first one is a classical fact in linear algebra, which will be used also in the next sections.

Lemma 36 *Consider a $k \times \kappa$ matrix A of maximal rank $k \leq \kappa$. For every vector $v \in \mathbb{R}^k$, the vector $\omega := A^T(AA^T)^{-1}v$ is a solution of the linear system $A\omega = v$. Indeed, ω gives the solution with smallest Euclidean norm.*

Remark 37 Note two big advantages of the solution ω determined through the formula $\omega = A^T(A \cdot A^T)^{-1}v$:

(a) ω depends smoothly upon A;
(b) ω goes to 0 when A is fixed and v goes to 0; indeed this statement remains true even if, while v goes to 0, the matrix A varies in a compact set over which $A \cdot A^T$ is invertible.

The second is a more sophisticated tool which is used indeed twice in this section.[14]

Lemma 38 *Consider a real analytic manifold $\mathcal{M}$ of dimension r and a real analytic map $F : \mathcal{M} \times \mathbb{R}^\kappa \to \mathbb{R}^k$. If, for each $q \in \mathcal{M}$, the set $\mathscr{Z}(q) := \{v : F(q, v) = 0\}$ has Hausdorff dimension at most d, then the set $\mathscr{Z} := \{v : \exists q \in \mathcal{M}$ with $F(q, v) = 0\}$ has dimension at most $r + d$.*

Proof (Proof of Theorem 29) Let ψ^r be the functions of Proposition 35 and set $\gamma := \sum_r d\psi^r \otimes d\psi^r$. After multiplying all the functions by a small factor, we can assume that $\gamma < g$. Using Theorem 20, we then find a C^1 embedding $w : \Sigma \to \mathbb{R}^{2n}$ such that $w^\sharp e = g - \gamma$. By density of C^∞ functions in C^1, we then get a smooth embedding v such that $\|v^\sharp e - (g - \gamma)\|_0 < \delta$, where $\delta > 0$ is a parameter which will be chosen later. Indeed, by the Whitney's theorem we can assume that $v(\Sigma)$ is

[14] It must be observed that Nash employs this fact without explicitly stating it and he does not prove it neither he gives a reference. He uses it twice, once in the proof of Theorem 29 and once in the proof of Proposition 35, and although in the first case one could appeal to a more elementary argument, I could not see an easier way in the second.

a real analytic subvariety, which will play an important role towards the end of the proof. Consider v as an embedding in the larger space $\mathbb{R}^{\bar{N}}$ with $\bar{N} = \frac{n(n+5)}{2}$. We will perturb v to a smooth free embedding $w_0 : \Sigma \to \mathbb{R}^{\bar{N}}$ with the property that $\|w_0^\sharp e - (g-\gamma)\|_0 < 2\delta$. Before coming to the proof of the existence of w_0, let us first see how we complete the argument.

First observe that the $(0,2)$ tensor $w_0^\sharp e - (g-\gamma)$ can be written as

$$w_0^\sharp e - (g-\gamma) = \sum_r b_r d\psi^r \otimes d\psi^r ,$$

where, thanks to Lemma 36, the coefficients b_r can be chosen smooth. In fact, notice that the coefficients become arbitrarily small as we decrease δ: for a suitable choice of δ we can thus assume $\|b_r\|_0 \le \frac{1}{2}$. This is the only requirement on δ: from now on we can consider that the smooth free embedding w_0 has been fixed, which in turn gives a positive threshold ε_0 for the applicability of Theorem 33. Next write

$$g - w_0^\sharp e = \gamma - \sum_r b_r d\psi^r \otimes d\psi^r = \sum_r (1-b_r) d\psi^r \otimes d\psi^r = \sum_r a_r^2 d\psi^r \otimes d\psi^r ,$$

for the smooth functions $a_r := \sqrt{1-b_r}$. Define $\bar{w} : \Sigma \to \mathbb{R}^{2N_0}$ setting

$$\bar{w}_{2(i-1)+1}(p) := \frac{a_r(p)}{\lambda} \sin \lambda\psi^r(p) , \qquad \bar{w}_{2i}(p) := \frac{a_r(p)}{\lambda} \cos \lambda\psi^r(p) .$$

A straightforward computation yields

$$\bar{w}^\sharp e = \sum_r a_r^2 d\psi^r \otimes d\psi^r + \frac{1}{\lambda^2} \sum_r da_r \otimes da_r .$$

In particular,

$$h := g - (w_0 \times \bar{w})^\sharp e = -\frac{1}{\lambda^2} \sum_r da_r \otimes da_r .$$

For λ sufficiently large we certainly have $\|h\|_3 \le \varepsilon_0$ and from Theorem 33 we achieve a C^k embedding $\bar{u} : \Sigma \to \mathbb{R}^{\bar{N}}$ such that $\bar{u}^\sharp e = w_0^\sharp e + h$. It turns out that $u := \bar{u} \times \bar{w}$ is a C^k embedding of Σ into $\mathbb{R}^N = \mathbb{R}^{\bar{N}} \times \mathbb{R}^{2N_0}$ and that $u^\sharp e = g$.

In order to complete the proof, we still need to perturb v to a free w_0. For any $\eta > 0$ we want to construct a free map $w_0 : \Sigma \to \mathbb{R}^{\bar{N}}$ such that $\|w_0 - v\|_1 \le \eta$. Clearly, for η sufficiently small w_0 is an embedding. In order to produce w_0 we consider the $2n + n(2n+1)$ functions given by

$$v_i\, , \; v_i v_j\, , \qquad j \le i \in \{1, \ldots, 2n\} ,$$

and those C^2 maps $w_0 : \Sigma \to \mathbb{R}^{\bar{N}}$ given by the formula

$$(w_0)_\alpha := \sum_i C^i_\alpha v_i + \sum_{j \le i} D^{ij}_\alpha v_i v_j ,$$

for constant coefficients $C^i_\alpha, D^{ij}_\alpha$. We claim that, for a generic choice of the constants C^i_α and D^{ij}_α, the map w_0 is free. Indeed, consider the set $\mathscr{G}$ of subspaces L of $\mathbb{R}^{n+\frac{n(n+1)}{2}}$ with dimension $n - 1 + \frac{n(n+1)}{2}$. For each $(p, L) \in \Sigma \times \mathscr{G}$, consider the set $\mathscr{C}(p, L)$ of coefficients $C^i_\alpha, D^\alpha_{ij}$ for which, in a local system of coordinates,

$$V_\alpha(p) := \left(\frac{\partial w_\alpha}{\partial x_1}(p), \ldots, \frac{\partial w_\alpha}{\partial x_n}(p), \frac{\partial^2 w_\alpha}{\partial x_1^2}(p), \frac{\partial^2 w_\alpha}{\partial x_1 \partial x_2}(p) \ldots, \frac{\partial^2 w_\alpha}{\partial x_n^2}(p)\right) \in L \tag{32}$$

for all $\alpha \in \{1, \ldots, \bar{N}\}$. This is a set of (linear) conditions which varies analytically as (p, L) varies in the $(2n - 1 + \frac{n(n+1)}{2}) = (\bar{N} - 1)$-dimensional manifold $\Sigma \times \mathscr{G}$. We next show that, if $\bar{d}$ is the dimension of the linear space of possible coefficients $C^i_\alpha, D^{ij}_\alpha$, then the dimension of each $\mathscr{C}(p, L)$ is at most $d = \bar{d} - \bar{N}$. In view of Lemma 38 this implies that the union of all $\mathscr{C}(p, L)$ has dimension at most $\bar{d} - 1$. Since the latter is indeed the closed set $\mathscr{B}$ of "bad coefficients" for which w is not free, we have conclude that $\mathscr{B}$ must have empty interior.

To complete the proof[15] it remains to bound the dimension of $\mathscr{C}(p, L)$. Hence fix p and, without loss of generality, assume that $(x_1, \ldots, x_n) = (v_1, \ldots, v_n)$ is a system of coordinates around p. Consider the $M = n + \frac{n(n+1)}{2}$ functions $f_1 = v_1, \ldots, f_n = v_n, f_{n+1} = v_1^2, f_{n+2} = v_1 v_2, \ldots, v_M = v_n^2$ and the corresponding vector valued map f. It is easy to check that the vectors $\frac{\partial f}{\partial x_1}(p), \ldots, \frac{\partial f}{\partial x_n}(p), \frac{\partial^2 f}{\partial x_1^2}(p), \frac{\partial^2 f}{\partial x_1 \partial x_2}(p), \ldots, \frac{\partial^2 f}{\partial x_n^2}(p)$ are linearly independent. But then it follows that the vectors

$$\bar{V}_j(p) := \left(\frac{\partial f_j}{\partial x_1}(p), \ldots, \frac{\partial f_j}{\partial x_n}(p), \frac{\partial^2 f_j}{\partial x_1^2}(p), \frac{\partial^2 f_j}{\partial x_1 \partial x_2}(p) \ldots, \frac{\partial^2 f_j}{\partial x_n^2}(p)\right)$$

are also linearly independent. Hence there is one of them which does not belong to L. For each $\alpha \in \{1, \ldots \bar{N}\}$ there is therefore at least one choice of the coefficients $C^i_\alpha, D^{ij}_\alpha$ for which the corresponding vector $V_\alpha(p)$ in (32) does not belong to L. Since α can be chosen in $\bar{N}$ different ways, the dimension of $\mathscr{C}(p, L)$ is at most $d = \bar{d} - \bar{N}$, which completes the proof.

[15] Indeed Nash does not give any argument and just refers to a similar reasoning that he uses in Proposition 35 below.

Proof (Proof of Proposition 35*)* The argument is very similar to the last part of the proof of Theorem 29 above. Consider again an embedding $v : \Sigma \to \mathbb{R}^{2n}$ which makes $v(\Sigma)$ a real analytic submanifold. Let then f_{ij} be the $n(2n+1)$ functions $v_i + v_j$ such that $i \leq j$ and consider

$$\psi^r := A^r_{ij} f_{ij}\,, \qquad \text{for } r \in \{1, \ldots, N_0\},$$

where the space of all possible constant coefficients A^r_{ij} has dimension $\bar{d}$. Our aim is to show that a generic choice of the coefficients give a system of functions ψ^r which satisfy the conclusions of the proposition.

Let therefore $\mathscr{B}$ be the closed subset of coefficients for which the conclusion fails, namely for each element in $\mathscr{B}$ there exists a point p at which the tensors $d\psi^r(p) \otimes d\psi^r(p)$ do not span the whole space $S_p := \mathrm{Sym}\,(T^*_p\Sigma \otimes T^*_p\Sigma)$. If we consider the set $\mathscr{G}_p$ of linear subspaces of S_p of codimension 1, the real analytic manifold $\mathscr{M} := \{(p, L) : L \in \mathscr{G}_p\}$ has dimension $n-1+\frac{n(n+1)}{2} = N_0 - 1$. For each (p, L) we let $\mathscr{C}(p, L)$ be the set of coefficients for which $d\psi^r(p)\otimes d\psi^r(p)$ belongs to L for every $r = 1, \ldots, N_0$: this is the zero set of a system of homogeneous quadratic polynomials in the coefficients A^r_{ij}. Moreover, in a real analytic atlas for $\mathscr{M}$ these quadratic polynomials depend analytically upon $(p, L) \in \mathscr{M}$. Set $\mathscr{B} = \cup_{(p,L)\in\mathscr{M}}\mathscr{C}(p, L)$. As above we can invoke Lemma 38: if we can bound the dimension of the each $\mathscr{C}(p, L)$ with $\bar{d} - N_0$, then the dimension of $\mathscr{B}$ is at most $\bar{d} - 1$.

Fix therefore (p, L) and for each r consider the linear space π_r of indices A^r_{ij}. Without loss of generality we can assume that $(v_1, \ldots, v_n) = (x_1, \ldots, x_n)$ is a system of coordinates around p. Therefore the set $\{df_{ij} \otimes df_{ij}$ with $i \leq j \leq n\}$ spans the whole space S_p and there is at least one element among them which does not belong to L. In turn this means that the subset $\mathscr{C}^r(p, L) \subset \pi_r$ of coefficients A^r_{ij} such that $d\psi^r \otimes d\psi^r$ belongs to L has codimension at least 1 in π_r. Therefore the dimension of $\mathscr{C}(p, L) = \mathscr{C}^1(p, L) \times \mathscr{C}^2(p, L) \times \ldots \times \mathscr{C}^{N_0}(p, L)$ is at most $d = \bar{d} - N_0$. This shows $d + N_0 - 1 < \bar{d}$ and completes the proof.[16]

Proof (Proof of Lemma 36*)* It is obvious that ω solves the desired linear system. Let now w be any solution of minimal Euclidean norm: w is uniquely determined by the property of being orthogonal to the kernel of A. However, the kernel of A consists of those vectors which are orthogonal to the image of A^T: since the ω of the lemma belongs to the image of A^T, this completes the proof.

[16]Nash suggests an alternative argument which avoids the discussion of the dimensions of $\mathscr{C}(p, L)$ and $\mathscr{B}$. One can apply his result on real algebraic varieties to find an embedding v which realizes $v(\Sigma)$ as a real algebraic submanifold, cf. Theorem 1. Then any set of coefficients A^r_{ij} which is algebraically independent over the minimal field $\mathbb{F}$ of definition of $v(\Sigma)$ (see Proposition 12) belongs to the complement of $\mathscr{B}$. Since $\mathbb{F}$ is finitely generated over the rationals (see Proposition 12), it has countable cardinality and the conclusion follows easily.

Proof (Proof of Lemma 38*)* Covering $\mathcal{M}$ with a real analytic atlas consisting of countably many charts, we can assume, without loss of generality, that $\mathcal{M}$ is the Euclidean r-dimensional ball B. Consider next $Z := \{(q, v) : F(q, v) = 0\} \subset B \times \mathbb{R}^\kappa \subset \mathbb{R}^r \times \mathbb{R}^\kappa$. If $\pi : \mathbb{R}^{r+\kappa} \to \mathbb{R}^\kappa$ is the projection on the second factor, then $\mathscr{Z} = \pi(Z)$ has at most the dimension of Z: it suffices therefore to show that $\dim(Z) \leq r + d$.

Now, Z is a real analytic subvariety in $\mathbb{R}^{r+\kappa}$ with the property that its slices $\{q\} \times \mathscr{Z}(q) := Z \cap (\{q\} \times \mathbb{R}^\kappa)$ all have dimension at most d. The dimension s of Z equals the dimension of its regular part Z^r and without loss of generality we can assume that Z^r is connected. Consider now standard coordinates $(x_1, \ldots, x_r)$ on $\mathbb{R}^r \times \{0\} \subset \mathbb{R}^{r+\kappa}$ and regard x_1 as a function over Z^r. By Sard's theorem almost every α is a regular value for x_1 on Z^r. If one such value α has nonempty preimage, then $Z^r \cap \{x_1 = \alpha\}$ is a submanifold of dimension $s - 1$. Otherwise it means that $x_1(Z^r)$ has measure 0: since however $x_1(Z)$ is connected, we must have $x_1(Z) = \{\alpha_0\}$ for some value α_0, that is, $Z^r \cap \{x_1 = \alpha_0\} = Z^r$. In both cases we have conclude that there is at least one value α_0 such that $Z^r \cap \{x_1 = \alpha_0\}$ is a smooth submanifold of dimension no smaller than $s-1$. Inductively repeating this argument, we conclude that there is a q such that $Z^r \cap (\{q\} \times \mathbb{R}^\kappa)$ is a regular submanifold of dimension at least $s - r$. Since $Z^r \cap (\{q\} \times \mathbb{R}^\kappa) \subset \{q\} \times \mathscr{Z}(q)$, we infer $s - r \leq d$, which concludes the proof of our claim.

4.4 Smoothing Operator

In order to show Theorem 33 we will need to smooth tensors efficiently and get sharp estimates on the $\|\cdot\|_k$ norms of the smoothing. This will be achieved, essentially, by convolution but, since we will need rather refined estimates, the convolution kernel must be chosen carefully. In the remaining sections the specific form of the regularizing operator will play no role: the only important ingredients are summarized in the following proposition.

Proposition 39 (Smoothing operator) *There is a family of smoothing operators* $\mathscr{S}_\varepsilon$ *with* $\varepsilon \in]0, 1[$ *such that*[17]

(a) $T \mapsto \mathscr{S}_\varepsilon T$ *is a linear map on the space of continuous* (i, j) *tensors; for each such* T $\mathscr{S}_\varepsilon T$ *is smooth and depends smoothly upon* ε.
(b) *For any integers* $r \geq s$ *and* i, j*, there is a constant* $C = C(r, s, i, j)$ *such that*

$$\|D^r(\mathscr{S}_\varepsilon T)\|_0 \leq C\varepsilon^{s-r}\|T\|_s \qquad \text{for every } C^s \ (i, j) \text{ tensor } T \text{ and } \varepsilon \leq 1; \tag{33}$$

[17]In Nash's paper the operator is called S_θ, where θ corresponds to ε^{-1}. Since it is nowadays rather unusual to parametrize a family of convolutions as Nash does, I have switched to a more modern convention.

(c) *If we denote by* $\mathscr{S}'_\varepsilon$ *the linear operator* $T \mapsto \frac{\partial}{\partial\varepsilon}\mathscr{S}_\varepsilon T$, *then for any integers* r, s, i, j, *there is a constant* $C = C(r, s, i, j)$ *such that*

$$\|D^r(\mathscr{S}'_\varepsilon T)\|_0 \le C\varepsilon^{s-r-1}\|T\|_s \qquad \text{for every } C^s \ (i, j) \text{ tensor } T \text{ and } \varepsilon \le 1; \tag{34}$$

(d) *For any integers* $s \ge r$ *and* i, j *there is a constant* $C = C(r, s, i, j)$ *such that*

$$\|D^r(T - \mathscr{S}_\varepsilon T)\|_0 \le C\varepsilon^{s-r}\|T\|_s \qquad \text{for every } C^s \ (i, j) \text{ tensor } T \text{ and } \varepsilon \le 1. \tag{35}$$

Proof As a first step we reduce the problem of smoothing tensors to that of smoothing functions. To achieve this, we fix a smooth embedding of Σ into $\mathbb{R}^{2n}$ (whose existence is guaranteed by the Whitney's embedding theorem), and we therefore regard Σ as a submanifold of $\mathbb{R}^{2n}$. We fix moreover a tubular neighborhood $V_{3\eta}$ of Σ and assume that the size 3η is sufficiently small so that the nearest point projection $\pi : V_{3\eta} \to \Sigma$ is well defined and C^∞. Consider now a coordinate patch U on Σ and a corresponding system of local coordinates $(u_1, \ldots, u_n)$. We then define the map $x : U \to \mathbb{R}^{2n}$ where $(x_1(u), \ldots, x_{2n}(u))$ gives the standard coordinates in $\mathbb{R}^{2n}$ of the point with coordinates u in U. If $\mathscr{N}(U) := \pi^{-1}(U)$, we then define $u : \mathscr{N}(U) \to U$ by letting $u(x)$ be the coordinates, in U, of $\pi(x)$. Clearly $u \circ x$ is the identity and $x \circ u$ becomes the identity when restricted on $U \subset \Sigma$. Then, given an (i, j) tensor T, which in the local coordinates on U can be expressed as

$$\sum_{\alpha_1,\ldots,\alpha_i,a_1,\ldots,a_j} T^{\alpha_1\ldots\alpha_i}_{a_1\ldots a_j}(u)\frac{\partial}{\partial u_{\alpha_1}}\cdots\frac{\partial}{\partial u_{\alpha_i}}du_{a_1}\cdots du_{a_j},$$

we define the functions

$$\mathscr{T}^{\beta_1\ldots\beta_i}_{b_1\ldots b_j}(x) = T^{\alpha_1\ldots\alpha_i}_{a_1\ldots a_j}(u(x))\frac{\partial x_{\beta_1}}{\partial u_{\alpha_1}}\cdots\frac{\partial x_{\beta_i}}{\partial u_{\alpha_i}}\frac{\partial u_{a_1}}{\partial x_{b_1}}\cdots\frac{\partial u_{a_j}}{\partial x_{b_j}}. \tag{36}$$

It is easy to check that the functions above do not depend on the chosen coordinates and thus can be defined globally on Σ. Conversely, if we have global functions $\mathscr{T}$ as above on Σ, we can "reconstruct a tensor" using, in local coordinates, the reverse formulae

$$T^{\alpha_1\ldots\alpha_i}_{a_1\ldots a_j}(u) = \mathscr{T}^{\beta_1\ldots\beta_i}_{b_1\ldots b_j}(x(u))\frac{\partial u_{\alpha_1}}{\partial x_{\beta_1}}\cdots\frac{\partial u_{\alpha_i}}{\partial x_{\beta_i}}\frac{\partial x_{b_1}}{\partial u_{a_1}}\cdots\frac{\partial x_{b_j}}{\partial u_{a_j}}. \tag{37}$$

Given these transformation rules and the smoothness of the maps $x \mapsto u(x)$ and $u \mapsto x(u)$, we easily conclude the estimates

$$\|D^k T\|_0 \le C \sum_{b_1,\ldots,b_j,\beta_1,\ldots,\beta_i} \|\mathscr{T}^{\beta_1\ldots\beta_i}_{b_1\ldots b_j}\|_k, \tag{38}$$

$$\|D^k \mathcal{T}^{\beta_1\ldots\beta_i}_{b_1\ldots b_j}\|_0 \leq C\|T\|_k\,, \tag{39}$$

for a constant $C = C(n, i, j, k)$ which is independent of the tensor T.

Thus, if we have defined a suitable family of smoothing operators $\mathcal{S}_\varepsilon$ on functions over Σ, we can extend them to tensors with the following algorithm: given a tensor T we produce the functions $\mathcal{T}^{\beta_1\ldots\beta_i}_{b_1\ldots b_j}$ using formula (36); we then apply the smoothing operator to each function, getting the functions $\mathcal{S}_\varepsilon \mathcal{T}^{\beta_1\ldots\beta_i}_{b_1\ldots b_j}$; we finally use the latter to define $\mathcal{S}_\varepsilon T$ through formula (37). Observe that each of these operations is linear in T.

As a second step we reduce the problem of regularizing functions over Σ to that of regularizing functions over $\mathbb{R}^{2n}$ by a simple extension argument. More precisely, consider a smooth cut-off function $\varphi : \mathbb{R}^+ \to \mathbb{R}$, which is identically 1 on $[0, \eta[$, strictly decreasing on $[\eta, 2\eta]$ and identically 0 on $[2\eta, \infty[$. Given a function f on Σ, we then extend it to a function $\tilde{f}$ on $V_{3\eta}$ setting $\tilde{f}(x) = \varphi(|x - \pi(x)|) f(\pi(x))$ and subsequently to $\mathbb{R}^{2n}$ by setting it identically 0 outside $V_{2\eta}$. Again, by the smoothness of π, it is easy to check that we have the estimate

$$\|D^k \tilde{f}\|_0 \leq C\|f\|_k$$

for some constant $C = C(k)$, where this time $D^k \tilde{f}$ denotes the usual (Euclidean) kth derivative and $\|\cdot\|_0$ is the usual maximum norm of a continuous compactly supported function on $\mathbb{R}^{2n}$. Conversely, if $\tilde{f} \in C^k_c(\mathbb{R}^{2n})$, we have

$$\|D^k(\tilde{f}|_\Sigma)\|_0 \leq C\|\tilde{f}\|_k = \sum_{i\leq k} \|D^i \tilde{f}\|_0\,.$$

Thus, if we can find a suitable regularization operator $\mathcal{R}_\varepsilon$ on $C^k_c(\mathbb{R}^{2n})$ which satisfies the properties analogous to (a), (b), (c), and (d), we achieve the corresponding desired operator on $C^k(\Sigma)$ via the rule $\mathcal{S}_\varepsilon f = (\mathcal{R}_\varepsilon \tilde{f})|_\Sigma$ (notice again that two points are crucial: the linearity of the maps $f \mapsto \tilde{f}$ and $\tilde{f} \mapsto \tilde{f}_\Sigma$ and the relation $f = \tilde{f}|_\Sigma$).

We now come to the operator $\mathcal{R}_\varepsilon$ regularizing functions on $\mathbb{R}^{2n}$, which is the convolution with a suitably chosen mollifier φ in the Schwartz class $\mathcal{S}$. More precisely, assuming that $m = 2n$ and that $\varphi \in \mathcal{S}(\mathbb{R}^m)$ has integral 1, we define $\varphi_\varepsilon(x) = \varepsilon^{-m}\varphi(\frac{x}{\varepsilon})$ and set

$$[\mathcal{R}_\varepsilon f](x) = f * \varphi_\varepsilon(x) = \int f(x-y)\varphi_\varepsilon(y)\,dy = \frac{1}{\varepsilon^m}\int f(x-y)\varphi\left(\frac{y}{\varepsilon}\right)dy\,.$$

The analog of property (a) is

$$\mathcal{R}_\varepsilon \text{ maps } C_c(\mathbb{R}^m) \text{ into } \mathcal{S}(\mathbb{R}^m) \text{ and depends smoothly on } \varepsilon. \tag{40}$$

The latter is, however, a very standard fact for convolutions. Estimate (b) is also a classical property. Indeed, given a multiindex $I = (i_1, \ldots, i_m) \in \mathbb{N}^m$, let $|I| = i_1 + \cdots + i_m$ and

$$\partial^I f = \frac{\partial^{|I|} f}{\partial x_1^{i_1} \partial x_2^{i_2} \cdots \partial x_m^{i_m}}.$$

If we fix natural numbers $r \geq s$ and consider a multiindex I with $|I| = r$, we can obviously write it as $I = I' + J$ where $|I'| = s$ and $|J| = r - s$. The usual properties of convolutions yield then the following estimate

$$\|\partial^I (\mathscr{R}_\varepsilon f)\|_0 = \|(\partial^{I'} f) * (\partial^J \varphi_\varepsilon)\|_0 \leq \|\partial^{I'} f\|_0 \|\partial^J \varphi_\varepsilon\|_{L^1} \leq \|D^s f\|_0 \varepsilon^{s-r} \|\partial^J \varphi\|_{L^1}.$$

Thus, if we define $C := \min_{|J|=r-s} \|\partial^J \varphi\|_{L^1}$, we achieve

$$\|\partial^I (\mathscr{R}_\varepsilon f)\|_0 \leq C \varepsilon^{r-s} \|D^s f\|_0 \qquad \text{when } s \leq r. \tag{41}$$

Coming to (c), we use elementary calculus to give a formula for $\mathscr{R}'_\varepsilon := \frac{\partial}{\partial \varepsilon} \mathscr{R}_\varepsilon$:

$$\mathscr{R}'_\varepsilon f(x) = \int f(x-y) \left[-\frac{m}{\varepsilon^{m+1}} \varphi\left(\frac{y}{\varepsilon}\right) - \frac{1}{\varepsilon^m} \nabla \varphi\left(\frac{y}{\varepsilon}\right) \cdot \frac{y}{\varepsilon^2} \right] dy.$$

If we set $\psi(y) := -m\varphi(y) - \nabla\varphi(y) \cdot y$ and $\psi_\varepsilon(y) = \varepsilon^{-m} \psi(\frac{y}{\varepsilon})$, we conclude the identity

$$\mathscr{R}'_\varepsilon f = \varepsilon^{-1} f * \psi_\varepsilon. \tag{42}$$

Note that even ψ belongs to the Schwartz class. Hence, by the argument given above, the following inequality

$$\|D^r (\mathscr{R}'_\varepsilon f)\|_0 \leq C \varepsilon^{s-r-1} \|D^s f\|_0 \tag{43}$$

is certainly valid for $r \geq s$. However, the crucial point of estimate (c) is its validity even in the range $r < s$! In order to achieve this stronger bound we need to choose a specific mollifier φ: more precisely we require that:

$$\forall k \in \mathbb{N} \qquad \exists \vartheta^{(k)} \in \mathscr{S} \quad \text{such that} \quad \frac{\partial^k \vartheta^{(k)}}{\partial x_1^k} = \psi. \tag{44}$$

With this property, for $s > r$ we can integrate by parts $k = s - r$ times to achieve the identity

$$\mathscr{R}'_\varepsilon f = \varepsilon^{s-r-1} \frac{\partial^{s-r} f}{\partial x_1^{s-r}} * \vartheta_\varepsilon^{(s-r)},$$

and, applying the same argument used for (41), we conclude (43).

In order to find a kernel φ such that (44) holds, we compute first the Fourier transform of ψ:

$$\hat{\psi}(\xi) = -m\hat{\varphi}(\psi) - \sum_j \left(-\frac{1}{i}\frac{\partial}{\partial \xi_j}\right)\left(i\xi_j\hat{\varphi}(\xi)\right) = \nabla\hat{\varphi}(\xi)\cdot\xi .$$

Assume $\hat{\varphi} \in C_c^\infty(\mathbb{R}^m)$ and equals $(2\pi)^{\frac{m}{2}}$ in a neighborhood of 0. Then φ belongs to $\mathscr{S}$ and has integral 1. Moreover $\hat{\psi}$ vanishes in a neighborhood of the origin, and thus $(i\xi_1)^{-k}\hat{\psi}$ belongs to $\mathscr{S}$. But then, if we let $\vartheta^{(k)}$ be the inverse Fourier transform of the latter function, we conclude that $\vartheta^{(k)} \in \mathscr{S}$ and that $\frac{\partial^k \vartheta^{(k)}}{\partial x_1^k} = \psi$.

To complete the proof, we finally show the analog of estimate (d), namely

$$\|D^r(f - \mathscr{R}_\varepsilon f)\|_0 \leq C\varepsilon^{s-r}\|D^s f\|_0 \qquad \text{when } s \geq r. \tag{45}$$

For $s = r$ it is an obvious outcome of (41). For $s > r$, we instead integrate (43) in ε:

$$\|D^r(f - \mathscr{R}_\varepsilon f)\|_0 \leq \int_0^\varepsilon \left\|D^r\left(\mathscr{R}'_\delta f\right)\right\|_0 d\delta \leq C\|D^s f\|_0 \int_0^\varepsilon \delta^{s-r-1}\, d\delta = C\varepsilon^{s-r}\|D^s f\|_0$$

(note that $s - r - 1 \geq 0$ under our assumptions!).

4.5 A Smooth Path to Prove the Perturbation Theorem

Recalling Sect. 4.2, we wish to construct

(i) a path $[t_0, \infty) \ni t \mapsto h(t)$ joining 0 to h
(ii) and a path $[t_0, \infty) \ni t \mapsto w(t)$ joining w_0 to $\bar{u}$

such that

$$\frac{d}{dt} w(t)^\sharp e = \dot{h}(t)\,. \tag{46}$$

Recall moreover that we have reduced (46) to solving (29) and (30) for the "velocity" $\dot{w}$ of w, at least in local coordinates. Assuming that $w(t)$ is a free map for every t, we can use Lemma 36 to find, in a given coordinate patch, a "canonical" solution of the linear system (29) and (28): more precisely we can write

$$\dot{w}_\alpha := \mathscr{L}_\alpha^{ij}(Dw, D^2 w)h_{ij} \tag{47}$$

where $\mathscr{L}^{ij}_\alpha(A, B)$ is a suitable collection of functions which depend smoothly (in fact analytically) upon the entries A and B. This defines a linear operator $\mathscr{L}(Dw, D^2w)$ from the space of $(0, 2)$ tensors over the coordinate patch U into the space of maps $\dot{w} : U \to \mathbb{R}^N$. Next, we wish to extend this operator to the whole manifold Σ: the crucial point is that, although derived in a coordinate patch, the formula above does not depend on the chosen coordinate patch.

Lemma 40 (Existence of the operator $\mathscr{L}$) *Assume $w : \Sigma \to \mathbb{R}^N$ is C^2 and free. Given any $(0, 2)$ tensor $\bar{h}$ and any coordinate patch, the map $\mathscr{L}(Dw, D^2w)\bar{h}$ defined above does not depend on the coordinates and the process defines, therefore, a global (linear) operator $\mathscr{L}(w)$ from the space of smooth symmetric $(0, 2)$ tensors over Σ into the space of smooth maps $C^\infty(\Sigma, \mathbb{R}^N)$.*

Proof Observe that, for each fixed $p \in \Sigma$, the linear space of vectors $z = \dot{w}(p)$ satisfying the system (29) and (30) is independent of the choice of coordinates (in other words, although the coefficients in the system might change, the solution set remains the same: this follows from straightforward computations!). Since, however, according to Lemma 36 the vector $[\mathscr{L}(Dw, D^2w)h](p)$ is the (unique) element of minimal norm in such vector space, it turns out that it is independent of the coordinates chosen to define $\mathscr{L}(Dw, D^2w)\bar{h}$.

Having defined the operator $\mathscr{L}(w)$ we can rewrite (46) as a "formal system of ordinary differential equations"

$$\begin{cases} \dot{w}(t) = \mathscr{L}(w(t))\dot{h}(t)\,, \\ \\ w(t_0) = w_0\,. \end{cases} \tag{48}$$

The problem with this approach is that the operator $\mathscr{L}$ "loses derivatives" in its nonlinear entry w, namely although it defines the velocity $\dot{w}$ at order 0, it depends on first and second derivatives of w. Hence, if $w, \dot{h} \in C^k$, then $\mathscr{L}(w)h$ is, a priori, only in C^{k-2}. There is therefore no classical functional analytic setting to solve (48) in the usual way, namely no Banach space where we can apply a Picard–Lindelöf or a Cauchy–Lipschitz iteration.

In order to get around this (very discouraging) issue, Nash considered the regularized problem

$$\begin{cases} \dot{w}(t) = \mathscr{L}(\mathscr{S}_{t^{-1}}w(t))\dot{h}(t) \\ \\ w(t_0) = w_0\,. \end{cases} \tag{49}$$

However, $h(t)$ must now be chosen carefully and, in fact, it will be chosen depending upon $w(t)$, so that the complete system will be given by the coupling of (49) with a second equation relating $w(t)$ and $h(t)$. In order to describe the latter, we introduce a function $\psi \in C^\infty(\mathbb{R})$ which is:

(a) identically equal to 0 on the negative real axis;
(b) identically equal to 1 on $[1, \infty)$;
(c) everywhere nondecreasing.

The path h is then linked to w through the relation

$$h(t) = \mathscr{S}_{t^{-1}} \left[\psi(t - t_0)h + \int_{t_0}^{t} [2d(\mathscr{S}_{\tau^{-1}} w(\tau) - w(\tau))] \odot d\dot{w}(\tau)\, \psi(t - \tau)\, d\tau \right]. \tag{50}$$

From now on the system (49) and (50) will be called *Nash's regularized flow equations*.

In order to gain some insight in the latter complicated relation, assume for the moment that we are able to find an initial value t_0 and a smooth curve $t \mapsto (w(t), h(t))$ in C^3 satisfying (49) and (50) over $[t_0, \infty)$. In particular, when we refer to a "smooth solution" of the regularized flow equations, we understand that $\mathscr{S}_{t^{-1}} w(t)$ is a free map for every t in the domain of definition.

Assume further that $w(t)$ converges in C^2 to some $\bar{u}$ for $t \uparrow \infty$ and that the integrands in the following computations all decay sufficiently fast, so that we can integrate over the whole halfline $[t_0, \infty)$. The relation (49) implies that

$$2d(\mathscr{S}_{t^{-1}} w(t)) \odot d\dot{w}(t) = \dot{h}(t)\,. \tag{51}$$

Integrating the latter identity between t_0 and ∞, we then get

$$\int_{t_0}^{\infty} 2d(\mathscr{S}_{\tau^{-1}} w(\tau)) \odot d\dot{w}(\tau)\, d\tau = h(\infty) - h(t_0) = h\,. \tag{52}$$

Letting $t \to \infty$ in (50) and using that $\mathscr{S}_{t^{-1}}$ converges to the identity, we conclude

$$h = h(\infty) = h + \int_{t_0}^{\infty} 2d(\mathscr{S}_{\tau^{-1}} w(\tau) - w(\tau)) \odot d\dot{w}(\tau) d\tau\,,$$

implying therefore

$$\int_{t_0}^{\infty} 2d(\mathscr{S}_{\tau^{-1}} w(\tau)) \odot d\dot{w}(\tau)\, d\tau = \int_{t_0}^{\infty} 2dw(\tau) \odot d\dot{w}(\tau) d\tau\,. \tag{53}$$

Combining the latter equality with (52) we achieve

$$\int_{t_0}^{\infty} 2dw(\tau) \odot d\dot{w}(\tau) d\tau = h\,. \tag{54}$$

On the other hand, the integrand in the left-hand side is precisely $\frac{d}{d\tau} w(\tau)^\sharp e$, and thus we immediately conclude

$$\bar{u}^\sharp e - w_0^\sharp e = w(\infty)^\sharp e - w(t_0)^\sharp e = h\,, \tag{55}$$

namely that $\bar{u}$ is the map in the conclusion of Theorem 33.

In order to carry out the program above, we obviously have to ensure that

(a) The regularized flow equations, namely the pair (49) and (50), is locally solvable; more precisely, if there is a solution in the interval $[t_0, t_1]$, it can be prolonged to some larger open interval $[t_0, t')$.
(b) We have uniform estimates ensuring the global solvability, namely any smooth solution on $[t_0, t')$ can be smoothly prolonged to the closed interval $[t_0, t']$.

The combination of (a) and (b) would then imply the existence of a global solution on $[t_0, \infty)$. We further have to ensure that

(c) The limit $\bar{u}$ of $w(t)$ for $t \to \infty$ exists in the strong C^3 topology, and we have the appropriate decay of the integrands needed to justify the "formal computations" (51), (52), (53), (54), and (55)

This last step will make the computations above rigorous and ensure that $\bar{u}$ is a C^3 isometric embedding. In order to complete the proof of Theorem 29, we will then only need to show that, when $h \in C^k$, then u is also in C^k.

The program above will be carried out in the subsequent sections under the assumption that t_0 is sufficiently large and $\|h\|_3$ sufficiently small, depending on the "initial value" w_0. Moreover, we will follow a somewhat different order. First we tackle a set of a priori estimates which are certainly powerful enough to conclude (b) and (c), cf. Proposition 41. We then examine the local existence of the solution, which combined with the estimates of Proposition 41 will immediately imply both global solvability and convergence to an isometry, cf. Proposition 44. Finally, the higher differentiability of $\bar{u}$ is achieved in Proposition 45.

4.6 A Priori Estimates for Solutions of Nash's Regularized Flow Equations

We start by fixing one important constant: $\varepsilon > 0$ will be chosen so that

$$\text{if } \|u - w_0\|_2 \le 4\varepsilon \quad \text{then } u \text{ is a free embedding.} \tag{56}$$

Our main a priori estimates are summarized in the following proposition, which is indeed the core of Nash's approach.

Proposition 41 (A priori estimates) *For any t_0 sufficiently large there is $\delta(t_0) > 0$ such that, if $\|h\|_3 \le \delta$, then the following holds. Consider any solution w of* (49) *and* (50) *over an interval I (with left endpoint t_0 and which might be closed, open or infinite). If*

$$\|w(t) - w_0\|_3 + t^{-1}\|w(t) - w_0\|_4 \leq 2\varepsilon\,, \tag{57}$$

$$t^4\|\dot{h}(t)\|_0 + \|\dot{h}(t)\|_4 \leq 2\,, \tag{58}$$

then indeed we have the improved bounds

$$\|w(t) - w_0\|_3 + t^{-1}\|w(t) - w_0\|_4 \leq \varepsilon\,, \tag{59}$$

$$t^4\|\dot{h}(t)\|_0 + \|\dot{h}(t)\|_4 \leq 1\,. \tag{60}$$

Moreover,

$$t^4\|\dot{w}(t)\|_0 + \|\dot{w}(t)\|_4 \leq C_0\,, \tag{61}$$

and, if $I = [t_0, \infty)$, then there is a function $\delta(s)$ with $\lim_{s\to\infty}\delta(s) = 0$ such that

$$\|w(t) - w(s)\|_3 \leq \delta(s) \qquad \text{for all } t \geq s \geq t_0. \tag{62}$$

Before coming to the proof we recall here a few useful estimates.

Lemma 42 *If T is a smooth (i, j) tensor on Σ and $r < \sigma < s$ are three natural numbers, then there is a constant $C = C(r, s, \sigma, i, j)$ such that*

$$\|T\|_\sigma \leq C\|T\|_r^\lambda\|T\|_s^{1-\lambda} \qquad \text{where } \sigma = \lambda r + (1-\lambda)s. \tag{63}$$

If $\Psi : \Gamma \to \mathbb{R}^k$ is a smooth map, with $\Gamma \subset \mathbb{R}^\kappa$ compact and r a natural number, then there is a constant $C(r, \Psi)$ such that

$$\|\Psi \circ v\|_r \leq C(1 + \|v\|_r) \qquad \text{for every smooth } v : \Sigma \to \Gamma. \tag{64}$$

For every $r \in \mathbb{R}$ there is a constant $C(r)$ such that

$$\|\varphi\psi\|_r \leq C\|\varphi\|_0\|\psi\|_r + C\|\varphi\|_r\|\psi\|_0 \qquad \text{for every } \varphi, \psi \in C^r(\Sigma). \tag{65}$$

The inequality extends as well to (tensor) product of tensors, where the constant will depend additionally only on the type of tensors involved.

The lemma above follows from rather standard and well-known arguments and we will give some explanations and references at the end of section. We underline here a crucial consequence, which will be used repeatedly in our arguments.

Remark 43 From (63) we easily conclude that, if $\|T(t)\|_k \leq \lambda t^j$ and $\|T\|_{k+i} \leq \lambda t^{j+i}$, then $\|T\|_{k+\kappa} \leq C\lambda t^{j+\kappa}$ for all intermediate $\kappa \in \{1, \ldots, j-1\}$.[18]

Proof (Proof of Proposition 41*)* First of all, if t_0 is chosen larger than a fixed constant, we can use (57) and Proposition 39(d) to conclude that $\|\mathscr{S}_{t^{-1}} w(t) - w_0\|_2 \leq 4\varepsilon$. In turn, by (56), this implies that, when computing the operator $\mathscr{L}$, the entries of $\mathscr{L}^{ij}_\alpha$ belong to a compact set where the corresponding functions are smooth. Observe moreover that $\|w(t)\|_3 \leq C$, for some constant C depending only upon the initial value w_0. We can thus apply (64) and Proposition 39 to conclude that

$$\|\mathscr{L}(\mathscr{S}_{t^{-1}} w(t))\|_\kappa \leq C(\kappa)(1 + t^{\kappa-1}) \tag{66}$$

where $C(\kappa)$ is a constant which depends only upon κ. In fact, for $\kappa \geq 1$ we have

$$\|\mathscr{L}(\mathscr{S}_{t^{-1}} w(t))\|_\kappa \overset{(64)}{\leq} C(\kappa)\|\mathscr{S}_{t^{-1}} w(t)\|_{\kappa+2} \leq C(\kappa)\|w(t)\|_3 t^{\kappa-1},$$

where the last inequality follows from Proposition 39(b). In the case of $\kappa = 0$, we use instead the estimate $\|\mathscr{S}_{t^{-1}} w(t)\|_2 \leq C\|w(t)\|_2$ (again cf. Proposition 39(b)).

Using now (65), from (49) we conclude that

$$\|\dot{w}(t)\|_0 \leq \|\mathscr{L}(\mathscr{S}_{t^{-1}} w(t))\|_0 \|\dot{h}(t)\|_0 \leq C t^{-4}, \tag{67}$$

$$\|\dot{w}(t)\|_4 \leq \|\mathscr{L}(\mathscr{S}_{t^{-1}} w(t))\|_4 \|\dot{h}(t)\|_0 + C\|\mathscr{L}(\mathscr{S}_{t^{-1}} w(t))\|_0 \|\dot{h}(t)\|_4 \leq C. \tag{68}$$

Indeed, this shows (61).

We next introduce some additional functions in order to make some expressions more manageable. More precisely

$$E(t) := 2d(\mathscr{S}_{t^{-1}} w(t) - w(t)) \odot d\dot{w}(t), \tag{69}$$

$$L(t) := \int_{t_0}^t E(\tau)\psi(t-\tau)\, d\tau. \tag{70}$$

Observe that with the introduction of these two quantities we can rewrite (50) as

$$h(t) = \mathscr{S}_{t^{-1}}[\psi(t-t_0)h + L(t)]. \tag{71}$$

Recalling Proposition 39, we have $\|\mathscr{S}_{t^{-1}} w(t) - w(t)\|_1 \leq C t^{-2}\|w(t)\|_3 \leq C t^{-2}$. Observe that $\|\dot{w}(t)\|_1 \leq C t^{-3}$, which follows from (67) and (68) because of Remark 43 (this is just one of several instances where such remark will be used!).

[18]Nash does not take advantage of this simple remark and introduces instead a rather unusual notation to keep track of all the estimates for the intermediate norms in the bounds corresponding to (59), (60) and (61).

Combining the latter estimate with (65), we then conclude $\|E(t)\|_0 \leq Ct^{-5}$. On the other hand,

$$\|\mathscr{S}_{t^{-1}} w(t) - w(t)\|_4 \leq Ct\,,$$

and hence again from (65) we conclude

$$\|E(t)\|_3 \leq C\|\mathscr{S}_{t^{-1}} w(t) - w(t)\|_4 \|\dot{w}(t)\|_1 + C\|\mathscr{S}_{t^{-1}} w(t) - w(t)\|_1 \|\dot{w}(t)\|_4 \leq Ct^{-2}\,. \tag{72}$$

The latter inequality yields

$$\|L(t)\|_3 \leq \int_{t_0}^{t} \|E(\tau)\|_3 \, d\tau \leq Ct_0^{-1}\,. \tag{73}$$

Next, we compute

$$\dot{h}(t) = \underbrace{\left(\tfrac{d}{dt}\mathscr{S}_{t^{-1}}\right)[\psi(t-t_0)h + L(t)]}_{=:P(t)} + \mathscr{S}_{t^{-1}}[\psi'(t-t_0)h + \dot{L}(t)]\,.$$

First, we observe that $\psi'(t-t_0)$ vanishes for $t > t_0 + 1$ and $t < t_0$. Hence

$$\|\psi'(t-t_0)\mathscr{S}_{t^{-1}} h\|_4 \leq \begin{cases} Ct_0\delta \text{ for } t \in [t_0, t_0+1], \\ 0 \qquad \text{otherwise.} \end{cases} \tag{74}$$

For the same reason (and because $\psi(0) = 0$) we can estimate

$$\|\dot{L}(t)\|_0 \leq \int_{\max\{t_0, t-1\}}^{t} \|E(\tau)\|_0 \, d\tau \leq Ct^{-5}\,, \tag{75}$$

$$\|\dot{L}(t)\|_3 \leq \int_{\max\{t_0, t-1\}}^{t} \|E(\tau)\|_3 \, d\tau \leq Ct^{-2}\,. \tag{76}$$

Next, recalling that $\mathscr{S}'_\varepsilon := \frac{d}{d\varepsilon}\mathscr{S}_\varepsilon$, we have

$$\tfrac{d}{dt}\mathscr{S}_{t^{-1}} = -t^{-2}\mathscr{S}'_{t^{-1}}\,.$$

Hence, using Proposition 39(c) and (73), it is straightforward to show that

$$t^4\|P(t)\|_0 + \|P(t)\|_4 \leq C(\|h(t)\|_3 + \|L(t)\|_3) \leq C\delta + Ct_0^{-1}\,, \tag{77}$$

where C is independent of δ. Combining (74), (75), (76), and (77) we get

$$t^4\|\dot{h}(t)\|_0 + \|\dot{h}(t)\|_4 \leq Ct^{-1} + C\delta(1+t_0^5) + Ct_0^{-1} \leq Ct_0^{-1} + C\delta t_0^5 \,. \tag{78}$$

Therefore, choosing first t_0 large enough and then $\delta \leq \delta_0(t_0)$ sufficiently small, we conclude a bound which is even stronger than (60): the left-hand side can be made smaller than any fixed $\eta > 0$.

The estimate on $\|w(t)-w_0\|_4$ in (59) is an obvious consequence of the one above on $\|\dot{h}(t)\|_4$ through integration of (49): it suffices to choose η smaller than a given constant. The proof of the remaining parts of (59) and (62) require instead a subtler argument. However, notice also that we just need to accomplish (62), since C_0 is a constant claimed to be independent of t_0.

In order to get (62) we integrate (49) and then integrate by parts:

$$\begin{aligned} w(t) - w(s) & \\ &= \int_s^t \mathscr{L}(\mathscr{S}_{\tau^{-1}}(w(\tau)))\dot{h}(\tau)\, d\tau \\ &= -\int_s^t \underbrace{\left[\frac{d}{d\tau}\mathscr{L}(\mathscr{S}_{\tau^{-1}}(w(\tau)))\right]}_{=:D(\tau)}(h(\tau)-h(t))\, d\tau + \mathscr{L}(\mathscr{S}_{t^{-1}}(w(s)))(h(t)-h(s))\,. \end{aligned} \tag{79}$$

First of all, integrating the bound (60) on $\dot{h}(t)$, we obviously conclude

$$\|h(t)-h(s)\|_0 \leq Cs^{-3} \qquad \text{for all } t \geq s \geq t_0. \tag{80}$$

Next, assuming that $t \geq s \geq t_0 + 1$, we have $\psi(s-t_0) = \psi(t-t_0) = 1$ and we can thus compute

$$\begin{aligned} h(t) - h(s) = (\mathscr{S}_{t^{-1}}h - \mathscr{S}_{s^{-1}}h) + \underbrace{\mathscr{S}_{t^{-1}}\int_s^t E(\tau)\,\psi(t-\tau)\, d\tau}_{\text{(I)}} & \\ + \underbrace{\mathscr{S}_{t^{-1}}\int_{s-1}^s E(\tau)\,(\psi(t-\tau)-\psi(s-\tau))\, d\tau}_{\text{(II)}} + \underbrace{(\mathscr{S}_{t^{-1}} - \mathscr{S}_{s^{-1}})L(s)}_{\text{(III)}}\,. & \end{aligned} \tag{81}$$

Note next that

$$\|\text{(I)}+\text{(II)}\|_3 \leq C\int_{s-1}^t \|E(\tau)\|_3\, d\tau \leq C\int_{s-1}^\infty \tau^{-2}\, d\tau \leq Cs^{-1}\,.$$

For what concerns (III) note that the bound (72) on $\|E(\tau)\|_3$ implies that

$$L(\infty) := \int_{t_0}^{\infty} E(\tau)\,d\tau$$

is well defined, it belongs to C^3, and it satisfies the following decay estimate:

$$\|L(\infty) - L(s)\|_3 \le C s^{-1}\,. \tag{82}$$

Thus we can bound

$$\|(\mathrm{III})\|_3 \le C s^{-1} + \|\mathcal{S}_{s^{-1}} L(\infty) - \mathcal{S}_{t^{-1}} L(\infty)\|_3\,,$$

which in turn leads to

$$\|h(t) - h(s)\|_3 \le C s^{-1} + \|\mathcal{S}_{s^{-1}} L(\infty) - \mathcal{S}_{t^{-1}} L(\infty)\|_3 \; + \|\mathcal{S}_{t^{-1}} h - \mathcal{S}_{s^{-1}} h\|_3\,. \tag{83}$$

Using the fact that $\mathcal{S}_{t^{-1}}$ converges to the identity for $t \to \infty$, we reach

$$\|h(t) - h(s)\|_3 \le \tilde{\delta}(s) \qquad \text{for all } t \ge s, \tag{84}$$

where $\tilde{\delta}(s)$ is a function such that $\lim_{s\to\infty} \tilde{\delta}(s) = 0$. Using (66), (80) and (83), we conclude

$$\|w(t) - w(s)\|_3 \le \bar{\delta}(s) + C \int_s^t (\|D(\tau)\|_3 \tau^{-3} + \|D(\tau)\|_0)\,d\tau\,, \tag{85}$$

for some function $\bar{\delta}(s)$ which converges to 0 as s goes to ∞.

In order to estimate carefully $D(t)$, we pass to local coordinates. Recalling the notation $\mathcal{L}^{ij}_\alpha = \mathcal{L}^{ij}_\alpha(A, B)$ of (47) we compute

$$\begin{aligned}
&\tfrac{d}{dt}\mathcal{L}^{ij}_\alpha (D\mathcal{S}_{t^{-1}} w(t), D^2\mathcal{S}_{t^{-1}} w(t)) \\
&\quad = \underbrace{D_A \mathcal{L}^{ij}_\alpha (D\mathcal{S}_{t^{-1}} w(t), D^2\mathcal{S}_{t^{-1}} w(t))}_{D'(t)} \circ \Big(-t^{-2} D\mathcal{S}'_{t^{-1}} w(t) + \mathcal{S}_{t^{-1}} D\dot{w}(t)\Big) \\
&\qquad + \underbrace{D_B \mathcal{L}^{ij}_\alpha (D\mathcal{S}_{t^{-1}} w(t), D^2\mathcal{S}_{t^{-1}} w(t))}_{D''(t)} \circ \Big(-t^{-2} D^2\mathcal{S}'_{t^{-1}} w(t) + \mathcal{S}_{t^{-1}} D^2\dot{w}(t)\Big)\,,
\end{aligned} \tag{86}$$

where $\circ$ denotes a suitable product structure. Now, as already argued for $\mathcal{L}(\mathcal{S}_{t^{-1}}(w(t))$, for any natural number κ we have

$$\|D'(t)\|_\kappa + \|D''(t)\|_\kappa \le C(\kappa)(1 + t^{\kappa-1})\,. \tag{87}$$

Moreover, having derived the bound $\|w(t)\|_4 \leq Ct$, we can take advantage of Proposition 39 to get

$$\|D(t)\|_0 \leq C\left(t^{-3}\|w(t)\|_4 + \|\dot{w}(t)\|_2\right) \leq Ct^{-2}. \tag{88}$$

In order to estimate the C^3 norm, we use (65), (87) and argue similarly to get:

$$\|D(t)\|_3 \leq Ct^2\left(\|\dot{w}(t)\|_2 + t^{-3}\|w(t)\|_4\right) + C\left(\|w(t)\|_4 + t\|\dot{w}(t)\|_4\right) \leq Ct. \tag{89}$$

Inserting the latter two inequalities in (85), we clearly conclude (62) and complete the proof.

Proof (Proof of Lemma 42*)* First of all, we observe that it suffices to prove all the claims for functions and in a local coordinate patch: hence, without loss of generality we can just prove the claim for functions on balls of $\mathbb{R}^n$.

Proof of (63)**.** By the classical extension theorems, it suffices to prove the inequality for functions defined on the whole $\mathbb{R}^n$ (under the assumptions that all norms are finite!). In such a case we will in fact have the stronger inequality

$$\|D^\sigma v\|_0 \leq C\|D^r v\|_0^\lambda \|D^s v\|_0^{1-\lambda}.$$

Clearly, it suffices to prove the inequality in the particular case where $r = 0 < \sigma < s$, where it takes the form

$$\|D^\sigma v\|_0 \leq C\|D^s v\|_0^{\sigma/s} \|v\|_0^{1-\sigma/s}.$$

If $v \equiv 0$, then there is nothing to prove. If $D^s v \equiv 0$, since the function is bounded, then we have $D^\sigma v \equiv 0$ and again the inequality is trivial. Otherwise, recall that we have the following elementary bound, with a constant C independent of v.

$$\|D^\sigma v\|_0 \leq C\|D^s v\|_0 + C\|v\|_0.$$

However, since we can rescale the function to $v_\varepsilon(r) = v(\varepsilon r)$, we also have the validity of

$$\|D^\sigma v\|_0 \leq C\varepsilon^{s-\sigma}\|D^s v\|_0 + C\varepsilon^{-\sigma}\|v\|_0,$$

with the very same constant C, i.e. independently of $\varepsilon > 0$. Choosing $\varepsilon = \|v\|_0^{1/s}\|D^s v\|_0^{-1/s}$ we conclude the proof.

Proof of (64)**.** Again we can assume that the domain of the function is $\mathbb{R}^n$. Denoting by D^j any partial derivative of order j, the chain rule can be written symbolically as

$$D^m(\Psi \circ v) = \sum_{l=1}^{m} (D^l \Psi) \circ v \sum_{\sigma} C_{l,\sigma} (Dv)^{\sigma_1} (D^2 v)^{\sigma_2} \dots (D^m v)^{\sigma_m} \tag{90}$$

for some constants $C_{l,\sigma}$, where the inner sum is over $\sigma = (\sigma_1, \dots, \sigma_m) \in \mathbb{N}^m$ such that

$$\sum_{j=1}^{m} \sigma_j = l, \quad \sum_{j=1}^{m} j\sigma_j = m.$$

From (63) we have

$$\|u\|_j \leq C_h \|u\|_0^{1-\frac{j}{m}} \|u\|_m^{\frac{j}{m}} \qquad \text{for } m \geq j \geq 0$$

(without loss of generality we assume both $\|u\|_0$ and $\|u\|_m$ nonzero, otherwise the inequality is trivial: thus we can use (63) also for the "extreme cases" $\sigma = r$ and $\sigma = s$!). Inserting the latter inequality in (90), we easily achieve (64).

Proof of (65)**.** Using the notation above we write the Leibniz rule as

$$D^m(\varphi\psi) = \sum_{j=0}^{m} \underbrace{C_{j,m} D^j \varphi D^{m-j} \psi}_{S_j} .$$

For each summand we use (63) and Young's inequality to write

$$\|S_j\|_0 \leq C\|\varphi\|_0^{1-j/m} \|\varphi\|_m^{j/m} \|\psi\|_0^{j/m} \|\psi\|_m^{1-j/m} \leq C\|\varphi\|_0 \|\psi\|_m + C\|\varphi\|_m \|\psi\|_0 .$$

□

4.7 Global Existence and Convergence to an Isometry

In this section we combine the bounds in Proposition 41 with a local solvability argument to show that there is a global solution to Nash's regularized flow equations.

Proposition 44 *There exist t_0 and δ such that, if $\|h\|_3 \leq \delta$, then there is a solution $t \mapsto w(t)$ of* (49) *and* (50) *on $[t_0, \infty)$ which satisfies the bounds* (59), (60), (61) *and* (62) *for every t. Moreover, for $t \to \infty$, $w(t)$ converges in C^3 to a free embedding $\bar{u}$ with $\bar{u}^\sharp e = w_0^\sharp e + h$.*

Proof The whole point lies in the following:

(Loc) assume $J = [t_0, t_1]$ is some closed interval (possibly trivial, namely, with $t_1 = t_0$) over which we have a solution of (49) and (50) satisfying the bounds (59), (60), (61) and (62). Then the solution can be prolonged on some open interval $[t_0, t_2[\supset [t_0, t_1]$ to a solution which satisfies the bounds (57) and (58).

The statement (Loc) and Proposition 41 easily imply the global existence claimed in the proposition. Indeed, if we let $[t_0, T)$ be the maximal interval over which there is a solution satisfying (59), (60), (61) and (62), the statement (Loc) with $t_1 = t_0$ and the a priori estimates in Proposition 41 imply that $T > t_0$, since for $t_1 = t_0$ we can simply set $w(t_0) = w$, $\dot h(t_0) = 0$ and all the bounds (59), (60), (61) and (62) would be trivially true. Moreover, if $T < \infty$, then the bounds in Proposition 41 imply that the solution can be smoothly extended to $[t_0, T]$ and (Loc) contradicts the maximality of T, establishing the global existence. The convergence to a C^3 $\bar u$ follows from the bound (62). In turn we have the bound

$$\|dw(t) \odot d\dot w(t)\|_0 + \|d(\mathscr{S}_{t^{-1}} w(t)) \odot d\dot w(t)\|_0 \le Ct^{-4},$$

so that all the integrals used in (52), (53), (54) and (55) converge in the uniform norm and define continuous functions. The computations in (52), (53), (54) and (55) are thus rigorous and yield $\bar u^\sharp e = w^\sharp e + h$.

Hence, in what follows we will focus on the proof of (Loc).

First of all, we rewrite (49) and (50) in terms of a fixed point for an integral operator on $(w, \lambda) := (w, \dot h)$. We start by writing

$$w(t) = w_0 + \int_{t_0}^t \mathscr{L}(\mathscr{S}_{\tau^{-1}} w(\tau))\lambda(\tau)\, d\tau =: w_0 + \int_{t_0}^t \mathscr{W}(w(\tau), \lambda(\tau))\, d\tau\,. \tag{91}$$

We then rewrite the function $E(t)$ of (69) as

$$E(t) = 2d(\mathscr{S}_{t^{-1}} w(t) - w(t)) \odot d(\mathscr{L}(\mathscr{S}_{t^{-1}} w(t))\lambda(t)) =: \mathscr{E}(w(t), \lambda(t))\,. \tag{92}$$

Finally,

$$\begin{aligned}
\lambda(t) &= \frac{d}{dt}\left\{\mathscr{S}_{t^{-1}}\left[\psi(t-t_0)h + \int_{t_0}^t \mathscr{E}(w(\tau), \lambda(\tau))\, \psi(t-\tau)\, d\tau\right]\right\}\\
&= \underbrace{\psi'(t-t_0)\mathscr{S}_{t^{-1}} h - t^{-2}\psi(t-t_0)\mathscr{S}'_{t^{-1}} h}_{=\mu(t)} - t^{-2}\mathscr{S}'_{t^{-1}} \int_{t_0}^t \mathscr{E}(w(\tau), \lambda(\tau))\, \psi(t-\tau)\, d\tau\\
&\quad + \mathscr{S}_{t^{-1}} \int_{t_0}^t \mathscr{E}(w(\tau), \lambda(\tau))\psi'(t-\tau) d\tau\,.
\end{aligned} \tag{93}$$

Observe now that the operator $\mathscr{W}$ is smooth on C^4, because of the regularization of $\mathscr{S}_t$ (cf. the proof of Proposition 41). The operator $\mathscr{E}$ is locally Lipschitz from C^4 to C^3 (cf. the proof of Proposition 41) because it loses one derivative, but on the other hand the operators $\mathscr{S}_t$ and $\mathscr{S}'_t$ in front of the integrals in the above expressions regularize again from C^3 to C^4. Hence the local existence in (Loc) follows from classical fixed point arguments.

We briefly sketch the details for the reader's convenience. We consider an interval $J = [t_0, t_1]$ as in (Loc) and $t_2 > t_1$, whose choice will be specified later. We consider a pair $(w, \lambda) \in C(J, C^4)$ which solves (91), (92) and (93) and satisfies

$$\|w(t) - w_0\|_3 + t^{-1}\|w(t) - w_0\|_4 \leq \varepsilon\,, \tag{94}$$

$$t^4\|\lambda(t)\|_0 + \|\lambda(t)\|_4 \leq 1 \tag{95}$$

(and in case $t_0 = t_1$ we simply set $w(t_0) = w_0$ and $\lambda(t_0) = 0$). We consider next the space X of pairs $(\underline{w}, \underline{\lambda}) \in C([t_0, t_2], C^4)$ such that

(a) $w = \underline{w}$ and $\lambda = \underline{\lambda}$ on the interval J;
(b) the following inequalities hold:

$$\|\underline{w}(t) - w_0\|_3 + t^{-1}\|\underline{w}(t) - w_0\|_4 \leq 2\varepsilon\,, \tag{96}$$

$$t^4\|\underline{\lambda}(t)\|_0 + \|\underline{\lambda}(t)\|_4 \leq 2\,. \tag{97}$$

On X we consider the norm $\|(\underline{w}, \underline{\lambda})\|_{4,0} := \max_{t\in[t_0,t_2]}(\|\underline{w}(t)\|_4 + \|\underline{\lambda}(t)\|_4)$. X is clearly a complete metric space. We then consider the transformation $\mathscr{A} : X \to C([t_0, t_2], C^4)$ given by $(\underline{w}, \underline{h}) \mapsto \mathscr{A}(\underline{w}, \underline{h}) = (\tilde{w}, \tilde{h})$ through the following formulas:

$$\tilde{w}(t) = w_0 + \int_{t_0}^{t} \mathscr{W}(\underline{w}(\tau), \underline{\lambda}(\tau))\, d\tau\,,$$

$$\tilde{\lambda}(t) = \mu(t) - t^{-2}\mathscr{S}'_{t^{-1}} \int_{t_0}^{t} \mathscr{E}(\underline{w}(\tau), \underline{\lambda}(\tau))\, \psi(t-\tau)\, d\tau + \mathscr{S}_{t^{-1}} \int_{t_0}^{t} \mathscr{E}(\underline{w}(\tau), \underline{\lambda}(\tau))\psi'(t-\tau) d\tau\,.$$

Now, if we assume $t_2 \leq t_1 + 1$, then $\max_t \|\mathscr{W}(\underline{w}(t), \underline{\lambda}(t)\|_4 \leq C$, because of the estimates (96) and (97). Hence we can estimate

$$\|\tilde{w}(t) - \underline{w}(t_1)\|_0 \leq \int_{t_1}^{t_2} \|\mathscr{W}(\underline{w}(\tau), \underline{\lambda}(\tau)\|_4\, d\tau \leq C(t_2 - t_1) \quad \forall t \geq t_1\,. \tag{98}$$

Similarly, since $\sup_t \|\mathscr{E}(\underline{w}(t), \underline{\lambda}(t))\|_3 \leq C$ and recalling the estimates of Proposition 39, we conclude that

$$\|\tilde{\lambda}(t) - \underline{\lambda}(t_1)\|_4 \leq \|\mu(t) - \mu(t_1)\|_4 + C(t_2 - t_1) \qquad \forall t \geq t_1\,.$$

From (94) and (95) and the smoothness of the map μ, it is easy to see that (96) and (97) is valid for the pair $(\tilde{w}, \tilde{\lambda})$ provided $t_2 - t_1$ is smaller than a certain threshold. In particular, for $t_2 - t_1$ small enough the operator $\mathscr{A}$ maps X into itself.

It remains to show the contraction property. Consider two pairs (w_1, λ_1), $(w_2, \lambda_2) \in X$ and $(\tilde{w}_i, \tilde{\lambda}_i) = \mathscr{A}(w_i, \lambda_i)$. Then, using the properties of the operators $\mathscr{S}_{t^{-1}}$ and $\mathscr{S}'_{t^{-1}}$ we easily conclude

$$\|\tilde{w}_1(t) - \tilde{w}_2(t)\|_{4,0} \leq \int_{t_1}^{t_2} \|\mathscr{W}(w_1(\tau), \lambda_1(\tau)) - \mathscr{W}(w_2(\tau), \lambda_2(\tau))\|_4 \, d\tau\,, \tag{99}$$

$$\|\tilde{\lambda}_1(t) - \tilde{\lambda}_2(t)\|_{4,0} \leq C \int_{t_1}^{t_2} \|\mathscr{E}(w_1(\tau), \lambda_1(\tau)) - \mathscr{E}(w_2(\tau), \lambda_2(\tau))\|_3 \, d\tau\,. \tag{100}$$

In turn, recalling the Lipschitz regularity of the operators $\mathscr{W}$ and $\mathscr{E}$, we easily achieve

$$\begin{aligned}\|\mathscr{A}(w_1, \lambda_1) - \mathscr{A}(w_2, \lambda_2)\|_{4,0} &= \|(\tilde{w}_1, \tilde{\lambda}_1) - (\tilde{w}_2, \tilde{\lambda}_2\|_{4,0} \\ &\leq C(t_2 - t_1)\|(w_1, \lambda_1) - (w_2, \lambda_2)\|_{4,0}\,.\end{aligned}$$

Again, it suffices to choose $t_2 - t_1$ smaller than a certain threshold to conclude that $\mathscr{A} : X \to X$ is a contraction.

4.8 *Higher Regularity of the Map* $\bar{u}$

Finally, in this section we complete the proof of Theorem 33 by showing the following result.

Proposition 45 *The map* $\bar{u}$ *of Proposition* 44 *belongs to* C^k *if* $h \in C^k$ *for* $k \geq 4$.

Proof The proof will be by induction on k. Assume that, under the assumption $h \in C^k$, we have shown that

$$\|w(t)\|_k + t^{-1}\|w(t)\|_{k+1} \leq C\,, \tag{101}$$

$$t^{k+1}\|\dot{h}(t)\|_0 + \|\dot{h}(t)\|_{k+1} \leq C\,, \tag{102}$$

for some constant C independent of t. We will then show that, under the assumption that $h \in C^{k+1}$, the same set of estimates hold with $k + 1$ in place of k, namely

$$\|w(t)\|_{k+1} + t^{-1}\|w(t)\|_{k+2} \leq C'\,, \tag{103}$$

$$t^{k+2}\|\dot{h}(t)\|_0 + \|\dot{h}(t)\|_{k+1} \leq C'\,, \tag{104}$$

with a constant C' which might be worse than C, but depends only on k and t_0 (the latter is, however, fixed in the statement of the proposition). Indeed the estimate for $\|w(t)\|_{k+1}$ will come from the following stronger claim: there is a function $\delta(s)$ which converges to 0 as $s \to \infty$ and such that

$$\|w(t) - w(s)\|_{k+1} \le \delta(s) \qquad \text{for all } t \ge s \ge t_0. \tag{105}$$

The claim obviously would complete the proof of the proposition, because it clearly shows that $w(t)$ converges in C^{k+1} as $t \uparrow \infty$. Hence, in the rest of the proof we will focus on showing (103), (104), and (105).

We start by estimating $\dot{w}(t)$ using (49) and recalling the same arguments of the proof of Proposition 41: from (101), (102), and Proposition 39 we conclude the bounds which are the analog of (67) and (68), namely

$$t^{k+1}\|\dot{w}(t)\|_0 + \|\dot{w}(t)\|_{k+1} \le C\,. \tag{106}$$

We next estimate the function $E(t)$ of (69), again using the arguments of Proposition 39. First, by Proposition 39(c) and (101) we get

$$t^k\|\mathscr{S}_{t-1}w(t) - w(t)\|_1 + \|\mathscr{S}_{t-1}w(t) - w(t)\|_{k+1} \le Ct\,. \tag{107}$$

Then, using (65) we conclude the bounds which are the analog of (72), namely

$$t^k\|E(t)\|_0 + \|E(t)\|_k \le Ct^{-2}\,. \tag{108}$$

We next recall the computation for $\dot{h}(t)$:

$$\begin{aligned}\dot{h}(t) = {} & \underbrace{-\frac{\psi(t-t_0)}{t^2}\mathscr{S}'_{t-1}h + \psi'(t-t_0)\mathscr{S}_{t-1}h}_{=:A(t)} \underbrace{-\frac{1}{t^2}\mathscr{S}'_{t-1}\overbrace{\int_{t_0}^{t} E(\tau)\psi(t-\tau)\,d\tau}^{=:L(t)}}_{=:B(t)} \\ & \underbrace{+\mathscr{S}_t\int_{\max\{t_0,t-1\}}^{t} E(\tau)\psi'(t-\tau)\,d\tau}_{=:C(t)}\,. \end{aligned} \tag{109}$$

Now, using that $h \in C^{k+1}$, Proposition 39(c), and the fact that $\psi'(t-t_0)$ vanishes for $t - t_0 > 1$, we easily conclude that

$$t^{k+2}\|A(t)\|_0 + \|A(t)\|_{k+2} \le C\,, \tag{110}$$

where the constant C depends on k and t_0 (which are both fixed). As for $C(t)$, we can use (108) and Proposition 39(b) to conclude

$$t^{k+2}\|C(t)\|_0 + \|C(t)\|_{k+2} \le C\,. \tag{111}$$

The estimate on $B(t)$ turns out to be more delicate. First notice that, by (108), we certainly conclude that $\|L(t)\|_k \le C$. Using now Proposition 39(c) we get however the weaker estimate

$$\|B(t)\|_{k+2} \le Ct\,. \tag{112}$$

We can now go back in the argument for (106) and recover $\|\dot{w}(t)\|_{k+2} \le Ct^2$. In turn, plugging this information in the derivation of (108) we get $\|E(t)\|_{k+1} \le Ct^{-1}$. The latter bound can be used to estimate $\|L(t)\|_{k+1} \le C \log t$ which in turn, using Proposition 39(c), improves (112) to

$$\|B(t)\|_{k+2} \le C \log t\,. \tag{113}$$

We can now iterate the whole process to reach, respectively,

$$\begin{aligned} \|\dot{h}(t)\|_{k+2} &\le C \log t\,, \\ \|\dot{w}(t)\|_{k+2} &\le C \log t\,, \\ \|w(t)\|_{k+2} &\le Ct \log t\,, \\ \|E(t)\|_{k+1} &\le Ct^{-2} \log t\,. \end{aligned}$$

Since however $t^{-2} \log t$ is integrable on $[t_0, \infty)$, we achieve the desired bound $\|B(t)\|_{k+2} \le C$ and indeed, using again Proposition 39(c),

$$t^{k+2}\|B(t)\|_0 + \|B(t)\|_{k+2} \le C\,. \tag{114}$$

Clearly (110), (114) and (111) yield (104). As already argued several times, we directly conclude $\|\dot{w}(t)\|_{k+2} \le C$ and $\|w(t)\|_{k+2} \le Ct$, namely (103). Besides, following the same reasoning as above we also conclude the following useful bound:

$$t^{k+1}\|E(t)\|_0 + \|E(t)\|_{k+1} \le Ct^{-2}\,. \tag{115}$$

Thus the only bound which remains to show is (105): the argument, however, follows almost verbatim the one for (62). We briefly sketch the details. First, we recall the computation in (79). Then, using the bound (104) we derive the analog of (80), namely

$$\|h(t) - h(s)\|_0 \le Cs^{-k-1} \qquad \text{for all } t \ge s \ge t_0. \tag{116}$$

Similarly, using (81) and (115) we derive

$$\begin{aligned} &\|h(t) - h(s)\|_{k+1} \\ &\quad \le Cs^{-1} + \|\mathscr{S}_{t^{-1}} h - \mathscr{S}_{s^{-1}} h\|_{k+1} + \|\mathscr{S}_{t^{-1}} L(\infty) - \mathscr{S}_{t^{-1}} L(\infty)\|_{k+1} \quad \forall t \ge s \ge t_0\,. \end{aligned} \tag{117}$$

Plugging these inequalities in (79) and using (66), we derive the existence of a function $\bar{\delta}(s)$ which converges to 0 as $s \to \infty$ and such that

$$\|w(t) - w(s)\|_{k+1} \le \bar{\delta}(s) + C \int_s^t (\|D(\tau)\|_{k+1}\tau^{-k-1} + \|D(\tau)\|_0)\, d\tau\,. \qquad (118)$$

This replaces the analogous estimate (85), where $D(t)$ is the quantity defined in (79). The estimate $\|D(\tau)\|_0 \le \tau^{-2}$ of (88) is certainly valid here as well. In order to estimate $\|D(t)\|_{k+1}$ we first recall the computations in (86) and the quantities $D'(t)$ and $D''(t)$ introduced there. Using the better bounds $\|w(t)\|_{k+2} \le Ct$ and (103), the estimate in (89) can in fact be improved to

$$\|D(t)\|_{k+1} \le Ct\,. \qquad (119)$$

Inserting the inequalities just found for $\|D(\tau)\|_0$ and $\|D(\tau)\|_{k+1}$ in (118), we immediately conclude (105), which completes our proof.

4.9 *The Nonclosed Case*

The proof of Corollary 30 uses a construction very similar to that employed Corollary 23 to show the existence of a short embedding of a noncompact manifold.

Proof (Proof of Corollary 30*)* Consider an open covering $\{U_\ell\}_\ell$ as in Lemma 25 and let $\mathscr{C}_i$ be the corresponding classes. As in the proof of Corollary 23, fix a family $\{\varphi_\ell\}_\ell$ of smooth functions with the properties that $\varphi_\ell \in C^\infty_c(U_\ell)$ and for every $p \in \Sigma$ there is at least one φ_ℓ which equals 1 on a neighborhood of p. Moreover, having ordered $\{U_\ell\}_\ell$ we fix a (strictly) decreasing number of parameters ε_ℓ, converging to 0.

Next consider the map $v^0 : \Sigma \to \mathbb{R}^{2(n+1)}$ defined in the following way: for each $i \in \{1, \dots, n+1\}$ and every $p \in \Sigma$, set

$$v^0_{2(i-1)+1}(p) = \varepsilon_\ell^2 \varphi_\ell(p) \quad \text{and} \quad v^0_{2i}(p) = \varepsilon_\ell \varphi_\ell(p)$$

if p is contained in some $U_\ell \in \mathscr{C}_i$, otherwise we set them equal to 0. As already shown in the proof of Corollary 23, the latter map is well-defined, and we let $h := (v^0)^\sharp e$. Provided we choose the ε_ℓ sufficiently small, we have $g - h > 0$.

For each U_ℓ fix a smooth map Φ_ℓ which maps U_ℓ diffeomorphically on the standard sphere $\mathbb{S}^n \setminus \{N\}$, where N denotes the north pole. We extend it to a smooth map on the whole manifold Σ by defining $\Phi_\ell \equiv N$ on $\Sigma \setminus U_\ell$. If σ denotes the standard metric on $\mathbb{S}^n$, we then select a sequence η_ℓ of sufficiently small positive numbers such that the tensor

$$\tilde{g} := g - h - \sum_\ell \eta_\ell \Phi_\ell^\sharp \sigma$$

is still positive definite. For each U_ℓ consider the tensor $g_\ell := \varphi_\ell^2 \left(\sum_\ell \varphi_\ell^2\right)^{-1} \tilde{g}$, so that

$$\sum_\ell g_\ell = \tilde{g}\,.$$

Observe that, since Φ_ℓ is a diffeomorphism on the support of g_ℓ, which in turn is contained in U_ℓ, the $(0,2)$ tensor $\bar{g}_\ell := (\Phi_\ell^{-1})^\sharp g_\ell$ is well-defined on $\mathbb{S}^n \setminus \{N\}$ and can be extended smoothly to $\mathbb{S}^n$ by setting it equal to 0. Thus there is an isometric embedding w^ℓ of $\mathbb{S}^n$ into $\mathbb{R}^{N_0}$ such that $(w^\ell)^\sharp e = \bar{g}_\ell + \eta_\ell \sigma$. By applying a translation we can assume that w^ℓ maps the north pole N in 0. Thus, $u^\ell := w^\ell \circ \Phi_\ell$ is a smooth map on Σ which vanishes identically outside U_ℓ and such that

$$(u^\ell)^\sharp e = g_\ell + \eta_\ell \Phi_\ell^\sharp \sigma\,.$$

Now, for each $i \in \{1, \ldots, n+1\}$ we define the map $v^i : \Sigma \to \mathbb{R}^{N_0}$ setting $v^i(p) = \varphi_\ell(p) u^\ell(p)$ if p belongs to some $U_\ell \in \mathscr{C}_i$ and 0 otherwise. Finally, let $u = v^0 \times v^1 \times \ldots \times v^{n+1}$. Then it is obvious from the construction and from Remark 34 that u is an isometry:

$$u^\sharp e = (v^0)^\sharp e + \sum_\ell g_\ell + \sum_\ell \eta_\ell \Phi_\ell^\sharp \sigma = h + \tilde{g} + \sum_\ell \eta_\ell \Phi_\ell^\sharp \sigma = g\,.$$

It follows therefore that u is necessarily an immersion. The argument of Corollary 23 finally shows that u is injective and completes the proof. Observe that, if we set instead

$$\tilde{g} := g - \sum_\ell \eta_\ell \Phi_\ell^\sharp \sigma\,,$$

and define analogously the maps w^ℓ, u^ℓ and v^i with $i \in \{1, \ldots, n+1\}$, the resulting map $\bar{u} = v^1 \times \ldots \times v^{n+1}$ is an isometric immersion of Σ: the only property which is lost compared to u is indeed the injectivity.

5 Continuity of Solutions of Parabolic Equations

5.1 Introduction

In 1958 Nash published his fourth masterpiece [76], a cornerstone in the theory of partial differential equations. His main theorem regarded bounded solutions of linear second-order parabolic equations with uniformly elliptic nonconstant coefficients.

More precisely, equations of the form

$$\partial_t u = \operatorname{div}_x(A(x,t)\nabla u)\,, \tag{120}$$

where:

(a) the unknown u is a function of time t and space $x \in \mathbb{R}^n$;
(b) $\partial_t u$ denotes the time partial derivative $\frac{\partial u}{\partial t}$;
(c) ∇u denotes the spatial gradient, namely the vector

$$\nabla u(x,t) = (\partial_1 u(x,t), \ldots, \partial_n u(x,t)) = \left(\frac{\partial u}{\partial x_1}(x,t), \ldots, \frac{\partial u}{\partial x_n}(x,t)\right),$$

(d) and $\operatorname{div}_x V$ denotes the (spatial) divergence of the vector field V, namely

$$\operatorname{div}_x V(x,t) = \partial_1 V_1(x,t) + \ldots + \partial_n V_n(x,t)\,.$$

Following Einstein's summation convention on repeated indices, we will often write

$$\operatorname{div}_x(A\nabla u) = \partial_i(A_{ij}\partial_j u)\,.$$

Assumption 46 In this section the coefficients A_{ij} will always satisfy the following requirements:

(S) Symmetry, namely $A_{ij} = A_{ji}$;
(M) Measurability, namely each $(x,t) \mapsto A_{ij}(x,t)$ is a (Lebesgue) measurable function;
(E) Uniform ellipticity, namely there is a $\lambda \geq 1$ such that

$$\lambda^{-1}|v|^2 \leq A_{ij}(x,t)v_i v_j \leq \lambda|v|^2 \qquad \forall (x,t) \in \mathbb{R}^n \times \mathbb{R} \text{ and } \forall v \in \mathbb{R}^n. \tag{121}$$

Clearly, since the coefficients A_{ij} are not assumed to be differentiable, we have to specify a suitable notion of solution for (120).

Definition 47 In what follows, the term *solution* of (120) in an open domain $\Omega \subset \mathbb{R}^n \times \mathbb{R}$ will denote a locally summable function u with locally square summable distributional derivatives $\partial_j u$ satisfying the identity

$$\int u(x,t)\partial_t\varphi(x,t)\,dx\,dt = \int \partial_i\varphi(x,t)A_{ij}(x,t)\partial_j u(x,t)\,dx\,dt \qquad \forall \varphi \in C^\infty_c(\Omega)\,. \tag{122}$$

The following is then Nash's celebrated Hölder continuity theorem. As usual we denote by $\|f\|_\infty$ the (essential) supremum of the measurable function f and, in case f coincides with a continuous function a.e., we state pointwise inequalities omitting the "almost everywhere" specification.

Theorem 48 (Nash's parabolic regularity theorem) *There are positive constants C and α depending only upon λ and n with the following property. If the matrix A satisfies Assumption* 46 *and u is a bounded distributional solution of* (120) *in $\mathbb{R}^n \times (0, \infty)$, then the following estimate holds for all $t_2 \geq t_1 > 0$ and all $x_1, x_2 \in \mathbb{R}^n$:*

$$|u(x_1, t_1) - u(x_2, t_2)| \leq C\|u\|_\infty \left[\frac{|x_1 - x_2|^\alpha}{t_1^{\alpha/2}} + \left(\frac{t_2 - t_1}{t_1} \right)^{\frac{\alpha}{2(1+\alpha)}} \right] . \tag{123}$$

From the above theorem, Nash derived a fundamental corollary in the case of second-order elliptic equations

$$\text{div}_x(A\nabla v) = 0 , \tag{124}$$

where the measurable coefficients A_{ij} do not depend on t.

Definition 49 If Ω is an open domain of $\mathbb{R}^n$, the term distributional solution v of (124) in Ω will denote a locally summable function v with locally square summable distributional derivatives $\partial_j u$ satisfying the identity

$$\int \partial_i v(x) A_{ij}(x) \partial_j \varphi(x) \, dx = 0 \qquad \forall \varphi \in C_c^\infty(\Omega) .$$

The following theorem is nowadays called De Giorgi–Nash theorem, since indeed De Giorgi proved it[19] independently of Nash in [22] (see [24] for the English translation).

Theorem 50 (De Giorgi–Nash) *There are positive constants C and β depending only upon λ and n with the following property. If the matrix A satisfies Assumption* 46 *and v is a bounded distributional solution of* (124) *in $B_{3r}(z) \subset \Omega$, then the following estimate holds for every $x, y \in B_r(z)$:*

$$|v(x) - v(y)| \leq C\|v\|_\infty r^{-\beta} |x - y|^\beta . \tag{125}$$

Theorem 50 was sufficient to give a positive answer to Hilbert's XIXth problem, namely the regularity of scalar minimizers of uniformly convex Lagrangians in any dimension, cf. [22, Teorema III]. The case $n = 2$ had been previously settled by Morrey in [66] and it was also known that the Hölder continuity of the first derivative

[19] In fact, De Giorgi's statement is stronger, since in his theorem $\|v\|_\infty$ in (125) is replaced by the L^2 norm of v (note that the power of r should be suitably adjusted: the reader can easily guess the correct exponent using the invariance of the statement under the transformation $u_r(x) = u(rx)$).

of the minimizer would suffice to conclude its full regularity, see [50, 67]. The De Giorgi–Nash theorem closed the gap.[20]

The De Giorgi–Nash Hölder continuity theorem is false for elliptic *systems*, as it was noticed by De Giorgi in [23]. In fact, for vectorial problems in the calculus of variations Nečas proved later the existence of nondifferentiable minimizers of smooth uniformly convex functionals when both the domain and the target have sufficiently large dimension. The methods of Nečas were refined further in [44] and [95], and recently the paper [65] used a different construction to show the existence of a nondifferentiable minimizer when the target is 2-dimensional and the domain 3-dimensional. Since Morrey's work shows the regularity for planar minimizers even in the vectorial case, the latter example is in the lowest possible dimensions. Finally, in [96] it was shown that if the domain is 5-dimensional, vectorial minimizers might even be unbounded!

Various authors rewrote, simplified and pushed further the De Giorgi–Nash theory. The two most important contributors are probably Moser [69] and Aronson [5]. Moser introduced the versatile Moser iteration, based on the study of the time-evolution of successive powers of the solution, which simplifies the proof (and avoids the explicit use of the entropy functional Q, see Definition 53). Moser further proved what is usually called Harnack inequality (although a more appropriate name in this case would probably be "Moser–Harnack"). For positive solutions v of (124), the inequality is the estimate

$$\sup_{B_r(x)} v \leq C \inf_{B_{2r}(x)} v,$$

where the constant C only depends on r, the dimension n and the ellipticity constant λ.

Aronson established a Gaussian type bound on the associated fundamental solution $S(x, t, \bar{x}, \bar{t})$ (cf. Theorem 52), more precisely he bounded the latter from above and from below with functions of the form

$$\frac{K}{(t-\bar{t})^{n/2}} e^{-B|x-\bar{x}|^2/(t-\bar{t})}$$

(Nash established the (weaker) upper bound with $K(t-\bar{t})^{-n/2}$, cf. Proposition 54).

These three results, namely the Hölder continuity, the Moser–Harnack inequality, and the Gaussian type bounds, are all connected and in some sense equivalent. Fine expositions of this, as well as clever rewritings/simplifications/improvements of the proofs, can be found in Bass [7, Ch. 7], [8] and Fabes and Stroock [32].

---

[20] Indeed, it was known that the first partial derivatives of the minimizer satisfy a uniformly elliptic partial differential equation with measurable coefficients. De Giorgi's stronger version of Theorem 50 would then directly imply the desired Hölder estimate. Nash's version was also sufficient, because a theorem of Stampacchia guaranteed the local boundedness of the first partial derivatives, cf. [93].

Most of the section will be dedicated to Nash's proof of Theorem 48, whereas Theorem 50 will be derived from Theorem 48 in the last section.

5.2 Preliminaries and Main Statements

Nash's approach to Theorem 48 follows initially the well-known path of proving "a priori estimates". More precisely, standard arguments reduce Theorem 48 to the following weaker version. In the rest of our discussion, we will use "smooth" to denote C^∞ functions. All the statements will indeed hold under much less restrictive regularity assumptions, namely the existence and continuity of a suitable number of derivatives needed to justify the computations contained in the arguments. On the other hand, since such precise results are not needed later, in order to keep the presentation less technical we will ignore the issue.

Theorem 51 (A priori estimate) *Theorem* 48 *holds under the additional assumptions that*

(A1) A_{ij} *is smooth on* $\mathbb{R}^n \times \mathbb{R}$ *for all* $i, j = 1, \ldots n$;
(A2) $A_{ij} = \delta_{ij}$ *outside of a compact set* $K \times [0, T]$;
(A3) u *is smooth.*

Observe a crucial point: it is well known (and it was well known at the time Nash wrote his note) that the assumptions (A1)–(A3) imply the smoothness of any solution of (120), but the crucial point in Theorem 51 is that the constants C and α of (123) are *independent* of A (more precisely, they depend only on the dimension n and the constant λ in (121)). We will focus on Theorem 51 for most of the subsequent sections and only at the end, in Sect. 5.8, we will show how to conclude Theorem 48 from it.[21]

Under the assumptions (A1)–(A3) of Theorem 51 we take advantage of the existence of fundamental solutions. More precisely, we recall the following theorem (see [35, Ch. 1.6]).

Theorem 52 *Under the assumptions of Theorem* 51 *there is a smooth map*

$$(x, t, \bar{x}, \bar{t}) \mapsto S(x, t, \bar{x}, \bar{t})$$

defined for $x, \bar{x} \in \mathbb{R}^n$ *and* $t > \bar{t}$ *with the following properties:*

(a) *The map* $(x, t) \mapsto S(x, t, \bar{x}, \bar{t}) = T(x, t)$ *is a classical solution of* (120) *on* $\mathbb{R}^n \times (\bar{t}, \infty)$.

[21] Nash does not provide any argument nor reference, he only briefly mentions that Theorem 48 follows from Theorem 51 using a regularization scheme and the maximum principle. Note that a derivation of the latter under the weak regularity assumptions of Theorem 48 is, however, not entirely trivial: in Sect. 5.8 we give an alternative argument based on a suitable energy estimate.

(b) *$T(\cdot, t)$ and $\partial_t^k T(\cdot, t)$ belong to the Schwartz space of rapidly decreasing smooth functions $\mathscr{S}(\mathbb{R}^n)$ and the corresponding seminorms can be bounded uniformly when t belongs to a compact subset of $(\bar{t}, \infty)$.*
(c) *$T > 0$ and $\int T(x,t)\,dx = 1$ for every $t > \bar{t}$.*
(d) *$T(\cdot, t)$ converges, in the sense of measures, to the Dirac mass $\delta_{\bar{x}}$ as $t \downarrow \bar{t}$, namely*

$$\lim_{t\downarrow\bar{t}} \int T(x,t)\varphi(x)\,dx = \varphi(\bar{x})$$

for any bounded continuous test function φ. Moreover, for any ball $B_r(\bar{x})$, the function $T(\cdot, t)$ converges to 0 on $\mathbb{R}^n \setminus B_r(\bar{x})$ with respect to all the seminorms of the Schwartz space.
(e) *For any u bounded smooth solution of (120) on $\mathbb{R}^n \times [\bar{t}, T[$ we have the representation formula*

$$u(x,t) = \int S(x,t,y,\bar{t})u(y,\bar{t})\,dy\,. \tag{126}$$

Vice versa, given a bounded smooth $u_0(y) =: u(y,\bar{t})$ the formula above gives the unique solution on $[\bar{t}, \infty[$ subject to the corresponding initial condition.
(f) *The properties above hold for the map $(\bar{x}, \bar{t}) \mapsto S(x,t,\bar{x},\bar{t}) = \bar{T}(\bar{x},\bar{t})$ on the domain $\mathbb{R}^n \times (-\infty, t)$, which therefore is a (backward in time) fundamental solution of the adjoint equation*

$$-\partial_{\bar{t}}\bar{T} = \partial_{\bar{x}_j}(A_{ij}\partial_{\bar{x}_i}\bar{T})\,. \tag{127}$$

Except for the smoothness, the existence of a map S with all the properties listed above is given in [35, Ch. 1] (note that point (f) is proved in [35, Th. 15]). The latter reference shows that S has continuous first-order derivatives (in time and space) and continuous second-order derivatives in space when the coefficients A_{ij} are C^2 (in fact $C^{1,\alpha}$, cf. [35, Th 10]). Decay properties for the function and its first-order space derivatives are then showed in [35, Th 11]. The higher regularity (and the decay of higher derivatives) when the coefficients A_{ij} are smooth and constant outside of a compact set, follows easily from the arguments given in [35], and we have stated it only for completeness: indeed the arguments of Nash do not really need this additional information.

In the remaining sections we will derive several bounds on the map S which will finally lead to a proof of Theorem 51 through the representation formula (126). Three very relevant quantities which we will compute on the fundamental solutions are the energy, the entropy and the first moment.

Definition 53 Under the assumptions of Theorem 51 let $T(x,t) := S(x,t,0,0)$, where S is the map of Theorem 52. We then introduce

(i) The *energy* $E(t) := \int T(x,t)^2\,dx$.

(ii) The *entropy* $Q(t) := -\int T(x,t)\log T(x,t)\,dx$.
(iii) The *first moment* $M(t) := \int T(x,t)|x|\,dx$.

On each of these quantities (which by Theorem 52 are smooth on $(0,\infty)$) Nash derives subtle crucial estimates, which we summarize in the following proposition.

Proposition 54 (Bounds on the energy, the entropy and the moment) *Under the assumptions of Theorem* 51 *there are positive constants* C_1, C_2, C_3 *and* C_4*, depending only upon* λ *and* n*, such that the following holds. If* T, E, Q *and* M *are as in Definition* 53*, then*

$$E(t) \le C_1 t^{-n/2}, \tag{128}$$

$$\|T(\cdot,t)\|_\infty \le C_2 t^{-n/2}, \tag{129}$$

$$Q(t) \ge -C_3 + \frac{n}{2}\log t, \tag{130}$$

$$C_4^{-1} t^{1/2} \le M(t) \le C_4 t^{1/2}. \tag{131}$$

The last bound is in fact the cornerstone of Nash's proof. With it he derives subsequently what he calls *G bound*.

Definition 55 Let T be as in Definition 53 and consider the "normalization" U of the fundamental solution: $U(y,t) := t^{n/2} T(t^{1/2} y, t)$. For any $\delta \in]0,1[$ the G_δ-functional is

$$G_\delta(t) = \int e^{-|y|^2} \log(U(y,t)+\delta)\,dy. \tag{132}$$

Proposition 56 (*G* bound) *Under the assumptions of Theorem* 51 *there are constants* C_5 *and* δ_0*, depending only upon* λ *and* n*, such that the following holds. If* G_δ *is as in Definition* 55*, then*

$$G_\delta(t) \ge -C_5(-\log\delta)^{1/2} \qquad \textit{for all } \delta < \delta_0. \tag{133}$$

In turn Proposition 56 will be used in an essential way to compare fundamental solutions for different source points. Observe in fact that the integrand defining G_δ is rather negative at those points ξ which are close to 0 (the "source" of the fundamental solution) and where at the same time the value of U is low. Our goal, namely bounding $G_\delta(t)$ from below by $-C(-\log\delta)^{1/2}$, is thus to gain control on such "bad points". In particular Proposition 56 allows to derive the central "overlap estimate" for fundamental solutions, namely the following result.

Proposition 57 (Overlap estimate) *Under the assumptions of Theorem* 51 *there are positive constants* C *and* α*, depending only upon* λ *and* n*, such that, if* S *is the* *map of Theorem 52, then*

$$\int |S(x,t,x_1,\bar{t}) - S(x,t,x_2,\bar{t})|\,dx \le C\left(\frac{|x_1-x_2|}{(t-\bar{t})^{1/2}}\right)^{\alpha} \qquad \text{for all } t > \bar{t}. \tag{134}$$

The Hölder estimate in space for a bounded solution u is a direct consequence of the overlap estimate and of (126), whereas the estimate in time will follow from additional considerations taking into account the other bounds derived above.

After collecting some elementary inequalities in the next section, we will proceed, in the subsequent three sections, to prove the three Propositions 54, 56, and 57. We will then show in Sect. 5.7 how Theorem 51 follows.

5.3 *Three Elementary Inequalities*

In deriving the estimates claimed in the previous section we will use three "elementary" inequalities on functions. All of them have been generalized in various ways in the subsequent literature and hold under less restrictive assumptions than those stated here: the statements given below are just sufficient for our purposes and I have tried to keep them as elementary as possible.

The first inequality is nowadays known as "Nash's inequality". In [76] Nash credits the proof to Elias Stein.

Lemma 58 (Nash's inequality) *There is a constant C, depending only upon n, such that the following inequality holds for any function $v \in \mathscr{S}(\mathbb{R}^n)$:*

$$\left(\int_{\mathbb{R}^n} |v(x)|^2\,dx\right)^{1+2/n} \le C\left(\int_{\mathbb{R}^n} |\nabla v(x)|^2\,dx\right)\left(\int_{\mathbb{R}^n} |v(x)|\right)^{4/n}. \tag{135}$$

The second is a Poincaré-type inequality in a "Gaussian-weighted" Sobolev space.

Lemma 59 (Gaussian Poincaré inequality) *The following inequality holds for any bounded C^1 function f on $\mathbb{R}^n$ with bounded derivatives and which satisfies the constraint $\int e^{-|\xi|^2} f(\xi)\,d\xi = 0$:*

$$2\int_{\mathbb{R}^n} e^{-|\xi|^2} f^2(\xi)\,d\xi \le \int_{\mathbb{R}^n} e^{-|\xi|^2} |\nabla f(\xi)|^2\,d\xi\,. \tag{136}$$

The proof of the final inequality in [76] is credited to Lennart Carleson:

Lemma 60 (Carleson's inequality) *There is a positive constant c, depending only on n, such that the following inequality holds for any positive function $T \in \mathscr{S}(\mathbb{R}^n)$*

with $\int_{\mathbb{R}^n} T(x)\,dx = 1$*:*

$$\int_{\mathbb{R}^n} |x| T(x)\,dx \geq c \exp\left[-\frac{1}{n}\int_{\mathbb{R}^n} T(x)\log T(x)\,dx\right]. \tag{137}$$

Proof (Proof of Lemma 58*)* Consider the Fourier transform[22] $\hat{v}$ of v:

$$\hat{v}(\xi) := (2\pi)^{-n/2}\int e^{ix\cdot\xi}\, v(x)\,dx\,.$$

Recalling the Plancherel identity and other standard properties of the Fourier transform we achieve

$$\int |v(x)|^2\,dx = \int |\hat{v}(\xi)|^2\,d\xi \tag{138}$$

$$\int |\nabla v(x)|^2\,dx = \int |\xi|^2 |\hat{v}(\xi)|^2\,d\xi \tag{139}$$

$$|\hat{v}(\xi)| \leq (2\pi)^{-n/2}\int |v(x)|\,dx \qquad \forall \xi \in \mathbb{R}^n\,. \tag{140}$$

Using (140) we obviously get

$$\int_{\{|\xi|\leq\rho\}} |\hat{v}(\xi)|^2\,d\xi \leq C\rho^n \left(\int |v(x)|\,dx\right)^2,$$

whereas using (139) we have

$$\int_{\{|\xi|\geq\rho\}} |\hat{v}(\xi)|^2\,d\xi \leq \int \frac{|\xi|^2}{\rho^2}|\hat{v}(\xi)|^2\,d\xi = \frac{1}{\rho^2}\int |\nabla v(x)|^2\,dx\,.$$

Equation (138) and the last two inequalities can be combined to reach

$$\int |v(x)|^2\,dx \leq C\rho^n \left(\int |v(x)|\,dx\right)^2 + \frac{1}{\rho^2}\int |\nabla v(x)|^2\,dx\,, \tag{141}$$

where the constant C is independent of ρ.

Next, the inequality (135) is trivial if v or ∇v vanishes identically. Hence, we can assume that both integrals in the right-hand side of (135) are nonzero. Under this

[22] In order to simplify the notation we omit the domain of integration when it is the entire space.

assumption (135) follows right away from (141) once we set

$$\rho = \left[\frac{\int |\nabla v(x)|^2 \, dx}{\left(\int |v(x)| \, dx \right)^2} \right]^{\frac{1}{n+2}} .$$

□

Proof (Proof of Lemma 59*)* Consider the Hilbert space H of measurable functions f such that $\int e^{-|\xi|^2} f^2(\xi) \, d\xi < \infty$, with the scalar product

$$\langle f, g \rangle := \int e^{-|\xi|^2} f(\xi) g(\xi) \, d\xi .$$

It is well known that a Hilbert basis of H is given by suitable products of the Hermite polynomials (cf. [4, Sec. 6.1]): if H_i denotes the Hermite polynomial of degree i in one variable, suitably normalized, we define, for any $I = (i_1, \dots, i_n) \in \mathbb{N}^n$

$$H_I(\xi) = H_{i_1}(\xi_1) H_{i_2}(\xi_2) \cdot \ldots \cdot H_{i_n}(\xi_n) .$$

We then have

$$\int e^{-|\xi|^2} f^2(\xi) \, d\xi = \sum_I \alpha_I^2 , \tag{142}$$

$$\int e^{-|\xi|^2} (\partial_{\xi_j} f)^2(\xi) \, d\xi = \sum_I \beta_{I,j}^2 , \tag{143}$$

where

$$\alpha_I = \int e^{-|\xi|^2} f(\xi) H_I(\xi) \, d\xi , \tag{144}$$

$$\beta_{I,j} = \int e^{-|\xi|^2} \partial_{\xi_j} f(\xi) H_I(\xi) \, d\xi . \tag{145}$$

Integrating by parts and using the relation

$$\partial_{\xi_j} (e^{-|\xi|^2} H_I(\xi)) = (2i_j)^{1/2} H_I(\xi)$$

we easily achieve the identity

$$\sum_{j=1}^n \beta_{I,j}^2 = 2|I| \alpha_I^2 .$$

Therefore we conclude

$$\int e^{-|\xi|^2} |\nabla f(\xi)|^2 \, d\xi = 2 \sum_I |I| \alpha_I^2 \, . \tag{146}$$

Note that $|I| \leq 1$ as soon as $I \neq (0, 0, \ldots 0)$. Thus, the inequality (136) is a trivial consequence of (142) and (146) provided $\alpha_{(0,0,\ldots,0)} = 0$. Since the Hermite polynomial H_0 is simply constant, the latter condition is equivalent to $\int e^{-|\xi|^2} f(\xi)\, d\xi = 0$.

Proof (Proof of Lemma 60*)* For every fixed $\lambda \in \mathbb{R}$, consider the function $\ell(\tau) = \tau \log \tau + \lambda \tau$ on $(0, \infty)$. Observe that the function is convex, it converges to 0 as $\tau \to \infty$ and converges to ∞ as $\tau \to \infty$. Its derivative $\ell'(\tau) = \log \tau + (1 + \lambda)$ vanishes if and only for $\tau_0 = e^{-1-\lambda}$ and moreover $\ell(\tau_0) = -e^{-\lambda-1} < 0$: the latter must thus be the minimum of the function and therefore

$$\tau \log \tau + \lambda \tau \geq -e^{-\lambda-1} \qquad \text{for every positive } \tau .$$

In particular, for any choice of the real numbers $a > 0$ and $b \in \mathbb{R}$ we have

$$\int (T(x) \log T(x) + (a|x| + b) T(x)) \, dx \geq -e^{-b-1} \int e^{-a|x|} \, dx \, . \tag{147}$$

In analogy with the quantities introduced in Definition 53, we consider the entropy and the moment, namely

$$Q := -\int T(x) \log T(x) \, dx \, , \tag{148}$$

$$M := \int |x| T(x) \, dx \, , \tag{149}$$

and we let $D(n)$ be the dimensional constant

$$D(n) := \int e^{-|x|} \, dx \, .$$

Then we can rewrite (147) as

$$-Q + aM + b \geq -e^{-b-1} a^{-n} D(n) \tag{150}$$

(where we have also used that $\int T(x)\, dx = 1$). Set $a := \frac{n}{M} > 0$ and $e^{-b} = \frac{e}{D(n)} a^n$. Then (150) turns into

$$-Q + n - \log\left(\frac{e}{D(n)} \left(\frac{n}{M}\right)^n\right) \geq -1 \, .$$

In turn, the latter is equivalent to

$$n - n \log n + \log D(n) + n \log M \geq Q\,.$$

Exponentiating the latter inequality we conclude $M \geq c(n) e^{Q/n}$ for some positive constant $c(n)$, which is precisely inequality (137).

5.4 Energy, Entropy and Moment Bounds

In this section we prove Proposition 54.

Proof (Proof of the energy estimate (128)*)* We differentiate E and compute

$$\begin{aligned}
E'(t) &= 2\int T(x,t)\partial_t T(t,x)\,dx = 2\int T(x,t)\partial_j(A_{ij}(x,t)\partial_i T(x,t))\,dx \\
&= -2\int \partial_j T(x,t) A_{ij}(x,t)\partial_j T(x,t)\,dx \leq -2\lambda^{-1}\int |\nabla T(x,t)|^2\,dx \\
&\overset{(135)}{\leq} -C\left(\int |T(x,t)|^2\,dx\right)^{1+2/n} = -CE^{1+2/n}\,,
\end{aligned}$$

where C is a positive constant depending only upon λ and n. Note moreover that in the last line we have used $\int T(x,t)\,dx = 1$. Since $E(t)$ is positive for every $t > 0$ we conclude that $\frac{d}{dt} E(t)^{-2/n} \geq C > 0$. By Theorem 52(d), $\lim_{t\downarrow 0} E(t)^{-1} = 0$ and thus we can integrate the differential inequality to conclude that

$$E(s)^{-2/n} = \int_0^s \frac{d}{dt} E(t)^{-2/n}\,dt \geq Cs\,,$$

which in turn implies $E(s) \leq C_1 s^{-n/2}$, where C_1 depends only upon λ and n.

Proof (Proof of the uniform bound (129)*)* By translation invariance, from the energy estimate we conclude

$$\int |S(x,t,\bar{x},\bar{t})|^2\,dx \leq C(t-\bar{t})^{-n/2}\,.$$

By Theorem 52(f) the above argument applies to the adjoint equation to derive also the bound

$$\int |S(x,t,\bar{x},\bar{t})|^2\,d\bar{x} \leq C(t-\bar{t})^{-n/2}\,.$$

On the other hand, using Theorem 52(e), we have

$$T(x,t) = \int S(x,t,\bar{x},\tfrac{t}{2})T(\bar{x},\tfrac{t}{2})\,d\bar{x}\,.$$

Using the Cauchy–Schwarz inequality, we then conclude

$$|T(x,t)|^2 \leq E(\tfrac{t}{2}) \textstyle\int |S(x,t,\bar{x},\tfrac{t}{2})|^2\,d\bar{x} \leq Ct^{-n}\,. \tag{151}$$

Proof (Proof of the entropy bound (130)*)* The L^∞ bound and the monotonicity of the logarithm gives easily

$$Q(t) \geq -\log\|T(\cdot,t)\|_\infty \int T(x,t)\,dx = -\log\|T(\cdot,t)\|_\infty \geq -C + \frac{n}{2}\log t\,.$$

Proof (Proof of the moment bound (131)*)* The first ingredient is Lemma 60, which gives $M(t) \geq Ce^{Q(t)/n}$. Next, differentiating the entropy we get

$$\begin{aligned}
Q'(t) &= -\int (1+\log T(x,t))\partial_t T(x,t)\,dx = -\int (1+\log T(x,t))\partial_j(A_{ij}(x,t)\partial_i T(x,t))\,dx\\
&= \int \partial_j \log T(x,t) A_{ij}(x,t)\partial_i T(x,t)\,dx\\
&= \int \left(\partial_j \log T(x,t) A_{ij}(x,t)\partial_i \log T(x,t)\right)\, T(x,t)\,dx\\
&\geq \lambda^{-1}\int |A(x,t)\nabla \log T(x,t)|^2 T(x,t)\,dx\,.
\end{aligned}$$

Recall that $\int T(x,t)\,dx = 1$ to estimate further

$$Q'(t) \geq \lambda^{-1}\left(\int |A(x,t)\nabla\log T(x,t)|T(x,t)\,dx\right)^2 = \lambda^{-1}\left(\int |A(x,t)\nabla T(x,t)|\,dx\right)^2.$$

Whereas, differentiating the momentum:

$$M'(t) = \int |x|\partial_j(A_{ij}(x,t)\partial_i T(x,t))\,dx = -\int \frac{x_j}{|x|}A_{ij}(x,t)\partial_i T(x,t)\,dx\,.$$

We thus conclude $|M'(t)|^2 \leq \lambda\, Q'(t)$.

Let us summarize the inequalities relevant for the rest of the argument, namely the entropy bound (130), Carleson's inequality, and the one just derived:

$$Q(t) \geq -C_3 + \frac{n}{2}\log t\,, \tag{152}$$

$$M(t) \geq Ce^{Q(t)/n}\,, \tag{153}$$

$$Q'(t)^{1/2} \geq \lambda^{-1/2}|M'(t)|\,. \tag{154}$$

Recall moreover that, from Theorem 52(d), $\lim_{t\downarrow 0} M(t) = 0$. We thus set $M(0) = 0$: this information and the three inequalities above will allow us to achieve the desired bound.

Define $nR(t) = Q(t) + C_3 - \frac{n}{2}\log t$. Observe that $Q'(t) = nR'(t) + \frac{n}{2t}$. Hence we can use (153) and integrate (154) to achieve

$$c_1 t^{1/2} e^{R(t)} \le M(t) \le c_2 \underbrace{\int_0^t \left(\tfrac{1}{2s} + R'(s)\right)^{1/2} ds}_{=:I(t)} . \tag{155}$$

Using the concavity of $\xi \mapsto (1+\xi)^{1/2}$ on $[-1,\infty)$, we conclude that $(1+\xi)^{1/2} \le 1 + \frac{\xi}{2}$ and thus

$$\left(\tfrac{1}{2s} + R'(s)\right)^{1/2} \le \left(\frac{1}{2s}\right)^{1/2}\left(1 + \frac{1}{2}R'(s)2s\right) = (2s)^{-1/2} + \left(\tfrac{s}{2}\right)^{1/2} R'(s) .$$

Hence

$$\begin{aligned} I(t) &\le \int_0^t (2s)^{-1/2}\, ds + \int_0^t \left(\tfrac{s}{2}\right)^{1/2} R'(s)\, ds = (2t)^{1/2} + \left(\tfrac{t}{2}\right)^{1/2} R(t) - \int_0^t (8s)^{-1/2} R(s)\, ds \\ &\le (2t)^{1/2} + \left(\tfrac{t}{2}\right)^{1/2} R(t) . \end{aligned}$$

Inserting the latter inequality in (155) and dividing by $t^{1/2}$ we conclude that

$$e^{R(t)} \le \frac{c_3 M(t)}{t^{1/2}} \le c_4 \left(1 + \frac{R(t)}{2}\right) , \tag{156}$$

where c_3 and c_4 are positive constants (depending only upon n and λ). Now, the map

$$\rho \mapsto e^{\rho} - c_4\left(1 + \frac{\rho}{2}\right)$$

converges to ∞ for $\rho \uparrow \infty$ and thus (156) implies that $R(t)$ is bounded by a constant which depends only upon λ and n. In turn, again from (156), we conclude (131).

5.5 G Bound

In this section we prove Proposition 56. We will use in an essential way the bounds of Proposition 54, especially the moment bound.

We begin by noting the obvious effect of the normalization $U(\xi,t) = t^{n/2} T(t^{1/2}\xi, t)$. All the estimates of Proposition 54 turn into corresponding "time-independent" bounds, which we collect here:

$$\int U(\xi,t)\,d\xi = 1\,, \tag{157}$$

$$\int |U(\xi,t)|^2\,d\xi \le C\,, \tag{158}$$

$$\|U(\cdot,t)\|_\infty \le C\,, \tag{159}$$

$$C^{-1} \le \int |\xi||U(\xi,t)|\,d\xi \le C\,, \tag{160}$$

for some constant C depending only on λ and n.

Moreover, the parabolic equation for T transforms into the equation

$$2t\partial_t U(\xi,t) = nU(\xi,t) + \xi_i\partial_i U(\xi,t) + 2\partial_j(A_{ij}(t^{1/2}\xi,t)\partial_i U(\xi,t))\,, \tag{161}$$

and observe that the "rescaled" coefficients $\bar{A}_{ij}(\xi,t) := A_{ij}(t^{1/2}\xi,t)$ satisfy the same ellipticity condition as A_{ij}, namely $\lambda^{-1}|v|^2 \le \bar{A}_{ij}v_iv_j \le \lambda|v|^2$.

Differentiating (132) we achieve

$$\begin{aligned}
2tG'_\delta(t) &= \int e^{-|\xi|^2}\frac{2t\partial_t U(\xi,t)}{U(\xi,t)+\delta}\,d\xi\\
&\overset{(161)}{=} \underbrace{n\int e^{-|\xi|^2}\frac{U(\xi,t)}{U(\xi,t)+\delta}\,d\xi}_{=:H_1(t)\ge 0} + \underbrace{\int e^{-|\xi|^2}\frac{\xi\cdot\nabla U(\xi,t)}{U(\xi,t)+\delta}\,d\xi}_{=:H_2(t)}\\
&\quad + \underbrace{2\int e^{-|\xi|^2}\frac{\partial_j(\bar{A}_{ij}(\xi,t)\partial_i U(\xi,t))}{U(\xi,t)+\delta}\,d\xi}_{=:H_3(t)}\,.
\end{aligned} \tag{162}$$

As for H_2, integrating by parts we get

$$\begin{aligned}
H_2(t) &= \int e^{-|\xi|^2}\xi\cdot\nabla(\log(U(\xi,t)+\delta)\,d\xi = -\int e^{-|\xi|^2}(n-2|\xi|^2)\log(U(\xi,t)+\delta)\,d\xi\\
&= -nG_\delta(t) + 2\int e^{-|\xi|^2}|\xi|^2\left(\log\delta + \log\left(1+\delta^{-1}U(\xi,t)\right)\right)d\xi\\
&\ge -nG_\delta(t) + 2\log\delta\int |\xi|^2 e^{-|\xi|^2}\,d\xi \ge -nG_\delta(t) + C\log\delta\,.
\end{aligned} \tag{163}$$

Finally, integrating by parts H_3:

$$\begin{aligned}
H_3(t) &= -2\int \partial_j\left(e^{-|\xi|^2}(U(\xi,t)+\delta)^{-1}\right)\bar{A}_{ij}(\xi,t)\partial_i U(\xi,t)\,d\xi\\
&= 4\int e^{-|\xi|^2}\xi_j\bar{A}_{ij}(\xi,t)\frac{\partial_i U(\xi,t)}{U(\xi,t)+\delta}\,d\xi + 2\int e^{-|\xi|^2}\frac{\partial_j U(\xi,t)\bar{A}_{ij}(\xi,t)\partial_i U(\xi,t)}{(U(\xi,t)+\delta)^2}\,d\xi
\end{aligned}$$

$$= \underbrace{4\int e^{-|\xi|^2}\xi_j \bar{A}_{ij}(\xi,t)\partial_i \log(U(\xi,t)+\delta)\,d\xi}_{:=H_4(t)}$$
$$+ \underbrace{2\int e^{-|\xi|^2}\partial_j \log(U(\xi,t)+\delta)\bar{A}_{ij}(\xi,t)\partial_i \log(U(\xi,t)+\delta)\,d\xi}_{=:H_5(t)}\,. \tag{164}$$

Note first that, by the ellipticity condition, the integrand of $H_5(t)$ is indeed nonnegative.

Next, for each (ξ, t) consider the quadratic form $\mathscr{A}(v, w) = \bar{A}_{ij}(\xi, t)v_i w_j$. The ellipticity condition guarantees that this is a scalar product. Hence, we have the corresponding Cauchy–Schwarz inequality $|\mathscr{A}(v, w)|^2 \leq \mathscr{A}(v, v)\mathscr{A}(w, w)$. Using this observation, $H_4(t)$ can be bounded by

$$\begin{aligned}
|H_4(t)| &\leq 4\int e^{-|\xi|^2}\left(\xi_i \bar{A}_{ij}(\xi,t)\xi_j\right)^{1/2}\left(\partial_h \log(U(\xi,t)+\delta)\bar{A}_{hk}(\xi,t)\partial_k \log(U(\xi,t)+\delta)\right)^{1/2} d\xi\\
&\leq 4\left(\int e^{-|\xi|^2}\xi_j \bar{A}_{ij}(\xi,t)\xi_j\,d\xi\right)^{1/2} H_5(t)^{1/2}\\
&\leq C H_5(t)^{1/2}\,.
\end{aligned} \tag{165}$$

Inserting (165), (164) and (163) in (162) we conclude the intermediate inequality

$$2tG'_\delta(t) \geq C\log\delta - nG_\delta(t) - CH_5(t)^{1/2} + H_5(t)\,. \tag{166}$$

The moment bound (160) will be used in a crucial way to prove the following

Lemma 61 *There are positive constants $\bar{G}$ and $\bar{c}$, both depending only upon λ and n, such that, if $\delta \leq 1$ and $G_\delta(t) \leq -\bar{G}$, then $H_5(t) \geq \bar{c}(1 - G_\delta(t))^2$.*

We postpone the proof of the lemma after showing how Proposition 56 follows easily from it and from the inequality (166). First of all observe that, under the assumption that $G_\delta(t) \geq -\tilde{G} \geq \bar{G}$, if the constant $\tilde{G}$ is chosen sufficiently large, then $H_5(t) - CH_5(t)^{1/2} \geq \bar{c}2G_\delta(t)^2$. Hence, we conclude the existence of positive constants $\tilde{G}$, $\tilde{c}$, C (depending only upon λ and n) such that

$$2tG'_\delta(t) \geq \tilde{c}G_\delta(t)^2 + C\log\delta \qquad \text{if } G_\delta(t) \leq -\tilde{G} \text{ and } \delta \leq 1. \tag{167}$$

Set therefore $C_5 := \left(\frac{C+1}{\tilde{c}}\right)^{1/2}$ and let $\delta_0 \leq 1$ be such that

$$C_5(-\log\delta_0)^{1/2} \geq \tilde{G}\,.$$

We now want to show that with these choices the estimate of Proposition 56 holds. In fact, assume that $\delta \leq \delta_0$ and that at some point $\tau > 0$ we have

$$G_\delta(\tau) < -C_5(-\log\delta)^{1/2}\,.$$

By our choice of δ_0 this would imply $G_\delta(\tau) < -\tilde{G}$, which in turn implies, by (167),

$$2\tau G_\delta'(\tau) \geq -\log\delta\,. \tag{168}$$

In particular, there is an $\varepsilon > 0$ such that G_δ is increasing on the interval $(\tau - \varepsilon, \tau)$. We then conclude that $G_\delta(\tau - \varepsilon) < -C_5(-\log\delta)^{1/2}$ and we can proceed further: it can only be that $G_\delta < -C_5(-\log\delta)^{1/2}$ on the whole interval $(0, \tau)$. But then (168) would be valid on $(0, \tau)$ and we would conclude that

$$\lim_{\tau\downarrow 0} G_\delta(\tau) = -\infty\,,$$

contradicting the trivial bound $G_\delta > \log\delta$.

In order to complete the proof of Proposition 56 it remains to show that Lemma 61 holds.

Proof (Proof of Lemma 61*)* Observe that, by the ellipticity condition,

$$H_5(t) \geq 2\lambda^{-1}\int e^{-|\xi|^2}|\nabla\log(U(\xi,t)+\delta)|^2\,d\xi\,. \tag{169}$$

We now wish to apply Lemma 59. We set for this reason

$$f(\xi) := \log(U(\xi,t)+\delta) - \pi^{-n/2}\int e^{-|\xi|^2}\log(U(\xi,t)+\delta)\,d\xi = \log(U(\xi,t)+\delta) - \pi^{-n/2}G_\delta(t)\,.$$

This choice achieves $\nabla f = \nabla\log(U+\delta)$ and $\int e^{-|\xi|^2} f(\xi)\,d\xi = 0$. We can thus apply Lemma 59 which, combined with (169), gives

$$H_5(t) \geq 4\lambda^{-1}\int e^{-|\xi|^2}\left(\log(U(\xi,t)+\delta) - \pi^{-n/2}G_\delta(t)\right)^2 d\xi\,. \tag{170}$$

Consider now the following function g on the positive real axis:

$$g(u) := u^{-1}(\log(u+\delta) - \pi^{-n/2}G_\delta(t))^2\,.$$

Since U is (strictly) positive, we have

$$\pi^{-n/2}G_\delta(t) > \pi^{-n/2}\log\delta\int e^{-|\xi|^2}d\xi = \log\delta\,. \tag{171}$$

Moreover g is nonnegative and vanishes only at the only positive point $\bar{u}$ such that

$$\log(\bar{u}+\delta) = \pi^{-n/2}G_\delta(t)\,.$$

Next, differentiating g we find

$$g'(u) = -u^{-2}(\log(u+\delta)-\pi^{-n/2}G_\delta(t))^2+2u^{-1}(u+\delta)^{-1}(\log(u+\delta)-\pi^{-n/2}G_\delta(t))\,.$$

Hence the derivative g' vanishes at $\bar{u}$ and at any other (positive) point u_m which solves

$$\underbrace{\log(u+\delta)-\pi^{-n/2}G_\delta(t)-2\frac{u}{u+\delta}}_{=:h(u)} = 0\,. \tag{172}$$

The function $h(u)$ is negative for $u \leq \bar{u}$ and thus any solution of the equation must be larger than $\bar{u}$. In fact

$$h(\delta) = \log 2 + \log\delta - \pi^{-n/2}G_\delta(t) - 1 \overset{(171)}{\leq} \log 2 - 1 < 0\,.$$

Since $\delta \leq 1$, we certainly conclude that any solution u_m of (172) must be larger than δ. On the other hand, differentiating h we find

$$h'(u) = \frac{2u}{(u+\delta)^2} - \frac{1}{u+\delta}\,,$$

which is strictly positive for $u \geq \delta$.

We conclude that there is a unique point $u_m > \bar{u}$ which satisfies (172). On the other hand

$$\lim_{u\uparrow\infty} g(u) = 0\,. \tag{173}$$

Hence u_m must be a local maximum for g, and g is strictly decreasing on $]u_m, \infty[$.

Observe next that

$$\log u_m < \log(u_m+\delta) \leq \pi^{-\frac{n}{2}}G_\delta(t) + 2\,.$$

We therefore conclude that

$$u_m < \exp(2+\pi^{-\frac{n}{2}}G_\delta(t)) =: U_0(t)\,.$$

Define

$$U^*(\xi,t) := \begin{cases} U(\xi,t) \text{ if } U(\xi,t) \geq U_0(t), \\ 0 \qquad\quad \text{otherwise.} \end{cases}$$

Summarizing we can bound

$$H_5(t) \geq c \int e^{-|\xi|^2} g(U^*(\xi,t))U^*(\xi,t)\, d\xi\,. \tag{174}$$

Recalling (159), we have $\|U^*(\cdot,t)\|_\infty \leq C$. If we set $\bar{C} = \max\{C, e^3\}$, we have $\|U^*(\cdot,t)\|_\infty \leq \bar{C}$ and, at the same time, $\bar{C} \geq e^3 \geq U_0(t) \geq u_m$, because for $G_\delta(t)$ we have the trivial bound

$$G_\delta(t) \leq \int \log(U(\xi,t)+\delta)\, d\xi \leq \int U(\xi,t)\, d\xi = 1\,. \tag{175}$$

Using the monotonicity of g on $]u_m, \infty[$ we then infer

$$H_5(t) \geq c \int e^{-|\xi|^2} (\log(\bar{C}+\delta) - \pi^{-n/2} G_\delta(t))^2 U^*(\xi,t)\, d\xi\,, \tag{176}$$

where c is a small but positive constant (depending only on λ and n) and $\bar{C}$ is a constant larger than e^3, also depending only on λ and n. In particular, the trivial bound (175) implies

$$\log(\bar{C}+\delta) - \pi^{-n/2} G_\delta(t) = \pi^{-n/2}\left(\pi^{n/2}\log(\bar{C}+\delta) - G_\delta(t)\right) \geq \pi^{-n/2}(1-G_\delta(t)) \geq 0\,,$$

and we therefore conclude

$$\begin{aligned} H_5(t) &\geq c_0(1-G_\delta(t))^2 \int e^{-|\xi|^2} U^*(\xi,t)\, d\xi \\ &= c_0(1-G_\delta(t))^2 \underbrace{\int_{|\xi|\geq \exp(2+G_\delta(t))} e^{-|\xi|^2} U(\xi,t)\, d\xi}_{=:I}\,. \end{aligned} \tag{177}$$

Clearly, in order to complete the proof of the lemma we just need to show the existence of positive constants $\bar{G}$ and $\bar{c}$ such that

$$G_\delta(t) \leq -\bar{G} \qquad \Longrightarrow \qquad I \geq \bar{c}\,.$$

Under the assumption $G_\delta(t) \leq -\bar{G}$, for any $\mu > 0$ we can write

$$I \geq e^{-\mu^2} \int_{\mu \geq |\xi| \geq \exp(2-\bar{G})} U(\xi,t)\, d\xi = e^{-\mu^2}\left(1 - \int_{|\xi|\leq \exp(2-\bar{G})} U(\xi,t)\, d\xi - \int_{|\xi|\geq\mu} U(\xi,t)\, d\xi\right).$$

Using (159) we have

$$\int_{|\xi|\leq \exp(2-\bar{G})} U(\xi,t)\, d\xi \leq C(\exp(2-\bar{G}))^n$$

for a constant C depending only on n and λ. In particular, if we choose $\bar{G}$ large enough we can assume that the integral above is bounded by $\frac{1}{4}$. Next, using (160) we get

$$\int_{|\xi|\geq\mu} U(\xi,t)\,d\xi \leq \frac{1}{\mu}\int U(\xi,t)|\xi|\,d\xi \leq \frac{C}{\mu}\,.$$

Thus, it suffices to fix μ large enough so that the latter integral is also smaller than $\frac{1}{4}$. With such choice, $G_\delta(t) \leq -\bar{G}$ implies $I \geq \frac{1}{2}e^{-\mu^2}$, which thus completes the proof.

5.6 Overlap Estimate

We are now ready to prove Proposition 57. First of all we notice that, without loss of generality, we can assume $\bar{t} = 0$. We thus consider two fundamental solutions $S(x,t,x_1,0)$ and $S(x,t,x_2,0)$. Fix for the moment a positive time t and set $\xi_i := x_i t^{-1/2}$ and

$$U_i(\xi) := t^{n/2} S(t^{1/2}\xi, t, x_i, 0)\,.$$

By Proposition 56 we have

$$\int e^{-|\xi-\xi_i|^2}\log(U_i(\xi)+\delta)\,d\xi \geq -C_5(-\log\delta)^{1/2} \tag{178}$$

for all $\delta \leq \delta_0$. In particular, in the rest of this paragraph we will certainly assume $\delta \leq 1$.

We then add the two inequalities above to get

$$\int \left[e^{-|\xi-\xi_1|^2}\log(U_1(\xi)+\delta) + e^{-|\xi-\xi_2|^2}\log(U_2(\xi)+\delta)\right] d\xi \geq -2C_5(-\log\delta)^{1/2} \quad \forall\delta\leq\delta_0\,. \tag{179}$$

Let

$$\begin{aligned}
U_+(\xi) &:= \max\{U_1(\xi), U_2(\xi)\}\,,\\
U_-(\xi) &:= \min\{U_1(\xi), U_2(\xi)\}\,,\\
f_+(\xi) &:= \max\{\exp(-|\xi-\xi_1|^2), \exp(-|\xi-\xi_2|^2)\}\,,\\
f_-(\xi) &:= \min\{\exp(-|\xi-\xi_1|^2), \exp(-|\xi-\xi_2|^2)\}\,.
\end{aligned}$$

Recalling the elementary bound $ac + bd \leq \max\{a, b\}\max\{c, d\} + \min\{a, b\}$ $\min\{c, d\}$ we then conclude

$$\int \left[f_+(\xi)\log(U_+(\xi)+\delta) + f_-(\xi)\log(U_-(\xi)+\delta)\right] d\xi \geq -2C_5(-\log\delta)^{1/2}. \tag{180}$$

Since $\delta \leq 1$, we have

$$\log(U_+(\xi)+\delta) \leq U_+(\xi) \leq U_1(\xi) + U_2(\xi),$$

and consequently we can bound

$$\int f_+(\xi)\log(U_+(\xi)+\delta)\, d\xi \leq \int (U_1(\xi)+U_2(\xi))\, d\xi \leq 2. \tag{181}$$

Next, we bound

$$\log(U_-(\xi)+\delta) = \log\delta + \log(1+\delta^{-1}U_-(\xi)) \leq \log\delta + \delta^{-1}U_-(\xi),$$

and thus

$$\int f_-(\xi)\log(U_-(\xi)+\delta)\, d\xi \leq \log\delta \int f_-(\xi)\, d\xi + \delta^{-1}\int U_-(\xi)\, d\xi. \tag{182}$$

Now, observe that $\int f_-(\xi)\, d\xi$ is simply a function w of $|\xi_1 - \xi_2|$, which is positive and decreasing. Thus, combining (180), (181), and (182) we achieve

$$\int U_-(\xi)\, d\xi \geq \max_{\delta\leq\delta_0} \delta\left[-2 - w(|\xi_1-\xi_2|)\log\delta - 2C_5(-\log\delta)^{1/2}\right] =: \phi(|\xi_1-\xi_2|). \tag{183}$$

The function ϕ is nonnegative and decreasing. Considering the rescaling which defined the U_i's we then conclude

$$\int \min\{S(x,t,x_1,0), S(x,t,x_2,0)\}\, dx = \int U_-(\xi)\, d\xi \geq \phi\left(\frac{|x_1-x_2|}{t^{1/2}}\right), \tag{184}$$

Next, recall the elementary identity

$$|\sigma - \tau| = \sigma + \tau - 2\min\{\sigma, \tau\},$$

valid for every positive σ and τ. In particular, we can combine it with (184) to conclude

$$\begin{aligned}\frac{1}{2}\int |S(x,t,x_1,0)-S(x,t,x_2,0)|\,dx &= 1-\int \min\{S(x,t,x_1,0),S(x,t,x_2,0)\}\,dx\\ &\leq 1-\phi\left(\frac{|x_1-x_2|}{t^{1/2}}\right) := \psi\left(\frac{|x_1-x_2|}{t^{1/2}}\right),\end{aligned} \tag{185}$$

where ψ is a positive increasing function strictly smaller than 1 everywhere. Observe, moreover, that with the same argument we easily achieve

$$\frac{1}{2}\int |S(x,t,x_1,\bar{t})-S(x,t,x_2,\bar{t})|\,dx \leq \psi\left(\frac{|x_1-x_2|}{(t-\bar{t})^{1/2}}\right), \tag{186}$$

whenever $t \geq \bar{t}$.

We will pass from (185) to (134) through an iterative argument. In order to implement such argument we introduce the functions

$$T_a(x,t) = \max\{S(x,t,x_1,0)-S(x,t,x_2,0),0\}, \tag{187}$$

$$T_b(x,t) = \max\{S(x,t,x_2,0)-S(x,t,x_1,0),0\}, \tag{188}$$

and

$$A(t) := \int T_a(x,t)\,dx = \int T_b(x,t)\,dx = \frac{1}{2}\int |S(x,t,x_1,0)-S(x,t,x_2,0)|\,dx\,.$$

Note, moreover, that although we have defined A only for $t > \bar{t}$, from the first identity in the derivation of (134) and the properties of the fundamental solution, it is easy to see that $\lim_{t\downarrow 0} A(t) = 1$.

Furthermore, let $T_a^*(x,t,\bar{t})$ and $T_b^*(x,t,\bar{t})$ be the solutions of (120) with respective initial data $T_a(x,\bar{t})$ and $T_b(x,\bar{t})$ at t. Note therefore the identities

$$T_a^*(x,t,\bar{t}) = \int S(x,t,y,\bar{t})T_a(y,\bar{t})\,dy = \int S(x,t,y,\bar{t})\underbrace{T_a(y,\bar{t})T_b(z,\bar{t})A(\bar{t})^{-1}}_{=:\chi(y,z,\bar{t})}\,dy\,dz\,, \tag{189}$$

$$T_b^*(x,t,\bar{t}) = \int S(x,t,z,\bar{t})T_b(z,\bar{t})\,dz = \int S(x,t,z,\bar{t})\chi(y,z,\bar{t})\,dy\,dz\,. \tag{190}$$

Moreover, $T_a^*(x,\bar{t},\bar{t}) - T_b^*(x,\bar{t},\bar{t}) = S(x,\bar{t},x_1,0) - S(x,\bar{t},x_2,0)$ and thus

$$T_a^*(x,t,\bar{t}) - T_b^*(x,t,\bar{t}) = S(x,t,x_1,0) - S(x,t,x_2,0) \qquad \text{for every } t \geq \bar{t}.$$

We therefore conclude the inequality

$$|S(x,t,x_1,0) - S(x,t,x_2,0)| \le \int |S(x,t,z,\bar t) - S(x,t,y,\bar t)|\chi(y,z,\bar t)\,dy\,dz\,. \tag{191}$$

Note that, in principle, $A(t,\bar t)$ is defined for $t > \bar t$. On the other hand, it follows easily from the first equality in (185), that $\lim_{t\downarrow \bar t} A(t,\bar t) = 1$. Integrating (191) we then obtain

$$A(t) \le \int \psi\left(\frac{|y-z|}{(t-\bar t)^{1/2}}\right)\chi(y,z,\bar t)\,dy\,dz \quad \forall t > \bar t\,. \tag{192}$$

Observe in particular that

$$A(t) < \int \chi(y,z,\bar t)\,dy\,dz = A(\bar t)\,. \qquad \forall t > \bar t\,, \tag{193}$$

namely A is strictly monotone decreasing.

Let $\varepsilon := \phi(1) = 1-\psi(1)$ and define $\sigma := 1-\frac{\varepsilon}{4}$. For each natural number $k\ge 1$ we let t_k be the first time such that $A(t_k)\le \sigma^k$, if such time exists. Since

$$A(|x_1-x_2|^2) \le \psi(1) = 1-\varepsilon < \sigma\,,$$

we have the inequality

$$t_1 \le |x_1-x_2|^2\,. \tag{194}$$

We wish to derive an iterative estimate upon $t_{k+1}-t_k$.

In order to do so, we let $x_0 := \frac{x_1+x_2}{2}$ and define the moments

$$M_a(t) := \int |x-x_0|T_a(x,t)\,dx\,, \tag{195}$$

$$M_b(t) := \int |x-x_0|T_b(x,t)\,dx\,, \tag{196}$$

$$M_k := \max\{M_b(t_k), M_a(t_k)\}\,. \tag{197}$$

Strictly speaking the moments are not defined for $t = 0$. However since the functions converge to 0 as $t\downarrow 0$, we set $M_a(0) = M_b(0) = 0$. Observe that

$$\int_{|y-x_0|\ge 2\sigma^{-k}M_k} T_a(y,t_k)\,dy \le \frac{\sigma^k}{2M_k}\int T_a(y,t_k)|y-x_0|\,dy \le \frac{\sigma^k}{2}\,.$$

Moreover, an analogous estimate is valid for T_b. Since the total integral of $T_a(y, t_k)$ (respectively $T_b(z, t_k)$) is in fact $A(t_k) = \sigma^k$, we conclude

$$\int_{|y-x_0|\leq 2\sigma^{-k}M_k} T_a(y)\,dy \geq \frac{\sigma^k}{2}, \tag{198}$$

$$\int_{|z-x_0|\leq 2\sigma^{-k}M_k} T_b(z)\,dz \geq \frac{\sigma^k}{2}. \tag{199}$$

Consider the domain $\Omega_k := \{(y, z) : |y - x_0| \leq 2\sigma^{-k}M_k, |z - x_0| \leq 2\sigma^{-k}M_k\}$ and its complement Ω_k^c. Observe that on Ω_k we have $|y - z| \leq 4\sigma^{-k}M_k$. Thus for $t' > t_k$ we can use (192) to estimate

$$\begin{aligned}
A(t') &\leq \int_{\Omega_k^c} \chi(y, z, t_k)\,dy\,dz + \psi\left(4\sigma^{-k}M_k(t' - t_k)^{-1/2}\right)\int_{\Omega_k} \chi(y, z, t_k)\,dy\,dz \\
&\leq \int \chi(y, z, t_k)\,dy\,dz - \left[1 - \psi\left(4\sigma^{-k}M_k(t' - t_k)^{-1/2}\right)\right]\int_{\Omega_k} \chi(y, z, t_k)\,dy\,dz \\
&\leq A(t_k) - \left[1 - \psi\left(4\sigma^{-k}M_k(t' - t_k)^{-1/2}\right)\right]A(t_k)^{-1}\left(\frac{\sigma^k}{2}\right)^2 \\
&= \sigma^k\left[\frac{3}{4} + \frac{1}{4}\psi\left(4\sigma^{-k}M_k(t' - t_k)^{-1/2}\right)\right].
\end{aligned} \tag{200}$$

If we set

$$t' := t_k + 16\sigma^{-2k}M_k^2,$$

then

$$\psi\left(4\sigma^{-k}M_k(t' - t_k)^{-1/2}\right) = \psi(1) = 1 - \varepsilon,$$

and (200) gives

$$A(t') \leq \sigma^k\left(1 - \frac{\varepsilon}{4}\right) = \sigma^{k+1}.$$

We thus infer the recursive estimate

$$t_{k+1} \leq t_k + 16\sigma^{-2k}M_k^2. \tag{201}$$

We wish next to estimate M_k. Observe that

$$\begin{aligned}
T_a(x, t') &= \max\{S(x, t', x_1, 0) - S(x, t', x_2, 0), 0\} = \max\{T_a^*(x, t', t) - T_b^*(x, t', t), 0\} \\
&\leq T_a^*(x, t', t) = \int S(x, t', y, t)T_a(y, t)\,dy.
\end{aligned}$$

Now,

$$M_a(t') = \int |x - x_0| T_a(x, t')\, dx \le \int (|x - y| + |y - x_0|) S(x, t', y, t) T_a(y, t)\, dy\, dx$$
$$= \int |y - x_0| T_a(y, t)\, dy + \int T_a(y, t) \int |x - y| S(x, t', y, t)\, dx\, dy\,.$$

Using the moment bound we then infer

$$M_a(t') \le M_a(t) + A(t) C_4 (t' - t)^{1/2}\,.$$

This, and the analogous bound on $M_b(t')$, leads to the recursive estimate

$$M_{k+1} \le M_k + \sigma^{k+1} C_4 (t_{k+1} - t_k)^{1/2} \le M_k (1 + 4C_4)\,.$$

Clearly, since $t_0 = 0$ and $M_0 = M_a(t_0) = M_b(t_0) = \frac{|x_1 - x_2|}{2}$, we have

$$M_k \le \frac{|x_1 - x_2|}{2} (1 + C_4)^k\,. \tag{202}$$

Thus the recursive bound (201) becomes

$$t_{k+1} \le t_k + 4|x_1 - x_2|^2 \big[\underbrace{\sigma^{-2}(1 + C_4)^2}_{B}\big]^k\,. \tag{203}$$

Summing (203) and taking into account that $t_1 \le |x_1 - x_2|^2$ we clearly reach

$$t_{k+1} \le 4|x_1 - x_2|^2 \frac{B^{k+1} - 1}{B - 1} \le 4|x_1 - x_2|^2 B^{k+1}\,, \tag{204}$$

where B is a constant larger than 2 which depends only on λ and n (if B as defined in (203) is smaller than 2, we can just enlarge it by setting it equal to 2).

We next set $t_0 = 0$ (and recall that $A(0) := \lim_{t \downarrow 0} A(t) = 1$). Hence, for any $t \ge 0$ there is a unique natural number k such that

$$t_k \le t < t_{k+1}\,.$$

We then conclude

$$\int |S(x, t, x_1, 0) - S(x, t, x_2, 0)|\, dx = A(t) \le A(t_k) \le \sigma^k \qquad \forall t \ge t_k\,. \tag{205}$$

Observe on the other hand that

$$k + 1 \ge -(\log B)^{-1} \log \frac{4|x_1 - x_2|^2}{t} \qquad \text{for all } t \ge t_k.$$

If we set $\alpha := -2(\log B)^{-1} \log \sigma$, which is a positive number depending therefore only upon λ and n, we reach the estimate

$$\int |S(x,t,x_1,0) - S(x,t,x_2,0)|\, dx \le \sigma^{-1} 4^{\alpha/2} \left(\frac{|x_1 - x_2|}{t^{1/2}} \right)^{\alpha} . \tag{206}$$

This is exactly the desired estimate, and hence the proof of Proposition 57 is finally complete.

5.7 *Proof of the a Priori Estimate*

First of all observe that, by Theorem 52(f), (134) can also be used to prove

$$\int |S(x_1,t,y,\bar{t}) - S(x_2,t,y,\bar{t})|\, dy \le C \left(\frac{|x_1 - x_2|}{(t-\bar{t})^{1/2}} \right)^{\alpha} \qquad \text{for all } t > \bar{t}. \tag{207}$$

This easily gives the Hölder continuity of any solution u through Theorem 52(e):

$$\begin{aligned} |u(x_1,t) - u(x_2,t)| &\le \int |S(x_1,t,y,0) - S(x_2,t,y,0)||u(y,0)|\, dy \\ &\le C\|u\|_\infty \left(\frac{|x_1 - x_2|}{t^{1/2}} \right)^{\alpha} . \end{aligned} \tag{208}$$

As for the time continuity, we use

$$u(x,t) - u(x,s) = \int S(x,t,y,s) u(y,s)\, dy - u(x,s) \int S(x,t,y,s)\, dy$$

to estimate

$$\begin{aligned} |u(x,s) - u(x,t)| &\le \int S(x,t,y,s)|u(y,s) - u(x,s)|\, dy \\ &\le \underbrace{\int_{|y-x|\le\rho} S(x,t,y,s)|u(y,s) - u(x,s)|\, dy}_{=I_1} \\ &\quad + \underbrace{\int_{|y-x|\ge\rho} S(x,t,y,s)|u(y,s) - u(x,s)|\, dy}_{=I_2} , \end{aligned} \tag{209}$$

where $\rho > 0$ will be chosen later. Using (208) (and the fact that the integral of the fundamental solution equals 1), we can estimate

$$I_1 \leq C\|u\|_\infty s^{-\alpha/2}\rho^\alpha\,. \tag{210}$$

For I_2 we use the moment bound (131):

$$I_2 \leq 2\rho^{-1}\|u\|_\infty \int |y-x|S(x,t,y,s)\,dy \leq C\|u\|_\infty \rho^{-1}(t-s)^{1/2}\,. \tag{211}$$

We thus get

$$|u(t,x)-u(s,x)| \leq C\|u\|_\infty \left(\rho^\alpha s^{-\alpha/2} + (t-s)^{1/2}\rho^{-1}\right).$$

Choosing $\rho^{1+\alpha} = s^{\alpha/2}(t-s)^{1/2}$ we conclude

$$|u(t,x)-u(s,x)| \leq C\|u\|_\infty \left(\frac{t-s}{s}\right)^{\frac{\alpha}{2(1+\alpha)}}\,. \tag{212}$$

The combination of (208) and (212) gives Theorem 51.

5.8 *Proof of Nash's Parabolic Regularity Theorem*

In order to conclude Theorem 48 from Theorem 51, fix measurable coefficients A_{ij} satisfying Assumption 46 and a bounded distributional solution u on $\mathbb{R}^n \times (0,\infty)$. Without loss of generality we can assume that the A_{ij} are defined also for negative times, for instance we can set $A_{ij}(x,-t) = A_{ij}(x,t)$ for every x and every $t > 0$. Next, we observe that, if φ is a smooth compactly supported nonnegative convolution kernel in $\mathbb{R}^n \times \mathbb{R}$, the regularized coefficients $B^\varepsilon_{ij} = A_{ij} * \varphi_\varepsilon$ satisfy Assumption 46 with the same constant λ in (121). Consider moreover a cutoff function ψ^ε which is nonnegative, compactly supported in $B_{2\varepsilon^{-1}} \times (-2\varepsilon, 2\varepsilon^{-1})$, identically equal to 1 on $B_{\varepsilon^{-1}} \times (-\varepsilon^{-1}, \varepsilon^{-1})$ and never larger than 1. If we set $A^\varepsilon_{ij} = \psi^\varepsilon B^\varepsilon_{ij} + (1-\psi^\varepsilon)\delta_{ij}$, again the matrix A^ε satisfies Assumption 46 with the same λ as the matrix A. Note also that

$$\lim_{\varepsilon\to 0}\|A^\varepsilon_{ij} - A_{ij}\|_{L^1(B_R(0)\times(-R,R))} = 0 \qquad \text{for every } R > 0. \tag{213}$$

We now wish to construct solutions u^ε to the "regularized" parabolic problem

$$\partial_t u^\varepsilon = \mathrm{div}_x(A^\varepsilon \nabla u^\varepsilon)\,, \tag{214}$$

which converge to our fixed solution u of the limiting Eq. (120). In order to do so, we fix a smooth mollifier χ and a family of cut-off functions β^ε *in space*. Such pair is the "spatial analog" of the pair $(\varphi, \psi^\varepsilon)$ used to regularize A. For every time s we define the regularized time-slice

$$\bar{u}^{\varepsilon,s}(x) := [u(\cdot, s) * \chi_\varepsilon](x)\beta^\varepsilon(x).$$

By classical parabolic theory, there is a unique smooth solution $u^{\varepsilon,s}$ of (214) on $\mathbb{R}^n \times [s, \infty[$ subject to the initial condition $u^{\varepsilon,s}(\cdot, s) = \bar{u}^{\varepsilon,s}$: in fact this statement follows easily from Theorem 52. Moreover, by the classical maximum principle (cf. for instance [35]) we have

$$\|u^{\varepsilon,s}\|_\infty \leq \|\bar{u}^{\varepsilon,s}\|_\infty \leq \|u\|_\infty. \tag{215}$$

The key to pass from Theorems 51 to 48 is then the following lemma.

Lemma 62 *For almost every $s > 0$, $u^{\varepsilon,s}$ converges weakly* in $L^\infty(\mathbb{R}^n \times (s, \infty))$ to u.*

We will turn to the lemma in a moment. With its aid Theorem 48 is a trivial corollary of Theorem 51 and of the estimate (215). Indeed the solutions $u^{\varepsilon,s}$ will satisfy the uniform estimate

$$|u^{\varepsilon,s}(x_1, t_1) - u^{\varepsilon,s}(x_2, t_2)| \leq C\|u\|_\infty \left[\frac{|x_1 - x_2|^\alpha}{(t_1 - s)^{\alpha/2}} + \left(\frac{t_2 - t_1}{t_1 - s}\right)^{\frac{\alpha}{2(1+\alpha)}}\right], \tag{216}$$

for all $t_2 \geq t_1 > s > 0$ and all $x_1, x_2 \in \mathbb{R}^n$. By the Ascoli–Arzelà Theorem the family $u^{\varepsilon,s}$ is precompact in C^0, and up to subsequences will then converge uniformly to a Hölder function u^s on any compact set $K \subset \mathbb{R}^n \times (s, \infty)$: by Lemma 62 u^s will coincide with u for almost every s and we will thus conclude

$$|u(x_1, t_1) - u(x_2, t_2)| \leq C\|u\|_\infty \left[\frac{|x_1 - x_2|^\alpha}{(t_1 - s)^{\alpha/2}} + \left(\frac{t_2 - t_1}{t_1 - s}\right)^{\frac{\alpha}{2(1+\alpha)}}\right]. \tag{217}$$

Letting now s go to 0 we achieve Theorem 48.

Proof (Proof of Lemma 62*)*
Step 1. First we will prove that (122) can in fact be upgraded to the following stronger statement for almost every pair of times $t > s$:

$$\begin{aligned}\int u(x,t)\varphi(x,t)\,dx = \int_s^t \int u(x,\tau)\partial_t\varphi(x,\tau)\,dx\,d\tau - \int_s^t \int \partial_i\varphi(x,\tau)A_{ij}(x,\tau)\partial_j u(x,\tau)\,dx\,d\tau \\ + \int u(x,s)\varphi(x,s)\,dx \qquad \forall\varphi \in C_c^\infty(\mathbb{R}^n \times (0,\infty)).\end{aligned} \tag{218}$$

The argument is standard, but we will include it for the reader's convenience. In particular we will prove that (218) holds for every pair $s < t$ satisfying the property

$$\lim_{\varepsilon \to 0} \frac{1}{\varepsilon} \left[\int_{s-\varepsilon}^{s} \int_{B_R} |u(x,\tau) - u(x,s)|\, dx\, d\tau + \int_{t}^{t+\varepsilon} \int_{B_R} |u(x,t) - u(x,\tau)|\, dx\, d\tau \right] = 0 \tag{219}$$

for all $R > 0$. By standard measure theory implies, any time that we fix $R \in \mathbb{N}$, (219) holds for almost every $s < t$.

On the other hand, to pass from (122) to (218) using (219) we just argue with the following classical procedure:

(i) We fix a monotone $\chi \in C^\infty(\mathbb{R})$ which is identically 1 on $]-\infty, 0]$ and identically 0 on $]1, \infty[$.
(ii) We test (122) with $\varphi(x,\tau)\chi(\frac{\tau-t}{\varepsilon})\chi(\frac{s-\tau}{\varepsilon})$.
(iii) We let ε go to 0.

Step 2. Next, using (215) and the weak* compactness of bounded sets in L^∞, we can assume the convergence of $u^{\varepsilon,s}$, up to subsequences, to some L^∞ function u^s. We wish to show that u^s has first-order distributional derivatives $\partial_j u^s$ which are locally square summable. In order to do so, we borrow some ideas from [6] and consider the function

$$h(x,t) := -\frac{\alpha |x|^2}{t},$$

where $\alpha > 0$ will be chosen in a moment. We use the Eq. (214) to derive the following equality:

$$\begin{aligned} &\int e^{h(x,t)} |u^{\varepsilon,s}(x,t)|^2\, dx + 2 \int_s^t \int e^{h(x,\tau)} \partial_j u^{\varepsilon,s}(x,\tau) A^\varepsilon_{ij}(x,\tau) \partial_i u^{\varepsilon,s}(x,\tau)\, dx\, d\tau \\ &\quad = \int_s^t \int e^{h(x,\tau)} \left[\partial_t h(x,\tau) |u^{\varepsilon,s}(x,\tau)|^2 - 2u^{\varepsilon,s}(x,\tau) \partial_j u^{\varepsilon,s}(x,\tau) A^\varepsilon_{ij}(x,\tau) \partial_i h(x,\tau) \right] dx\, d\tau \\ &\qquad + \int e^{h(x,s)} |u^{\varepsilon,s}(x,s)|^2\, dx\,. \end{aligned} \tag{220}$$

Note that, for each fixed ε the solution $u^{\varepsilon,s}$ is smooth and all derivatives are bounded, by standard regularity theory for linear parabolic differential equations, see for instance [30, Sec. 7.2.3]. Thus all the integrals above are finite and the equality above follows from usual calculus formulae.

Now, observe that the last integral in (220) is bounded by $C\|u\|_\infty^2$ for some constant $C = C(\alpha, s)$. Using the ellipticity of A^ε_{ij} we can thus estimate

$$\begin{aligned}\int e^{h(x,t)}|u^{\varepsilon,s}(x,t)|^2\,dx + 2\lambda^{-1}\int_s^t\int e^{h(x,\tau)}|\nabla u^{\varepsilon,s}(x,\tau)|^2\,dx\,d\tau \\ \leq \int_s^t\int e^{h(x,\tau)}\left[\partial_t h(x,\tau)|u^{\varepsilon,s}(x,\tau)|^2 + 2\lambda|u^{\varepsilon,s}(x,\tau)||\nabla u^{\varepsilon,s}(x,\tau)||\nabla h(x,\tau)|\right]dx\,d\tau \\ + C\|u\|_\infty^2\,.\end{aligned}$$

The weight h has the following fundamental property:

$$\partial_t h = -\frac{1}{4\alpha}|\nabla h|^2\,. \tag{221}$$

Thus, it suffices to choose α small, depending only upon λ, to conclude, via Young's inequality,

$$\begin{aligned}\int e^{h(x,t)}|u^{\varepsilon,s}(x,t)|^2\,dx + 2\lambda^{-1}\int_s^t\int e^{h(x,\tau)}|\nabla u^{\varepsilon,s}(x,\tau)|\,dx\,d\tau \\ \leq \lambda\int_s^t\int e^{h(x,\tau)}|\nabla u^{\varepsilon,s}(x,\tau)|^2\,dx\,d\tau + C\|u\|_\infty^2\,.\end{aligned}$$

The latter inequality gives an upper bound on

$$\int_s^t\int e^{h(x,\tau)}|\nabla u^{\varepsilon,s}(x,\tau)|^2\,dx\,d\tau$$

which depends upon $\|u\|_\infty$ and λ, but not upon ε. We thus infer a uniform bound for $\|\nabla u^{\varepsilon,s}\|_{L^2(B_R(0)\times(s,\infty))}$ for every positive R. In turn such bound implies that the partial derivatives $\partial_j u^s$ are locally square summable and that $\partial_j u^{\varepsilon,s}$ converge (locally) weakly in L^2 to $\partial_j u^s$ (again up to subsequences, which we do not label for notational convenience).

Step 3. Passing to the limit in the weak formulation of (214) and using that the initial data $u^{\varepsilon,s}(\cdot,s)$ converges (locally in L^1) to $u(\cdot,s)$, we then infer the corresponding of (218) for every $t > s$ (in this case we need no restriction upon t because we know that u^s converges locally uniformly!), namely, the validity of

$$\begin{aligned}\int u^s(x,t)\varphi(x,t)\,dx = \int u^s(x,s)\varphi(x,s)\,dx + \int_s^t\int u^s(x,\tau)\partial_\tau\varphi(x,\tau)\,dx\,d\tau \\ - \int_s^t\int \partial_i\varphi(x,\tau)A_{ij}(x,\tau)\partial_j u^s(x,\tau)\,dx\,d\tau\end{aligned} \tag{222}$$

for every test function $\varphi \in C_c^\infty(\mathbb{R}^n \times (0,\infty))$. If we consider $w := u - u^s$ we then subtract (222) from (218) to conclude the following identity for almost every pair $t \geq s$ and for every test $\varphi \in C_c^\infty(\mathbb{R}^n \times (0,\infty))$:

$$\int w(x,t)\varphi(x,t)\,dx = \int_s^t \int w(x,\tau)\partial_\tau \varphi(x,\tau)\,dx\,d\tau \\ - \int_s^t \int \partial_i \varphi(x,\tau) A_{ij}(x,\tau)\partial_j w(x,\tau)\,dx\,d\tau\,. \tag{223}$$

Our goal is to use the latter integral identity, which is a weak form of (120) with initial data $w(\cdot,s) = 0$, to derive that $w = 0$ almost everywhere: this would imply that $u = u^s$ almost everywhere and thus complete the proof of the lemma.

Step 4. In order to carry on the above program we wish to test (223) with $\varphi = e^h w$, but we must face two difficulties:

(i) w is not smooth enough. Indeed the first-order partial derivatives in space are locally square summable and pose no big difficulties, but note that in (223) there is a term with a partial derivative in time, which for $e^h w$ is not even a summable function.
(ii) $e^h w$ is not compactly supported in space (the assumption of being compactly supported in time can be ignored, since all domains of integration are bounded in time).

In order to remove these two problems we fix a cutoff function $\chi \in C_c^\infty(\mathbb{R}^n)$ and a compactly supported smooth kernel *in space only*, namely, a nonnegative $\gamma \in C_c^\infty(\mathbb{R}^n)$ with integral 1. We then consider the spatial regularization

$$w * \gamma_\varepsilon(x,\tau) = \int w(y,\tau)\gamma\left(\frac{x-y}{\varepsilon}\right)dy\,,$$

and define the test function $\varphi := \chi^2 e^h w * \gamma_\varepsilon$. The map $x \mapsto w * \gamma_\varepsilon(x,t)$ is smooth for every fixed t and moreover $\|\nabla(w*\gamma_\varepsilon)(\cdot,t)\|_\infty \leq C\|w\|_\infty \varepsilon^{-1}$. To gain regularity in time we can use the weak form of the equation to show that, in the sense of distributions,

$$\partial_t(w * \gamma_\varepsilon) = (\mathrm{div}_x(A\nabla w)) * \gamma_\varepsilon = (A_{ij}\partial_j w) * \partial_i \gamma_\varepsilon\,. \tag{224}$$

Since $\partial_t w$ is locally square summable, we conclude that $\partial_t(w * \gamma_\varepsilon)$ is a locally bounded measurable function and thus that $w * \gamma_\varepsilon$ is locally Lipschitz in the space-time domain $\mathbb{R}^n \times (0,\infty)$. Hence the test function $\varphi := \chi^2 e^h w * \gamma_\varepsilon$ is Lipschitz and compactly supported and, although the test function in our definition of distributional solution is assumed to be smooth, it is easy check that, nonetheless, (223) holds for our (possibly less regular) choice. Inserting such φ

in (223), and using (224), we then achieve

$$
\begin{aligned}
\int e^{h(x,t)} w(x,t) w * \gamma_\varepsilon(x,t) \chi^2(x)\, dx
&= \int_s^t \int e^{h(x,\tau)} \partial_t h(x,\tau) w(x,\tau) w * \gamma_\varepsilon(x,\tau) \chi^2(x)\, dx\, d\tau \\
&\quad + \underbrace{\int_s^t \int e^{h(x,\tau)} w(x,\tau)[(A_{ij}\partial_j w) * \partial_i \gamma_\varepsilon](x,\tau) \chi^2(x)\, dx\, d\tau}_{=:(I)} \\
&\quad - \int_s^t \int e^{h(x,\tau)} \partial_i w(x,\tau) A_{ij}(x,\tau) \chi(x) \cdot \\
&\qquad \cdot [\partial_j w * \gamma_\varepsilon(x,\tau)\chi(x) + w * \gamma_\varepsilon(x,\tau)(\partial_j h(x,\tau)\chi(x) + 2\partial_j \chi(x))]\, dx\, d\tau\,.
\end{aligned}
$$

Next, assuming that γ is a symmetric kernel, we can use the standard identity

$$
\int (f * \gamma)(x) g(x)\, dx = \int f(x)(g * \gamma)(x)\, dx
$$

to conclude

$$
(I) = -\int_s^t \int e^{h(x,\tau)} \partial_j w(x,\tau) A_{ij}(x,\tau)[(\chi^2 \partial_i w + \chi^2 w \partial_i h + 2w\chi\partial_i \chi) * \gamma_\varepsilon](x,\tau)\, dx\, d\tau\,.
$$

Letting ε go to 0 we then conclude

$$
\begin{aligned}
\int e^{h(x,t)} w^2(x,t) \chi^2(x)\, dx
&= -2 \int_s^t \int e^{h(x,\tau)} \chi^2(x) \partial_i w(x,\tau) A_{ij}(x,\tau) \partial_j w(x,\tau)\, dx\, d\tau \\
&\quad + \int_s^t \int e^{h(x,\tau)} \chi^2(x) w^2(x,\tau) \partial_t h(x,\tau)\, dx\, d\tau \\
&\quad - 2 \int_s^t \int e^{h(x,\tau)} w(x,\tau) \chi(x) \partial_i w(x,\tau) A_{ij}(x,\tau)(2\partial_j \chi(x) + \chi(x)\partial_j h(x,\tau))\, dx\, d\tau\,.
\end{aligned}
$$

Using now the ellipticity of A_{ij} and (221) we achieve

$$
\begin{aligned}
\int e^{h(x,t)} w^2(x,t) \chi^2(x)\, dx
&\le -2\lambda^{-1} \int_s^t \int e^{h(x,\tau)} \chi^2(x) |\nabla w(x,\tau)|^2\, dx\, d\tau
\end{aligned}
$$

$$
\begin{aligned}
&- (4\alpha)^{-1} \int_s^t \int e^{h(x,\tau)} \chi^2(x) w^2(x,\tau) |\nabla h(x,\tau)|^2 \, dx \, d\tau \\
&+ 2\lambda \int_s^t \int e^{h(x,\tau)} |w(x,\tau)| |\nabla w(x,\tau)| (\chi^2(x) |\nabla h(x,\tau)| + 2|\chi(x)| |\nabla \chi(x)|) \, dx \, d\tau \, .
\end{aligned}
$$

From the latter we recover, using Young's inequality,

$$
\begin{aligned}
&\int e^{h(x,t)} w^2(x,t) \chi^2(x) \, dx \\
&\quad \le -(4\alpha)^{-1} \int_s^t \int e^{h(x,\tau)} \chi^2(x) w^2(x,\tau) |\nabla h(x,\tau)|^2 \, dx \, d\tau \\
&\quad + C(\lambda) \int_s^t \int e^{h(x,\tau)} \chi^2(x) w^2(x,\tau) |\nabla h(x,\tau)|^2 \, dx \, d\tau \\
&\quad + C(\lambda) \int_s^t \int e^{h(x,\tau)} w^2(x,\tau) |\nabla \chi(x)|^2 \, dx \, d\tau \, ,
\end{aligned}
$$

where $C(\lambda)$ is a constant which only depends on λ. Hence, choosing α sufficiently small, depending only on λ, we conclude

$$
\int e^{h(x,t)} w^2(x,t) \chi^2(x) \, dx \le C(\lambda) \int_s^t \int e^{h(x,\tau)} w^2(x,\tau) |\nabla \chi(x)|^2 \, dx \, d\tau \, . \tag{225}
$$

Next, consider a cut-off function $\beta \in C_c^\infty(B_2)$ which is identically 1 on B_1 and, for any $R > 0$, set $\chi(x) := \beta(\frac{x}{R})$. Insert the latter in (225). Using that $|\nabla \chi(x)| \le CR^{-1}$ and the fact that $e^h w^2$ is integrable, when we let $R \uparrow \infty$ we conclude

$$
\int e^{h(x,t)} w^2(x,t) \, dx \le 0 \, .
$$

This implies that $w(\cdot, t) \equiv 0$ for almost every $t \ge s$ and thus concludes the proof.

5.9 *Proof of the De Giorgi–Nash Theorem*

By standard Sobolev space theory, cf. [30, Sec. 7.2&7.3], $v|_{B_{3r}(z)}$ is the unique minimum of the energy functional

$$
\mathscr{E}(w) := \int_{B_{3r}(z)} \partial_i w(x) A_{ij}(x) \partial_j w(x) \, dx \tag{226}
$$

among those functions $w \in W^{1,2}(B_{3r}(z))$ such that $w - v \in W_0^{1,2}(B_{3r}(z))$. If we first extend A and v and we then regularize them by convolution to A^ε and v^ε, we can consider the corresponding solutions of the regularized elliptic equations, using the same arguments of the last section (a proof of the regularity of the solution can be found, for instance, in [30, Sec. 6.3]. By the maximum principle (cf. again [30, Sec. 6.4]), we will have $\|v^\varepsilon\|_\infty \leq \|v\|_\infty$ and v^ε will be a minimizer of the corresponding regularized energy functional. Since $\|v^\varepsilon\|_{W^{1,2}(B_{3r}(z))}$ would be uniformly bounded, we can assume, after extraction of a convergent subsequence, that v^ε converges weakly in $W^{1,2}(B_{3r}(z))$ to some $\bar{v}$, which in turn is a distributional solution of (124) subject to the constraint $\bar{v}-v \in W^{1,2}_0(B_{3r}(z))$. As such, $\bar{v}$ must be a minimizer of the same variational problem as $v|_{B_{3r}(z)}$, which we already know to be unique. Thus $\bar{v} = v|_{B_{3r}(z)}$ and so it suffices to prove Theorem 50 under the a priori assumption that A and v are smooth. Moreover, by rescaling v to $\bar{v}(x) := v(rx+z)$, we can assume that $r = 1$ and $z = 0$.

Under these additional assumptions, we can consider $v(x,t) := v(x)$ as a stationary smooth solution of the parabolic problem

$$\partial_t u(x,t) = \partial_j(A_{ij}(x)\partial_i u(x,t)) \tag{227}$$

on $C_3 := B_3 \times (0,\infty)$. Theorem 50 is then a simple corollary of Theorem 51 and the following proposition, which is a direct outcome of the theory developed by Nash.

Proposition 63 **(L^∞ estimate for the initial-boundary value problem)** *There is a constant C depending only upon n and λ with the following property. Assume that $A_{ij}(x,t)$ satisfies the assumptions of Theorem* 51 *and $w : \overline{B}_2 \times [0,\infty[\to \mathbb{R}$ is a smooth bounded solution of* (120) *with $w(x,0) = 0$ for every x. Then*

$$\|w(\cdot,t)\|_{L^\infty(B_1)} \leq C\|w\|_\infty t^{1/2}\,. \tag{228}$$

With Proposition 63 at hand, it is easy to conclude Theorem 50. Indeed, multiply v by a smooth cut-off function $\varphi \in C^\infty_c(B_3)$ taking values in $[0,1]$ and identically 1 on B_2. Extend φv smoothly on $\mathbb{R}^n$ by setting it equal to 0 on $\mathbb{R}^n \setminus B_3$. Let z be the solution on $\mathbb{R}^n \times [0,\infty)$ of (227) with $z(\cdot,0) = \varphi v$. Note that $\|z\|_\infty \leq \|v\|_\infty$ by the maximum principle. We can apply Proposition 63 to $w(\cdot,t) := z(\cdot,t) - v(\cdot)$ to conclude

$$\|v - z(\cdot,t)\|_{L^\infty(B_1)} \leq 2C\|v\|_\infty t^{1/2}\,. \tag{229}$$

On the other hand, by Theorem 51 we have

$$|z(x_1,t) - z(x_2,t)| \leq C\|v\|_\infty \frac{|x_1-x_2|^\alpha}{t^{\alpha/2}}\,. \tag{230}$$

In particular, for $x_1, x_2 \in B_1$, we can combine the last two inequalities to conclude

$$|v(x_1) - v(x_2)| \leq C\|v\|_\infty \left(t^{1/2} + \frac{|x_1-x_2|^\alpha}{t^{\alpha/2}}\right) \qquad \forall t > 0\,. \tag{231}$$

Choose now $t^{1/2+\alpha/2} = |x_1 - x_2|^\alpha$ to conclude that

$$|v(x_1) - v(x_2)| \le C\|v\|_\infty |x_1 - x_2|^{\alpha/(1+\alpha)}. \tag{232}$$

So, to complete the proof of Theorem 51 we only need to show Proposition 63.

Proof (Proof of Proposition 63*)* Consider any smooth solution u of (120) in $C_2 := \bar{B}_2 \times [0, \infty[$. The boundary values on ∂C_2 determine then the solution through a representation formula of the form

$$u(x,t) := \int_{\partial C_2} u(\xi)\rho(x,t,\xi)\,d\xi\,,$$

where the integral is taken with respect to the standard surface measure on the boundary ∂C_2, cf. [35, Sec. 1.4]. If we set $\xi = (y(\xi), \tau(\xi))$, then the kernel $\rho(x,t,\xi)$ satisfies the conditions

(i) $\int \rho(x,t,\xi)\,d\xi = 1$;
(ii) $\rho(x,t,\xi) \ge 0$;
(iii) $\rho(x,t,\xi) = 0$ if $t \le \tau(\xi)$.

Since the fundamental solutions $S(x,t,x_0,t_0)$ with $t_0 < 0$ are also smooth solutions of the parabolic equation in the cylinder C_2, we reach the identity

$$S(x,t,x_0,t_0) = \int_{\partial C_2} S(y(\xi), \tau(\xi), x_0, t_0)\rho(x,t,\xi)\,d\xi\,. \tag{233}$$

Multiplying by $|x - x_0|$ and integrating we then have

$$\int |x - x_0| S(x,t,x_0,t_0)\,dx_0 = \int\int_{\partial C_2} |x - x_0| S(y(\xi), \tau(\xi), x_0, t_0)\rho(x,t,\xi)\,d\xi\,dx_0\,. \tag{234}$$

In particular, using the moment bound (131) we conclude

$$\int\int_{\partial C_2} (|x - y(\xi)| - |x_0 - y(\xi)|) S(y(\xi), \tau(\xi), x_0, t_0)\rho(x,t,\xi)\,d\xi\,dx_0 \le C_4(t - t_0)^{1/2}\,. \tag{235}$$

From the latter inequality, using again the moment bound, we achieve

$$\begin{aligned}
&\int_{\partial C_2} |x - y(\xi)|\rho(x,t,\xi)\,d\xi \\
&\quad = \int\int_{\partial C_2} |x - y(\xi)| S(y(\xi), \tau(\xi), x_0, t_0)\rho(x,t,\xi)\,d\xi\,dx_0
\end{aligned}$$

$$\leq C_4(t-t_0)^{1/2} + \int_{\partial C_2} \int |x_0 - y(\xi)| S(y(\xi), \tau(\xi), x_0, t_0)\, dx_0\, \rho(x,t,\xi)\, d\xi$$

$$\stackrel{(131)}{\leq} 2C_4(t-t_0)^{1/2} . \tag{236}$$

Letting t_0 go to 0, we thus conclude

$$\int_{\partial C_2} |x - y(\xi)| \rho(\xi, x, t)\, d\xi \leq 2C_4 t^{1/2} . \tag{237}$$

Let $\mathscr{L} := \partial C_2 \setminus B_2(0) \times \{0\}$ and observe that $|x - y(\xi)| \geq 2 - |x|$ if $\xi \in \mathscr{L}$. Thus, using (237) and the fact that $\rho \geq 0$, we conclude

$$2C_4 t^{1/2} \geq \int_{\mathscr{L}} |x - y(\xi)|\, \rho(x,t,\xi)\, d\xi \geq (2 - |x|) \int_{\mathscr{L}} \rho(x,t,\xi)\, d\xi . \tag{238}$$

Consider now a solution w as in the proposition. Since $w = 0$ on $B_2(0) \times \{0\}$, for any (x, t) we have

$$|w(x,t)| \leq \int_{\mathscr{L}} \rho(x,t,\xi) |w(\xi)|\, d\xi \stackrel{(238)}{\leq} \frac{Ct^{1/2}}{2 - |x|} \|w\|_\infty . \tag{239}$$

The latter inequality for $x \in B_1(0)$ obviously implies (228).

6 The Other Papers in Pure Mathematics

6.1 A Path Space and Stiefel–Whitney Classes

In 1955 Whitney communicated to the Proceedings of the National Academy of Sciences a two pages note of Nash, [74], where he gives a very direct proof of the topological invariance of the Stiefel–Whitney classes of smooth manifolds, a theorem proved 3 years before by Thom (cf. [98]). For the definition of Stiefel–Whitney classes of a smooth vector bundle we refer to [64]: given a differentiable manifold its Stiefel–Whitney classes are then the corresponding classes of the tangent bundle and the theorem of Thom shows that such classes are a topological invariant. In fact, Thom derived this consequence from a stronger theorem, namely that the homotopy type of a tangent bundle as fiber space over a topological manifold M is the same for any differentiable structure on M. Nash shows that this conclusion can be inferred from the definition of an appropriate path space X of the topological manifold M, where, loosely speaking, the tangent bundles can be embedded.

Definition 64 Given a topological manifold M, X is the space of continuous mappings $\gamma : [0,1] \to M$ which do not "recross" the starting point $\gamma(0)$. X is endowed with the topology induced by uniform convergence and with a natural projection map $\pi : X \to M$ defined by $\pi(\gamma) := \gamma(0)$.

Given a differentiable structure on M, we can define on its tangent bundle a smooth Riemann tensor g and use it to "embed the tangent bundle in X" (more precisely, we will embed the sphere bundle in X, see below). To this aim, first of all we assume, by suitably modifying g, that

(I) any pair of points in the Riemannian manifold (M, g) with geodesic distance no larger than 1 can be joined by a unique geodesic segment of length 1.

Hence we can consider the subset G of X consisting of those paths which are geodesic segments with length 1 parametrized with arc-length. Of course, the sphere bundle on M given by the tangent vectors v with unit length is isomorphic to G as fiber bundle over M. Nash's main observation can then be stated as follows.

Theorem 65 *If (I) holds, then G is a fiber deformation retract of X, i.e., there is a continuous map $\Phi : [0,1] \times X \to X$ such that*

(a) *$\Phi(0, \gamma) = \gamma$ for every $\gamma \in X$;*
(b) *$\Phi(1, \gamma) \in G$ for every $\gamma \in X$;*
(c) *$\Phi(1, \eta) = \eta$ for every $\eta \in G$;*
(d) *$\pi(\Phi(s, \gamma)) = \pi(\gamma)$ for every $\gamma \in X$ and every $s \in [0,1]$.*

The proof, which Nash sketches very briefly, is an elementary exercise.

6.2 *Le problème de Cauchy pour les équations différentielles d'un fluide général*

In 1962, 4 years after his last masterpiece on the continuity of solutions to parabolic equations, Nash published a 12 pages paper in French, whose aim was to prove the short-time existence of smooth solutions to the compressible Navier–Stokes equations for a viscous heat-conducting fluid. More precisely he considers the following system of five partial differential equations, in the unknowns ρ, v and T which represent, respectively, the density, the velocity and the temperature of the fluid and are therefore functions of the time t and the space $x \in \mathbb{R}^3$:

$$\begin{cases} \partial_t \rho + \mathrm{div}_x (\rho v) = 0\,, \\[2ex] \rho \partial_t v_i + \rho \left[v_j \partial_j v_i \right] + \partial_i p = \partial_j \sigma_{ij} + \rho F_i\,, \\[2ex] \partial_t T + v_j \partial_j T = \frac{1}{\rho T S_T} \left[\mathrm{div}\,(\varkappa \nabla T) + \rho^2 T S_\rho \mathrm{div}\, v \right] + \frac{2\eta}{\rho T S_T} \mathscr{S}(v)_{ij} \mathscr{S}(v)_{ij} + \frac{\zeta}{\rho T S_T} (\mathrm{div}\, v)^2\,. \end{cases} \tag{240}$$

In the system above[23]:

(i) We use Einstein's convention on repeated indices;
(ii) The pressure p is a function of the density ρ and the temperature T;
(iii) σ_{ij} is the Cauchy stress tensor, given by the formula

$$\sigma_{ij} = \eta\left(\partial_i v_j + \partial_j v_i\right) + \left(\zeta - \tfrac{2}{3}\eta\right) \operatorname{div} v\, \delta_{ij}\,, \tag{241}$$

with η and ζ (the viscosity coefficients) which are functions of ρ and T;
(iv) $F = (F_1, F_2, F_3)$ is the external force acting on the fluid;
(v) $\varkappa$, the heat conductivity, is a function of the temperature T and the density ρ;
(vi) The entropy S is a function of ρ and T, whereas S_T and S_ρ are the corresponding partial derivatives with respect to T and ρ;
(vii) $\mathscr{S}(v)$ is the traceless part of the symmetrized derivative of v, more precisely

$$\mathscr{S}(v)_{ij} = \frac{1}{2}\left[\partial_i v_j + \partial_j v_i - \tfrac{2}{3}\operatorname{div} v\, \delta_{ij}\right]. \tag{242}$$

The functions $\eta, \zeta, \varkappa, S$ and p are thus known and determined by the thermodynamical properties of the fluid. They display a rather general behavior, although they must obey some restrictions: we refer to the classical textbook [61] for their physical meaning and for the derivation of the equations. In his paper Nash assumes that all the functions $\eta, \zeta, \varkappa, p, S$ and S_T are real analytic and positive.[24]

Similarly, the external force F is given. Nash considers then the Cauchy problem for (240) in the whole threedimensional space, namely he assumes that the density, the velocity and the pressure are known at a certain time, which without loss of generality we can assume to be the time 0. This problem has received a lot of attention in the last 30 years and we refer to the books [33, 63] for an account of the latest developments in the mathematical treatment of (240).

---

[23]The first two equations are the first two equations from [77, p. 487, (1)] whereas the third should correspond to [77, p. 488, (1c)]. The latter is derived by Nash from the third equation in [77, p. 487, (1)], which in turn corresponds to the classical conservation law for the entropy, see, for instance, [61, (49.5)]. The third equation of [77, p. 487, (1)] contains two typos, which disappear in [77, p. 488, (1c)]. The latter however contains another error: Nash has η and ζ in place of $\frac{\eta}{\rho T S_T}$ and $\frac{\zeta}{\rho T S_T}$, but it is easy to see that this would not be consistent with the way he describes its derivation. Nash's error has no real consequence for the rest of the note, since he treats the coefficients in front of $\mathscr{S}(v)_{ij}\mathscr{S}(v)_{ij}$ and $(\operatorname{div} v)^2$ as arbitrary real analytic functions of ρ and T and the same holds for $\frac{\eta}{\rho T S_T}$ and $\frac{\zeta}{\rho T S_T}$ under the assumption $S_T \neq 0$. The latter inequality is needed in any case even to treat Nash's "wrong" equation for T.

[24]Indeed Nash does not mention the positivity of S_T, although this is certainly required by his argument when he reduces the existence of solutions of (240) to the existence of a solutions of a suitable parabolic system, cf. [77, (6) and (7)]: the equation in T is parabolic if and only if $\frac{\varkappa}{\rho T S_T}$ is positive.

I also have the impression that his argument does not really need the positivity of S and p, although these are quite natural assumptions from the thermodynamical point of view.

In order to give his existence result, Nash first passes to the Lagrangian formulation of (240) and he then eliminates the density ρ. Subsequently he shows the existence, for a finite time, of a (sufficiently) smooth solution of the resulting system of equations under the assumption that the initial data and the external force are (sufficiently) smooth. In particular, he writes the system as a second-order parabolic linear system of partial differential equations with variable coefficients, where the latter depend upon the unknowns (it must be noted that such dependence involves first-order spatial derivatives of the unknowns and their time integrals). The existence result is therefore achieved through a fixed point argument, taking advantage of classical estimates for second-order linear parabolic systems.

6.3 Analyticity of the Solutions of Implicit Function Problems with Analytic Data

In 1966 Nash turned again one last time to the isometric embedding problem, addressing the real analytic case. More precisely, his aim was to prove that, if in Theorem 29 we assume that the metric g is real analytic, then there is a real analytic isometric embedding of (Σ, g) in a sufficiently large Euclidean space. The most important obstacle in extending the proof of [75] to the real analytic case is the existence of a suitable smoothing operator which replaces the one in Sect. 4.4 in the real analytic context.

In his 12 pages paper Nash gives indeed two solutions to the problem. Most of the paper is devoted to prove the existence of a suitable (real) analytic smoothing operator on a general compact real analytic manifold. But he also remarks that the real analytic case of the isometric embedding problem for compact Riemannian manifolds Σ can be reduced to the existence of real analytic isometric embeddings for real analytic Riemannian manifolds which are tori, at the price of enlarging the dimension of the Euclidean target: it simply suffices to take a real analytic immersion of Σ into $\mathbb{T}^{2n+1}$ using Whitney's theorem and then to extend the real analytic Riemannian metric g on Σ to the whole torus (a problem which can be solved using Cartan's work [16]). On the other hand the existence of a suitable regularizing analytic operator on the torus is an elementary consequence of the Fourier series expansion.

Nash leaves the existence of real analytic embeddings for noncompact real analytic Riemannian manifolds as open and it points out that "...The case of non-compact manifolds seem to call for a non-trivial generalization of the methods". The noncompact case was indeed settled later by Gromov (cf. [39]).

6.4 Arc Structure of Singularities

In 1968 Nash wrote his last paper in pure mathematics. Although it was published 28 years later (see [79]), its content was promoted by Hironaka and later by

Lejeune-Jalabert (cf. [28]): thus the content of Nash's work became known very much before it was finally published. Nash's idea is to use the space of complex analytic arcs in a complex algebraic variety as a tool to study its singularities and in particular their resolutions (whose existence had been established only 4 years before Nash's paper in the celebrated work of Hironaka, [48]). In his paper he formulated a question which became known in algebraic geometry as Nash's problem. A complete solution of the problem has not yet been reached although a lot of progress has been made in recent years (we refer the reader to the very recent survey [21]).

Nash's problem (and his ideas) are nowadays formulated for varieties (in fact, schemes) on a general algebraically closed field of any characteristic. However [79] is concerned with complex varieties and in this brief description we will stick to the latter case. Take therefore a complex variety V. The space X of arcs in V is then given by the jets of holomorphic maps $x : \Omega \to V$ where Ω is an arbitrary open subset of $\mathbb{C}$ containing the origin.[25] An interesting case is that where $W = V_s$ is the set of singularities of V: $X(V_s)$ consists of those arcs which "pass through" a singularity. In [79] Nash realized that this space has, roughly speaking, the structure of an "infinite dimensional complex variety" (for a precise formulation we refer to [79, p. 32] or to [21, Th. 2.6]; see also the earlier work of Greenberg [37]) which has finitely many irreducible components, cf. [79, Prop. 1]. Nash calls such components *arc families*.

The main idea of Nash is to establish a relation between the arc families of $X(V_s)$ and the irreducible components of the image of V_s through a resolution of the singularities of V. More precisely, having fixed a resolution of the singularities $V^* \to V$ (namely a *smooth* algebraic variety V^* together with a proper birational map $V^* \to V$), we can look at the components $W_1^*, \dots W_L^*$ of the image W^* of V_s in V^*. Nash lifts almost every arc in $X(V_s)$ to a unique arc of $X(W^*)$ and through this procedure establishes the existence of an injective map from the arc families of $X(V_s)$ to the components of W^*, cf. [79, Prop. 2].[26] As a corollary, given two different resolutions V^* and V^{**}, and the corresponding components $W_1^*, \dots, W_{L^*}^*$, $W_1^{**}, \dots, W_{L^{**}}^{**}$ of the preimage of V_s in V^* and V^{**}, Nash establishes the existence of a birational correspondence $W_j^* \to W_k^{**}$ between those pairs which correspond to the same arc family (cf. [79, Cor., p. 38]).

As a consequence of his considerations, such components are *essential*, i.e., they must appear in any resolution of the singularities of V. He then raised the question whether all essential components must correspond to an arc family: this is what algebraic geometers call, nowadays, Nash's problem. In high dimension the answer is known to be negative since the work [52] and it has been shown very recently

[25] In the modern literature it is customary to take an equivalent definition of X through formal power series; we refer to [58] for the latter and for several important subtleties related to variants of the Nash arc space.

[26] In fact, Nash claims the proposition with *any* algebraic subset W of V in place of V_s but, although the proposition does hold for $W = V_s$, it turns out to be false for a general algebraic subset W; cf. [21, Ex. 3.7] for a simple explicit counterexample.

that in fact the answer is negative already for some threedimensional varieties, cf. [20, 56]. It must be noticed that Nash was indeed rather careful with the higher dimensional case of his question: quoting [79, p. 31] "...We do not know how complete is the representation of essential components by arc families". However in the twodimensional case, i.e. the case of algebraic surfaces, it is a classical fact that there is a unique minimal resolution, namely containing only essential components, and Nash conjectured that each essential component is indeed related to an arc family. The conjecture has been proved only recently in [34].

Nonetheless the studies on Nash's problem are very far from being exhausted. Indeed the answer has been proved to be affirmative in a variety of interesting cases (see the survey articles [21, 84]) and several mathematicians are looking for the "correct formulation" of the question (see, for instance, [56]), possibly leading to a complete understanding of the relations between resolutions of the singularities and the arc space.

6.5 The Nash Blow-Up

In algebraic geometry the term "Nash blow-up" refers to a procedure with which, roughly speaking, the singular points of an algebraic variety are replaced by all the limits of the tangent spaces to the regular points. If X is an algebraic subvariety of $\mathbb{C}^n$ of pure dimension r, the Nash blow-up is then the (closure of the) graph of the Gauss map: more precisely, if we denote by $\mathbf{Gr}\,(r, n)$ the Grassmanian of r-dimensional complex linear subspaces of $\mathbb{C}^n$, then the Nash blow-up of X is the closure of the set of pairs $(x, T_x X) \in \mathbb{C}^n \times \mathbf{Gr}\,(r, n)$, where x varies among all regular points of X and $T_x X$ denotes the tangent space to X at x. Although such definition is given in terms of the embedding, it can be shown that in fact the Nash blow-up of X depends only upon X. Since there is in general no canonical ideal to blow up to realize the transformation, some authors prefer the term "Nash modification".

A long standing open problem is whether after a finite number of Nash blow-ups every singular variety becomes smooth (indeed, in characteristic p the answer is negative and one needs to state the problem in terms of "normalized Nash blow-ups", cf. [83]). According to [92], such question was posed by Nash to Hironaka in a private communication in the early 1960s and the term "Nash blow-up" was first used by Nobile a decade later in [83], where he proved that the answer to Nash's question is affirmative for curves in characteristic 0. Building upon the work of Hironaka [48], Spivakovsky proved in the late 1980s that the answer is affirmative for surfaces in characteristic 0 for the normalized Nash blow-up (cf. [92]). In general the question of Nash is still widely open and constitutes an active area of research.

Lejeune-Jalabert noticed some years ago that the same problem was posed by Semple a decade or so before Nash, in [90]. For this reason some authors (cf. [36, 100]) use the term "Semple–Nash modification".

Acknowledgements I am very grateful to Helge and Ragni for entrusting to me this portion of the Nash volume, a wonderful occasion to deepen my understanding of the mathematics of a true genius, who has had a tremendous influence in my own work.

Most of the manuscript has been written while I was visiting the CMSA at Harvard and I wish to thank Shing-Tung Yau and the staff at CMSA for giving me the opportunity to carry on my work in such a stimulating environment.

Several friends and colleagues have offered me kind and invaluable help with various portions of this note. In particular I wish to thank Davide Vittone for giving me several precious suggestions with the Sects. 3 and 4 and reading very carefully all the manuscript; Gabriele Di Cerbo, Riccardo Ghiloni and János Kollár for clarifying several important points concerning Sect. 2 and pointing out a few embarassing mistakes; Tommaso de Fernex and János Kollar for kindly reviewing a first rather approximate version of Sect. 6.4; Eduard Feireisl for his suggestions on Sect. 6.3; Cedric Villani for allowing me to steal a couple of paragraphs from his beautiful review of [76] in the Nash memorial article [26]; Francois Costantino for helping me with a delicate topological issue; Jonas Hirsch and Govind Menon for proofreading several portions of the manuscript; Helge Holden for going through all the manuscript with extreme care.

This work has been supported by the grant agreement 154903 of the Swiss National Foundation.

References

1. S. Akbulut and H. King. On approximating submanifolds by algebraic sets and a solution to the Nash conjecture. *Invent. Math.*, 107(1):87–98, 1992.
2. A. G. Akritas. Sylvester's forgotten form of the resultant. *Fibonacci Quart.*, 31(4):325–332, 1993.
3. A. D. Alexandrov. *Intrinsic geometry of convex surfaces*. OGIZ, Moscow-Leningrad, 1948.
4. G. E. Andrews, R. Askey, and R. Roy. *Special functions*. Cambridge: Cambridge University Press, 1999.
5. D. G. Aronson. Bounds for the fundamental solution of a parabolic equation. *Bull. Amer. Math. Soc.*, 73:890–896, 1967.
6. D. G. Aronson. Non-negative solutions of linear parabolic equations. *Ann. Scuola Norm. Sup. Pisa (3)*, 22:607–694, 1968.
7. R. F. Bass. *Diffusions and elliptic operators*. Springer-Verlag, New York, 1998.
8. R. F. Bass. On Aronson's upper bounds for heat kernels. *Bull. London Math. Soc.*, 34(4):415–419, 2002.
9. J. Bochnak, M. Coste, and M.-F. Roy. *Real Algebraic Geometry*. Springer-Verlag, Berlin, 1998.
10. J. F. Borisov. The parallel translation on a smooth surface. IV. *Vestnik Leningrad. Univ.*, 14(13):83–92, 1959.
11. J. F. Borisov. $C^{1,\alpha}$-isometric immersions of Riemannian spaces. *Doklady*, 163:869–871, 1965.
12. T. Buckmaster, C. De Lellis, P. Isett, and L. Székelyhidi, Jr. Anomalous dissipation for 1/5-Hölder Euler flows. *Ann. of Math. (2)*, 182(1):127–172, 2015.
13. T. Buckmaster, C. De Lellis, L. Székelyhidi, Jr., and V. Vicol. Onsager's conjecture for admissible weak solutions. *Comm. Pure Appl. Math.*, to appear.
14. C. Burstin. Ein Beitrag zum Problem der Einbettung der Riemannschen Räume in euklidischen Räumen. *Rec. Math. Moscou*, 38(3–4):74–85, 1931.
15. E. Cartan. Sur la possibilité de plonger un espace Riemannien donné dans un espace Euclidien. *Ann. Soc. Polon. Math.*, 6:1–7, 1928.
16. H. Cartan. Variétés analytiques réelles et variétés analytiques complexes. *Bull. Soc. Math. France*, 85:77–99, 1957.

17. R. L. Cohen. The immersion conjecture for differentiable manifolds. *Ann. of Math. (2)*, 122(2):237–328, 1985.
18. S. Cohn-Vossen. Zwei Sätze über die Starrheit der Eiflächen. *Nachrichten Göttingen*, 1927:125–137, 1927.
19. S. Conti, C. De Lellis, and L. Székelyhidi, Jr. h-principle and rigidity for $C^{1,\alpha}$ isometric embeddings. In *Nonlinear partial differential equations*, volume 7 of *Abel Symp.*, pages 83–116. Springer, Heidelberg, 2012.
20. T. de Fernex. Three-dimensional counter-examples to the Nash problem. *Compos. Math.*, 149(9):1519–1534, 2013.
21. T. de Fernex. The space of arcs of an algebraic variety. In *Algebraic geometry: Salt Lake City 2015*, Proc. Sympos. Pure Math., Vol. 97.1, pp. 169–197, 2015.
22. E. De Giorgi. Sulla differenziabilità e l'analiticità delle estremali degli integrali multipli regolari. *Mem. Accad. Sci. Torino. Cl. Sci. Fis. Mat. Nat. (3)*, 3:25–43, 1957.
23. E. De Giorgi. Un esempio di estremali discontinue per un problema variazionale di tipo ellittico. *Boll. Un. Mat. Ital. (4)*, 1:135–137, 1968.
24. E. De Giorgi. *Selected papers*. Springer-Verlag, Berlin, 2006. Edited by L. Ambrosio, G. Dal Maso, M. Forti, M. Miranda and S. Spagnolo.
25. C. De Lellis, D. Inauen, and L. Székelyhidi, Jr. A Nash–Kuiper theorem for $C^{1,\frac{1}{5}-\delta}$ immersions of surfaces in 3 dimensions. *Rev. Mat. Iberoam.*, 34(3):1119–1152, 2018.
26. C. De Lellis, H. King, J. Milnor, Nachbar J., L. Székelyhidi, Jr., C. Villani, and J. Weinstein. John Forbes Nash, Jr. 1928–2015. *Notices of the AMS*, 63(5):492–506, 2017.
27. C. De Lellis and L. Székelyhidi, Jr. Dissipative continuous Euler flows. *Invent. Math.*, 193(2):377–407, 2013.
28. M. Demazure, H. Pinkham, and B. Teissier, editors. *Séminaire sur les Singularités des Surfaces*, volume 777 of *Lecture Notes in Mathematics*. Springer, Berlin, 1980. Held at the Centre de Mathématiques de l'École Polytechnique, Palaiseau, 1976–1977.
29. Y. Eliashberg and N. Mishachev. *Introduction to the h-principle*. American Mathematical Society, Providence, RI, 2002.
30. L. C. Evans. *Partial differential equations*. American Mathematical Society, Providence (R.I.), 1998.
31. L. C. Evans and R. F. Gariepy. *Measure theory and fine properties of functions*. CRC Press, Boca Raton, FL, 1992.
32. E. B. Fabes and D. W. Stroock. A new proof of Moser's parabolic Harnack inequality using the old ideas of Nash. *Arch. Rational Mech. Anal.*, 96(4):327–338, 1986.
33. E. Feireisl. *Dynamics of viscous compressible fluids*. Oxford University Press, Oxford, 2004.
34. J. Fernández de Bobadilla and M. P. Pereira. The Nash problem for surfaces. *Ann. of Math. (2)*, 176(3):2003–2029, 2012.
35. A. Friedman. *Partial differential equations of parabolic type*. Prentice-Hall, 1964.
36. P. D. González Pérez and B. Teissier. Toric geometry and the Semple-Nash modification. *Rev. R. Acad. Cienc. Exactas Fís. Nat. Ser. A Math. RACSAM*, 108(1):1–48, 2014.
37. M. J. Greenberg. Rational points in Henselian discrete valuation rings. *Inst. Hautes Études Sci. Publ. Math.*, (31):59–64, 1966.
38. M. L. Gromov. Isometric imbeddings and immersions. *Dokl. Akad. Nauk SSSR*, 192:1206–1209, 1970.
39. M. L. Gromov. *Partial differential relations*. Springer-Verlag, Berlin, 1986.
40. M. L. Gromov and V. A. Rohlin. Imbeddings and immersions in Riemannian geometry. *Uspehi Mat. Nauk*, 25(5 (155)):3–62, 1970.
41. M. Günther. Zum Einbettungssatz von J. Nash. *Math. Nachr.*, 144:165–187, 1989.
42. M. Günther. Isometric embeddings of Riemannian manifolds. In *Proceedings of the International Congress of Mathematicians, Vol. I, II (Kyoto, 1990)*, pages 1137–1143. Math. Soc. Japan, Tokyo, 1991.
43. R. S. Hamilton. The inverse function theorem of Nash and Moser. *Bull. Amer. Math. Soc. (N.S.)*, 7(1):65–222, 1982.

44. W. Hao, S. Leonardi, and J. Nečas. An example of irregular solution to a nonlinear Euler-Lagrange elliptic system with real analytic coefficients. *Ann. Scuola Norm. Sup. Pisa Cl. Sci. (4)*, 23(1):57–67, 1996.
45. R. Hartshorne. *Algebraic geometry.* Springer-Verlag, New York-Heidelberg, 1977.
46. A. Hatcher. *Algebraic Topology.* Cambridge University Press, Cambridge, 2002.
47. G. Herglotz. Über die Starrheit der Eiflächen. *Abh. Math. Semin. Hansische Univ.*, 15:127–129, 1943.
48. H. Hironaka. Resolution of singularities of an algebraic variety over a field of characteristic zero. I, II. *Ann. of Math. (2)* ***79*** *(1964), 109–203; ibid. (2)*, 79:205–326, 1964.
49. M. W. Hirsch. *Differential Topology.* Springer-Verlag, New York-Heidelberg, 1976.
50. E. Hopf. Zum analytischen Charakter der Lösungen regulärer zweidimensionaler Variationsprobleme. *Math. Z.*, 30:404–413, 1929.
51. P. Isett. A Proof of Onsager's Conjecture. *Ann. Math. (2)*, 188:871–963, 2018.
52. S. Ishii and J. Kollár. The Nash problem on arc families of singularities. *Duke Math. J.*, 120(3):601–620, 2003.
53. H. Jacobowitz. Implicit function theorems and isometric embeddings. *Ann. of Math. (2)*, 95:191–225, 1972.
54. M. Janet. Sur la possibilité de plonger un espace Riemannien donné à n dimensions dans un espace Euclidien à $\frac{n(n+1)}{2}$ dimensions. *C. R. Acad. Sci., Paris*, 183:942–943, 1926.
55. Z. Jelonek. On the extension of real regular embedding. *Bull. Lond. Math. Soc.*, 40(5):801–806, 2008.
56. J. M. Johnson and J. Kollár. Arc spaces of cA-type singularities. *J. Singul.*, 7:238–252, 2013.
57. A. Källén. Isometric embedding of a smooth compact manifold with a metric of low regularity. *Ark. Mat.*, 16(1):29–50, 1978.
58. J. Kollár and A. Némethi. Holomorphic arcs on singularities. *Invent. Math.*, 200(1):97–147, 2015.
59. S. G. Krantz and H. R. Parks. *A Primer of Real Analytic Functions.* Birkhäuser Verlag, Basel, 1992.
60. N. H. Kuiper. On C^1-isometric imbeddings. I, II. *Nederl. Akad. Wetensch. Proc. Ser. A.* ***58*** *= Indag. Math.*, 17:545–556, 683–689, 1955.
61. L. D. Landau and E. M. Lifshitz. *Course of theoretical physics. Vol. 6, Fluid dynamics.* Pergamon Press, Oxford, second edition, 1987.
62. H. Lewy. On the existence of a closed convex surface realizing a given Riemannian metric. *Proc. Natl. Acad. Sci. USA*, 24:104–106, 1938.
63. P.-L. Lions. *Mathematical topics in fluid mechanics. Vol. 2.* The Clarendon Press, Oxford University Press, New York, 1998. Compressible models, Oxford Science Publications.
64. J. W. Milnor and J. D. Stasheff. *Characteristic classes.* Princeton University Press, Princeton, N. J., 1974.
65. C. Mooney and O. Savin. Some Singular Minimizers in Low Dimensions in the Calculus of Variations. *Arch. Ration. Mech. Anal.*, 221(1):1–22, 2016.
66. C. B. Morrey, Jr. On the solutions of quasi-linear elliptic partial differential equations. *Trans. Am. Math. Soc.*, 43:126–166, 1938.
67. C. B. Morrey, Jr. Second-order elliptic systems of differential equations. In *Contributions to the theory of partial differential equations*, Annals of Mathematics Studies, no. 33, pages 101–159. Princeton University Press, Princeton, N. J., 1954.
68. J. Moser. A new technique for the construction of solutions of non-linear differential equations. *Proc. Nat. Acad. Sci. USA*, 47:1824–1831, 1961.
69. J. Moser. On Harnack's theorem for elliptic differential equations. *Comm. Pure Appl. Math.*, 14:577–591, 1961.
70. J. Moser. A rapidly convergent iteration method and non-linear differential equations. II. *Ann. Scuola Norm. Sup. Pisa (3)*, 20:499–535, 1966.
71. J. Moser. A rapidly convergent iteration method and non-linear partial differential equations. I. *Ann. Scuola Norm. Sup. Pisa (3)*, 20:265–315, 1966.
72. J. Nash. Real algebraic manifolds. *Ann. of Math. (2)*, 56:405–421, 1952.

73. J. Nash. C^1 isometric imbeddings. *Ann. of Math. (2)*, 60:383–396, 1954.
74. J. Nash. A path space and the Stiefel-Whitney classes. *Proc. Nat. Acad. Sci. U.S.A.*, 41:320–321, 1955.
75. J. Nash. The imbedding problem for Riemannian manifolds. *Ann. of Math. (2)*, 63:20–63, 1956.
76. J. Nash. Continuity of solutions of parabolic and elliptic equations. *Amer. J. Math.*, 80:931–954, 1958.
77. J. Nash. Le problème de Cauchy pour les équations différentielles d'un fluide général. *Bull. Soc. Math. France*, 90:487–497, 1962.
78. J. Nash. Analyticity of the solutions of implicit function problems with analytic data. *Ann. of Math. (2)*, 84:345–355, 1966.
79. J. Nash. Arc structure of singularities. *Duke Math. J.*, 81(1):31–38 (1996), 1995.
80. J. Nash. *The essential John Nash*. Princeton University Press, Princeton, NJ, 2002. Edited by H. W. Kuhn and S. Nasar.
81. L. Nirenberg. *The determination of a closed convex surface having given line element*. ProQuest LLC, Ann Arbor, MI, 1949. Thesis (Ph.D.)–New York University.
82. L. Nirenberg. The Weyl and Minkowski problems in differential geometry in the large. *Comm. Pure Appl. Math.*, 6:337–394, 1953.
83. A. Nobile. Some properties of the Nash blowing-up. *Pacific J. Math.*, 60(1):297–305, 1975.
84. C. Plénat and M. Spivakovsky. The Nash problem and its solution: a survey. *J. Singul.*, 13:229–244, 2015.
85. A. V. Pogorelov. *Izgibanie vypuklyh poverhnosteĭ*. Gosudarstv. Izdat. Tehn.-Teor. Lit., Moscow-Leningrad, 1951.
86. A. V. Pogorelov. *Extrinsic geometry of convex surfaces*. American Mathematical Society, Providence, R.I., 1973.
87. W. Rudin. *Principles of mathematical analysis*. McGraw-Hill Book Co., New York-Auckland-Düsseldorf, third edition, 1976.
88. L. Schläfli. Nota alla memoria del sig. Beltrami, "Sugli spazii di curvatura costante". *Annali di Mat. (2)*, 5:178–193, 1871.
89. J. Schwartz. On Nash's implicit functional theorem. *Comm. Pure Appl. Math.*, 13:509–530, 1960.
90. J. G. Semple. Some investigations in the geometry of curve and surface elements. *Proc. London Math. Soc. (3)*, 4:24–49, 1954.
91. M. Shiota. *Nash manifolds*, volume 1269 of *Lecture Notes in Mathematics*. Springer-Verlag, Berlin, 1987.
92. M. Spivakovsky. Sandwiched singularities and desingularization of surfaces by normalized Nash transformations. *Ann. of Math. (2)*, 131(3):411–491, 1990.
93. G. Stampacchia. Sistemi di equazioni di tipo ellittico a derivate parziali del primo ordine e proprietà delle estremali degli integrali multipli. *Ricerche Mat.*, 1:200–226, 1952.
94. N. Steenrod. *The Topology of Fibre Bundles*. Princeton Mathematical Series, vol. 14. Princeton University Press, Princeton, N. J., 1951.
95. V. Šverák and X. Yan. A singular minimizer of a smooth strongly convex functional in three dimensions. *Calc. Var. Partial Differential Equations*, 10(3):213–221, 2000.
96. V. Šverák and X. Yan. Non-Lipschitz minimizers of smooth uniformly convex functionals. *Proc. Natl. Acad. Sci. USA*, 99(24):15269–15276, 2002.
97. L. Székelyhidi, Jr. The h-principle and turbulence. *ICM 2014 Proceedings Volume*, 2014.
98. R. Thom. Espaces fibrés en sphères et carrés de Steenrod. *Ann. Sci. Ecole Norm. Sup. (3)*, 69:109–182, 1952.
99. A. Tognoli. Su una congettura di Nash. *Ann. Scuola Norm. Sup. Pisa (3)*, 27:167–185, 1973.
100. L. D. Tráng and B. Teissier. On the mathematical work of Professor Heisuke Hironaka. *Publ. Res. Inst. Math. Sci.*, 44(2):165–177, 2008.
101. A. H. Wallace. Algebraic approximation of manifolds. *Proc. London Math. Soc. (3)*, 7:196–210, 1957.

102. A. Weil. *Foundations of Algebraic Geometry*. American Mathematical Society, New York, 1946.
103. H. Weyl. Über die Bestimmung einer geschlossenen konvexen Fläche durch ihr Linienelement. Zürich. Naturf. Ges. 61, 40–72, 1916.
104. J. H. C. Whitehead. On C^1-complexes. *Ann. of Math. (2)*, 41:809–824, 1940.
105. H. Whitney. Differentiable manifolds. *Ann. of Math. (2)*, 37(3):645–680, 1936.
106. H. Whitney. The self-intersections of a smooth n-manifold in $2n$-space. *Ann. of Math. (2)*, 45:220–246, 1944.
107. S.-T. Yau. Open problems in geometry. In *Differential geometry: partial differential equations on manifolds (Los Angeles, CA, 1990)*, volume 54 of *Proc. Sympos. Pure Math.*, pages 1–28. Amer. Math. Soc., Providence, RI, 1993.

A Few of Louis Nirenberg's Many Contributions to the Theory of Partial Differential Equations

Robert V. Kohn

1 Introduction

Mathematics is the language of science, and partial differential equations are a crucial component: they provide the language we use to describe—and the tools we use to understand—phenomena in many areas including geometry, engineering, and physics.

Louis Nirenberg's contributions to this field have been hugely influential. His impact includes the solution of many important problems, and—more importantly—the introduction of many fundamentally new ideas.

The depth, variety, and extent of his work make it difficult to synthesize. That challenge has nevertheless been undertaken twice, by YanYan Li [50] and by Tristan Rivière [73], with admirable success. Rather than attempt another synthesis, I shall focus here on six specific topics:

- his early work on the Weyl and Minkowski problems;
- his results with Shmuel Agmon and Avron Douglis on elliptic regularity;
- his paper with Fritz John on functions with bounded mean oscillation;
- his work with Luis Caffarelli and me on the Navier–Stokes equations;
- his results with Haim Brezis on nonlinear elliptic equations with critical exponents; and
- his work with Basilis Gidas, Wei-Ming Ni, and Henri Berestycki on the "method of moving planes" and the "sliding method."

The author gratefully acknowledges support from NSF through grant DMS-1311833.

R. V. Kohn (✉)
Department of Mathematics, Courant Institute of Mathematical Sciences, New York University, New York, NY, USA
e-mail: kohn@cims.nyu.edu

H. Holden, R. Piene (eds.), *The Abel Prize 2013–2017*,
https://doi.org/10.1007/978-3-319-99028-6_20

My goal is to capture—to the extent possible in a few pages—the character of these contributions. I shall point to some related and/or subsequent work; however my discussions are necessarily incomplete, since a comprehensive review of even one of these topics would be a gargantuan task.

In focusing on these topics, I am necessarily omitting many important accomplishments; fortunately quite a few have been summarized elsewhere. For example, I do not touch his work at the interface between PDE and several complex variables—but these have been discussed by Joseph Kohn[1] and by Simon Donaldson.[2] The articles just cited also discuss other aspects of his work, and the surveys [50, 73] touch almost everything. Another rich source is [80], where leading researchers discuss five of his themes in the context of recent, related work of their own.

Louis is a friend, colleague, and role model to an entire community of mathematicians (myself included). A thoughtful and dedicated mentor, he has advised 46 PhD students (starting with Walter Littman in 1956, and ending with Kanishka Perera in 1997, according to the Mathematical Genealogy website), while also having a formative influence on countless postdocs and collaborators. His influence has been amplified by Louis' outstanding ability as an expositor: he writes in a way that invites the reader's participation, with detailed introductions that put his work in context and explain its main ideas. In addition to many research articles he has also written influential survey articles, including one on elliptic theory [69] and another on variational & topological methods [70]. His book *Topics in Nonlinear Functional Analysis*, written in 1974 and reprinted in 2001 [71], is still widely used today.

A cross-cutting theme in Louis' research is his exquisite taste in problems. One very successful mode has been to recognize, through specific challenges, the need for new PDE tools or estimates. His uncanny ability to identify such challenges—and to find the required tools or estimates—has been a major driver of his impact. The early work on the Weyl and Minkowski problems (Sect. 2) and the work with Brezis on nonlinear elliptic equations with critical exponents (Sect. 6) are examples of such work; additional examples include his paper with Newlander on the integrability of almost-complex structures [65] and his introduction (with Joseph Kohn) of the class of pseudodifferential operators [45].

A very different, also very successful mode has been to identify tools that are clearly important, and explore their scope systematically. His work with Agmon and Douglis on elliptic regularity (Sect. 3) and that with Gidas and Ni on the method of moving planes (Sect. 7) have this character. Another favorite example is his systematic treatment of interpolation inequalities (known as Gagliardo–Nirenberg inequalities, since they were found independently by E. Gagliardo [33] and by Nirenberg, who announced them at the 1958 International Congress of Mathematicians and published them as Section 2 of [69]).

---

[1]Louis Nirenberg receives the National Medal of Science, *Notices Amer. Math. Soc.* **43**(10), 1111–1116 (1996) (includes "Nirenberg's work in partial differential equations" by L. Caffarelli, and "Nirenberg's work in complex analysis" by J. J. Kohn).

[2]Donaldson, S.: On the work of Louis Nirenberg. *Notices Amer. Math. Soc.* **58**(3), 469–472 (2011).

But this tidy framework is too narrow to accommodate all Louis' work. In particular, he has always loved puzzles—especially ones involving estimates or inequalities—and this has led to many successful collaborations. The work with Fritz John on functions of bounded mean oscillation (Sect. 4) is, in my view, an example of that type.

Louis' vision, leadership, and accomplishments have been recognized by many awards over the years; being selective, the list includes (besides the 2015 Abel Prize) the 1959 Bôcher prize, the 1982 Crafoord Prize, the 1994 Leroy P. Steele Prize for Lifetime Achievement, the 1995 National Medal of Science, and the 2010 Chern Medal.

His stature has led to many interviews[3] as well as video available at the Simons Foundation's Science Lives site.[4] These delightful resources capture (among other things) Louis' engaging wit, generosity, and taste.

2 The Weyl Problem, the Minkowski Problem, and Fully Nonlinear PDE in Two Space Dimensions

Nirenberg's PhD thesis, completed in 1949, was entitled *The determination of a closed convex surface having given line element* [66]. The corresponding papers, published in 1953, are entitled *The Weyl and Minkowski problems in differential geometry in the large* and *On nonlinear elliptic partial differential equations and Hölder continuity* [67, 68]. This work proved two long-standing conjectures in differential geometry, and fundamentally advanced our understanding of fully-nonlinear PDE in two space dimensions. Not many PhD theses achieve so much!

The environment in which he did this work was rather unusual. The research group established at New York University by Richard Courant was still very small; its leaders (besides Courant) were Kurt Friedrichs, James J. Stoker and Fritz John (who arrived in 1946, shortly after Nirenberg's arrival as a graduate student). Government funding permitted substantial expansion after the war, and Courant and his colleagues had a remarkable eye for talent. As a result, Nirenberg's fellow PhD students were a truly remarkable group—including Avron Douglis, Harold Grad, Eugene Isaacson, Joseph Keller, Martin Kruskal, Peter Lax, and Cathleen Morawetz. (It was also a relatively large group: according to the Mathematical Genealogy website, NYU granted 37 mathematics PhD's in the four-year period 1948–1951.)

[3]Interview with Louis Nirenberg, interviewed by A. Jackson. *Notices Amer. Math. Soc.* **49**(4), 441–449 (2002), and Interview with Louis Nirenberg, interviewed by M. Raussen and C. Skau. *Newsletter of the European Mathematical Society*, Dec 2015, 33–38; reprinted in *Notices Amer. Math. Soc.* **63**(2), 135–140 (2016).

[4]Louis Nirenberg, interviewed by Jalal Shatah, on the Simons Foundation's Science Lives website https://www.simonsfoundation.org/2014/04/21/louis-nirenberg/. (Accessed 8 March 2018).

His thesis work provides an outstanding example of how specific challenges can lead to the development of fundamentally new tools. The challenges, in this case, were the Weyl and Minkowski problems—two easy-to-believe conjectures about two-dimensional surfaces in three-dimensional space, which had been open for many years. A framework for viewing them as nonlinear PDE problems was already well-established, and Hans Lewy had used it to obtain solutions when the data are analytic [48, 49]. But the analytic category is very rigid! Each problem's natural formulation involves data that are a few times differentiable. Solving the problems in that setting required a new a priori estimate for fully-nonlinear PDE in two space dimensions. Nirenberg's fundamental contribution was to obtain that estimate.

The crucial estimate says that if u solves a PDE of the form $F(D^2u, Du, u, x) = 0$ in a two-dimensional domain,

(i) u, Du, and D^2u are continuous, with L^∞ norm at most K, and
(ii) the equation is elliptic with a positive ellipticity bound λ,

then in any subdomain D^2u is actually *Hölder continuous* (with a uniform bound depending only on K, λ, the C^1 norm of F, and the choice of subdomain). The key point, of course, is that while D^2u was only assumed to be bounded and continuous, the PDE assures that it is significantly better: Hölder continuous. Higher regularity follows by differentiating the equation and using linear PDE estimates (provided the regularity of F permits). Nirenberg's proof of this regularity theorem was related to the theory of quasiconformal mappings, drawing inspiration from Morrey's proof that 2D quasiconconformal mappings with bounded distortion are Hölder continuous [57].

As noted above, the specific challenges that led Nirenberg to consider this regularity issue were questions from differential geometry, raised by Weyl in 1916 and Minkowski in 1903. The Weyl problem has its roots in the fact that a convex surface in $\mathbb{R}^3$ has nonnegative Gaussian curvature. It seeks a sort of converse:

> Given a Riemannian metric g on the two-dimensional sphere S^2 with positive Gaussian curvature, can it be realized by a convex two-dimensional surface in $\mathbb{R}^3$? In other words, is there a map $H: S^2 \to \mathbb{R}^3$ such that $\|DH(x)v\|^2_{\mathbb{R}^3} = \|v\|^2_{g(x)}$ for every $x \in S^2$ and every $v \in T_xS^2$?

The Minkowski problem has its roots in the fact that if M is a strictly convex surface in $\mathbb{R}^3$, K_M is its Gaussian curvature, and $\nu_M : M \to S^2$ is its Gauss map (taking $x \in M$ to the outward unit normal to M at x), then (by elementary arguments) one has $\int_{S^2} \frac{x}{K_M(\nu_M^{-1}(x))}\, dA = 0$, where the variable of integration is $x = (x_1, x_2, x_3) \in S^2 \subset \mathbb{R}^3$ and the integral is with respect to surface area on S^2. The Minkowski problem seeks a sort of converse:

> Given a positive function K on S^2 satisfying $\int_{S^2} \frac{x}{K(x)}\, dA = 0$, is there a strictly convex surface M such that $K(x) = K_M(\nu_M^{-1}(x))$?

The suggestion to look at these problems came from James Stoker. This is not surprising in view of Stoker's longstanding interest in differential geometry (in fact,

Stoker gave a new, simple proof in 1950 that a solution of the Minkowski problem is necessarily unique [81]). However Nirenberg has said that as a PhD student he worked most closely with Kurt Friedrichs.[3]

Nirenberg's solution of each problem used what was known even then as "the method of continuity." Focusing (for simplicity of language) on the Weyl problem, the method consists of

(i) showing that the given metric (call it g_1) can be joined to the standard metric (call it g_0) by continuous path in the space of matrics with positive curvature (call it g_t, $0 \leq t \leq 1$);
(ii) showing that the set of t for which g_t is realizable is an open subset of $[0, 1]$; and
(iii) showing that the set of t for which g_t is realizable is a closed subset of $[0, 1]$.

The essence of this program was already present in Weyl's work; in fact, his 1916 paper [84] identified the fundamental issues and obtained several key estimates, though he lacked the PDE tools to complete the program. Nirenberg's treatment of (i) followed Weyl's. The proof of (ii) required solving a degenerate system of PDE's; Nirenberg's treatment used an iteration scheme, whose convergence was proved using estimates for certain 2nd order linear PDE (this was in large part a modern implementation of Weyl's ideas). Weyl had reduced the proof of (iii) to the study of a fully nonlinear PDE in two space dimensions, and he had shown that the solution was C^2, but this was not enough to conclude the argument. Nirenberg's regularity result—showing that the solution was actually $C^{2,\alpha}$ for some α—was the crucial ingredient permitting completion of the program.

His solution of the Minkowski problem followed a similar strategy. There, too, the argument used the method of continuity, and relied on prior work (in this case a 1938 paper by Lewy [49] and a 1939 paper by Miranda [56]) for identification of a suitable PDE-based framework. The prior work had reduced the analogue of (iii) to the study of a fully nonlinear PDE in two space dimensions, and Miranda had shown that the solution was C^2. Nirenberg's regularity result (showing that the solution was actually $C^{2,\alpha}$) was again the crucial ingredient permitting completion of the program.

In 1949—the year Nirenberg completed his PhD—another solution of the Weyl problem was published by the Soviet mathematician A.V. Pogorelov, using methods completely different from Nirenberg's. (Briefly: A.D. Alexandroff had shown the existence of a sort of weak solution, obtained by taking a limit of polyhedra; Pogorelov proved the regularity of those weak solutions.) Pogorelov also published a solution of the Minkowski problem in 1952. A discussion of Pogorelov's work and its relation to Nirenberg's can be found in the Math Reviews entry for [67], which is MR0058265. Pogorelov too was an outstanding mathematician, who did this work at the very beginning of his career. The independent solutions by Nirenberg and Pogorelov provide a reminder that while Soviet mathematics was remarkably strong in the post World War II period, communication with the West was quite limited.

In attacking the Weyl and Minkowski problems, Nirenberg was solving problems that others had claimed before. Indeed, a 1940 paper by Caccioppoli addressed the Weyl problem using the method of continuity. However, as Nirenberg wrote, in establishing point (iii) Cacciopoli "refers to previous publications on nonlinear second order elliptic equations (see [18] for references). These papers contain only sketches of proofs—details are not presented—and it is not clear that all the results mentioned there are fully established." Concerning the Minkowski problem: Miranda's 1939 paper [56] claimed a full solution, but it relied on Cacciopoli's not-fully-established results. By the time Nirenberg and Pogorelov worked on these problems, there seems to have been a consensus that the previous "solutions" were incomplete.

Nirenberg's proof of $C^{2,\alpha}$ regularity for solutions of fully-nonlinear elliptic equations was limited to two space dimensions. This was sufficient for the Weyl and Minkowski problems, since they involve two-dimensional surfaces in $\mathbb{R}^3$. It is natural, however, to ask what happens in higher dimensions: is a C^2 solution of a uniformly elliptic, fully nonlinear equation $F(D^2u, Du, u, x) = 0$ necessarily $C^{2,\alpha}$ in space dimension $n \geq 3$? The answer is yes, but the proof requires methods entirely different from those of Nirenberg's 1953 paper. (I thank N. Nadirashvili for input on this topic.) Briefly: if u solves such an equation, then for any i the partial derivative $v = \partial u/\partial x_i$ is a *viscosity solution* of the linear elliptic PDE obtained by formally differentiating the original equation (see, e.g., Corollary 1.3.2 of [61]). Since the leading-order term of this equation has the form $\sum a_{kl} \frac{\partial^2 v}{\partial x_k \partial x_l}$ with $a_{kl}(x)$ continuous, the regularity theory for viscosity solutions of linear elliptic equations is applicable, and it shows that v is $C^{1,\alpha}$ for some $\alpha > 0$ [19]. Interestingly, if the condition $u \in C^2$ is replaced by $u \in C^{1,1}$ then the argument breaks and higher regularity becomes false: a recent paper by Nadirashvili, Tkachev, and Vlăduţ [60] identified a nonlinear elliptic PDE of the form $F(D^2u) = 0$ in $\mathbb{R}^5$ with an (explicit) viscosity solution of the form $u(x) = p(x)/|x|$, where p is a homogeneous polynomial in x of degree 3. Since u is homogeneous of degree 2, it is C^2 except at $x = 0$, with bounded but discontinuous second derivatives at the origin.

3 Elliptic Regularity for Boundary Value Problems: The Agmon–Douglis–Nirenberg Estimates

The 1950s was a period of rapid development in our understanding of elliptic PDE, and Nirenberg was a major player. The following discussion will focus on linear PDE with variable coefficients, since this is the heart of the matter. It should be understood, however, that these results are also crucial for the study of nonlinear PDE (for example, permitting existence theorems to be proved using fixed-point theorems or iteration arguments).

As background: by 1950 there was a rather comprehensive understanding of second-order elliptic PDE for a scalar-valued unknown: a treatment involving L^2-type estimates using Hilbert space methods was presented, for example, in volume 2 of Courant and Hilbert's *Methoden der Mathematischen Physik* [24], and estimates involving Hölder norms were established by Schauder in 1934. However a similarly general understanding of higher-order equations and elliptic systems was not yet available. Progress in those directions began in the early 1950s with work by Vishik [83], Browder [15], and Gårding [34] among others. Another key development was the work of Calderón and Zygmund on singular integral operators [22], which provided the crucial tools needed for L^p-type estimates.

Nirenberg's contributions in the 1950s included the following key advances:

- His 1955 paper with Avron Douglis, *Interior estimates for elliptic systems of partial differential equations* [26], extended elliptic theory to a much more general class of systems than had been considered before, obtaining Schauder-type interior estimates involving Hölder norms. Roughly speaking, this work identified what it should *mean* for a system to be elliptic. An important feature of the definition is that the system need not be of the same order in each unknown.
- His 1959 paper with Shmuel Agmon and Avron Douglis, *Estimates near the boundary for solutions of elliptic partial differential equations satisfying general boundary conditions. I* [1], provided estimates up to the boundary, for any elliptic boundary value problem involving a scalar-valued unknown. This work was notable both for its scope and for its method. Concerning the scope: while previous work provided a full understanding of problems with Dirichlet-type boundary conditions, the 1959 paper achieved something similar for *any* boundary condition satisfying the "complementing condition." (Roughly speaking: these are boundary conditions for which, in the homogeneous constant-coefficient case for a half-space, separation of variables reveals that a solution which is periodic on the boundary must decay exponentially toward the interior of the domain.) Concerning the method: the paper's starting point was a study of the constant coefficient case in a half-space, obtaining an explicit solution analogous to the Poisson kernel representation of a harmonic function. These representations were then used to obtain estimates for the solution (up to the boundary, even for PDE's with variable coefficients in domains with curved boundaries), by applying tools from potential theory and the then-recently-developed theory of singular integral operators. This produced both estimates of Schauder type (estimating Hölder-type norms of the solution in terms of those of the data) and also analogous estimates of L^p-type. Related estimates were obtained by Felix Browder, in work done independently around the same time [16].

This ground-breaking work was done during a period of dramatic progress, to which many others contributed. The introductions of Nirenberg's papers are notable not only for their transparent discussions of the papers' methods and achievements, but also for their richly detailed discussions of related work by others.

The 1955 paper dealt with systems but obtained only interior estimates. The 1959 paper dealt with boundary estimates but was restricted to scalar-valued unknowns. It was of course a natural idea to combine the papers' methods, to obtain estimates up to the boundary for elliptic systems with general boundary conditions. Such results were already within view by 1959: the Introduction of [1] says "In this paper we shall derive 'estimates near the boundary' for elliptic equations of arbitrary order under general boundary conditions, not merely Dirichlet boundary conditions. We have obtained these results for general elliptic systems, but for simplicity, we treat here in detail the theory of a single equation for one function. Systems will be treated in a forthcoming paper." It took a few years to wrap things up (which is not surprising, considering the generality of the outcome):

- Nirenberg's 1964 paper with Shmuel Agmon and Avron Douglis, *Estimates near the boundary for solutions of elliptic partial differential equations satisfying general boundary conditions. II* [2], provided Schauder-type and L^p estimates up to the boundary, for boundary value problems involving the full range of systems considered in [26]. As in [1], the boundary conditions considered are essentially the most general ones permitting such estimates. (An indirect characterization is that for the homogeneous constant-coefficient case in a half-space, a solution which is periodic on the boundary must decay exponentially toward the interior of the domain; a more algebraic characterization is included in the paper. Such boundary conditions are said to satisfy the "complementing condition.") Like the earlier work [1] on scalar-valued unknowns, the analysis combines a thorough understanding of half-space problems with tools from potential theory and singular integral operators. However the paper's focus on general systems made the analysis of the half-space problems quite different from what was done in [1].

It is interesting to compare the scientific style of Nirenberg's earlier work on the Weyl and Minkowski problems with that of the papers just discussed. One could say that the earlier work was problem-driven, while that with Agmon and Douglis was method-driven. Indeed, the starting point of the earlier work was to solve the Weyl and Minkowski problems, while that of the later was to identify the full power and scope of certain methods. And yet, upon reflection the contrast is not so sharp: once he saw that the key to the Weyl and Minkowski problems was a regularity theorem for a 2nd order, fully-nonlinear elliptic PDE in two space dimensions, Nirenberg proved a rather general result of this type—capturing the full power of his method—and explored additional applications, for example to the existence of solutions to quasilinear PDE [68]. As for the work with Agmon and Douglis: no specific challenge was needed, since by the 1950s the importance of a priori estimates for elliptic equations and systems was well-established.

The Agmon–Douglis–Nirenberg estimates helped establish a sound foundation for the theory of elliptic PDE. Since the strength of this work lies partly in its generality, no example can capture its full importance. Let me nevertheless mention a favorite example, namely the applicability of this theory to linear elasticity. In the early days our understanding of elastostatics relied heavily on Korn's inequality (whose early proofs for traction-type boundary conditions were complicated and relied heavily on the special form of the problem). From a modern perspective, Korn's inequality is not completely irrelevant—it assures us, for example, that solutions to traction problems are unique up to rigid motions. But as far as elliptic estimates are concerned, the equations of elastostatics are just another example of an elliptic system to which the Agmon–Douglis–Nirenberg theory applies.

I also have a favorite example concerning the importance of permitting elliptic systems to be of different orders in different unknowns: if Ω is a bounded domain in $\mathbb{R}^n$, consider the generalized Stokes system

$$-\Delta u_i + \nabla_i p = f_i, \quad \operatorname{div} u = g \quad \text{in } \Omega$$

with boundary condition

$$\sum_{j=1}^{n} e_{ij}(u) n_j - p n_i = h_i \quad \text{at } \partial\Omega$$

(using the notation $e_{ij}(u) = \frac{1}{2}[\nabla_i u_j + \nabla_j u_i]$). When $n = 2$ or 3 and $g = 0$, problems of this form arise both in elasticity (when the material is incompressible) and in fluid dynamics (Stokes flow). The system is second-order in u and first-order in p, but it meets the requirements of the Agmon–Douglis–Nirenberg theory.

4 Functions with Bounded Mean Oscillation

Research is unpredictable: tools and results developed in a particular context often have impact in other contexts, leading to entirely unanticipated consequences. The focus of this section—Nirenberg's 1961 paper *On functions of bounded mean oscillation* with Fritz John [43]—provides a fine example.

This paper addressed the question: suppose a function u has *bounded mean oscillation* on a cube $Q_0 \subset \mathbb{R}^n$, in the sense that its mean oscillation on sub-cubes is finite:

$$\sup_{Q \subset Q_0} \frac{1}{|Q|} \int_Q |u - u_Q| \, dx =: \|u\|_{\mathrm{BMO}(Q_0)} < \infty$$

(here Q ranges over cubes contained in Q_0, $|Q|$ is the volume of Q, and u_Q is the average of u on Q). Elementary examples (for example $\log |x|$) show that u need not be L^∞, but suggest that u can be large only on very small sets. The John–Nirenberg paper quantified this; its main result was that if $\|u\|_{\mathrm{BMO}(Q_0)} \leq K$ then

$$|\{x : |u - u_{Q_0}| > \sigma\}| \leq B e^{-b\sigma/K} |Q_0|$$

for some constants B and b depending only on the dimension n. This yields, by elementary arguments, control of various norms of $u - u_{Q_0}$; in particular

$$\frac{1}{|Q_0|} \int_{Q_0} e^{\beta K^{-1} |u - u_{Q_0}|} \, dx \leq C$$

for constants β and C depending only on n, and

$$\frac{1}{|Q_0|}\int |u - u_{Q_0}|^p\,dx \le C_{p,n}K^p \tag{1}$$

for any $p < \infty$. Focusing on the latter: while $u - u_{Q_0}$ is not uniformly of order K, its L^p norms are controlled for any $p < \infty$ as if that were the case.

The immediate motivation came from Fritz John's work on elasticity [41]. In nonlinear elasticity the deformation of an elastic body $\Omega \subset \mathbb{R}^3$ is a map $f\colon \Omega \to \mathbb{R}^3$. Writing $Df(x) = R(x)E(x)$ where $R(x)$ is a rotation and $E(x) = \left[(Df(x))^T Df(x)\right]^{1/2}$, the nonlinear elastic energy controls the nonlinear strain $|E(x) - I|$ but not the local rotation $R(x)$. In linear elasticity, Korn's inequality provides L^2 control of the infinitesimal rotation in terms of the L^2 norm of the linear strain; John's goal was a fully nonlinear analogue of this result. He found a proof that if the nonlinear strain is uniformly small on a cube, then the BMO norm of Df is also small:

$$\|E(x) - I\|_{L^\infty(Q)} \le \epsilon \quad \text{implies that} \quad \|Df\|_{\mathrm{BMO}(Q)} \le C\epsilon \tag{2}$$

provided ϵ is sufficiently small. Since E stays close to I by hypothesis, this is really an estimate on the oscillation of $R(x)$. Knowing Nirenberg's analytical power—and his love of inequalities—John drew Nirenberg into exploring the implications of (2). This was the origin of the John–Nirenberg paper; note that (1) with $p = 2$ shows that $R(x)$ stays close in L^2 to its average on Q, turning (2) into the nonlinear Korn-like inequality

$$\frac{1}{|Q|}\int_Q |Df - (Df)_Q|^2\,dx \le C \sup_{x\in Q} |E(x) - I|^2. \tag{3}$$

It was clear from the start that their estimates on BMO functions would have implications far beyond elasticity. Indeed the John–Nirenberg paper includes, as an application, a new proof of a result due to M. Weiss and A. Zygmund (namely: if G is periodic and $G(x+h) + G(x-h) - 2G(x) = O\left(h/|\log h|^\beta\right)$ for some $\beta > 1/2$ then G is the indefinite integral of some function g belonging to every L^p). A more dramatic application was provided by Jürgen Moser in the very same issue of Comm. Pure Appl. Math.: he used the John–Nirenberg theory to prove a Harnack inequality for the solution of a divergence-form elliptic equation

$$\sum_{i,j=1}^{n} \partial_i(a_{ij}(x)\partial_j u) = 0,$$

when the matrix-valued function $a_{ij}(x)$ is merely L^∞ and uniformly elliptic [59]. The Hölder regularity of such u was a landmark result proved in 1957 by Ennio De Giorgi [25] and John Nash [62]. Moser had given a third proof in 1960 [58].

Harnack's inequality implies Hölder continuity by an elementary argument (this is Section 5 of [59]), so Moser's application of the John–Nirenberg estimate provided a fourth proof of the celebrated De Giorgi–Nash–Moser regularity theorem.

A different perspective on BMO began to emerge in the mid-1960s, when J. Peetre, S. Spanne, and E. Stein observed (independently) that while singular integral operators (such as Riesz transforms) are not bounded linear operators on $L^\infty(\mathbb{R}^n)$, they *are* bounded linear operators on BMO$(\mathbb{R}^n)$, the space of functions on all $\mathbb{R}^n$ such that

$$\sup_{\text{cubes Q}} \frac{1}{|Q|} \int_Q |u - u_Q|\, dx =: \|u\|_{\mathrm{BMO}(\mathbb{R}^n)} < \infty.$$

This was the first indication that BMO$(\mathbb{R}^n)$ deserved attention as a function space, and would be the "right" substitute for L^∞ in many results of harmonic analysis. The correctness of this viewpoint became clear in the early 1970s, when C. Fefferman showed [29, 30] that

(a) $f \in \mathrm{BMO}(\mathbb{R}^n)$ exactly if $f = g_0 + \sum_{j=1}^n R_j g_j$ where $g_0, \dots g_n$ are L^∞ and R_j is the jth Riesz transform (which acts in Fourier space as $\xi_j/|\xi|$); and
(b) BMO$(\mathbb{R}^n)$ is dual (using the L^2 inner product) to the Hardy space $H^1(\mathbb{R}^n)$ (which consists, by definition, of functions in L^1 whose Riesz transforms are also in L^1).

Returning to elasticity, it is natural to ask: is the John–Nirenberg theory of any use for the analysis of nonlinear elastic boundary value problems? A 1972 paper by John provides an attractive answer, by proving the uniqueness of nonlinear elastic equilibria when the boundary displacement is fixed and only deformations with uniformly small strain are considered [42]. The idea is relatively simple: writing the deformation as $f(x) = x + u(x)$ and proceeding as one would for linear elasticity (with u as the elastic displacement), one needs to show that the higher-order terms neglected in the linear theory are truly unimportant. This is done using a consequence of the John–Nirenberg theory slightly different from those displayed above: if a function g has small BMO norm and average value 0 on a (nice enough) domain Ω then

$$\int_\Omega |g|^3\, dx \le C\|g\|_{\mathrm{BMO}(\Omega)} \int |g|^2\, dx.$$

In the proof of the uniqueness theorem, this is applied with $g = Df_1 - Df_2$ where f_1 and f_2 are two uniformly-small-strain elastic equilibria.

Fritz John's arguments required uniform bounds on the strain. This is a serious handicap, since one rarely knows in advance that the solution of a nonlinear elasticity problem has uniformly small strain. Forty years after the work of John and Nirenberg, the relationship between nonlinear strain and rotation was revisited by G. Friesecke, R.D. James, and S. Mü [31]. They improved (3) by showing that

$$\int_Q |Df - (Df)_Q|^2\,dx \le C \int_Q |E(x) - I|^2\,dx, \tag{4}$$

and used this estimate to explore the connection between 3D elasticity and various plate theories [31, 32].

I started by noting the unpredictability of research progress. In 1961 John and Nirenberg anticipated connections to elasticity (this was after all their starting point), and they also anticipated connections to analysis (this is clear from their new proof of the Weiss–Zygmund result). But they could not have anticipated the deep links to harmonic analysis that emerged a decade later, and I don't think they anticipated that (4) would be true without assuming uniformly small strain.

5 Partial Regularity for the 3D Navier–Stokes Equations

I had the privilege of collaborating with Louis Nirenberg and Luis Caffarelli around 1981 on partial regularity for the incompressible Navier–Stokes equations. I was a 2nd-year postdoc at Courant in 1980–1981, and Luis had just joined the faculty. It was Louis' suggestion that we look together at Vladimir Scheffer's work on Navier–Stokes [75, 76], which none of us had read before. The discussions that followed were an incredible learning experience! Their outcome was our paper *Partial Regularity of Suitable Weak Solutions of the Navier Stokes Equations* [20].

The incompressible Navier–Stokes equations describe the flow of a viscous, Newtonian fluid (such as water). Focusing for simplicity on the problem in all $\mathbb{R}^3$ with unit viscosity and no forcing, the equations say that the velocity u and pressure p solve the initial value problem

$$\begin{aligned} u_t + u \cdot \nabla u - \Delta u + \nabla p &= 0 \\ \nabla \cdot u &= 0 \\ u(x, 0) &= u_0(x). \end{aligned} \tag{5}$$

For this to be adequate as a description of the fluid, there should be a unique solution of (5) for any (sufficiently smooth) initial data u_0 with suitable decay as $|x| \to \infty$. We still don't know whether this is true or not. Indeed: if u_0 is smooth enough (and decays at infinity) there is a unique classical solution for a while at least, but for large initial data we cannot rule out the development of singularities in finite time. The solution can be continued for all time as a Leray–Hopf weak solution, but we do not know that such weak solutions are unique. (Nonuniqueness of Leray–Hopf weak solutions seems a real possibility, in view of recent progress including [17, 38, 40].)

The program that Scheffer began in the late 1970s seems natural in hindsight, but at the time it was revolutionary. There was by then a well-established literature on the partial regularity of minimizers for problems from geometry and the calculus of variations. It was Scheffer's idea to study the partial regularity of weak solutions to the Navier–Stokes equations using similar methods. His main result was that for a suitably constructed weak solution, the singular set has 5/3-dimensional parabolic Hausdorff measure zero in space-time. Our paper [20] obtained a similar result with 5/3 replaced by 1. The improved result places substantial restrictions upon the geometry of the singular set; for example, in an axisymmetric solution the only possible location of a singularity is on the axis. (The definition of parabolic Hausdorff measure is similar to that of ordinary Hausdorff measure, except that it uses coverings not by balls but rather by parabolic cylinders Q_r having radius r in space and extent r^2 in time.)

Can a solution with smooth initial data really develop a singularity? We still don't know. Leray suggested looking for self-similar singular solutions, i.e., ones of the form

$$u(x,t) = (T-t)^{-1/2} w\left(x/\sqrt{T-t}\right), \tag{6}$$

but we now know there are no such solutions with locally finite energy [63, 82]. Leray's ansatz can be generalized by looking for a solution that remains "bounded in similarity variables," i.e., such that

$$u(x,t) = (T-t)^{-1/2} w(y,s) \quad \text{where } y = x/\sqrt{T-t} \text{ and } s = -\ln(T-t).$$

This leads to an autonomous evolution for $w(y,s)$, namely

$$w_s + w \cdot \nabla w - \Delta w + \frac{1}{2} w + \frac{1}{2} y \cdot \nabla w + \nabla q = 0, \tag{7}$$

to be solved in all $\mathbb{R}^3$ and all sufficiently large s, with $\nabla \cdot w = 0$ and a suitable decay condition as $|y| \to \infty$. Leray's proposal was to look for a *stationary* solution of (7), but to give an example of a singular solution it would suffice to find *any* solution of (7) that exists for all $s > s_0$ and doesn't decay to 0 as $s \to \infty$. Alas, we have no idea whether such a w exists or not.

In looking for possible examples of singular solutions, it is natural to focus on solutions with special symmetry. Since the partial regularity theory does not rule out an axially symmetric solution developing a singularity along its axis, considerable attention has been devoted to the axially symmetric setting. The main result there is that if blowup occurs, then it must be "type II" in both space and time, in the sense that the functions $(T-t)^{1/2}|u(x,t)|$ and $(x_1^2 + x_2^2)^{1/2}|u(x,t)|$ must both be unbounded as t approaches the singular time T. Paraphrasing the first of these estimates: in the axially symmetric setting (with symmetry around the x_3 axis), if a solution blows up at time T then its L^∞ norm must grow faster than $(T-t)^{-1/2}$, and the associated solution of (7) must have $\|w\|_{L^\infty} \to \infty$ as $s \to \infty$ [23, 44, 78].

Returning for a moment to Scheffer's program, it is natural to hope for a proof that the parabolic Hausdorff dimension of the singular set is strictly less than 1. Alas, it seems that this would require an entirely new approach. Indeed, Scheffer's results and ours rely mainly on a "generalized energy inequality" (Eq. (11) below). The generalized energy inequality permits a nonzero forcing term f on the right-hand side of the Navier–Stokes equation provided that $u \cdot f \leq 0$, and it permits u to be discontinuous in time provided that $|u|^2$ only jumps downward. Using observations such as these, Scheffer has shown that the generalized energy inequality is consistent with u being singular on a set of parabolic Hausdorff dimension α for any $\alpha < 1$ [77]. Thus, the result of [20] seems to be more or less optimal, if the generalized energy inequality is to be used as the main tool and parabolic Hausdorff measure is used to measure the size of the singular set. (There are other ways to measure the size of the singular set; for some results using "box-counting dimension" see [46] the references cited there.)

The rest of this section provides a little more detail concerning the contributions of [20]. The main ingredients of a partial regularity theorem are:

(a) a weak solution, with some global estimates;
(b) a result of the form "locally sufficiently small implies regular;" and
(c) a covering argument.

Concerning (a): multiplying the Navier–Stokes equation by u, integrating in space, and integrating by parts leads formally to $\frac{d}{dt}\int |u|^2\,dx + 2\int |\nabla u|^2\,dx = 0$. For Leray–Hopf weak solutions the formal argument breaks down but we have still have an energy inequality:

$$\int_{\mathbb{R}^3\times\{t\}} |u|^2\,dx + 2\iint_{\mathbb{R}^3\times(0,t)} |\nabla u|^2\,dx\,d\tau \leq \int_{\mathbb{R}^3} |u_0|^2\,dx \tag{8}$$

where u_0 is the initial data and we focus for simplicity only on the whole-space problem. This clearly implies

$$\int_{\mathbb{R}^3\times\{t\}} |u|^2\,dx \leq M \quad \text{and} \quad \iint_{\mathbb{R}^3\times(0,t)} |\nabla u|^2\,dx\,dt \leq M/2 \tag{9}$$

for all t, where $M = \int_{\mathbb{R}^3} |u_0|^2\,dx$ is fixed by the initial data. It also implies that

$$\iint_{\mathbb{R}^3\times(0,t)} |u|^{10/3} + |p|^{5/3}\,dx\,d\tau \leq CM^{5/3} \tag{10}$$

for all t. (The estimate for u follows from (9) using the Gagliardo–Nirenberg estimate $\int_{\mathbb{R}^3} |u|^{10/3}\,dx \leq C\left(\int_{\mathbb{R}^3} |\nabla u|^2\,dx\right)\left(\int_{\mathbb{R}^3} |u|^2\right)^{2/3}$ and integration in time. The estimate for p follows from that for u, since we are discussing the whole-space problem: taking the divergence of the equation gives $\Delta p = -\sum_{i,j=1}^3 \nabla_i\nabla_j(u_iu_j)$, and for each i, j the singular integral operator $\Delta^{-1}\nabla_i\nabla_j$ is a bounded linear map from $L^{5/3}$ to itself.)

The energy inequality (8) is global, but partial regularity is a local matter. Therefore we need something similar but more local—a *generalized energy inequality*. For a smooth, compactly supported, scalar-valued function $\phi(x,t)$, multiplying the Navier–Stokes equation by $u\phi$, integrating in space, and integrating by parts leads formally to $\frac{d}{dt}\int |u|^2\phi\,dx + 2\int |\nabla u|^2\phi\,dx = \int |u|^2(\phi_t+\Delta\phi) + (|u|^2+2p)u\cdot\nabla\phi\,dx$; for weak solutions the formal argument breaks down, but (for suitably-constructed weak solutions) one gets the *generalized energy inequality*

$$2\iint |\nabla u|^2\phi\,dx\,dt \le \iint |u|^2(\phi_t+\Delta\phi) + (|u|^2+2p)u\cdot\nabla\phi\,dx\,dt \tag{11}$$

for smooth, compactly-supported functions ϕ such that $\phi \ge 0$.

Concerning (b): The Navier–Stokes equation has the following scale invariance: if $u(x,t)$ and $p(x,t)$ solve (5) then so does

$$u_\lambda(x,t) = \lambda u(\lambda x, \lambda^2 t), \quad p_\lambda(x,t) = \lambda^2 p(\lambda x, \lambda^2 t) \tag{12}$$

for any $\lambda > 0$. A result of the form "locally small implies regular" should be scale-invariant; in other words its hypothesis should have "dimension zero" under the convention that each spatial dimension x_i has dimension 1, time t has dimension 2, each velocity component u_i has dimension -1, the pressure p has dimension -2, and $\partial/\partial x_i$ has dimension -1. Note that in this parabolic setting, a local estimate should involve an integral over a parabolic cylinder

$$Q_r(x_0,t_0) = \{(x,t) \,:\, |x-x_0| < r,\ t_0 - r^2 < t < t_0\}.$$

The heart of the partial regularity theory in [20] is the following "locally small implies regular" result: there is a constant $\epsilon_0 > 0$ such that

$$\limsup_{r\to 0} r^{-1}\iint_{Q_r(x_0,t_0)} |\nabla u|^2\,dx\,dt < \epsilon_0 \quad \text{implies that } u \text{ is regular at } (x_0,t_0). \tag{13}$$

The proof makes use of a rather different "locally small implies regular" result: there is a constant $\epsilon_1 > 0$ such that

$$r^{-2}\iint_{Q_r(x_0,t_0)} |u|^3 + |p|^{3/2}\,dx\,dt < \epsilon_1 \quad \text{implies that } u \text{ is regular on } Q_{r/2}(x_0,t_0). \tag{14}$$

(This is a simplified version of Proposition 1 and Corollary 1 of [20]. The result there was more complicated, because it was not known at the time that for the solution of Navier–Stokes in a bounded domain the pressure was in $L^{3/2}$.) The latter estimate (and its proof) are quite close to what Scheffer had done before.

I will not attempt to discuss the proofs of these results, except to remark upon the relation between them: the proof of (13) in [20] proceeds by showing, roughly speaking, that if $r^{-1} \iint_{Q_r(x_0,t_0)} |\nabla u|^2 \, dx\, dt$ is small enough then $r^{-2} \iint_{Q_r(x_0,t_0)} |u|^3 + |p|^{3/2} \, dx\, dt$ decays as r decreases, becoming eventually less than ϵ_1. Alternative proofs of these "locally small implies regular" results have since been given by others [47, 53]. A well-organized and modern exposition is available in [74].

Concerning (c): the covering arguments used to estimate the size of the singular set are quite standard. Using (10) and (14) one can show that the singular set has 5/3-dimensional parabolic Hausdorff measure zero. Indeed, by (14) and Hölder's inequality, if (x_0, t_0) is a singular point then for any $r > 0$ the parabolic cylinder Q_r centered at (x_0, t_0) has

$$r^{-5/3} \iint_{Q_r} |u|^{10/3} + |p|^{5/3} \, dx\, dt \geq \epsilon_1'$$

for some fixed positive constant ϵ_1'. By a parabolic variant of the Vitali covering lemma, one concludes that for any $\delta > 0$ the singular set is contained in a union of parabolic cylinders Q_j whose radii $r_j < \delta$ satisfy

$$\sum r_i^{5/3} \leq C \iint_{\cup_j Q_j} |u|^{10/3} + |p|^{5/3} \, dx\, dt.$$

As $\delta \to 0$ this shows that the singular set has Lebesgue measure 0; since $\cup_j Q_j$ is contained in a δ-neighborhood of the singular set, the right hand side of the preceding estimate tends to 0 as $\delta \to 0$. So the singular set has 5/3-dimensional parabolic Hausdorff measure 0. (This argument is close to what Scheffer did in [75, 76].)

The proof that the singular set has one-dimensional parabolic measure zero proceeds similarly, except that it combines the small-implies-regular result (13) with the global estimate on $\iint |\nabla u|^2 \, dx\, dt$. It estimates the one-dimensional measure whereas the previous argument estimated the 5/3-dimensional measure, because it relies on a global estimate for $\iint |\nabla u|^2 \, dx\, dt$ (which has scaling dimension 1) whereas the previous argument relied on a global estimate for $\iint |u|^{10/3} + |p|^{5/3} \, dx\, dt$ (which has scaling dimension 5/3).

Evidently, the outcome of the argument requires a suitable synergy between the form of the small-implies-regular result and the global estimate being used. Our paper [20] obtained additional results by considering global energy-type estimates with weighted norms, associated with formal calculations of $\frac{d}{dt} \int |u|^2 |x| \, dx + 2 \int |\nabla u|^2 |x| \, dx$ and $\frac{d}{dt} \int |u|^2 |x|^{-1} \, dx + 2 \int |\nabla u|^2 |x|^{-1} \, dx$. In doing so, we needed some analogues of the Gagliardo–Nirenberg interpolation inequalities in norms weighted by powers of $|x|$. Convinced that such estimates would have other uses as well, we wrote a separate paper on this topic [21]. The estimates proved there have indeed been used in many settings, and they have been generalized in various ways—for example to interpolation estimates involving weighted Hölder norms [51] and fractional derivatives [64]. The extremals and sharp constants for these estimates have also attracted considerable attention (see, e.g., [27, 52]).

6 Nonlinear Elliptic Equations Involving Critical Exponents

Nirenberg's 1983 paper with Haim Brezis, *Positive solutions of nonlinear elliptic equations involving critical Sobolev exponents* [14], was a landmark development in our understanding of semilinear PDE involving critical exponents. Its focus was the existence of solutions to

$$\begin{aligned} -\Delta u &= u^p + f(x,u) && \text{in } \Omega \\ u &> 0 && \text{in } \Omega \\ u &= 0 && \text{at } \partial\Omega \end{aligned} \tag{15}$$

in a bounded domain $\Omega \subset \mathbb{R}^n$ when $n \geq 3$, p is the "critical exponent"

$$p = \frac{n+2}{n-2},$$

and $f(x,u)$ grows slower than u^p at infinity.

To explain the issues, it is convenient to focus on the special case $f(x,u) = a(x)u$, when the PDE becomes

$$-\Delta u = u^p + a(x)u. \tag{16}$$

A necessary condition for the existence of a positive solution is that

$$\text{the first Dirichlet eigenvalue of } -\Delta - a \text{ is positive,} \tag{17}$$

as one easily verifies by multiplying (16) by the associated eigenfunction and integrating by parts. This condition is definitely not sufficient, since by Pohozhaev's identity there is no solution when $a(x) = 0$ and Ω is star-shaped.

To understand why $p = (n+2)/(n-2)$ is special, note that $p = (n+2)/(n-2)$ is equivalent to $p+1 = 2n/(n-2)$, the exponent that appears in the scale-invariant Sobolev inequality

$$\|u\|_{L^{2n/(n-2)}(\Omega)} \leq C\|\nabla u\|_{L^2(\Omega)} \quad \text{for } u \in H_0^1(\Omega). \tag{18}$$

A key point is that bounded sequences in $H_0^1(\Omega)$ are precompact in L^{q+1} for $q < (n+2)/(n-2)$, but not in $L^{2n/(n-2)}$. This is relevant to the problem at hand because when the exponent is subcritical (i.e., when p is replaced by q such that $1 < q < (n+2)/(n-2)$) there are straightforward variational approaches, either

(i) seeking a positive critical point of

$$\int_\Omega \frac{1}{2}|\nabla u|^2 - \frac{1}{q+1}|u|^{q+1} - \frac{1}{2}a(x)u^2\,dx, \tag{19}$$

or else

(ii) solving the variational problem

$$\inf_{\substack{\int_\Omega |u|^{q+1}\,dx=1 \\ u=0 \text{ at } \partial\Omega}} \int_\Omega |\nabla u|^2 - a(x)u^2\,dx \tag{20}$$

(for which the Euler–Lagrange equation is $-\Delta u = a(x)u + \mu u^q$ with μ constant; the eigenvalue condition (17) assures that $\mu > 0$, so that a well-chosen scalar multiple of u solves (16)).

When q is subcritical approach (i) is tractable since the functional (19) satisfies the Palais–Smale condition, and approach (ii) also works since the direct method of the calculus of variations applies straightforwardly to (20). For the critical exponent, however, neither approach works (at least, not straightforwardly): (i) is dubious since the analogue of (19) doesn't satisfy the Palais–Smale condition; and (ii) is dubious since the constraint $\int_\Omega |u|^{p+1}\,dx = 1$ is not preserved under weak convergence in $H_0^1(\Omega)$. Moreover this is not just a technical issue, since (as noted above) there is in fact no positive solution in a star-shaped domain when $p = (n+2)/(n-2)$ and $a(x) = 0$.

The essential phenomenon here is the study of a variational problem involving both $\int |\nabla u|^2\,dx$ and $\int |u|^{p+1}\,dx$, for the special value of p where the latter is controlled by the former but with a lack of compactness. What intrigued Brezis and Nirenberg was the observation that the existence or nonexistence of a solution can depend, for such problems, on the presence (and form of) "lower order terms" such as $\int a(x)u^2\,dx$. This observation had already been made in a special case by Thierry Aubin, in a 1976 paper on the Yamabe problem [4] (a problem from geometry which is easily reduced to solving a PDE quite similar to (16) but on a Riemannian manifold without boundary rather than a domain in $\mathbb{R}^n$). Indeed it was Aubin's work that attracted their attention to this area.

Actually, variational problems involving a lack of compactness arise in a great variety of settings. The work of P.L. Lions on concentration compactness [54, 55] is a rich source of examples; in a different direction, the recent paper [35] by Ghoussoub and Robert discusses many examples involving the existence of extremals for Sobolev-type inequalities involving weighted norms.

The goal of the Brezis–Nirenberg paper [14] was to understand when and how the presence of a "lower order term" permits the variational approaches (i) or (ii) to succeed even when the exponent is critical. Their results include a rather complete understanding about when approach (ii)—based on a minimization analogous to (20)—suffices to solve (16) in space dimension $n \geq 4$. Focusing on this part of the story for a moment, let

$$S = \inf_{\substack{\int_\Omega |u|^{p+1}\,dx=1 \\ u=0 \text{ at } \partial\Omega}} \int_\Omega |\nabla u|^2\,dx \tag{21}$$

with $p = (n+2)/(n-2)$. (Clearly $S^{-1/2}$ is the best constant for the scale-invariant Sobolev estimate (18); the value of this constant doesn't depend on Ω and is the same as the best constant for the analogous scale invariant inequality in all $\mathbb{R}^n$; in particular, the value of S is known.) Now let J be the minimum value of (20) with q replaced by p:

$$J = \inf_{\substack{\int_\Omega |w|^{p+1}\,dx=1 \\ w=0 \text{ at } \partial\Omega}} \int_\Omega |\nabla u|^2 - a(x)u^2\,dx. \tag{22}$$

Brezis and Nirenberg showed that when $n \geq 4$ and the eigenvalue condition (17) holds, the following are equivalent:

(a) $a(x) > 0$ somewhere in Ω
(b) $J < S$
(c) the minimum defining J is achieved.

(This statement combines several of the results in [14], following the lead of [12].) The proofs that $(b) \to (c)$ and $(c) \to (a)$ are relatively elementary, and they work even when $n = 3$. The assertion $(a) \to (b)$—proved using a well-chosen test function for J—is what restricts the result to $n \geq 4$.

The case $n = 3$ is surprisingly different, and the treatment in [14] was limited to the case when Ω is a ball and $a(x)$ is constant. A full understanding was achieved only in 2002 by Druet [28]; the $n = 3$ analogue of assertion (a) turns out to be that $g(x, x) > 0$ somewhere in Ω, where $g(x, y)$ is the regular part of the Green's function for $-\Delta - a$ on Ω.

I have focused thus far on approach (ii), which minimizes a suitable functional subject to the constraint $\int_\Omega |u|^{p+1}\,dx = 1$. Brezis and Nirenberg also studied approach (i), which is more useful when $f(x, u)$ is nonlinear in u, for example when the PDE is

$$-\Delta u = u^p + \mu u^q \tag{23}$$

where $q < p = (n+2)/(n-2)$ and $\mu > 0$ is constant. If p were subcritical it would be standard to find a critical point using the mountain-pass lemma. When p is critical, they show this still works (despite the failure of the Palais–Smale condition) when the min-max value of the functional (the critical value, so to speak) is strictly less than $\frac{1}{n}S^{n/2}$. Using this result, they show (for example) that when $n \geq 4$ Eq. (23) has a positive solution for any $\mu > 0$ and any bounded Ω.

I have already mentioned Pierre-Louis Lions' work on concentration compactness, which was roughly contemporaneous with [14]. While its focus was very similar—namely, variational problems with a lack of compactness—its method was rather different. Briefly: Lions focused on classifying the mechanisms by which compactness can be lost (and developing methods for ruling them out in specific examples), while Brezis and Nirenberg focused more sharply on a particular class of problems. The two investigations complement each other nicely: conditions for existence analogous to $J < S$ show up also in Lions' work, but Brezis and Nirenberg achieved a more complete understanding for the particular problems they addressed.

We have thus far discussed two particular methods for finding solutions of (15). Their failure does not necessarily imply nonexistence, as [14] makes clear by pointing to examples (such as $-\Delta u = u^p$ in the shell $\{1 < |x| < 2\}$). The nonexistence theorems in [14] are mainly for star-shaped domains, proved using Pohozhaev's identity or something similar. The fact that $-\Delta u = u^p$ has a positive solution in a shell but not in a star-shaped domain suggests that the *topology* of Ω might be relevant—and this was confirmed by Bahri and Coron in a 1988 paper [5], which developed an approach to existence theorems that takes advantage of nontrivial topology.

7 The Method of Moving Planes and the Sliding Method

Nirenberg's 1979 paper with Basilis Gidas and Wei-Ming Ni, *Symmetry and related properties via the maximum principle* [36], began the development of a powerful and flexible toolkit for showing that the solutions of certain nonlinear elliptic PDE respect the symmetry suggested by their boundary conditions. Their approach, which soon became known as the *method of moving planes*, drew inspiration from work by Alexandroff on problems from geometry (for which their citation was [39]) and work by Serrin on PDE's with overdetermined boundary conditions [79]. The essential contribution of [36] was to show that far from being a trick that solves a few specific problems, the method of moving planes provides an intuitive and flexible approach for proving the symmetry of positive solutions, for a broad class of nonlinear PDE. While the 1979 paper [36] focused mainly on problems in bounded domains, it also considered some problems in all $\mathbb{R}^n$, and a followup paper [37] obtained additional results in that setting.

The method of moving planes is particularly well-suited to the study of positive solutions of equations of the form $\Delta u + f(u) = 0$ (as I'll discuss in some detail below). The introduction of [36] points briefly to the equations $\Delta u + u^{(n+2)/(n-2)} = 0$ and $\Delta u - u^{(n+2)/(n-2)} = 0$ in space dimension $n > 2$ as motivating examples, explaining their relevance to Yang–Mills field theory and geometry. However the paper is method-oriented not application-oriented, written with confidence that the method of moving planes would in due course find many applications. And indeed it has! While a survey is beyond the scope of this article (and beyond the expertise of this author), let me mention one recent thread. The methods of Gidas, Ni, and Nirenberg have been extended to positive solutions of some *nonlocal* problems, by considering equivalent local problems in one more space dimension; for discussion and selected references see the segment of [80] by Xavier Cabré.

In the late 1980s Nirenberg returned to this area in a fruitful collaboration with Henri Berestycki. Their focus was on certain nonlinear PDE's in infinite cylinders, whose solutions describe moving combustion fronts. In this setting, a key goal is to prove monotonicity of the solution (with respect to the cylinder's axial variable). To achieve this goal, they introduced a maximum-principle-based approach to monotonicity, known as the "sliding method," whose spirit is similar to the method of moving planes [8, 9].

The early 1990s saw another important development, of a methodological character. Since the method of moving planes and the sliding method rely on versions of the maximum principle, the early papers had to exercise considerable care to be sure the required versions of the maximum principle were true. Besides complicating the analysis, this limited the statements of the theorems, for example by not permitting domains with corners. However it was understood in the early 1990s that a uniformly elliptic operator of the form $Lu = \sum a_{ij}(x)\partial_{ij}u + \sum b_i(x)\partial_i u + c(x)u$ with bounded, measurable coefficients satisfies a maximum principle in a domain Ω ($Lu \geq 0$ in Ω and $u \leq 0$ at $\partial\Omega$ implies $u \leq 0$ in Ω) provided only that Ω has sufficiently small measure. Nirenberg's 1991 paper with Berestycki, *On the method of moving planes and the sliding method* [10], shows how this version of the maximum principle permits dramatic simplification of the proofs of many results, and extends their validity to more general domains (e.g., ones with corners). (For an expository account of these developments with much more detail than given here, see [13].)

The preceding paragraphs are at best an incomplete survey of Nirenberg's work in this area. In his paper with Berestycki on problems in cylinders [9], a subtlety quite distinct from the sliding method involves understanding the solution's asymptotics at $\pm\infty$; this is analyzed using Nirenberg's 1963 results with Agmon [3] and related results by Pazy [72]. While [9] obtains qualitative results about solutions that are assumed to exist, a 1992 paper with Berestycki obtains rather complete information about the existence and uniqueness of traveling fronts for cylinder analogues of the most-studied one-dimensional models [11]. Later, Nirenberg wrote two papers with Berestycki and Caffarelli [6, 7] applying the method of moving planes or the sliding method to the monotonicity and symmetry of some problems in unbounded domains.

But achieving completeness is a hopeless task. Rather, let me try to communicate the elegant simplicity of the method of moving planes and the sliding method, by discussing two examples from Introduction of [10]. The first uses the method of moving planes:

Let Ω be a bounded domain in $\mathbb{R}^n$ which is convex in the x_1 direction and symmetric about $x_1 = 0$. Suppose $u \in C(\overline{\Omega}) \cap C^2(\Omega)$ solves

$$\begin{aligned} -\Delta u &= f(u) && \text{in } \Omega \\ u &> 0 && \text{in } \Omega \\ u &= 0 && \text{at } \partial\Omega \end{aligned} \tag{24}$$

where f is locally Lipschitz. Then u is symmetric with respect to x_1, and $\partial_1 u < 0$ for $x_1 > 0$.

The hypothesis that u be positive is crucial; for example, when Ω is a ball centered at 0 there are plenty of Dirichlet eigenfunctions that are not symmetric in x_1—but they are not positive. The hypothesis that Ω be convex is also crucial; for example, when Ω is the shell $\{1 < |x| < 2\}$ the equation $-\Delta u = u^q$ has a non-radial positive solution with $u = 0$ at $\partial\Omega$ when the space dimension is $n > 2$ and q is below but sufficiently close to the critical value $(n+2)/(n-2)$ [14].

The following argument is general, but we visualize it in Fig. 1 by taking Ω to be a diamond. Writing $x = (x_1, y)$ for points in $\mathbb{R}^n$, let $-a = \inf_{x \in \Omega} x_1$. For $-a < \lambda < 0$ let T_λ be the hyperplane $x_1 = \lambda$, let Σ_λ be the part of Ω where x_1 is less than λ, and observe that the function $(x_1, y) \mapsto u(2\lambda - x_1, y)$ is the reflection of u about the hyperplane T_λ. The key idea is to compare u with its reflection, by considering the function

$$w_\lambda(x_1, y) = u(2\lambda - x_1, y) - u(x_1, y).$$

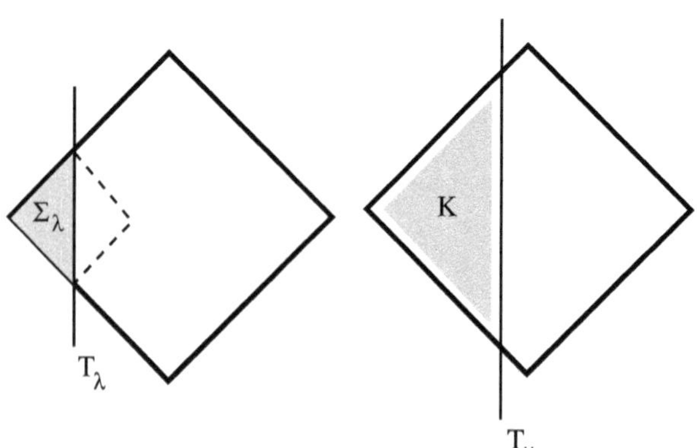

Fig. 1 The method of moving planes, when Ω is a diamond. Left: the hyperplane T_λ and the region Σ_λ (shaded); the broken line shows the boundary of the reflection of Σ_λ. Right: in the argument by contradiction, the set K (shaded) occupies most of Σ_μ

Since f is locally Lipschitz and u is bounded, w solves an equation of the form

$$-\Delta w_\lambda + c_\lambda(x) w_\lambda = 0 \tag{25}$$

in Σ_λ, with $c_\lambda(x)$ bounded. Moreover $w_\lambda \geq 0$ on $\partial\Sigma_\lambda$ (in fact, it vanishes on the part of the boundary where $x_1 = \lambda$ and it is strictly positive on the part of the boundary that belongs to $\partial\Omega$). The main task in the method of moving planes lies in proving that

$$w_\lambda > 0 \text{ on } \Sigma_\lambda \text{ whenever } -a < \lambda < 0. \tag{26}$$

The symmetry of u follows from (26) by elementary arguments combined with relatively standard applications of the maximum principle (for full details see [10] or [13]).

We now sketch the proof of (26), using the fact that the PDE (25) has a maximum principle on a domain of sufficiently small volume. When λ is close to $-a$ the set Σ_λ is thin in the x_1 direction; so the maximum principle applies and $w_\lambda > 0$ in Σ_λ. Now let $\mu \leq 0$ be the largest value such that $w_\lambda > 0$ on Σ_λ for $\lambda \in (-a, \mu)$. If $\mu = 0$ we're done, so we assume $\mu < 0$ and seek a contradiction. By continuity we have $w_\mu \geq 0$ in Σ_μ, and it follows (using a version of the usual maximum principle) that in fact $w_\mu > 0$ in Σ_μ. Now let K be a compact subset of Σ_μ such that $\Sigma_\mu \setminus K$ has small measure. Evidently w_μ is bounded away from 0 on K. Therefore (by continuity) $w_{\mu+\epsilon}$ is strictly positive on K when ϵ is sufficiently small. Since $\Sigma_{\mu+\epsilon} \setminus K$ has small volume, the maximum principle on sets with small volume shows that $w_{\mu+\epsilon} > 0$ on $\Sigma_{\mu+\epsilon}$. It follows that $w_{\mu+\epsilon} > 0$ on the entire set $\Sigma_{\mu+\epsilon}$, contradicting the definition of μ. Thus $\mu = 0$ and the argument is complete.

Turning to the sliding method: the following example is again from the Introduction of [10] (though the statement there is a bit more general).

> Let Ω be a bounded domain in $\mathbb{R}^n$ which is convex in the x_1 direction, and assume $\partial\Omega$ contains no segment parallel to the x_1 axis. Suppose $u \in C(\overline{\Omega}) \cap C^2(\Omega)$ solves
>
> $$-\Delta u = f(u) \quad \text{in } \Omega \tag{27}$$
>
> where f is locally Lipschitz, with boundary data such that
>
> $$u = \phi \quad \text{at } \partial\Omega.$$
>
> Assume that for any three points $x' = (x_1', y), x = (x_1, y), x'' = (x_1'', y)$ with $x', x'' \in \partial\Omega$, we have
>
> $$\phi(x') \leq u(x) \leq \phi(x''). \tag{28}$$

Then u is strictly monotone in x_1, in the sense that

$$u(x_1 + \tau, y) > u(x_1, y) \text{ when } \tau > 0, \text{ if } (x_1, y) \text{ and } (x_1 + \tau, y) \text{ are both in } \Omega.$$

Furthermore, if f is differentiable then $\partial_1 u > 0$. Finally, u is the unique solution of the given boundary value problem satisfying (28).

The proof uses translation rather than reflection: for $\tau > 0$, it compares u with its translate by τ, by considering the difference

$$w_\tau(x_1, y) = u(x_1 + \tau, y) - u(x_1, y).$$

This function is defined in the domain D_τ obtained by intersecting Ω with its translation $\Omega - \tau e_1$. It once again solves a PDE of the form (25), and the (28) assures that $w_\tau \geq 0$ at ∂D_τ. The main task is to show that

$$w_\tau > 0 \text{ on } D_\tau \text{ for all } \tau > 0 \text{ such that } D_\tau \text{ is nonempty.} \tag{29}$$

The argument is parallel to that used in the first example. Briefly: when τ is large the domain D_τ is small and the maximum principle for (25) in small domains gives $w_\tau > 0$ on D_τ. On the other hand if $w_\tau > 0$ on D_τ for all $\tau > \tau_1 > 0$, an argument similar the one given before (relying once again on the maximum principle for domains with small volume) shows that we also have $w_\tau > 0$ on D_τ for $\tau = \tau_1 - \epsilon$ when ϵ is sufficiently small.

The monotonicity of u and the other conclusions follow from (29) by elementary arguments combined with relatively standard applications of the maximum principle (for full details see [10]).

8 Conclusion

As indicated by my title, I have discussed just a few of Louis Nirenberg's many contributions. The topics I have selected are important, but many topics I have omitted are also very important. Writing about—indeed, thinking about—Louis' impact is truly a humbling experience. It was a great pleasure to see his contributions recognized by the 2015 Abel Prize.

References

1. Agmon, S., Douglis, A., Nirenberg, L.: Estimates near the boundary for solutions of elliptic partial differential equations satisfying general boundary conditions. I. *Comm. Pure Appl. Math.* **12**, 623–727 (1959).
2. Agmon, S., Douglis, A., Nirenberg, L.: Estimates near the boundary for solutions of elliptic partial differential equations satisfying general boundary conditions. II. *Comm. Pure Appl. Math.* **17**, 35–92 (1964).
3. Agmon, S., Nirenberg, L.: Properties of solutions of ordinary differential equations in Banach space. *Comm. Pure Appl. Math.* **16**, 121–239 (1963).
4. Aubin, Th.: Equations différentielles non linéaires et problème de Yamabe concernant la courbure scalaire. *J. Math. Pures et Appl.* **55**, 269–293 (1976).
5. Bahri, A., Coron, J.M.: On a nonlinear elliptic equation involving the critical Sobolev exponent: The effect of the topology of the domain. *Comm. Pure Appl. Math.* **41**, 253–294 (1988).
6. Berestycki, H., Caffarelli, L., Nirenberg, L.: Monotonicity for elliptic equations in unbounded Lipschitz domains. *Comm. Pure Appl. Math.* **50**, 1089–1111 (1997).
7. Berestycki, H., Caffarelli, L., Nirenberg, L.: Further qualitative properties for elliptic equations in unbounded domains. *Ann. Scuola Norm. Sup. Pisa Cl. Sci. (4)* **25**(1–2), 69–94 (1997).
8. Berestycki, H., Nirenberg, L.: Monotonicity, symmetry, and antisymmetry of solutions of semilinear elliptic equations. *J. Geom. Phys.* **5**, 237–275 (1988).
9. Berestycki, H., Nirenberg, L.: Some qualitative properties of solutions of semilinear elliptic equations in cylindrical domains, In: Rabinowitz, P., Zehnder, E. (Eds.), *Analysis, Et Cetera*, pp. 115–164, Academic Press (1990).
10. Berestycki, H., Nirenberg, L.: On the method of moving planes and the sliding method. *Bol. Soc. Bras. Mat. (N.S.)* **22** 1–37 (1991).
11. Berestycki, H., Nirenberg, L.: Travelling fronts in cylinders. *Ann. Inst. H. Poincaré Anal. Nonlin.* **9**, 497–572 (1992).
12. Brezis, H.: Elliptic equations with limiting Sobolev exponents—the impact of topology. *Comm. Pure Appl. Math.* **39**, S17–S39 (1986).
13. Brezis, H.: Symmetry in nonlinear PDE's. In: Giaquinta, M., Shatah, J., Varadhan, S.R.S. (eds.) Differential Equations—La Pietra 1996. *AMS Proc. Symp. Pure Math.* **65**, 1–12 (1999).
14. Brezis, H., Nirenberg, L.: Positive solutions of nonlinear elliptic equations involving critical exponents. *Comm. Pure Appl. Math.* **36**, 437–477 (1983).
15. Browder, F.E.: The Dirichlet problem for linear elliptic equations of arbitrary even order with variable coefficients. *Proc. Natl. Acad. Sci. U.S.A.* **38**, 230–235 (1952).
16. Browder, F.E.: Estimates and existence theorems for elliptic boundary value problems: *Proc. Natl. Acad. Sci. U.S.A.* **45**, 365–372 (1959).
17. Buckmaster, T., Vicol, V.: Nonuniqueness of weak solutions to the Navier–Stokes equation. *Ann. of Math.*, in press.
18. Caccioppoli, R.: Ovaloidi di metrica assegnata. *Pontificia Academia Scientiarum. Commentationes.* **4**(1) 1–20 (1940).
19. Caffarelli, L.: Elliptic second order equations. *Rend. Sem. Mat. Fis. Milano* **58**. 253–284 (1988).
20. Caffarelli, L., Kohn, R.V., Nirenberg, L.: Partial regularity of suitable weak solutions of the Navier–Stokes equations. *Comm. Pure Appl. Math.* **35**, 771–831 (1982).
21. Caffarelli, L., Kohn, R.V., Nirenberg, L.: First order interpolation inequalities with weights. *Compos. Math.* **53**, 259–275 (1984).
22. Calderón, A.P., Zygmund, A.: On the existence of singular integrals. *Acta. Math.* **88**, 85–139 (1952).
23. Chen, C.-C., Strain, R.M., Tsai, T.-P., Yau, H.-T.: Lower bounds on the blow-up rate of the axisymmetric Navier–Stokes equations II. *Comm. Partial Diff. Eqns.* **34**, 203–232 (2009).
24. Courant R., Hilbert, D.: *Methoden der Mathematische Physik*, Vol. 2. Springer, Berlin (1937).

25. De Giorgi, E.: Sulla differenziabilità e lanaliticità delle estremali degli integrali multipli regolari. *Mem. Accad. Sci. Torino Cl. Sci. Fis. Mat. Nat. (3)*, **3**, 25–43 (1957).
26. Douglis, A., Nirenberg, L.: Interior estimates for elliptic systems of partial differential equations. *Comm. Pure Appl. Math.* **8**, 503–538 (1955).
27. Dolbeault, J., Esteban, M.J., Loss, M.: Rigidity versus symmetry breaking via nonlinear flows on cylinders and Euclidean spaces. *Invent. Math.* **206**, 397–440 (2016).
28. Druet, O.: Elliptic equations with critical Sobolev exponents in dimension 3. *Ann. Inst. H. Poincaré Anal. Nonlin.* **19**, 125–142 (2002).
29. Fefferman, C.: Characterizations of bounded mean oscillation. *Bull. Amer. Math. Soc.* **77**, 587–588 (1971).
30. Fefferman, C., Stein, E.: H^p spaces of several variables. *Acta Math.* **129**, 137–193 (1972).
31. Friesecke, G., James, R.D., Müller, S.: A theorem on geometric rigidity and the derivation of nonlinear plate theory from three-dimensional elasticity. *Comm. Pure Appl. Math.* **55**, 1461–1506 (2002).
32. Friesecke, G., James, R.D., Müller, S.: A hierarchy of plate models derived from nonlinear elasticity by Gamma-convergence. *Arch. Rational Mech. Anal.* 180 (2006) 183–236.
33. Gagliardo, E.: Proprietà di alcune classi di funzioni in più variabili. *Ricerche Mat.* **7**, 102–137 (1958).
34. Gårding, L.: Dirichlet's problem for linear elliptic partial differential equations. *Math. Scandinavica* **1**, 55–72 (1953).
35. Ghoussoub, N., Robert, F.: Sobolev inequalities for the Hardy–Schrödinger operator: extremals and critical dimensions. *Bull. Math. Sci.* **6**, 89–144 (2016).
36. Gidas, B., Ni, W.-M., Nirenberg, L.: Symmetry and related properties via the maximum principle. *Commun. Math. Phys.* **68**, 209–243 (1979).
37. Gidas, B., Ni, W.-M., Nirenberg, L.: Symmetry of positive solutions of nonlinear elliptic equations in $\mathbb{R}^n$. In: *Mathematical Analysis and Applications, Part A*, pp. 369–402, *Adv. in Math. Suppl. Stud.*, 7A, Academic Press (1981).
38. Guillod, J., Šverák, V.: Numerical investigations of non-uniqueness for the Navier–Stokes initial value problem in borderline spaces. arXiv:1704.00560 (2017).
39. Hopf, H.: *Lectures on Differential Geometry in the Large*, Stanford University (1956).
40. Jia, H., Šverák, V.: Are the incompressible 3D Navier–Stokes equations locally ill-posed in the natural energy space? *J. Funct. Anal.* **268**, 3734–3766 (2015).
41. John, F.: Rotation and strain. *Comm. Pure Appl. Math.* **14**. 391–413 (1961).
42. John, F.: Uniqueness of non-linear elastic equilibrium for prescribed boundary displacements and sufficiently small strains. *Comm. Pure Appl. Math.* **25**, 617–634 (1972).
43. John. F., Nirenberg, L.: On functions of bounded mean oscillation. *Comm. Pure Appl. Math.* **14**, 415–426, (1961).
44. Koch, G., Nadirashvili, N., Seregin, G.A., Šverák, V.: Liouville theorems for the Navier–Stokes equations and applications. *Acta Math.* **203**, 83–105 (2009).
45. Kohn, J.J., Nirenberg, L.: An algebra of pseudo-differential operators. *Comm. Pure Appl. Math.* **18**, 269–305 (1965).
46. Kukavica, I., Pei, Y.: An estimate on the parabolic fractal dimension of the singular set for solutions of the Navier–Stokes system. *Nonlinearity* **25**, 2775–2783 (2012).
47. Ladyzhenskaya, O.A., Seregin, G.A.: On partial regularity of suitable weak solutions to the three-dimensional Navier–Stokes equations. *J. Math. Fluid Mech.* **1**, 356–387 (1999).
48. Lewy, H.: On the existence of a closed convex surface realizing a given Riemannian metric. *Proc. Natl. Acad. Sci. U.S.A.* **24**, 104–106 (1938).
49. Lewy, H.: On differential geometry in the large. I. Minkowski's problem. *Trans. Amer. Math. Soc.* **43**, 258–270 (1938).
50. Li, YanYan: The work of Louis Nirenberg. In: *Proceedings of the International Congress of Mathematicians 2010*,vol. I, 127–137. Hindustan Book Agency, New Delhi (2010). Also available at http://www.wias-berlin.de/imu/archive/ICM2010/www.icm2010.in/wp-content/icmfiles/laudaions/chern.pdf. (Accessed 8 March 2018.)

51. Lin, C.-S.: Interpolation inequalities with weights. *Comm. Partial Diff. Eqns.* **11**, 1515–1538 (1986).
52. Lin, C.-S., Wang, Z.-Q.: Symmetry of extremal functions for the Caffarelli–Kohn–Nirenberg inequalities. *Proc. Amer. Math. Soc.* **132**, 1686–1691 (2004).
53. Lin, F.: A new proof of the Caffarelli-Kohn-Nirenberg theorem. *Comm. Pure Appl. Math.* **51**, 241–257 (1998).
54. Lions, P.-L.: The concentration-compactness principle in the calculus of variations. The locally compact case. Part I and Part II. *Ann. Inst. H. Poincaré Anal. Nonlin.* **1**, 109–145 and 223–283 (1984).
55. Lions, P.-L.: The concentration-compactness principle in the calculus of variations. The limit case. Part I and Part II. *Rev. Mat. Iberoamericana* **1**(1), 145–201 and **1**(2), 45–121 (1985).
56. Miranda, C.: Su un problema di Minkowski. *Rend. Sem. Mat. Roma* **3**, 96–108 (1939).
57. Morrey, C.B.: On the solutions of quasilinear elliptic partial differential equations. *Trans. Amer. Math. Soc.* **43**, 126–166 (1938).
58. Moser, J.: A new proof of De Giorgi's theorem concerning the regularity problem for elliptic differential equations. *Comm. Pure Appl. Math.* **13**, 457–468 (1960).
59. Moser, J.: On Harnack's theorem for elliptic differential equations. *Comm. Pure Appl. Math.* **14**, 577–591 (1961).
60. Nadirashvili, N., Tkachev, V., Vlăduţ, S.: A non-classical solution to a Hessian equation from Cartan isoparametric cubic. *Adv. Math.* **231**, 1589–1597 (2012).
61. Nadirashvili, N., Tkachev, V., Vlăduţ, S.: *Nonlinear Elliptic Equations and Nonassociative Algebras*. Mathematical Surveys & Monographs **200**, American Mathematical Society (2014).
62. Nash, J.: Parabolic equations. *Proc. Natl. Acad. Sci. U.S.A.* **43**. 754–758 (1957).
63. Nečas, J., Růžička, M., Šverák, V.: On Leray's self-similar solutions of the Navier–Stokes equations. *Acta Math.* **176**, 283–294 (1996).
64. Nguyen, H.-M., Squassina, M.: Fractional Caffarelli–Kohn–Nirenberg inequalities. *J. Funct. Anal.* **274**(9), 2661–2672 (2018).
65. Newlander, A., Nirenberg, L.: Complex analytic coordinates in almost complex manifolds. *Ann. of Math.* **65**, 391–404 (1957).
66. Nirenberg, L.: *The Determination of a Closed Convex Surface Having Given Line Element*. PhD Thesis, New York University (1949).
67. Nirenberg, L.: The Weyl and Minkowski problems in differential geometry in the large. *Comm. Pure Appl. Math.* **6**, 337–394 (1953).
68. Nirenberg, L.: On nonlinear elliptic partial differential equations and Hölder continuity. *Comm. Pure Appl. Math.* **6**, 103–156 (1953).
69. Nirenberg, L.: On elliptic partial differential equations. *Ann. Scuola Norm. Sup. Pisa (3)* **13**, 115–162 (1959).
70. Nirenberg, L.: Variational and topological methods in nonlinear problems. *Bull. Amer. Math. Soc. (N.S.)* **4**, 267–302 (1981).
71. Nirenberg, L.: *Topics in Nonlinear Functional Analysis*, Courant Lecture Notes Vol. 6, Courant Institute of Mathematical Sciences, New York, and American Mathematical Society, Providence (2001).
72. Pazy, A.: Asymptotic expansions of solutions of ordinary differential equations in Hilbert space. *Arch. Rational Mech. Anal.* **24**, 193–218 (1967).
73. Rivière, T.: Exploring the unknown: the work of Louis Nirenberg on partial differential equations. *Notices Amer. Math. Soc.* **63**(2), 120–125 (2016). A longer, more detailed version is available as arXiv:1505.04930.
74. Robinson, J.C., Rodrigo, J.L., Sadowski, W.: *The Three-Dimensional Navier–Stokes Equations: Classical Theory*. Cambridge University Press (2016).
75. Scheffer, V.: Hausdorff measure and the Navier–Stokes equations. *Comm. Math. Phys.* **55**, 97–112 (1977).
76. Scheffer, V.: The Navier–Stokes equations on a bounded domain. *Comm. Math. Phys.* **73**, 1–42 (1980).

77. Scheffer, V.: Nearly one-dimensional singularities of solutions to the Navier–Stokes inequality. *Comm. Math. Phys.* **110**, 525–551 (1987).
78. Seregin, G., Šverák, V.: On type I singularities of the local axisymmetric solutions of the Navier–Stokes equations. *Comm. Partial Diff. Eqns.* **34**, 171–201 (2009).
79. Serrin, J.: A symmetry problem in potential theory. *Arch. Rational Mech. Anal.* **43**, 304–318 (1971).
80. Sormani, C.: Recent applications of Nirenberg's classical ideas. *Notices Amer. Math. Soc.* **63**(2), 126–134 (2016) (includes "The Gidas–Ni–Nirenberg theorem" by X. Cabré; "Nirenberg's problem" by S.-Y. A. Chang; "Caffarelli–Kohn–Nirenberg and the Navier–Stokes problem" by G. Seregin; "Stability of the Gagliardo–Nirenberg Sobolev inequality" by E. Carlen and A. Figalli; and "Isometric embeddings of surfaces" by M.-T. Wang and S.-T. Yau).
81. Stoker, J.J.: On the uniqueness theorems for the embedding of convex surfaces in three-dimensional space. *Comm. Pure Appl. Math.* **3**, 231–257 (1950).
82. Tsai, T.-P.: On Leray's self-similar solutions of the Navier–Stokes equations satisfying local energy estimates. *Arch. Rational Mech. Anal.* **143**, 29–51 (1998).
83. Vishik, M.I.: On strongly elliptic systems of differential equations. *Mat. Sbornik, N.S.* **29**(71), 615–676 (1951).
84. Weyl, H.: Über die bestimmung einer geschlossenen konvexen fläche durch ihr linienelement. *Vierteljahrsschrift der Naturforschenden Gessellschaft*, Zürich **61**, 40–72 (1916).

List of Publications for John Forbes Nash, Jr.

1945

[1] (with J.F. Nash, Sr.). Sag and tension calculus for wire spans using catenary formulas. *Electrical Engineering Transactions*, 64:685–692.

1950

[2] Non-cooperative games, PhD thesis, Princeton University. [EJN, pp. 51–84].[1]

[3] Equilibrium points in n-person games. *Proc. Nat. Acad. Sci. U. S. A.*, 36:48–49. [EJN, pp. 49–50].

[4] The bargaining problem. *Econometrica*, 18:155–162. [EJN, pp. 37–46]. Also available in *Bargaining and the theory of cooperative games: John Nash and beyond*, pages 5–12. Edward Elgar, Cheltenham, 2010.

[5] (with L.S. Shapley). A simple three-person poker game. In *Contributions to the Theory of Games*, Annals of Mathematics Studies, no. 24, pages 105–116. Princeton University Press, Princeton, N. J.

1951

[6] Non-cooperative games. *Ann. of Math. (2)*, 54:286–295. [EJN, pp. 85–98].

1952

[7] Real algebraic manifolds. *Ann. of Math. (2)*, 56:405–421. [EJN, pp. 127–150].

1953

[8] Two-person cooperative games. *Econometrica*, 21:128–140. [EJN, pp. 99–114]. Also available in *Bargaining and the theory of cooperative games: John Nash and beyond*, pages 589–601. Edward Elgar, Cheltenham, 2010.

[1] Here [EJN] stands for "The Essential John Nash". See [24].

H. Holden, R. Piene (eds.), *The Abel Prize 2013–2017*,
https://doi.org/10.1007/978-3-319-99028-6_21

[9] (with J.P. Mayberry and M. Shubik). A comparison of treatments of a duopoly situation. *Econometrica*, 21:141–154.

1954

[10] C^1 isometric imbeddings. *Ann. of Math. (2)*, 60:383–396.
[11] Parallel control. RAND/RM-1361, Santa Monica, CA: RAND, 8-27-54. [EJN, pp. 117–125].
[12] (with G. Kalish, J. Milnor, and J. Nering). Some experimental n-person games. In *Decision processes*, pages 301–327. Wiley, New York.
[13] Results on continuation and uniqueness of fluid flow. *Bull. Amer. Math. Soc.*, 60(2):165–166.

1955

[14] A path space and the Stiefel–Whitney classes. *Proc. Nat. Acad. Sci. U.S.A.*, 41:320–321.

1956

[15] The imbedding problem for Riemannian manifolds. *Ann. of Math. (2)*, 63:20–63. Also available in *Uspehi Mat. Nauk*, 26(4(160)):173–216, 1971 (in Russian). [EJN, pp. 151–209, including author's note].

1957

[16] Parabolic equations. *Proc. Nat. Acad. Sci. U.S.A.*, 43:754–758.

1958

[17] Continuity of solutions of parabolic and elliptic equations. *Amer. J. Math.*, 80:931–954. [EJN, pp. 211–240].

1962

[18] Le problème de Cauchy pour les équations différentielles d'un fluide général. *Bull. Soc. Math. France*, 90:487–497.

1966

[19] Analyticity of the solutions of implicit function problems with analytic data. *Ann. of Math. (2)*, 84:345–355. Also available in *Uspehi Mat. Nauk*, 26(4(160)):217–226, 1971 (in Russian).

1995

[20] (with H.W. Kuhn, J.C. Harsanyi, R. Selten, J.W. Weibull, E. van Damme, and P. Hammerstein). The work of John F. Nash Jr. in game theory. *Duke Math. J.*, 81(1):1–29. Also available in *J. Econ. Theory*, 69(1):153–185, 1996.
[21] Arc structure of singularities. *Duke Math. J.*, 81(1):31–38.
[22] John Forbes Nash, Jr.—Autobiography. In *Les prix Nobel 1994*, pages 275–279. Nordstedts Tryckeri, Stockholm. [EJN, pp. 5–11].

1996

[23] *Essays on Game Theory*. Edward Elgar Publishing, Cheltenham, UK.

2002

[24] *The Essential John Nash*. Princeton University Press, Princeton, NJ. Edited by Harold W. Kuhn and Sylvia Nasar.

[25] Ideal money. *Southern Economic Journal*, 69(1):4–11.

2008

[26] The agencies method for modeling coalitions and cooperation in games. *Int. Game Theory Rev.*, 10(4):539–564.

2009

[27] Gale at Princeton. *Games Econ. Behav.*, 66(2):627.

[28] Ideal money and asymptotically ideal money. In *Contributions to game theory and management. Vol. II*, pages 281–293. Grad. Sch. Manag., St. Petersb. Univ., St. Petersburg.

[29] Studying cooperative games using the method of agencies. In *Contributions to game theory and management. Vol. II*, pages 294–306. Grad. Sch. Manag., St. Petersb. Univ., St. Petersburg. Also available in *Int. J. Math. Game Theory Algebra*, 18(4–5):413–426 and in *Game theory and applications. Vol. 14*, volume 14 of *Game Theory Appl.*, pages 88–103. Nova Sci. Publ., Hauppauge, NY.

2012

[30] (with R. Nagel, A. Ockenfels and R. Selten). The agencies method for coalition formation in experimental games. *Proceedings of the National Academy of Sciences*, 109(50):20358–20363.

2016

[31] (with M. T. Rassias, editor) *Open Problems in Mathematics*. Springer, Switzerland, pp. 543.

List of Publications for Louis Nirenberg

1953

[1] A strong maximum principle for parabolic equations. *Comm. Pure Appl. Math.*, 6:167–177.

[2] The Weyl and Minkowski problems in differential geometry in the large. *Comm. Pure Appl. Math.*, 6:337–394.

[3] (with S. Agmon and M. H. Protter). A maximum principle for a class of hyperbolic equations and applications to equations of mixed elliptic-hyperbolic type. *Comm. Pure Appl. Math.*, 6:455–470.

[4] On nonlinear elliptic partial differential equations and Hölder continuity. *Comm. Pure Appl. Math.*, 6:103–156; addendum, 395.

1954

[5] On a generalization of quasi-conformal mappings and its application to elliptic partial differential equations. In *Contributions to the theory of partial differential equations*, Annals of Mathematics Studies, no. 33, pages 95–100. Princeton University Press, Princeton, N. J.

1955

[6] Remarks on strongly elliptic partial differential equations. *Comm. Pure Appl. Math.*, 8:649–675.

[7] (with A. Douglis). Interior estimates for elliptic systems of partial differential equations. *Comm. Pure Appl. Math.*, 8:503–538.

[8] (with L. Bers). On a representation theorem for linear elliptic systems with discontinuous coefficients and its applications. In *Convegno Internazionale sulle Equazioni Lineari alle Derivate Parziali, Trieste, 1954*, pages 111–140. Edizioni Cremonese, Roma.

[9] (with L. Bers). On linear and non-linear elliptic boundary value problems in the plane. In *Convegno Internazionale sulle Equazioni Lineari alle Derivate Parziali, Trieste, 1954*, pages 141–167. Edizioni Cremonese, Roma.

H. Holden, R. Piene (eds.), *The Abel Prize 2013–2017*,
https://doi.org/10.1007/978-3-319-99028-6_22

1956

[10] Estimates and existence of solutions of elliptic equations. *Comm. Pure Appl. Math.*, 9:509–529.

1957

[11] Uniqueness in Cauchy problems for differential equations with constant leading coefficients. *Comm. Pure Appl. Math.*, 10:89–105.

[12] (with A. Newlander). Complex analytic coordinates in almost complex manifolds. *Ann. of Math. (2)*, 65:391–404.

[13] (with C. B. Morrey, Jr.). On the analyticity of the solutions of linear elliptic systems of partial differential equations. *Comm. Pure Appl. Math.*, 10:271–290.

1958

[14] (with K. Kodaira and D. C. Spencer). On the existence of deformations of complex analytic structures. *Ann. of Math. (2)*, 68:450–459.

1959

[15] On elliptic partial differential equations. *Ann. Scuola Norm. Sup. Pisa (3)*, 13:115–162.

[16] (with S. Agmon and A. Douglis). Estimates near the boundary for solutions of elliptic partial differential equations satisfying general boundary conditions. I. *Comm. Pure Appl. Math.*, 12:623–727.

[17] (with P. Hartman). On spherical image maps whose Jacobians do not change sign. *Amer. J. Math.*, 81:901–920.

1960

[18] (with D. C. Spencer). On rigidity of holomorphic imbeddings. In *Contributions to function theory (Internat. Colloq. Function Theory, Bombay, 1960)*, pages 133–137. Tata Institute of Fundamental Research, Bombay.

1961

[19] Elementary remarks on surfaces with curvature of fixed sign. In *Proc. Sympos. Pure Math., Vol. III*, pages 181–185. American Mathematical Society, Providence, R.I.

[20] (with F. John). On functions of bounded mean oscillation. *Comm. Pure Appl. Math.*, 14:415–426.

[21] Inequalities in boundary value problems for elliptic differential equations. In *Proc. Internat. Sympos. Linear Spaces (Jerusalem, 1960)*, pages 351–356. Jerusalem Academic Press, Jerusalem; Pergamon, Oxford.

[22] Comments on elliptic partial differential equations. In *Proc. Sympos. Pure Math., Vol. IV*, pages 101–108. American Mathematical Society, Providence, R.I.

1963

[23] Rigidity of a class of closed surfaces. In *Nonlinear Problems (Proc. Sympos., Madison, Wis., 1962)*, pages 177–193. Univ. of Wisconsin Press, Madison, Wis.

[24] (with S. Agmon). Properties of solutions of ordinary differential equations in Banach space. *Comm. Pure Appl. Math.*, 16:121–239.

[25] Comportement à l'infini pour des équations différentielles ordinaires das un espace de Banach. In *Les Équations aux Dérivées Partielles (Paris, 1962)*, pages 167–173. Editions du Centre National de la Recherche Scientifique, Paris.

[26] (with F. Treves). Solvability of a first order linear partial differential equation. *Comm. Pure Appl. Math.*, 16:331–351.

[27] Elliptic partial differential equations and ordinary differential equations in Banach space. In *Differential Equations and Their Applications (Proc. Conf., Prague, 1962)*, pages 121–122. Publ. House Czechoslovak Acad. Sci., Prague; Academic Press, New York.

[28] Some aspects of linear and nonlinear partial differential equations. In *Proc. Internat. Congr. Mathematicians (Stockholm, 1962)*, pages 147–162. Inst. Mittag-Leffler, Djursholm.

[29] Some problems in the theory of linear and non-linear partial differential equations. *Uspehi Mat. Nauk*, 18(4 (112)):101–118.

[30] On solvability of first order partial differential equations. In *Putlines Joint Sympos. Partial Differential Equations (Novosibirsk, 1963)*, pages 191–194. Acad. Sci. USSR Siberian Branch, Moscow.

1964

[31] (with S. Agmon and A. Douglis). Estimates near the boundary for solutions of elliptic partial differential equations satisfying general boundary conditions. II. *Comm. Pure Appl. Math.*, 17:35–92.

[32] Partial differential equations with applications in geometry. In *Lectures on Modern Mathematics, Vol. II*, pages 1–41. Wiley, New York.

1965

[33] (with J. J. Kohn). An algebra of pseudo-differential operators. *Comm. Pure Appl. Math.*, 18:269–305.

[34] (with J. J. Kohn). Non-coercive boundary value problems. *Comm. Pure Appl. Math.*, 18:443–492.

1966

[35] (with P.D. Lax). On stability for difference schemes: A sharp form of Gårding's inequality. *Comm. Pure Appl. Math.*, 19:473–492.

[36] An extended interpolation inequality. *Ann. Scuola Norm. Sup. Pisa (3)*, 20:733–737.

1967

[37] (with S. Agmon). Lower bounds and uniqueness theorems for solutions of differential equations in a Hilbert space. *Comm. Pure Appl. Math.*, 20:207–229.

[38] (with P.D. Lax). A sharp inequality for pseudo-differential and difference operators. In *Singular Integrals (Proc. Sympos. Pure Math., Chicago, Ill., 1966)*, pages 213–217. Amer. Math. Soc., Providence, R.I.

[39] (with J. J. Kohn). Degenerate elliptic-parabolic equations of second order. *Comm. Pure Appl. Math.*, 20:797–872.

1969

[40] On pseudo-differential operators. In *Lectures in Differential Equations, Vol. I*, pages 179–199. Van Nostrand, New York.

[41] (with S. S. Chern and H. I. Levine). Intrinsic norms on a complex manifold. In *Global Analysis (Papers in Honor of K. Kodaira)*, pages 119–139. Univ. Tokyo Press, Tokyo.

[42] (with F. Trèves). Conditions nécessaires de résolubilité locale des équations pseudo-différentielles. *C. R. Acad. Sci. Paris Sér. A-B*, 269:A774–A777.

[43] (with F. Trèves). Conditions suffisantes de résolubilité locale des équations aux dérivées partielles linéaires. *C. R. Acad. Sci. Paris Sér. A-B*, 269:A853–A856.

1970

[44] (with F. Trèves). On local solvability of linear partial differential equations. I. Necessary conditions. *Comm. Pure Appl. Math.*, 23:1–38.

[45] (with F. Trèves). On local solvability of linear partial differential equations. II. Sufficient conditions. *Comm. Pure Appl. Math.*, 23:459–509. Correction, *Comm. Pure Appl. Math.*, 24(2):279–288, 1971.

[46] Pseudo-differential operators. In *Global Analysis (Proc. Sympos. Pure Math., Vol. XVI, Berkeley, Calif., 1968)*, pages 149–167. Amer. Math. Soc., Providence, R.I.

[47] A characterization of convex bodies. *J. Fac. Sci. Univ. Tokyo Sect. I*, 17:397–402.

1971

[48] (with J. F. Treves). Remarks on the solvability of linear equations of evolution. In *Symposia Mathematica, Vol. VII (Convegno sulle Problemi di Evoluzione, INDAM, Rome, 1970)*, pages 325–338. Academic Press, London.

[49] Generalized degree and nonlinear problems. In *Contributions to nonlinear functional analysis (Proc. Sympos., Math. Res. Center, Univ. Wisconsin, Madison, Wis., 1971)*, pages 1–9. Academic Press, New York.

[50] A proof of the Malgrange preparation theorem. In *Proceedings of Liverpool Singularities—Symposium, I (1969/70)*, pages 97–105. Lecture Notes in Mathematics, Vol. 192. Springer, Berlin.

[51] An application of generalized degree to a class of nonlinear problems. In *Troisième Colloque sur l'Analyse Fonctionnelle (Liège, 1970)*, pages 57–74. Vander, Louvain.

1972

[52] An abstract form of the nonlinear Cauchy–Kowalewski theorem. *J. Differential Geometry*, 6:561–576. Collection of articles dedicated to S. S. Chern and D. C. Spencer on their sixtieth birthdays.

[53] (with H. Brézis and G. Stampacchia). A remark on Ky Fan's minimax principle. *Boll. Un. Mat. Ital. (4)*, 6:293–300. Reprinted as *Boll. Unione Mat. Ital. (9)*, 1(2):257–264, 2008.

1973

[54] (with H. F. Walker). The null spaces of elliptic partial differential operators in $\mathbf{R}^n$. *J. Math. Anal. Appl.*, 42:271–301. Collection of articles dedicated to Salomon Bochner.

[55] (with J. J. Kohn). A pseudo-convex domain not admitting a holomorphic support function. *Math. Ann.*, 201:265–268.

[56] *Lectures on linear partial differential equations.* American Mathematical Society, Providence, R.I. Also available in *Uspehi Mat. Nauk*, 30(4(184)):147–204, 1975 (in Russian).

1974

[57] (with C. Loewner). Partial differential equations invariant under conformal or projective transformations. In *Contributions to analysis (a collection of papers dedicated to Lipman Bers)*, pages 245–272. Academic Press, New York.

[58] *Topics in nonlinear functional analysis.* Courant Institute of Mathematical Sciences, New York University, New York. Also available in Russian, Izdat, "Mir", Moscow, 1977. Revised reprint, American Mathematical Society, Providence, RI, 2001.

1975

[59] On a problem of Hans Lewy. In *Fourier integral operators and partial differential equations (Colloq. Internat., Univ. Nice, Nice, 1974)*, pages 224–234. Lecture Notes in Math., Vol. 459. Springer, Berlin.

[60] Monge–Ampère equations and some associated problems in geometry. In *Proceedings of the International Congress of Mathematicians (Vancouver, B. C., 1974), Vol. 2*, pages 275–279. Canad. Math. Congress, Montreal, Que.

1976

[61] Nonlinear differential equations invariant under certain geometric transformations. In *Symposia Mathematica, Vol. XVIII (Convegno sulle Trasformazioni Quasiconformi e Questioni Connesse, INDAM, Rome, 1974)*, pages 399–405. Academic Press, London.

[62] Propagation of singularities for linear partial differential equations and reflections at a boundary. In *Séminaire Goulaouic–Schwartz (1975/1976), Équations aux dérivées partielles et analyse fonctionnelle*, pages Exp. No. 26, 19. Centre Math., École Polytech., Palaiseau. Paper presented at Séminaire Goulaouic–Lions–Schwartz 1974–1975.

1977

[63] (with D. Kinderlehrer). Regularity in free boundary problems. *Ann. Scuola Norm. Sup. Pisa Cl. Sci. (4)*, 4(2):373–391.
[64] (with H. Brézis). Image d'une somme d'opérateurs non linéaires et applications. *C. R. Acad. Sci. Paris Sér. A-B*, 284(21):A1365–A1368.
[65] (with H. Brézis). Some first-order nonlinear equations on a torus. *Comm. Pure Appl. Math.*, 30(1):1–11.

1978

[66] (with D. Kinderlehrer). Analyticity at the boundary of solutions of nonlinear second-order parabolic equations. *Comm. Pure Appl. Math.*, 31(3):283–338.
[67] (with D. Kinderlehrer). Hodograph methods and the smoothness of the free boundary in the one phase Stefan problem. In *Moving boundary problems (Proc. Sympos. and Workshop, Gatlinburg, Tenn., 1977)*, pages 57–69. Academic Press, New York.
[68] (with H. Brézis). Characterizations of the ranges of some nonlinear operators and applications to boundary value problems. *Ann. Scuola Norm. Sup. Pisa Cl. Sci. (4)*, 5(2):225–326.
[69] (with D. Kinderlehrer and J. Spruck). Régularité dans les problèmes elliptiques à frontière libre. *C. R. Acad. Sci. Paris Sér. A-B*, 286(24):A1187–A1190.
[70] (with H. Brézis). Forced vibrations for a nonlinear wave equation. *Comm. Pure Appl. Math.*, 31(1):1–30.
[71] (with D. Kinderlehrer). The smoothness of the free boundary in the one phase Stefan problem. *Comm. Pure Appl. Math.*, 31(3):257–282.
[72] (with D. Kinderlehrer and J. Spruck). Regularity in elliptic free boundary problems. *J. Analyse Math.*, 34:86–119.

1979

[73] (with B. Gidas and W. M. Ni). Symmetry and related properties via the maximum principle. *Comm. Math. Phys.*, 68(3):209–243.
[74] The use of topological, functional analytic, and variational methods in nonlinear problems. *Confer. Sem. Mat. Univ. Bari*, (163–168):391–398.
[75] (with D. Kinderlehrer and J. Spruck). Regularity in elliptic free boundary problems. II. Equations of higher order. *Ann. Scuola Norm. Sup. Pisa Cl. Sci. (4)*, 6(4):637–683.

1980

[76] (with S. Webster and P. Yang). Local boundary regularity of holomorphic mappings. *Comm. Pure Appl. Math.*, 33(3):305–338.

[77] (with H. Brézis and J.-M. Coron). Free vibrations for a nonlinear wave equation and a theorem of P. Rabinowitz. *Comm. Pure Appl. Math.*, 33(5):667–684.

[78] Remarks on nonlinear problems. In *The Chern Symposium 1979 (Proc. Internat. Sympos., Berkeley, Calif., 1979)*, pages 189–197. Springer, New York-Berlin.

1981

[79] Variational and topological methods in nonlinear problems. *Bull. Amer. Math. Soc. (N.S.)*, 4(3):267–302.

[80] (with B. Gidas and W. M. Ni). Symmetry of positive solutions of nonlinear elliptic equations in $\mathbf{R}^n$. In *Mathematical analysis and applications, Part A*, volume 7 of *Adv. in Math. Suppl. Stud.*, pages 369–402. Academic Press, New York-London.

[81] Comments on nonlinear problems. *Matematiche (Catania)*, 36(1):109–119 (1983). With an appendix by Chang Shou Lin.

1982

[82] (with L. Caffarelli and R. Kohn). Partial regularity of suitable weak solutions of the Navier-Stokes equations. *Comm. Pure Appl. Math.*, 35(6):771–831.

[83] Elliptic equations with critical nonlinear exponent. *Rend. Semin. Mat. Fis. Milano*, 52:187–191.

1983

[84] (with H. Brézis). Positive solutions of nonlinear elliptic equations involving critical Sobolev exponents. *Comm. Pure Appl. Math.*, 36(4):437–477.

[85] The Dirichlet problem for the Monge–Ampère equation. In *Methods of functional analysis and theory of elliptic equations (Naples, 1982)*, pages 193–198. Liguori, Naples.

1984

[86] (with L. Caffarelli and J. Spruck). The Dirichlet problem for nonlinear second-order elliptic equations. I. Monge–Ampère equation. *Comm. Pure Appl. Math.*, 37(3):369–402. Correction, *Comm. Pure Appl. Math.*, 40(5):659–662, 1987.

[87] (with L. Caffarelli and R. Kohn). First order interpolation inequalities with weights. *Compositio Math.*, 53(3):259–275.

[88] The work of Yau Shing-Tung. In *Proceedings of the International Congress of Mathematicians, Vol. 1, 2 (Warsaw, 1983)*, pages 15–19. PWN, Warsaw.

1985

[89] Uniqueness in the Cauchy problem for a degenerate elliptic second order equation. In *Differential geometry and complex analysis*, pages 213–218. Springer, Berlin.

[90] (with L. Caffarelli, J. J. Kohn, and J. Spruck). The Dirichlet problem for nonlinear second-order elliptic equations. II. Complex Monge–Ampère, and uniformly elliptic, equations. *Comm. Pure Appl. Math.*, 38(2):209–252.

[91] (with L. Caffarelli and J. Spruck). The Dirichlet problem for nonlinear second-order elliptic equations. III. Functions of the eigenvalues of the Hessian. *Acta Math.*, 155(3–4):261–301.

1986

[92] (with L. Caffarelli and J. Spruck). The Dirichlet problem for the degenerate Monge–Ampère equation. *Rev. Mat. Iberoamericana*, 2(1–2):19–27.

[93] (with L. Caffarelli and J. Spruck). Nonlinear second order elliptic equations. IV. Starshaped compact Weingarten hypersurfaces. In *Current topics in partial differential equations*, pages 1–26. Kinokuniya, Tokyo.

1988

[94] (with L. Caffarelli and J. Spruck). Nonlinear second-order elliptic equations. V. The Dirichlet problem for Weingarten hypersurfaces. *Comm. Pure Appl. Math.*, 41(1):47–70.

[95] (with L. Caffarelli and J. Spruck). On a form of Bernstein's theorem. In *Analyse mathématique et applications*, pages 55–66. Gauthier-Villars, Montrouge.

[96] Fully nonlinear elliptic equations. In *The mathematical heritage of Hermann Weyl (Durham, NC, 1987)*, volume 48 of *Proc. Sympos. Pure Math.*, pages 217–225. Amer. Math. Soc., Providence, RI.

[97] Fully nonlinear second order elliptic equations. In *Calculus of variations and partial differential equations (Trento, 1986)*, volume 1340 of *Lecture Notes in Math.*, pages 239–247. Springer, Berlin.

[98] Mathematical methods in nonlinear problems. *Fiz.-Mat. Spis. B'' lgar. Akad. Nauk.*, 30(63)(4):266–271 (in Bulgarian).

[99] (with H. Berestycki and B. Larrouturou). A nonlinear elliptic problem describing the propagation of a curved premixed flame. In *Mathematical modeling in combustion and related topics (Lyon, 1987)*, volume 140 of *NATO Adv. Sci. Inst. Ser. E Appl. Sci.*, pages 11–28. Nijhoff, Dordrecht.

[100] (with H. Berestycki). Monotonicity, symmetry and antisymmetry of solutions of semilinear elliptic equations. *J. Geom. Phys.*, 5(2):237–275.

1989

[101] Variational methods in nonlinear problems. In *Topics in calculus of variations (Montecatini Terme, 1987)*, volume 1365 of *Lecture Notes in Math.*, pages 100–119. Springer, Berlin.

[102] On fully nonlinear elliptic equations of second order. In *Séminaire sur les Équations aux Dérivées Partielles, 1988–1989*, pages Exp. No. XVI, 6. École Polytech., Palaiseau.

1990

[103] (with H. Berestycki). Some qualitative properties of solutions of semilinear elliptic equations in cylindrical domains. In *Analysis, et cetera*, pages 115–164. Academic Press, Boston, MA.

[104] (with H. Berestycki and L. A. Caffarelli). Uniform estimates for regularization of free boundary problems. In *Analysis and partial differential equations*, volume 122 of *Lecture Notes in Pure and Appl. Math.*, pages 567–619. Dekker, New York.

1991

[105] (with H. Berestycki). Travelling front solutions of semilinear equations in n dimensions. In *Frontiers in pure and applied mathematics*, pages 31–41. North-Holland, Amsterdam.

[106] (with H. Brezis). Remarks on finding critical points. *Comm. Pure Appl. Math.*, 44(8–9):939–963.

[107] (with H. Berestycki). On the method of moving planes and the sliding method. *Bol. Soc. Brasil. Mat. (N.S.)*, 22(1):1–37.

[108] (with H. Berestycki). Asymptotic behaviour via the Harnack inequality. In *Nonlinear analysis*, Sc. Norm. Super. di Pisa Quaderni, pages 135–144. Scuola Norm. Sup., Pisa.

1992

[109] (with H. Berestycki). Travelling fronts in cylinders. *Ann. Inst. H. Poincaré Anal. Non Linéaire*, 9(5):497–572.

[110] *On the maximum principle*. AMS-MAA Joint Lecture Series. American Mathematical Society, Providence, RI. A joint AMS-MAA lecture presented in Orono, Maine, August 1991.

[111] Some personal remarks about S. S. Chern. In *Chern—a great geometer of the twentieth century*, pages 98–101. Int. Press, Hong Kong.

1993

[112] (with H. Berestycki and S. Varadhan). État fondamental et principe du maximum pour les opérateurs elliptiques du second ordre dans des domaines généraux. *C. R. Acad. Sci. Paris Sér. I Math.*, 317(1):51–56.

[113] (with H. Brezis). H^1 versus C^1 local minimizers. *C. R. Acad. Sci. Paris Sér. I Math.*, 317(5):465–472.

[114] The maximum principle and principal eigenvalue for second order elliptic equations in general bounded domains. In *Current problems of analysis and mathematical physics (Italian) (Taormina, 1992)*, pages 189–194. Univ. Roma "La Sapienza", Rome.

[115] (with H. Berestycki and I. Capuzzo-Dolcetta). Problèmes elliptiques indéfinis et théorèmes de Liouville non linéaires. *C. R. Acad. Sci. Paris Sér. I Math.*, 317(10):945–950.

[116] (with H. Berestycki and L. A. Caffarelli). Symmetry for elliptic equations in a half space. In *Boundary value problems for partial differential equations and applications*, volume 29 of *RMA Res. Notes Appl. Math.*, pages 27–42. Masson, Paris.

1994

[117] (with H. Berestycki and S. R. S. Varadhan). The principal eigenvalue and maximum principle for second-order elliptic operators in general domains. *Comm. Pure Appl. Math.*, 47(1):47–92.

[118] Partial differential equations in the first half of the century. In *Development of mathematics 1900–1950 (Luxembourg, 1992)*, pages 479–515. Birkhäuser, Basel.

[119] (with H. Berestycki and I. Capuzzo-Dolcetta). Superlinear indefinite elliptic problems and nonlinear Liouville theorems. *Topol. Methods Nonlinear Anal.*, 4(1):59–78.

1995

[120] (with H. Brezis). Degree theory and BMO. I. Compact manifolds without boundaries. *Selecta Math. (N.S.)*, 1(2):197–263.

[121] (with H. Berestycki and I. Capuzzo-Dolcetta). Variational methods for indefinite superlinear homogeneous elliptic problems. *NoDEA Nonlinear Differential Equations Appl.*, 2(4):553–572.

[122] The maximum principle and related topics. In *Modern methods in complex analysis (Princeton, NJ, 1992)*, volume 137 of *Ann. of Math. Stud.*, pages 283–289. Princeton Univ. Press, Princeton, NJ.

1996

[123] Degree theory beyond continuous maps. In *Partial differential equations and mathematical physics (Copenhagen, 1995; Lund, 1995)*, volume 21 of *Progr. Nonlinear Differential Equations Appl.*, pages 262–263. Birkhäuser Boston, Boston, MA.

[124] (with H. Berestycki and L. A. Caffarelli). Inequalities for second-order elliptic equations with applications to unbounded domains. I. *Duke Math. J.*, 81(2):467–494. A celebration of John F. Nash, Jr.

[125] Degree theory beyond continuous maps. *CWI Quarterly*, 9(1–2):113–120. SMC 50 jubilee (Amsterdam, 1996).

[126] (with H. Brezis). Degree theory and BMO. II. Compact manifolds with boundaries. *Selecta Math. (N.S.)*, 2(3):309–368. With an appendix by the authors and Petru Mironescu.

1997

[127] (with H. Brezis). A Lyapounov–Schmidt procedure involving a nonlinear projection. In *Multidimensional complex analysis and partial differential equations (São Carlos, 1995)*, volume 205 of *Contemp. Math.*, pages 25–32. Amer. Math. Soc., Providence, RI.

[128] (with H. Berestycki and L. A. Caffarelli). Monotonicity for elliptic equations in unbounded Lipschitz domains. *Comm. Pure Appl. Math.*, 50(11):1089–1111.

[129] Lipman Bers and partial differential equations. In *Lipa's legacy (New York, 1995)*, volume 211 of *Contemp. Math.*, pages 455–461. Amer. Math. Soc., Providence, RI.

[130] (with H. Brezis). Removable singularities for nonlinear elliptic equations. *Topol. Methods Nonlinear Anal.*, 9(2):201–219.

[131] (with H. Berestycki and L. Caffarelli). Further qualitative properties for elliptic equations in unbounded domains. *Ann. Scuola Norm. Sup. Pisa Cl. Sci. (4)*, 25(1–2):69–94.

1998

[132] (with Y. Li). The Dirichlet problem for singularly perturbed elliptic equations. *Comm. Pure Appl. Math.*, 51(11–12):1445–1490.

1999

[133] (with H. Brezis, Y. Li, and P. Mironescu). Degree and Sobolev spaces. *Topol. Methods Nonlinear Anal.*, 13(2):181–190.

[134] Estimates for elliptic equations in unbounded domains and applications to symmetry and monotonicity. In *Harmonic analysis and partial differential equations (Chicago, IL, 1996)*, Chicago Lectures in Math., pages 263–274. Univ. Chicago Press, Chicago, IL.

2000

[135] (with J. N. Mather, H. P. McKean, and P. H. Rabinowitz). Jürgen K. Moser (1928–1999). *Notices Amer. Math. Soc.*, 47(11):1392–1405.

[136] Variational methods in nonlinear problems. In *The mathematical sciences after the year 2000 (Beirut, 1999)*, pages 116–122. World Sci. Publ., River Edge, NJ.

[137] (with Y. Li). A variational result in a domain with boundary. *Methods Appl. Anal.*, 7(3):489–494. Cathleen Morawetz: a great mathematician.

2002

[138] (with I. Ekeland). A convex Darboux theorem. *Methods Appl. Anal.*, 9(3):329–344.

2003

[139] (with Y. Li). Estimates for elliptic systems from composite material. *Comm. Pure Appl. Math.*, 56(7):892–925. Comment, *Abstract and applied analysis*, pages 249–255. World Sci. Publ., River Edge, NJ, 2004.

[140] Laurent Schwartz and some of his mathematical work. *Gaz. Math.*, (98, suppl.):97–103. Laurent Schwartz (1915–2002).

[141] Ol′ ga Arsen′ evna Oleĭnik. *Tr. Semin. im. I. G. Petrovskogo*, (23):5–6 (in Russian). Also available in *J. Math. Sci. (N. Y.)*, 120(3):1241, 2004 (in English).

[142] In memoriam: Natascha Artin Brunswick, Editor Emeritus. *Teor. Veroyatnost. i Primenen.*, 48(3):638.

2004

[143] (with J. J. Kohn, P. A. Griffiths, H. Goldschmidt, E. Bombieri, B. Cenkl and P. Garabedian). Donald C. Spencer (1912–2001). *Notices Amer. Math. Soc.*, 51(1):17–29, 2004.

[144] (with S. Friedlander, P. Lax, C. Morawetz, G. Seregin, N. Ural′tseva, and M. Vishik). Olga Alexandrovna Ladyzhenskaya (1922–2004). *Notices Amer. Math. Soc.*, 51(11):1320–1331.

2005

[145] (with Y. Li). The distance function to the boundary, Finsler geometry, and the singular set of viscosity solutions of some Hamilton–Jacobi equations. *Comm. Pure Appl. Math.*, 58(1):85–146.

[146] Some recollections of working with François Treves. In *Geometric analysis of PDE and several complex variables*, volume 368 of *Contemp. Math.*, pages 371–373. Amer. Math. Soc., Providence, RI.

[147] Memories of Guido Stampacchia. In *Variational analysis and applications*, volume 79 of *Nonconvex Optim. Appl.*, pages 79–80. Springer, New York.

[148] The distance function to the boundary and singular set of viscosity solutions of Hamilton–Jacobi equation. In *Variational analysis and applications*, volume 79 of *Nonconvex Optim. Appl.*, pages 765–772. Springer, New York.

[149] (with I. Ekeland). Regularity in an unusual variational problem. *J. Math. Fluid Mech.*, 7(suppl. 3):S332–S348.

[150] (with Y. Li). Regularity of the distance function to the boundary. *Rend. Accad. Naz. Sci. XL Mem. Mat. Appl. (5)*, 29:257–264.

2006

[151] (with Y. Y. Li). Generalization of a well-known inequality. In *Contributions to nonlinear analysis*, volume 66 of *Progr. Nonlinear Differential Equations Appl.*, pages 365–370. Birkhäuser, Basel.

[152] (with Y. Y. Li). A geometric problem and the Hopf lemma. I. *J. Eur. Math. Soc. (JEMS)*, 8(2):317–339.

[153] (with Y. Y. Li). A geometric problem and the Hopf lemma. II. *Chinese Ann. Math. Ser. B*, 27(2):193–218.

2008

[154] Comments on some papers of Alberto P. Calderón. In *Selected papers of Alberto P. Calderón*, pages 607–621. Amer. Math. Soc., Providence, RI.

2009

[155] (with L. Caffarelli and Y. Y. Li). Some remarks on singular solutions of nonlinear elliptic equations. I. *J. Fixed Point Theory Appl.*, 5(2):353–395.

[156] (with Y. Y. Li). Partial results on extending the Hopf lemma. *Rend. Mat. Appl. (7)*, 29(1):97–115. Errata, *Rend. Mat. Appl. (7)*, 29(3–4):365.

[157] The Dirichlet problem for some fully nonlinear elliptic equations. In *Colloquium De Giorgi 2007 and 2008*, volume 2 of *Colloquia*, pages 91–98. Ed. Norm., Pisa.

2011

[158] On singular solutions of nonlinear elliptic and parabolic equations. *Milan J. Math.*, 79(1):3–12.

[159] Ordinary differential equations in Banach spaces. In *Abstract differential equations (Italian)*, volume 29 of *C.I.M.E. Summer Sch.*, pages 123–170. Springer, Heidelberg.

[160] Louis Nirenberg, editor. *Pseudo-differential operators. Lectures given at the summer school of the Centro Internazionale Matematico Estivo (C.I.M.E.), Stresa, Italy, August 26–September 3, 1968. Reprint of the 1969 original published by Edizioni Cremonese.* Berlin: Springer; Firenze: Fondazione CIME Roberto Conti, reprint of the 1969 original published by edizioni cremonese edition.

2012

[161] (with L. Caffarelli and Y. Y. Li). Some remarks on singular solutions of nonlinear elliptic equations. II. Symmetry and monotonicity via moving planes. In *Advances in geometric analysis*, volume 21 of *Adv. Lect. Math. (ALM)*, pages 97–105. Int. Press, Somerville, MA.

2013

[162] (with L. Caffarelli and Y. Li). Some remarks on singular solutions of nonlinear elliptic equations III: viscosity solutions including parabolic operators. *Comm. Pure Appl. Math.*, 66(1):109–143.

2015

[163] (with S. Anastasio, R. G. Douglas, C. Foias, W.-M. Ching, M. Davis, M. Sharir, M. Wigler, J. A. Fisher, P. D. Lax, P. M. Willig). In memory of Jacob Schwartz. *Notices Amer. Math. Soc.*, 62(5):473–490.

2017

[164] (with A. Córdoba, E. Stein, T. Tao, J. J. Kohn, S.-Y. A. Chang, C. R. Graham, D. Córdoba, B. Klartag, J. Fröhlich, L. Seco, M. Weinstein). Ad honorem Charles Fefferman. *Notices Amer. Math. Soc.*, 64(11):1254–1273.

Curriculum Vitae for John Forbes Nash, Jr.

Born:	June 13, 1928 in Bluefield, West Virginia, USA
Died:	May 23, 2015 in Monroe Township, New Jersey, USA
Degrees/education:	Bachelor of Science, Carnegie Institute of Technology, 1948
	Master of Science, Carnegie Institute of Technology, 1948
	PhD, Princeton University, 1951

H. Holden, R. Piene (eds.), *The Abel Prize 2013–2017*,
https://doi.org/10.1007/978-3-319-99028-6_23

Positions:	Consultant, RAND Corporation, 1950, 1952, 1953, 1954
	C.L.E. Moore Instructor, Massachusetts Institute of Technology, 1951–1953
	Assistant Professor, Massachusetts Institute of Technology, 1953–1957
	Associate Professor, Massachusetts Institute of Technology, 1957–1959
	Senior Research Mathematician, Princeton University, 1995–2015
Visiting positions:	Institute for Advanced Study, 1956–1957 and 1961–1962
Memberships:	American Academy of Arts and Sciences, 1995
	National Academy of Sciences, 1996
	American Philosophical Society, 2006
	American Mathematical Society, Fellow, 2013
	Norwegian Academy of Science and Letters, 2015
Awards and prizes:	John von Neumann Theory Prize, 1978
	Nobel Memorial Prize in Economic Sciences, 1994
	Leroy P. Steele Prize for Seminal Contribution to Research, 1999
	Abel Prize, 2015
Honorary degrees:	Carnegie Mellon University, 1999
	University of Naples Federico II, 2003
	University of Antwerp, 2007
	City University of Hong Kong, 2011

Curriculum Vitae for Louis Nirenberg

Born:	February 28, 1925 in Hamilton, Canada
Degrees/education:	Bachelor of Science, McGill University, 1945
	Master of Science, New York University, 1947
	PhD, New York University, 1949
Positions:	Research Associate, New York University, 1949–1951
	Assistant Professor, New York University, 1951–1954
	Associate Professor, New York University, 1954–1957

H. Holden, R. Piene (eds.), *The Abel Prize 2013–2017*,
https://doi.org/10.1007/978-3-319-99028-6_24

	Professor, New York University, 1957–1999
	Professor Emeritus, New York University, 1999–
	Director, Courant Institute of Mathematical Sciences, 1970–1972
Visiting positions:	ETH, 1951–1952
	University of Göttingen, one month in 1952
	Institute for Advanced Study, spring 1958
Memberships:	American Academy of Arts and Sciences, 1965
	National Academy of Sciences, 1969
	Accademia dei Lincei, 1978
	Accademia Mediterranea della Scienza, 1982
	American Philosophical Society, 1987
	Académie des Sciences, Membre associé étranger, 1989
	Istituto Lombardo, Accademia Scienze e Lettere, 1991
	Ukrainian Academy of Sciences, 1994
	American Mathematical Society, Fellow, 2013
	Norwegian Academy of Science and Letters, 2015
Awards and prizes:	Sloan Fellow, 1958–1960
	Bôcher Memorial Prize, 1959
	Guggenheim Fellow, 1966–1967, 1975–1976
	Colloquium Lecture, American Mathematical Society, 1974
	Fermi Lecture, Scuola Normale Superiore di Pisa, Fermi medal 1974
	Hermann Weyl Lectures at Institute for Advanced Study, 1980
	Crafoord Prize, 1982
	Jeffery–Williams Prize, 1987
	Leroy P. Steele Prize for Lifetime Achievement, 1994
	National Medal of Science, 1995
	Chern Medal, 2010
	Leroy P. Steele Prize for Seminal Contribution to Research, 2014
	Abel Prize, 2015
Honorary degrees:	McGill University, 1986
	Honorary Professorship, Nankai University, 1987
	Honorary Professorship at Zhejiang University, 1988
	University of Pisa, 1990
	Université de Paris IX Paris-Dauphine, 1990
	McMaster University, 2000
	University of British Columbia, 2010
	Honorary Professorship, Peking University, 2016

Part IV
2016 Sir Andrew J. Wiles

"for his stunning proof of Fermat's Last Theorem by way of the modularity conjecture for semistable elliptic curves, opening a new era in number theory"

ABEL
PRISEN

Citation

The Norwegian Academy of Science and Letters has decided to award the Abel Prize for 2016 to **Sir Andrew J. Wiles**, University of Oxford,

> for his stunning proof of Fermat's Last Theorem by way of the modularity conjecture for semistable elliptic curves, opening a new era in number theory

Number theory, an old and beautiful branch of mathematics, is concerned with the study of arithmetic properties of the integers. In its modern form the subject is fundamentally connected to complex analysis, algebraic geometry, and representation theory. Number theoretic results play an important role in our everyday lives through encryption algorithms for communications, financial transactions, and digital security. Fermat's Last Theorem, first formulated by Pierre de Fermat in the seventeenth century, is the assertion that the equation $x^n + y^n = z^n$ has no solutions in positive integers for $n > 2$. Fermat proved his claim for $n = 4$, Leonhard Euler found a proof for $n = 3$, and Sophie Germain proved the first general result that applies to infinitely many prime exponents. Ernst Kummer's study of the problem unveiled several basic notions in algebraic number theory, such as ideal numbers and the subtleties of unique factorization. The complete proof found by Andrew Wiles relies on three further concepts in number theory, namely elliptic curves, modular forms, and Galois representations. Elliptic curves are defined by cubic equations in two variables. They are the natural domains of definition of the elliptic functions introduced by Niels Henrik Abel. Modular forms are highly symmetric analytic functions defined on the upper half of the complex plane, and naturally factor through shapes known as modular curves. An elliptic curve is said to be modular if it can be parametrized by a map from one of these modular curves. The modularity conjecture, proposed by Goro Shimura, Yutaka Taniyama, and André Weil in the 1950s and 1960s, claims that every elliptic curve defined over the rational numbers is modular. In 1984, Gerhard Frey associated a semistable elliptic curve to any hypothetical counterexample to Fermat's Last Theorem, and strongly suspected that this elliptic curve would not be modular. Frey's non-modularity was proven via Jean-Pierre Serre's epsilon conjecture by Kenneth Ribet in 1986. Hence, a proof of the Shimura–Taniyama–Weil modularity conjecture for semistable elliptic curves would also yield a proof of Fermat's Last Theorem. However, at the time the modularity conjecture was widely believed to be completely inaccessible. It was therefore a stunning advance when Andrew Wiles, in a breakthrough paper published in 1995, introduced his modularity lifting technique and proved the semistable case of the modularity conjecture. The modularity lifting technique of Wiles concerns the Galois symmetries of the points of finite order in the abelian group structure on an elliptic curve. Building upon Barry Mazur's deformation theory for such Galois representations, Wiles identified a numerical criterion which ensures that modularity for points of order p can be lifted to modularity for points of order any power of p, where p is an odd prime. This lifted modularity is then sufficient to prove that the elliptic curve is modular. The numerical criterion was confirmed in the semistable case by using an important companion paper written jointly with Richard Taylor. Theorems of Robert Langlands and Jerrold Tunnell show that in many cases the Galois representation given by the points of order three is modular. By an ingenious switch from one prime to another, Wiles showed that in the remaining cases the Galois representation given by the points of order five is modular. This completed his proof of the modularity conjecture, and thus also of Fermat's Last Theorem. The new ideas introduced by Wiles were crucial to many subsequent developments, including the proof in 2001 of the general case of the modularity conjecture by Christophe Breuil, Brian Conrad, Fred Diamond, and Richard Taylor. As recently as 2015, Nuno Freitas, Bao V. Le Hung, and Samir Siksek proved the analogous modularity statement over real quadratic number fields. Few results have as rich a mathematical history and as dramatic a proof as Fermat's Last Theorem.

First Steps

Andrew Wiles

I was born in a college room in Cambridge so my choice of profession was perhaps inevitable. My childhood was also spent in Cambridge except for the period from age two to six during which we lived in Nigeria. Apparently I resisted education at first, refusing to go to school for a whole term, but finally succumbed. Holidays then and later were spent on the estate and farm of a wonderful school that my maternal grandfather had created from nothing in Sussex.

My earliest memories of problem solving are not purely mathematical. My father and I had devised a game in which we had to prove the most unlikely propositions. For example I remember his proving that he was the pope. The last line of the proof went 'So I and the pope are two, and two is one, so I and the pope are one'. I can no longer remember how he proved that two equals one, but the proof was probably linguistic rather than mathematical. It was surely not the argument a teacher later gave me (as a warning about induction) that all numbers were equal. For assume it is true for any large set of n numbers. Then we will prove it for $n + 1$. Consider the first n numbers in the set. They are all equal. Similarly the last n are. So since n is large the two sets overlap and they are all equal.

My father, although by then a theologian, had previously worked with the codebreakers at Bletchley Park during the war and regularly did the crossword puzzles. I joined in the easier ones, but I was drawn more to mathematical problems. My mother had turned down a place at Cambridge to read mathematics because she wanted to read physics, but I never enjoyed applied subjects even in mathematics.

Electronic Supplementary Material The online version of this chapter (https://doi.org/10.1007/978-3-319-99028-6_25) contains supplementary material, which is available to authorized users.

A. Wiles (✉)
Mathematical Institute, University of Oxford, Oxford, UK
e-mail: wiles@maths.ox.ac.uk

H. Holden, R. Piene (eds.), *The Abel Prize 2013–2017*,
https://doi.org/10.1007/978-3-319-99028-6_25

Beach work, early 1960's

I enjoyed doing mathematics as far back as I can remember but the first time I recall getting really excited was ironically due to a rather weak teacher at school. The class would love to distract him with his stories of mountain climbing but I was noticing patterns and discovering the excitement of new ideas. For example I remember realizing that there was a formula for the sum of the angles in a polygon. What worked really well with this teacher was that he never gave me any notion that such a result might be well known.

Searching for more stimulation at the public library I found E.T. Bell's book on Fermat's Last Theorem. Its yellow plastic cover told very briefly the story of the problem and of the prize for it. I was immediately hooked. I searched around for more information about congruences and started trying to prove it. As a twelve and thirteen year old I had a much more inspiring teacher, Mrs Briggs. One of her favourite lessons was to give a proof of Pythagoras's theorem (due I believe to Yaglom) in which she left out a crucial step. We were supposed to find the gap. I remember finding this very difficult, more difficult than the proof itself.

In high school I was very lucky to have another teacher, Graham Townsend, who had been a number theorist. He introduced me to some beautiful mathematics and in particular gave me a copy of Hardy and Wright. This together with Davenport's 'The higher arithmetic' became my favourite mathematical reading. In Davenport's book there is an elementary description of the class number formula for certain imaginary quadratic fields where it can be described as the difference between a sum of non-residues and a sum of residues. This was the first time that I was overwhelmed by the beauty of our subject. Its allure was only increased by the fact that the proof was analytic and at that time still beyond my reach. But I had already been fascinated by class numbers as they were the basis of the classical approach to FLT and this taught me the important lesson that one cannot do number theory in isolation as the methods are drawn from many fields.

When I arrived at Oxford as an undergraduate I was still trying to prove FLT and thought I had a solution. I told my tutor, Dominic Welsh, who was all set to arrange for someone with some expertise to listen to me when I found the mistake. Dominic was very supportive in my undergraduate studies and would seek out tutors that he (or I) thought would be helpful, but the curriculum was not flexible enough to allow more than a brief course in number theory. Nevertheless I would retreat into it when I could. I tried to read what I found of Fermat in the library but sadly it was in Latin and my Latin was not up to it. However in terms of real research I had to bide my time.

Moving to Cambridge for my graduate studies I was very lucky to be assigned John Coates as a supervisor who arrived in Cambridge just as my preliminary course work year was over. Beginning research was like emerging into daylight. My encounter with professional research started in the summer when attending a conference on algebraic number theory at Durham where Coates, Iwasawa, Ribet, Serre, Tate and others lectured. John introduced me to Iwasawa theory and after doing some initial work on explicit reciprocity laws he was generous to let me work with him on elliptic curves and in particular on the Birch Swinnerton-Dyer conjecture. I loved doing research rather than attending courses and it turned into a very successful collaboration. None had been able to prove any part of the conjecture for more than a finite number of curves. We were able to prove one direction of the original part of the conjecture for a natural family of rational elliptic curves E, those with complex multiplication. The result was that if the L-function of the curve $L(E, s)$ did not vanish at $s = 1$ then the curve had only finitely many rational points.

After two years working with Coates he left Cambridge and so I moved to Harvard. This was more of a change than I expected and was in many ways the perfect complement to Cambridge. At Cambridge I had been very focussed on research and had learned what I needed as I went along. It was extremely productive but also somewhat insular. At Harvard there was an abundance of professors, students and visitors all talking about number theory and related fields and the atmosphere was much more intense than anything I had encountered before. When I arrived I started to work on Iwasawa's main conjecture and to do this I started studying Barry Mazur's paper Modular Curves and the Eisenstein Ideal. This was still a preprint but I saw that it had techniques that were stronger than those used in Ribet's work which I was trying to generalize. In particular I spent those two years getting familiar with the language of modern arithmetic geometry and also, because it was the subject of seminars and courses, with the rudiments of automorphic form theory. Prior to going to Harvard I had not known what a modular curve was. At some point during this period I also had my PhD oral as I had left Cambridge too early to be allowed to take it. Cambridge kindly appointed Barry Mazur and John Tate as my examiners and as a consequence I had a rather relaxed and informal exam. At the end of the two years I was able to take Ribet's work a bit further. I then joined forces with Mazur and in the following two years we solved Iwasawa's conjecture.

After my Harvard experience and brief trips to Paris and Bonn I settled in Princeton in 1982. Again it was a great centre for number theory, with Shimura, Iwasawa, Dwork and Katz on the faculty not to speak of Deligne and Langlands at the IAS. Although my motivation for studying Iwasawa's conjecture at Harvard had been to then try to prove the analogue for elliptic curves (first those with complex multiplication) I felt that the approach that had worked for modular curves would require some much deeper geometry than was yet available so I opted instead for generalizing Iwasawa's conjecture to the case of totally real fields. This was probably a fortunate decision as the proof in the complex multiplication case when it did come (due to Rubin) came from a quite different source, the Euler system methods of Kolyvagin, which were introduced somewhat later. The proofs in the general case, which did rely more on the approach of Mazur and myself, as well as techniques from the totally real case and from the theory of Euler systems, were not successful for many years until the work of Skinner, Urban and Kato. Moreover it did use more sophisticated geometry as well as a much deeper understanding of automorphic forms than I had at the time.

The Spring of 1985–1986 I spent in Paris. While I was trying to complete my work on the Iwasawa conjecture for totally real fields there was a flurry of excitement which had been initiated by a lecture of Frey. He had suggested a link between the modularity conjecture for elliptic curves and Fermat's Last Theorem. His suggestion had been incomplete but it led to Serre formulating his conjectures on the modularity of residual Galois representations. Perhaps surprisingly I was not caught up in this excitement while I was in Paris as they had seemed to me to be too optimistic. It was not until I returned from my sabbatical, in the late summer of 1986, that I learned that Ribet had actually proved a key part of them and that now Fermat's Last Theorem was a consequence of the modularity conjecture.

The opportunity to go back to work on elliptic curves and at the same time being able to work on FLT was irresistible. I had no intuition how to approach the modularity conjecture (known then as the Weil conjecture) but I quickly started trying to prove all Galois representations of a very special type were modular using techniques derived from the paper of Mazur I had studied at Harvard, together with the techniques I had introduced for the Iwasawa conjecture in the totally real case. These seemed to give a toe hold. For a while I considered only reducible residual representations, i.e. those that would correspond to elliptic curves with an isogeny, but after some months I realized that Langlands and Tunnell's results on solvable base change would give a method of studying the 3-adic representations and so of considering all elliptic curves. A more detailed description of the work of the next few years is given in the introduction to the paper 'Modular Elliptic Curves and Fermat's Last Theorem'.

Wedding day, Oxford, August 1988

During this period of work on FLT between 1986–1994 I was not quite the recluse in the attic that some accounts have portrayed. I married my wife Nada Canaan in the summer of 1988 and spent the next year in Oxford and Cambridge as a Royal Society professor. Nada had been a doctoral student in molecular biology at Princeton and we had met there. It was not until our honeymoon that I owned up to being obsessed with a very old and intractable problem. Fortunately perhaps she had never heard of it. Returning to Princeton at the end of 1989 we moved house and then had our three daughters between 1990 and 1994. Having small children was actually a perfect balance to working on the problem because both required a hundred per cent of one's attention when engaged with them.

During these early years at Harvard and Princeton I had a number of students, some of whom became collaborators. One happy outcome of this is that they often carried on with problems that I had left behind, or that had left me behind, but

Cambridge, June 23. 1993. (Photo: Peter Goddard/Isaac Newton Institute)

usually working in a language close enough to mine that I could follow the progress. Much of this has been described in Chris Skinner's accompanying account of my work. I am indebted to him for this essay, not only for its lucidity and scope, but also for presenting it in a style that so perfectly reflects my own preferences.

The Mathematical Works of Andrew Wiles

Christopher Skinner

Abstract This paper surveys the published mathematical works of Andrew Wiles up through the time he was awarded the Abel Prize.

1 Introduction

Andrew Wiles was awarded the Abel Prize for 2016 for

> his stunning proof of Fermat's Last Theorem by way of the modularity conjecture for semistable elliptic curves, opening a new era in number theory,

as noted in the citation by the Abel Prize Committee of the Norwegian Academy of Science and Letters. The proof of the modularity conjecture and the final resolution of Fermat's Last Theorem was a towering achievement, but it stands as perhaps only the highest pinnacle in a range of landmark theorems. As the citation for his 1989 election as a Fellow of the Royal Society reads:

> Andrew Wiles is almost unique amongst number-theorists in his ability to bring to bear new tools and new ideas on some of the most intractable problems of number theory. His finest achievement to date has been his proof, in joint work with Mazur, of the "main conjecture" of Iwasawa theory for cyclotomic extensions of the rational field. This work settles many of the basic problems on cyclotomic fields which go back to Kummer, and is unquestionably one of the major advances in number theory in our times. Earlier he did deep work on the conjecture of Birch and Swinnerton-Dyer for elliptic curves with complex multiplication – one offshoot of this was his proof of an unexpected and beautiful generalisation of the classical explicit reciprocity laws of Artin–Hasse–Iwasawa. Most recently, he has made new progress on the construction of l-adic representations attached to Hilbert modular forms, and has applied these to prove the "main conjecture" for cyclotomic extensions of totally real fields – again a remarkable result since none of the classical tools of cyclotomic fields applied to these problems.

C. Skinner (✉)
Department of Mathematics, Princeton University, Princeton, NJ, USA
e-mail: cmcls@princeton.edu

H. Holden, R. Piene (eds.), *The Abel Prize 2013–2017*,
https://doi.org/10.1007/978-3-319-99028-6_26

This paper surveys the published mathematical works of Andrew Wiles up through the time he was awarded the Abel Prize. It is an attempt to describe the important theorems highlighted in the above citations and to also give some sense of the new ideas and new tools Wiles introduced. Our hope is that this can serve as both an introduction to these results as well as a brief guide to the those seeking to navigate their proofs, especially the number-theorist or graduate student.

As noted especially in his Abel Prize citation, many of Wiles's proofs have significantly shaped the development of algebraic number theory. This is especially true of his proof of the modularity of semistable elliptic curves, which opened the door to previously unimagined progress on a host of problems related to modular forms and automorphic forms far beyond elliptic curves and GL_2 – progress that continues unabated as this article is written (yet, all the proofs bear the clear stamp of Wiles's seminal ideas). To give some idea of the impact of Wiles's work, at the end of some of the sections we have included brief remarks about subsequent developments. However, these only touch upon some of the highlights. To have tried to be comprehensive would have meant writing a substantially longer article!

A reader seeking a description of Wiles's work aimed at a broad mathematical audience can also consult [85] and [43], which include personal reflections on the influence of Wiles's work by some who have worked with him and by others whose work has followed on his.

2 Explicit Reciprocity Laws

Class field theory, the intrinsic classification of the abelian Galois extensions of a local or global field in terms of the arithmetic of the field, is one of the great achievements of algebraic number theory. However, both the formulation of the reciprocity maps of class field theory and the proofs of the reciprocity laws left open many problems and questions about making the theory explicit. For local fields, Lubin–Tate formal groups and Lubin–Tate extensions yielded an explicit construction of the abelian extensions that closely parallels the local case of the Kronecker–Weber theorem (which describes all abelian extensions of $\mathbb{Q}_p$ as subfields of cyclotomic extensions $\mathbb{Q}_p(\zeta_n)$). However, this still left open the problem of making other aspects of the theory explicit, especially the computation of the Hilbert symbol.

Artin, Hasse, and Iwasawa all proved some important explicit formulas for the Hilbert symbols for cyclotomic extensions of $\mathbb{Q}_p$. Because of the role that the Hilbert symbol plays in defining and proving properties of the power residue symbol, such explicit formulas for the Hilbert symbol are often referred to as *explicit reciprocity laws.*

One of Wiles's earliest published results, proved while a graduate student at the University of Cambridge, is a generalization of the reciprocity laws of Artin, Hasse, and Iwasawa with $\mathbb{Q}_p$ replaced by an arbitrary p-adic local field K, the cyclotomic extensions of $\mathbb{Q}_p$ replaced with the Lubin–Tate extensions of K, and the classical

Hilbert symbol replaced with the Hilbert symbols for Lubin–Tate formal groups. These reciprocity laws were used in Coates's and Wiles's proof of their groundbreaking work toward the Birch–Swinnerton-Dyer Conjecture for elliptic curves with complex multiplication and especially in Rubin's subsequent strengthening of this result.

In the following we first recall the definition of the classical Hilbert symbol and some of its properties and then the explicit reciprocity laws of Artin, Hasse, and Iwasawa. We then recall Lubin–Tate formal groups and Lubin–Tate extensions (these also appear in the subsequent discussion of Coates–Wiles homomorphisms and the Coates–Wiles Theorem) and their Hilbert symbols. This is followed by the statements of Wiles's higher explicit reciprocity laws and then the example of the formal multiplicative group over $\mathbb{Z}_p$, which recovers the explicit laws of Artin, Hasse, and Iwasawa as a special case.

Let K be a p-adic local field, that is, a finite extension of $\mathbb{Q}_p$. Let $\mathcal{O}$ be its ring of integers and let $\mathfrak{m}$ be the maximal ideal of $\mathcal{O}$. Let $\overline{K}$ be a separable algebraic closure of K, and let $\mathrm{Gal}(\overline{K}/K)^{\mathrm{ab}}$ be the quotient of the absolute Galois group $\mathrm{Gal}(\overline{K}/K)$ by the closure of its commutator. Let $\mathrm{Rec}_K : K^\times \to \mathrm{Gal}(\overline{K}/K)^{\mathrm{ab}}$ be the reciprocity map of local class field theory, normalized so that uniformizers map to arithmetic Frobenius elements.

2.1 *The Hilbert Symbol*

Let $n > 0$ be a positive integer such that K contains the group μ_n of nth roots of unity. The *Hilbert symbol* associated with K and n is the bilinear pairing

$$(\cdot,\cdot)_n : K^\times \times K^\times \to \mu_n$$

defined by

$$(\alpha,\beta)_n = \frac{\mathrm{Rec}_K(\beta)(\alpha^{1/n})}{\alpha^{1/n}}.$$

It follows from Kummer theory that $(K^\times)^n$ is the kernel of both sides of this pairing. In addition to being bilinear, the Hilbert symbol also has the following fundamental properties:

(H1) $(\alpha,\beta)_n = 1$ if and only if β is a norm from $K(\alpha^{1/n})$.
(H2) $(\alpha,-\alpha)_n = 1 = (\alpha, 1-\alpha)_n$.
(H3) $(\alpha,\beta)_n = (\beta,\alpha)_n^{-1}$.

Because of (H1), some authors refer to the Hilbert symbol as the *norm residue symbol*.

In some instances, it is possible to give relatively simple expressions for the Hilbert symbol.

$n = 2$

It follows from (H1) that $(\cdot,\cdot)_2$ has a very concrete definition:

$$(\alpha,\beta)_2 = 1 \iff \alpha x^2 + \beta y^2 = z^2 \text{ has a solution } 0 \neq (x,y,z) \in K^3.$$

$p \nmid n$

This is the case of the *tame Hilbert symbol*. Let q be the order of the residue field $\mathcal{O}/\mathfrak{m}$. Then $\mathcal{O}^\times = \mu_{q-1} \times 1+\mathfrak{m}$. In particular, every unit $x \in \mathcal{O}^\times$ can be uniquely written as $x = \omega(x)\langle x\rangle$ with $\omega(x) \in \mu_{q-1}$ and $\langle x\rangle \in 1+\mathfrak{m}$. If $p \nmid n$, then $n|q-1$ and

$$(\alpha,\beta)_n = \omega\left((-1)^{\mathrm{ord}_K(\alpha)\mathrm{ord}_K(\beta)}\frac{\alpha^{\mathrm{ord}_K(\beta)}}{\beta^{\mathrm{ord}_K(\alpha)}}\right)^{(q-1)/n}.$$

This relation is used to define the nth power residue symbol and to explain many of its properties.

2.2 The Explicit Reciprocity Laws of Artin, Hasse, and Iwasawa

Very generally, an explicit reciprocity law for K is an expression for the Hilbert symbol $(\alpha,\beta)_n$ in terms of relatively simple (often analytic) functions of α and β. In light of the formulas in the preceding section, the case of most interest is when $p \mid n$ and especially when $n = p^m$ is a power of p.

Let p be an odd prime. For each non-negative integer $m \geq 0$, let ζ_m be a primitive p^{m+1}th root of unity; we assume these are chosen so that $\zeta_m = \zeta_{m+1}^p$. Let $\Phi_m = \mathbb{Q}_p(\zeta_m)$. Let $\mathfrak{p}_m$ be the maximal ideal of the ring of integers of Φ_m. Then $\pi_m = 1-\zeta_m$ is a uniformizer of $\mathfrak{p}_m$. Let T_m denote the trace map from Φ_m to $\mathbb{Q}_p$, and for $n \geq m$ let $N_{n,m}$ be the norm map from Φ_n to Φ_m. Let $\Phi_m' = \cap_{n>m} N_{n,m}(\Phi_n^\times) \subset \Phi_m^\times$. Let $(\cdot,\cdot)_{p^{m+1}}$ be the Hilbert symbol for Φ_m. We then have the explicit reciprocity laws of Artin, Hasse, and Iwasawa:

(R1) For $\beta \in 1+\mathfrak{p}_m$, $(\zeta_m,\beta)_{p^{m+1}} = \zeta_m^{-T_m(\log\beta)/p^{m+1}}$.

(R2) For $\beta \in 1+\mathfrak{p}_m$, $(\pi_m,\beta)_{p^{m+1}} = \zeta_m^{T_m(\frac{\zeta_m}{\pi_m}\log\beta)/p^{m+1}}$.

(R3) For $\alpha \in 1+\mathfrak{p}$ and $\beta \in \Phi_m'$, $(\alpha,\beta)_{p^{m+1}} = \zeta_m^{-T_n(\zeta_n\log\alpha\cdot\frac{1}{\beta'}\frac{d\beta'}{d\pi_n})/p^{n+1}}$ for any $n \geq 2m+1$ and any $\beta' \in \Phi_n'$ such that $\beta = N_{n,m}(\beta')$.

(R4) For $\alpha \in 1+\mathfrak{p}_m^{2p^m}$ and $\beta \in \Phi_m^\times$, $(\alpha,\beta)_{p^{m+1}} = \zeta_m^{-T_m(\zeta_m\log\alpha\cdot\frac{1}{\beta}\frac{d\beta}{d\pi_m})/p^{m+1}}$.

Here log is the usual p-adic logarithm, and $\frac{d\beta}{d\pi_m} = \sum_{r=-N}^{\infty} ra_r\pi_m^{r-1}$ for some choice of expansion $\beta = \sum_{r=-N}^{\infty} a_r\pi_m^r$ with $a_r \in \mathbb{Z}_p$ (note that this is not uniquely-defined, but the values of the expressions in (R3) and (R4) do not depend on the choices of expansions). The reciprocity laws (R1) and (R2) were proved by Artin and Hasse, as was (R4) in the special case $m = 0$. The general cases of (R3) and (R4) were proved by Iwasawa.

2.3 Wiles's Higher Explicit Reciprocity Laws

Wiles proved an important generalization of the reciprocity laws of Artin, Hasse, and Iwasawa with $\mathbb{Q}_p$ replaced by an arbitrary p-adic local field K and the fields Φ_m replaced with Lubin–Tate extensions. In particular, Wiles proves the analog of (R3) and the analog of the $m = 0$ case of (R4). The impetus for considering such generalizations arose out of his work with John Coates on the Birch–Swinnerton-Dyer conjecture for elliptic curves with complex multiplication, though the problem is a very natural one.

Let K be a p-adic local field. A Lubin–Tate extension of K arises from a one-parameter formal group law over $\mathcal{O}$. The latter is a power series $F \in \mathcal{O}[\![X, Y]\!]$ satisfying $F(X, Y) = X + Y +$ (higher order terms), $F(X, F(Y, Z)) = F(F(X, Y), Z)$ (associativity), and $F(X, Y) = F(Y, X)$ (commutativity). It is straightforward to deduce that there is then a unique $i_F \in X\mathcal{O}[\![X]\!]$ such that $F(X, i_F(X)) = 0$ (existence of an inverse) and that $F(X, 0) = X$ and $F(0, Y) = Y$. Let L be any algebraic extension of K and let $\mathfrak{m}_L$ be the maximal ideal of the integral closure $\mathcal{O}_L$ of $\mathcal{O}$ in L. Then for any positive integer $r > 0$, F defines an abelian group structure on $\mathfrak{m}_L^r$ that we denote by $\oplus_F$: $x \oplus_F y = F(x, y)$. The inverse of an element $x \in \mathfrak{m}_L^r$ with respect to this group structure is $i_F(x)$, and we write $x \ominus_F y$ to mean $x \oplus_F i_F(y)$ (subtraction for this group structure). The convention is to write $F(\mathfrak{m}_L^r)$ for $\mathfrak{m}_L^r$ with this abelian group structure. A homomorphism of two such group laws $F, G \in \mathcal{O}[\![X, Y]\!]$ is an $f \in X\mathcal{O}[\![X]\!]$ such that $f(F(X, Y)) = G(f(X), f(Y))$. Clearly, $f(x \oplus_F y) = f(x) \oplus_G f(y)$, so $x \mapsto f(x)$ is a homomorphism from $F(\mathfrak{m}_L^r)$ to $G(\mathfrak{m}_L^r)$.

Let π be a uniformizer of $\mathfrak{m}$ and let q be the order of the residue field $\mathcal{O}/\mathfrak{m}$. Let $f \in \mathcal{O}[\![X]\!]$ be a power series such that $f(X) = \pi X +$ (higher order terms) and $f(X) \equiv X^q \bmod \mathfrak{m}$. Lubin and Tate proved that there is a unique one-parameter formal group law F such that f is an endomorphism of F, and furthermore there is a unique injective homomorphism $\mathcal{O} \hookrightarrow \mathrm{End}(F)$, written $a \mapsto [a]_F$, such that $[\pi]_F = f$ and $[a]_F = aX +$ (higher order terms). Then $F(\mathfrak{m}_L^r)$ is an $\mathcal{O}$-module for the action $a \odot_F x = [a]_F(x)$ (note that $\pi \odot_F x = f(x)$). Such an F is often called a Lubin–Tate formal group law. For each positive integer $r > 0$, let

$$M_r = \{x \in \mathfrak{m}_{\overline{K}} : \pi^r \odot_F x = 0\}.$$

This is a cyclic $\mathcal{O}$-module isomorphic to $\mathcal{O}/\mathfrak{m}^r$. For each non-negative integer $m \geq 0$ let

$$\Phi_m = K[M_{m+1}].$$

This is a Lubin–Tate extension of K. Lubin and Tate also proved that Φ_m is a totally ramified extension of K of degree $(q-1)q^m$ such that:

(LT1) The kernel of the surjection $K^\times \twoheadrightarrow \mathrm{Gal}(\Phi_m/K)$ determined by the reciprocity map Rec_K is $\pi^{\mathbb{Z}} \times 1 + \mathfrak{m}^{m+1}$ (so the norm group corresponding to the extension Φ_m is $N_{\Phi_m/K}(\Phi_m^\times) = \pi^{\mathbb{Z}} \times 1 + \mathfrak{m}^{m+1}$).
(LT2) For $u \in \mathcal{O}^\times$ and $x \in M_{m+1}$, $\mathrm{Rec}_K(u)(x) = u^{-1} \odot_F x$.

It follows from (LT1) that the extensions Φ_m depend only on the choice of uniformizer π and not on the formal group law F. In fact, all Lubin–Tate formal group laws over $\mathcal{O}$ associated with a fixed uniformizer π are isomorphic.

Let $\mathfrak{p}_m$ be the maximal ideal of the ring of integers Φ_m. The Hilbert symbol for F and m is the bilinear pairing

$$(\cdot,\cdot)_{F,m} : F(\mathfrak{p}_m) \times \Phi_m^\times \to M_{m+1}$$

defined by

$$(\alpha,\beta)_{F,m} = \mathrm{Rec}_{\Phi_m}(\beta)(\alpha') \ominus_F \alpha',$$

for any $\alpha' \in F(\mathfrak{m}_{\overline{K}})$ such that $\pi^{m+1} \odot_F \alpha' = \alpha$. Note that this pairing is $\mathcal{O}$-linear in α (for the $\mathcal{O}$-action $\odot_F$). The Hilbert symbols $(\cdot,\cdot)_{F,m}$ also have the following properties:

(H1)$_F$ $(\alpha,\beta)_{F,m} = 0$ if and only if β is a norm from $\Phi_m(\alpha')$, where $\pi^{m+1} \odot_F \alpha' = \alpha$.
(H2)$_F$ For $n \geq m$, $\pi^{n-m} \odot_F (\alpha,\beta)_{F,n} = (\alpha, N_{n,m}(\beta))_{F,m}$ for any $\alpha \in F(\mathfrak{p}_m)$ and $\beta \in \Phi_n^\times$.

Here $N_{n,m}$ is the norm map from Φ_n to Φ_m.

It is for the Hilbert symbols $(\cdot,\cdot)_{F,m}$ that Wiles proved explicit reciprocity laws. For each $m \geq 0$, let $v_m \in M_{m+1}$ be an $\mathcal{O}$-generator, chosen so that $\pi \odot_F v_{m+1} = v_m$. Each v_m is a uniformizer of $\mathfrak{p}_m$. Let T_m and N_m be, respectively, the trace and norm maps from Φ_m to K. Let $\Phi_m' = \cap_{n>m} N_{n,m}(\Phi_n^\times) \subset \Phi_m^\times$.

Theorem 1 (Wiles's higher explicit reciprocity laws) *Suppose p is odd.*

(W1) [4, (10)] *For $\beta \in 1 + \mathfrak{p}_m$,*

$$(v_m,\beta)_{F,m} = \left(\frac{1}{\pi^{m+1}}(N_m\beta^{-1} - 1)\right) \odot_F v_m.$$

(W2) [4, Thm. 1] *For* $\alpha \in F(\mathfrak{p}_m)$ *and* $\beta \in \Phi'_m$,

$$(\alpha, \beta)_{F,m} = \left(\frac{1}{\pi^{n+1}} T_n \left(\frac{1}{\lambda'_F(v_n)} \frac{1}{\beta'} \frac{d\beta'}{dv_n} \lambda_F(\alpha)\right)\right) \odot_F v_m$$

for any $n \geq 2m+1$ *and any* $\beta' \in \Phi'_n$ *such that* $N_{n,m}(\beta') = \beta$.

(W3) [4, Thm. 23] *For* $\alpha \in F(\mathfrak{p}_0^2)$ *and* $\beta \in \Phi_0^\times$,

$$(\alpha, \beta)_{F,0} = \left(T_0(\lambda_F\left(\alpha\right) \frac{1}{\lambda'_F(v_0)} \frac{1}{\beta} \frac{d\beta}{dv_0}\right)\right) \odot_F v_0.$$

Here, $\lambda_F \in K[\![X]\!]$ is the formal logarithm of F. This is the unique isomorphism λ_F of F with the formal additive group (as formal group laws over K) such that $\lambda'_F(0) = 1$. The power series $\lambda_F(X)$ converges on $\mathfrak{m}_{\overline{K}}$, and the formal derivative λ'_F of λ_F has coefficients in $\mathcal{O}$.

The reciprocity law (R1) follows from (W1), while (W2) is a generalization of (R3), and (W3) generalizes the $m = 0$ case of (R4). Wiles showed that (R2) follows from (W2).

The expression in (W1) is an immediate consequence of the definition of $(\cdot, \cdot)_{F,m}$ and the description of the action of $\mathrm{Rec}_{\Phi_m}(\beta)$ in (LT2). Wiles's proof of (W2) closely follows Iwasawa's proof of his reciprocity law (R3), substituting Lubin–Tate theory for the cyclotomic theory. One place the proofs differ is that where Iwasawa's proof rests essentially on the reciprocity law (R2) of Artin and Hasse, Wiles's proof gets by with just (W1). For a fixed choice of π all Lubin–Tate formal groups F are isomorphic, which allowed Wiles to make calculations for a particularly nice Lubin–Tate group (that corresponding to $f = \pi X + X^q$); this replaces the explicitness of cyclotomic fields in Iwasawa's proof. We note that (W3) for a height one formal group (so $q = p$) had already been proved by Coates and Wiles in [2].

2.4 Example: The Formal Multiplicative Group Over $\mathbb{Z}_p$

In this example we explain the relation between Wiles's reciprocity laws and those of Artin–Hasse and Iwasawa. Let $K = \mathbb{Q}_p$ and let $f(X) = (X+1)^p - 1 = pX + \cdots + X^p$ (so $\pi = p$). Then $F(X, Y) = X + Y + XY$ is just the formal multiplicative group, and $M_r = \{\zeta - 1 : \zeta \in \mu_{p^r}\}$, so $\Phi_m = \mathbb{Q}_p(\zeta_m - 1) = \mathbb{Q}_p(\zeta_m)$. From the definitions of the various Hilbert symbols we find that in this case

$$(\alpha, \beta)_{F,m} = (1+\alpha, \beta)_{p^{m+1}} - 1,$$

where $(\alpha, \beta)_{p^{m+1}}$ is the Hilbert symbol for Φ_m. Let $v_m = \zeta_m - 1 = -\pi_m$. It follows that

$$(\alpha, \beta)_{F,m} = a \odot_F v_m \iff (1+\alpha, \beta)_{p^{m+1}} = \zeta_m^a.$$

To see that (W1) implies (R1) we need only observe that by local class field theory $N_m(1+\mathfrak{p}_m)$ is contained in $1+p^{m+1}$, so

$$N_m(\beta^{-1})-1\equiv\log(N_m(\beta)^{-1})\equiv -T_m(\log\beta)\bmod p^{2m+2}.$$

To see that (W2) implies (R3) we just note that in this case $\lambda_F(\alpha)=\log(1+\alpha)$, $\lambda_F'(v_m)=1/\zeta_m$, and $\frac{d\beta}{dv_m}=-\frac{d\beta}{d\pi_m}$ (which is the source of the minus sign in (R3)). This also shows that (W3) implies the $m=0$ case of (R4). Noting that $N_{n,m}(\zeta_n-1)=\zeta_m-1$, so $1-\zeta_m\in\Phi_m'$, we take $\beta=\zeta_m-1$ and $\beta'=\zeta_n-1$ in (W2). As $\frac{d\beta'}{d\pi_n}=1$ we have

$$T_n\left(\frac{1}{\lambda'(v_n)}\frac{1}{\beta'}\frac{d\beta'}{dv_n}\lambda_F(\alpha)\right)=T_n\left(\frac{\zeta_n}{\zeta_n-1}\log(1+\alpha)\right).$$

A direct computation shows that the trace of $\frac{\zeta_n}{\zeta_n-1}$ from Φ_n to Φ_m is just $p^{n-m}(\frac{\zeta_m}{\zeta_m-1})$, which then implies that

$$T_n\left(\frac{1}{\lambda'(v_n)}\frac{1}{\beta'}\frac{d\beta'}{dv_n}\lambda_F(\alpha)\right)\Big/\,p^{n+1}=T_m\left(\frac{\zeta_m}{\pi_m}\log(1+\alpha)\right)\Big/\,p^{m+1},$$

and so (R2) follows from (W2).

2.5 Further Developments

Not long after Wiles proved his reciprocity laws, Coleman [38] placed some of the constructions in a more general context (including the ‘Coleman isomorphism,’ an important generalization of the Lemma in Sect. 3.1 below) and also formulated an even more general reciprocity for Lubin–Tate extensions [39]. This conjecture was subsequently proved by de Shalit [44]. Wiles's explicit reciprocity laws were used in the first proof of the Coates–Wiles theorem about the Birch–Swinnerton-Dyer Conjecture for CM elliptic curves [3] (see also Sect. 4 below) and particularly in Rubin's proof [74] of his generalization of the Coates–Wiles theorem. Some very general explicit reciprocity laws were given by Perrin-Riou [69, 70], Kato [55, 56], and Colmez [40], as part of their approaches to the Iwasawa theory of elliptic curves and more general motives.

3 The Coates–Wiles Homomorphisms and p-adic L-Functions

The proofs by Coates and Wiles of their ground-breaking results on the arithmetic of elliptic curves with complex multiplication rely on some special homomorphisms. These homomorphisms, which are defined on groups of local units and take values in the ring of integers of a p-adic field, are well-adapted to recognizing when groups of global units have non-trivial p-power index in the groups of local units and – miraculously – to relating the indices to values of Hecke L-functions. Though the idea behind these homomorphisms originates in work of Kummer, they were first introduced by Coates and Wiles, and these homomorphisms are now generally called *Coates–Wiles homomorphisms.* These homomorphisms enabled Coates and Wiles to generalize to imaginary quadratic fields a theorem of Iwasawa on the structure of certain quotients of groups of norm-compatible local units by special subgroups of norm compatible global units and to relate this structure to p-adic L-functions. The argument of Coates and Wiles even provided a new, streamlined proof of Iwasawa's theorem.

In the following we describe the Coates–Wiles homomorphisms in the context of Lubin–Tate formal groups and Lubin–Tate extensions. We then describe the new proof of Iwasawa's theorem and its generalization. The application to the Birch–Swinnerton Dyer conjecture is described in the following separate section.

We keep with the notation introduced in the preceding discussion of Wiles's higher explicit reciprocity laws and in particular the notation in Sect. 2.3. So p is an odd prime, K is a finite extension of $\mathbb{Q}_p$, π is fixed uniformizer of the maximal ideal $\mathfrak{m}$ of O, and $F \in O[\![X, Y]\!]$ is a Lubin–Tate formal group law associated with some $f(X) = \pi X + \text{(higher order terms)} \in XO[\![X]\!]$.

3.1 The Coates–Wiles Homomorphisms

The Tate module of F is the inverse limit

$$T_\pi F = \varprojlim_n M_n,$$

where the limit is taken with respect to the transition maps $\pi^{n-m} \odot_F : M_n \to M_m$, $n \geq m$. The Tate module $T_\pi F$ is a free O-module of rank one. An O-basis of $T_\pi F$ is just a sequence $v = (v_n)$, v_n an O-generator of M_n, such that $\pi \odot_F v_{n+1} = v_n$. Let $G_n = \mathrm{Gal}(\Phi_n/K)$ and $G = \varprojlim_n G_n = \mathrm{Gal}(\Phi_\infty/K)$, where $\Phi_\infty = \cup_n \Phi_n$. The group G acts on $T_\pi F$ via an isomorphism $\chi : G \xrightarrow{\sim} O^\times$ (so $\sigma(v) = \chi(\sigma) \odot_F v_n$). The character χ is independent of the choice of F.

Let $U_n = 1 + \mathfrak{p}_n$, and let $U = \varprojlim_n U_n$ be the projective limit with respect to the norm maps $N_{n,m} : U_n \to U_m$. It is easily deduced from (LT1) that $U = \varprojlim_n U'_n$, where $U'_n \subset U_n$ is the subgroup of units whose norm to K is 1. Furthermore, the image of the natural projection of U to U_n is U'_n. The group G also acts on U.

The definition of the Coates–Wiles homomorphisms for F begin with the following lemma.

Lemma 2 ([5, Thm. 5]) *Let $v = (v_n)$ be an $\mathcal{O}$-basis of $T_\pi F$ and $u = (u_n) \in U$. There exists a unique power series $g_{v,u} \in \mathcal{O}[\![T]\!]^\times$ such that $u_n = g_{v,u}(v_n)$ for all $n \geq 0$.*

Let v be a fixed $\mathcal{O}$-basis of $T_\pi F$. For $k \geq 0$, the kth Coates–Wiles homomorphism is

$$\delta_k : U \to \mathcal{O}$$

$$\delta_k(u) = \left(\frac{1}{\lambda'_F(T)}\frac{d}{dT}\right)^k \log g_{v,u}(T)\bigg|_{T=0} = \left(\frac{d}{dZ}\right)^k \log h_{v,u}(Z)\bigg|_{Z=0}, \tag{CW1}$$

where $h_{v,u} \in K[\![Z]\!]$ is the power series such that $h_{v,u}(\lambda_F(T)) = g_{v,u}(T)$. The far right-hand side of (CW1) is the definition of $\delta_k(u)$ given in [5, p. 15]. While λ_F does not have coefficients in $\mathcal{O}$ in general, $\lambda'_F(T) = \frac{\partial}{\partial X}F(X,T)|_{X=0} \in \mathcal{O}[\![T]\!]^\times$, and $\frac{d}{dT}\log g_{v,u}(T) = \frac{g'_{v,u}(T)}{g_{v,u}(T)} \in \mathcal{O}[\![T]\!]$, so δ_k is $\mathcal{O}$-valued. The uniqueness of $g_{v,u}$ implies δ_k is a homomorphism (since $g_{v,uu'} = g_{v,u}g_{v,u'}$). Noting that $\lambda([\chi(\sigma)]_F(T)) = \chi(\sigma)\lambda_F(T)$, it follows that

$$\delta_k(\sigma(u)) = \chi^k(\sigma)\delta_k(u). \tag{CW2}$$

It is easy to see that the Coates–Wiles homomorphisms are independent of the choice of v, and even the group F, up to $\mathcal{O}^\times$-multiple.

Remark 3 In [3] Coates and Wiles introduced a precursor to the homomorphisms δ_k. These are homomorphisms $\varphi_k : U_0 \to \mathcal{O}/\mathfrak{m}$. The definition given in [3, (8)] is simple: Suppose F is the Lubin–Tate formal group associated with the basic polynomial $f(X) = \pi X + X^q$. Given $u_0 \in U_0$, let $g(T) \in \mathcal{O}[\![T]\!]$ such that $g(v_0) = u_0$. Then $\varphi_k(u) = c_k$, where $T\frac{d}{dT}\log g(T) = \sum_{i=1}^\infty c_i T^i$. It turns out that $\varphi_k(u_0) = \delta_k(u) \bmod \mathfrak{m}$ for $u = (u_n) \in U$, given a suitable compatibility among the various generators v_0 chosen in the definitions of these homomorphisms (and this relation is always true up to multiplication by a non-zero scalar in $(\mathcal{O}/\mathfrak{m})^\times$).

3.2 Groups of Local Units

For simplicity we now assume that $K = \mathbb{Q}_p$, so $q = p$ but it may be that $\pi \neq p$. Since $K = \mathbb{Q}_p$, $\chi : G \xrightarrow{\sim} \mathbb{Z}_p^\times$ and so $G = \Delta \times \Gamma$, where Δ is the torsion subgroup (identified via χ with $\mu_{p-1} \subset \mathbb{Z}_p^\times$) and Γ is the subgroup identified via χ with $1 + p\mathbb{Z}_p$. For $1 \leq i \leq p-1$, let $\epsilon_i = \frac{1}{p-1}\sum_{\sigma\in\Delta}\chi^{-i}(\sigma)\sigma \in \mathbb{Z}_p[\Delta]$. The ϵ_i are orthogonal idempotents satisfying $1 = \sum_{i=1}^{p-1}\epsilon_i$, so there are decompositions $U_n' = \oplus_{i=1}^{p-1}U_n^{(i)}$, $U_n^{(i)} = \epsilon_i U_n'$, and

$$U = \oplus_{i=1}^{p-1}U^{(i)}, \quad U^{(i)} = \epsilon_i U = \varprojlim_n U_n^{(i)}.$$

It follows from (CW2) that the Coates–Wiles homomorphism δ_k factors through the projection to $U^{(i)}$ for $i \equiv k \bmod (p-1)$.

Let $\Lambda = \mathbb{Z}_p[[\Gamma]]$. Fixing a topological generator $\gamma \in \Gamma$ (such as the element satisfying $\chi(\gamma) = 1 + p$) identifies Λ with the power series ring $\mathbb{Z}[[T]]$ via $\gamma \mapsto 1 + T$. The group U is naturally a Λ-module, and it follows from (CW2) that for $h \in \Lambda$,

$$\delta_k(h(T)\cdot u) = h(\chi(\gamma)^k - 1)\delta_k(u). \tag{CW3}$$

Of course, each of the summands $U^{(i)}$ is also a Λ-module. In fact:

Lemma 4 ([5, Lem. 2, Cor. 3]) *Suppose $K = \mathbb{Q}_p$. If Φ_0 does not contain a primitive pth root of unity, then each $U^{(i)}$, $1 \leq i < p-1$, is a free Λ-module of rank one satisfying*

$$U^{(i)}/\omega_n U^{(i)} \xrightarrow{\sim} U_n^{(i)}, \quad \omega_n = \gamma^{n+1} - 1 \in \Lambda,$$

and $U^{(i)}$ is generated by an element ε_i such that $\delta_k(\varepsilon_i) \in \mathcal{O}^\times$ for all $k \equiv i \bmod (p-1)$. If Φ_0 contains a primitive pth root of unity then the conclusions hold for $1 < i < p-1$.

Let $\widehat{\mathbb{Z}}_p^{\mathrm{ur}}$ be the p-adic completion of the ring of integers of the maximal unramified extension of $\mathbb{Q}_p$. One of the important features of the Coates–Wiles homomorphisms is that they can be interpolated by elements in $\widehat{\mathbb{Z}}_p^{\mathrm{ur}}[[\Gamma]] = \widehat{\mathbb{Z}}_p^{\mathrm{ur}}[[T]]$ (where the identification is again given by $\gamma \mapsto 1 + T$).

Theorem 5 ([5, Thm. 16]) *Suppose $K = \mathbb{Q}_p$. Let $u \in U$. There is a $\widehat{\mathbb{Z}}_p^{\mathrm{ur}}$-valued measure μ_u on G such that for each $\chi \geq 1$,*

$$\int_G \chi^k(\sigma)\, d\mu_u = \Omega_F^k\left(1 - \frac{\pi^k}{p}\right)\delta_k(u).$$

Here $\Omega_F \in (\widehat{\mathbb{Z}}_p^{\mathrm{ur}})^\times$ is a constant that depends on F.

In general, Ω_F satisfies $\Omega_F^{\mathrm{Frob}_p}/\Omega_F = \pi/p$ and is only uniquely defined up to $\mathbb{Z}_p^\times$-multiple. If $\pi = p$, then Ω_F can be taken to be 1 and each μ_u is a $\mathbb{Z}_p$-valued measure.

The $\widehat{\mathbb{Z}}_p^{\mathrm{ur}}$-valued measures on G are naturally identified with the elements of the completed group ring $\widehat{\mathbb{Z}}_p^{\mathrm{ur}}[\![G]\!] = \oplus_{i=1}^{p-1}\widehat{\mathbb{Z}}_p^{\mathrm{ur}}[\![\Gamma]\!]e_i$. If $H_u = \sum_{i=1}^{p-1} H_{u,i}e_i$ is the element identified with μ_u, then $H_{u,i} \in \widehat{\mathbb{Z}}_p^{\mathrm{ur}}[\![T]\!]$ has the property that

$$H_{u,i}(\chi^k(\gamma) - 1) = \Omega_F^k\left(1 - \frac{\pi^k}{p}\right)\delta_k(u), \quad k \equiv i \bmod (p-1). \tag{CW4}$$

It follows from this and (CW3) that

$$H_{h\cdot u,i}(T) = h(T)H_{u,i}(T), \quad h \in \Lambda. \tag{CW5}$$

In particular, the map $U^{(i)} \to \widehat{\mathbb{Z}}_p^{\mathrm{ur}}[\![\Gamma]\!]$, $u \mapsto H_{u,i}$, is a homomorphism of Λ-modules. This induces an isomorphism $U\hat{\otimes}_{\mathbb{Z}_p}\widehat{\mathbb{Z}}_p^{\mathrm{ur}} \xrightarrow{\sim} \widehat{\mathbb{Z}}_p^{\mathrm{ur}}[\![T]\!]$ for $1 < i < p-1$, and even for $i = 1$ if Φ_0 does not contain a primitive pth root of unity.

3.3 Example: Cyclotomic Units

Let $K = \mathbb{Q}_p$ and $F(X,Y) = X + Y + XY$ be the formal multiplicative group law. This is the Lubin–Tate group law associated with the power series $f(T) = (1+T)^p - 1$. As before, for each integer $n \geq 0$ let ζ_n be a primitive p^{n+1}th root of unity, chosen so that $\zeta_{n+1}^p = \zeta_n$. Then $v = (v_n) = (\zeta_n - 1)$ is a $\mathbb{Z}_p$-generator of the Tate module of F.

Fix an integer $a \neq \pm 1$ prime to p and let

$$u_{a,n} = \left(\frac{\zeta_n^a - 1}{\zeta_n - 1}\right)^{p-1}.$$

Then $u_{a,n} \in U_n$ and $u_a = (u_{a,n}) \in U$. By inspection we see that

$$g_{v,u_a}(T) = \left(\frac{(1+T)^a - 1}{T}\right)^{p-1}.$$

Making the formal change of variable $1 + T = e^Z = 1 + Z + Z/2 + \cdots$ yields

$$\delta_k(u_a) = (p-1)\left(\frac{d}{dZ}\right)^{k-1}\left(\frac{ae^{aZ}}{e^{aZ}-1} - \frac{e^Z}{e^Z - 1}\right)\Bigg|_{Z=0}.$$

Recalling that $\frac{te^t}{e^t-1} = \sum_{n=0}^{\infty} B_n \frac{t^n}{n!}$, where B_k is the kth Bernoulli number, one then finds that for an even integer $k > 0$,

$$\delta_k(u_a) = (p-1)(a^k - 1)\frac{B_k}{k} = (p-1)(1-a^k)\zeta(1-k),$$

where $\zeta(s)$ is the Riemann zeta function. (This can all be justified by working over $\mathbb{C}$ and restricting to regions of absolute convergence.)

Let $H_{u_a,i} \in \Lambda = \mathbb{Z}_p[\![T]\!]$ be the power series associated with u_a as in (CW4). It then follows that for an even integer $1 < i \leq p-1$ and any positive integer $k \equiv i \bmod (p-1)$,

$$H_{u_a,i}(\chi^k(\gamma) - 1) = (p-1)(1-a^k)(1-p^{k-1})\zeta(1-k).$$

Let $r_a \in \mathbb{Z}_p$ be such that $a = \omega(a)^{-1}\chi(\gamma)^{r_a}$ (so $r_a = \log_p a / \log_p \chi(\gamma)$), where $\omega(a) \in \mu_{p-1} \subset \mathbb{Z}_p^\times$ is the $(p-1)$th root of unity such that $\omega(a) \equiv a \bmod p$. Let $R_{a,i}(T) = (1 - \omega(a)^{-i}(1+T)^{r_a}) \in \mathbb{Z}_p[\![T]\!]$. Then $R_{a,i}(\chi^k(\gamma) - 1) = (1 - a^k)$ for all $k \equiv i \bmod (p-1)$. Furthermore, $R_{a,i}(0) = 1 - \omega(a)^{-i}$, so if a is a primitive root mod p and $i < p-1$, then $R_{a,i} \in \mathbb{Z}_p[\![T]\!]^\times$. Choosing $a \neq \pm 1$ to be a primitive root mod p we then put

$$G_i(T) = \frac{1}{p-1} R_{a,i}(T)^{-1} H_{u_a,i}(T) \in \mathbb{Z}_p[\![T]\!], \quad i \text{ even}, \ 1 < i < p-1.$$

This satisfies

$$G_i(\chi^k(\gamma) - 1) = (1 - p^{1-k})\zeta(1-k), \quad k > 0, k \equiv i \bmod (p-1).$$

From this one recognizes G_i to be the Kubota–Leopoldt p-adic L-function $L_p(\omega^i, s)$ for the even Dirichlet character ω^i, in the sense that

$$G_i(\chi(\gamma)^{1-s} - 1) = L_p(\omega^i, s).$$

One also sees that for any a prime to p, $H_{u_a,i}$ is a multiple of G_i (a unit multiple if a is a primitive root mod p).

Let $\mathcal{C} \subset U$ be the subgroup generated by the units u_a as a varies over all integers not equal to ± 1 and prime to p. This is the group of cyclotomic units. Let $\overline{\mathcal{C}}$ be the closure of $\mathcal{C}$ in U. Suppose $1 < i < p-1$ is even. Then $U^{(i)}$ is a free Λ-module of rank 1 and the map $u \mapsto H_{u,i}$ is a Λ-module isomorphism $U^{(i)} \xrightarrow{\sim} \Lambda$ that maps $\overline{\mathcal{C}}^{(i)} = \epsilon_i \overline{\mathcal{C}}$ onto $G_i \cdot \Lambda$. In particular, this recovers the following theorem of Iwasawa:

Theorem 6 (Iwasawa) *Let $1 < i < p-1$ be an even integer. There is a Λ-module isomorphism $U^{(i)}/\overline{\mathcal{C}}^{(i)} \cong \Lambda/(G_i)$ with $G_i(T) \in \Lambda$ the power series satisfying $G_i(\chi(\gamma)^{1-s} - 1) = L_p(\omega^i, s)$.*

Remark 7 Much earlier, Kummer had proved that if $(U_0/\overline{\mathcal{C}}_0)^{(i)}$ is non-zero, then $p|(2\pi\sqrt{-1})^i\zeta(i)$. Here $1 < i < p-1$ is an even integer, and $\mathcal{C}_0$ is the group of cyclotomic units in Φ_0, that is, the group generated by the $u_{a,0}$. In Kummer's proof, the role of the Coates–Wiles homomorphisms is played by the homomorphisms φ_i (see the remark at the end of Sect. 3.1).

3.4 Example: Elliptic Units

Let E be an elliptic curve with complex multiplication by the ring of integers $O_{\mathcal{K}}$ of an imaginary quadratic field $\mathcal{K} \subset \mathbb{C}$. For simplicity we assume that $\mathcal{K}$ has class number 1 and that E is defined over $\mathcal{K}$. We may then assume that E has a Weierstrass model given by an equation

$$E : y^2 = 4x^3 - g_2x - g_3$$

with $g_2, g_3 \in O_{\mathcal{K}}$ and with discriminant Δ divisible only by the primes of $\mathcal{K}$ of bad reduction for E and possibly those dividing 2 and 3. The curve E then has a complex uniformization $\xi : \mathbb{C}/\Lambda \to E(\mathbb{C})$, $\xi(z) = (\wp(z), \wp'(z))$, where Λ is a lattice and $\wp$ is the Weierstrass $\wp$-function for this lattice. We identify $O_{\mathcal{K}}$ with a subring of $\mathrm{End}(E)$ so that $a \cdot \xi(z) = \xi(az)$ for $a \in O_{\mathcal{K}}$.

The theory of complex multiplication associates with E a Hecke character $\psi : \mathbb{A}_{\mathcal{K}}^\times \to \mathbb{C}^\times$ of $\mathcal{K}$ such that: (1) $\psi(z) = z^{-1}$ for $z \in \mathbb{C}^\times$, (2) ψ is unramified at a finite place v of $\mathcal{K}$ if and only if E has good reduction at v, (3) if E has good reduction at a finite place v of $\mathcal{K}$, then for any uniformizer $\varpi_v \in \mathcal{K}_v$, $\psi(\varpi_v) \in \mathcal{K}^\times$ is the generator of the prime $\mathfrak{p}_v$ corresponding to v such that the reduction modulo $\mathfrak{p}_v$ of multiplication by $\psi(\varpi_v)$ is the Frobenius homomorphism, and (4) $L(E,s) = L(\psi,s)L(\overline{\psi},s)$. Let $\mathfrak{f}$ be the conductor of ψ; this is divisible only by primes of bad reduction for E. We also write ψ for the corresponding Hecke character of modulus $\mathfrak{f}$. Then for any fractional ideal $\mathfrak{a} \in \mathbb{I}_{\mathcal{K}}^{\mathfrak{f}}$, $\psi(\mathfrak{a}) \in \mathcal{K}^\times$; if $\mathfrak{a}$ is integral, then $\psi(\mathfrak{a}) \in O_{\mathcal{K}}$.

Let $p > 5$ be a prime that splits in $\mathcal{K}$, say $p = \mathfrak{p}\overline{\mathfrak{p}}$, and such that E has good reduction at $\mathfrak{p}$ and $\overline{\mathfrak{p}}$. Let $K = \mathcal{K}_{\mathfrak{p}} = \mathbb{Q}_p$ and $O = O_{\mathcal{K}_{\mathfrak{p}}} = \mathbb{Z}_p$. Let $t = -2x/y$. This is a local uniformizer at the origin of E. There exists a power series $w(t) = t^3 +$(higher order terms) $\in O[\![t]\!]$ such that $x = x(t) = t/w(t) \in t^{-2}O[\![t]\!]^\times$ and $y = y(t) = -2/w(t) \in t^{-3}O[\![t]\!]^\times$. Then $(x(t), y(t)) \in E(O(\!(t)\!))$. The formal group of E over O is given by a formal group law $F(X,Y) = X + Y +$ (higher order terms) $\in O[\![X,Y]\!]$ that has the property that $(x(t_1), y(t_1)) + (x(t_2), y(t_2)) = (x(F(t_1,t_2)), y(F(t_1,t_2)))$, where the addition is of points on E. This is in fact a Lubin–Tate formal group law over O associated with some $f(X) = \pi X +$ (higher order terms) with $\pi \in O_{\mathcal{K}}$ the generator of $\mathfrak{p}$ such that the reduction modulo $\mathfrak{p}$ of multiplication by π is the Frobenius homomorphism (in particular, $\pi = \psi(\varpi)$ for any uniformizer ϖ of K).

Following standard convention, for any algebraic extension L/K we write $\widehat{E}(\mathfrak{m}_L)$ for the group with underlying set the maximal ideal $\mathfrak{m}_L$ of the ring of integers of L and group law $\oplus_F$ defined by F. There is then a Galois equivariant isomorphism $\widehat{E}(\mathfrak{m}_{\overline{K}}) \xrightarrow{\sim} E_1(\overline{K})$, $t \mapsto (x(t), y(t))$, where E_1 is the kernel of the reduction modulo $\mathfrak{p}$ map, that is compatible with $\mathcal{O}_{\mathcal{K}}$-actions. This isomorphism induces a Galois equivariant isomorphism of M_n (the kernel of multiplication by π^{n+1} in $\widehat{E}(\mathfrak{m}_{\overline{K}})$) and $E[\pi^{n+1}]$ and hence an isomorphism of Tate modules

$$T_\pi \widehat{E} = \varprojlim_n M_n \xrightarrow{\sim} T_\pi E = \varprojlim_n E[\pi^{n+1}].$$

In particular, an $\mathcal{O}$-basis $v = (v_n)$ of the left-hand side is identified with a $\mathcal{O}$-basis (Q_n) of the right-hand side (so for all $n \geq 0$, $Q_n \in E[\pi^{n+1}]$ is an $\mathcal{O}$-generator satisfying $\pi \cdot Q_{n+1} = Q_n$). The prime $\mathfrak{p}$ is totally ramified in the field $F_n = \mathcal{K}(E[\pi^{n+1}])$, and the Lubin–Tate extension Φ_n is just the completion of F_n at the unique prime above $\mathfrak{p}$.

Let $\mathfrak{a} \subset \mathcal{O}_{\mathcal{K}}$ be an integral ideal that is prime to $\mathfrak{f}$ and $\mathfrak{p}$. Let α be a generator of $\mathfrak{a}$. We define a rational function $\theta_{\mathfrak{a}} \in \mathcal{K}(E) = \mathcal{K}(x, y)$:

$$\theta_{\mathfrak{a}} = \alpha^{-12} \Delta^{N(\mathfrak{a})-1} \prod_{P \in E[\mathfrak{a}]-0} (x - x(P))^{-6}.$$

Here $N(\mathfrak{a})$ is the norm of $\mathfrak{a}$. A remarkable feature of $\theta_{\mathfrak{a}}$ is that if $\mathfrak{c} \subset \mathcal{O}_{\mathcal{K}}$ is an integral ideal prime to $\mathfrak{a}$ and not a power of a prime ideal and if $Q \in E[\mathfrak{c}]$ is an $\mathcal{O}_{\mathcal{K}}$-generator, then $\theta_{\mathfrak{a}}(Q)$ is a unit of the ring of integers of the ray class field $\mathcal{K}_{\mathfrak{c}}$ of conductor $\mathfrak{c}$. Let B be a finite set of integral ideals $\mathfrak{b}$ prime to $\mathfrak{f}$ and $\mathfrak{p}$ and such that $B \leftrightarrow \mathrm{Gal}(\mathcal{K}_{\mathfrak{f}}/\mathcal{K})$, $\mathfrak{b} \mapsto \mathrm{Rec}_{\mathcal{K}}(\mathfrak{b})$. Let $S \in E[\mathfrak{f}]$ be an $\mathcal{O}_{\mathcal{K}}$-generator. We define another rational function in $\mathcal{K}(E)$ by

$$\Lambda_{\mathfrak{a}}(P) = \prod_{\sigma \in \mathrm{Gal}(\mathcal{K}_{\mathfrak{f}}/\mathcal{K})} \theta_{\mathfrak{a}}(P + \sigma(S)) = \prod_{\mathfrak{b} \in B} \theta_{\mathfrak{a}}(P + \psi(\mathfrak{b}) \cdot S).$$

As $Q_n + \sigma(S)$ is an $\mathcal{O}_{\mathcal{K}}$-generator of $E[\mathfrak{p}^{n+1}\mathfrak{f}]$, it follows that $\Lambda_{\mathfrak{a}}(Q_n)$ is a unit in F_n. Let $u_{\mathfrak{a},n} = \Lambda_{\mathfrak{a}}(Q_n)^{p-1} \in U_n$. Then $u_{\mathfrak{a}} = (u_{\mathfrak{a},n}) \in U$ is a norm-compatible collection of global units.

Via the Laurent series expansions $x = x(t)$ and $y = y(t)$, the function field $\mathcal{K}(E) = \mathcal{K}(x, y)$ is embedded in $K((t))$ as the subfield $K(x(t), y(t))$. Let $\Lambda_{\mathfrak{a},p}(t)$ be the image of $\Lambda_{\mathfrak{a}}$ under this embedding. It is easily checked using the definition of $\Lambda_{\mathfrak{a}}$ and the formula for $x(P + \psi(\mathfrak{b})S)$ in terms of $x(P)$ and $y(P)$ that $\Lambda_{\mathfrak{a},p}(t) \in \mathcal{O}[[t]]^\times$. As $\Lambda_{\mathfrak{a},p}(v_n) = \Lambda_{\mathfrak{a}}(Q_n)$, it follows that

$$g_{v,u_{\mathfrak{a}}}(t) = \Lambda_{\mathfrak{a},p}(t)^{p-1} \in \mathcal{O}[[t]]^\times$$

is the unique power series satisfying $g_{v,u_\mathfrak{a}}(v_n) = u_{\mathfrak{a},n}$ for all $n \geq 0$. Similarly, the map ξ embeds $\mathcal{K}(E)$ in $\mathbb{C}((z))$ as the subfield $\mathcal{K}(\wp(z), \wp'(z))$; let $\Lambda_{\mathfrak{a},\infty}(z) = \Lambda_\mathfrak{a} \circ \xi$. The field $\mathcal{K}(E)$ is stable under the derivations $\frac{1}{\lambda'(t)}\frac{d}{dt}$ of $K((t))$ and $\frac{d}{dz}$ of $\mathbb{C}((z))$ and these are equal on $\mathcal{K}(E)$.

Since E has complex multiplication by $\mathcal{O}_\mathcal{K}$ and $\mathcal{K}$ has class number 1, the lattice Λ is a free $\mathcal{O}_\mathcal{K}$-module of rank one generated by some period Ω_∞. Then the point $S \in E[\mathfrak{f}]$ in the definition of $\Lambda_\mathfrak{a}$ is just $\xi(u)$ for some $u = \Omega_\infty/f$ with f some generator of $\mathfrak{f}$. We then have

$$\delta_k(u_\mathfrak{a}) = (p-1)\left(\frac{1}{\lambda'(t)}\frac{d}{dt}\right)^k \log \Lambda_{\mathfrak{a},p}(T)\Big|_{T=0} = (p-1)\left(\frac{d}{dz}\right)^k \log \Lambda_{\mathfrak{a},\infty}(z)\Big|_{z=0},$$

and a computation (cf. [3, Lem. 21]) shows that the right-hand equals

$$(p-1)12(-1)^k(k-1)!f^k(N(\mathfrak{a})-\psi(\mathfrak{a})^k)\Omega_\infty^{-k}L_\mathfrak{f}(\overline{\psi}^k, k).$$

For $1 \leq i \leq p-1$, let $H_{u_\mathfrak{a},i} \in \widehat{\mathbb{Z}}_p^{\mathrm{ur}}[[T]]$ be the power series associated with $u_\mathfrak{a}$ as in (CW4), and let $\Omega_p = \Omega_F$. Then for any positive integer $k \equiv i \bmod (p-1)$,

$$\begin{aligned} &H_{u_\mathfrak{a},i}(\chi^k(\gamma)-1) \\ &\quad = (p-1)12(-1)^k(k-1)!f^k(N(\mathfrak{a})-\psi(\mathfrak{a})^k)\left(\frac{\Omega_p}{\Omega_\infty}\right)^k(1-\frac{\pi^k}{p})L_\mathfrak{f}(\overline{\psi}^k, k), \end{aligned}$$

where χ is the character as in Sect. 3.1. If $\mathfrak{a}$ is chosen so that $N(\mathfrak{a}) \not\equiv \psi(\mathfrak{a})^i \bmod p$, then there is a unit power series $R_\mathfrak{a} \in \widehat{\mathbb{Z}}_p^{\mathrm{ur}}[[T]]^\times$ such that $G_i = R_\mathfrak{a}^{-1}H_{u_\mathfrak{a},i}$ satisfies

$$\begin{aligned} &G_i(\chi^k(\gamma)-1) \\ &\quad = (-1)^k(k-1)!f^k\left(\frac{\Omega_p}{\Omega_\infty}\right)^k(1-\frac{\psi(\mathfrak{p})^k}{p})L_\mathfrak{f}(\overline{\psi}^k, k), \quad k>0, k \equiv i \bmod (p-1) \end{aligned}$$

(cf. [5, Thm. 18]). This is one of the p-adic L-functions for $\mathcal{K}$ also constructed by Katz.

Let $\mathcal{E} \subset U$ be the subgroup generated by the units $u_\mathfrak{a}$ as $\mathfrak{a}$ varies over all proper integral ideals of $\mathcal{K}$ that are prime to $\mathfrak{f}$ and $\mathfrak{p}$. This is the group of elliptic units. Let $\overline{\mathcal{E}}$ be the closure of $\mathcal{E}$ in U. If $1 < i < p-1$, then $U^{(i)}$ is a free Λ-module of rank 1. The hypothesis that $a_\mathfrak{p}(E) \not\equiv 1 \bmod p$ is equivalent to Φ_0 not containing a primitive pth root of unity, in which case $U^{(1)}$ is also a free Λ-module of rank 1. In particular, the map $u \mapsto H_{u,i}$ is a Λ-module isomorphism $U^{(i)} \otimes_{\mathbb{Z}_p} \widehat{\mathbb{Z}}_p^{\mathrm{ur}}[[T]] \xrightarrow{\sim} \widehat{\mathbb{Z}}_p^{\mathrm{ur}}[[T]]$.

Theorem 8 ([5, Thm. 22]) *Let $1 \leq i < p-1$, and suppose $a_\mathfrak{p}(E) \not\equiv 1 \bmod p$ if $i = 1$. There is a Λ-module isomorphism $(U/\overline{\mathcal{E}})^{(i)} = U^{(i)}/\overline{\mathcal{E}}^{(i)} \cong \Lambda/(\mathcal{G}_i)$ with $\mathcal{G}_i(T) \in \Lambda$ such that $\mathcal{G}_i$ generates the ideal $G_i \cdot \widehat{\mathbb{Z}}_p^{\mathrm{ur}}[[T]]$ in $\widehat{\mathbb{Z}}_p^{\mathrm{ur}}[[T]] = \Lambda\widehat{\otimes}_{\mathbb{Z}_p}\widehat{\mathbb{Z}}_p^{\mathrm{ur}}$.*

One important consequence of this theorem is:

Corollary 9 *Suppose* $a_{\mathfrak{p}}(E) \not\equiv 1 \bmod p$. *If there exists a non-zero* Λ-*homomorphism* $U/\overline{\mathcal{E}} \to \mathbb{Z}_p(\chi)$, *then* $L_{\mathfrak{f}}(\overline{\psi}, 1) = 0$.

We explain how the corollary follows from the theorem: A Λ-homomorphism $U/\overline{\mathcal{E}} \to \mathbb{Z}_p(\chi)$ must factor through the projection to $(U/\overline{\mathcal{E}})^{(1)}$ and even to

$$(U/\overline{\mathcal{E}})^{(1)}/(\gamma - \chi(\gamma))(U/\overline{\mathcal{E}})^{(1)} \cong \Lambda/(\mathcal{G}_1, \gamma - \chi(\gamma)) \cong \mathbb{Z}_p/(\mathcal{G}_1(\chi(\gamma) - 1)),$$

where the first isomorphism follows from the theorem. If the homomorphism is non-zero, then it must be that $\mathcal{G}_1(\chi(\gamma) - 1) = 0$ and hence that $0 = G_1(\chi(\gamma) - 1) = -f(\Omega_p/\Omega_\infty)(1 - \psi(\mathfrak{p})/p)L_{\mathfrak{f}}(\overline{\psi}, 1)$. Since $\pi = \psi(\mathfrak{p}) \neq p$, it must be that $L_{\mathfrak{f}}(\overline{\psi}, 1) = 0$.

Remark 10 In [3], Coates and Wiles proved a preliminary version of this result that is very much in the spirit of Kummer's theorem for cyclotomic units (see the remark at the end of Sect. 3.3). In their proof, which shows that if $(U_0/\overline{\mathcal{E}}_0)^{(1)}$ is non-zero, where $\mathcal{E}_0$ is the subgroup generated by the $u_{\mathfrak{a},0}$, then $\mathfrak{p} \mid L_{\mathfrak{f}}(\overline{\psi}, 1)/\Omega_\infty$, the role of the homomorphisms δ_k is played by the Kummer homomorphisms φ_k (see the remark at the end of Sect. 3.1).

Remark 11 The hypothesis that $a_{\mathfrak{p}}(E) \not\equiv 1 \bmod p$ is only used to conclude that Φ_0 has no non-trivial pth-root of unity and so $U^{(1)}$ is a free Λ-module of rank 1. A closer analysis of $(U/\overline{\mathcal{E}})^{(1)}$ can be made, along the lines that Iwasawa did for cyclotomic extensions, that yields a similar conclusion as in the corollary in the case $i = 1$ without the hypothesis on $a_{\mathfrak{p}}(E)$ modulo p. This hypothesis, however, is needed for the result of Coates and Wiles described in the preceding remark: it ensures that $\mathfrak{p} \nmid (1 - \frac{\pi}{p})$.

3.5 Further Developments

Shortly after this work of Coates and Wiles, Coleman [38] proved a generalization of the lemma in Sect. 3.1. In the context of the cyclotomic units, this yields a map $U \to \mathbb{Z}_p[[G]]$, $u \mapsto \mu_u$ where μ_u is the measure in the theorem in Sect. 3.2, that is almost an isomorphism (see the end of Sect. 3.2). In general this is referred to as 'Coleman's isomorphism.' This was generalized in Perrin-Riou [69, 70] and Kato [55, 56] to maps on norm compatible collections of local cohomology groups of a p-adic Galois representation over a p-adic field K:

$$Exp^* : H^1_{\mathrm{Iw}}(K, V) = \varprojlim_n H^1(K(\zeta_{p^{n+1}}), V) = H^1(K, \Lambda_K \otimes V) \to D(V),$$

where $\Lambda_K = \mathbb{Z}_p[[\Gamma_K]]$, with $\Gamma_K = \mathrm{Gal}(K(\zeta_{p^\infty})/K)$, and $D(V)$ is the (φ, Γ_K)-module associated to the p-adic representation V. The Coates–Wiles homomorphisms become a very explicit expression for how these maps interpolate the Bloch–Kato exponential. If $V = \mathbb{Q}_p(1)$, then this recovers the Coates–Wiles homomorphisms for cyclotomic units. Recent work of Schneider and Venjakob [80] have attempted to define a parallel theory with the cyclotomic extension $K(\zeta_{p^\infty})/K$ replaced with the local Coates–Wiles extensions for elliptic curves with complex multiplication. These generalizations of the Coleman isomorphism and the Coates–Wiles homomorphisms have played an important role in recent progress on the Iwasawa theory of elliptic curves and modular forms, especially [57] and [61].

4 The Birch–Swinnerton-Dyer Conjecture

Elliptic curves are a seemingly inexhaustible source of intriguing problems for the number-theorist. As witness to the attraction of the arithmetic of elliptic curves, in his introduction to a volume of papers dedicated to John Coates on the occasion of Coates's 60th birthday Wiles wrote [24]:

> Needless to say for those who have devoted some time to this subject, it is so full of fascinating problems that it is hard to turn from this to anything else. The conjecture of Birch and Swinnerton-Dyer. . . had made the old subject irresistible.

Wiles surely knew of what he wrote. While a graduate student at the University of Cambridge, Wiles – together with John Coates – made the first progress toward the Birch–Swinnerton-Dyer Conjecture that went beyond computational examples. And if not always front and center, elliptic curves have never been out of sight in all of Wiles's mathematical works.

In the following we recall the Birch–Swinnerton-Dyer Conjecture and the Coates–Wiles Theorem. We include a brief overview of their proof of this theorem, especially describing the central role played by the Coates–Wiles homomorphisms.

4.1 Elliptic Curves and the BSD Conjecture

Let E be an elliptic curve over F, that is, a smooth, proper, geometrically connected curve of genus one together with a distinguished F-rational point. The curve E has the structure of a commutative algebraic group over F with the distinguished point being the identity element. In particular, the set $E(F)$ of F-rational points on E has the structure of an abelian group. A fundamental result of Mordell, extended by Weil, is that

$$E(F) \text{ is a finitely generated abelian group.}$$

Let

$$r(F) = \mathrm{rank}_{\mathbb{Z}} E(F)$$

be the rank of this group.

In the late 1950s Birch and Swinnerton-Dyer, led in part by computational evidence, made a remarkable conjecture about $r(F)$:

Conjecture 12 (The Birch–Swinnerton-Dyer Conjecture) *Let E be an elliptic curve over a number field F. Then*

$$r(F) = \mathrm{ord}_{s=1} L(E/F, s).$$

Here $L(E/F, s)$ is the Hasse–Weil L-function of E over F. This is defined first as an Euler product

$$L(E/F, s) = \prod_{v} L_v(E/F; q_v^{-s})^{-1},$$

where v runs over the finite places of F, q_v is the order of the residue field $\mathbb{F}_v$ at v, and $L_v(E/F; X) \in \mathbb{Z}[X]$ is a polynomial of degree at most 2. For all but finitely many places v, $L_v(E/F; X) = 1 - a_v(E)X + q_v X^2$ with $1 - a_v(E) + q_v$ being the number of points on E over the residue field $\mathbb{F}_v$. It follows from the Riemann hypothesis for E over the $\mathbb{F}_v$'s that this product converges absolutely for $\mathrm{Re}(s) > \frac{3}{2}$ and so defines a holomorphic function on this half-plane. At the time that Birch and Swinnerton-Dyer formulated their conjecture, it was an accepted conjecture – though far from proved ! – that each $L(E/F, s)$ has an analytic continuation to the entire complex plane. However, it was known by results of Deuring from the 1950s that this was true for elliptic curves with complex multiplication. Such curves include those with Weierstrass models of the form $y^2 = x^3 - Dx$, $D \in \mathbb{Q}^\times$, and hence those curves that were originally studied by Birch and Swinnerton-Dyer.

Proving the Birch–Swinnerton-Dyer Conjecture is one of the Clay Mathematics Institute's seven *Millennium Prize Problems* announced in 2000. The official problem description was written by Andrew Wiles and later published as [23]. His description of the problem includes additional context and history.

4.2 The Coates–Wiles Theorem

In 1977 Coates and Wiles published the first theoretical evidence toward the Birch–Swinnerton-Dyer Conjecture:

Theorem 13 ([3, Thm. 1]) *Let E be an elliptic curve with complex multiplication by the maximal order of an imaginary quadratic field $\mathcal{K}$ with class number one.*

Let $F = \mathbb{Q}$ or $\mathcal{K}$ and suppose E is defined over F. If $E(F)$ is infinite, then $L(E/F, 1) = 0$.

Another way of stating the conclusion: $r(F) > 0 \implies \mathrm{ord}_{s=1} L(E/F, s) > 0$.

Remark 14 The hypothesis that E has complex multiplication by a *maximal* order of $\mathcal{K}$ is not essential. If E has complex multiplication by an order in $\mathcal{K}$, then E is isogeneous over $\mathcal{K}$ to an elliptic curve with complex multiplication by a maximal order. Both the rank of $E(\mathcal{K})$ and the L-function $L(E/\mathcal{K}, s)$ are invariants of the $\mathcal{K}$-isogeny class of E, so the general case reduces to the case of multiplication by the maximal order, at least when $F = \mathcal{K}$. If $F = \mathbb{Q}$, then $E(\mathbb{Q})$ being infinite implies $E(\mathcal{K})$ is infinite, and $L(E/\mathcal{K}, s) = L(E/\mathbb{Q}, s)^2$, so this case reduces to the case over $\mathcal{K}$.

4.3 Idea of the Proof: Bird's Eye View

As indicated in the preceding remark, it suffices to prove the theorem in the case $F = \mathcal{K}$. Here is a bird's eye view of the proof in [3] for this case: Let p be a prime of good reduction for E that splits in $\mathcal{K}$: $p = \mathfrak{p}\bar{\mathfrak{p}}$. Starting with a point $P \in E(\mathcal{K})$ of infinite order, Coates and Wiles construct an abelien extension L_n of $F_n = K[E[\mathfrak{p}^{n+1}]]$. For p suitably nice, using class field theory and the arithmetic of the elliptic curve, they deduce from the existence of the L_n that the groups $(U_0/\mathcal{E}_0)^{(1)}$ are non-zero, and from the connection of these groups with the L-function of the Hecke character ψ of $\mathcal{K}$ associated with E they then conclude that $L(\bar{\psi}, 1)/\Omega_\infty$ is divisible by $\mathfrak{p}$ (see the remark at the end of Sect. 3.4). As there are infinitely many suitably nice p, it must be that $L(\bar{\psi}, 1) = 0$. As $L(E/\mathcal{K}, 1) = L(\psi, 1)L(\bar{\psi}, 1)$, the theorem follows.

For the interested reader, in the next section we provide more details for a slight variation of the above argument. This proof makes use of the later developments in the paper [5], which were described in Sect. 3.1 and particularly in the example in Sect. 3.4. For even more details about the proofs of the Coates–Wiles theorem as well as these other results, the interested reader should consult Rubin's notes [77].

Remark 15 In an earlier work [1] Coates and Wiles used similar arguments to prove an analog of Kummer's criterion. The latter states that the class number of $\mathbb{Q}(\zeta_p)$ is divisible by p (p an odd prime) if and only if at least one of the numbers $\zeta^*(k) = (k-1)!(2\pi)^{-k}\zeta(k) = (-1)^{1+k/2}B_k/2k$, $0 < k < p-1$ an even integer, is divisible by p. Let w be the number of roots of unity in $\mathcal{K}$, and let p be an odd prime of good reduction for E such that p splits in $\mathcal{K}$: $p = \mathfrak{p}\bar{\mathfrak{p}}$. Coates and Wiles proved [1, Thm. 1] that for p suitably nice there exists a $\mathbb{Z}/p\mathbb{Z}$ extension of the ray class field $\mathcal{K}[\mathfrak{p}]$ of $\mathcal{K}$ of modulus $\mathfrak{p}$ that is unramified away from the prime above $\mathfrak{p}$ and distinct from the ray class field $\mathcal{K}[\mathfrak{p}^2]$ of modulus $\mathfrak{p}^2$ if and only if one of the numbers

$$L^*(\psi^k, k) = w(k-1)!\Omega_\infty^{-k} L(\psi^k, k), \quad 1 \le k \le p-1, \ k \equiv 0 \bmod w,$$

is divisible by p. The proof described above is essentially a refinement of this criterion in the spirit of Herbrand's theorem.

4.4 A Variation on the Proof in [3]

We begin by choosing a nice prime p. Let p be a prime such that

(p1) $p > 5$,
(p2) p splits in $\mathcal{K}$: $p = \mathfrak{p}\bar{\mathfrak{p}}$,
(p3) E has good reduction at both $\mathfrak{p}$ and $\mathfrak{p}$,
(p4) $a_{\mathfrak{p}}(E) \not\equiv 1 \bmod p$.

These are the primes satisfying the conditions imposed in Sect. 3.4. There are infinitely many such primes: The primes satisfying (p2) are a set of density $\frac{1}{2}$, and (p1) and (p3) exclude only finitely many of these. Suppose E has good reduction at $\mathfrak{p}$. If $a_{\mathfrak{p}}(E) \equiv 1 \bmod p$, then, since $p > 5$, the Riemann hypothesis for the reduction of E mod $\mathfrak{p}$ forces $a_{\mathfrak{p}}(E) = 1$. Since E has complex multiplication by $\mathcal{K}$, the roots of $x^2 - x + p$ must then belong to $\mathcal{K}$. In particular, $1 - 4p = -Da^2$ for some integer a, where $-D$ is the discriminant of $\mathcal{K}$. Let q be an odd prime not dividing D. Then $1 + Da^2$ can belong to only half of the non-zero residue classes modulo q. So if we additionally require that $4p$ not belong to one of these residue classes, then (p4) will also be satisfied. Therefore, there is a set of primes of density $\frac{1}{2} \times \frac{2}{q-1} = \frac{1}{q-1}$ such that (p1)–(p4) hold.

Let $K = \mathcal{K}_{\mathfrak{p}} = \mathbb{Q}_p$ and $\mathcal{O} = \mathcal{O}_{\mathcal{K}_{\mathfrak{p}}} = \mathbb{Z}_p$. As noted in Sect. 3.4, since p satisfies (p2) and (p3) the formal group law $\hat{E}$ of E over $\mathcal{O}$ is a Lubin–Tate formal group law associated with some $f(X) = \pi X +$ (higher order terms) with π a generator of $\mathfrak{p}$ such that the reduction modulo $\mathfrak{p}$ of $E \xrightarrow{\times\pi} E$ is just the Frobenius homomorphism. Under the identification of $\hat{E}(\mathfrak{m}_{\overline{K}})$ with the kernel $E_1(\overline{K})$ of the mod $\mathfrak{p}$ reduction map, the $\mathcal{O}$-module M_n is identified ($\mathrm{Gal}(\overline{K}/K)$-equivariantly) with $E[\pi^{n+1}]$. For $n \geq 0$, let $F_n = \mathcal{K}(E[\pi^{n+1}])$. Then $F_n/\mathcal{K}$ is totally ramified at the prime $\mathfrak{p}$ and the completion of F_n at the unique prime above $\mathfrak{p}$ is just the Lubin–Tate extension Φ_n of K. Let $F_\infty = \cup F_n$ and $\Phi_\infty = \cup\Phi_n$. Then $G = \mathrm{Gal}(\Phi_\infty/K) \xrightarrow{\sim} \mathrm{Gal}(F_\infty/\mathcal{K})$. The π-adic Tate module $T_\pi E$ of E is a free $\mathcal{O}$-module of rank one, and so $\mathrm{Gal}(F_\infty/\mathcal{K})$ acts on $T_\pi E$ via a $\mathcal{O}^\times$ valued character. As a character of G this is just the character denoted χ in Sect. 3.1. It is also the $\mathfrak{p}$-adic Galois character of the Hecke character ψ of $\mathcal{K}$ associated with E by the theory of complex multiplication.

Let $P \in E(\mathcal{K})$ be a point of infinite order. For each integer $n \geq 0$, choose a point $P_n \in E(\overline{\mathcal{K}})$ such that $\pi^{n+1}P_n = P$. Let $L_n = F_n[P_n]$. Then L_n is an abelian extension of F_n of p-power order and a Galois extension of $\mathcal{K}$. The extension L_n/F_n is unramified at all primes not dividing $\mathfrak{p}$, and, if n is sufficiently large, it is non-trivial and ramified at the unique prime of F_n above $\mathfrak{p}$ (see [3, Lems. 33, 35]). Let $L_\infty = \cup L_n$ and let $\mathfrak{X}_\infty = \mathrm{Gal}(L_\infty/F_\infty)$. The group G acts on $\mathfrak{X}_\infty$ by conjugation, and $\mathfrak{X}_\infty$ admits a G-invariant, injective homomorphism $\phi : \mathfrak{X}_\infty \hookrightarrow T_\pi E$ given by

$\phi(\tau) = (\tau(P_n) - P_n) \in T_\pi E$. In particular, $\mathfrak{X}_\infty$ is a free $\mathbb{Z}_p$-module of rank one and for $\sigma \in G$ we have $\sigma\tau\sigma^{-1} = \chi(\sigma)\tau$.

Let $\mathfrak{U}_n = \mathcal{O}_{F_n}^\times \cap U_n$ be the group of units in the ring of integers of F_n that are congruent to 1 modulo the unique prime above $\mathfrak{p}$. Let $\overline{\mathfrak{U}}_n$ be its closure in U_n. Since L_n/F_n is unramified outside the unique prime over $\mathfrak{p}$ and is an abelian extension of p-power order, the reciprocity map of global class field theory induces a homomorphism $U_n/\overline{\mathfrak{U}}_n \xrightarrow{\mathrm{Rec}_{F_n}} \mathrm{Gal}(L_n/F_n)$ whose image is the inertia group at the prime above $\mathfrak{p}$ (which is non-trival if n is sufficiently large). Let $\overline{\mathfrak{U}} \subset U$ be the inverse limit $\overline{\mathfrak{U}} = \varprojlim \overline{\mathfrak{U}}_n$ with respect to the norm maps. Then

$$\varphi = \varprojlim_n \mathrm{Rec}_{F_n} : U/\overline{\mathfrak{U}} \to \mathfrak{X}_\infty = \varprojlim \mathrm{Gal}(L_n/F_n)$$

is a G-equivariant homomorphism with non-trivial image. As the group $\overline{\mathcal{E}}$ of elliptic units is contained in $\overline{\mathfrak{U}}$, upon fixing a $\mathbb{Z}_p$-basis of $\mathfrak{X}_\infty$ the homomorphism φ induces a non-zero G-equivariant homomorphism $U/\overline{\mathcal{E}} \to \mathbb{Z}_p(\chi)$. As explained in Sect. 3.4, it follows from Coates's and Wiles's analysis of the structure of $U/\overline{\mathcal{E}}$ in [5] that the existence of such a homomorphism implies that $L(\overline{\psi}, 1) = 0$ and hence $L(E/\mathcal{K}, 1) = 0$; see the Corollary near the end of Sect. 3.4.

Remark 16 Property (p4) is only included for convenience, and the theorem can be proved without it (see the remark at the end of Sect. 3.4). This is contained in Rubin's extension [74] of the Coates–Wiles theorem.

4.5 Further Developments

Extensions of the Coates–Wiles theorem followed quickly after the publication of [3]. Arthaud, in [28] and her PhD thesis, extended the theorem to the situation where F is an abelian extension of $\mathcal{K}$ such that $F(E_{\mathrm{tor}})$, the field obtained by adjoining to F the coordinates of all the torsion points of E, is also abelian over $\mathcal{K}$. Rubin [74] proved a strengthening of Arthaud's result, which included the following consequence: Suppose $F = \mathcal{K}$ and $M/\mathcal{K}$ is any finite abelian extension. If ξ is a character of $\mathrm{Gal}(M/\mathcal{K})$ such that the ξ-isotypical subspace of $E(M) \otimes \mathbb{C}$ is non-zero, then $L(\overline{\psi}\xi, 1) = 0$. Here we have identified the Galois character ξ with a Hecke character of $\mathcal{K}$ in the usual manner via class field theory. Rubin and Wiles [7] combined this result with results about p-parts of class groups of $\mathbb{Z}_{\ell_1} \times \cdots \times \mathbb{Z}_{\ell_s}$-extensions ($\ell_i \neq p$) and congruences between L-values of Hecke characters and Bernoulli numbers to prove: Suppose $\mathcal{K} = \mathbb{Q}(\sqrt{-p})$, p an odd prime, and $E_{/\mathbb{Q}}$ has complex multiplication by an order in $\mathcal{K}$. Let $N \geq 1$ be any integer such that $p \nmid N$. Then the Mordell–Weil group $E(\mathcal{K} \cdot \mathbb{Q}(\zeta_{N^\infty})^+)$ is finitely-generated. Greenberg [51] proved a partial converse to the Coates–Wiles theorem, showing that if $\mathrm{ord}_{s=1} L(E, s)$ is odd, then either $E(\mathbb{Q})$ is infinite or the Tate–Shafarevich group of E is infinite. Rubin [75] combined many of the ingredients from the proof of the Coates–Wiles theorem with ideas of Thaine to provide the first examples of (CM) elliptic curves with proven finite Tate–Shafarevich groups. These ideas eventually led to a proof of the Iwasawa Main Conjectures for imaginary quadratic fields [76] and for CM elliptic curves (at ordinary primes). Rubin and Pollack [73] later proved the Main Conjecture for CM curves at supersingular primes. These results encompass the Coates–Wiles theorem and even establish the p-part of the conjectured formula of Birch and Swinnerton-Dyer when $L(E, 1) \neq 0$.

5 The Main Conjecture

Most of Wiles's early works were closely intertwined with Iwasawa theory. Iwasawa's legacy is clearly visible in Wiles's higher explicit reciprocity laws and especially in the applications of the Coates–Wiles homomorphisms to p-adic L-functions and the arithmetic of $\mathbb{Z}_p$-extensions of imaginary quadratic fields (including the Coates–Wiles Theorem). It seems fitting that Wiles would then go on to prove Iwasawa's Main Conjecture, first for $\mathbb{Q}$ in joint work with Barry Mazur, and then for all totally real fields.

In the following we recall some of the main features of Iwasawa's Main Conjecture, including some of its arithmetic consequences. We then briefly describe Mazur's and Wiles's proof of the Main Conjecture for $\mathbb{Q}$. We follow this with a description of Wiles's *second* proof of the Main Conjecture, which was the template for his proof of the Main Conjecture for totally real fields. It was also the prototype for subsequent proofs of other 'main conjectures,' including for most elliptic curves over $\mathbb{Q}$.

5.1 *Iwasawa's Main Conjecture*

Let p be a fixed prime. For simplicity we consider only the case $p > 2$ in all that follows.

In order to view algebraic numbers as both complex numbers and p-adic numbers, we fix embeddings $\overline{\mathbb{Q}} \hookrightarrow \mathbb{C}$ and $\overline{\mathbb{Q}} \hookrightarrow \overline{\mathbb{Q}}_p$. The latter embedding identifies $G_{\mathbb{Q}_p} = \mathrm{Gal}(\overline{\mathbb{Q}}_p/\mathbb{Q}_p)$ as a subgroup of $G_{\mathbb{Q}} = \mathrm{Gal}(\overline{\mathbb{Q}}/\mathbb{Q})$.

For each $n \geq 1$ let ζ_{p^n} be a primitive p^nth root of unity. The field $\mathbb{Q}(\zeta_{p^{n+1}})$ is a Galois extension of $\mathbb{Q}$ with Galois group $G_n = \mathrm{Gal}(\mathbb{Q}(\zeta_{p^{n+1}})/\mathbb{Q})$ canonically isomorphic to $(\mathbb{Z}/p^{n+1}\mathbb{Z})^\times$, the isomorphism being $\sigma \mapsto a \bmod p^n$ for $\sigma(\zeta_{p^n}) = \zeta_{p^n}^a$. Put $\mathbb{Q}(\zeta_{p^\infty}) = \cup_{n=0}^\infty \mathbb{Q}(\zeta_{p^{n+1}})$. Then there is a canonical isomorphism of profinite groups:

$$G = \mathrm{Gal}(\mathbb{Q}(\zeta_{p^\infty})/\mathbb{Q}) = \varprojlim_n G_n \xrightarrow{\sim} \varprojlim_n (\mathbb{Z}/p^{n+1}\mathbb{Z})^\times = \mathbb{Z}_p^\times.$$

The resulting character

$$\varepsilon : \mathrm{Gal}(\overline{\mathbb{Q}}/\mathbb{Q}) \twoheadrightarrow \mathrm{Gal}(\mathbb{Q}(\zeta_{p^\infty})/\mathbb{Q}) = G \xrightarrow{\sim} \mathbb{Z}_p^\times$$

is the p-adic cyclotomic character. Corresponding to the decomposition $\mathbb{Z}_p^\times = \mu_{p-1} \times 1 + p\mathbb{Z}_p$ we write $\varepsilon(\sigma) = \omega(\sigma)\langle\sigma\rangle$. In particular, $\omega : G_{\mathbb{Q}} \twoheadrightarrow \mu_{p-1}$ is a surjective character.

The groups G_n decompose compatibly as $G_n = \Delta \times \Gamma_n$ with Δ cyclic of order $p-1$ and Γ_n cyclic of order p^n. The subgroup Δ projects isomorphically onto $\mathrm{Gal}(\mathbb{Q}(\zeta_p)/\mathbb{Q}) \xrightarrow{\sim} (\mathbb{Z}/p\mathbb{Z})^\times$. The Galois group G then decomposes as $G = \Delta \times \Gamma$ with $\Gamma = \varprojlim_n \Gamma_n$. The group Γ is a cyclic pro-p-group and identified with $1+p\mathbb{Z}_p$. A convenient topological generator of Γ is the element $\gamma = 1 + p$ (since $p > 2$, $1 + p\mathbb{Z}_p = (1+p)^{\mathbb{Z}_p}$). The choice of a generator γ determines an isomorphism of Γ with $\mathbb{Z}_p$ and of Γ_n with $\mathbb{Z}/p^n\mathbb{Z}$.

The character ω factors as

$$\omega : G_{\mathbb{Q}} \twoheadrightarrow \Delta \xrightarrow{\sim} \mathrm{Gal}(\mathbb{Q}(\zeta_p)/\mathbb{Q}) \xrightarrow{\sim} \mu_{p-1},$$

and so can be viewed as a character of Δ. In particular, every character of Δ is just a power of ω.

The cyclotomic $\mathbb{Z}_p$-extension of $\mathbb{Q}$ is the field $\mathbb{Q}_\infty = \mathbb{Q}(\zeta_{p^\infty})^\Delta$. The Galois group $\mathrm{Gal}(\mathbb{Q}_\infty/\mathbb{Q})$ is canonically identified with Γ and so with $\mathbb{Z}_p$ (hence the terminology). The field $\mathbb{Q}_\infty$ is the union of the fields $\mathbb{Q}_n = \mathbb{Q}(\zeta_{p^{n+1}})^\Delta$, each of which is Galois over $\mathbb{Q}$ with Galois group $\mathrm{Gal}(\mathbb{Q}_n/\mathbb{Q}) = \Gamma_n \xrightarrow{\sim} \mathbb{Z}/p^n\mathbb{Z}$.

For any number field F we write F_∞ for the composite field $F \cdot \mathbb{Q}_\infty$. This is the cyclotomic $\mathbb{Z}_p$-extension of F.

The Main Conjecture of Iwasawa Theory for Totally Real Fields

Let F be a totally real number field and K/F a cyclic Galois extension with K a CM field such that $K \cap F_\infty = F$. Let $\chi : \mathrm{Gal}(K/F) \to \overline{\mathbb{Q}}_p^\times$ be an odd character (that is, χ is non-trivial on complex conjugation). Let $H = K(\zeta_p)$. The Galois group $\mathrm{Gal}(H_\infty/F)$ admits a decomposition

$$\mathrm{Gal}(H_\infty/F) \xrightarrow{\sim} \mathrm{Gal}(H/F) \times \mathrm{Gal}(F_\infty/F) = \Delta_F \times \Gamma_F,$$

where the map is the natural projection (restriction of the Galois action) to each factor. In particular, $\Delta_F = \mathrm{Gal}(H/F)$ is a finite group, and χ can be viewed as a character of Δ_F. Note that Δ_F projects canonically to $\mathrm{Gal}(\mathbb{Q}(\zeta_p)/\mathbb{Q}) \cong \Delta$, so any character of Δ (such as ω) determines a character of Δ_F. The group Γ_F is identified with a closed subgroup of $\Gamma = \mathrm{Gal}(\mathbb{Q}_\infty/\mathbb{Q})$ of finite-index via the canonical projection, so Γ_F is isomorphic to $\mathbb{Z}_p$. Let $\gamma_F \in \Gamma_F$ be a fixed topological generator. The Main Conjecture identifies two polynomials associated with χ and γ_F, one having an algebraic origin and the other an analytic origin.

The algebraic polynomial is defined as follows. Let $X = \mathrm{Gal}(L_\infty/H_\infty)$ with L_∞/H_∞ being the maximal unramified abelian pro-p-extension of H_∞. Let $V = X \otimes_{\mathbb{Z}_p} \overline{\mathbb{Q}}_p$. Iwasawa proved that V is a finite-dimensional $\overline{\mathbb{Q}}_p$-vector space. Let V^χ be the χ-isotopical subspace, that is, the subspace on which the natural action of Δ_F is via the character χ. The group Γ_F acts naturally on V^χ, and we let $f_\chi(T) \in \overline{\mathbb{Q}}_p[T]$ be the characteristic polynomial of the action of $\gamma_F - 1$ on V^χ.

The analytic polynomial arises from the p-adic L-function $L_p(\psi, s)$ associated to the even character $\psi = \chi^{-1}\omega$ of Δ_F. The p-adic L-functions for such characters are generalizations of the Kubota–Leopoldt p-adic L-functions for even Dirichlet characters. In this generality they were first constructed by Deligne and Ribet. Let $\mathcal{O}_\psi$ be the ring of integers of the finite extension of $\mathbb{Q}_p$ obtained by adjoining the values of ψ. Let $H_\psi(T) = \varepsilon(\gamma_F)(1+T)-1$ if $\psi = 1$ (i.e., $\chi = \omega$) and otherwise let $H_\psi(T) = 1$. Deligne and Ribet proved that there is a power series $G_\psi(T) \in \mathcal{O}_\psi[\![T]\!]$ such that

$$L_p(\psi, 1-s) = \frac{G_\psi(u^s - 1)}{H_\psi(u^s - 1)}, \quad u = \varepsilon(\gamma_F).$$

In particular, if $n \geq 1$ is any positive integer, then

$$L^{(p)}(\omega^{-n}\psi, 1-n) = \frac{G_\psi(u^n - 1)}{H_\psi(u^n - 1)},$$

where the left-hand side is the value at $s = 1-n$ of the usual Hecke L-function for the character $\omega^{-n}\psi$ (the superscript '(p)' denotes the omission of Euler factors at the primes above p).

Let $\varpi \in \mathcal{O}_\psi$ be a uniformizer. The Weierstrass preparation theorem for $\mathcal{O}_\psi[\![T]\!]$ asserts that any power series $P(T) \in \mathcal{O}_\psi[\![T]\!]$ can be uniquely expressed as $F(T) = \varpi^\mu \cdot f(T) \cdot u(T)$ with $\mu \geq 0$ an integer, $f(T)$ a monic polynomial such that $f(T) \equiv T^{\deg(f)} \bmod \varpi$, and $u \in \mathcal{O}_\psi[\![T]\!]^\times$ a unit. The exponent μ (the 'μ-invariant') and the polynomial $f(T)$ (the 'Weierstrass polynomial of $F(T)$') do not depend on the choice of ϖ. Let $g_\psi(T) \in \mathcal{O}_\psi[T]$ be the Weierstrass polynomial of the power series $G_\psi(T)$.

Conjecture 17 (The Main Conjecture of Iwasawa Theory for F) *For K/F and χ as above and for $u = \varepsilon(\gamma_F)$,*

$$f_\chi(T) = g_{\chi^{-1}\omega}(u(1+T)^{-1} - 1).$$

That is, the eigenvalues of the action of $\gamma_F - 1$ on V^χ are exactly the zeros of $G_{\chi^{-1}\omega}(u(1+T)^{-1} - 1)$, when both are counted with multiplicity.

For $F = \mathbb{Q}$ this conjecture was made by Iwasawa. The extension of the conjecture to totally real fields was made by Coates, Greenberg, and Oesterlé. The case $F = \mathbb{Q}$ of this conjecture was proved by Mazur and Wiles [9]. The general conjecture was proved by Wiles [13] following partial results in [11].

The Main Conjecture with μ-Invariant

If χ has order prime to p then there is an a priori stronger version of the Main Conjecture. We suppose that the degree of K/F is prime to p (for example, K could be the splitting field of χ). Then $\Delta_F \cong \mathrm{Gal}(K(\zeta_p)/F)$ also has order prime to p, and we can define an idempotent $e_\chi = \frac{1}{\#\Delta_F}\sum_{\sigma\in\Delta_F}\chi^{-1}(\sigma)\sigma \in \mathcal{O}_\chi[\Delta_F]$. We set $X^\chi = e_\chi(X\otimes_{\mathbb{Z}_p}\mathcal{O}_\chi)$. Iwasawa proved that that under the natural action of Γ_F on X, X is a finitely-generated torsion $\mathbb{Z}_p[\![\Gamma_F]\!]$-module. Hence X^χ is a finitely-generated torsion Λ_χ-module, where $\Lambda_\chi = \mathcal{O}_\chi[\![\Gamma_F]\!]$. The choice of the generator $\gamma_F \in \Gamma_F$ identifies Λ_χ with the power-series ring $\mathcal{O}_\chi[\![T]\!]$ via $\gamma_F \mapsto 1+T$.

Finitely-generated Λ_χ-modules have a structure theory reminiscent of that of finitely-generated modules over a PID. In particular, X^χ admits a Λ_χ-homomorphism to a product of cyclic Λ_χ-modules

$$X^\chi \to \prod_{i=1}^{r_\chi} \Lambda_\chi/(f_{\chi,i}) \tag{1}$$

with finite-order kernel and cokernel. The ideal

$$\mathrm{char}_{\Lambda_\chi}(X^\chi) = \left(\prod_{i=1}^{r_\chi} f_{\chi,i}\right) \subset \Lambda_\chi$$

is well-defined and called the characteristic ideal of X^χ.

Conjecture 18 (The Main Conjecture of Iwasawa Theory for F (with μ-invariant)) *For K/F and χ as above with K/F of degree prime to p, and for $u = \varepsilon(\gamma)$,*

$$\mathrm{char}_{\Lambda_\chi}(X^\chi) = (G_{\chi^{-1}\omega}(u(1+T)^{-1}-1)).$$

For $F = \mathbb{Q}$ this is a consequence of the work of Mazur and Wiles [9]. For general totally real F, Wiles also proved this conjecture in [13].

Let $\varpi \in \mathcal{O}_\chi$ be a uniformizer. Then there is an integer $\mu_\chi \geq 0$ (the μ-invariant) such that $\mathrm{char}_{\Lambda_\chi}(X^\chi) = (\varpi^{\mu_\chi} f_\chi(T))$. Similarly, by Weierstrass preparation, $G_\psi(T) = \varpi^{\mu_\psi} g_\psi(T)u(T)$ with $u(T) \in \Lambda_\chi^\times$. So the conjecture '(with μ-invariant)' strengthens the previous one by asserting that in addition to $f_\chi(T) = g_{\chi^{-1}\omega}(u(1+T)^{-1}-1)$, when K/F has degree prime to p one has $\mu_\chi = \mu_{\chi^{-1}\omega}$. It has been conjectured that both μ_χ and $\mu_{\chi^{-1}\omega}$ are always zero, but this has only been proved for $F = \mathbb{Q}$ (first by Ferraro and Washington).

A Refined Class Number Formula

One important consequence of the proof of the Main Conjecture is a refined class number formula for certain Hecke L-series. We explain this in a special case.

We now take $F = \mathbb{Q}$, $K = \mathbb{Q}(\zeta_p)$, and $\chi = \omega^i$ for some odd integer $1 < i < p-1$. Then $H = K$ and $H_\infty = \mathbb{Q}(\zeta_{p^\infty})$, so $\mathrm{Gal}(H_\infty/F) = G = \Delta \times \Gamma$. The character $\psi = \chi^{-1}\omega$ is just ω^j for the even integer $0 < j < p-1$ such that $j \equiv 1 - i \bmod p-1$. Also, $\Lambda_\chi = \mathbb{Z}_p[\![T]\!] = \Lambda$. The power series $G_{\omega^j}(T)$ is just the power series $G_j(T)$ that appeared in the Example 3.3.

Let $X_n = \mathrm{Gal}(L_n/\mathbb{Q}(\zeta_{p^{n+1}}))$ be the Galois group of the maximal unramified abelian pro-p-extension L_n of $\mathbb{Q}(\zeta_{p^{n+1}})$. By class field theory there is a G_n-equivariant isomorphism of X_n with the p-part A_n of the class group of $\mathbb{Q}(\zeta_{p^{n+1}})$. Consequently, there are isomorphisms $X_n^\chi = e_\chi(X_n) \cong e_\chi(A_n) = A_n^\chi$. Iwasawa proved that $X^\chi/(\gamma^{p^n} - 1)X^\chi \xrightarrow{\sim} X_n^\chi \cong A_n^\chi$, and that X^χ has no non-zero finite-order Λ-submodule. Together with the Main Conjecture, these imply that

$$\begin{aligned}\#A_n^\chi = \#X^\chi/(\gamma^{p^n} - 1)X^\chi &= \#\Lambda/(\mathrm{char}_\Lambda(X^\chi), \gamma^{p^n} - 1)\\ &= \#\mathbb{Z}_p[\![T]\!]/(G_{\chi^{-1}\omega}(u(1+T)^{-1} - 1), (1+T)^{p^n} - 1).\end{aligned}$$

Specializing to $n = 0$ yields a refined class number formula proposed by Iwasawa and Leopoldt:

A Refined Class Number Formula *Let $\chi = \omega^i$ with $1 < i < p-1$ an odd integer. Then*

$$\#A_0^\chi = \#\mathbb{Z}_p/(L(\chi^{-1}, 0)). \tag{RCF}$$

There is an analogous formula for odd Hecke characters χ of totally real fields if the order of χ is prime to p.

Significant progress toward this refined class number formula (RCF) had preceded [9]. Using the arithmetic of cyclotomic fields, and especially the cyclotomic units appearing in Sect. 3.3, Herbrand had proved that the non-triviality of the left-hand side of (RCF) implies the non-triviality of the right-hand side. Ribet, using deep results about the geometry of the modular curve $X_1(p)$, later proved the opposite implication. In [6] Wiles greatly extended both Ribet's techniques and ideas from Mazur's Eisenstein ideal paper to show that if A_0^χ is a cyclic $\mathbb{Z}_p$-module, then the refined class number formula (RCF) holds. This paper laid the foundation for the subsequent proof by Mazur and Wiles of the main conjecture (and hence the refined class number formula).

Remark 19 In our statement of the refined class number formula (RCF) we omitted the case $i = 1$. This was solely for convenience, and the formula can be easily extended to include this case: the residue formula for the p-adic ζ-function shows that $G_{\omega^0}(T)$ is a unit, while it follows from results of Stickelberger that $X^\omega = 0$.

Some additional arithmetic applications of the Main Conjecture can be found in [14].

The Birch–Tate Conjecture

Another important consequence of the proof of the Main Conjecture is the (prime-to-2) Birch–Tate formula for the order of the K_2-group of the ring of integers of a totally real number field:

The Birch–Tate Formula *Let F be a totally real number field and $\mathcal{O}_F$ its ring of integers. Then*

$$\#K_2\mathcal{O}_F \sim w_2(F)\zeta_F(-1), \tag{BTF}$$

where '$\sim$' denotes equality up to a unit in $\mathbb{Z}[\frac{1}{2}]$.

Coates showed how (BTF) follows from Tate's description of K_2 and the Main Conjecture.

Other Formulations

There are other equivalent formulations of the main conjecture for $\mathbb{Q}$. We explain one in the case where $F = \mathbb{Q}$, $K/F = \mathbb{Q}(\zeta_p)/\mathbb{Q}$, and $\chi = \omega^i$, $1 \leq i < p-1$ an odd integer. Let $\psi = \chi^{-1}\omega$.

Let $\mathcal{E}_n$ be the group of units in $\mathbb{Z}[\zeta_p]$ and let $\mathcal{E} = \varprojlim_n \mathcal{E}_n$ be the projective limit with respect to the norm maps from $\mathbb{Q}(\zeta_{p^{n+1}})$ to $\mathbb{Q}(\zeta_{p^n})$. Then $\mathcal{E} \subset U$, where U is the group of local units so denoted in Sect. 3.3. In particular, the group $\mathcal{C}$ of cyclotomic units also described there is a subgroup of $\mathcal{E}$. Let $\overline{\mathcal{E}}$ be the closure of $\mathcal{E}$ in U. An alternative formulation of the Main Conjecture involves $(\overline{\mathcal{E}}/\overline{\mathcal{C}})^\psi = e_\psi(\overline{\mathcal{E}}/\overline{\mathcal{C}})$.

Conjecture 20 (The Main Conjecture for ω^i (alternative formulation)) *Let $\chi = \omega^i$ with $1 \leq i < p-1$ an odd integer, and let $\psi = \chi^{-1}\omega$. Then*

$$\mathrm{char}_\Lambda(X^\psi) = \mathrm{char}_\Lambda((\overline{\mathcal{E}}/\overline{\mathcal{C}})^\psi).$$

Note that this formulation does not involve p-adic L-functions. The equivalence with the statement given before comes via the theorem stated in Sect. 3.3, as we explain.

Let $Y = \mathrm{Gal}(M_\infty/K_\infty)$, where M_∞ is the maximal abelian pro-p-extension of K_∞ that is unramified away from the primes above p. Global class field theory then gives an exact sequence

$$0 \to (\overline{\mathcal{E}}/\overline{\mathcal{C}})^\psi \to (U/\overline{\mathcal{C}})^\psi \to Y^\psi \to X^\psi \to 0.$$

A duality argument from global class field theory shows that $\mathrm{char}_\Lambda(Y^\psi)$ is the image of $\mathrm{char}_\Lambda(X^\chi)$ under the involution on Λ that sends T to $u(1+T)^{-1}-1$. In particular, the Main Conjecture for χ is true if and only if $\mathrm{char}_\Lambda(Y^\psi) = (G_\psi(T))$. By the theorem in Sect. 3.3, $\mathrm{char}_\Lambda((U/\overline{\mathcal{C}})^\psi) = (G_\psi(T))$. Using this, the equivalence of the Main Conjecture with the alternative formulation then follows from the exact sequence above and basic properties of characteristic ideals.

5.2 The Proof by Mazur and Wiles

In [9] Mazur and Wiles proved the Main Conjecture of Iwasawa Theory for $\mathbb{Q}$:

Theorem 21 ([9, Thm., p. 211]) *The Main Conjecture of Iwasawa Theory for $\mathbb{Q}$ is true.*

The proof in [9] builds on the groundwork laid in [6] (and in Mazur's work on the Eisenstein ideal). The main ingredients are (1) the work of Kubert and Lang connecting the Stickelberger ideal – and hence the p-adic L-functions of even Dirichlet characters – to the cuspidal subgroups in the Jacobians of modular curves, and (2) a deep analysis of the geometry of modular curves and their Jacobians, especially their (bad) reduction at p, permitting the construction of suitable extensions of the (p-parts) of the cuspidal subgroups; the splitting fields of these extensions over the various $K(\zeta_{p^n})$ correspond to 'large' quotients of X^χ. These quotients are then shown to be large enough that the Main Conjecture follows. In the following we attempt a more detailed sketch of these arguments in a special case.

We now focus on the case where $F = \mathbb{Q}$, $K/F = \mathbb{Q}(\zeta_p)/\mathbb{Q}$, and $\chi = \omega^i$ for some odd integer $1 \leq i < p-1$. Then $\psi = \chi^{-1}\omega = \omega^j$ for some even integer $0 \leq j < p-2$. Let $\Lambda = \mathbb{Z}_p[\![\Gamma]\!]$, which is identified with $\mathbb{Z}_p[\![T]\!]$ via $\gamma \mapsto 1+T$.

An important observation in [9] is that it suffices to prove the following for all n and all odd χ:

$$\mathrm{Fitt}_\Lambda(X^\chi/((\gamma u)^{p^n}-1)X^\chi) \subset (G_\psi(u(1+T)^{-1}-1), (\gamma u)^{p^n}-1) \qquad \text{(MW1)}$$

Here, for a finite R-module M, $\mathrm{Fitt}_R(M)$ denotes the 0th Fitting ideal of M. The reduction to (MW1) goes as follows: By properties of Fitting ideals,

$$\mathrm{Fitt}_\Lambda(X^\chi) \bmod ((\gamma u)^{p^n}-1) = \mathrm{Fitt}_\Lambda(X^\chi/((\gamma u)^{p^n}-1)X^\chi),$$

so knowing (MW1) for all n implies that $\mathrm{Fitt}_\Lambda(X^\chi) \subset (G_\psi(u(1+T)^{-1}-1))$. Since the characteristic ideal of X^χ is contained in $\mathrm{Fitt}_\Lambda(X^\chi)$ (with finite index), it then follows that

$$\mathrm{char}_\Lambda(X^\chi) \subseteq (G_\psi(u(1+T)^{-1}-1)). \qquad \text{(MW2)}$$

As previously explained by Iwasawa, the analytic class number formula implies that the products over the odd characters χ of the left- and right-hand sides of (MW2) are equal. Combined with the inclusions (MW2) for all odd χ, it follows that the inclusion in (MW2) must be an equality for each χ. That is, the Main Conjecture is true. Note that this proves the Main Conjecture for all odd characters of Δ at once – it is an all-or-none strategy.

To give an idea of the arguments in [9] establishing (MW1), we begin with the case $n = 0$, which is essentially done in [6]. First we note that $X^{\omega} = 0$ and $X^{\omega^{-1}} = 0$, by classical results of Stickelberger and Herbrand, hence the inclusion (MW1) is immediate in these cases (for all n). So we focus on $\chi \neq \omega^{\pm 1}$.

The Cuspidal Subgroup

The modular curve $X_1(p)$ has a canonical model over $\mathbb{Q}$ such that the cusp ∞ is defined over $\mathbb{Q}(\zeta_p)^+$ and the cusp 0 is defined over $\mathbb{Q}$. In this model the action of $\mathrm{Gal}(\overline{\mathbb{Q}}/\mathbb{Q})$ on ∞ is via the diamond Hecke operators and factors through the surjection to $\mathrm{Gal}(\mathbb{Q}(\zeta_p)/\mathbb{Q})$: If $\sigma \in \mathrm{Gal}(\overline{\mathbb{Q}}/\mathbb{Q})$ is such that its image in $\mathrm{Gal}(\mathbb{Q}(\zeta_p)/\mathbb{Q}) \cong (\mathbb{Z}/p\mathbb{Z})^\times$ corresponds to $a \in (\mathbb{Z}/p\mathbb{Z})^\times$, then $\sigma \cdot \infty = \langle a \rangle \cdot \infty$. In particular, this gives a well-defined action of $\mathrm{Gal}(\mathbb{Q}(\zeta_{p^\infty})/\mathbb{Q}) = \Delta \times \Gamma$ on (the Galois orbit of) ∞; Γ acts trivially. Let $\mathbb{Z}[\Delta]^0$ be the degree 0-elements in $\mathbb{Z}[\Delta]$ (the kernel of the augmentation map), and let $\mathcal{C} \subset J_1(p)$ be the image of $\mathbb{Z}[\Delta]^0 \cdot \infty$; this is a finite subgroup (by the Manin–Drinfeld theorem). Let $\theta = \chi\omega$, and put

$$C(\theta) = e_\theta(\mathcal{C} \otimes \mathbb{Z}_p) \subset J_1(p)[p^\infty].$$

Here $e_\theta \in \mathbb{Z}_p[\Delta]$ is the idempotent associated with θ. So $C(\theta)$ is just the θ-isotypical piece of the p-primary part of $\mathcal{C}$. As $\theta \neq \omega^2$, the theory of modular units developed by Kubert and Lang shows that there is an isomorphism of Λ-modules

$$C(\theta) \cong \Lambda/(G_\psi(u(1+T)^{-1} - 1), u(1+T) - 1) \cong \mathbb{Z}_p/(L(\theta^{-1}, -1)). \quad \text{(MW3)}$$

The Good Quotient

The modular curve $X_0(p)$ also has a canonical model over $\mathbb{Q}$ and the standard projection map $X_1(p) \to X_0(p)$ of modular curves (given by $\tau \mapsto \tau$ on the usual complex uniformizations by the complex upper half-plane) is defined over $\mathbb{Q}$ and induces a map of Jacobians $J_0(p) \to J_1(p)$ by Picard functoriality. Let A be the quotient of $J_1(p)$ by the image of this map. Then A has good reduction over $\mathbb{Q}(\zeta_p)^+$. As the action of the diamond operators on $J_0(p)$ is trivial and $\theta \neq 1$, the group $C(\theta)$ injects into A. Let $\Phi \subset A[p^\infty]$ be the maximal subgroup of μ-type, that is, the maximal $\mathbb{Q}$-subgroup whose irreducible $\mathbb{Z}_p[\mathrm{Gal}(\overline{\mathbb{Q}}/\mathbb{Q})]$-subquotients are all isomorphic to μ_p; this is a finite group. Let $B = A/\Phi$. Then B also has good reduction over $\mathbb{Q}(\zeta_p)^+$, and $C(\theta)$ injects into B since $\theta \neq \omega$. The abelian variety B is the so-called 'good quotient' of $J_1(p)$.

The Extension

Let $\mathbb{T}$ be the commutative subalgebra of the endomorphism ring of $J_1(p)$ generated by the usual Hecke operators (T_n and $\langle n \rangle$ for $(n, p) = 1$ and U_p). Then $\mathbb{T}$ acts on $C(\theta)$. Let $\mathbb{I} \subset \mathbb{T}$ be the annihilator of $C(\theta)$. Since the image of $J_0(p)$ is $\mathbb{T}$-stable, $\mathbb{T}$ also acts on A and even on B since Φ is also $\mathbb{T}$-stable by maximality. Mazur and Wiles consider the extension

$$0 \to C(\theta) \to B[\mathbb{I}] \to M \to 0$$

of $\mathbb{Z}_p[\mathrm{Gal}(\overline{\mathbb{Q}}/\mathbb{Q})]$-modules. Here M is defined to be the cokernel of the inclusion $C(\theta) \subset B[\mathbb{I}]$. By analyzing the geometry of the abelian variety B, Mazur and Wiles show that

$$\begin{aligned} &\text{(i) } B[\mathbb{I}] \cong C(\theta) \oplus M \text{ as a } \mathrm{Gal}(\overline{\mathbb{Q}}_p/\mathbb{Q}_p)\text{-module,} \\ &\text{(ii) the action of } \mathrm{Gal}(\overline{\mathbb{Q}}/\mathbb{Q}) \text{ on } M \text{ is via the cyclotomic character } \varepsilon, \\ &\text{(iii) } \mathrm{length}_{\mathbb{Z}_p}(M) \geq \mathrm{length}_{\mathbb{Z}_p}(\mathbb{T}/\mathbb{I}) = \mathrm{length}_{\mathbb{Z}_p}(C(\theta)). \end{aligned} \tag{MW4}$$

Let $r \geq 0$ be so large that $F_r = \mathbb{Q}(\zeta_{p^{r+1}})$ contains the splitting field of M. Let L/F_r be the splitting field over F_r of the extension $B[\mathbb{I}]$. As the Galois action on $B[p^\infty]$ is unramified at each prime $\ell \neq p$, it follows from (MW4)(i) that L/F_r is an unramified extension of p-power order. Consider the pairing

$$(\cdot, \cdot) : \mathrm{Gal}(L/F_r) \times M \to C(\theta), \quad (\sigma, m) = \sigma(\tilde{m}) - \tilde{m},$$

where $\tilde{m} \in B[\mathbb{I}]$ is any element projecting to m. This is a $\mathrm{Gal}(\overline{\mathbb{Q}}/\mathbb{Q})$-equivariant pairing. The kernel in $\mathrm{Gal}(L/F_r)$ is trivial by the definition of L, while the kernel in M is trivial since B has no non-zero μ-type subgroups by construction. In particular, there is a $\mathrm{Gal}(\overline{\mathbb{Q}}/\mathbb{Q})$-equivariant isomorphism

$$\mathrm{Gal}(L/F_r) \xrightarrow{\sim} \mathrm{Hom}_{\mathbb{Z}_p}(M, C(\theta)).$$

The action of $\mathrm{Gal}(\overline{\mathbb{Q}}/\mathbb{Q})$ on the right-hand side factors through $G = \Delta \times \Gamma$, and the action of Δ is by $\theta\omega^{-1} = \chi$ while the action of Γ is via ϵ^{-1} (so γ acts as u^{-1}); this follows from (MW4)(ii). It follows that $\mathrm{Gal}(L/F_r)$ is a quotient of $X^\chi/(\gamma u - 1)X^\chi$. In particular, there is a Λ-module surjection

$$X^\chi/(\gamma u - 1)X^\chi \twoheadrightarrow \mathrm{Hom}_{\mathbb{Z}_p}(M, C(\theta)).$$

The inclusion (MW1) (for $n = 0$) now follows from (MW3), (MW4)(iii), and basic properties of Fitting ideals.

The proof of (MW1) for $n > 0$ follows along similar lines. The role of the abelian variety A is played by an abelian variety quotient $J_1(p^n) \twoheadrightarrow A_n$ that is defined inductively on n. The abelian variety A_n has good reduction over $\mathbb{Q}(\zeta_{p^n})^+$. Analyzing the geometry of the reduction of A_n, Mazur and Wiles used the ideas of Kubert and Lang to construct a cuspidal divisor subgroup $C_n(\theta) \subset A_n[p^\infty]$ that is isomorphic to $\Lambda/(G_\psi(u(1+T)^{-1}-1), (u(1+T))^{p^n}-1)$ as a Λ-module. This replaces (MW3). The construction of the analog of the extension $B[\mathbb{I}]$ proceeds similarly, though additional complications arise from $C_n(\theta)$ not being a cyclic $\mathbb{Z}_p$-module.

A more detailed description of the results of Mazur and Wiles can be found in the Bourbaki seminar by Coates [37].

Remark 22 Iwasawa was led to his Main Conjecture in large part by analogy with Weil's theory of the action of Frobenius on the p-power torsion in the Jacobian of a smooth, geometrically connected, projective curve X over a finite field $\mathbb{F}_\ell$, $\ell \neq p$. Mazur and Wiles [8] uncovered another such analogy in the course of proving the Main Conjecture for $\mathbb{Q}$.

5.3 *Wiles's Proof for Totally Real Fields*

In the paper [11], Wiles made substantial progress towards a proof of the Main Conjecture for totally real fields. This included a new proof of the Main Conjecture for $\mathbb{Q}$. These methods were refined in [13], culminating in a proof of the complete Main Conjecture for totally real fields:

Theorem 23 ([13, Thms. 1.2,1.4]) *Let F be a totally real number field. The Main Conjecture for F (including the case with μ-invariant) is true.*

We note that we are continuing to assume that p is odd. However, Wiles's results in [13] cover some cases for $p = 2$, including the Main Conjecture for $\mathbb{Q}$ for $p = 2$ (without μ-invariant).

In Wiles's new proof, the detailed analysis of the cuspidal divisor group $C_n(\theta)$ in [9], which relied on explicit modular units and a study of the mod p reduction of Jacobians of modular curves, is replaced with an analysis of congruences between Eisenstein series and modular forms. The good quotient abelian variety B_n is replaced with a suitable lattice in a sum of p-adic Galois representations associated with some 'good' cuspidal eigenforms. The reduction of this lattice modulo the analog of the ideal $\mathbb{I}$ – the Eisenstein ideal – becomes an extension of the sought-for type. This is a vast generalization of the method used by Ribet [71] to prove the converse to Herbrand's theorem, an early step toward the refined class number formula (RCF).

We try to give some further idea of these new methods in the context of the $n = 0$ case of (MW1).

Let $\mathbb{T}_{\mathbb{Z}_p} = \mathbb{T} \otimes \mathbb{Z}_p$. Recall that $\theta = \chi\omega$ and let $\mathbb{I}_\theta \subset \mathbb{T}_{\mathbb{Z}_p}$ be the ideal generated by $U_p - 1$ and $T_\ell - 1 - \theta^{-1}(\ell)\ell$ and $\langle \ell \rangle - \theta^{-1}(\ell)$ for all primes $\ell \neq p$. The ideal $\mathbb{I}_\theta$ is sometimes called the Eisenstein ideal since the (abstract) Hecke operators generating it annihilate the weight 2 Eisenstein series with q-expansion

$$E_2(\theta^{-1}) = \frac{L(\theta^{-1}, -1)}{2} + \sum_{n=1}^{\infty} \left(\sum_{d|n} \theta^{-1}(d)d \right) q^n.$$

One key step in Wiles's approach is to prove that there is a $\mathbb{Z}_p$-algebra surjection:

$$\mathbb{T}_{\mathbb{Z}_p}/\mathbb{I}_\theta \twoheadrightarrow \mathbb{Z}_p/(L(\theta^{-1}, -1)), \tag{EC1}$$

which necessarily maps each T_ℓ to $1 + \theta^{-1}(\ell)\ell \bmod L(\theta^{-1}, -1)$. This is essentially proved by showing that there is a modular form $G = 1 + \sum_{n=1} a_n q^n$ of level p and Nebentypus θ^{-1} with $\mathbb{Z}_p$-integral q-expansion. Then $F = E_2(\theta^{-1}) - \frac{L(\theta^{-1},-1)}{2} \cdot G = \sum_{n=0}^{\infty} b_n q^n$ is essentially a cusp form with $\mathbb{Z}_p$-integral q-expansion coefficients b_n with the property that $b_\ell \equiv 1 + \theta^{-1}(\ell)\ell \bmod L(\theta^{-1}, -1)$ for all $\ell \neq p$. And the map in (EC1) is just given by $T_\ell \mapsto b_\ell \bmod L(\theta^{-1}, -1)$. To make this precise, one needs to apply Hida's ordinary projector $e = \varinjlim_n U_p^{n!}$ to F (in part because of a possible constant term at the cusp 0.)

The next key step is to construct a suitable lattice in $T_p J_1(p) \otimes \mathbb{Q}_p$. Let $T = e(T_p J_1(p))$. This is a module for the ordinary Hecke algebra $\mathbb{T}_p = e\mathbb{T}_{\mathbb{Z}_p}$. Let $V = T \otimes_{\mathbb{Z}_p} \mathbb{Q}_p$. Then

(i) V is a free $\mathbb{T}_p \otimes_{\mathbb{Z}_p} \mathbb{Q}_p$-module of rank 2;

(ii) V is an irreducible $\mathbb{T}_p \otimes_{\mathbb{Z}_p} \mathbb{Q}_p[\mathrm{Gal}(\overline{\mathbb{Q}}/\mathbb{Q})]$-module;

(iii) there is a unique $\mathrm{Gal}(\overline{\mathbb{Q}}_p/\mathbb{Q}_p)$-stable $\mathbb{T}_p \otimes_{\mathbb{Z}_p} \mathbb{Q}_p$-submodule $V^+ \subset V$ that is free of rank one; the quotient V/V^+ is unramified at p.

(EC2)

The space V can be identified as the sum of the p-adic Galois representations associated with certain newforms of weight 2 and level p. Properties (i)–(iii) can be deduced from corresponding properties of these representations (the results of the work of many mathematicians, especially Deligne, Langlands, Mazur and Wiles). Let $v^+ \in V^+$ be a $\mathbb{T}_p \otimes_{\mathbb{Z}_p} \mathbb{Q}_p$-generator, and let $L = \mathbb{T}_p[\mathrm{Gal}(\overline{\mathbb{Q}}/\mathbb{Q})] \cdot v^+ \subset V$. This is a full lattice in V by (ii). In particular, it is a faithful $\mathbb{T}_p$-module. One then considers the quotient $L/\mathbb{I}_\theta L$. Using the Eichler–Shimura congruence relation (which describes the trace of a Frobenius at ℓ on V as the operator T_ℓ for $\ell \neq p$), one can deduce that $\overline{L}$ is a $\mathrm{Gal}(\overline{\mathbb{Q}}/\mathbb{Q})$ extension

$$0 \to L_1 \to L/\mathbb{I}_\theta L \to L_2 \to 0 \qquad \text{(EC3)}$$

with $\mathrm{Gal}(\overline{\mathbb{Q}}/\mathbb{Q})$ acting trivially on L_1 and by $\theta^{-1}\varepsilon$ on L_2. Furthermore, as a $\mathbb{T}_p$-module, $L_2 \cong \mathbb{T}_p/\mathbb{I}_\theta\mathbb{T}_p = \mathbb{T}_{\mathbb{Z}_p}/\mathbb{I}_\theta$ (the last equality follows since $U_p - 1 \in \mathbb{I}_\theta$), and the faithfulness of L as a $\mathbb{T}_p$-module together with (EC2)(i) implies that $\mathrm{length}_{\mathbb{Z}_p}(L_1) \geq \mathrm{length}_{\mathbb{Z}_p}(\mathbb{T}_p/\mathbb{I}_\theta\mathbb{T}_p)$. The Galois action on L is unramified away from p, while (EC2)(iii) and the choice of v^+ implies that (EC3) splits as a $\mathrm{Gal}(\overline{\mathbb{Q}}_p/\mathbb{Q}_p)$-extension. Using (EC1), the inclusion (MW1) (for $n = 0$) can now be deduced exactly as before.

In [13] Wiles used a variant of this approach to establish (MW2) directly, using Hida families of modular forms over the Iwasawa algebra Λ (for Hida families see Sect. 6.3 below). But the structure of the argument was the same as sketched above.

The advantage of this approach is that the geometry of the modular curves and their Jacobians has receded into the background: it only plays a role in establishing the properties of the Galois representations associated with newforms. This was a crucial shift that allowed the arguments to extend to Hilbert modular forms and so to a proof of the Main Conjecture for totally real fields. Suppose $F \neq \mathbb{Q}$. Eisenstein series exist on the Hilbert modular variety and not on any Shimura curve, but the two-dimensional p-adic Galois representations one seeks to use are found (with a few exceptions) in the cohomology of Shimura curves (which have no cusps) and not in the cohomology of the Hilbert modular variety.

Not surprisingly, there were technical challenges to making this argument work for general totally real fields. Among these were (a) our ignorance of the truth of Leopoldt's conjecture for general totally real fields (that F_∞ should be the unique $\mathbb{Z}_p$-extension of F), (b) the possibility of exceptional zeros of the forms $\zeta - 1$, $\zeta \in \mu_{p^\infty}$, of $f_\chi(T)$, and (c) that certain Hilbert modular forms do not have Jacquet–Langlands transfers to quaternion algebras with an associated Shimura curve (their two-dimensional Galois representations are not found in the cohomology of any Shimura curve). Wiles succeeded in overcoming (a) and especially (b) by a limiting argument, working with characters $\chi\rho$ with ρ of p-power order but such that $f_{\chi\rho}(T)$ does not have the exceptional zero of interest. In hindsight this argument presages the 'patching' argument that was to be so crucial to the proof of the modularity of elliptic curves. The existence of the Galois representations not found in the cohomology of Shimura curves was established by Wiles in [12] (see Sect. 6.4 below).

5.4 *Further Developments*

Ideas of Thaine and Kolyvagin eventually led to a proof of Iwasawa's Main Conjecture for $\mathbb{Q}$ using the Euler system of cyclotomic units by Rubin (see [78] and Rubin's appendix in [64]). This argument actually proved the alternative formulation described in section "Other Formulations". But Wiles's proof remains the only proof to date of the Iwasawa Main Conjecture for general totally real fields. The structure of his proof – combining special (usually Eisenstein) congruences for Hida families with lattice constructions in Galois representations to construct 'large' pieces of Selmer groups – has proven to be robust and has been applied to make progress on other 'main conjectures.' Mazur and Tilouine [67] and then Hida and Tilouine [54] used this strategy to prove results in the direction of the anticyclotomic main conjecture for CM fields. Eric Urban and the author [84] followed this same strategy, but with classical modular forms replaced with modular forms for $U(2, 2)$, to prove one divisibility (p-adic L-function divides characteristic ideal) in the main conjecture for many elliptic curves. Kato had already proven the opposite divisibility [57], and in combination this resulted in the main conjecture for (most) elliptic curves at ordinary primes. The analog of the refined class number formula (RCF) in this case is the p-part of the conjectured formula for Birch and Swinnerton-Dyer when $L(E, 1) \neq 0$, as well as the consequence that $L(E, 1) = 0$ if and only if the Selmer group of $E/\mathbb{Q}$ has infinite order (which is consistent with the Birch–Swinnerton-Dyer Conjecture).

6 Galois Representations

The p-adic Galois representations associated with Hecke characters, elliptic curves, and modular forms have played prominent roles in the proofs of all of Wiles's theorems. In most instances, fundamental properties of these representations (e.g., ramification, characteristic polynomials of Frobenius elements) had already been established. But in a few cases, Wiles had to first prove the necessary properties or even prove that the desired Galois representations exist. Such results, for example, were required to complete the proof of the Main Conjecture of Iwasawa theory for totally real fields.

In the following we recall the Galois representations associated with modular forms and Hilbert modular forms and describe some of Wiles's results pertaining to them. These results include Wiles's proof that the p-adic Galois representation associated with a p-ordinary (Hilbert) modular eigenform is ordinary at p (that is, the restriction to a decomposition group at p is upper-triangular) as well as his proof of the existence of the 'missing' representations associated to p-ordinary Hilbert modular forms when the degree of the totally real field is even.

Let p be a prime. In this section we do not assume that p is odd unless explicitly stated.

For a number field F, let $\overline{F}$ be a separable algebraic closure of F and let $G_F = \mathrm{Gal}(\overline{F}/F)$. For each place v of F, let $\overline{F}_v$ be a separable algebraic closure of the completion F_v and let $G_{F_v} = \mathrm{Gal}(\overline{F}_v/F)$. Let $I_v \subset G_{F_v}$ be the inertia subgroup. We write Frob_v for a (arithmetic) Frobenius element in G_{F_v}/I_v. Fix an embedding $\overline{F} \hookrightarrow \overline{F}_v$. This identifies G_{F_v} with a decomposition subgroup in G_F.

6.1 Galois Representations of Modular Forms

Let $f \in S_k(N,\chi)$ be a newform of weight k, level N, and Nebentypus χ. Let $f(\tau) = \sum_{n=1}^{n} a_n q^n$, $q = e^{2\pi i\tau}$, be the usual Fourier expansion of f. The L-function of f is just

$$L(f,s) = \sum_{n=1}^{\infty} a_n n^{-s} = \prod_{\text{primes } \ell} L_\ell(f;\ell^{-s})^{-1},$$

where $L_\ell(f;X) = 1 - a_\ell X + \chi(\ell)\ell^{k-1}X^2$ if $\ell \nmid N$, and $L_\ell(f;X) = 1 - a_\ell X$ if $\ell \mid N$. The a_n are the eigenvalues of Hecke operators acting on f and generate a finite extension $\mathbb{Q}(f) \subset \mathbb{C}$ of $\mathbb{Q}$. The a_n are in fact algebraic integers and so belong to the ring of integers $\mathbb{Z}(f)$ of $\mathbb{Q}(f)$. Let $\mathfrak{p} \subset \mathbb{Z}(f)$ be a prime above p and let $K = \mathbb{Q}(f)_\mathfrak{p}$ and $\mathcal{O} = \mathbb{Z}(f)_\mathfrak{p}$. Deligne showed that there is a two-dimensional continuous semisimple K-linear representation V_f of $G_\mathbb{Q}$,

$$\rho_f : G_\mathbb{Q} \to \mathrm{Aut}_K(V_f),$$

with the following properties: (1) $\det \rho_f = \chi\varepsilon^{k-1}$ is the product of the finite Galois character associated with χ and the $(k-1)$-th power of the p-adic cyclotomic character, (2) V_f is unramified at the primes $\ell \nmid pN$, and (3) for $\ell \nmid pN$, $\operatorname{trace} \rho_f(\mathrm{Frob}_\ell) = a_\ell$. We note that by the Chebotarev Density Theorem and the Brauer–Nesbitt Theorem, (2) and (3) are enough to characterize ρ_f. From (1) together with the fact that $\chi(-1) = (-1)^k$ we conclude that (4) the eigenvalues of a complex conjugation are $+1$ and -1 (ρ_f is odd). Also, refinements of Deligne's work by Langlands and Carayol improved on (4), showing that: (5) the level N of f equals the prime-to-p Artin conductor of V_f, and the Euler factors of the L-function $L(f,s) = \sum_{n=1}^{\infty} a_n n^{-s} = \prod_\ell (1 - a_\ell \ell^{-s} + \chi(\ell)\ell^{k-1-2s})^{-1}$ can be recovered from V_f:

$$L_\ell(f;X) = \det(1 - X\,\mathrm{Frob}_\ell \,|(V_f)_{I_\ell}), \quad \ell \neq p,$$

where $(-)_{I_\ell}$ denotes the I_ℓ-coinvariants (i.e., the maximal unramified quotient). Properties (1)-(5) can be more succinctly expressed in a (slightly stronger) statement of 'local–global compatibility' with respect to the local Langlands correspondence, but we do not go into this here.

A modular form is said to be ordinary at $\mathfrak{p}$ if a_p is a $\mathfrak{p}$-adic unit (a unit in $\mathbb{Z}(f)_\mathfrak{p}$). In this case, it is a consequence of results of Mazur and Wiles [10] and results of Hida that:

Theorem 24 ([10, Prop. 2], [12, Thm. 2]) *Suppose f is ordinary at $\mathfrak{p}$. Then*

$$\rho_f|G_{\mathbb{Q}_p} \cong \begin{pmatrix} \alpha^{-1}\chi\varepsilon^{k-1} & * \\ 0 & \alpha \end{pmatrix},$$

where α is the unique unramified character of $G_{\mathbb{Q}_p}$ such that $\alpha(\mathrm{Frob}_p) = \alpha_p$, where $\alpha_p = a_p$ if $p \mid N$, and otherwise α_p is a unit root of $X^2 - a_pX + \chi(p)p^{k-1}$.

Note that α_p is uniquely determined if $k \geq 2$ or χ is ramified at p. But if $k = 1$ and χ is unramified, there are two unit roots. However, in that case the restriction of the representation to $G_{\mathbb{Q}_p}$ is split.

The result in [10] imposes some restrictions on the χ and the level N. These were lifted in [11] and [12]. Below we provide a brief sketch of the argument from the latter two papers as well as include a few words about the results for Hilbert modular forms. But before doing this we recall an important tool for studying Galois representations that was introduced in [12].

6.2 Pseudorepresentations

In [12] Wiles introduced the notion of a pseudorepresentation, which has ever since been an extremely useful tool in the hands of those studying Galois representations.

Let G be a group and R a commutative ring. As introduced in [12, pp. 563–564], a pseudorepresentation of G in R is a tuple of maps (a, d, x), $a, d : G \to R$ and $x : G \times G \to R$, satisfying certain relations. These relations are those one would expect if $a(g)$ and $d(g)$ were the diagonal entries of a two-dimensional representation $\rho : G \to \mathrm{GL}_2(R)$, $\rho(g) = \begin{pmatrix} a(g) & b(g) \\ c(g) & d(g) \end{pmatrix}$, and x was given by $x(g, g') = b(g)c(g')$. In particular, for such a representation ρ, the maps $a(g)$, $d(g)$, and $x(g, g') = b(g)c(g')$ form a pseudorepresentation. If G and R are a topological group and ring, respectively, then we say a pseudorepresentation (a, d, x) of G in R is continuous if a, d, and x are continuous maps. Two important properties of pseudorepresentations are:

(PR1) If some x_{g_0,h_0} is invertible in R or if x is identically 0, then there exists a representation $\rho : G \to \mathrm{GL}_2(R)$ such that $\operatorname{trace}\rho(g) = a(g) + d(g)$ and $\det\rho(\sigma) = a(g)d(g) - x(g, g)$.

(PR2) If $\{\mathfrak{a}_i\}_{i\in I}$ is a collection of ideals of R and $\{(a_i, b_i, x_i)\}_{i\in I}$ are pseudorepresentations in $R/\mathfrak{a}_i$ such that (a_i, b_i, x_i) and (a_j, b_j, x_j) agree modulo $\mathfrak{a}_i + \mathfrak{a}_j$ for all $i, j \in I$, then there is a pseudorepresentation (a, d, x) in $R/\bigcap_{i\in I}\mathfrak{a}_i$ such that for each $i \in I$, (a_i, d_i, x_i) is the reduction of (a, d, x) modulo $\mathfrak{a}_i$.

An important example of a continuous pseudorepresentation is that associated with a newform f. Returning to the notation from the previous section, fix a basis of V with respect to which complex conjugation c has image $\rho_f(c) = \begin{pmatrix} 1 & 0 \\ 0 & -1 \end{pmatrix}$. The matrix entries $a(\sigma)$, $d(\sigma)$, and $x(\sigma, \tau) = b(\sigma)c(\sigma)$ with respect to this basis are a continuous pseudorepresentation of $G_{\mathbb{Q}}$ in $\mathbb{Z}(f)_{\mathfrak{p}}$ and even in the $\mathbb{Z}_p$-subalgebra generated by the a_ℓ for all but finitely many primes ℓ. (Note that $a(\sigma) = \frac{1}{2}(\operatorname{trace}\rho(\sigma) + \operatorname{trace}\rho(c\sigma))$, etc.)

6.3 Ordinarity of Galois Representations

We return to the sketch of the proof of the theorem stated in Sect. 6.1.

Suppose first that f has weight 2. A construction of Shimura associates to f an abelian variety A over $\mathbb{Q}$ of dimension equal to $[\mathbb{Q}(f) : \mathbb{Q}]$ with an injection $\mathbb{Z}(f) \hookrightarrow \mathrm{End}_{\mathbb{Q}}(A)$ and such that $V_f \cong T_pA \otimes_{\mathbb{Z}(f)\otimes\mathbb{Z}_p} K$. If $p \nmid N$ or χ is ramified at p, then A has potentially good reduction at p; it acquires good reduction over any extension that splits the ramification at p of the character χ. Working over such an extension if necessary one then has that the trace of any power of the Frobenius acting on $D(A[p^\infty]) \otimes_{\mathbb{Z}(f)} \mathbb{Z}(f)_{\mathfrak{p}}$, $D(A[\mathfrak{p}^\infty])$ being the Dieudonné module of the p-divisible group $A[p^\infty]$, agrees with the trace of the power of the Frobenius at p on any ℓ-adic representation associated to f, $\ell \neq p$. The theorem in these cases then follows easily. To deduce the theorem in the remaining cases from these, Wiles makes use of certain p-adic analytic families of modular forms – Hida families.

We will say that an eigenform $f \in S_k(N,\chi)$ is a p-stablilized newform if $p \mid N$ and f is an eigenform for the U_p-operator and either f is a newform or $f(\tau) = f_0(\tau) - \beta_p f_0(p\tau)$ for some newform f_0 of level $N_0 = N/p$ with $p \nmid N_0$. For example, if f is a newform that is ordinary at $\mathfrak{p}$, then either $f_\alpha = f$ is p-stabilized (when $p \mid N$) or $f_\alpha(\tau) = f(\tau) - \beta_p f(p\tau)$ is p-stabilized, where β_p is the non-unit root of $X^2 - a_pX + \chi(p)p^{k-1}$ if $k \geq 2$. In both cases the eigenvalue of U_p acting on f_α is the $\mathfrak{p}$-adic unit α_p. We call f_α the ordinary p-stabilization of f.

Let $f \in S_k(N,\chi)$ be a newform that is ordinary at $\mathfrak{p}$. Let $\mathcal{O} = \mathbb{Z}(f)_{\mathfrak{p}}$ and $\Lambda_{\mathcal{O}} = \mathcal{O}[\![\Gamma]\!]$, where Γ is as in Sect. 5.1. Let $\gamma \in \Gamma$ be a topological generator and $u = \varepsilon(\gamma)$. There is a finite, normal extension $\mathbb{I}$ of $\Lambda_{\mathcal{O}}$ and a formal q-expansion

$$\mathbf{f} = \sum_{n=1}^{\infty} \mathbf{a}_n q^n \in \mathbb{I}[\![q]\!]$$

with the following properties: (a) for all but finitely many $\mathcal{O}$-algebra homomorphisms $\phi : \mathbb{I} \to \overline{\mathbb{Q}}_p$ such that $\phi(\gamma) = \zeta u^{k_\phi - 2}$, $k_\phi \geq 2$ an integer and ζ a primitive p^{r_ϕ}th root of unity,

$$f_\phi = \sum_{n=1}^{\infty} \phi(\mathbf{a}_n) q^n$$

is the q-expansion of a p-stabilized newform of level dividing $Np^{r_\phi+1}$ and Nebentypus $\chi\omega^{2-k_\phi}\chi_\zeta$, where $\chi_\zeta : (\mathbb{Z}/p^{r_\phi+1}\mathbb{Z})^\times \to \mu_{p^\infty}$ is the character of order p^{r_ϕ} such that $\chi_\zeta(u) = \zeta^{-1}$, and (b) there is some such ϕ_0 with $k_{\phi_0} = k$ and $r_{\phi_0} = 0$ for which $f_{\phi_0} = f_\alpha$, the ordinary p-stabilization of f. The formal series $\mathbf{f}$ satisfying (a) is often referred to as a Hida family. Part of the assertion here is that there is a Hida family specializing to f. The Hida family $\mathbf{f}$ is used as follows.

For ϕ as in (a), let $P_\phi = \ker(\phi)$. Applying property (PR2) to the set of ideals P_ϕ and the pseudo-representations associated to the f_ϕ, it follows that there is a continuous pseudorepresentation $(\mathbb{a}, \mathbb{d}, \mathbb{x})$ of $G_{\mathbb{Q}}$ in $\mathbb{I}$ that specializes under each ϕ to the pseudorepresentation of f_ϕ. Let $F_{\mathbb{I}}$ be the field of fractions of $\mathbb{I}$. By property (PR1), there is a semisimple representation $\rho_{\mathfrak{f}} : G_{\mathbb{Q}} \to \mathrm{GL}_2(F_{\mathbb{I}})$ such that trace $\rho_{\mathfrak{f}} = \mathbb{a} + \mathbb{d}$ and $\det \rho_{\mathfrak{f}} = \chi\varepsilon\Phi$, where $\Phi : G_{\mathbb{Q}} \twoheadrightarrow \Gamma \hookrightarrow \Lambda_{\mathcal{O}}^{\times}$ is the canonical character. Let $V_{\mathfrak{f}}$ be the representation space of ρ_f. As each localization $\mathbb{I}_{P_\phi}$ is a DVR, for each ϕ there exists a $G_{\mathbb{Q}}$-stable rank two $\mathbb{I}_{P_\phi}$-lattice $M_\phi \subset V_{\mathfrak{f}}$ such that the semisimplification of $M_{\mathfrak{f}}$ mod P_ϕ is isomorphic to V_{f_ϕ} as a $G_{\mathbb{Q}}$-representation. To prove the theorem it then suffices to prove that there is a $F_{\mathbb{I}}$-line in $V_{\mathfrak{f}}$ on which $G_{\mathbb{Q}_p}$ acts as $\alpha_{\mathfrak{f}}^{-1}\chi\varepsilon\Phi$, where $\alpha_{\mathfrak{f}} : G_{\mathbb{Q}_p} \to \mathbb{I}^{\times}$ is the continuous unramified character such that $\alpha_{\mathfrak{f}}(\mathrm{Frob}_p) = \mathbb{a}_p$. For then the intersection of this line with M_{ϕ_0} reduces modulo P_{ϕ_0} to a K-line on which $G_{\mathbb{Q}_p}$ acts as $\alpha^{-1}\chi\varepsilon^{k-1}$.

Let $\mathcal{X}$ be the set of homomorphisms ϕ such that $k_\phi = 2$ and $\chi\chi_\zeta$ is ramified at p. The set of primes $P_\phi = \ker(\phi)$, $\phi \in \mathcal{X}$, is Zariski dense in $\mathbb{I}$. Since the conclusion of the above theorem has been observed to hold for each ρ_{f_ϕ}, $\phi \in \mathcal{X}$, it readily follows that on $G_{\mathbb{Q}_p}$, $\mathbb{a} + \mathbb{d} = \alpha_{\mathfrak{f}} + \alpha_{\mathfrak{f}}^{-1}\chi\varepsilon\Phi$ and $\mathbb{x} = 0$.

Let $\tau \in G_{\mathbb{Q}_p}$ such that $\alpha_{\mathfrak{f}}(\tau) \neq \alpha_{\mathfrak{f}}^{-1}\chi\Phi(\tau)$. Then there exists a basis of $V_{\mathfrak{f}}$ with respect to which $\rho_{\mathfrak{f}}(\tau) = \begin{pmatrix} \alpha_{\mathfrak{f}}^{-1}\chi\varepsilon\Phi(\tau) & 0 \\ 0 & \alpha_{\mathfrak{f}}(\tau) \end{pmatrix}$. Then either $\rho|_{G_{\mathbb{Q}_p}} = \begin{pmatrix} \alpha_{\mathfrak{f}}^{-1}\chi\varepsilon\Phi & * \\ 0 & \alpha_{\mathfrak{f}} \end{pmatrix}$ or $\rho|_{G_{\mathbb{Q}_p}} = \begin{pmatrix} \alpha_{\mathfrak{f}}^{-1}\chi\varepsilon\Phi & 0 \\ * & \alpha_{\mathfrak{f}} \end{pmatrix}$. As observed above, if the former holds then we are done, so assume the latter holds. For all but finitely many $\phi \in \mathcal{X}$, the matrix entries of $\rho_{\mathfrak{f}}$ with respect to this basis belong to $\mathbb{I}_{P_\phi}$; let $\mathcal{X}' \subset \mathcal{X}$ be the subset of such ϕ. Since the conclusion of the theorem holds for f_ϕ, $\phi \in \mathcal{X}'$, it follows that the reduction of $*$ modulo P_ϕ is zero for all $\phi \in \mathcal{X}'$. But the set of such P_ϕ is Zariski dense in $\cap_{\phi\in\mathcal{X}'}\mathbb{I}_{P_\phi}$, hence $* = 0$.

6.4 Galois Representations of Hilbert Modular Forms

In [12] Wiles showed that the Theorem in Sect. 6.1 generalizes to include Hilbert modular forms over any totally real field F. However, for this to even make sense, Wiles had to also prove the existence of the Galois representations for some Hilbert modular forms. We content ourselves with just indicating the broad outlines of what is proved in [12].

Let f be a Hilbert modular newform over a totally real field F. As in [12], suppose f has (parallel) weight $k \geq 1$ and level $\mathfrak{n}$ (an integral ideal of F). If $k \geq 2$ and

(G) either $[F : \mathbb{Q}]$ is odd or f is 'square-integrable' at some prime $\mathfrak{q} \mid \mathfrak{n}$ (more precisely, the local constituent at some $\mathfrak{q}$ of the cuspidal automorphic representation associated with f is supercuspidal or a twist of the Steinberg representation),

then the two-dimensional p-adic Galois representations that one expects to be associated with f occur in the cohomology of Shimura curves over F. But otherwise they do not. Part of [12] is the construction of the missing representations – for those f that are ordinary at all primes above p – and a proof that they satisfy the analog of property (5) of Sect. 6.1 (cf. [12, Thms. 2.1.2 & 2.1.3]).

Wiles developed enough of an analog of Hida's theory for Hilbert modular forms to prove that if f is ordinary, then it is the specialization of an analytic family $\mathfrak{f}$ over some $\mathbb{I}$ as in Sect. 6.3. With this in hand, the arguments of Sect. 6.3 carry over if (G) holds. When (G) does not hold, Wiles again makes use of pseudorepresentations: By analyzing congruences between Hida families, he proves that there are infinitely many height one primes $P \subset \mathbb{I}$ such that there is a prime $\mathfrak{q} \nmid p\mathfrak{n}$ such that $\mathbb{I}/P \cong \mathbb{T}(\mathfrak{n}\mathfrak{q})_{\mathbb{I}}^{\mathfrak{q}-\mathrm{new}}/\mathfrak{a}_{\mathfrak{q}}$. Here $\mathbb{T}(\mathfrak{n}\mathfrak{q})_{\mathbb{I}}^{\mathfrak{q}-\mathrm{new}}$ is the Hecke algebra over $\mathbb{I}$ for the space of Hida families over $\mathbb{I}$ that specialize to Hilbert modular forms that are ordinary at the primes above p (but are not necessarily eigenforms) and that are new at the prime $\mathfrak{q}$ and $\mathfrak{a}_{\mathfrak{q}}$ is an ideal. As (G) holds for the eigenforms that are new at $\mathfrak{q}$, Wiles deduces that their pseudorepresentations are specializations of a pseudorepresentation of G_F in $\mathbb{T}(\mathfrak{n}\mathfrak{q})_{\mathbb{I}}^{\mathfrak{q}-\mathrm{new}}$, which yields a pseudorepresentation into $\mathbb{I}/P$. From property (PR2) of pseudorepresentations, it then follows that there is a pseudorepresentation in $\mathbb{I}$ that is unramified. The Galois representation $\rho_{\mathfrak{f}} : G_F \to \mathrm{GL}_2(F_{\mathbb{I}})$ that then comes from property (PR1) can be easily seen to have all the desired properties. The existence and properties of ρ_f follow.

6.5 Further Developments

Taylor [86] used congruences and pseudorepresentations much as in Wiles's argument to construct the expected p-adic Galois representations for all Hilbert modular newforms of weights at least 2. Blasius and Rogawski [31] exploited the existence of endoscopic lifts from $U(2)$ to $U(3)$ to find most of the Galois representations associated with Hilbert modular forms in the cohomology of Picard modular surfaces. This extended the class of such representations that have a geometric realization, but forms of parallel weight 2 over an even degree field and forms that have (partial) weight one are not covered by this result. These results left open the full local-global compatibility of these representations at ramified primes. For primes not dividing p this was completed by Blasius [32], while the general analog of Wiles's ordinarity result at the prime p was established by Saito [79] and Liu [65] and in [83]. The latter two papers crucially use congruences and p-adic families much as in Wiles's arguments. Taylor extended Wiles's notion of pseudorepresentations to higher dimensions, using them to construct Galois representations for Siegel modular forms of low weight [87] and even some cohomological cuspidal represenations for GL_2 over an imaginary quadratic field [88]. The use of p-adic families of pseudorepresentations to construct missing Galois representations and prove their local-global properties has permeated the field, as evidenced, for example, in the work of Chenevier and Harris on Galois representations for unitary groups [34]. Pseudorepresentations are an indispensible tool in the construction [53] of Galois representations for cohomological automorphic representations of GL_n

over totally real and CM fields and in Scholze's remarkable construction [81] of Galois representations for torsion classes in the cohomology of these same groups.

7 Modularity of Elliptic Curves and Fermat's Last Theorem

Wiles's most celebrated breakthrough has been the proof that all semistable elliptic curves over $\mathbb{Q}$ are modular, one consequence of which was the truth of Fermat's Last Theorem (finally!).

A history of the resolution of this problem in Wiles's own words can be found in the introduction to [16] and his address [15] to the 1994 International Congress of Mathematicians. An additional discussion of this in the context of other developments and problems in number theory can be found in the article [19], which is essentially the text of Wiles's talk at the 1998 ICM.

The impact of this modularity theorem has been far-reaching, going well beyond a proof of Fermat's Last Theorem. For example, many tools, techniques, and results have been developed for *modular* elliptic curves (e.g., Heegner points, the Gross–Zagier formula, the BSD conjecture in cases of analytic rank at most one, p-adic L-functions, etc.). Thanks to Wiles's theorem, and its subsequent extension to all elliptic curves over $\mathbb{Q}$, the previously common hypothesis that "E is a modular elliptic curve" has been dropped. But this was just the start. The methods pioneered in the proof of this modularity theorem have been remarkably robust and adaptable, leading to resolutions of a host of problems within the circle of the Langlands Program, including:

- the modularity of all elliptic curves over $\mathbb{Q}$,
- the modularity of all elliptic curves over real quadratic fields and the meromorphic continuation of the Hasse–Weil L-functions of elliptic curves over all totally real fields;
- a proof of Serre's Conjecture,
- a proof of the Artin Conjecture for odd two-dimensional representations of $\mathrm{Gal}(\overline{\mathbb{Q}}/\mathbb{Q})$,
- the automorphy of small symmetric powers of modular newforms, and the potential automorphy of all symmetric powers,
- a proof of the Sato–Tate Conjecture for elliptic curves and even for all Hilbert modular newforms (of weights at least two),
- the potential automorphy of a pure, regular, totally odd, polarizable, weakly compatible system of ℓ-adic Galois representations,

to list just a few. More are sure to come!

In the following we recall the modularity conjecture for elliptic curves and Wiles's modularity lifting theorem. We explain how the latter yielded a proof of the former, at least for semistable curves, and we describe some of the key features of Wiles's proof.

7.1 Elliptic Curves and Modular Forms

Let E be an elliptic curve over a number field F. There is an associated L-series $L(E/F, s)$, the Hasse–Weil L-function of E over F. As recalled in Sect. 4.1, this L-series is first defined as an Euler product $L(E/F, s) = \prod_v L_v(E/F; q_v^{-s})^{-1}$, where v runs over all finite places of F. The product converges absolutely for $\mathrm{Re}(s) > \frac{3}{2}$ and defines a holomorphic function in this region. Hasse conjectured that $L(E/F, s)$ has a meromorphic continuation to all of $\mathbb{C}$ and satisfies a functional equation relating $L(E/F, s)$ to $L(E/F, 2 - s)$. For elliptic curves with complex multiplication this was proved by Deuring, who identified $L(E/F, s)$ as the L-function of a Hecke character. Shimura verified Hasse's conjecture for those elliptic curves over $\mathbb{Q}$ that admit a parameterization by a modular curve $X_0(N)$.

The converse theorems of Hecke and Weil show that if one assumes a holomorphic continuation and functional equation for all the twists of $L(E/F, s)$ by Hecke characters of F as well as boundedness in vertical strips, then the series $L(E/F, s)$ is a Mellin transform of a cuspidal modular newform. For $F = \mathbb{Q}$ this suggested the following conjecture, a precise version of a remarkable idea of Shimura and Taniyama:

Conjecture 25 (The Modularity Conjecture for Elliptic Curves) *Let E be an elliptic curve over $\mathbb{Q}$ of conductor N. Let $L(E, s) = \sum_{n=1}^{\infty} a_n n^{-s}$ be the Hasse–Weil L-series of $E/\mathbb{Q}$, and let $f(\tau) = \sum_{n=1}^{\infty} a_n e^{2\pi i \tau}$.*

(a) *$f(\tau)$ is a cuspidal newform of weight 2 for the congruence subgroup $\Gamma_0(N)$.*
(b) *There exists a non-constant $\mathbb{Q}$-rational morphism $\phi : X_0(N) \to E$ such that the pullback $\phi^* \omega_E$ of any non-zero invariant differential form on E is a multiple of the differential form on $X_0(N)$ defined by $f(\tau)\, d\tau$.*

In particular, $L(E, s)$ has a holomorphic continuation, and $\Lambda_E(s) = N^{s/2}(2\pi)^{-s}\Gamma(s)L(E, s)$ satisfies $\Lambda_E(s) = w_E \Lambda(E, 2 - s)$, $w_E = \pm 1$.

It is a consequence of a construction of Shimura and of the isogeny theorem of Faltings that parts (a) and (b) of this conjecture are equivalent.

The Modularity Conjecture was verified by Shimura for elliptic curves having complex multiplication. It is now known to hold in full generality as a consequence of Wiles's groundbreaking work [16, 17] and subsequent developments, especially by Breuil, Conrad, Diamond, and Taylor.

The modularity of elliptic curves, which is a priori stronger than Hasse's conjecture, has proved to be the more useful concept. In particular, the morphisms $\phi : X_0(N) \to E$ have provided an indispensible tool for studying the arithmetic of elliptic curves.

7.2 *Modularity in Terms of Galois Representations*

While the converse theorems of Hecke and Weil provide a satisfactory characterization of the Dirichlet series arising from modular forms, they are generally of little use in directly establishing modularity. To apply the converse theorem one needs to have already proved that a Dirichlet series (and its twists) has good analytic properties – and essentially the only known way to establish these properties is to know that the Dirichlet series comes from a modular form! However, The Modularity Conjecture has a Galois-theoretic reformulation that suggests another approach.

Let $E/\mathbb{Q}$ be an elliptic curve. For a prime p, the subgroup $E[p^n]$ of p^n-torsion points is a free $\mathbb{Z}/p^n\mathbb{Z}$-module of rank two equipped with a continuous action of $G_{\mathbb{Q}} = \mathrm{Gal}(\overline{\mathbb{Q}}/\mathbb{Q})$. The p-adic Tate module

$$T_pE = \varprojlim_n E[p^n]$$

is therefore a free $\mathbb{Z}_p$-module of rank two with a continuous $\mathbb{Z}_p$-linear $G_{\mathbb{Q}}$-action. The Weil pairing yields a non-degenerate $G_{\mathbb{Q}}$-invariant alternating $\mathbb{Z}_p$-pairing $T_pE \times T_pE \rightarrow \mathbb{Z}_p(1)$, and so the determinant of the $\mathbb{Z}_p$-linear representation of $G_{\mathbb{Q}}$ on T_pE is just the p-adic cyclotomic character $\varepsilon : G_{\mathbb{Q}} \rightarrow \mathbb{Z}_p^{\times}$. By the criterion of Néron–Ogg–Shafarevich, T_pE is unramified at all primes $\ell \nmid pN$. More generally, the conductor N of E is just the prime-to-p Artin conductor of the Galois representation $V_pE = T_pE \otimes_{\mathbb{Z}_p} \mathbb{Q}_p$. Furthermore, the Euler factors of the L-function $L(E, s)$ can recovered from V_pE:

$$L_\ell(E; X) = \det(1 - \mathrm{Frob}_\ell \,|V_pE_{I_\ell}), \quad \ell \neq p,$$

where $I_\ell \subset G_{\mathbb{Q}_\ell} = \mathrm{Gal}(\overline{\mathbb{Q}}_\ell/\mathbb{Q}_\ell)$ is the inertia subgroup, $\mathrm{Frob}_\ell \in G_{\mathbb{Q}_\ell}/I_\ell$ is the arithmetic Frobenius element, and $V_pE_{I_\ell}$ denotes the inertia coinvariants. Additionally, p-adic Hodge theory allows the Euler factor at p to also be recovered from V_pE.

There is a similar story for holomorphic newforms $f \in S_2(\Gamma_0(N))$, as explained in Sect. 6.1. We use the notation introduced in Sect. 6.1: so $\rho_f : G_{\mathbb{Q}} \rightarrow \mathrm{Aut}_K(V_f)$, $K = \mathbb{Q}(f)_{\mathfrak{p}}$ and $\dim_K V_f = 2$, is the p-adic Galois representation associated to f and some prime $\mathfrak{p} \mid p$ of the field $\mathbb{Q}(f)$ generated by the Hecke eigenvalues of f.

This leads to the following Galois-theoretic reformulation of The Modularity Conjecture:

Conjecture 26 (The Modularity Conjecture for Elliptic Curves (Galois formulation)) *Let $E/\mathbb{Q}$ be an elliptic curve and p a prime. There is a cuspidal newform f and a prime above p in $\mathbb{Q}(f)$ such that the continuous $G_{\mathbb{Q}}$-representations V_pE and V_f are isomorphic.*

Note that the f in this conjecture, if it exists, must necessarily be the f in the first version of The Modularity Conjecture. Furthermore, if this Galois version holds for one prime p then it holds for all primes p.

So to prove the modularity of elliptic curves, this Galois-theoretic formulation of The Modularity Conjecture suggests that one attempt to identify the set of Galois representations V_pE as a subset of the V_f. But how can one do this? The key, it turns out, is to attempt something more ambitious: prove that any two-dimensional p-adic Galois representations that looks like the Galois representation attached to a newform is indeed isomorphic to some V_f (after possibly extending scalars). Such a conjecture had certainly been in the air for some time, but it was finally written down by Fontaine and Mazur. But why should this prove any more tractable? And where does one start?

A place to start is suggested by a conjecture of Serre, essentially a 'mod p' version of the modularity conjecture of Fontaine and Mazur. Let $\rho : G_{\mathbb{Q}} \to \mathrm{Aut}_K(V)$ be a continuous representation with K a finite extension of $\mathbb{Q}_p$ and V a two-dimensional K-space. Let O be the ring of integers of K. Then V contains a Galois-stable O-lattice, say T; this is a free O-module of rank 2 with a continuous O-linear $G_{\mathbb{Q}}$-action. Let $\mathfrak{p} \subset O$ denote the maximal ideal and let $\mathbb{F} = O/\mathfrak{p}$ be the residue field, a finite extension of $\mathbb{F}_p$. Let $\overline{V} = T/\mathfrak{p}T$ be the mod $\mathfrak{p}$-reduction of T. So $\overline{V}$ is a two-dimensional $\mathbb{F}$-space with a continuous $\mathbb{F}$-linear action $\bar{\rho} : G_{\mathbb{Q}} \to \mathrm{Aut}_{\mathbb{F}}(\overline{V})$. If $\overline{V}$ is irreducible then its isomorphism class does not depend on the choice of T. The representation $\bar{\rho}$ inherits many properties from ρ (e.g., if ρ is odd, unramified outside a finite set of primes, or its determinant is a power of the cyclotomic character, then the same holds for $\bar{\rho}$).

Conjecture 27 (Serre's Conjecture) *Let $\mathbb{F}$ be a finite extension of $\mathbb{F}_p$ and $\bar{\rho} : G_{\mathbb{Q}} \to Aut_{\mathbb{F}}(\overline{V})$ a continuous 2-dimensional $\mathbb{F}$-representation of $G_{\mathbb{Q}}$. Suppose $\bar{\rho}$ is odd, irreducible, and unramified outside a finite set of primes. Then there is a newform of level $N(\bar{\rho})$, weight $k(\bar{\rho})$, and Nebentypus $\epsilon_{\bar{\rho}}$ and a prime above p in $\mathbb{Q}(f)$ such that $\bar{\rho}_f$ is isomorphic to $\bar{\rho}$ (after possibly extending scalars).*

In this conjecture $N(\bar{\rho})$ and $k(\bar{\rho})$ are well-defined integers depending only on $\overline{V}$. In particular, $N(\bar{\rho})$ is just the prime-to-p Artin conductor of the representation $\overline{V}$. The character $\epsilon_{\bar{\rho}}$ satisfies $\det \bar{\rho} = \epsilon_{\bar{\rho}}\omega^{k(\bar{\rho})-1}$.

Let $\rho : G_{\mathbb{Q}} \to \mathrm{Aut}_K(V)$ be as above. We say that ρ is modular if it is isomorphic to some ρ_f (after possibly extending scalars). Similarly, we say that $\bar{\rho}$ is modular if it is isomorphic to some $\bar{\rho}_f$ (again, after possibly extending scalars).

Accepting Serre's Conjecture, one then might the ask the question: If $\bar{\rho}$ is modular, is ρ also modular? This is precisely the kind of question addressed by the modularity lifting theorems in [16, 17].

7.3 Wiles's Modularity Lifing Theorem

In [16] Wiles restricts the focus to $p \geq 3$ and $\bar{\rho}$ such that

(r1) $\bar{\rho}|_{G_{\mathbb{Q}_p}}$ is either ordinary and $G_{\mathbb{Q}_p}$-distinguished (meaning that there is a $G_{\mathbb{Q}_p}$-stable line $\overline{V}^+ \subset \overline{V}$ such that I_p acts trivially on $\overline{V}/\overline{V}^+$ and $G_{\mathbb{Q}_p}$ acts by distinct characters on $\overline{V}^+$ and $\overline{V}/\overline{V}^+$) or flat (meaning that it is the Galois representation on the $\overline{\mathbb{Q}}_p$-points of a finite flat group scheme over $\mathbb{Z}_p$);
(r2) if $\bar{\rho}$ is ramified at a prime $\ell \equiv -1 \bmod p$, then either $\bar{\rho}|_{G_{\mathbb{Q}_\ell}}$ is reducible (after possibly extending scalars) or $\bar{\rho}|_{I_\ell}$ is absolutely irreducible;
(r3) $\bar{\rho}|_{\mathrm{Gal}(\overline{\mathbb{Q}}/\mathbb{Q}(\sqrt{p^*}))}$ is absolutely irreducible, where $p^* = (-1)^{\frac{p-1}{2}} p$.

One of the main results in [16] is:

Theorem 28 ([16, Thm. 0.2]) *If $\rho : G_{\mathbb{Q}} \to \mathrm{Aut}_K(V)$ is such that*

(a) *ρ is continuous, unramified outside a finite set of primes, and odd;*
(b) *$\bar{\rho}$ is modular and satisfies (r1)-(r3);*
(c) *either (i) $\bar{\rho}$ is ordinary, $\det \rho = \chi \varepsilon^{k-1}$ for an integer $k \geq 2$ and χ a finite order character, and V has a $G_{\mathbb{Q}_p}$-stable line $V^+ \subset V$ such that V/V^+ is unramified at p, or (ii) $\bar{\rho}$ is flat and $\rho|_{G_{\mathbb{Q}_p}}$ is equivalent to the representation on the Tate module of a p-divisible group,*

then ρ is modular.

At first glance, such a theorem seems to fall short of what is wanted since it still has the modularity of $\bar{\rho}$ as a hypothesis. But that it has real teeth – especially when combined with the existence of compatible families of p-adic Galois representations (for varying p) – can be seen from some of its consequences.

The proof of this modularity lifting theorem galvanized the algebraic number theory world. It went far beyond what anyone else had imagined could be proved at the time. As Barry Mazur remarked in the BBC/Horizon documentary about Wiles's proof: *"What was unique about those lectures [by Wiles, on his proof] were the glorious ideas, how many new ideas were presented...."*

7.4 Modularity of Semistable Elliptic Curves and FLT

One consequence of the modularity lifting theorem of the preceding section is:

Theorem 29 ([16, Thm. 0.4]) *Let E be a semistable elliptic curve over $\mathbb{Q}$. Then E is modular.*

Let $\rho_{E,p} : G_{\mathbb{Q}} \to \mathrm{Aut}_{\mathbb{Z}_p}(T_p E) \cong \mathrm{GL}_2(\mathbb{Z}_p)$ be the Galois representation on the p-adic Tate module of E. Then $\bar{\rho}_{E,p}$ is just the representation of $G_{\mathbb{Q}}$ on $E[p]$. The deduction of the modularity of E from Wiles's modularity lifting theorem follows from a remarkable interplay of the representations $\bar{\rho}_{E,3}$ and $\bar{\rho}_{E,5}$.

The 3 − 5 switch.

Suppose first that $\bar{\rho}_{E,3}$ is irreducible. As E is semistable, the properties (r1)-(r3) are satisfied. And fortuitously, $\bar{\rho}_{E,3}$ is modular: this follows from deep results of Langlands and Tunnell proving cases of the Artin Conjecture. Since E is semistable, all the hypotheses of the modularity lifiting theorem are satisfied: E is modular.

Next, suppose that $\bar{\rho}_{E,3}$ is reducible. As E is semistable, it turns out that $\bar{\rho}_{E,5}$ must be irreducible. And as E is semistable, $\bar{\rho}_{E,5}$ satisfies (r1)-(r3). But the results of Langlands and Tunnell do not apply to $\bar{\rho}_{E,5}$. Wiles proves that there is another elliptic curve E' such that $\bar{\rho}_{E',5} \cong \bar{\rho}_{E,5}$ and $\bar{\rho}_{E',3}$ is irreducible and either E' or a quadratic twist of E' is semistable. Applying the modularity lifting theorem to $\rho_{E',3}$ (or a quadratic twist – modularity is stable under twisting) we then have that E', and hence $\rho_{E',5}$, is modular. This implies that $\bar{\rho}_{E,5} \cong \bar{\rho}_{E',5}$ is modular, and so – applying the modularity lifting theorem again, this time to $\rho_{E,5}$ – E is modular!

7.5 Fermat's Last Theorem is True

The most celebrated consequence of the modularity of semistable curves is:

Theorem 30 ([16, Thm. 0.5]) *Let $p \geq 3$ be a prime. If $u^p + v^p + w^p = 0$ with $u, v, w \in \mathbb{Q}$. Then $uvw = 0$.*

In particular, Fermat's Last Theorem is true. The proof is based on a remarkable idea of Frey, made precise by Serre in [82]: If there were a solution with $uvw \neq 0$, then there would exist a semistable elliptic curve E (constructed from u, v, and w) such that $N(\bar{\rho}_{E,p}) = 2$ and $k(\bar{\rho}_{E,p}) = 2$. Ribet [72] proved that if $\bar{\rho}_{E,p}$ is modular, then it is isomorphic to some $\bar{\rho}_f$ for an f of level $N(\bar{\rho}_{E,p})$ and weight $k(\bar{\rho}_{E,p})$. But there are no cuspidal newforms of level 2 and weight 2. And Wiles proved that such an E, and hence $\bar{\rho}_{E,p}$, must be modular!

7.6 A Brief Overview of the Proof of Modularity Lifting

Let $\bar{\rho} : G_{\mathbb{Q}} \to \mathrm{GL}_2(\mathbb{F})$ be a representation as in the theorem in Sect. 7.3. The proof of the modularity lifting theorem begins by reformulating it in terms of rings.

A Ring-Theoretic Reformulation

This starts by considering deformations of $\bar{\rho}$.

Let A be a complete local Noetherian ring with maximal ideal $\mathfrak{m}_A$ and residue field $\mathbb{F}$. A deformation of $\bar{\rho}$ over A is an equivalence classes of representations ρ :

$G_{\mathbb{Q}} \to \mathrm{GL}_2(A)$ such that $\rho \bmod \mathfrak{m}_A = \bar{\rho}$. It is natural to impose various conditions on these deformations (e.g., restricting the ramification or imposing restrictions on $\rho|_{G_{\mathbb{Q}_p}}$). In particular, Wiles considers deformation conditions ('deformation data') $\mathcal{D} = (?, \Sigma, \mathcal{O}, \mathcal{M})$, where $? \in \{\text{flat}, \text{ord}, \text{Sel}, \text{str}\}$, Σ is a finite set of primes containing p and all those at which $\bar{\rho}$ is ramified, $\mathcal{O}$ is the ring of integers of some p-adic field with residue field $\mathbb{F}$; $\mathcal{M}$ is a subset of those primes different from p at which $\bar{\rho}$ is ramified. A deformation ρ is of type $\mathcal{D}$ if (i) A is an $\mathcal{O}$-algebra; (ii) ρ is unramified outside Σ; (iii) at each prime $q \in \mathcal{M}$, $\rho|_{G_{\mathbb{Q}_q}}$ is 'minimal' (essentially the same as $\bar{\rho}|_{G_{\mathbb{Q}_q}}$); and (iv) $\rho|_{G_{\mathbb{Q}_p}}$ is of type ?, which is a restriction on $\rho|_{G_{\mathbb{Q}_p}}$ (e.g., if ? = flat then each $\rho \bmod \mathfrak{m}_A^n$ arises from a finite flat group scheme). When $? \in \{\text{flat}, \text{Sel}, \text{str}\}$, then the determinant of ρ is just the product of the p-adic cyclotomic character ε and a finite-order character.

It turns out that there is a universal deformation $\rho_{\mathcal{D}} : G_{\mathbb{Q}} \to \mathrm{GL}_2(R_{\mathcal{D}})$ of type $\mathcal{D}$; this goes back to work of Mazur [66], who introduced the study of deformations of Galois representations. The ring $R_{\mathcal{D}}$ is a complete local Noetherian $\mathcal{O}$-algebra and so a quotient of a power series ring $\mathcal{O}[\![x_1, \ldots, x_g]\!]$. The number g of variables x_i can be taken to equal $\dim_{\mathbb{F}} \mathfrak{m}_{R_{\mathcal{D}}}/(\varpi, \mathfrak{m}_{R_{\mathcal{D}}}^2)$, where ϖ is a uniformizer of $\mathcal{O}$. By the universality property of $R_{\mathcal{D}}$, $\mathrm{Hom}_{\mathbb{F}}(\mathfrak{m}_{R_{\mathcal{D}}}/(\varpi, \mathfrak{m}_{R_{\mathcal{D}}}^2), \mathbb{F})$ is identified with the deformations over the dual numbers $\mathbb{F}[\epsilon]/(\epsilon^2)$ that are of type $\mathcal{D}$. The latter can also be identified with the elements of a Galois cohomology group $H^1_{\mathcal{D}}(\mathbb{Q}, \mathrm{ad}\bar{\rho}) \subset H^1(\mathbb{Q}, \mathrm{ad}\bar{\rho})$, which is defined by local conditions. So

$$g = \dim_{\mathbb{F}} \mathrm{Hom}_{\mathbb{F}}(\mathfrak{m}_{R_{\mathcal{D}}}/(\varpi, \mathfrak{m}_{R_{\mathcal{D}}}^2), \mathbb{F}) = \dim_F H^1_{\mathcal{D}}(\mathbb{Q}, \mathrm{ad}\bar{\rho}). \qquad \text{(ML1)}$$

This gives an initial arithmetic handle on $R_{\mathcal{D}}$.

Suppose $\rho : G_{\mathbb{Q}} \to \mathrm{GL}_2(\mathcal{O})$ is a type $\mathcal{D}$ deformation of $\bar{\rho}$, such as might arise from the Tate module of an elliptic curve or be associated with a newform. Then there is an $\mathcal{O}$-algebra homomorphism $\pi : R_{\mathcal{D}} \twoheadrightarrow \mathcal{O}$ by universality. Let $\mathfrak{p} = \ker \pi$ (so $\rho_{\mathcal{D}} \bmod \mathfrak{p}$ is in the deformation class of ρ). By considering the type $\mathcal{D}$ deformations over $\mathcal{O}[\epsilon]/(\varpi^n\epsilon, \epsilon^2)$, one finds that

$$\mathrm{Hom}_{\mathcal{O}}(\mathfrak{p}/\mathfrak{p}^2, K/\mathcal{O}) = H^1_{\mathcal{D}}(\mathbb{Q}, \mathrm{ad}\rho \otimes_{\mathcal{O}} K/\mathcal{O}). \qquad \text{(ML2)}$$

Here K is the field of fractions of $\mathcal{O}$ and the subscript '$\mathcal{D}$' denotes a subgroup defined by local conditions depending on $\mathcal{D}$. This observation played a key role in Wiles's proof of his modularity lifting theorem. The cohomology group $H^1_{\mathcal{D}}(\mathbb{Q}, \mathrm{ad}\rho \otimes_{\mathcal{O}} K/\mathcal{O})$ is closely related to the Selmer groups defined by Greenberg and Bloch and Kato and others, and its order is expected to be related to a special value of a certain L-function.

As $\bar{\rho}$ is assumed to be irreducible and modular, by work of Ribet and others [45, 72] there is an eigenform with an associated p-adic Galois representation that is a deformation of type $\mathcal{D}$. From this it follows that there is a universal modular deformation $\rho^{\mathrm{mod}}_{\mathcal{D}} : G_{\mathbb{Q}} \to \mathrm{GL}_2(\mathbb{T}_{\mathcal{D}})$ of type $\mathcal{D}$. If ? $\neq$ ord, then $\mathbb{T}_{\mathcal{D}}$ is a localization of a weight two Hecke algebra $\mathbb{T}_2(N_{\mathcal{D}}; \mathcal{O})$ at a particular maximal ideal

$\mathfrak{m}_{\mathcal{D}}$. If ? = ord, then $\mathbb{T}_{\mathcal{D}}$ is a projective limit of such algebras with the p-part of the level allowed to vary. The representation $\rho_{\mathcal{D}}$ captures all the modular forms whose representations are deformations of type $\mathcal{D}$.

By universality there is a (surjective) map

$$\varphi_{\mathcal{D}} : R_{\mathcal{D}} \twoheadrightarrow \mathbb{T}_{\mathcal{D}}.$$

If $\varphi_{\mathcal{D}}$ were an isomorphism, then every deformation $\rho : G_{\mathbb{Q}} \to \mathrm{GL}_2(O)$ of type $\mathcal{D}$ would be modular – it would come from an O-homomorphism $\mathbb{T}_{\mathcal{D}} \to O$. If ? $\neq$ ord, then this would imply that ρ is the Galois representation associated with a newform. If ? = ord, then this would follow provided $\det \rho = \chi \varepsilon^{k-1}$ for an integer $k \geq 2$ and χ a finite order character.

So an ambitious version of the modularity lifting theorem would be: *$\varphi_{\mathcal{D}}$ is an isomorphism.* And this is exactly what Wiles proved. But why would one expect this? And where could one start to prove it?

If ? $\neq$ ord, then $\mathbb{T}_{\mathcal{D}}$ is a finite free O-module. Let $\pi : \mathbb{T}_{\mathcal{D}} \to O$ be an O-homomorphism associated with an eigenform. Let $\mathfrak{p} = \ker(\pi \circ \varphi_{\mathcal{D}}) \subset R_{\mathcal{D}}$ and let $\eta = \pi(\mathrm{Ann}_{\mathbb{T}_{\mathcal{D}}}(\ker \pi)) \subset O$ (this is non-zero). Wiles proves a remarkable isomorphism criterion:

$$\varphi_{\mathcal{D}} \text{ is an isomorphism of complete intersections over } O \iff \#\mathfrak{p}/\mathfrak{p}^2 = \#O/\eta. \tag{ML3}$$

So to prove that $\varphi_{\mathcal{D}}$ is an isomorphism it suffices to verify the equality on the right-hand side. Wiles's proof, however, takes a slightly different path.

Wiles's proof of the modularity lifting theorem proceeds in two large steps, making use of both implications in (ML3). He first proves that $\varphi_{\mathcal{D}_{\min}}$ is an isomorphism of complete intersections over O for a minimal deformation problem $\mathcal{D}_{\min}$. It follows from the $\Rightarrow$ implication of (ML3) that the equality on the right-hand side of (ML3) holds for the minimal deformation problem $\mathcal{D}_{\min}$. He then combines this equality with a close analysis of how η changes with the deformation problem to conclude that the right-hand equality of (ML3) also holds for non-minimal $\mathcal{D}$. It then follows from the $\Leftarrow$ implication of (ML3) that $\varphi_{\mathcal{D}}$ is also an isomorphism of complete intersections over O for non-minimal $\mathcal{D}$.

Remark 31 The equality on the right-hand side of (ML3) is suggested by the Bloch–Kato conjectures for the symmetric square L-function of the eigenform f. Using that $H^1(X_1(N_{\mathcal{D}}), O)_{\mathfrak{m}_{\mathcal{D}}}$ is a free $\mathbb{T}_{\mathcal{D}}$-module of rank 2 (which Wiles proves, extending and generalizing earlier arguments of Mazur), it then follows from a result of Hida that

$$\#O/\eta = \#O/(N_{\mathcal{D}}\phi(N_{\mathcal{D}})L_{\mathcal{D}}(\mathrm{Sym}^2 f, 2)/4\pi \Omega_f^+ \Omega_f^-),$$

where $\Omega_f^{\pm}$ are the canonical periods associated to f (these are well-defined up to units in O) and the superscript '$\mathcal{D}$' denotes an incomplete symmetric square L-

function (the missing Euler factors depend on $\mathcal{D}$). On the other hand, the Bloch–Kato conjectures predict that

$$\#\mathcal{O}/(N_{\mathcal{D}}\phi(N_{\mathcal{D}})L_{\mathcal{D}}(\mathrm{Sym}^2 f, 2)/4\pi\Omega_f^+\Omega_f^-) = \#H^1_{\mathcal{D}}(\mathbb{Q}, \mathrm{ad}\rho_f \otimes_{\mathcal{O}} K/\mathcal{O}).$$

Combined with (ML2) this would imply the equality on the right-hand side of (ML3).

$R_{\mathcal{D}} = \mathbb{T}_{\mathcal{D}}$: the minimal case

There is a minimal deformation problem $\mathcal{D}_{\min}$ for $\bar{\rho}$: essentially, ? is as restrictive as possible, Σ is the set of primes at which $\bar{\rho}$ is ramified together with p, and $\mathcal{M} = \Sigma\backslash\{p\}$. That $\varphi_{\mathcal{D}_{\min}}$ is an isomorphism of complete intersections over $\mathcal{O}$ is the main result of the article [17] by Wiles and Richard Taylor. This result is proved by what has come to be called 'Taylor–Wiles patching.' It is in this argument that the condition (r3) is crucially used. Another key ingredient here is the work of Ribet *et al.* proving that if $\bar{\rho}$ is modular (hypothesis (b) in the theorem) then it has a modular lift of minimal type $\mathcal{D}_{\min}$, so $\mathbb{T}_{\mathcal{D}_{\min}}$ exists. The patching proceeds more-or-less as follows.

Let $g = \dim_{\mathbb{F}} H^1_{\mathcal{D}_{\min}}(\mathbb{Q}, \mathrm{ad}\bar{\rho})$. For each integer $m > 0$, Wiles proved that there are infinitely many g-tuples $Q = (q_1, \ldots, q_g)$ of distinct primes $q_i \equiv 1 \bmod p^m$ for which there is an auxiliary deformation problem $\mathcal{D}_Q$ such that: (i) $\dim_{\mathbb{F}} H^1_{\mathcal{D}_Q}(\mathbb{Q}, \mathrm{ad}\bar{\rho}) = g$; (ii) $R_{\mathcal{D}_Q}$ and $\mathbb{T}_{\mathcal{D}_Q}$ are naturally $\mathcal{O}[\Delta_Q]$-modules, Δ_Q being the p-Sylow subgroup of $\prod_{q_i \in Q}(\mathbb{Z}/q_i\mathbb{Z})^\times$; (iii) $R_{\mathcal{D}_Q}/I_Q R_{\mathcal{D}_Q} \xrightarrow{\sim} R_{\mathcal{D}_{\min}}$ and $\mathbb{T}_{\mathcal{D}_Q}/I_Q\mathbb{T}_{\mathcal{D}_Q} \xrightarrow{\sim} \mathbb{T}_{\mathcal{D}_{\min}}$ for $I_Q \subset \mathcal{O}[\Delta_Q]$ the augmentation ideal; and (iv) $\mathbb{T}_{\mathcal{D}_Q}$ is a free $\mathcal{O}[\Delta_Q]$-module of finite rank. Both the minimality of $\mathcal{D}_{\min}$ and the condition (r3) are used to find these sets Q.

By property (i) of the sets Q, each $R_{\mathcal{D}_Q}$ (and hence $\mathbb{T}_{\mathcal{D}_Q}$) can be presented as a quotient of the power series ring $P = \mathcal{O}[[x_1, \ldots, x_g]]$ (see (ML1)). The ring $\mathcal{O}[\Delta_Q]$ can be expressed as a quotient of the power series ring $S = \mathcal{O}[[y_1, \ldots, y_g]]$ by mapping each $1 + y_i$ to a generator δ_i of the p-Sylow subgroup of $(\mathbb{Z}/q_i\mathbb{Z})^\times$. So by (ii) there is a homomorphism of $\mathcal{O}$-algebras $S \to R_{\mathcal{D}_Q}$ such that $1 + y_i$ maps to $\delta_i \in R_{\mathcal{D}_Q}$, and by (iii) $R_{\mathcal{D}_Q}/(y_1, \ldots, y_g)R_{\mathcal{D}_Q} = R_{\mathcal{D}_{\min}}$ and $\mathbb{T}_{\mathcal{D}_Q}/(y_1, \ldots, y_g)\mathbb{T}_{\mathcal{D}_Q} = \mathbb{T}_{\mathcal{D}_{\min}}$. Intuitively, property (iv) shows that $\mathbb{T}_{\mathcal{D}_Q}$ (and hence also $R_{\mathcal{D}_Q}$ and P) comes 'closer and closer' to being a free S-module of finite rank as m increases. If it were possible to make sense of the case $m = \infty$, then the corresponding Hecke ring would be a free S-module of finite rank and so have Krull dimension at least $g + 1$. As P is a regular local ring of Krull dimension $g + 1$, it follows that the surjective homomorphism from P to the Hecke algebra would have to be an isomorphism. As this map would factor through the (putative) deformation ring for this case, it would then also follow that this deformation ring would have to be isomorphic to the Hecke ring, and hence – looking modulo $(y_1, \ldots, y_g)$ – that $\varphi_{\mathcal{D}_{\min}}$ is an isomorphism of complete intersections over $\mathcal{O}$. The patching argument makes up for not having such a deformation problem for $m = \infty$.

Recall that ϖ is a uniformizer of $\mathcal{O}$. Replace $\mathbb{T}_{\mathcal{D}_{\min}}$ and $R_{\mathcal{D}_{\min}}$ with their respective quotients $\mathbb{T} = \mathbb{T}_{\mathcal{D}_{\min}}/\varpi\mathbb{T}_{\mathcal{D}_{\min}}$ and $R = R_{\mathcal{D}_{\min}}/(\varpi, \mathfrak{p}\ker(\varphi_{\mathcal{D}_{\min}}))$. As $\mathbb{T}_{\mathcal{D}_{\min}}$ is a finite free $\mathcal{O}$-module, to prove that $\varphi_{\mathcal{D}_{\min}}$ is an isomorphism of complete intersections over $\mathcal{O}$ it suffices to prove that the induced map $\varphi : R \twoheadrightarrow T$ is an isomorphism of complete intersections over $\mathbb{F}$.

Let $I_m = ((1+y_1)^{p^m} - 1, \ldots, (1+y_g)^{p^m} - 1) \subset S$. To finish the 'patching' and the proof that φ is an isomorphism of complete intersections, for each $m \geq 1$ consider the quotient rings $\mathbb{T}_{Q,m} = \mathbb{T}_{\mathcal{D}_Q}/(\varpi, I_m)\mathbb{T}_{\mathcal{D}_Q}$ and $R_{Q,m} = \mathrm{im}\{R_{\mathcal{D}_Q} \to \mathbb{T}_{Q,m} \times R\}$ and the induced surjection $\varphi_{Q,m}$. A key point here is that all these rings are finite $\mathbb{F}$-spaces with $\mathbb{F}$-dimension bounded in terms of m and $\dim_{\mathbb{F}} R$. The result is an infinite collection of commutative diagrams

$$\begin{array}{ccccc}
P \otimes_{\mathcal{O}} \mathbb{F} = \mathbb{F}[[x_1, \ldots, x_g]] & \twoheadrightarrow & R_{Q,m} & \xrightarrow{\varphi_{Q,m}} & \mathbb{T}_{Q,m} \\
& \nearrow & \downarrow{\scriptstyle /(y_i)} & & \downarrow{\scriptstyle /(y_i)} \\
S \otimes_{\mathcal{O}} \mathbb{F} = \mathbb{F}[[y_1, \ldots, y_g]] & & R & \xrightarrow{\varphi} & \mathbb{T}
\end{array}$$

with $\mathbb{T}_{Q,m}$ a free module over $\mathbb{F}[[y_1, \ldots, y_g]]/I_m$. For a fixed m there are only finitely many isomorphism classes of such diagrams. So appealing to a pigeon hole principle argument yields a sequence of sets Q_m, $m \geq 1$, whose associated diagrams are compatible as m varies. Taking limits over m one ends up with isomorphisms $\mathbb{F}[[x_1, \ldots, x_g]] \xrightarrow{\sim} \varprojlim_m R_{Q_m,m} \xrightarrow{\sim} \varprojlim_m \mathbb{T}_{Q_m,m}$ and $\mathbb{F}[[x_1, \ldots, x_g]]/(y_1, \ldots, y_g) \xrightarrow{\sim} R \xrightarrow{\sim} \mathbb{T}$, essentially by the argument sketched earlier.

$R_{\mathcal{D}} = \mathbb{T}_{\mathcal{D}}$: Relaxing Restrictions

Wiles showed that $\varphi_{\mathcal{D}}$ is an isomorphism of complete intersections over $\mathcal{O}$ if $\varphi_{\mathcal{D}_{\min}}$ is an isomorphism of complete intersections over $\mathcal{O}$. The proof made use of the numerical criterion (ML3). Let $\pi_{\mathcal{D}_{\min}} : \mathbb{T}_{\mathcal{D}_{\min}} \to \mathcal{O}$ be a homomorphism (this is associated with an eigenform). For each deformation problem $\mathcal{D}$ this induces a homomorphism $\pi_{\mathcal{D}} : \mathbb{T}_{\mathcal{D}} \to \mathcal{O}$. Let $\mathfrak{p}_{\mathcal{D}} = \ker(\pi_{\mathcal{D}} \circ \varphi_{\mathcal{D}})$ and $\eta_{\mathcal{D}} = \pi_{\mathcal{D}}(\mathrm{Ann}_{\mathbb{T}_{\mathcal{D}}}(\ker \pi_{\mathcal{D}}))$. By (ML2),

$$\#\mathfrak{p}_{\mathcal{D}}/\mathfrak{p}_{\mathcal{D}}^2 = \#H^1_{\mathcal{D}}(\mathbb{Q}, \mathrm{ad}\rho \otimes_{\mathcal{O}} K/\mathcal{O}) \leq \#H^1_{\mathcal{D}_{\min}}(\mathbb{Q}, \mathrm{ad}\rho \otimes_{\mathcal{O}} K/\mathcal{O}) \cdot \prod_{\ell \in \Sigma} c_{\mathcal{D},\ell},$$

where each $c_{\mathcal{D},\ell}$ is the order of a subquotient of $H^1(\mathbb{Q}_\ell, \mathrm{ad}\rho \otimes_{\mathcal{O}} K/\mathcal{O})$ depending on $\mathcal{D}$. It follows that

$$\#\mathfrak{p}_{\mathcal{D}}/\mathfrak{p}_{\mathcal{D}}^2 \leq \#\mathfrak{p}_{\mathcal{D}_{\min}}/\mathfrak{p}_{\mathcal{D}_{\min}}^2 \cdot \prod_{\ell \in \Sigma} c_{\mathcal{D},\ell}.$$

On the other hand, by a careful analysis of how the congruence ideals $\eta_{\mathcal{D}}$ change as $\mathcal{D}$ varies, Wiles also proves – generalizing work of Ihara, Ribet, and Mazur – that

$$\#\mathcal{O}/(\eta_{\mathcal{D}}) \geq \#\mathcal{O}/(\eta_{\mathcal{D}_{\min}}) \cdot \prod_{\ell \in \Sigma} c_{\mathcal{D},\ell}.$$

As $\varphi_{\mathcal{D}_{\min}}$ is an isomorphism of complete intersections over $\mathcal{O}$, $\#\mathfrak{p}_{\mathcal{D}_{\min}}/\mathfrak{p}^2_{\mathcal{D}_{\min}} = \#\mathcal{O}/(\eta_{\mathcal{D}_{\min}})$ by (ML3). Combining this with the two preceding inequalities yields $\#\mathfrak{p}_{\mathcal{D}}/\mathfrak{p}^2_{\mathcal{D}} \leq \#\mathcal{O}/(\eta_{\mathcal{D}})$. As the opposite equality always holds, it follows that $\#\mathfrak{p}_{\mathcal{D}}/\mathfrak{p}^2_{\mathcal{D}} = \#\mathcal{O}/(\eta_{\mathcal{D}})$ and hence – appealing to (ML3) once more – that $\varphi_{\mathcal{D}}$ is an isomorphism of complete intersections over $\mathcal{O}$.

A complete exposition of much of the arguments in [16, 17], focusing on the proof of modularity of semistable curves (and hence Fermat's Last Theorem), is given in the article [42] by Darmon, Diamond, and Taylor.

Remark 32 As recalled in the remark at the end of Sect. 7.6, the equality in (ML3) can be restated as an instance of the Bloch–Kato conjecture for the symmetric square L-function of a modular form. In particular, cases of this conjecture are a *corollary* of Wiles's proof of the modularity lifting theorem. The proof of cases of the Bloch–Kato conjecture for the symmetric square L-functions via such '$R = \mathbb{T}$' results was further developed by Diamond, Flach, and Guo in [48].

7.7 Modularity of Residually Reducible Representations

While the '3 – 5 switch' showed that Wiles's modularity lifting theorem was sufficient to prove modularity of all semistable elliptic curves, it seems unlikely that one will always have similar tricks at one's disposal. And so it is desirable to remove some of the conditions imposed in (r1)-(r3). The papers [18, 20, 22] made progress toward removing the irreducibility required in (r3). For example, one of the main results of [20] is:

Theorem 33 ([20, Thm.]) *Suppose that $\rho : G_{\mathbb{Q}} \to \mathrm{GL}_2(E)$ is a continuous representation, irreducible and unramified outside a finite set of primes, where E is a finite extension of $\mathbb{Q}_p$. Suppose also that $\bar{\rho}^{ss} \cong 1 \oplus \chi$ and that*

(i) $\chi|_{G_{\mathbb{Q}_p}} \neq 1$;
(ii) $\rho|_{I_p}$ *is equivalent to a subgroup of* $\begin{pmatrix} * & * \\ 0 & 1 \end{pmatrix}$;
(iii) $\det \rho = \psi \varepsilon^{k-1}$ *for some* $k \geq 2$ *and is odd;*

where ψ is of finite order. Then ρ is the representation associated with a modular form.

The proof of this theorem combined some of the elements of the proof of the main conjecture with some generalizations of the arguments in [16, 17]. Another feature of this proof was the crucial use of solvable base change for modular forms.

A similar but generally weaker theorem is proved for representations of G_F for a totally real field F [20, Thms. A & B].

These theorems are proved by considering deformations rings for all the residual representations corresponding to extensions of χ by 1. These deformation rings are less well-behaved than in the residually irreducible case. In particular, they may not be equidimensional: the dimension of the part corresponding to reducible deformations will grow with the set Σ at which ramification is allowed, while one expects the part corresponding to irreducible deformations to have finite dimension (equal to $d + 2 + \delta_F$ for d the degree of the totally real field F and $0 \leq \delta_F \leq d/2$ the possible defect in Leopoldt's conjecture for F). Matters are further complicated by the fact that in general there is no representation over the associated Hecke algebras (which are localizations of Hecke rings at *Eisenstein* maximal ideals) but only pseudorepresentations. So there is no natural map from the deformation rings to the Hecke algebras.

The argument in [20] proceeds roughly as follows. One first considers certain deformation problems $\mathcal{D} = (\mathcal{O}, \Sigma, c, \mathcal{M})$ over a totally real field F, with $\mathcal{O}$ the ring of integers of a local field with finite residue field $\mathbb{F}$ containing the values of χ; Σ is a finite set of primes of F containing all those over p and those at which χ is ramified, $c \in H^1(F, \mathbb{F}(\chi^{-1}))$ is a class unramified outside Σ and split at all primes of F above p, and $\mathcal{M} \subset \Sigma$ is a set of primes of F not dividing p but at which either c or χ is ramified. Much as in the irreducible case, there is a universal deformation ring $R_{\mathcal{D}}$ classifying certain deformations of the residual representation

$$\rho_c : \mathrm{Gal}(F_\Sigma/F) \to \left(\begin{smallmatrix} 1 & * \\ 0 & \chi \end{smallmatrix}\right),$$

where the implied extension is given by c. A prime $\mathfrak{p} \subset R_{\mathcal{D}}$ is said to be pro-modular if the pseudorepresentation associated with $\rho_{\mathcal{D}}$ mod $\mathfrak{p}$ is induced from the pseudorepresentation over the related Hecke ring $\mathbb{T}_{\mathcal{D}}$ via an $\mathcal{O}$-algebra homomorphism $\mathbb{T}_{\mathcal{D}} \to R_{\mathcal{D}}/\mathfrak{p}$. The aim is then to show that all primes of $R_{\mathcal{D}}$ are pro-modular.

Under suitable hypotheses – which include that the degree d of F is very large compared to the size of Σ, the number of primes of F above p, and the dimension of the space of possible c's – the pro-modularity of all the primes of $R_{\mathcal{D}}$ is shown in essentially three steps: (I) prove that for a 'nice' prime $\mathfrak{p} \subset R_{\mathcal{D}}$ every minimal prime contained in $\mathfrak{p}$ is pro-modular (and hence all primes on the components of Spec $R_{\mathcal{D}}$ corresponding to such minimal primes are pro-modular); (II) prove that $R_{\mathcal{D}}$ has at least one nice prime; and (III) conclude that all primes of $R_{\mathcal{D}}$ are pro-modular. Included in the definition of a 'nice' prime $\mathfrak{p}$ is that it is pro-modular and that $R_{\mathcal{D}}/\mathfrak{p}$ has characteristic p.

The proof of (I) is modeled on the proof of the modularity lifting theorem. However, complications arise as the residual representation is now taken to be the representation $\rho_{\mathfrak{p}} = \rho_{\mathcal{D}}$ mod $\mathfrak{p}$ over the (infinite, characteristic p) fraction field of $R_{\mathcal{D}}/\mathfrak{p}$. As there is no natural map from $R_{\mathcal{D}}$ to $\mathbb{T}_{\mathcal{D}}$ the proof relies on the link provided by a universal *pseudo-deformation* ring $R_{\mathcal{D}}^{ps}$. There are then natural homomorphisms $R_{\mathcal{D}}^{ps} \to R_{\mathcal{D}}$ and $R_{\mathcal{D}}^{ps} \to \mathbb{T}_{\mathcal{D}}$. The patching and congruence arguments of [16, 17] are adapted to show that these maps become isomorphisms after localization and completion at a nice prime.

The proof of (II) uses the Eisenstein congruences as exploited in the proof of the main conjecture for totally real fields [13] to show that there is a nice prime $\mathfrak{p}_0$ for *some* extension c_0 of χ by 1. Then a ring-theoretic argument relying on the hypotheses about the size of d is used to show that there are primes in the subring of traces of a $R_{\mathcal{D}_0}$, for $\mathcal{D}_0 = (\mathcal{O}, \Sigma, c_0, \mathcal{M})$, which correspond to representations with reductions related to other classes c; in fact all c's can be obtained this way and so one finds that there are nice primes for all classes c (this argument makes use of (I) for the nice prime $\mathfrak{p}_0$).

The proof of (III) exploits a connectivity result of Raynaud (which ultimately applies because of the hypothesis on the size of d). This result essentially allows one to deduce that the irreducible components of Spec $R_{\mathcal{D}}$ are connected by nice primes. Effectively, given any partition $\mathcal{C} = \mathcal{C}_1 \sqcup \mathcal{C}_2$ of the set $\mathcal{C}$ of connected components of Spec $R_{\mathcal{D}}$ with each $\mathcal{C}_i \neq \emptyset$, there must exist $C_i \in \mathcal{C}_i$, $i = 1, 2$, such that $C_1 \cap C_2$ contains a prime $\mathfrak{p}$ that satisfies all the hypotheses of being nice except possibly being pro-modular – and even the latter holds, say, if the primes on the component C_1 are all a priori pro-modular. Applying this with $\mathcal{C}_1$ being the set of all pro-modular components – which is non-empty by (I) and (II) – it then follows from (I) that $\mathcal{C}_2$ must be empty: all primes of $R_{\mathcal{D}}$ are pro-modular.

The final step in the proof of the above modularity theorem for residually reducible representations makes use of solvable base change to move to a situation where the hypotheses on the size of the degree of the totally real field hold. A two-dimensional p-adic Galois representation ρ of G_F is known to be associated with a Hilbert modular form over F if for some totally real solvable extension F'/F the restriction $\rho|_{G_{F'}}$ is associated with a Hilbert modular forms over F'. This is a consequence of results about base change for automorphic form on GL_2. For the purposes of the modularity theorem for residually reducible representations, one moves from the base totally real field F to a large solvable extension F' with prescribed local behavior at a finite number primes and such that the p-part of the class field of $F'(\chi)$ (the splitting field of χ over F') is controlled relative to that of $F(\chi)$. When $F(\chi)$ is abelian over $\mathbb{Q}$ the existence of such a field F' can be deduced from a theorem of Washington on the behavior of the p-part of the class group of $\mathbb{Z}_\ell$-extensions ($\ell \neq p$). However, for more general totally real fields the existence of a suitable F' remains an open question.

Remark 34 The paper [18] examined some special cases of residually reducible deformations where the deformation rings could be identified with Hecke algebras. These essentially amount to the cases where the dimension of the space of possible extensions of 1 by χ is one-dimensional (note the reversal of the ordering of the characters). The proof in [18] of the isomorphism between the deformation rings and Hecke rings amounts to a direct verification of the numerical criterion on the right-hand side of (ML3). Essentially, taking $\mathfrak{p}$ to correspond to a suitable reducible deformation, the refined class number formula RCF implies that the order of $\mathfrak{p}/\mathfrak{p}^2$

is bounded by an L-value while the techniques used to prove the main conjecture show that the size of the corresponding $\mathcal{O}/(\eta)$ is bounded below by the same L-value. The paper [22] showed that the methods in [20] also applies to the situation where $\bar{\rho}|_{\mathrm{Gal}(\overline{\mathbb{Q}}/\mathbb{Q}(\sqrt{p^*}))}$ is reducible.

7.8 Further Developments

Wiles's proof of his modularity lifting theorem opened the floodgates: Diamond [46] extended Wiles's methods to show that all elliptic curves over $\mathbb{Q}$ that are semistable at 3 and 5 are modular. This was followed by the crucial observation, made independently by Diamond [47] and Fujiwara, that the patching arguments of Step 1 actually provide the Gorenstein-ness and multiplicity one results about Hecke algebras that were used as inputs in Wiles's arguments; this allowed for subsequent applications of the methods beyond modular curves, to situations where the structure of the cohomology as a module for the Hecke algebra was less well-understood. A closer analysis of the local deformations at the prime p led to the proof by Breuil, Conrad, Diamond, and Taylor that all elliptic curves over $\mathbb{Q}$ are modular [33, 41]. As already noted, the papers [18, 20], and [22] made some progress in removing the hypothesis that $\bar{\rho}$ is irreducible (hypothesis (r1)), while [21] showed that an application of solvable base change could replace the delicate analysis of Ribet *et al* that was used in Step 1 (to prove the existence of a modular lift of minimal type). These additional techniques rendered the methods very robust, allowing for some generalizations to totally real ground fields F as well as to higher-dimensional Galois representations. Taylor [89, 90] essentially proved a potential version of both Serre's Conjecture and the Fontaine–Mazur Conjecture ('potential' means modular over a totally real extension E/F). Clozel, Harris, and Taylor [35] made progress on extending Wiles methods to higher-dimensional essentially (conjugate) self-dual representations of the Galois group of totally real or CM fields, essentially extending Step 1 but running into obstacles to generalizing Step 2. These obstacles were circumvented in [91], building on ideas of Kisin [62, 63], especially the idea to consider the global deformation problem relative to local deformation problems. This led to a proof of the Sato–Tate Conjecture for elliptic curves with non-integral j-invariants [52]. These methods were further refined and strengthened resulting in a proof by Barnet-Lamb, Gee, and Geraghty of the Sato–Tate Conjecture for Hilbert modular forms over any totally real field [29]. The issue of weakening the analog of the hypothesis on $\rho|_{\mathrm{Gal}(\overline{\mathbb{Q}}/\mathbb{Q}(\sqrt{(-1)^{(p-1)/2}p}))}$ ('bigness') was addressed by Snowden and Wiles [27] and by Thorne [92]. Techniques for proving potential automorphy have advanced quickly. As of the preparation of this paper, the state of the art seems to be [30] and [68], but that will almost surely have changed by the time it has gone to press! Significant progress toward the full Fontaine–Mazur Conjecture for two-dimensional representations of $G_{\mathbb{Q}}$ has been made, especially by Emerton [49] and Kisin [63]. Khare and Wintenberger [58–60] combined many of these modular lifting theorems with an extensive generalization of 'the 3 − 5 switch' to prove Serre's Conjecture in full. Freitas, Le Hung, and Siksek [50] were able to use these modular lifting results to prove that all elliptic curves over real quadratic fields are modular. Thorne [93] has generalized many of the arguments in [20] to higher-dimensional representations, and Clozel and Thorne [36] have used some of these results to prove new cases of functoriality for symmetric powers of holomorphic modular forms.

8 Solvable Points on Genus One Curves

Galois theory has provided both an indispensable tool for the algebraic number-theorist as well as an important framework in which to couch problems. Of course one of the earliest successes of Galois theory was the proof that the general polynomial of degree five or higher cannot be solved in radicals (there is no 'quintic formula,' etc., generalizing the quadratic formula or Cardano's formulas for cubic polynomials). One Galois-theoretic statement of this is: The solutions to a polynomial of degree five or higher with coefficients in a field F need not lie in a solvable extension of F, that is, a field extension whose normal closure over F has solvable Galois group. For polynomials of degrees less than five, the solutions always lie in solvable extensions.

From this Galois-theoretic perspective, there is a very natural question generalizing from polynomials to higher dimensional varieties: When does a variety X over a field F have a solvable point, that is, a point in a solvable extension of F? Perhaps the most natural case to consider is that of a smooth, geometrically connected, projective curve X. This is the problem that Wiles picked up, jointly with his then PhD student Mirela Çiperiani in the mid-2000s. Previously, Ambus Pál had proved that such a curve X of genus $g = 0, 2, 3$, or 4 always has a solvable point. But for curves over a number field F, the problem was open for all other g. Wiles, together with Çiperiani, considered the case of genus one curves.

Genus one curves are particularly rich from an arithmetic perspective. The Jacobian of a genus one curve is an elliptic curve, which means that one can frequently bring to bear on a problem all the tools one has for studying elliptic curves. And this is what Çiperiani and Wiles do, proving:

Theorem 35 ([25, Thms. 0.0.1 & 0.0.2]) *Let X be a smooth, geometrically connected, projective curve of genus one over $\mathbb{Q}$, and let E be the Jacobian of X. If*

(a) $\mathrm{ord}_{s=1}L(E,s) \leq 1$ *or E has semistable reduction;*
(b) $X(\mathbb{Q}_p) \neq \emptyset$ *for all primes p,*

then X has a point in a solvable extension of $\mathbb{Q}$.

The proof of this theorem brings together the modularity of elliptic curves, the results of Gross–Zagier and Kolyvagin, work of Vatsal and Cornut on the existence of anticyclotomic twists $L(E, \chi, s)$ of order 1 at $s = 1$, and Galois cohomology arguments reminiscent of the patching arguments from the proof of the modularity lifting theorems.

8.1 *An Idea of the Proof*

Let X and E be as in the theorem. Note that the curve X is a homogeous space for E and is isomorphic to the trivial homogeneous space over any field in which X has a point.

Let F be a number field. The Galois cohomology group $H^1(F, E)$ classifies homogeneous spaces over F that are trivial over some finite extension of F. So the problem is to show that the class $[X]$ of X in $H^1(\mathbb{Q}, E)$ is in the kernel of the restriction map $H^1(\mathbb{Q}, E) \stackrel{\text{res}}{\to} H^1(F, E)$ for some *solvable* extension F of $\mathbb{Q}$. The hypothesis that $X(\mathbb{Q}_p) \neq \emptyset$ for all primes p means that $[X]$ lies in the kernel

$$Ш(E/\mathbb{Q}) = \ker\{H^1(\mathbb{Q}, E) \to \prod_{v \leq \infty} H^1(\mathbb{Q}_v, E)\},$$

the Tate–Shafaravich group of E. Note that X, being of genus one, always has point over $\mathbb{Q}_\infty = \mathbb{R}$. The idea of the proof is to show that each element of $Ш(E/\mathbb{Q})$ becomes trivial over some solvable extension.

The Case of Analytic Rank at Most One

Suppose that $\mathrm{ord}_{s=1} L(E, s) \leq 1$. (Note that even making sense of this order requires analytic continuation of $L(E, s)$ to $s = 1$, which was only proved through the modularity of E!) It is known that in this case there is an imaginary quadratic field K satisfying the Heegner hypothesis for N (all prime divisors of N split and K) and such that $\mathrm{ord}_{s=1} L(E/K, s) = 1$. Then by the Gross–Zagier theorem, the Heegner point over K on E has infinite order. It then follows from the results of Kolyvagin that $E(K)$ has rank one and $Ш(E/K)$ is finite. But Kolyvagin also proved a structure theorem for $Ш(E/K)$ in this case. Roughly, this says: The group $Ш(E/K)$ is the direct sum of finite cyclic subgroups, each generated by an element whose image in $H^1(K[n], E)$, for $K[n]$ some ring class group of K (with n the squarefree product of 'Kolyvagin primes'), is trivial. This means that every element of $Ш(E/K)$ is trivial over some abelian extension of K and therefore over some solvable extension of $\mathbb{Q}$. The same is then also true of every element in $Ш(E/\mathbb{Q})$ and in particular of $[X]$.

The Case of Analytic Rank Greater than One

Suppose that $\text{ord}_{s=1} L(E, s) > 1$. Kolyvagin has conjectured that the Selmer group of E over an imaginary quadratic field K for which the Heegner hypothesis holds always has a structure much as in the analytic rank one case. This would again imply the triviality of $[X]$ over an abelian extension of K. However, Kolyvagin's conjecture is not known in general.

Çiperiani and Wiles instead proceed along the following lines. Let p be a prime. For any number field F, the Selmer group $H^1_{\text{Sel}}(F, E[p^N])$ surjects onto the p^N-torsion $Ш(E/F)[p^N]$ of the Tate–Shafarevich group, with the kernel being the image of $E(F)/p^N E(F)$. So it suffices to show that there is a solvable extension F of K such that the image of $H^1_{\text{Sel}}(K, E[p^N])$ in $H^1_{\text{Sel}}(F, E[p^N])$ lies in the subgroup $E(F)/p^N E(F)$. They then exploit:

The Unramified Under Ramified Principle

This essentially says that the classes in $H^1_{\text{Sel}}(F, E[p^N])$ are contained in the subgroup of $H^1(F, E[p^n])$ generated by classes that are ramified as some 'nice' primes of F.

Generic Analytic Rank One

Results of Cornut and Vatsal show that $\text{ord}_{s=1} L(E, \chi, s) = 1$ for χ an anticyclotomic character of K of p-power order and sufficienly large conductor p^n. Then Kolyvagin's construction produces many classes in $H^1(K[p^n], E[p^N])$ that are ramified at nice primes and whose restriction to a solvable extension come from a point on E.

Making use of the $\mathbb{Z}/p^N[\text{Gal}(K[p^n]/K)]$-module structure on the various groups of classes in $H^1(K_n, E[p^N])$, they then use an ingenious patching argument (as n, N, and the sets of nice primes vary) to show that the classes coming from the generic rank one property fill up enough of the classes ramified at nice primes to capture all of the Selmer classes.

9 Class Groups of Quadratic Fields

While there has been remarkable progress on understanding the arithmetic of elliptic curves in the fifty or so years since Birch and Swinnerton-Dyer made their conjecture, a great deal of which traces back to the groundbreaking works of Wiles, there is still much that remains unknown about even the arithmetic of number fields.

It can be sobering to recall how little is understood about class groups of even quadratic fields.

Wiles addressed the natural problem of the existence of an imaginary quadratic field with class number indivisible by a given prime ℓ and having prescribed factorizations at a finite set of primes:

Theorem 36 ([26, Thm. A]) *Let $\ell \geq 3$ be a prime. Let S_-, S_+, and S_0 be disjoint finite sets of primes such that*

(a) *S_- contains no prime q with $q \equiv 1 \bmod \ell$ and $q \equiv -1 \bmod 4$;*
(b) *S_+ contains no prime q with $q \equiv -1 \bmod \ell$;*
(c) *S_0 contains no prime q with $q \equiv 1 \bmod \ell$.*

Then there exists an imaginary quadratic field L satisfying

(i) *ℓ does not divide the class number h_L of L;*
(ii) *L is inert at each prime in S_-, split at each prime in S_+, and ramified at each prime in S_0.*

This theorem can be restated in terms of Dirichlet characters as: There exists an odd quadratic character χ such that (i) $\ell \nmid L(0, \chi)$, and (ii) $\chi(q) = -1$ if $q \in S_-$, $\chi(q) = +1$ if $q \in S_+$, and $\chi(q) = 0$ if $q \in S_0$.

Previous results, by Hartung, Horie, and Brunier, had either not imposed any factorization conditions, assumed ℓ to be much larger than the primes in S, or imposed more restrictions on the primes in S.

The proof makes use of the trace formula on Shimura curves. A key observation is that the trace of a Hecke operator on a suitable space of modular forms for a quarternion algebra (the holomorphic differentials on the Shimura curve) can be expressed as a sum of class numbers of orders in certain imaginary quadratic fields. The heart of the argument is a good choice of the Shimura curve so that the resulting sum is of class numbers of orders in fields with the prescribed factorizations at the primes in $S = S_- \cup S_+ \cup S_0$ while being able to chose the Hecke operator to have trace a unit modulo ℓ. The latter is achieved by making a clever choice of auxiliary level structures at some additional primes and exploiting the relation of traces of Hecke operators with traces of Frobenius elements in the Galois representations associated with modular forms.

Remark 37 One motivation for considering the problem of the existence of such quadratic extensions, especially over a general totally real field, arises from the results in [20], where it is explained that having such extensions would yield much stronger modularity theorems for residually reducible Galois representation of general totally real fields (see also the last part of the discussion of the theorem in Sect. 7.7).

Acknowledgements It is a pleasure to thank Manjul Bhargava, John Coates, Karl Rubin, Peter Sarnak, and Richard Taylor for their comments on an earlier version of this article. During the preparation of this article the author was partly supported by the grant DMS-1301842 from the

National Science Foundation and by the Simons Investigator Grant #376203 from the Simons Foundation.

References

1. J. Coates and A. Wiles, *Kummer's criterion for Hurwitz numbers*, Algebraic number theory (Kyoto Internat. Sympos., Res. Inst. Math. Sci., Univ Kyoto, Kyoto, 1976), Japan Soc. Promotion Sci., Tokyo, 1977, pp. 9–23.
2. ——, *Explicit reciprocity laws*, Journées Arithmétiques de Caen (Univ Caen, Caen, 1976), Soc. Math. France, Paris, 1977, pp. 7–17. Astérisque No. 41–42.
3. ——, *On the conjecture of Birch and Swinnerton-Dyer*, Invent. Math. **39** (1977), no. 3, 223–251.
4. A. Wiles, *Higher explicit reciprocity laws*, Ann. Math. (2) **107** (1978), no. 2, 235–254.
5. J. Coates and A. Wiles, *On p-adic L-functions and elliptic units*, J. Austral. Math. Soc. Ser. A **26** (1978), no. 1, 1–25.
6. A. Wiles, *Modular curves and the class group of* $\mathbf{Q}(\zeta_p)$, Invent. Math. **58** (1980), no. 1, 1–35.
7. K. Rubin and A. Wiles, *Mordell-Weil groups of elliptic curves over cyclotomic fields*, Number theory related to Fermat's last theorem (Cambridge, Mass., 1981), Progr. Math., vol. 26, Birkhäuser, Boston, Mass., 1982, pp. 237–254.
8. B. Mazur and A. Wiles, *Analogies between function fields and number fields*, Amer. J. Math. **105** (1983), no. 2, 507–521.
9. ——, *Class fields of abelian extensions of* **Q**, Invent. Math. **76** (1984), no. 2, 179–330.
10. ——, *On p-adic analytic families of Galois representations*, Compositio Math. **59** (1986), no. 2, 231–264.
11. A. Wiles, *On p-adic representations for totally real fields*, Ann. of Math. (2) **123** (1986), no. 3, 407–456.
12. ——, *On ordinary* λ*-adic representations associated to modular forms*, Invent. Math. **94** (1988), no. 3, 529–573.
13. ——, *The Iwasawa conjecture for totally real fields*, Ann. of Math. (2) **131** (1990), no. 3, 493–540.
14. ——, *On a conjecture of Brumer*, Ann. of Math. (2) **131** (1990), no. 3, 555–565.
15. ——, *Modular forms, elliptic curves, and Fermat's last theorem*, Proceedings of the International Congress of Mathematicians, Vol. 1, 2 (Zürich, 1994), Birkhäuser, Basel, 1995, pp. 243–245.
16. ——, *Modular elliptic curves and Fermat's last theorem*, Ann. of Math. (2) **141** (1995), no. 3, 443–551.
17. R. Taylor and A. Wiles, *Ring-theoretic properties of certain Hecke algebras*, Ann. of Math. (2) **141** (1995), no. 3, 553–572.
18. C. Skinner and A. Wiles, *Ordinary representations and modular forms*, Proc. Nat. Acad. Sci. U.S.A. **94** (1997), no. 20, 10520–10527.
19. A. Wiles, *Twenty years of number theory*, Mathematics: frontiers and perspectives, Amer. Math. Soc., Providence, RI, 2000, pp. 329–342. MR1754786
20. C. Skinner and A. Wiles, *Residually reducible representations and modular forms*, Inst. Hautes Études Sci. Publ. Math. **89** (1999), 5–126 (2000).
21. ——, *Base change and a problem of Serre*, Duke Math. J. **107** (2001), no. 1, 15–25.
22. ——, *Nearly ordinary deformations of irreducible residual representations*, Ann. Fac. Sci. Toulouse Math. (6) **10** (2001), no. 1, 185–215.
23. A. Wiles, *The Birch and Swinnerton-Dyer conjecture*, The millennium prize problems, Clay Math. Inst., Cambridge, MA, 2006, pp. 31–41.
24. ——, *Foreword [In honour of John H. Coates on the occasion of his sixtieth birthday]*, Doc. Math. **Extra Vol.** (2006), 3–4.

25. M. Çiperiani and A. Wiles, *Solvable points on genus one curves*, Duke Math. J. **142** (2008), no. 3, 381–464.
26. A. Wiles, *On class groups of imaginary quadratic fields*, J. Lond. Math. Soc. (2) **92** (2015), no. 2, 411–426.
27. A. Snowden and A. Wiles, *Bigness in compatible systems*, Elliptic curves, modular forms and Iwasawa theory, Springer Proc. Math. Stat., vol. 188, Springer Cham, 2016, pp. 469–492.

Other References

28. N. Arthaud, *On Birch and Swinnerton-Dyer's conjecture for elliptic curves with complex multiplication. I*, Compositio Math. **37** (1978), no. 2, 209–232.
29. T. Barnet-Lamb, T. Gee, and D. Geraghty, *The Sato-Tate conjecture for Hilbert modular forms*, J. Amer. Math. Soc. **24** (2011), no. 2, 411–469.
30. T. Barnet-Lamb, T. Gee, D. Geraghty, and R. Taylor, *Potential automorphy and change of weight*, Ann. of Math. (2) **179** (2014), no. 2, 501–609.
31. D. Blasius and J. Rogawski, *Motives for Hilbert modular forms*, Invent. Math. **114** (1993), no. 1, 55–87.
32. D. Blasius, *Hilbert modular forms and the Ramanujan conjecture*, Noncommutative geometry and number theory, Aspects Math., E37, Friedr. Vieweg, Wiesbaden, 2006, pp. 35–56.
33. C. Breuil, B. Conrad, F. Diamond, and R. Taylor, *On the modularity of elliptic curves over* **Q***: wild 3-adic exercises*, J. Amer. Math. Soc. **14** (2001), no. 4, 843–939.
34. G. Chenevier and M. Harris, *Construction of automorphic Galois representations, II*, Camb. J. Math. **1** (2013), no. 1, 53–73.
35. L. Clozel, M. Harris, and R. Taylor, *Automorphy for some l-adic lifts of automorphic mod l Galois representations*, Publ. Math. Inst. Hautes Études Sci. **108** (2008), 1–181. With Appendix A, summarizing unpublished work of Russ Mann, and Appendix B by Marie-France Vignéras.
36. L. Clozel and J. Thorne, *Level-raising and symmetric power functoriality, III*, Duke Math. J. **166** (2017), no. 2, 325–402.
37. J. Coates, *The work of Mazur and Wiles on cyclotomic fields*, Bourbaki Seminar, Vol. 1980/81, Lecture Notes in Math., vol. 901, Springer, Berlin-New York, 1981, pp. 220–242.
38. R. Coleman, *Division values in local fields*, Invent. Math. **53** (1979), no. 2, 91–116.
39. ——, *The arithmetic of Lubin-Tate division towers*, Duke Math. J. **48** (1981), no. 2, 449–466.
40. P. Colmez, *Théorie d'Iwasawa des représentations de de Rham d'un corps local*, Ann. of Math. (2) **148** (1998), no. 2, 485–571 (French).
41. B. Conrad, F. Diamond, and R. Taylor, *Modularity of certain potentially Barsotti-Tate Galois representations*, J. Amer. Math. Soc. **12** (1999), no. 2, 521–567.
42. H. Darmon, F. Diamond, and R. Taylor, *Fermat's last theorem*, Current developments in mathematics, 1995 (Cambridge, MA), Int. Press, Cambridge, MA, 1994, pp. 1–154.
43. H. Darmon, *Andrew Wiles's marvelous proof*, Notices Amer. Math. Soc. **64** (2017), no. 3, 209–216.
44. E. de Shalit, *The explicit reciprocity law in local class field theory*, Duke Math. J. **53** (1986), no. 1, 163–176.
45. F. Diamond, *The refined conjecture of Serre*, Elliptic curves, modular forms, & Fermat's last theorem (Hong Kong, 1993), Ser. Number Theory, I, Int. Press, Cambridge, MA, 1995, pp. 22–37.
46. ——, *On deformation rings and Hecke rings*, Ann. of Math. (2) **144** (1996), no. 1, 137–166.
47. ——, *The Taylor-Wiles construction and multiplicity one*, Invent. Math. **128** (1997), no. 2, 379–391.

48. F. Diamond, M. Flach, and L. Guo, *The Tamagawa number conjecture of adjoint motives of modular forms*, Annales scientifiques de l'École Normale Supérieure **37** (2004), no. 5, 663–727.
49. M. Emerton, *A local-global compatibility conjecture in the p-adic Langlands programme for* $GL_{2/\mathbf{Q}}$, Pure Appl. Math. Q. **2** (2006), no. 2, Special Issue: In honor of John H. Coates., 279–393.
50. N. Freitas, B. Le Hung, and S. Siksek, *Elliptic curves over real quadratic fields are modular*, Invent. Math. **201** (2015), no. 1, 159–206.
51. R. Greenberg, *On the Birch and Swinnerton-Dyer conjecture*, Invent. Math. **72** (1983), no. 2, 241–265.
52. M. Harris, N. Shepherd-Barron, and R. Taylor, *A family of Calabi-Yau varieties and potential automorphy*, Ann. of Math. (2) **171** (2010), no. 2, 779–813.
53. M. Harris, K.-W. Lan, R. Taylor, and J. Thorne, *On the rigid cohomology of certain Shimura varieties*, Res. Math. Sci. **3** (2016), Paper No. 37, 308pp.
54. H. Hida and J. Tilouine, *On the anticyclotomic main conjecture for CM fields*, Invent. Math. **117** (1994), no. 1, 89–147.
55. K. Kato, *Generalized explicit reciprocity laws*, Adv. Stud. Contemp. Math. (Pusan) **1** (1999), 57–126. Algebraic number theory (Hapcheon/Saga, 1996).
56. ——, *Generalized explicit reciprocity laws*, Adv. Stud. Contemp. Math. (Pusan) **1** (1999), 57–126. Algebraic number theory (Hapcheon/Saga, 1996).
57. ——, *p-adic Hodge theory and values of zeta functions of modular forms*, Astérisque **295** (2004), ix, 117–290 (English, with English and French summaries). Cohomologies p-adiques et applications arithmétiques. III.
58. C. Khare, *Serre's modularity conjecture: the level one case*, Duke Math. J. **134** (2006), no. 3, 557–589.
59. C. Khare and J.-P. Wintenberger, *Serre's modularity conjecture. I*, Invent. Math. **178** (2009), no. 3, 485–504.
60. ——, *Serre's modularity conjecture. II*, Invent. Math. **178** (2009), no. 3, 505–586.
61. G. Kings, D. Loeffler, and S. Zerbes, *Rankin-Eisenstein classes and explicit reciprocity laws*, Camb. J. Math. **5** (2017), no. 1, 1–122.
62. M. Kisin, *Moduli of finite flat group schemes, and modularity*, Ann. of Math. (2) **170** (2009), no. 3, 1085–1180.
63. ——, *The Fontaine-Mazur conjecture for* GL_2, J. Amer. Math. Soc. **22** (2009), no. 3, 641–690.
64. S. Lang, *Cyclotomic fields I and II*, 2nd ed., Graduate Texts in Mathematics, vol. 121, Springer-Verlag, New York, 1990. With an appendix by Karl Rubin. MR1029028
65. T. Liu, *Lattices in filtered* (φ, N)*-modules*, J. Inst. Math. Jussieu **11** (2012), no. 3, 659–693.
66. B. Mazur, *Deforming Galois representations*, Galois groups over **Q** (Berkeley, CA, 1987), Math. Sci. Res. Inst. Publ., vol. 16, Springer, New York, 1989, pp. 385–437.
67. B. Mazur and J. Tilouine, *Représentations galoisiennes, différentielles de Kähler et "conjectures principales"*, Inst. Hautes Études Sci. Publ. Math. **71** (1990) (French).
68. S. Patrikis and R. Taylor, *Automorphy and irreducibility of some l-adic representations*, Compos. Math. **151** (2015), no. 2, 207–229.
69. B. Perrin-Riou, *Théorie d'Iwasawa des représentations p-adiques sur un corps local*, Invent. Math. **115** (1994), no. 1, 81–161 (French). With an appendix by Jean-Marc Fontaine.
70. ——, *Théorie d'Iwasawa et loi explicite de réciprocité*, Doc. Math. **4** (1999), 219–273 (French, with English summary).
71. K. Ribet, *A modular construction of unramified p-extensions of* $\mathbf{Q}(\mu_p)$, Invent. Math. **34** (1976), no. 3, 151–162.
72. ——, *On modular representations of* $\mathrm{Gal}(\overline{\mathbf{Q}}/\mathbf{Q})$ *arising from modular forms*, Invent. Math. **100** (1990), no. 2, 431–476.
73. R. Pollack and K. Rubin, *The main conjecture for CM elliptic curves at supersingular primes*, Ann. of Math. (2) **159** (2004), no. 1, 447–464.
74. K. Rubin, *Elliptic curves with complex multiplication and the conjecture of Birch and Swinnerton-Dyer*, Invent. Math. **64** (1981), no. 3, 455–470.

75. ——, *Tate-Shafarevich groups and L-functions of elliptic curves with complex multiplication*, Invent. Math. **89** (1987), no. 3, 527–559.
76. ——, *On the main conjecture of Iwasawa theory for imaginary quadratic fields*, Invent. Math. **93** (1988), no. 3, 701–713.
77. ——, *Elliptic curves with complex multiplication and the conjecture of Birch and Swinnerton-Dyer*, Arithmetic theory of elliptic curves (Cetraro, 1997), Lecture Notes in Math., vol. 1716, Springer, Berlin, 1999, pp. 167–234.
78. ——, *Euler systems*, Annals of Mathematics Studies, vol. 147, Princeton University Press, Princeton, NJ, 2000. Hermann Weyl Lectures. The Institute for Advanced Study.
79. T. Saito, *Hilbert modular forms and p-adic Hodge theory*, Compos. Math. **145** (2009), no. 5, 1081–1113.
80. P. Schneider and O. Venjakob, *Coates-Wiles homomorphisms and Iwasawa cohomology for Lubin-Tate extensions*, Elliptic curves, modular forms and Iwasawa theory, Springer Proc. Math. Stat., vol. 188, Springer, Cham, 2016, pp. 401–468.
81. P. Scholze, *On torsion in the cohomology of locally symmetric varieties*, Ann. of Math. (2) **182** (2015), no. 3, 945–1066.
82. J.-P. Serre, *Sur les représentations modulaires de degré* 2 *de* $\mathrm{Gal}(\overline{\mathbf{Q}}/\mathbf{Q})$, Duke Math. J. **54** (1987), no. 1 (French).
83. C. Skinner, *A note on the p-adic Galois representations attached to Hilbert modular forms*, Doc. Math. **14** (2009), 241–258.
84. C. Skinner and E. Urban, *The Iwasawa main conjectures for* GL_2, Invent. Math. **195** (2014), no. 1, 1–277.
85. C. Skinner, K. Rubin, B. Mazur, M. Çiperiani, C. Khare, and J. Thorne, *The mathematical works of Andrew Wiles*, Notices Amer. Math. Soc. **64** (2017), no. 3, 217–227.
86. R. Taylor, *On Galois representations associated to Hilbert modular forms*, Invent. Math. **98** (1989), no. 2, 265–280.
87. ——, *l-adic representations associated to modular forms over imaginary quadratic fields. II*, Invent. Math. **116** (1994), no. 1–3, 619–643.
88. ——, *l-adic representations associated to modular forms over imaginary quadratic fields. II*, Invent. Math. **116** (1994), no. 1–3, 619–643.
89. ——, *Remarks on a conjecture of Fontaine and Mazur*, J. Inst. Math. Jussieu **1** (2002), no. 1, 125–143.
90. ——, *On the meromorphic continuation of degree two L-functions*, Doc. Math. **Extra Vol.** (2006), 729–779.
91. ——, *Automorphy for some l-adic lifts of automorphic mod l Galois representations. II*, Publ. Math. Inst. Hautes Études Sci. **108** (2008), 183–239.
92. J. Thorne, *On the automorphy of l-adic Galois representations with small residual image*, J. Inst. Math. Jussieu **11** (2012), no. 4, 855–920. With an appendix by Robert Guralnick, Florian Herzig, Richard Taylor and Thorne.
93. ——, *Automorphy lifting for residually reducible l-adic Galois representations*, J. Amer. Math. Soc. **28** (2015), no. 3, 785–870.

List of Publications for Sir Andrew J. Wiles

1977

[1] (with J. Coates). Kummer's criterion for Hurwitz numbers. In *Algebraic number theory (Kyoto Internat. Sympos., Res. Inst. Math. Sci., Univ. Kyoto, Kyoto, 1976)*, 9–23. Japan Soc. Promotion Sci., Tokyo.

[2] (with J. Coates). Explicit reciprocity laws. In *Journées Arithmétiques de Caen (Univ. Caen, Caen, 1976)*, Astérisque, 41–42:7–17. Soc. Math. France, Paris.

[3] (with J. Coates). On the conjecture of Birch and Swinnerton-Dyer. *Invent. Math.*, 39(3):223–251, 1977.

1978

[4] Higher explicit reciprocity laws. *Ann. Math. (2)*, 107(2):235–254.

[5] (with J. Coates). On p-adic L-functions and elliptic units. *J. Austral. Math. Soc. Ser. A*, 26(1):1–25.

1980

[6] Modular curves and the class group of $\mathbf{Q}(\zeta_p)$. *Invent. Math.*, 58(1):1–35.

1982

[7] (with K. Rubin). Mordell–Weil groups of elliptic curves over cyclotomic fields. In *Number theory related to Fermat's last theorem (Cambridge, Mass., 1981)*, 237–254, *Progr. Math.*, 26. Birkhäuser, Boston, Mass.

1983

[8] (with B. Mazur). Analogies between function fields and number fields. *Amer. J. Math.*, 105(2):507–521.

1984

[9] (with B. Mazur). Class fields of abelian extensions of $\mathbf{Q}$. *Invent. Math.*, 76(2):179–330.

H. Holden, R. Piene (eds.), *The Abel Prize 2013–2017*,
https://doi.org/10.1007/978-3-319-99028-6_27

1986

[10] On p-adic representations for totally real fields. *Ann. of Math. (2)*, 123(3):407–456.

[11] (with B. Mazur). On p-adic analytic families of Galois representations. *Compositio Math.*, 59(2):231–264.

1988

[12] On ordinary λ-adic representations associated to modular forms. *Invent. Math.*, 94(3):529–573.

1990

[13] The Iwasawa conjecture for totally real fields. *Ann. of Math. (2)*, 131(3):493–540.

[14] On a conjecture of Brumer. *Ann. of Math. (2)*, 131(3):555–565.

1995

[15] Modular elliptic curves and Fermat's last theorem. *Ann. of Math. (2)*, 141(3):443–551.

[16] (with R. Taylor). Ring-theoretic properties of certain Hecke algebras. *Ann. of Math. (2)*, 141(3):553–572.

[17] Modular forms, elliptic curves, and Fermat's last theorem. In *Proceedings of the International Congress of Mathematicians, Vol. 1, 2 (Zürich, 1994)*, 243–245. Birkhäuser, Basel.

1997

[18] (with C.M. Skinner). Ordinary representations and modular forms. *Proc. Nat. Acad. Sci. U.S.A.*, 94(20):10520–10527.

1999

[19] (with C.M. Skinner). Residually reducible representations and modular forms. *Inst. Hautes Études Sci. Publ. Math.*, (89):5–126.

2000

[20] Twenty years of number theory. In *Mathematics: frontiers and perspectives*, 329–342. Amer. Math. Soc., Providence, RI.

2001

[21] (with C.M. Skinner). Base change and a problem of Serre. *Duke Math. J.*, 107(1):15–25.

[22] (with C.M. Skinner). Nearly ordinary deformations of irreducible residual representations. *Ann. Fac. Sci. Toulouse Math. (6)*, 10(1):185–215.

2006

[23] The Birch and Swinnerton-Dyer conjecture. In *The millennium prize problems*, 31–41. Clay Math. Inst., Cambridge, MA.

[24] Foreword [In honour of John H. Coates on the occasion of his sixtieth birthday]. *Doc. Math.*, (Extra Vol.):3–4 (electronic).

2008

[25] (with M. Çiperiani). Solvable points on genus one curves. *Duke Math. J.*, 142(3):381–464.

2015

[26] On class groups of imaginary quadratic fields. *J. Lond. Math. Soc. (2)*, 92(2):411–426.

2016

[27] (with A. Snowden). Bigness in compatible systems. In *Elliptic curves, modular forms and Iwasawa theory*, 469–492, Springer Proc. Math. Stat., 188, Springer, Cham.

2018

[28] (with S. Patrikis and A. Snowden). Residual Irreducibility of Compatible Systems. *Int. Math. Res. Not. IMRN*, 2:571–587.

Curriculum Vitae for Sir Andrew John Wiles, KBE, FRS

Born:	April 11, 1953 in Cambridge, UK
Degrees/education:	Bachelor of Science, Oxford University, 1974
	PhD, Cambridge University, 1980
Positions:	Junior Research Fellow, Cambridge University, 1977–1980
	Benjamin Peirce Assistant Professor, Harvard University, 1977–1980
	Professor, Princeton University, 1982–2011
	Royal Society Professor, Oxford University, 1988–1990

H. Holden, R. Piene (eds.), *The Abel Prize 2013–2017*,
https://doi.org/10.1007/978-3-319-99028-6_28

	Eugene Higgins Professor of Mathematics, Princeton University, 1994–2011
	Royal Society Research Professor, Oxford University, 2011–
	Regius Professor of Mathematics, Oxford University, 2018–
Visiting positions:	Visiting Professor, Sonderforschungsbereich Theoretische Mathematik, Bonn, spring 1981
	Member, Institute for Advanced Study, fall 1981, fall 1992, 1995–1996
	Visiting professor, University of Paris, Orsay, spring 1982
	Guggenheim Fellow, Institut des Hautes Études Scientifiques and École Normale Supérieure, 1985–1986
	Visiting Professor, Isaac Newton Institute and Department of Pure Mathematics, Cambridge
Memberships:	Fellow of the Royal Society, 1989
	American Academy of Arts and Sciences, foreign honorary member, 1994
	National Academy of Sciences, foreign associate, 1996
	American Philosophical Society, 1997
	Académie des Sciences, membre associé étranger, 1998
	Academia Europaea, 2015
	Norwegian Academy of Science and Letters, 2016
Awards and prizes:	Whitehead Prize, 1988
	Rolf Schock Prize, 1995
	Ostrowski Prize, 1995
	Fermat Prize, 1995
	Wolf Prize, 1995/1996
	Royal Medal, 1996
	National Academy of Sciences Award in Mathematics, 1996
	Cole Prize, 1997
	Wolfskehl Prize, 1997
	IMU Silver Plaque, 1998
	King Faisal International Prize in Science, 1998
	Clay Research Award, 1999
	KBE, 2000
	Pythagoras Award, 2004
	Shaw Prize, 2005
	Abel Prize, 2016
	Copley Medal, 2017
Honorary degrees:	University of Warwick, 1998
	Oxford University, 1999
	University of Nottingham, 1999
	Columbia University, 2003
	Yale University, 2005
	University of Cambridge, 2010
	University of Bristol, 2018

Part V
2017 Yves Meyer

"for his pivotal role in the development of the mathematical theory of wavelets"

ABEL
PRISEN

Citation

The Norwegian Academy of Science and Letters has decided to award the Abel Prize for 2017 to **Yves Meyer** of the École normale supérieure Paris–Saclay, France

for his pivotal role in the development of the mathematical theory of wavelets.

Fourier analysis provides a useful way of decomposing a signal or function into simply-structured pieces such as sine and cosine waves. These pieces have a concentrated frequency spectrum, but are very spread out in space. Wavelet analysis provides a way of cutting up functions into pieces that are localised in both frequency and space. Yves Meyer was the visionary leader in the modern development of this theory, at the intersection of mathematics, information technology and computational science. The history of wavelets goes back over a hundred years, to an early construction by Alfréd Haar. In the late 1970s the seismologist Jean Morlet analysed reflection data obtained for oil prospecting, and empirically introduced a new class of functions, now called "ondelettes" or "wavelets", obtained by both dilating and translating a fixed function. In the spring of 1985, Yves Meyer recognised that a recovery formula found by Morlet and Alex Grossmann was an identity previously discovered by Alberto Calderón. At that time, Yves Meyer was already a leading figure in the Calderón–Zygmund theory of singular integral operators. Thus began Meyer's study of wavelets, which in less than ten years would develop into a coherent and widely applicable theory. The first crucial contribution by Meyer was the construction of a smooth orthonormal wavelet basis. The existence of such a basis had been in doubt. As in Morlet's construction, all of the functions in Meyer's basis arise by translating and dilating a single smooth "mother wavelet", which can be specified quite explicitly. Its construction, though essentially elementary, appears rather miraculous. Stéphane Mallat and Yves Meyer then systematically developed multiresolution analysis, a flexible and general framework for constructing wavelet bases, which places many of the earlier constructions on a more conceptual footing. Roughly speaking, multiresolution analysis allows one to explicitly construct an orthonormal wavelet basis from any bi-infinite sequence of nested subspaces of $L^2(R)$ that satisfy a few additional invariance properties. This work paved the way for the construction by Ingrid Daubechies of orthonormal bases of compactly supported wavelets. In the following decades, wavelet analysis has been applied in a wide variety of arenas as diverse as applied and computational harmonic analysis, data compression, noise reduction, medical imaging, archiving, digital cinema, deconvolution of the Hubble space telescope images, and the recent LIGO detection of gravitational waves created by the collision of two black holes. Yves Meyer has also made fundamental contributions to problems in number theory, harmonic analysis and partial differential equations, on topics such as quasi-crystals, singular integral operators and the Navier–Stokes equations. The crowning achievement of his pre-wavelets work is his proof, with Ronald Coifman and Alan McIntosh, of the L^2-boundedness of the Cauchy integral on Lipschitz curves, thus resolving the major open question in Calderón's program. The methods developed by Meyer have had a long-lasting impact in both harmonic analysis and partial differential equations. Moreover, it was Meyer's expertise in the mathematics of the Calderón–Zygmund school that opened the way for the development of wavelet theory, providing a remarkably fruitful link between a problem set squarely in pure mathematics and a theory with wide applicability in the real world.

For My Mother

Yves Meyer

My mother died thirty years ago from a stroke. Her life had been sad. Her marriage was a failure. My father never wished to live with us, even when I was a baby. I was born in 1939 and my sister, Danièle, in 1938. My father sold the pharmacy he was running at 17 boulevard du Temple, Paris, and he enrolled in the Army in 1944. He was assigned to the Hospital Pharmacy Department at Rabat, Morocco. My mother stayed a full year with us in Paris. She finally decided to take us to Rabat against the wish of my father. In Rabat, the three of us were living in a single room of a low-cost hotel. My father was working, living, and sleeping at the hospital. We were used to his absence.

I remember the fierce beauty of the Ocean. I remember the delight of the mint tea with pastries at the terrace of les Oudayas, an old castle from the fifteenth century that has been preserved with the greatest care. In the seventeenth century this fortress was protecting Rabat from most of the attacks from the sea. It is the place where the river Bouregreg meets the Ocean.

After spending two years in Rabat, my father received a position at Hospital Pharmacy Department at Tunis. This time we traveled together. The journey was fabulous. It lasted almost a week with one main stop in Algiers. I remember the beauty of the parks of Algiers. This summer had been dreadfully hot. We were

---

Electronic Supplementary Material The online version of this chapter (https://doi.org/10.1007/978-3-319-99028-6_29) contains supplementary material, which is available to authorized users.

Y. Meyer (✉)
CMLA, Centre de Mathématiques et de Leurs Applications, École normale supérieure de Cachan, Cachan Cedex, France
e-mail: Yves.Meyer@cmla.ens-cachan.fr

H. Holden, R. Piene (eds.), *The Abel Prize 2013–2017*,
https://doi.org/10.1007/978-3-319-99028-6_29

Oudayas, Rabat. (Photo: private)

Left: My mother and me. Middle: My mother, Dany and me. Right: My mother with Dany in Rabat. (Photo: private)

traveling in a slow train without air conditioning. We finally arrived to Tunis. I remember a hot and dusty city. At Rabat the heat was moderated by a blessed sea breeze.

In Tunis we lived for two years in a tiny room of a hotel. Afterwards we obtained housing in a suburb of Tunis. Then my father enrolled as a volunteer to Indochina. My mother stayed in Tunis. The French war in Indochina was a tragedy and ended with the Diên Biên Phu disaster. My father returned to Tunis in 1953 and decided finally to try to live with us. I was fourteen.

I was educated by my mother and by my teachers of Lycée Carnot (now *Lycée pilote Bourguiba de Tunis*). My mother demanded that my eventual successes be a revenge for her broken life. She put the strongest expectations upon me. For her success meant power and money. In these postwar years it was fashionable to take Greek and Latin at high school. My mother imposed this choice that unfortunately had the effect of no sciences, with the exception of mathematics. Mathematics was taught at a good level in the humanities.

I loved ancient Greek. I worshiped Socrates, as described by Plato. I loved his ironical lack of respect for the mighty people. In the Gorgias, Socrates is advised by Callicles to learn the cheap tricks of rhetoric that are needed for winning in the political disputes of Athens. Socrates refuses and advocates a life devoted to the search for truth.

Most of my schoolmates belonged to the Jewish community, or I should rather speak of many Jewish communities. Indeed Jewish people from many countries emigrated to Tunis. The first diaspora came during the Roman era. These first immigrants adopted Arabic names, as Taïeb, Haïat, and were speaking an Arabic dialect mixed with many Hebrew words (Judeo-Tunisian Arabic). They ultimately became extremely poor, as it is told in "La statue de sel" by Albert Memmi. The Spanish Jews who had been expelled from Spain on July 31, 1492, by the Catholic Kings found a safe exile in North African and in the Ottoman Empire. Finally the "Grana" came from Livorno in the eighteenth century. Besides Taïeb, Haïat,... my friends were Forti, Modigliani, Houri,... In the streets of Tunis one could hear people speaking Italian, Arabic, French, and even Russian. Tunis was a colorful and peaceful melting pot. Distinct religions coexisted. Tunis was a paradise, as Trieste or Sarajevo have been. Tunisia was not a French colony. It was a foreign country enjoying a temporary agreement with France. This agreement ("protectorat") paved the road to full independence. Independence was reached peacefully in 1955. Habib Bourguiba and Pierre Mendès-France were smart people and could understand each other.

In Tunisia beauty was present everywhere. In the landscape, in the architecture, and even in the dresses of women walking in the streets. Roman ruins had been preserved. A beautiful aqueduct built twenty centuries ago by the Roman emperor Hadrian was still standing high near the place we were living. This aqueduct had been restored in 1267 by El-Mostancir after a partial destruction by the Vandals. I loved this aqueduct. I was admiring it every day on my way to high school.

The culture of Tunisia results from the influence of a large number of civilizations. We can list Carthaginian, Roman, Vandal, Jewish, Christian, Arab, Islamic, Turkish, and French (and the native Amazigh should not be forgotten). Tunisians are quite tolerant after having been exposed to so many cultures and religions. But the Lycée Carnot of my childhood was far from being perfect, since it was focused on the European culture. There were only a few Muslims among the students. Muslims were mostly studying in the famous Sadiki College, founded in 1875 by the Grand Vizier Pasha Kheireddine. Not only the Arabic language but also the Tunisian culture was absent from Lycée Carnot.

Lycèe Carnot. (Photo: private)

For instance we never heard of Ibn Khaldun who was born in Tunis, lived in the fourteenth century and was one of the greatest historian of all times. What is the point of studying ancient Greek and being blind to the culture of the people around us. I am still ashamed of not being able to speak Arabic. I was shaped that way.

My family left Tunis in August 1956. I was seventeen. My father who was living with us since his return from Indochina received a position in Strasbourg, the city where he was born. Compared to Tunis Strasbourg was hell. In 1956 the German influence was still strong. People were speaking a German dialect and had kept some traditional German values of the nineteen century. I was quite unhappy there but this lasted only one year.

I was admitted at Ecole Normale Supérieure in July 1957. My mother strongly objected my decision to choose Ecole Normale Supérieure instead of Ecole Polytechnique. Ecole Normale Supérieure leads to a professorship either in high school where the alumni are given a position immediately after graduating, or for those who are gifted in research and willing to start a Ph.D. program, eventually to a position at the University. Ecole Normale Supérieure did not offer a graduate program. If you enter Ecole Polytechnique you will likely become a manager, an important person. This was not my goal. I was still influenced by Socrates. I did not want to be involved in the industrial development of my country. Today I cannot defend this viewpoint. In Ecole Normale Supérieure we were housed together with students in humanities. I loved discussing literature with Jacques English (who was a philosopher) and Yoshio Abe (a specialist of Baudelaire). Y. Abe wrote an essay on the years he spent at Ecole Normale Supérieure. Here is what he remembered of our discussions:

> Among the mathematics majors, Meyer was the best and brightest. He loves literature almost the same level as he loves mathematics. We often discussed Marivaux and Proust, but I forgot the details of our discussions. What I now remember about Meyer is the following: on our way back from the concert of Ms. Clara Haskil just before she passed away, he asked me "The general public do not really understand the state that genius musicians like her attained through their painstaking efforts; they are left lonely even after they got a big round of applause from the audience. If that is the case, what is the point of holding a concert and playing the instrument despite her illness? What do you think?"

After graduating from Ecole Normale Supérieure, I started my professional life as a high school teacher. For three years I taught mathematics at Prytanée militaire de La Flèche. Let me explain this decision and what is Prytanée. The King Henri IV established the Collège Royal in 1604. René Descartes had been a student there. Napoléon transferred this high school to the Army and gave it the name of Prytanée militaire de La Flèche. While his officers were fighting in Napoléon's campaigns, their children were being educated in the Prytanée. In 1960 the Algerian war of Independence was still raging. I had the good fortune of doing my military service as a professor in the Prytanée. Teaching in Prytanée militaire de La Flèche was better than fighting a bad fight in an unjust war. The Army offered this alternative. My military service lasted two years and ended in November 1962. Some of my students of these years are now my colleagues. Let me single out François Ledrappier and Paul-Jean Cahen.

In 1963 I married Anne Limpaler. I obtained a teaching assistant position at the University of Strasbourg. In these times Ph.D. did not exist in France. Instead we had "thèse d'État", which could last ten years or more. One could be a teaching assistant for ever. My dedication to harmonic analysis originated from reading the first edition of Trigonometric Series by Antoni Zygmund. I fell in love with this book, both with the material and the style of the exposition. A few years later I met Zygmund. He was the deep, human, simple, and sensitive person I guessed from his book. Harmonic analysis did not exist in Strasbourg. In the beginning of my mathematical career, I was working completely alone. I wrote my Ph.D. dissertation in complete isolation. I constructed non trivial multipliers of the Hardy space $H^1(\mathbb{R})$. I found interesting examples but I was unable to prove what I conjectured, the fact that Hörmander's condition suffices. This was achieved by Elias Stein. I was competing with Stein without knowing it. When my thesis was completely written and typed by my wife I asked for Pierre Cartier's opinion. He advised me to bring it to Jean-Pierre Kahane. In these years the role of the supervisor was mostly to tell you if what you were aiming to do had been done before.

Kahane thought that my findings were sufficient for a "thèse d'État". Elias Stein was on the way to publishing a much better theorem than the results of my thesis. I was destroyed. When I presented my dissertation I could not stop saying "I obtained this theorem but Elias Stein proved a much better result…" Finally the probabilist

Paul André Meyer who belonged to the committee said "Yves, it is not your role to criticize your results, it is the role of the committee." My defense was a disaster.

In 1966 I obtained a temporary position at the mathematics department in Orsay. At that time this department belonged to the University of Paris. Then I read the extraordinary book "Ensembles parfaits et Séries trigonométriques" by J.P. Kahane and R. Salem. I was enthusiastic. The role played by Pisot numbers in the problem of uniqueness for trigonometric series had been discovered by Salem and Zygmund and was detailed in this book. It was truly fascinating. Here is the theorem. Let $\theta > 2$ and $E_\theta \subset \mathbb{R}$ be the Cantor set with dissection ratio $1/\theta$. More precisely E_θ is the compact set consisting of all sums $\sum_0^\infty \epsilon_k \theta^{-k}$ where $\epsilon_k \in \{0, 1\}$. Then E_θ is a set of uniqueness for the trigonometric expansion if and only if θ is a Pisot number. Pisot numbers were discovered by Axel Thue in 1912. I was fascinated by the interplay between number theory and harmonic analysis which is so profound and beautiful in this result.

I decided to extend the work of Salem and Zygmund to the problem of spectral synthesis. More precisely I wanted to prove that E_θ is a set of spectral synthesis when θ is a Pisot number. Carl Herz did it in 1957 when $\theta = 3$. It took me five years of intense work to complete this program. Let me describe my approach. Spectral synthesis holds for a compact set E if every function in L^∞ whose Fourier transform $\hat{f}$ is supported by E is weak star limit of a sequence of trigonometric polynomials whose frequencies belong to E. For achieving this goal one samples the dilated Fourier transform $\widehat{f}(\theta^{-N}x)$ on the discrete set $\Lambda_{\theta,N}$ consisting of the 2^N sums $\sum_0^{N-1} \epsilon_k \theta^k$ where $\epsilon_k \in \{0, 1\}$. Let Λ_θ be the increasing union of these $\Lambda_{\theta,N}$. In Herz' proof $\Lambda_3 \subset \mathbb{N}$ and $\widehat{f}(3^{-N}x)$ can be sampled on a regular grid with a good control of the L^∞ norms. The proof ends there. When θ is not an integer much more work is needed. The following question shall be answered. What are the discrete sets Λ of real numbers that have the property that any *mean periodic function* f whose spectrum is contained in Λ is an *almost periodic function*? If $\Lambda = \Lambda_\theta$ the answer is positive if and only if θ is a Pisot number. More generally I proved that *model sets* have this property. These splendid patterns will be rediscovered four years later as Penrose pavings. Fifteen years later they will be known as quasi-crystals. Number theory plays a key role in this investigation. I presented this work on spectral synthesis and Pisot numbers as an invited speaker at the ICM 1970, which took place in Nice. During these years I often discussed with Charles Pisot. The book [2] I wrote on the subject in 1970 did not find an audience and was eventually pulped by Elsevier. Dan Shechtman did not benefit from my findings.

Penrose pavings (discovered by Roger Penrose in 1974) mimic the beauty of the Roman mosaics in Tunisia. Penrose pavings are quasi-crystals. As illustrated by the following picture the medieval Islamic artists already found these splendid pavings.

In this part of my mathematical life, I was obsessed by the search of beauty. I viewed mathematics as an art where only rigor and beauty matter. Was I simply irresponsible?

In 1974 my work in number theory did not find an audience. Even the small group of students of Pisot viewed my research as heretical. I was too much ahead of time. Fifteen years later my book would be a classic. I was quite discouraged, but I had the good fortune to have Raphy Coifman provide some guidance. At that time Coifman was a professor at Washington University in Saint Louis. The first day of my visit there Raphy said we should solve Calderón's conjectures. In the early sixties A. Calderón was intrigued by a remarkable discovery achieved by E. de Giorgi in 1956. De Giorgi proved Hölder estimates on solutions of some elliptic PDEs without any smoothness assumption on the coefficients. A. Calderón wanted to recover De Giorgi's estimates through an improved pseudo-differential calculus where the smoothness assumptions on the coefficients are minimal. This program was so ambitious that one might have feared some lengthy preliminary work. This was not A. Calderón's style. Instead he pointed to a concise question that happened to be the magic key opening all doors. Moreover the same key was expected to yield striking discoveries in real and complex analysis. Here is Calderón's conjecture. Let $A(x)\colon \mathbb{R} \mapsto \mathbb{C}^m$ be a Lipschitz function of the real variable x. We have $|A(x') - A(x)| \leq C|x - x'|, x, x' \in \mathbb{R}$. The Lipschitz norm of A is the optimal C and is also given by $\|a\|_\infty$ where $a(x) = \frac{d}{dx}A(x)$. Let $B \subset \mathbb{C}^m$ be the smallest compact set such that

$$\frac{A(x) - A(y)}{x - y} \in B, \quad x, y \in \mathbb{R}.$$

Let us assume that Φ is analytic on a neighborhood of this compact set B and let $T_A f(x) = pv \int K_A(x, y) f(y)\, dy$ be the singular integral operator defined by the singular kernel

$$K_A(x, y) = \Phi\left(\frac{A(x) - A(y)}{x - y}\right) \frac{1}{x - y}.$$

A. Calderón made the following claim: There exist a constant $C = C(B, \Phi)$ such that for $f \in L^2$ and A as above, one has

$$\|T_A(f)\|_2 \leq C \|f\|_2.$$

This conjecture is the magic key opening new chapters in complex analysis, linear PDEs, and nonlinear PDEs. If $m = 1$ and $\Phi(t) = (1 + it)^{-1}$, T_A is the *Cauchy integral on a Lipschitz curve* and the boundedness of the Cauchy integral has far reaching applications to complex analysis. The extension to higher dimensions of these new singular integral operators is provided by the *method of rotations*. The boundedness of the Cauchy integral for all Lipschitz curves implies that the method of layer potentials can be used to solve the Dirichlet and Neumann problems in Lipschitz domains. Moreover Calderón's program is seminal in the modern theory of multi-linear operators.

It took us seven years to prove it. On the way Raphy and I developed a new theory of multi-linear operators $T = L^{p_1} \times L^{p_1} \times \cdots \times L^{p_m} \mapsto L^p$ where $1/p = 1/p_1 + \cdots + 1/p_m$. During my visits to the University of Chicago I often discussed with Alberto Calderón and Antoni Zygmund, which was inspiring and delightful. What Raphy and I eventually achieved (with the collaboration of Alan McIntosh in 1981) was immediately applauded and used by all the mathematicians working in the field. I was invited to speak on these results at ICM 1982 (held in 1983 at Warsaw). Soon afterwards my students Guy David and Jean-Lin Journé discovered the fabulous $T(1)$ theorem [1] and most of what Raphy and I have achieved with much effort and pain was suddenly trivial. For the first time in my life I gained international recognition. At that time I was still a pure mathematician.

While I was solving some hard problems in pure mathematics Jacques-Louis Lions was boosting French industry by developing applied mathematics at the highest level. It took me many years to understand the deepness of his fight. I changed my views on my research during the Fall of 1983. I was a professor at Ecole Polytechnique. My friend and colleague Charles Goulaouic was dying of cancer. My mathematical talent could not even improve the ultrasound examination of his liver. I was an invalid in front of such human suffering and distress.

In 1984 J-L. Lions gave me an important problem in control theory. This problem was a key issue in the construction of the International Space Station. The issue was to control and attenuate some vibrations that could damage the structure. The control would be achieved by firing a tiny rocket on the station. I fortunately solved the problem raised by Lions [3]. Then I emerged from depression, and understood for the first time in my life that my skills in pure mathematics could be used in real-life problems. In my research I then abolished the frontier between pure and applied mathematics. A few months later Louis Nirenberg found a simpler proof. Then Jacques-Louis Lions found a third proof that led to the HUM strategy.

One year later I was joining the wavelet group led by Ingrid Daubechies, Alexander Grossmann and Jean Morlet. Then for the first time in my life I understood some of the problems raised in physics and signal processing. Stimulating discussions with Ingrid Daubechies played a pivotal role in my construction of orthonormal wavelet bases. Soon after I had the incredible chance of working with Stéphane Mallat. That happened in 1987 at the University of Chicago. Mallat and I spent three days in the office of Antoni Zygmund. During these three days Mallat unveiled a spectacular discovery. He proved that my newborn orthonormal wavelet bases were nothing other than the quadrature mirror filters that were already used by the signal processing community. More precisely every orthonormal wavelet basis $\mathscr{B}$ is rooted in a quadrature mirror filters. The converse is not true and some quadrature mirror filters are unstable. They do not provide us with an orthonormal wavelet basis. This point was later on clarified by Albert Cohen [5]. Mallat's discovery explains why the signal processing community immediately adopted orthonormal wavelet bases. During my wavelet decade (1984–1994) I could abolish the frontier between mathematics and the other sciences. I was extremely happy. Moreover I enrolled Raphy Coifman in this endeavor.

My interest in Navier–Stokes equations arose from the wavelet revolution. My source of inspiration was (1) a series of talks and preprints by Marie Farge and (2) an intriguing paper by Guy Battle and Paul Federbush entitled "Navier and Stokes meet the wavelets". Following the views of Marie Farge, Guy Battle and Paul Federbush, it was reasonable to believe that wavelet based Galerkin schemes could overcome pseudo-spectral algorithms that are acknowledged as being the best solvers for Navier–Stokes equations. Indeed turbulent flows are active over a full range of scales and one is tempted to decouple Navier–Stokes equations into (1) a sequence E_j; $j \in \mathbb{Z}$ of equations where the evolution is confined to

Raphy Coifman and me. (Photo: private)

a given scale 2^j and (2) some description of the nonlinear interactions between scales or of the energy transfers across scales. The only existing algorithm that permit to travel across scales while keeping an eye on the frequency contents is the Littlewood-Paley expansion (or the wavelet analysis). Furthermore micro-local analysis and Littlewood-Paley expansions have been successfully applied to Navier–Stokes or Euler equations by Jean-Yves Chemin and his students. Times were ripe for replacing Littlewood–Paley analysis by fast numerical schemes that have the same scientific contents, i.e. by wavelet analysis. It is surprising that this endeavor was not a success story. As often in science, something else was found. My student Marco Cannone made two main discoveries. He proved that Littlewood–Paley expansions were more effective than wavelet expansions in the Battle–Federbush paper. He then observed that a strategy due to Fujita and Kato (but also used by Cazenave and Weissler) was even more effective. Once this was clarified, Marco Cannone and Fabrice Planchon improved on the Fujita–Kato theorem. Indeed they proved global existence for solutions $u(x;t) \in \mathscr{C}([0;\infty); L^3(\mathbb{R}^3))$ whenever the initial condition u_0 is oscillating. Uniqueness was proved by Pierre-Gilles Lemarié a few years later. The oscillating character of u_0 is defined by the smallness of a Besov norm in a suitable Besov space. An equivalent condition is given by simple size estimates of the wavelet coefficients. These methods did not yield the limiting case. The best result in this direction was finally obtained by Herbert Koch and Daniel Tataru [4].

Nine years ago I returned to pure mathematics. In collaboration with my student Basarab Matei I unveiled the magic properties of irregular sampling on quasi-crystals. One can take advantages of holes in the spectrum of a band-limited signal f to down-sample f below the Nyquist rate. The Lebesgue measure of the spectrum of f replaces the Nyquist rate one samples on a quasi-crystal. At that time I was unaware of the remarkable work of Nir Lev and Alexander Olevskii on the subject. Now Alexander Olevskii and I are also studying quasi-crystalline measures. A discrete measure μ is quasi-crystalline if (a) the support of μ is a locally finite set, (b) μ is a tempered distribution, and (c) the support of the distributional Fourier transform $\widehat{\mu}$ of μ is also locally finite. Olevskii and I completed some missing proofs in an old and forgotten paper by Andrew Guinand that was published in 1959 in Acta Matematica.

The success of my research is mostly due to my friends. Let me single out Raphy Coifman, and praise a friendship over more than forty years. Working with Alexander Olevskii is a blessing. The success of my research is also due to my incredible students. I shared so much with them. We are a family. Alberto Calderón was my spiritual father and my love and gratitude have no bounds. My colleague and friend Robert Ryan kindly improved this manuscript. Let me thank Bob once more for his help.

References

1. David, Guy; Journé, Jean-Lin. A boundedness criterion for generalized Calderón-Zygmund operators, Annals of Mathematics. Second Series, (1984) Vol. 120 (2): 371–397.
2. Y.Meyer. Algebraic numbers and harmonic analysis, North Holland, New York (1972).
3. Y.Meyer. Etude d'un modèle mathématique issu du contrôle des structures déformables. Nonlinear Partial Differential Equations and their applications. Collège de France, (1985) Vol. VII (234–242) Brezis, H and Lions, J-L. eds, Pitman.
4. Y.Meyer. Oscillating patterns in image processing and in some nonlinear evolution equations, (Lewis Memorial Lectures) AMS (2001).
5. Y.Meyer. Wavelets and operators, Vol. 1 and 2, Cambridge Univ. Press (1997).

A Journey Through the Mathematics of Yves Meyer

Albert Cohen

Abstract The mathematics of Yves Meyer cover a wide range of fields, as various as number theory, harmonic analysis, operator theory, partial differential equations, control theory, signal and image processing.This survey gives an introductory and self-contained overview of Yves Meyer's contributions to these areas of research.

Foreword

The mathematics of Yves Meyer cover a wide range of fields, as various as number theory, harmonic analysis, operator theory, partial differential equations, control theory, signal and image processing.

We have chosen to illustrate this unusual variety by a selection of representative highlights. Each of them strikes the imagination either by its depth and elegance, far reaching vision, or its potential for applications in science and technology, sometimes all such features combined together. The main themes, that are treated in separate sections, are the following:

1. Problems in harmonic analysis and number theory, that paved the way to the mathematical theory of quasicrystals.
2. The study of singular integral multilinear operators, which culminated with the treatment of the Cauchy integral on a Lipschitz graph.
3. The construction of wavelet bases and other time-frequency systems, and their application to the analysis of global and local smoothness.
4. Topics in partial differential equations, such as control of waves, compensated compactness, oscillations in nonlinear evolution systems.

Our objective is to offer a limited but self-contained survey of Yves Meyer's contributions and to evoke the broad research programs in which these contributions

A. Cohen (✉)
Laboratoire Jacques-Louis Lions, Sorbonne Université, Paris, France
e-mail: cohen@ann.jussieu.fr

H. Holden, R. Piene (eds.), *The Abel Prize 2013–2017*,
https://doi.org/10.1007/978-3-319-99028-6_30

take place. We have tried to reach a unified notation in the statement of the various results, which sometimes leads to formulations that slightly differ from the original ones. The reader interested in more detailed treatments is invited to consult the beautiful monographs and books by Yves Meyer which are referred to throughout the discussion, according to their relevance to each topic.

The author is deeply indebted to several colleagues for discussions, comments and suggestions. He would like to thank in particular Jean-Paul Allouche, Aline Bonami, Wolfgang Dahmen, Ronald DeVore, Guy David, Alain Haraux, Stéphane Jaffard, François Murat and Fabrice Planchon.

1 Harmonic Analysis, Quasicrystals and Number Theory

In 1982, Dan Shechtman found out by electron microscopy that certain metallic alloys have X-ray diffraction spectra with seemingly contradictory properties [87]. On the one hand, these spectra exhibit sharp spots which indicate some long range structure, similar to crystals. On the other hand, these spots obey angular symmetries of $\pi/5$ which are in contradiction with a periodic organization of the atoms. Quasicrystals had been discovered, leading scientists to reconsider some of their conception of matter after few years of controversy. The Nobel prize in chemistry was awarded to Shechtman in 2011 for his achievement.

Aside from their experimental discovery, particular examples of quasicrystals had been provided in the beautiful tilings constructed in 1974 by Roger Penrose, however a consistent mathematical description of quasicrystal was seemingly missing. Amazingly, the theoretical model was already available, introduced a few years earlier in foundational works by Yves Meyer on harmonic analysis and number theory.

As we discuss further, one initial motivation was the study of the properties of certain sets obtained by a dissection process similar to the generation of the Cantor triadic set. The same concepts that led to the mathematical theory of quasicrystals played a crucial role in studying the spectral synthesis problem for such fractal sets. They were also used to revisit the proof of a celebrated result of Raphaël Salem and Antoni Zygmund on the uniqueness problem for trigonometric series. Yves Meyer was awarded the Salem prize in 1970 for these outstanding results, which are, among many others, detailed in his books [64, 65] and monograph [67].

The relations between number theory and harmonic analysis remain an object of high interest for Yves Meyer, as illustrated by two recent contributions that we evoke at the end of this section. The first one shows that certain quasicrystals constitute natural grids for sampling band-limited functions with unknown Fourier support of given measure. The second one establishes new Poisson summation formulas for atomic measures supported on non-uniform grids, reviving the work of Andrew Guinand.

1.1 Spectral Synthesis

From its early days, the theory of Fourier series has been a source of problems which stimulated the development of entire fields of mathematics. In particular the theory of sets developed by Georg Cantor after 1870 has been motivated by questions such as the following: if a trigonometric series, written in modern form as

$$\sum_{n\in\mathbb{Z}} c_n e^{inx} \tag{1}$$

converges towards 0 everywhere outside of a set S, is it necessarily the null series? If such a property holds, S is called a *set of uniqueness*, otherwise a *set of multiplicity*. These properties are obviously preserved by translation of S and it can be proved that they are also preserved by dilation of S. Countable sets are known to be of uniqueness type, while sets with positive Lebesgue measure are obviously of multiplicity type. This leads to focus on uncountable sets with null Lebesgue measure.

Natural candidates for this investigation are the fractal sets obtained by a dissection process: for a given real number $\theta > 2$, starting from the interval $[0, 1]$, one removes the interval $]\kappa, 1-\kappa[$ where $\kappa = \theta^{-1}$. A similar removal process is applied at the new scale on the two remaining intervals, and iterated. The particular case $\theta = 3$ leads to the Cantor triadic set. For a general $\theta > 2$, up to a rescaling by $\theta - 1$, the limit set is of the form

$$E_\theta = \Big\{ \sum_{k\geq 1} \varepsilon_k \theta^{-k} \ : \ \varepsilon_k \in \{0, 1\} \Big\}. \tag{2}$$

In 1955, Raphaël Salem and Antoni Zygmund gave a beautiful characterization of sets of the above form that are of uniqueness type [84] in terms of a number theoretic property. We recall that a number $\theta \in \mathbb{R}$ is algebraic if it is a solution of an equation $a_0 + a_1 x + \cdots + a_n x^n = 0$ with $a_j \in \mathbb{Z}$. Its conjugates are the other roots when considering the equation of minimal degree n, also called the degree of θ. Non-algebraic real numbers are called transcendentals.

Definition 1 A *Pisot–Vijayaraghavan number* is an algebraic number $\theta > 1$ such that all its conjugates have moduli strictly less than 1. A *Salem number* is an algebraic number $\theta > 1$ such that all its conjugates have moduli less than or equal to 1 with at least one case of equality.

Pisot–Vijayaraghavan numbers were discovered by Axel Thue in 1912 and studied separately by Charles Pisot and Tirukkannapuram Vijayaraghavan. Integers larger than 2 or the golden ratio $\frac{1+\sqrt{5}}{2}$ are examples of Pisot–Vijayaraghavan numbers, while square roots of integers larger than 2 are not. Charles Pisot showed that such numbers are also characterized by the following property: there exists a real $a > 0$ such that, for all $j \geq 0$,

$$a\theta^j = m_j + \varepsilon_j, \tag{3}$$

where the m_j are natural numbers and $(\varepsilon_j)_{j\geq 0} \in \ell^2$. In words, up to rescaling, the iterated powers of θ are asymptotically close to the set of integers, as opposed to iterated powers of a generic real number. The characterization by Salem and Zygmund of dissection sets which have the property of uniqueness is the following.

Theorem 2 *The dissection set E_θ is a set of uniqueness if and only if θ is a Pisot–Vijayaraghavan number strictly larger than* 2.

The fascinating interplay between fractal geometry, number theory and harmonic analysis is explored in depth in the book of Jean-Pierre Kahane and Raphaël Salem [48]. After completing his doctoral thesis, Yves Meyer embraced this line of research. One of his first contributions to number theory concerns the concept of *normal set* that was studied at that period by François Dress, Michel Mendès France and Gérard Rauzy. A set $E \subset \mathbb{R}$ is called a normal set if there exists a sequence $(\lambda_n)_{n\geq 1}$ of real numbers such that E is the set of all x for which the sequence $(x\lambda_n)_{n\geq 1}$ is equidistributed modulo 1. The following result obtained by Yves Meyer in [63] implies that the set of transcendental numbers is a normal set.

Theorem 3 *For all $\varepsilon > 0$, there exists a sequence $(\lambda_n)_{n\geq 1}$ of real numbers such that $|\lambda_n - n| \leq \varepsilon$ and such that the sequence $(x\lambda_n)$ is equidistributed modulo* 1 *if and only if x is a transcendental number.*

In the end of the 1960s, Yves Meyer decided to work on the spectral synthesis of bounded continuous functions. Given a compact set $E \subset \mathbb{R}$, one considers the space $\mathscr{B}_E$ of bounded continuous functions on $\mathbb{R}$ whose spectrum is contained in E, and its subset $\mathscr{S}_E$ that consists of all finite combinations of complex exponential functions

$$e_\lambda(x) := \exp(i2\pi\lambda x), \tag{4}$$

for $\lambda \in E$. The spectrum of a continuous function f can be defined as the set of all λ such that e_λ belongs to the closure of the translates $\{f(\cdot - y)\}_{y\in\mathbb{R}}$ for the topology of uniform convergence on compact sets. For a bounded continuous function this set is the support of its distributional Fourier transform. Note that throughout this section, the Fourier transform is defined by

$$\hat{f}(\omega) = \int_{\mathbb{R}^d} f(x)e^{-i2\pi\langle\omega,x\rangle}dx, \tag{5}$$

while the factor 2π is omitted in the definition used in further Sects. 2, 3 and 4.

One wants to know if $\mathscr{S}_E$ is dense in $\mathscr{B}_E$ in the weak* topology $\sigma(L^\infty, L^1)$. If the answer is positive, E is called a *set of spectral synthesis*. A countable set E is always a set of spectral synthesis, in this case the weak* topology can be replaced by the strong one.

In addition, one is interested in concrete approximations schemes through linear operators of finite rank

$$L_k : \mathscr{B}_E \mapsto \mathscr{S}_{F_k}, \tag{6}$$

where $(F_k)_{k\geq 1}$ is a sequence of finite subsets of E such that $\#(F_k) \to \infty$. These operators should thus satisfy

$$\lim_{k\to\infty} \int_{\mathbb{R}} L_k f(x) g(x) dx = \int_{\mathbb{R}} f(x) g(x) dx, \quad f \in \mathscr{B}_E, \ g \in L^1(\mathbb{R}). \tag{7}$$

It can be checked that this implies the uniform stability property

$$\|L_k f\|_{L^\infty} \leq C \|f\|_{L^\infty}, \quad f \in \mathscr{B}_E, \ k \geq 0, \tag{8}$$

for some fixed constant C, as well as uniform convergence of $L_k f$ towards f on any compact set.

Raphaël Salem wanted to understand if the answer to the problem of spectral synthesis for the dissection sets E_θ can also be given in terms of number theoretic properties, similar to sets of uniqueness. In the particular case of the triadic Cantor set E_3, a positive answer was given by Carl Herz in [41].

Yves Meyer attacked the spectral synthesis problem for more general dissection sets E_θ by studying a specific approximation procedure: observing that

$$E_\theta = F_k + \theta^{-k} E_\theta, \quad F_k := \Big\{ \sum_{j=1}^{k} \varepsilon_j \theta^{-j} \ : \ \varepsilon_j \in \{0, 1\} \Big\}, \tag{9}$$

one finds that any $f \in \mathscr{B}_E$ can be decomposed into

$$f(x) = \sum_{\lambda \in F_k} a_\lambda(x) e_\lambda(x), \tag{10}$$

where each function a_λ belongs to $\mathscr{B}_{\theta^{-k} E_\theta}$. The functions a_λ therefore have slower variation as $k \to \infty$, which suggests the approximation

$$L_k f(x) := \sum_{\lambda \in F_k} a_\lambda(0) e_\lambda(x). \tag{11}$$

Yves Meyer gave a characterization of the dissection sets for which this approximation process is successful [67].

Theorem 4 *For any real number $\theta > 2$, the following properties are equivalent:*

- *E_θ is a set of spectral synthesis and the linear approximation scheme defined by* (11) *satisfies the uniform bound* (8).
- *θ is a Pisot–Vijayaraghavan number such that for any $k \geq 0$ and with n the degree of θ, the identity*

$$\varepsilon_0 + \varepsilon_1\theta + \cdots + \varepsilon_k\theta^k = (1 - \theta^{k+1})(q_0 + q_1\theta + \cdots + q_n\theta^n), \tag{12}$$

with $\varepsilon_j \in \{-1, 0, 1\}$ and $q_j \in \mathbb{Q}$, implies that $\varepsilon_j = q_j = 0$ for all j.

In particular, all integers $\theta > 2$ fall in the above described category. While the above approximation process (11) may fail for general Pisot–Vijayaraghavan numbers, Yves Meyer proves in [65] that it is still possible to construct a sequence of operators $(L_k)_{k\geq 1}$ which has the required properties, thereby establishing the following result.

Theorem 5 *If $\theta > 2$ is a Pisot–Vijayaraghavan number, then E_θ is a set of spectral synthesis.*

1.2 Coherent Sets of Frequencies

The proof by Yves Meyer of the above results relies in good part on the study of the discrete grid Λ_θ that consists of all finite sums of the form

$$\sum_{k\geq 0} \varepsilon_k\theta^k, \quad \varepsilon_k \in \{0, 1\}. \tag{13}$$

These sums are the frequencies that appear after properly rescaling the functions $L_k f$ in (11). The study of these grids also led Meyer to revisit the work of Salem and Zygmund, providing a new and simpler proof of Theorem 2.

One crucial property is encapsulated in the following definition proposed in [65]. We now use the multivariate notation

$$e_\lambda(x) = \exp(i2\pi\langle\lambda, x\rangle), \tag{14}$$

for $\lambda \in \mathbb{R}^d$

Definition 6 A set $\Lambda \subset \mathbb{R}^d$ is a *coherent set of frequencies* if there exist a compact set $K \subset \mathbb{R}^d$ and a constant C such that every finite trigonometric sum $g := \sum_{\lambda\in\Lambda} c_\lambda e_\lambda$ with frequencies in Λ satisfies

$$\|g\|_{L^\infty(\mathbb{R})} \leq C\|g\|_{L^\infty(K)}. \tag{15}$$

The above concept was already contained in a fundamental work of Jean-Pierre Kahane [47] that studies the links between the two well-known variants of periodic functions:

- Almost periodic functions, introduced by Harald Bohr in 1923, are the continuous functions which are uniform limits on $\mathbb{R}^d$ of finite linear combinations of the functions $(e_\lambda)_{\lambda\in\mathbb{R}^d}$.
- Mean periodic functions, introduced by Jean Delsarte in 1934, are the continuous functions f such that the set $\tau(f)$ spanned by the translates $\{f(\cdot - y)\}_{y\in\mathbb{R}^d}$ in the topology of uniform convergence on compact sets is a strict subspace of $C(\mathbb{R}^d)$.

In 1947, Laurent Schwartz proved in [85] that any univariate mean-periodic function f of spectrum Λ can be uniformly approximated on every compact set by finite linear combinations of

$$e_{\lambda,p}(x) := x^p \exp(i2\pi\lambda x), \quad p \in \mathbb{N},\ \lambda \in \Lambda. \tag{16}$$

This result was extended to the multivariate case in 1955 by Bernard Malgrange.

In the case where f is also bounded, one shows that a similar approximation result holds when only using the trigonometric functions e_λ, which suggests that it could be an almost periodic function. However, Jean-Pierre Kahane shows that there exist bounded mean-periodic functions that are not uniformly continuous and therefore not almost periodic. This leads him to the following question: what condition on $\Lambda \subset \mathbb{R}$ characterizes the existence of a bounded mean-periodic function f with spectrum contained in Λ that is not almost periodic? One main result from [47] is that a necessary and sufficient condition is that Λ should be a coherent set of frequencies, with K in Definition 6 being an interval.

One trivial instance of a coherent set of frequencies is given by the regular lattice $\Lambda = \mathbb{Z}^d$, or any of its subsets, for which one can take $K = [0, 1]^d$. The particular sets discussed in the seminal work of Kahane are lacunary sequences of Sidon type for which the interval K satisfying (15) can be chosen with arbitrarily small measure.

Examples of coherent sets of frequencies that are not subsets of a regular lattice were provided by Yves Meyer. As to the grids Λ_θ given by (13), the following result is proved in [68].

Theorem 7 *For $\theta > 2$, the set Λ_θ is a coherent set of frequencies if and only if θ is a Pisot–Vijayaraghavan number.*

Other examples, that are discussed further, paved the way to the mathematical theory of quasicrystals. The relevance of coherent sets of frequencies to the spectral synthesis problem lies in the following result.

Theorem 8 *Let Λ be a coherent set of frequencies. Then, there exist a neighbourhood V of the origin and a constant $C > 1$, such that if $(a_\lambda)_{\lambda\in\Lambda}$ is a collection of* *functions in $\mathscr{B}_V$, one has*

$$\sup_{x,y\in\mathbb{R}^d}\Big|\sum_{\lambda\in\Lambda}a_\lambda(y)e_\lambda(x)\Big|\le C\sup_{x\in\mathbb{R}^d}\Big|\sum_{\lambda\in\Lambda}a_\lambda(x)e_\lambda(x)\Big|. \tag{17}$$

Conversely, this property implies that Λ is a coherent set of frequencies.

In the above inequality, the function of the variable x appearing on the right side has spectrum in $\Lambda+V$, while the function of the variables x and y on the left side has spectrum in $\Lambda\times V$ and coincides with f on the diagonal $x=y$. This construction was termed in French as *principe des soucoupes*, since $\Lambda+V$ may be visualized as a sequence of saucers distributed on a table which are "piled up" in $\Lambda\times V$. When applied to the set Λ_θ, this principle is used by Meyer to establish the stability of approximation schemes such as (11) and therefore the spectral synthesis property for the sets E_θ.

1.3 Mathematical Models of Quasicrystals

The mathematical model for a perfect crystal is a set $\Lambda\subset\mathbb{R}^d$ of the form

$$\Lambda := F+L, \tag{18}$$

where F is a finite set and L is a full rank lattice, that is,

$$L=B\mathbb{Z}^d, \tag{19}$$

where B is a $d\times d$ invertible matrix. Let us denote by $|L|:=|\det(B)|$ the measure of the fundamental volume of L. The Poisson summation formula, that plays a central role in crystallography, may be written as the distributional identity

$$|L|\sum_{\lambda\in L}\widehat{\delta_\lambda}=\sum_{\lambda^*\in L^*}\delta_{\lambda^*}, \tag{20}$$

where the dual lattice L^* is the set of points $\lambda^*\in\mathbb{R}^d$ such that $\langle\lambda,\lambda^*\rangle\in\mathbb{Z}$ for all $\lambda\in\Lambda$ or equivalently $L^*=(B^t)^{-1}\mathbb{Z}^d$.

Let us now turn to the mathematical models of quasicrystals that emerged from the work of Yves Meyer [64, 65, 67]. All of them are instances of sets of coherent frequencies. We begin with a preliminary definition that describes a certain uniformity in the placement of points.

Definition 9 A set $\Lambda\subset\mathbb{R}^d$ is a *Delone set* if it is both

- Uniformly discrete: there exists $r>0$ such that any ball $B(x,r)$ contains at most one point of Λ.

- Relatively dense: there exists $R > 0$ such that any ball $B(x, R)$ contains at least one point of Λ.

Obviously, lattices and perfect crystals obey this property, but many other sets also do. In particular, Delone sets are generally *not* sets of coherent frequencies. Mathematical quasicrystals are Delone sets with additional structural properties that are described by Yves Meyer in various ways.

The first description is obtained by the *cut and project* scheme which was implicit in earlier work on algebraic number theory: the set of interest is obtained by projecting a "slice" cut from a higher dimensional lattice in general position.

Definition 10 Let L be a full rank lattice L of $\mathbb{R}^{d+m}$ for some $d, m > 0$. Denoting by $p_1(x) \in \mathbb{R}^d$ and $p_2(x) \in \mathbb{R}^m$ the components of $x \in \mathbb{R}^{d+m}$ such that $x = (p_1(x), p_2(x))$, we assume that p_1 is a bijection between L and $p_1(L)$ with dense image. A similar property is assumed for p_2. Let $K \subset \mathbb{R}^m$ be a Riemann integrable compact set of positive measure. The associated *model set* $\Lambda = \Lambda(L, K) \in \mathbb{R}^d$ is defined by

$$\Lambda := \{p_1(x) \ : \ x \in L, \ p_2(x) \in K\}. \tag{21}$$

Model sets are particular cases of more general notions of quasicrystals which are discussed further. Their link with the grids Λ_θ given by (13) is expressed by the following result.

Theorem 11 *If θ is a Pisot–Vijayaraghavan number, then Λ_θ is contained in a model set Λ.*

Yves Meyer observes that the Poisson formula (20) on the underlying lattice structure L induces a similar formula for the model set $\Lambda(L, K)$ however involving weights: for any smooth function φ compactly supported in K, one has

$$\sum_{x\in L} \varphi(p_2(x))\hat{\delta}_{p_1(x)} = \sum_{x^*\in L^*} \psi(p_2(x^*))\delta_{p_1(x^*)}, \tag{22}$$

where

$$\psi(\omega) := |L|^{-1}(2\pi)^d \hat{\varphi}(-\omega). \tag{23}$$

Formula (22) shows that the Fourier transform of an atomic measure of the form $\sum_{\gamma\in\Lambda} w(\gamma)\delta_\Lambda$, thus supported by the model set Λ, is an atomic measure on a dense set of points, however with higher weights on certain points induced by the fast decay of ψ. Antonio Cordoba proved that such Poisson summation formulas always require weights unless Λ is a lattice [23].

The second mathematical description of quasicrystals involves a concept of duality motivated by the study of almost periodic functions.

Definition 12 Let $\Lambda \subset \mathbb{R}^d$ and $\varepsilon > 0$. The *ε-dual* of Λ is the set

$$\Lambda^\varepsilon := \{\nu \in \mathbb{R}^d \ : \ |\exp(i2\pi\langle \nu, \lambda\rangle) - 1| \leq \varepsilon, \ \lambda \in \Lambda\}. \tag{24}$$

Obviously if Λ is a full rank lattice, one has $\Lambda^* \subset \Lambda^\varepsilon$ for all $\varepsilon > 0$, and equality holds for ε sufficiently small. On the other hand, the ε-dual of most Delone sets is reduced to $\{0\}$. Structural properties may therefore be described by imposing that the set Λ^ε consists of many elements.

Definition 13 A set Λ is *ε-harmonious* if Λ^ε is relatively dense, and *harmonious* if this holds for all $\varepsilon > 0$.

The third description, now commonly used as the mathematical definition of a quasicrystal and termed as *Meyer sets* by Robert Moody [78], is based on the study of the difference set

$$\Lambda - \Lambda := \{\lambda - \nu \ : \ \lambda, \nu \in \Lambda\}. \tag{25}$$

The set $\Lambda - \Lambda$ is relevant to crystallography since diffraction patterns are related to the Fourier transform of the autocorrelation of the density function, therefore involving the interatomic distance vectors $\lambda - \nu$.

Definition 14 A set $\Lambda \subset \mathbb{R}^d$ is a *Meyer set* if it is a Delone set such that

$$\Lambda - \Lambda \subset \Lambda + F, \tag{26}$$

where F is a finite set.

Jeffrey Lagarias shows in [49] that this property has an even simpler equivalent expression.

Theorem 15 *A set $\Lambda \subset \mathbb{R}^d$ is a Meyer set if and only if it is a Delone set such that $\Lambda - \Lambda$ is also a Delone set.*

The relations between the above descriptions of quasicrystals have been established by Yves Meyer and can be summarized as follows.

Theorem 16 *Let $\Lambda \subset \mathbb{R}^d$ be a Delone set. The following properties are equivalent:*

- *Λ is a Meyer set.*
- *Λ is a harmonious set.*
- *There exists a model set $\Lambda(L, K)$ and a finite set F such that $\Lambda \subset F + \Lambda(L, K)$.*

Meyer sets, and therefore harmonious and models sets, are sets of coherent frequencies.

Among the many connections between the above models of quasicrystals and number theory, a striking one appears when asking which dilation factors leave such sets invariant. These factors are obviously the integers in the case of a full rank lattice. The answer for more general quasicrystals is provided in a beautiful result of Yves Meyer.

Theorem 17 *Let Λ be a Meyer set. If $\theta > 1$ is such that $\theta\Lambda \subset \Lambda$, then it is a Pisot–Vijayaraghavan or Salem number. Conversely, for any Pisot–Vijayaraghavan or Salem number θ, there exists a Meyer set Λ such that $\theta\Lambda \subset \Lambda$.*

1.4 Universal Sampling Sets

Yves Meyer did not have in mind the mathematical description of quasicrystals when introducing and studying the previously described discrete sets. One of his intuitions was that such sets could serve as natural *sampling* grids for relevant classes of functions.

Sampling theory has been motivated since the 1960s by the development of discrete telecommunications. It is well known, since the foundational work of Claude Shannon and Harry Nyquist, that regular grids are particularly suitable for the sampling of certain band-limited functions. This may be seen as a direct consequence of the Poisson summation formula (20) which yields, for any sufficiently nice function f,

$$|L| \sum_{\lambda \in L} f(\lambda) e^{i2\pi\langle\lambda,\omega\rangle} = \sum_{\lambda^* \in L^*} \hat{f}(\omega + \lambda^*). \tag{27}$$

This last formula shows that, if $E \subset \mathbb{R}^d$ is a compact set with translates $(E + \lambda^*)_{\lambda^* \in L^*}$ having intersections of null measure, functions with Fourier transform supported in E are then stably determined by their sampling on Γ. Such sets E should in particular satisfy

$$|E| \le \frac{1}{|L|}. \tag{28}$$

One elementary example, for which equality holds in the above, is the fundamental volume of the lattice L^*, that is,

$$E_{L^*} = (B^*)^{-1}([0, 1]^d), \tag{29}$$

or any of its translates. A theory of stable sampling on more general discrete sets was developed in the 1960s by Henry Landau and Arne Beurling, and can be summarized as follows. If $E \subset \mathbb{R}^d$ is a compact set, one considers the Paley–Wiener space $\mathscr{F}_E$ that consists of all functions in $L^2(\mathbb{R}^d)$ with Fourier transform supported on E.

Definition 18 A set $\Lambda \subset \mathbb{R}^d$ has the property of *stable interpolation* for $\mathscr{F}_E$ if there exists a constant C such that

$$\sum_{\lambda\in\Lambda} |f(\lambda)|^2 \leq C\|f\|_{L^2}^2, \quad f \in \mathscr{F}_E. \tag{30}$$

It has the property of *stable sampling* for $\mathscr{F}_E$ if there exists a constant C such that

$$\|f\|_{L^2}^2 \leq C\sum_{\lambda\in\Lambda} |f(\lambda)|^2, \quad f \in \mathscr{F}_E. \tag{31}$$

Necessary conditions for stable interpolation and sampling can be stated in terms of certain density notions introduced by Beurling for discrete sets. The upper density of $\Lambda \subset \mathbb{R}^d$ is defined by

$$\overline{\text{dens}}(\Lambda) = \limsup_{R\to\infty} \sup_{x\in\mathbb{R}^d} \frac{\#(\Lambda \cap B(x,R))}{|B(x,R)|} \tag{32}$$

and its lower density by

$$\underline{\text{dens}}(\Lambda) = \liminf_{R\to\infty} \inf_{x\in\mathbb{R}^d} \frac{\#(\Lambda \cap B(x,R))}{|B(x,R)|} \tag{33}$$

The set is said to have uniform density if $\overline{\text{dens}}(\Lambda) = \underline{\text{dens}}(\Lambda) := \text{dens}(\Lambda)$. The following result was obtained by Henri Landau in [55].

Theorem 19 *If Λ is a set of stable sampling for $\mathscr{F}_E$, then $\underline{\text{dens}}(\Lambda) \geq |E|$. If Λ is a set of stable interpolation for $\mathscr{F}_E$, then $\overline{\text{dens}}(\Lambda) \leq |E|$.*

Sufficient conditions cannot however be stated in terms of the sole density, unless E has some simple structure. For example, when E is a univariate interval, Beurling showed that sets with density strictly larger than $|E|$ have the stable sampling property. The case of a lattice L is instructive: on the one hand, the set E_{L^*} has measure $|E_{L^*}| = |L|^{-1} = \text{dens}(L)$ and satisfies the stable sampling and interpolation properties in view of (27). On the other hand, other sets E with the same or even smaller measure could have their translates by Λ^* overlapping with non-zero measure, which is a principal obstruction to these properties. This phenomenon is well known in electrical engineering as *aliasing*.

A natural question is therefore: does there exist sets Λ that are suitable for the sampling of band-limited functions with Fourier support contained in *any arbitrary* set E of prescribed measure? Such sets Λ are called *universal sampling sets*.

In view of the previous remarks, regular lattices cannot be universal sampling sets. In [80, 81], Alexander Olevskii and Alexander Ulanovskii gave the first construction of a set Λ of uniform density that has the stable sampling property for any E such that

$$|E| < \text{dens}(\Lambda). \tag{34}$$

Yves Meyer had the intuition that the model sets given by (21) could offer a natural alternative solution to this problem. The density of a model set $\Lambda = \Lambda(L, K)$ is uniform and given by

$$\mathrm{dens}(\Lambda) = \frac{|K|}{|L|}. \tag{35}$$

The following result was established together with Bassarab Matei [59, 60], for model sets $\Lambda := \Lambda(L, K) \subset \mathbb{R}^d$ such that K is a univariate interval. Such model sets are called *simple quasicrystals*.

Theorem 20 *A simple quasicrystal Λ is a set of stable sampling for any E such that* (34) *holds.*

In [75], Yves Meyer showed that the above theorem could be derived from a duality principle established in his early work on model sets [68], and that this new approach is intimately linked with the problem studied by Jean-Pierre Kahane in [47].

There exists an interesting parallel between this result and the theory of compressed sensing developed by Emmanuel Candès and Terence Tao which deals with finite discrete signals represented by large vectors. Compressed sensing reveals in particular that, with high probability, vectors with unknown support of a prescribed size s can be reconstructed by Fourier measurements at m random frequencies when m is larger than s by a logarithmic factor [9]. Theorem 20 may thus be viewed as a deterministic analog to compressed sensing, where sparse vectors are replaced by band-limited functions.

1.5 New Poisson Summation Formulas

In 2015, Yves Meyer returned to topics combining harmonic analysis and number theory, by considering the following general problem: which measures μ supported on a discrete set Λ have distributional Fourier transform $\hat{\mu}$ also supported on a discrete set $\widehat{\Lambda}$?

In view of the classical Poisson formula (20), trivial examples are the Dirac comb $\sum_{\lambda \in L} \delta_\lambda$ on a lattice L, as well as its translated and modulated versions

$$\sum_{\theta \in F} \sum_{\lambda \in L} P_\theta(\lambda) \delta_{\lambda+\theta}, \tag{36}$$

where F is a finite set and each P_θ is a finite combination of complex exponential functions e_a with $a \in \mathbb{R}^d$. A striking result due to Nir Lev and Alexander Olevskii [52] shows that these examples are the only ones if it is required that the supports are also uniformly discrete.

Theorem 21 *If the support of a univariate measure μ and of its Fourier transform $\hat{\mu}$ are both uniformly discrete, then μ is of the form* (36). *The same holds for a multivariate measure under the assumption that μ or $\hat{\mu}$ is positive.*

Non-trivial examples can be obtained if one replaces the uniformly discrete condition by the weaker condition that Λ and $\widehat{\Lambda}$ are locally finite, that is, their intersection with any compact set is finite. Measures (assumed to be tempered distributions) that are supported on locally finite sets as well as their Fourier transform are termed by Yves Meyer as *crystalline measures*. Examples of univariate crystalline measure have been proposed by Lev and Olevskii [53], but older examples turned out to be already provided in a paper from 1959 by Andrew Guinand [38].

The distribution considered by Guinand is of the form

$$\sigma := -2\delta_0' + \sum_{n=1}^{\infty} \frac{r_3(n)}{\sqrt{n}} (\delta_{\sqrt{n}} - \delta_{-\sqrt{n}}) \tag{37}$$

with $r_3(n)$ being defined as the number of points $k \in \mathbb{Z}^3$ such that $|k|^2 = n$. Guinand proved that it satisfies

$$\hat{\sigma} = i\sigma. \tag{38}$$

Yves Meyer shows in [74] how this construction can be modified in order to get rid of the Dirac derivative δ_0'. Defining the measure

$$\mu := \sum_{n=1}^{\infty} \chi(n) \frac{r_3(n)}{\sqrt{n}} (\delta_{\sqrt{n}/2} - \delta_{-\sqrt{n}/2}), \tag{39}$$

where $\chi(n) = -1/2$ if $n \in \mathbb{N} \setminus 4\mathbb{N}$, $\chi(n) = 4$ if $n \in 4\mathbb{N} \setminus 16\mathbb{N}$, and $\chi(n) = 0$ if $n \in 16\mathbb{N}$, Meyer proves that it satisfies

$$\hat{\mu} = -i\mu. \tag{40}$$

Other examples have been provided since then, also with distributions of points of the form $\pm\sqrt{n}$, which therefore become denser at infinity.

Jesús Ildefonso Diaz had the intuition that the constructions of Guinand and Meyer should be linked with the wave equation and the application of Huygens principle on the three-dimensional torus. This was confirmed by the following result obtained jointly with Yves Meyer [30].

Theorem 22 *Let $\mathbb{T}$ be a three-dimensional torus. Let v be a finitely supported measure on V such that $\int_{\mathbb{T}} dv = 0$, and let $u : \mathbb{T} \times \mathbb{R} \to \mathbb{R}$ be the solution to the Cauchy problem*

$$\frac{\partial^2 u}{\partial t^2} - \Delta u = 0, \quad u(\cdot, 0) = 0, \quad \frac{\partial u}{\partial t}(\cdot, 0) = v. \tag{41}$$

Then $u(x_0, \cdot)$ is a crystalline measure for every $x_0 \in \mathbb{T}$ which does not belong to the support of v.

The above result gives a new interpretation to the constructions of Guinand and Meyer. For example, the measure μ in (39) is obtained by taking

$$v = \sum_{k \in \frac{1}{2}\mathbb{Z}^3 \setminus \mathbb{Z}^3} \frac{1}{2}\delta_k - \sum_{k \in \frac{1}{4}\mathbb{Z}^3 \setminus \frac{1}{2}\mathbb{Z}^3} \frac{1}{16}\delta_k, \tag{42}$$

for the initial velocity in (41), having identified $\mathbb{T}$ with $\mathbb{R}^3/\mathbb{Z}^3$.

The non-standard Poisson formulas obtained by Guinand, Lev, Olevskii, Meyer and others are fascinating, and the topic is still at its infancy. No one knows yet if these formulas will prove to be useful, either in the description of physical phenomena or in concrete applications. As we have seen, both turned out to be the case for the model sets, a few years after their introduction by Yves Meyer.

2 The Calderón Program

The mathematical analysis of linear and nonlinear partial differential equations has motivated the systematic study of certain classes of operators that generalize differential operators with constant coefficients.

Classical instances of such classes, pushed into the forefront by Lars Hörmander in the 1960s, are the *pseudo-differential operators* which have the general form

$$Tu(x) = \frac{1}{(2\pi)^d} \int_{\mathbb{R}^d} e^{ix\cdot\omega} \sigma(x,\omega)\hat{u}(\omega)d\omega. \tag{43}$$

where

$$\hat{u}(\omega) := \int_{\mathbb{R}^d} u(x)e^{-ix\cdot\omega}dx, \tag{44}$$

is the Fourier transform of u and where σ is a given function defined on $\mathbb{R}^{2d}$. The function σ is called the symbol of T, and reduces to a polynomial in ω in the case of differential operator with constant coefficients. Pseudo-differential calculus relies on the fact that several properties and operations on such operators have simple expressions at the level of their symbol. However this typically comes at the price that σ should be infinitely differentiable. Consequently, operations such as multiplication by non-smooth functions are excluded.

Alberto Calderón wanted to overcome this limitation, motivated by the study of linear PDEs involving non-smooth coefficients or domains with non-smooth boundaries, which are encountered in many realistic situations. He also had in mind nonlinear PDEs as he wrote: *"the aim of this greater generality is to obtain stronger estimates and to prepare the ground for applications to the theory of quasilinear and nonlinear differential equations"*. These objectives led Calderón to consider other types of operators that are defined by means of singular integral kernels, and to formulate far reaching conjectures on their boundedness, as we detail further.

In 1974, Yves Meyer started an intense and long term collaboration with Ronald Coifman which led to the complete proof of Calderón's conjectures. This came together with the introduction and analysis of multilinear singular operators. One highly celebrated result on which we focus our attention is the boundedness in L^2 of the operator associated with the Cauchy integral on a Lipschitz curve, which was established in 1981 in collaboration with Alan McIntosh [19]. Several beautiful offsprings came out from this intense scientific activity, often obtained by students of Yves Meyer, and we mention a few of them in the end of this section.

2.1 Calderón–Zygmund Operators and Calderón's Conjectures

In the 1950s Alberto Calderón and Antoni Zygmund addressed the problem of finding minimal conditions for the boundedness of integral operators in L^p spaces. One objective was to go beyond standard convolution operators and allow for non-integrable kernels which arise naturally in the studies of PDEs, as discussed further. The most commonly used class which emerged from their work is the following.

Definition 23 A *Calderón–Zygmund operator* is an integral operator

$$Tu(x) = \int_{\mathbb{R}^d} K(x, y)u(y)dy, \tag{45}$$

which acts boundedly in $L^2(\mathbb{R}^d)$ and such that, for some constant C, its kernel satisfies the estimates

$$|K(x, y)| \le C|x - y|^{-d}, \tag{46}$$

and

$$|\nabla_x K(x, y)| + |\nabla_y K(x, y)| \le C|x - y|^{-(d+1)}, \tag{47}$$

for all $x \neq y$.

Since $K(x, \cdot)$ can be non-integrable, the integral defining T needs, in such case, to be defined by an appropriate limiting process. The simplest example of such an operator is the univariate Hilbert transform

$$Hu(x) = \lim_{\varepsilon \to 0} \frac{1}{\pi} \int_{|y|>\varepsilon} \frac{u(x - y)}{y} dy, \tag{48}$$

which is the convolution operator with the distribution $x \mapsto \frac{1}{\pi}\mathrm{pv}(\frac{1}{x})$.

One nice feature of Calderón–Zygmund operators is that their boundedness in L^p spaces for $1 < p < \infty$ follows for free from the L^2 boundedness assumption combined with the estimates (46) and (47). The proof of this fact is based on a subtle decomposition of the function u, also due to Calderón and Zygmund, see [22, 86] or [76] for a general introduction. In the case of the Hilbert transform, the L^2 boundedness follows trivially from Parseval's equality, since the Fourier transform of $x \mapsto \frac{1}{\pi}\mathrm{pv}(\frac{1}{x})$ is the uniformly bounded function $\omega \mapsto -i\,\mathrm{sgn}(\omega)$. This is not anymore applicable for operators which do not have a convolution structure, and for which L^2 boundedness becomes a central topic of investigation.

Calderón considered the operators T_A associated with kernels of the form

$$K_A(x,y) = \Phi\Big(\frac{A(x)-A(y)}{x-y}\Big)\frac{1}{x-y}, \qquad x,y \in \mathbb{R}, \quad x \neq y, \tag{49}$$

where $A : \mathbb{R} \to \mathbb{C}^m$ is a Lipschitz function, that is, such that $\|A'\|_{L^\infty} = L < \infty$. Here, Φ is a given function that is assumed to be analytic in a neighbourhood of the smallest compact set B such that $\frac{A(x)-A(y)}{x-y} \in B$ for all $x, y \in \mathbb{R}^d$. Similar to the Hilbert transform, such operators are defined by a proper limiting process. Calderón's conjectures can be summarized as follows.

Claim The operators T_A act boundedly in L^2 with $\|T_A\|_{L^2\to L^2} \le C(B,\Phi)$.

A case of particular interest is met when $\Phi(z) = (1+iz)^{-1}$ and A takes its values in $\mathbb{R}$. The operator T_A is then

$$T_A u(x) = \int_{\mathbb{R}} \frac{u(y)}{(x-y)+i(A(x)-A(y))}dy, \tag{50}$$

which, up to multiplication by $1+iA'(y)$ in the numerator, is the Cauchy integral on the Lipschitz graph $\{z = z(x) = x + iA(x) \ : \ x \in \mathbb{R}\}$. Since any closed Lipschitz curve $\Gamma \subset \mathbb{C}$ can be locally parameterized by such graphs up to proper rotations, the L^2 boundedness of operators of the type (50) implies that of the Cauchy integral operator

$$T_\Gamma u(z) = \frac{1}{2i\pi}\int_\Gamma \frac{u(z')}{z-z'}dz' \tag{51}$$

from $L^2(\Gamma)$ into itself.

The boundedness in $L^2(\Gamma)$ of the Cauchy integral has far reaching applications in complex analysis, where it can be rephrased as follows: if $\Omega \subset \mathbb{C}$ is the domain that admits Γ as its boundary and Ω^c denotes its complement, one has the direct sum decomposition

$$L^2(\Gamma) = \mathscr{H}^2(\Omega) \oplus \mathscr{H}^2(\Omega^c). \tag{52}$$

The Hardy space $\mathscr{H}^2(\Omega)$ in the above identity is the closure in $L^2(\Gamma)$ of the polynomials, while $\mathscr{H}^2(\Omega^c)$ is the closure in $L^2(\Gamma)$ of the rational functions with poles contained in Ω and vanishing at infinity.

From the perspective of PDEs, it relates to the treatment of boundary value problems by integral formulations. If $\Omega \subset \mathbb{R}^d$ is a bounded domain, consider the two problems

$$\text{(i)}\ -\Delta v = 0 \text{ on } \Omega \text{ (interior)} \quad \text{or} \quad \text{(ii)}\ -\Delta v = 0 \text{ on } \Omega^c \text{ (exterior)}, \tag{53}$$

with boundary conditions on $\Gamma = \partial D$ of either type $v = f$ (Dirichlet) or $\frac{\partial v}{\partial n} = \nabla v \cdot \nu = g$ (Neumann) where ν is the outer normal vector. The approach introduced by Carl Friedrich Gauss and subsequently studied by Carl Neumann gives integral representations of the solutions to these four problems, all based on the fundamental solution Φ of the Laplace equation on $\mathbb{R}^d$ and its gradient $\nabla\Phi$. They are expressed as single layer potential integrals

$$v_{\mathrm{d},\varepsilon}(x) = \int_\Gamma \varphi_\varepsilon(y)\Phi(x-y)dy, \tag{54}$$

for the Dirichlet problems, and double layer potential integrals

$$v_{\mathrm{n},\varepsilon}(x) = \int_\Gamma \psi_\varepsilon(y)\nabla\Phi(x-y)\cdot\nu(y)dy, \tag{55}$$

for the Neumann problems, where $\varepsilon = 1$ for the interior problem (i) and $\varepsilon = -1$ for the exterior problem (ii). The functions φ_ε and ψ_ε are solutions to

$$\Big(\frac{\varepsilon}{2}I - \mathscr{K}\Big)\varphi_\varepsilon = f \quad \text{and} \quad \Big(-\frac{\varepsilon}{2}I - \mathscr{K}\Big)\psi_\varepsilon = g, \quad \text{on } \Gamma, \tag{56}$$

where

$$\mathscr{K}u(x) := \int_\Gamma u(y)\nabla\Phi(x-y)\cdot\nu(y)dy = -\frac{1}{\omega_d}\int_\Gamma \frac{(x-y)\cdot\nu(y)}{|x-y|^d}u(y)dy, \tag{57}$$

is the double layer potential operator on Γ, with ω_d the volume of the unit ball of $\mathbb{R}^d$.

In the bidimensional case $d = 2$, it is readily seen that $\mathscr{K}$ is equivalent to the Cauchy integral T_Γ when identifying $x = (x_1, x_2)$ with $z = x_1 + ix_2$. In turn, the L^2 boundedness of the Cauchy integral appears as a crucial step in the resolution of (56) for Lipschitz domains. This analysis can be extended to the higher dimensional case $d > 2$ by means of the so-called method of rotations introduced by Calderón in [7]. Using such ingredients, the integral treatment of the above boundary value problems on Lipschitz domains with f or g in $L^2(\Gamma)$ was completed by Gregory Verchota in [91], based on the L^2 boundedness result proved in [19] by Coifman, McIntosh and Meyer.

Calderón's research program reveals a fascinating interplay between real and complex analysis. It was the view of Calderón and Zygmund that tools from complex analysis such as the Cauchy integral could benefit from being investigated by real analysis tools. This road was followed by Yves Meyer and his collaborators, while significant advances were also being achieved by Alberto Calderón using complex variable techniques.

2.2 Commutators and the Cauchy Integral

One natural angle of attack to the Calderón conjectures is to consider developments of the function Φ into power series, which naturally leads to study the operators Γ_k defined, with an appropriate limiting process, by

$$\Gamma_k u(x) = \int_{\mathbb{R}} \frac{(A(x) - A(y))^k}{(x-y)^{k+1}} u(y) dy, \tag{58}$$

for all integers $k \geq 0$. Note that these operators only depend on the function

$$a := A', \tag{59}$$

which belongs to L^∞ when A is Lipschitz continuous.

In the case $k = 0$, one has $\Gamma_0 = \pi H$, where H is the Hilbert transform (48) for which L^2 boundedness is trivial. The next operator Γ_1 identifies with a commutator:

$$\Gamma_1 = [\Lambda, A] = \Lambda A - A\Lambda, \tag{60}$$

where (with an abuse of notation) A is the operator of pointwise multiplication by A and Λ is the homogeneous operator of order 1 defined by

$$\widehat{\Lambda u}(\omega) = \pi|\omega|\hat{u}(\omega). \tag{61}$$

Equivalently, $\Lambda = \pi DH$, where D is the differentiation operator. This identification gives the intuition of why Lipschitz functions constitute the natural class for picking A: if D was put in place of Λ, the resulting commutator would identify by Leibniz rule with the pointwise multiplication by a, which is bounded in L^p spaces if and only if a is uniformly bounded.

The boundedness of Γ_1 in L^2 was established in 1965 by Alberto Calderón in [6], together with similar results for commutators between the multiplication operator A and more general homogeneous operators of order 1 that commute with translation. The proof relies on a result of complex analysis concerning the Hardy space $\mathcal{H}^p$, that consists of the functions F that are analytic on $\{\Im(z) > 0\}$ and such that $F(\cdot + iy)$ has a limit $f \in L^p(\mathbb{R})$ as $y > 0$ tends to 0. The norm of this space is $\|F\|_{\mathcal{H}^p} :=$

$\|f\|_{L^p(\mathbb{R})}$ and Calderón proves that for $p < \infty$ this L^p norm is equivalent to that of the Lusin area function

$$Sf(t) = \Big(\int_{|x-t|\le y} |F'(x+iy)|^2 dxdy\Big)^{1/2}. \tag{62}$$

The boundedness of the commutators of interest is then established by introducing an appropriate analytic function F and making use of the above equivalence.

All attempts to prove the boundedness of Γ_1 in L^2 by real variable methods were unsuccessful until Ronald Coifman and Yves Meyer attacked the systematic treatment of the operators Γ_k. These operators can be identified, up to normalization by $k!$, as higher order commutators $[\dots[[\Lambda_k, A], A], \dots, A]$ where $\Lambda_k := \pi D^k H$. One may write $\Gamma_k u = M_k(a, \dots, a, u)$, where M_k is the multilinear operator defined by

$$M_k(a_1, \dots, a_k, u) := \int_{\mathbb{R}} \frac{\prod_{i=1}^k (A_i(x) - A_i(y))}{(x-y)^{k+1}} u(y)dy, \tag{63}$$

with $a_i = A_i'$. This led Coifman and Meyer to introduce and study classes of multilinear singular operators, searching for boundedess results from $L^{p_1} \times \cdots \times L^{p_{k+1}} \to L^p$ where $\frac{1}{p_1} + \cdots + \frac{1}{p_{k+1}} = \frac{1}{p}$. The boundedness of Γ_k in L^2 for Lipschitz functions A corresponds to the particular values $p_1 = \cdots = p_k = \infty$ and $p_{k+1} = p = 2$.

The first and second commutators Γ_1 and Γ_2 were successfully treated in [20] and the boundedness in L^2 of Γ_k for all $k \ge 1$ was finally established a couple of years later in [21]. In this work Coifman and Meyer use real variable techniques, fully exploiting the Calderón–Zygmund properties of the integral kernel.

One natural strategy for the treatment of the Cauchy integral through the operator T_A in (50) is to study the convergence of the series

$$\sum_{k\ge 0} (-i)^k \Gamma_k, \tag{64}$$

that results from the expansion $(1+it)^{-1} = \sum_{k\ge 0}(-i)^k t^k$ when $|t| < 1$. The first results by Coifman and Meyer on the boundedness in L^2 of the high order commutators however did not provide exploitable estimates for the norms $\|\Gamma_k\|_{L^2\to L^2}$. This approach was by-passed by Calderón in 1977 who established by means of complex analysis the boundedness of T_A in L^2 for functions A with sufficiently small Lipschitz constant [6].

Theorem 24 *There exists a constant $\alpha > 0$ such that for all A such that $\|a\|_{L^\infty} \le \alpha$, with $a = A'$, the operator T_A is bounded in L^2.*

In 1980, Alan McIntosh suggested to Ronald Coifman and Yves Meyer a new and powerful approach to study the multilinear operators M_k, which was inspired by techniques developed by Tosio Kato. One principal ingredient lies in an integral representation formula of the operator $L_k : u \mapsto M_k(a_1, \ldots, a_k, u)$ expressed by

$$L_k = \frac{1}{k!} \sum_{\sigma \in S_k} \int_{-\infty}^{+\infty} R_t P_{a_{\sigma(1)}} R_t P_{a_{\sigma(2)}} \cdots R_t P_{a_{\sigma(k)}} R_t \frac{dt}{t}, \tag{65}$$

where $R_t = (I - itD)^{-1}$ and P_{a_i} is the operator of pointwise multiplication by a_i, and where S_k is the set of permutations of $\{1, \ldots, k\}$. The study of each term in (65) by real variable techniques led Yves Meyer and his collaborators to the following estimates [19].

Theorem 25 *There exists a constant C such that, with $a_i = A_i'$, one has*

$$\|M_k(a_1, \ldots, a_k, u)\|_{L^2} \le C(1+k)^4 \|a_1\|_{L^\infty} \ldots \|a_k\|_{L^\infty} \|u\|_{L^2}, \quad k \ge 0, \tag{66}$$

In particular, with $a_j = a = A'$,

$$\|\Gamma_k\|_{L^2 \to L^2} \le C(1+k)^4 \|a\|_{L^\infty}^k, \quad k \ge 0. \tag{67}$$

These estimates immediately yield the convergence of (64) when $\|a\|_{L^\infty} \le \delta < 1$, with $\|T_A\|_{L^2 \to L^2}$ bounded by $C(1-\delta)^{-5}$ for some constant C, however this limitation on a can now be seen to be unnecessary. Indeed, if $\|a\|_{L^\infty} = M$ with M not necessarily smaller than 1, one introduces $\tilde{A}$ such that $A(x) = (M^2 + 1)\tilde{A}(x) - iM^2 x$, so that the kernel of T_A is that of $T_{\tilde{A}}$ multiplied by $(M^2+1)^{-1}$. Since $\|\tilde{a}\|_{L^\infty}^2 \le \delta := (M^2 + M^4)(M^2+1)^{-2} < 1$, the following celebrated result follows as a direct consequence.

Theorem 26 *There exists a constant C such that for any Lipschitz function A, one has*

$$\|T_A\|_{L^2 \to L^2} \le C(1+M)^9, \tag{68}$$

where $M = \|a\|_{L^\infty}$ with $a = A'$.

The techniques proposed for the treatment of the Cauchy integral also led to the complete proof of Calderón's conjectures for general integral operators with kernels of the form (49), and the holomorphy assumption on Φ could be replaced by weaker smoothness assumptions [15].

2.3 Offsprings and Related Results

Several developments came out from the join efforts of Coifman and Meyer, offering new insight on Calderón's conjectures and opening new doors towards the resolution of deep problems.

One remarkable such offspring is the $T(1)$ theorem, established in 1984 by Guy David and Jean-Lin Journé, which gives a simple necessary and sufficient condition for the L^2 boundedness of an integral operator T whose kernel K satisfies the Calderón–Zygmund off-diagonal estimates (46) and (47). The criterion involves the space $\mathrm{BMO}(\mathbb{R}^d)$ introduced by Fritz John and Louis Nirenberg, which consists of functions that have mean oscillation

$$\mathrm{osc}_Q(f) := |Q|^{-1} \int_Q \left| f(x) - |Q|^{-1} \int_Q f(y)dy \right| dx, \tag{69}$$

uniformly bounded over all curves $Q \subset \mathbb{R}^d$. This space has its norm given by

$$\|f\|_{\mathrm{BMO}} := \sup_Q \mathrm{osc}_Q(f), \tag{70}$$

where the supremum is taken over all cubes. The space $\mathrm{BMO}(\mathbb{R}^d)$ is "slightly larger" than $L^\infty(\mathbb{R}^d)$: for example, the function $x \mapsto \log(|x|)$ belongs to $\mathrm{BMO}(\mathbb{R})$. One preliminary requirement is that T should be weakly continuous in $L^2(\mathbb{R}^d)$ which means that

$$|\langle Tf, g\rangle_{L^2}| \le CR^d(\|f\|_{L^\infty} + R\|\nabla f\|_{L^\infty})(\|g\|_{L^\infty} + R\|\nabla g\|_{L^\infty}), \tag{71}$$

for all $f, g \in \mathscr{D}(\mathbb{R}^d)$ whose support is contained in a ball of radius R.

Theorem 27 *Let T be an integral operator that is weakly continuous on $L^2(\mathbb{R}^d)$ and whose kernel K satisfies* (46) *and* (47)*. Then T defines a bounded operator on $L^2(\mathbb{R}^d)$ if and only if the images by T and T^* of the constant function with value* 1 *satisfy*

$$T(1),\ T^*(1) \in \mathrm{BMO}(\mathbb{R}^d). \tag{72}$$

A bound for $\|T\|_{L^2 \to L^2}$ depends only on the constants in (46) *and* (47) *and on the* BMO *norms of $T(1)$ and $T^*(1)$.*

Intuitively, the condition (72) means that if the kernel is singular, it should satisfy some cancellation properties as reflected for example in the case of the Hilbert transform (48). Note that $T(1)$ and $T^*(1)$ do not make sense a priori and should be defined by a proper limiting process.

Several earlier results from the Calderón program, such as the boundedness of the first commutator Γ_1 in L^2, can be viewed as direct consequences of the $T(1)$

theorem. Indeed, if $\Gamma = [\Lambda, A]$ where Λ is an homogeneous pseudo-differential of order 1, one has

$$\Gamma(1) = \Lambda(A). \tag{73}$$

Thus $\Gamma(1)$ is the image of $a = A' \in L^\infty$ by a homogeneous operator of order 0, which is known to belong to BMO.

The $T(1)$ theorem may not be well adapted for the treatment of certain operators, which motivates the study of criterions that generalize (72). Yves Meyer and Alan McIntosh proposed to study the action of T on *accretive* functions, that is, functions b such that $\Re(b) \geq \beta$ for some $\beta > 0$. They established boundedness of T in L^2 under the condition that $T(b_1) = T^*(b_2) = 0$ where b_1 and b_2 are two such functions [56]. The $T(b)$ theorem subsequently obtained by Guy David, Jean-Lin Journé and Stephen Semmes [28], requires that

$$T(b_1),\ T^*(b_2) \in \mathrm{BMO}(\mathbb{R}^d). \tag{74}$$

In particular, the boundedness of the Cauchy integral in L^2 may be viewed as a direct consequence of the $T(b)$ theorem by a suitable choice of the accretive function related to $a = A'$.

Several alternative proofs of the boundedness in L^2 of the Cauchy integral were provided after the seminal work of Coifman, McIntosh and Meyer. Two of them are proposed by Ronald Coifman, Peter Jones and Stephen Semmes in [17], one of which uses matrix representations based on appropriate Haar systems. One notable proof was established in 1995 by Mark Melnikov and Joan Verdera [61], using a discrete geometric notion of curvature $c(z_1, z_2, z_3)$, originally due to Karl Menger, that is defined as the inverse radius of the circle passing through distinct points (z_1, z_2, z_3) in the complex plane. The Menger curvature also writes

$$c(z_1, z_2, z_3) = \frac{4S(z_1, z_2, z_3)}{|z_1 - z_2|\,|z_2 - z_3|\,|z_3 - z_1|}, \tag{75}$$

where $S(z_1, z_2, z_3)$ is the area of the triangle with vertices (z_1, z_2, z_3). The right side quantity appears naturally when estimating the images of characteristic functions χ_I by the Cauchy integral, and these estimates lead to the boundedness result.

The new ideas introduced by Melnikov and Verdera played a key role in the resolution of the long standing Painlevé problem, that is, characterizing the geometry of removable sets of singularities of bounded analytic functions. The locus of such singularities are the compact sets $K \subset \mathbb{C}$ that have analytic capacity $\gamma(K) = 0$, that is, such that functions which are bounded and analytic on $\mathbb{C} \setminus K$ are necessarily constant. Isolated points are instances of such sets, while pieces of smooth curves are not. Melnikov found a connexion between the analytic capacity and a notion of Menger curvature for a measure μ defined by

$$c^2(\mu) := \int_{\mathbb{C}^3} c^2(x, y, z)\, d\mu(x)\, d\mu(y)\, d\mu(z). \tag{76}$$

A first step was achieved in 1998 by Guy David who gave a proof of the Vitushkin conjecture for sets of finite measure [27].

Theorem 28 *If K has finite one dimensional Hausdorff measure, then $\gamma(K) = 0$ if and only if K is totally unrectifiable, that is, its intersection with any rectifiable curve has null Hausdorff measure.*

Finally, the Painlevé problem was solved in 2003 by Xavier Tolsa who gave the following characterization [90].

Theorem 29 *A compact set K is not removable if and only if there exists a non-trivial positive Radon measure μ supported on K with finite Menger curvature.*

Another long standing problem that was stimulated by the work of Coifman and Meyer concerns the study of operators T defined on a Hilbert spaces through certain sesquilinear forms. If (H_0, H_1) are Hilbert spaces such that H_1 is dense in H_0 with continuous embedding, one says that a sesquilinear form B defined and bi-continuous on H_1 is β-accretive for some $\beta > 0$ if and only if it satisfies

$$\Re(B(u,u)) \geq 0 \quad \text{and} \quad \Re(B(u,u)) + \|u\|_{H_0}^2 \geq \beta\|u\|_{H_1}^2, \quad u \in H_1. \tag{77}$$

Such sesquilinear forms yield an operator T acting from some domain H_2 into H_0, where

$$\langle Tu, v\rangle_{H_0} = B(u,v), \quad u \in H_2, v \in H_1, \tag{78}$$

where H_2 is the set of $u \in H_1$ such that $|B(u,v)| \leq C(u)\|v\|_{H_0}$ for all $v \in H_1$. The operator T is called maximal accretive and there exists a unique maximal accretive square root operator S such that $S^2 = T$. This operator may be written as an integral

$$S = \frac{2}{\pi}\int_0^\infty (1+\lambda^2 T)^{-1} T d\lambda. \tag{79}$$

The conjecture raised by Tosio Kato is the following:

Claim *The domain of the square root S coincides with the space H_1.*

The above claim trivially holds in the case where B is symmetric, corresponding to self-adjoint operators. However, a counterexample was found by Alan McIntosh in the non-symmetric case for the above general Hilbert space formulation. The problem remained to prove or disprove the conjecture in the relevant case that motivated Kato: second order elliptic operators acting on functions of d variables according to

$$u \mapsto Tu := -\text{div}(a\nabla u), \tag{80}$$

where $a : \mathbb{R}^d \mapsto \mathcal{M}(\mathbb{C})$ is a complex matrix valued map such that

$$\beta|z|^2 \leq \Re(\langle a(x)z, z\rangle) \leq C|z|^2, \quad x \in \mathbb{R}^d,\ z \in \mathbb{C}^d. \tag{81}$$

In this case the natural spaces are $H_0 = L^2(\mathbb{R}^d)$ and $H_1 = H^1(\mathbb{R}^d)$, the usual Sobolev space.

Alan McIntosh noticed a similarity between the efforts pursued by Coifman and Meyer in studying the Cauchy integral through the Calderón commutators and his attempts to attack Kato's conjecture. In turn, the techniques introduced by Coifman, McIntosh and Meyer in [19] for obtaining the multilinear estimates (66) were used in the same work to prove the validity of Kato's conjecture for the above elliptic operators in the univariate case $d = 1$. The approach was based on a suitable development into Neumann series inside the integral (79) and could be generalized to the multivariate case $d > 1$, however assuming a is sufficiently close to the constant identity matrix in L^∞ norm. A positive answer to Kato's conjecture for second order elliptic operators in higher dimension was finally obtained in 2002 by Pascal Auscher, Steve Hofmann, Michael Lacey, Alan McIntosh and Philippe Tchamitchian [1]. One key aspect of this work, which has impact on other problems such as boundary value problems for second order elliptic operators, is a suitable adaptation of the $T(b)$ theorem that avoids the recourse to the Neumann series. A nice perspectival account on this groundbreaking contribution is given by Yves Meyer in [73].

3 Wavelets and Time-Frequency Analysis

In 1984, Yves Meyer began to work on wavelets, joining forces with Jean Morlet, Alexander Grossmann, Ingrid Daubechies, and many others, in the development of this new area of interdisciplinary research. The construction of wavelet theory took place during the decade of the 1980s. It benefited greatly from ideas coming from various (and sometimes completely disjoint) sources: theoretical harmonic analysis, approximation theory, computer vision and image analysis, computer aided geometric design, digital signal processing. One of the fundamental contributions of Yves Meyer was to recognize and organize these separate developments into a unified and elegant theory.

One major stimulus was the vision of powerful applications in areas as diverse as signal and image processing, statistics, or fast numerical simulation. This perspective was confirmed in the following decades. Wavelets have since then become a powerful computational tool, at the heart of hundreds of algorithms and industrial patents.

Here again, Yves Meyer played a key role in identifying the mathematical properties that are of critical use in such applications. He was also one of the first to point out some intrinsic limitations of wavelets and promote alternative analysis strategies. As discussed in the end of this section, one recent and spectacular illustration is the detection in 2016 of gravitational waves, which uses one such variant. The main references for the results presented in this section are the books [46, 70] by Yves Meyer and his collaborators, as well as [25] by Ingrid Daubechies.

3.1 From Fourier Transforms to Wavelet Transforms

The process of *analyzing* and *representing* an arbitrary function f by means of elementary functions has been at the heart of fundamental and applied advances in science and technology for several centuries. In more recent decades, implementation of this process on computers by fast algorithms has become of ubiquitous use in scientific computing.

In the foundational example of the univariate Fourier transform, the elementary building blocks, sometimes called *atoms*, are the complex exponentials e_ω defined by $e_\omega(t) = e^{i\omega t}$: we may write the Fourier transform of a function f as

$$\hat{f}(\omega) = \int_{-\infty}^{+\infty} f(\omega)\overline{e_\omega(t)}dt. \tag{82}$$

and the inverse Fourier transform as

$$f(t) = \frac{1}{2\pi}\int_{-\infty}^{+\infty} \hat{f}(\omega)e_\omega(t)dt. \tag{83}$$

In a similar way, Fourier series $\sum_{n\in\mathbb{Z}} c_n(f)e_n$ decompose functions of period 2π into atoms e_n defined by $e_n(t) := e^{int}$ for $n \in \mathbb{Z}$, with coefficients given by

$$c_n(f) = \frac{1}{2\pi}\int_{-\pi}^{\pi} f(t)\overline{e_n(t)}dt \tag{84}$$

From the computational side, these transforms can be discretized and implemented via the fast Fourier transform algorithm in $\mathcal{O}(N \log N)$ computational time, where N is the size of discretization.

The above atoms e_ω and e_n have no localization since their modulus is constant and equal to 1 independently of t. This property constitutes a major defect when trying to efficiently detect the local frequency content of functions by means of Fourier analysis. It also makes Fourier representations numerically ineffective for functions that are not smooth everywhere. For example, the Fourier coefficients $c_n(f)$ of a 2π periodic piecewise smooth function f with a jump discontinuity at a single point $t_0 \in [-\pi, \pi]$ decay like $|n|^{-1}$, which affects the convergence of the Fourier series on the whole of $\mathbb{R}$.

Time-frequency analysis aims to provide representations which better capture the local frequency content of a function. Musical scores may be viewed as a caricatural sketch for such representations: the atoms are represented by the notes which represent a pure frequency occurring during a given time interval.

One first approach consists of pre-multiplying the function f by a smooth and well localized non-negative function g as well as its translates $g(\cdot - \tau)$ before applying the Fourier transform. The resulting *short time* Fourier transform

$$Gf(\omega,\tau) := \int_{-\infty}^{+\infty} f(t)g(t-\tau)e^{-i\omega t}dt, \tag{85}$$

was studied by Denis Gabor in 1945 with the Gaussian function e^{-t^2} as a specific choice for g. This transform can also be written as

$$Gf(\omega,\tau) = \langle f, g_{\omega,\tau}\rangle, \quad g_{\omega,\tau}(t) := g(t-\tau)e^{i\omega t}, \tag{86}$$

where $\langle f, g\rangle := \int_{\mathbb{R}} f(t)\overline{g(t)}dt$ is the L^2 inner product. Since $\hat{g}_{\omega,\tau}(\xi) = e^{-i\tau\xi}\hat{g}(\xi - \omega)$, and $\hat{g}$ is also a Gaussian, the functions $g_{\omega,\tau}$ are well localized both in time around τ and in frequency around ω, and are sometimes called time-frequency atoms. The reconstruction of f from Gf is given by

$$f(t) = C\int_{-\infty}^{+\infty}\int_{-\infty}^{+\infty} Gf(\omega,\tau)e^{i\omega t}d\omega d\tau, \quad C := \left(2\pi\int_{-\infty}^{+\infty} g(t)dt\right)^{-1}, \tag{87}$$

which is a straightforward consequence of the Fourier inversion formula.

The temporal resolution of the above short time Fourier transform is inherently limited in scale by the width of the window described by g. This led Jean Morlet, who was an engineer in geoseismic, to propose a different approach: starting from a given function ψ which is well localized both in time and frequency and such that $\int_{\mathbb{R}} \psi = 0$, one generates a continuous family of functions $\psi_{a,b}$ called *wavelets*, by translation and dilations

$$\psi_{a,b}(t) = a^{-1/2}\psi\Big(\frac{t-b}{a}\Big), \quad a > 0, \quad b \in \mathbb{R}, \tag{88}$$

where $a^{-1/2}$ is an L^2 normalization factor. The wavelet transform of f is then defined by

$$Wf(a,b) := \langle f, \psi_{a,b}\rangle = a^{-1/2}\int_{-\infty}^{+\infty} f(t)\psi\Big(\frac{t-b}{a}\Big)dt. \tag{89}$$

The generating function ψ is sometimes called the *mother wavelet*. In contrast to the short time Fourier transform, the possibility of letting the scale parameter a tend to 0 gives access to the analysis of arbitrarily localized features. Since $\hat{\psi}_{a,b}(\omega) = e^{-ib\omega}a^{1/2}\hat{\psi}(a\omega)$, this comes at the price of a loss of frequency localization due to the inverse scaling factor a^{-1}. Jean Morlet also proposed the reconstruction formula

$$f(t) = C_\psi^{-1}\int_0^{+\infty}\int_{-\infty}^{+\infty} Wf(a,b)\psi_{a,b}(t)\frac{db\,da}{a^2}, \tag{90}$$

where $C_\psi := \int_{-\infty}^{+\infty} \frac{|\hat{\psi}(\omega)|^2}{|\omega|} d\omega$ is assumed to be non-zero and finite. This formula was first recognized by Alex Grossmann, a theoretical physicist, as a particular case of "reproducing formulas" from group representation theory that are commonly used in quantum mechanics.

Yves Meyer then recognized it as an instance of a reproducing formula introduced by Alberto Calderón in the 1960s which has the following general form for functions f defined on $\mathbb{R}^d$:

$$f(x) = C_{\psi,\tilde{\psi}}^{-1} \int_0^\infty f * \psi_t * \tilde{\psi}_t \, dt, \tag{91}$$

with the notation $g_t = t^{-d} g(t^{-1} \cdot)$. This formula holds for $f \in L^2(\mathbb{R}^d)$ provided that ψ and $\tilde{\psi}$ are radial functions with sufficient smoothness and localization properties, and the constant $C_{\psi,\tilde{\psi}}$ is the integral $\int_0^\infty \hat{\psi}(t\omega)\hat{\tilde{\psi}}(t\omega)\frac{dt}{t}$ for any non-zero $\omega \in \mathbb{R}^d$, which is assumed to be non-zero and finite.

3.2 *Orthonormal Wavelet Bases*

In numerical computation, one is naturally led to *sample* the above defined transforms Gf and Wf. For the short time Fourier transform, one easily checks that this can be done in such way that no information is lost on f, assuming for simplicity that the localizing function g is supported on an interval of length T_0, is non-negative and does not vanish on an interval of length t_0. Namely, then f can be reconstructed from the family of functions $f_m = g(\cdot - mt_0)f$ for $m \in \mathbb{Z}$, and each of these functions is characterized by its Fourier coefficients which are given by

$$c_n(f_m) = Gf(mt_0, n\omega_0), \quad n \in \mathbb{Z}. \tag{92}$$

where $\omega_0 := 2\pi/T_0$. The natural sampling of Gf is therefore on the lattice $(mt_0, n\omega_0)_{m,n\in\mathbb{Z}}$, which amounts in tiling the time-frequency plane by the rectangles associated with the atoms

$$g_{n,m} := g_{n\omega_0, mt_0}, \quad n, m \in \mathbb{Z}. \tag{93}$$

In the case of wavelets, such a tiling is more naturally linked to a lattice of the form $(a_0^{-n}, mb_0a_0^{-n})_{n,m\in\mathbb{Z}}$, for some fixed $a_0 > 1$ and $b_0 > 0$, leading therefore to the discrete family of wavelets

$$\psi_{n,m} := a_0^{n/2}\psi(a_0^n \cdot - mb_0), \quad n, m \in \mathbb{Z}. \tag{94}$$

The problem of understanding if the sampling $Wf(a_0^{-n}, mb_0a_0^{-n}) = \langle f, \psi_{n,m}\rangle$ permits the characterization and stable reconstruction of f, is related to the following concept: a sequence $(f_n)_{n\geq 0}$ in a Hilbert space H is called a *frame*, if there exists constants $0 < A \leq B$ such that for all $f \in H$,

$$A\|f\|^2 \leq \sum_{n\geq 0} |\langle f, f_n\rangle|^2 \leq B\|f\|^2, \tag{95}$$

where $\|\cdot\|$ and $\langle\cdot,\cdot\rangle$ are the norm and inner product of H. Such a stability property implies the existence of a dual frame $(e_n)_{n\geq 0}$, such that any $f \in H$ can be reconstructed according to

$$f = \sum_{n\geq 0} \langle f, f_n\rangle e_n. \tag{96}$$

In the case of the wavelet family (94), Ingrid Daubechies showed that, for any mother wavelet ψ satisfying some mild assumptions, it constitutes a frame whenever $a_0 b_0 < K$ for some $K = K(\psi)$. This was a first step towards stable numerical algorithms.

Frame decompositions of the above type are generally redundant: the systems $(e_n)_{n\geq 0}$ and $(f_n)_{n\geq 0}$ are complete but linearly dependent, in contrast to bases. Two elementary examples of orthonormal wavelet bases were known since long, both of the form

$$\psi_{j,k}(t) = 2^{j/2}\psi(2^j t - k), \quad j,k \in \mathbb{Z}, \tag{97}$$

corresponding to the dilation factor $a_0 = 2$ and translation step $b_0 = 1$. The first one, introduced by Alfred Haar in 1910, uses for ψ the piecewise constant function

$$\psi(t) := \chi_{[0,1/2[} - \chi_{[1/2,1[}, \tag{98}$$

where χ_E stands for the characteristic function of a set E. The L^2-orthonormality of the family (97) is then straightforward. Their completeness follows by observing that the contribution

$$Q_j f = \sum_{k\in\mathbb{Z}} \langle f, \psi_{j,k}\rangle \psi_{j,k}, \tag{99}$$

of the single scale level j is the difference $P_{j+1}f - P_j f$, where $P_j f$ is the L^2-projection of f onto piecewise constant functions over the dyadic intervals $I_{j,k} := [2^{-j}k, 2^{-j}(k+1)[, k \in \mathbb{Z}$. The second elementary example is the so-called Shannon wavelet, which is obtained by taking ψ such that

$$\hat{\psi}(\omega) = \chi_{]-2\pi,-\pi]}(\omega) + \chi_{[\pi,2\pi[}(\omega). \tag{100}$$

In this example, the above contribution $Q_j f$ now amounts to a filtered version of f on the frequency band $|\omega| \in [2^j\pi, 2^{j+1}\pi[$, that is,

$$\widehat{Q_j f}(\omega) = \hat{\psi}(2^{-j\omega})\hat{f}(\omega). \tag{101}$$

The wavelet coefficients $\langle f, \psi_{j,k}\rangle$ are, up to a normalizing factor, the samples of the band-limited function $Q_j f$ at the points $2^{-j}k$, which uniquely characterizes it according to the classical sampling theorem of Claude Shannon.

While the Shannon wavelet is well localized in the frequency domain, it has poor decay $\mathcal{O}(t^{-1})$ in the time domain due to the fact that $\hat{\psi}$ has a jump discontinuity. The opposite situation holds for the Haar wavelet, and a natural question was therefore if orthonormal wavelet bases can be constructed with a function ψ that is well localized in both domains. Yves Meyer's initial intuition was that such a result could not hold, by analogy with the time-frequency atoms of (93) for which a negative answer is known in the form of the following result established in 1981 by Roger Balian and Francis Low.

Theorem 30 *Let g be such that the time-frequency system* (93) *forms an orthonormal basis. Then*

$$\int_{-\infty}^{+\infty} |x|^2|g(x)|^2dx + \int_{-\infty}^{+\infty} |\omega|^2|\hat{g}(\omega)|^2d\omega = \infty. \tag{102}$$

After some attempts to disprove their existence, Yves Meyer turned the table in 1985 and gave a beautiful construction of orthonormal wavelet bases that belong to the Schwartz class

$$\mathcal{S}(\mathbb{R}) := \{f \in C^\infty(\mathbb{R}) \ : \ \sup_{x\in\mathbb{R}} |x|^k|f^{(l)}(x)| < \infty, \ k, l \geq 0\}, \tag{103}$$

and are therefore well localized both in time and frequency. The construction is based on a subtle regularization of $\hat{\psi}$ in the definition of the Shannon wavelet by (100).

The idea of splitting a function f into dyadic frequency bands by smooth filtering operations has been ubiquitous in harmonic analysis, since its introduction by John Littlewood and Raymond Paley in the 1930s. The Littlewood–Paley decomposition can be described in the following way: starting with a non-negative and even function $\kappa \in C^\infty(\mathbb{R})$ that is compactly supported on $[-4\pi/3, 4\pi/3]$ and takes value 1 on $[-2\pi/3, 2\pi/3]$, one defines the function

$$\sigma(\omega) = \kappa(\omega/2) - \kappa(\omega). \tag{104}$$

which is supported on $[-8\pi/3, -2\pi/3] \cup [2\pi/3, 8\pi/3]$. The low-pass and band-pass components of f at each scale level j are defined by

$$\widehat{S_j f}(\omega) = \kappa(2^{-j}\omega)\hat{f}(\omega) \ \text{ and } \ \widehat{\Delta_j f}(\omega) = \widehat{S_{j+1} f}(\omega) - \widehat{S_j f}(\omega) = \sigma(2^{-j}\omega)\hat{f}(\omega). \tag{105}$$

The resulting decompositions of f according to

$$f = S_0 f + \sum_{j\geq 0} \Delta_j f = \sum_{j\in\mathbb{Z}} \Delta_j f, \tag{106}$$

come as useful substitutes to orthonormal decompositions when working in L^p spaces for $p \neq 2$. In particular, for $1 < p < \infty$ the L^p norm of a function f is equivalent to that of the square-function $Sf := \left(\sum_{j\in\mathbb{Z}} |\Delta_j f|^2\right)^{1/2}$.

The wavelet construction proposed by Yves Meyer obeys similar principles but requires some finer properties. First, one assumes a symmetry in the transition region, namely

$$\kappa(\pi + \omega) + \kappa(\pi - \omega) = 1, \quad \omega \in [0, \pi/3], \tag{107}$$

which implies in particular the partition of the unity property

$$\sum_{j\in\mathbb{Z}} \sigma(2^{-j}\omega) = 1. \tag{108}$$

Then, the function $\hat{\psi}$ is taken such that $|\hat{\psi}|^2 = \sigma$ but with a phase choice that makes it symmetric around the point $x = \frac{1}{2}$.

Theorem 31 *Let κ and σ be defined as above and let $\hat{\psi}(\omega) := e^{i\omega/2}\sqrt{\sigma(\omega)}$. Then, the system* (97) *is an orthonormal basis of $L^2(\mathbb{R})$.*

Another orthonormal wavelet basis with good smoothness and localization properties had been obtained earlier in the work of Jan-Olov Strömberg [88]. The technique used by Strömberg was Gram–Schmidt orthonormalization applied to the Schauder basis, the primitive of the Haar system, leading to a piecewise affine wavelet ψ with exponential spatial decay but limited smoothness. Strömberg also constructed arbitrarily smooth wavelets following the same principle but his work was unnoticed at that time. By its elegant simplicity, Meyer's construction was celebrated as a major milestone, standing at the crossroad between several research programs.

3.3 *Multiresolution Analysis*

In the area of numerical signal and image processing, multiscale approximations and decompositions have been considered by engineers since the 1970s, with certain features analogous to Littlewood–Paley analysis. For example, starting from an image that is discretized on a pixel grid, one may consider its approximations obtained by recursively averaging the value of the light intensity over squares of 2×2, 4×4, 8×8 pixels, etc. Stéphane Mallat understood that such numerical procedures were naturally linked with the construction of wavelet bases. His decisive meeting with Yves Meyer led him to formalize the following concept which became central in wavelet theory [57].

Definition 32 A *multiresolution analysis* of $L^2(\mathbb{R})$ is a sequence $(V_j)_{j\in\mathbb{Z}}$ of subspaces of $L^2(\mathbb{R})$ which satisfies the following properties:

- $V_j \subset V_{j+1}$.
- $\overline{\bigcup_{j\in\mathbb{Z}} V_j} = L^2(\mathbb{R})$.
- $\bigcap_{j\in\mathbb{Z}} V_j = \{0\}$.
- $f \in V_j \iff f(2\cdot) \in V_{j+1}$.
- There exists a function $\varphi \in V_0$ such that $(\varphi(\cdot - k))_{k\in\mathbb{Z}}$ is a Riesz basis of V_0.

For any countable set $\mathcal{F}$, a family $(e_k)_{k\in\mathcal{F}}$ in a Hilbert space V is called a Riesz basis if it is complete and there exists constants $0 < c \le C < \infty$ such that

$$c \sum_{k\in\mathcal{F}} |x_k|^2 \le \| \sum_{k\in\mathcal{F}} c_k e_k \|_V^2 \le C \sum_{k\in\mathcal{F}} |x_k|^2, \tag{109}$$

holds for any finitely supported coefficient sequence $(x_k)_{k\in\mathcal{F}}$, and therefore by density for any sequence in $\ell^2(\mathcal{F})$. The second and third properties in the above definition mean that $\lim_{j\to+\infty} \|f - P_j f\|_{L^2} = 0$ and $\lim_{j\to-\infty} \|P_j f\|_{L^2} = 0$, for all $f \in L^2(\mathbb{R})$, where P_j is the orthogonal projector onto V_j. The fourth property readily implies that a Riesz basis for V_j consists of the rescaled functions

$$\varphi_{j,k}(t) = 2^{j/2}\varphi(2^j t - k), \quad k \in \mathbb{Z}. \tag{110}$$

The function φ is called a *scaling function*. In the most simple case where V_j are the space of piecewise constant functions over the dyadic intervals $[2^{-j}k, 2^{-k}(k+1)[$, $k \in \mathbb{Z}$, one simply has $\varphi = \chi_{[0,1]}$ and (110) is an orthonormal basis of V_j. We have seen that this particular multiresolution analysis is associated with the Haar system. Other simple examples of multiresolution analysis spaces are given by spline functions: the functions of V_j are piecewise polynomial functions of some given degree $m \ge 1$ over the same dyadic intervals, with global C^{m-1} smoothness. The scaling function is then given by the $m+1$-fold convolution $\varphi := (*)^{m+1}\chi_{[0,1]}$ known as B-spline of degree m.

Stéphane Mallat and Yves Meyer gave a simple recipe for deriving the wavelet basis associated with a given multiresolution analysis. A key observation is that since $\varphi \in V_0 \subset V_1$, it should satisfy a *two-scale difference equation* of the form

$$\varphi(t) = \sum_{n\in\mathbb{Z}} h_n \varphi(2t - n), \tag{111}$$

for some coefficient sequence $(h_n)_{n\in\mathbb{Z}}$. Then the construction of the wavelet basis is summarized by the following result.

Theorem 33 *Assume that $(\varphi(\cdot - k))_{k\in\mathbb{Z}}$ is an orthonormal basis of V_0. Let $\psi \in V_1$ be defined by*

$$\psi(t) = \sum_{n\in\mathbb{Z}} g_n \varphi(2t-n), \quad g_n := (-1)^n h_{1-n}. \tag{112}$$

Then $(\psi_{j,k})_{k\in\mathbb{Z}}$ is an orthonormal basis of the space W_j defined as the orthogonal complement of V_j in V_{j+1}. In turn $(\psi_{j,k})_{j,k\in\mathbb{Z}}$ is an orthonormal basis of $L^2(\mathbb{R})$.

When the integer translates of the scaling function φ only constitute a Riesz basis, such as for splines of degree $m \geq 1$, a proper recombination of these translates leads to a new scaling function with orthonormal translates, so that the above construction readily applies. A drawback of this approach is that the new scaling function might lose certain desirable properties such as compact support.

This problem was circumvented in 1988 by Ingrid Daubechies in her celebrated construction of compactly supported orthonormal wavelets [24]. In this construction, the scaling function φ is not defined explicitly but instead as a solution to the scaling equation (111). For all integers $m \geq 0$, a finitely supported sequence $(h_n)_{n=0,\dots,2m+1}$ is carefully designed, such that the solution to (111) is compactly supported on $[0, 2m+1]$, has orthonormal integer translates, similar approximation properties as the splines of degree m, and Hölder smoothness that grows linearly with m. Aside from the case $m = 0$ that corresponds to the Haar system, the functions φ and ψ do not have explicit expressions, while the coefficients h_n and g_n are explicitly given. The orthonormality requirement in the above construction can be relaxed, as shown in 1992 by Albert Cohen, Ingrid Daubechies and Jean-Christophe Feauveau [13]. The resulting *biorthogonal wavelet bases* offer additional flexibility, allowing for example for symmetric or piecewise polynomial generating functions, which are not conciliable with compactly supported orthonormal wavelets. This has led engineers and numerical analysts to adopt them in most practical applications (in particular for the still image compression standard JPEG 2000).

The multiresolution analysis framework was immediately extended by Mallat and Meyer to multivariate functions, by tensorizing the spaces V_j in the different variables. This leads to multivariate wavelet bases of the form

$$\psi^{\varepsilon}_{\varepsilon,j,k} = 2^{dj/2}\psi_\varepsilon(2^j \cdot -k), \quad j \in \mathbb{Z},\ k \in \mathbb{Z}^d, \tag{113}$$

for $\varepsilon = (\varepsilon_1, \dots, \varepsilon_d) \in \{0,1\}^d \setminus \{(0,\dots,0)\}$, where

$$\psi_\varepsilon(x_1,\dots,x_d) := \psi_{\varepsilon_1}(x_1)\cdots\psi_{\varepsilon_d}(x_x), \quad \psi_0 := \varphi,\ \psi_1 := \psi. \tag{114}$$

Adaptation of these bases to more general bounded domains of $\mathbb{R}^d$ as well as to various types of manifolds came in the following years, again based on the multiresolution concept.

In addition to this geometric flexibility, the hierarchical structure of the multiresolution framework is the key to fast algorithms for decomposing or reconstructing a functions into wavelet components. These algorithms are based on the sequences (h_n) and (g_n) which connect the successive scale levels. They were familiar to electrical engineers as *iterated filter bank* decompositions. The reconstruction algorithms are also of the same type as *subdivision schemes* proposed in geometric modeling for the fast design of curves and surfaces by iterated refinements.

3.4 Smoothness Spaces and Sparsity

When expanding a function f into a given basis $(e_n)_{n\geq 0}$, a desirable feature is that the resulting decomposition $f = \sum_{n\geq 0} x_n e_n$ is numerically stable: operations such as perturbations, thresholding or truncation of the coefficients x_n should effect the norm of f in a well-controlled manner. Such prescriptions can be encapsulated in the following classical property.

Definition 34 A sequence $(e_n)_{n\geq 0}$ in a separable Banach space X is an *unconditional basis*, if the following properties hold.

(i) It is a *Schauder basis*: every $f \in X$ admits a unique expansion $\sum_{n\geq 0} x_n e_n$ that converges towards f in X.
(ii) There exists a finite constant $C \geq 1$ such that for any finite set $F \subset \mathbb{N}$,

$$|x_n| \leq |y_n|, \quad n \in F \implies \Big\|\sum_{n\in F} x_n e_n\Big\|_X \leq C\Big\|\sum_{n\in F} y_n e_n\Big\|_X \tag{115}$$

The property (115) means that membership of f in X only depends on the moduli of its coordinates $|x_n|$. In other words, multiplier operators of the form

$$T : \sum_{n\geq 0} x_n e_n \to \sum_{n\geq 0} c_n x_n e_n, \tag{116}$$

should act boundedly in X if $(c_n)_{n\geq 0}$ is a bounded sequence. Orthonormal and Riesz bases are obvious examples of unconditional bases in Hilbert spaces.

While the trigonometric system is a Schauder bases in $L^p(]-\pi,\pi[)$ when $1 < p < \infty$, it does not constitute an unconditional basis when $p \neq 2$, and it is thus not possible to characterize the space L^p through a property of the moduli of the Fourier coefficients. The same situation is met for classical smoothness spaces, such as the Sobolev spaces $W^{m,p}_{\text{per}}(]-\pi,\pi[)$ that consist of 2π-periodic functions having distributional derivatives up to order m in L^p_{loc}: apart from the Hilbertian case $p = 2$, for which one has

$$f \in W^{m,2}_{\text{per}}(]-\pi,\pi[) \iff \sum_{n\in\mathbb{Z}} (1 + |n|^{2m})|c_n(f)|^2 < \infty, \tag{117}$$

no such characterization is available when $p \neq 2$.

Yves Meyer showed that, in contrast to the trigonometric system, wavelet bases are unconditional bases for most classical function spaces that are known to possess one. The case of L^p spaces for $1 < p < \infty$ is treated by the following observation: if the mother wavelet ψ has C^1 smoothness, the multiplier operator (116) by a bounded sequence can be identified to a Calderón–Zygmund operator as introduced in Definition 23, therefore acting boundedly in $L^p(\mathbb{R}^d)$. Conversely, Yves Meyer showed that Calderón–Zygmund operators are "almost diagonalized" by wavelet bases in the sense that the resulting matrices have fast off-diagonal decay. This property plays a key role in the numerical treatment of partial differential and integral equations by wavelet methods, as discussed for instance in [3, 11].

The characterization of more general function spaces by the size properties of wavelet coefficients is particularly simple for an important class of smoothness spaces introduced by Oleg Besov. There exist several equivalent definitions of Besov spaces. The original one uses the mth order L^p-modulus of smoothness

$$\omega_m(f,t)_p := \sup_{|h|\le t} \|\Delta_h^m f\|_{L^p}, \tag{118}$$

where Δ_h^m is the m-th power of the finite difference operator $\Delta_h : f \mapsto f(\cdot+h)-f$. For $s > 0$, any integer $m > s$, and $0 < p, q < \infty$, a function $f \in L^p(\mathbb{R}^d)$ belongs to the space $B_q^{s,p}(\mathbb{R}^d)$ if and only if the function $g : t \to t^{-s}\omega_m(f,t)_p$ belongs to $L^q([0,\infty[, \frac{dt}{t})$. One may use

$$\|f\|_{B_q^{s,p}} := \|f\|_{L^p} + |f|_{B_q^{s,p}}, \quad \text{with} \quad |f|_{B_q^{s,p}} := \|g\|_{L^q([0,\infty[,\frac{dt}{t})}, \tag{119}$$

as a norm for such spaces, also sometimes denoted by $B_q^s(L^p(\mathbb{R}^d))$. Roughly speaking, functions in $B_q^{s,p}(\mathbb{R}^d)$ have up to s (integer or not) derivatives L^p. The third index q may be viewed as a fine tuning parameter, which appears naturally when viewing Besov spaces as real interpolation spaces between Sobolev space [4]: for example, with $0 < s < m$,

$$B_q^s(L^p) = [L^p, W^{m,p}]_{\theta,q}, \quad s = \theta m. \tag{120}$$

Particular instances are the Hölder spaces $B_\infty^{s,\infty} = C^s$ and Sobolev spaces $B_p^{s,p} = W^{s,p}$, when s is not an integer or when $p = 2$ for all values of s.

Let (ψ_λ) denote a multivariate wavelet basis of the type (113), where for simplicity λ denotes the three indices (e,j,k) in (113). Denoting by $|\lambda| := j = j(\lambda)$ the scale level of $\lambda = (e,j,k)$, we consider the expansion

$$f = \sum_{|\lambda|\ge 0} d_\lambda \psi_\lambda, \tag{121}$$

where the coarsest scale level $|\lambda| = 0$ also includes the translated scaling functions that decompose $P_0 f$.

The characterization of $B_q^{s,p}(\mathbb{R}^d)$ established by Yves Meyer for such expansions requires some minimal prescriptions: one assumes that for an integer $r > s$ the univariate mother wavelet ψ and scaling functions φ that appear in (113) have derivatives up to order r that decay sufficiently fast at infinity, for instance faster than any polynomial rate, and that $\int_{-\infty}^{+\infty} t^k \psi(t)dt = 0$ for all $k = 0, 1, \dots, r-1$.

Theorem 35 *Let (ψ_λ) be a wavelet basis satisfying the above assumptions. Then, one has the norm equivalence*

$$\|f\|_{B_q^{s,p}} \sim \|\varepsilon\|_{\ell^q}, \tag{122}$$

where the sequence $\varepsilon = (\varepsilon_j)_{j\geq 0}$ is defined by

$$\varepsilon_j := 2^{sj} 2^{(\frac{d}{2}-\frac{d}{p})j} \|(d_\lambda)_{|\lambda|=j}\|_{\ell^p}. \tag{123}$$

A closely related characterization of Besov spaces uses Littlewood–Paley analysis and has a form similar to the above with $\varepsilon_j := \|\Delta_j f\|_{L^p}$, where Δ_j are the dyadic blocks of (105). In the wavelet characterization, these blocks are replaced by the components $Q_j f = P_{j+1} f - P_j f$ which are further discretized into the local components $d_\lambda \psi_\lambda$. Similar results have been obtained for Besov spaces defined on general bounded Lipschitz domains $\Omega \subset \mathbb{R}^d$ with wavelet bases adapted to such domains.

The norm equivalence (122) shows that membership of f in Besov spaces is characterized by simple weighted summability properties of its wavelet coefficients. In the particular case $q = p$, this equivalence takes the very simple form

$$\|f\|_{B_p^{s,p}} \sim \|(2^{(s+\frac{d}{2}-\frac{d}{p})|\lambda|} d_\lambda)\|_{\ell^p}. \tag{124}$$

As an immediate consequence, classical results such as the critical Sobolev embedding $B_p^{s,p} \subset L^2$ for $s = \frac{d}{p} - \frac{d}{2}$ take the trivial form of the embedding $\ell^p \subset \ell^2$ for $p < 2$. While this embedding is not compact, an interesting approximation property still holds: when retaining only the n largest coefficients in the wavelet decomposition of f, the resulting approximation f_n satisfies

$$\|f - f_n\|_{L^2} \leq C n^{-s/d} \|f\|_{B_p^{s,p}}. \tag{125}$$

This follows immediately from the fact that, for $p < 2$, the decreasing rearrangement of $(d_k)_{k\geq 1}$ of a sequence $(d_\lambda) \in \ell^p$ satisfies the tail bound

$$\Big(\sum_{k\geq n} d_k^2\Big)^{1/2} \leq n^{\frac{1}{2}-\frac{1}{p}} \|(d_\lambda)\|_{\ell^p}. \tag{126}$$

This last estimate shows that ℓ^p summability governs the compressibility of a sequence, in the sense of how fast it can be approximated by *n-sparse* sequences.

The theory of best n-term wavelet approximation, generalizing the above remarks, has been developed by Ronald DeVore and his collaborators, in close relation with other nonlinear approximation procedures such as free knot splines or rational approximation, see [29] for a detailed survey.

A particularly useful feature of nonlinear wavelet approximation is that piecewise smooth signals, such as images, can be efficiently captured since the large coefficients are only those of the few wavelets whose supports contain the singularities. This is an instance of *sparse approximation* which aims at accurately capturing functions by a small number of well chosen coefficients in a basis or dictionary expansion. Sparse approximation in unconditional bases was identified in [31] by David Donoho as a key ingredient for powerful applications in signal and image compression and statistical estimation. A detailed exposition of these applications can be found in [58].

Pushed into the forefront by the work of Yves Meyer, David Donoho and Ronald DeVore, sparse approximation became within a few years a prominent concept in scientific computing. The theory of *compressed sensing* developed by Emmanuel Candès, Justin Romberg and Terence Tao [8], and by David Donoho [32], shows in particular that sparsity is highly beneficial when solving ill-posed linear inverse problems: an n-sparse vector of size $N \gg n$ and unknown support can be stably reconstructed from m well chosen linear measurements, where m is of the order of n up to logarithmic factors. Moreover, such a reconstruction can be performed by searching for the solution of minimal ℓ^1 norm, therefore relying on simple techniques from convex optimization.

3.5 *Chirps, Time-Frequency Bases and Gravitational Waves*

The existence of gravitational waves was first predicted by Albert Einstein in 1916, as small perturbations of space-time that propagate at the speed of light. While intensively studied, both theoretically and numerically, these perturbations had never been observed due to their extremely small amplitude (about 10^{-6} of the size of the hydrogene atom nucleus), until the breakthrough of the Ligo–Virgo collaborative detection programs. The Nobel prize of physics was attributed in 2017 to Rainer Weiss, Kip S. Thorne, and Barry C. Barish for their pioneering work in the conception of these experiments.

The gravitational wave first detected in 2016 is attributed to the fusion of two black holes and has for a large part the *chirp* behaviour

$$f(t) \sim |t - t_0|^{-1/4} \cos(|t - t_0|^{5/8} + \varphi_0). \tag{127}$$

Loosely speaking a chirp is a signal whose "instantaneous frequency" evolves with time in some controlled manner. This can be formalized by conditions of the form

$$f(t) = Re\left(a(t)e^{i\varphi(t)}\right), \tag{128}$$

where

$$\left|\frac{a'(t)}{a(t)}\right| \ll |\varphi'(t)| \quad \text{and} \quad |\varphi''(t)| \ll |\varphi'(t)|^2. \tag{129}$$

The first condition says that the amplitude has little variation over the pseudo-period $\frac{2\pi}{|\varphi'(t)|}$, and the second one that the pseudo-period itself varies slowly. Typical examples are ultrasounds emitted by bats and recordings of voice signals, as well the above gravitational wave.

Time-frequency analysis such as the short-time Fourier transform provides the natural tools for the study of chirps. However, their precise detection in a noisy environment requires non-redundant decompositions in which they are as sparse as possible. Theorem 30 expresses a principal obstruction to the construction of orthonormal time-frequency bases. Yves Meyer played a major role in the development of alternate approaches in order to circumvent this difficulty.

The first such approach was originally suggested by Kenneth Wilson and formalized by Ingrid Daubechies, Stéphane Jaffard and Jean-Lin Journé [26]: an orthonormal basis of $L^2(\mathbb{R})$ is constructed by allowing Fourier localization around two frequencies of the same amplitude and opposite signs, taking for all $n \in \mathbb{Z}$ the functions $\varphi_{0,n}(t) = \varphi(t-n)$ and

$$\varphi_{l,n}(t) = \begin{cases} \sqrt{2}\varphi\left(t - \frac{n}{2}\right)\cos(2\pi l t) \; l \geq 0, \; l+n \in 2\mathbb{Z}, \\ \\ \sqrt{2}\varphi\left(t - \frac{n}{2}\right)\sin(2\pi l t) \; l > 0, \; l+n \in 2\mathbb{Z}+1. \end{cases} \tag{130}$$

The generating function φ should satisfy certain symmetry properties. One possible choice is the scaling function associated with the orthonormal wavelet basis of Yves Meyer, which is defined by $\hat{\varphi} = \sqrt{\kappa}$ where κ is the symmetric and smooth cut-off that satisfies (107). A variant of this system, where the family is made redundant by additional dilations, was proposed in the papers of Sergei Klimenko and his collaborators for the sparse representation of gravitational waves and used for their detection. Compact support in the Fourier domain is mandatory here, due to very strong noise components that lie away from the main Fourier zone of interest.

There is an interesting parallel between the development of time-scale and time-frequency analysis: starting on both sides with continuous algorithms, a computational breakthrough is brought by the introduction of orthonormal bases leading to discrete decompositions, together with fast algorithms. In both cases, function spaces are naturally associated (Besov spaces in the case of wavelets and the so-called modulation spaces introduced by Hans Feichtinger and Karlheinz Gröchenig in the case of Wilson bases), and applications cover classes of signals that are sparse in such bases (piecewise regular functions for wavelets and chirps for Wilson bases). In both setting, it is possible to further improve on sparse representations, by introducing libraries of orthonormal bases which allow for the selection of a particular one for a given signal.

In the case of wavelets, such libraries are known as *wavelet packets* and were introduced by Yves Meyer, in collaboration with Ronald Coifman and Victor Wickerhauser. They are based on applying the numerical filters given by the sequences (h_n) and (g_n) in (111) and (112) in order to adaptively split the frequency domain.

In the case of time-frequency bases, these libraries are based on a variant of Wilson bases constructed by Henrique Malvar. The Malvar basis is of the form

$$\varphi_{j,k}(t) = \varphi(t-j)\cos\Big(\pi\Big(k+\frac{1}{2}\Big)(t-j)\Big), \quad j \in \mathbb{Z},\ k \in \mathbb{N}, \tag{131}$$

now referred to as the MDCT (Modified Discrete Cosinus Transform) and used in audio compression formats, e.g. MP3 or MPEG2 AAC. Martin Vetterli pointed out the analogy between Wilson and Malvar bases, and as a result, adaptive Malvar bases were introduced by Yves Meyer and Ronald Coifman. These bases are of the form

$$\varphi_{j,k}(t) = \sqrt{\frac{2}{l_j}}\varphi_j(t)\cos\Big(\frac{\pi}{l_j}\Big(k+\frac{1}{2}\Big)(t-a_j)\Big), \quad j \in \mathbb{Z},\ k \in \mathbb{N}, \tag{132}$$

where the support of the function φ_j is localized around the interval $[a_j, a_{j+1}]$ of length l_j. They were used in speech segmentation by Victor Wickerhauser and Eva Wesfreid, where the lengths l_j automatically adapt to the changes in the signal through an entropy minimization criterium and thus perform automatic segmentation.

3.6 *Pointwise Smoothness and Multifractal Analysis*

Let us finally return to wavelet bases. In addition to the characterization of smoothness classes given by Theorem 35, their localization properties give access to the more refined analysis of the smooth or singular behaviour at a given point x. There exist various notions of pointwise smoothness, the most intuitive one being that a function f from $\mathbb{R}^d$ to $\mathbb{R}$ has Hölder smoothness α at a point $x_0 \in \mathbb{R}^d$ if an only if, there exists a polynomial π of degree strictly less than α such that

$$|f(x) - \pi(x)| \leq C|x - x_0|^\alpha, \tag{133}$$

for all x in a neighbourhood of x_0. The Hölder exponent $\alpha_f(x_0)$ of f at x_0 is the supremum of those α such that (133) holds.

Since global Hölder smoothness of order α is characterized by the decay property $|d_\lambda| \leq C2^{-(\alpha+\frac{d}{2})|\lambda|}$ of the wavelet coefficients, an often used heuristics is that (133) is equivalent to the same decay property for only those $\lambda = (j, k, \varepsilon)$ such that $|2^j x_0 - k|$ remains uniformly bounded.

Such a statement is actually wrong, but Stéphane Jaffard and Yves Meyer showed that it can be modified in various ways into a correct one. In particular Yves Meyer showed in [71] that a slightly more precise version of this local decay condition characterizes a new notion of pointwise smoothness, the *weak scaling exponent*, which has the remarkable property of being covariant with respect to fractional primitives or derivatives.

The ability of wavelets to characterize pointwise smoothness plays a key role in multifractal analysis, an area of research that started from the study of fully developed turbulence. One objective is to study the size of the sets of points at which a given function f has as certain amount of smoothness. This can be quantified by the *spectrum of singularity* of f, which is defined as the function $\alpha \mapsto d_f(\alpha)$ where $d_f(\alpha)$ is the Hausdorff dimension of the set of points x where $\alpha_f(x) = \alpha$. A function is said to be multifractal if its spectrum of singularity is not concentrated at a single smoothness α. One example of a multifractal function, for which the spectrum of singularity was completely characterized with the help of wavelets, is the Riemann function

$$R(x) := \sum_{n=1}^{\infty} \frac{\sin(\pi n^2 x)}{n^2}. \tag{134}$$

In practice, the estimation of d_f from discretized data is subject to numerical difficulties. This motivated Uriel Frisch and Giorgio Parisi to instead study a related quantity $s_f(p)$ that can be defines as the supremum of those s such that $f \in B^{s,p}_{\infty}(\mathbb{R}^d)$. Their conjecture, known as *multifractal formalism*, was that the functions d_f and s_f are related by the Legendre transform

$$d_f(\alpha) = \inf_p \{p\alpha - s_f(p) + d\}. \tag{135}$$

Using wavelet bases, Stéphane Jaffard proved that this conjecture is false for certain functions, but generically true in the sense of Baire's categories inside Besov and Sobolev spaces [44].

The classification of pointwise singularities by means of wavelets was pushed one step further by Yves Meyer, who considered chirps where the "instantaneous frequency" in (128) diverges at a point x_0. This leads to pointwise singularities that are typically of the form

$$f_{\alpha,\beta}(x) = |x - x_0|^{\alpha} \sin\left(\frac{1}{|x - x_0|^{\beta}}\right). \tag{136}$$

Based on the heuristic supplied by such toy-examples, a general framework for such behaviors is developed in [45] where the sine function is replaced by a fairly arbitrary oscillating function, which yields a characterization by precise estimates on the wavelet coefficients. An application is completely worked out: it concerns the above mentioned Riemann function which is shown to have an explicit asymptotic development for the chirp behavior of near rational points of the form $x_0 = (2p+1)/(2q+1)$. For example, at $x_0 = 1$,

$$R(1+x) = -\frac{x}{2} + \sum_{k\geq 1} |x|^{k+1/2} g_k\Big(\frac{1}{x}\Big), \tag{137}$$

where g_k is essentially a primitive of order k of the Riemann function itself.

4 Partial Differential Equations

We discuss in this final section several striking contributions by Yves Meyer to the field of partial differential equations. A common feature in these contributions is that techniques from harmonic analysis play a central role.

The behaviour of the solutions to the linear wave equation is studied in relation with non-harmonic trigonometric series. The phenomenon of compensated compactness in nonlinear quantities is described by mean of Hardy spaces and Littlewood–Paley analysis. The latter is also used to establish improved Sobolev embeddings and understand the role of oscillations in the existence and uniqueness theory of nonlinear PDEs such as Navier–Stokes equations.

These last aspects are among the topics covered in the beautiful monograph [72], which also surveys in depth the successes as well as shortcomings of multiscale methods such as wavelets in nonlinear PDEs and image processing.

4.1 The Wave Equation and Control Problems

One instance of PDE studied in the early works of Yves Meyer is the linear wave equation,

$$\frac{\partial^2 u}{\partial t^2} - \Delta u = 0, \tag{138}$$

set on bounded domains $\Omega \subset \mathbb{R}^d$, which has intimate connexions with non-harmonic trigonometric series. Imposing for instance the homogeneous boundary conditions $u(x,t) = 0$ for $x \in \partial\Omega$, one searches for an expansion of u of the form

$$u(x,t) = \sum_{n\geq 1} c_n(t)\varphi_n(x), \tag{139}$$

where $(\varphi_n)_{n\geq 1}$ is an L^2-orthonormal basis of eigenfunctions of the Laplace operator with the same boundary condition. If $0 \leq \lambda_1 \leq \lambda_2 \leq \cdots$ is the associated sequence of eigenvalues, and if the initial conditions are

$$u(x,0) = u_0(x) = \sum_{n\geq 1} c_{n,0}\varphi_n(x), \tag{140}$$

and

$$\frac{\partial u}{\partial t}(x,0) = u_1(x) = \sum_{n\geq 1} c_{n,1}\varphi_n(x), \tag{141}$$

one finds that the coefficients in (139) have the form

$$c_n(t) = c_{n,0}\cos(\alpha_n t) + c_{n,1}\alpha_n^{-1}\sin(\alpha_n t), \tag{142}$$

where $\alpha_n = \sqrt{\lambda_n}$. The behaviour of the solution at each given point $x \in \Omega$ is thus described by a trigonometric sum with frequencies $(\alpha_n)_{n\geq 1}$. Such sums are called *non-harmonic* Fourier series since the α_n are generally not the integer multiples of a fixed number.

From the study of these series, Yves Meyer derived several deep results on the behaviour of solutions to the wave equation. We discuss three of them, all concerned with the case where Ω is a domain in $\mathbb{R}^2$ or $\mathbb{R}^3$. Other results dealing with the wave equation on the sphere S^{d-1} are given in [67].

The first one from [66] shows that solutions with rather smooth initial conditions may surprisingly become unbounded at any point as t grows.

Theorem 36 *Let Ω be a square. There exist functions φ that belong to $C^1(\Omega)$ such that the wave equation with this initial data $(u_0, u_1) = (\varphi, 0)$ satisfies $\limsup_{t\to\infty} |u(x,t)| = \infty$ for almost every $x \in \Omega$.*

A variant of this result, also established by Yves Meyer, shows that the maximal value of $u(\cdot,t)$ may blow up in an even more brutal way: there exist initial data in $C^1(\Omega)$ such that $\|u(\cdot,t)\|_{L^\infty} = \infty$ for any $t > 0$. In other words, the membrane modeled by the wave equation can be instantaneously broken. This is in sharp contrast with the behaviour of the Dirichlet energy $\int_\Omega |\nabla u(x,t)|^2 dx$ of these solutions, which is given by $\sum_{n\geq 0} \lambda_n |c_n(t)|^2$ and therefore remains uniformly bounded as t grows. By similar energy reasoning, one finds that if the initial condition u_0 has finite H^2 norm, the solution would remain uniformly bounded by the continuous embedding of $H^2(\Omega)$ into $C(\overline{\Omega})$.

The second result concerns the following inverse problem: is it possible to recover the initial conditions (u_0, u_1) to the wave equation from the observation of $t \mapsto u(x,t)$ at a given point x on some given time interval $[0,T]$? This important question lies at the interface of exact controllability and pointwise oscillation theory, see [40] for an easier but related result of *non-oscillation* in a rectangle.

The answer to the question is easily seen to be negative if an eigenfunction φ of the Laplace operator vanishes at the given point x, since the initial condition

$(\varphi, 0)$ then produces a solution such that $u(x,t) = 0$ for all $t \in \mathbb{R}$. One therefore considers only *strategic points* which are those $x \in \Omega$ such that no eigenfunction vanishes at x. It is easily seen that such points do not exist if the spectrum has multiplicities, since in such a case one can always recombine two eigenfunctions with the same eigenvalue in such a way that they vanish at an arbitrarily given point. On the other hand, since the set of zeroes of any eigenfunction has measure 0, and the spectrum of the Laplace operator is countable, as soon as all eigenvalues are simple, the set of non-strategic points is negligible. In particular, for rectangles $\Omega = [a,b] \times [c,d]$, strategic points exist and have full measure in Ω if, and only if, the quantity $(b-a)^2/(d-c)^2$ is irrational.

Yves Meyer gave the following negative answer to the observation problem: there exist strategic points $x \in \Omega$ such that, for any $T > 0$, one can find a non-trivial smooth solution to the wave equation that vanishes at x for all $t \in [0,T]$. This result was later extended to *all* points, without irrationality condition, by Stéphane Jaffard [42], relying on the theory of frames.

Theorem 37 *Let Ω be a rectangle. For any $x \in \Omega$ and $T > 0$, there exists a non-trivial and C^∞ solution to the wave equation such that*

$$u(x,t) = 0, \quad t \in [0,T]. \tag{143}$$

A third result deals with an optimal control problem raised by Jacques-Louis Lions. One studies on a regular bounded domain $\Omega \subset \mathbb{R}^3$ the wave equation

$$\frac{\partial^2 u}{\partial t^2} - \Delta u = f, \quad f(x,t) = v(t)\delta_z(x), \tag{144}$$

with initial conditions $(u_0, u_1) = (0,0)$ and homogeneous boundary conditions $u(x,t) = 0$ for $x \in \partial\Omega$. Here $z \in \Omega$ is a given point and $v \in L^2([0,T])$ is the function representing the control at this point. For a given target profile $u^* \in L^2(\Omega)$ at time $T > 0$, one seeks to minimize, for some $\kappa > 0$,

$$J(v) = \int_\Omega |u(x,T) - u^*(x)|^2 dx + \kappa \int_0^T |v(t)|^2 dt. \tag{145}$$

One crucial prerequisite for solving this problem was the existence of a solution u such that $u(\cdot,t) \in L^2(\Omega)$ for all $t \in [0,T]$. Yves Meyer gave in [69] the following positive answer.

Theorem 38 *For any control function $v \in L^2([0,T])$, the Eq.* (144) *has a unique solution $u \in C([0,T], L^2(\Omega))$, with $\frac{\partial u}{\partial t} \in C([0,T], H^{-1}(\Omega))$ and $\frac{\partial^2 u}{\partial t^2} \in C([0,T], H^{-2}(\Omega))$.*

Alternate proofs of this results were subsequently proposed, first by Louis Nirenberg, then by Jacques-Louis Lions who used an approach that led him to the Hilbert Uniqueness Method (HUM).

The proof of Yves Meyer combines results on non-harmonic trigonometric series with estimates by Pham The Lai improving a theorem of Lars Hörmander on the spectral function $e_\lambda(x) := \sum_{\lambda_n \le \lambda} |\varphi_n(x)|^2$ of the Laplace operator on the domain $\Omega \subset \mathbb{R}^3$. These estimates are of the form

$$e_\lambda(x) - \frac{1}{2\pi^2}\lambda^{3/2} = \mathcal{O}(\lambda \text{dist}(z, \partial\Omega)^{-1}), \tag{146}$$

and were used by Yves Meyer to quantify the difficulty of controlling the solution as the point z is near to the boundary: if $\|v\|_{L^2} \le 1$, one has

$$\sup_{t\in[0,T]} \|u(\cdot, t)\|_{L^2} \le C_\Omega \sqrt{1+T}\, \text{dist}(z, \partial\Omega)^{-1/2}, \tag{147}$$

where the constant C_Ω only depends on Ω.

It seems that this proof is one of the rare circumstances where such a result can be established through harmonic analysis arguments for a *general* domain. A similar result for the control of bending plates on arbitrary domains was obtained by Alain Haraux and Stéphane Jaffard in [39]. Aside from the result itself, Yves Meyer's argument demonstrates the astonishing power of Hörmander–Pham estimates, which may be useful in completely different settings that we are unable to imagine today.

4.2 Compensated Compactness

In general, nonlinear operations that are continuous with respect to strong topologies in L^p spaces fail to be continuous with respect to weak topologies. One elementary example is the product mapping

$$(u, v) \mapsto uv. \tag{148}$$

This map is continuous from $L^2(\Omega) \times L^2(\Omega)$ to $L^1(\Omega)$ for any domain $\Omega \subset \mathbb{R}^d$, but there exist sequences $(u_n)_{n\ge0}$ and $(v_n)_{n\ge0}$ that converge weakly in L^2 towards some u and v, while $u_n v_n$ does not converge towards uv in the distributional sense. This behaviour is typically produced by resonating oscillations, as can be seen for example when taking $u_n(x) = v_n(x) = \sin(nx)$ on a unit interval.

The theory of *compensated compactness*, introduced by Luc Tartar and François Murat [77, 89], studies instances where non-linear operations may recover weak continuity due to certain cancellation phenomenons. Such a study is particularly relevant for the issue of the existence of solutions to many nonlinear systems of PDEs or variational problems. One groundbreaking result proved by Tartar and Murat, sometimes referred to as the div-curl lemma, takes the following form.

Theorem 39 *Let $(E_n)_{n\ge0}$ and $(B_n)_{n\ge0}$ be sequences of vector fields that weakly converge towards E and B in $L^2(\mathbb{R}^d)^d$ and such that* $(\text{div}E_n)_{n\ge0}$ *and all compo-*

nents of $(\mathrm{curl} B_n)_{n\ge 0}$ *are bounded in* $L^2(\mathbb{R}^d)$. *Then* $E_n \cdot B_n$ *converges towards* $E \cdot B$ *in the distributional sense.*

The additional condition on the sequences $(\mathrm{div} E_n)_{n\ge 0}$ and $(\mathrm{curl} B_n)_{n\ge 0}$ could be further weakened into compactness of these sequences in $H^{-1}(\mathbb{R}^d)$. Several spectacular applications of compensated compactness are given in [89], in particular to homogenization of elliptic PDEs and to viscous approximations of entropy solutions to univariate hyperbolic conservation laws. Tartar and Murat have also shown that the inner product is the only nonlinear map enjoying the weak continuity property as described in Theorem 39.

A related question is whether this behaviour could be related to a certain gain of integrability of the product $E \cdot B$ when such control on the quantities $\mathrm{div} E$ and $\mathrm{curl} B$ hold. One first such result has been obtained by Stefan Müller [79] for the Jacobian $J(u) = \det(\nabla u)$ of a vector field $u \in W^{1,d}(\mathbb{R}^d)^d$, when this Jacobian is non-negative. The gain is expressed by the fact that the $J(u)$ not only belongs to $L^1(\mathbb{R}^d)$ but also to the space $L \log L_{loc}(\mathbb{R}^d)$ that consists of the functions f such that $f \log(2+f) \in L^1_{loc}(\mathbb{R}^d)$.

Theorem 40 *If* $u \in W^{1,d}(\mathbb{R}^d)^d$ *and* $J(u) \ge 0$, *then* $J(u) \in L \log L_{loc}$.

The relation between the Jacobian and the div-curl nonlinearities clearly appears if one expands $\det(\nabla u)$ with respect to its first column, which gives

$$J(u) = \sum_{i=1}^{d} \frac{\partial u_1}{\partial x_i} m_{1,i}(u), \tag{149}$$

where $m_{1,i}(u)$ is the corresponding cofactor. One then observes that the vector fields E and B with components defined by $E_i = m_{1,i}(u)$ and $B_i = \frac{\partial u_1}{\partial x_i}$ are divergence and curl free, respectively.

The intuition of Pierre-Louis Lions was that this gain could be better described in terms of the Hardy space $\mathscr{H}^1(\mathbb{R}^d)$. The latter (not to be confused with the Sobolev space $H^1(\mathbb{R}^d) = W^{1,2}(\mathbb{R}^d)$) consists of all $f \in L^1(\mathbb{R}^d)$ such that the maximal function

$$x \mapsto \sup_{t>0} |f * \varphi_t(x)|, \quad \varphi_t = t^{-d}\varphi(t^{-1}\cdot), \tag{150}$$

belongs to $L^1(\mathbb{R}^d)$, where φ is an arbitrary non-negative function in $\mathscr{D}(\mathbb{R}^d)$ such that $\int_{\mathbb{R}^d} \varphi = 1$. This intuition proved to be correct, as expressed by the following results obtained by Ronald Coifman, Pierre-Louis Lions, Yves Meyer and Stephen Semmes [18].

Theorem 41 *If* $u \in W^{1,d}(\mathbb{R}^d)^d$, *then its Jacobian* $J(u) = \det(\nabla u)$ *belongs to* $\mathscr{H}^1(\mathbb{R}^d)$.

Since it is known that $f \in L \log L_{loc}(\mathbb{R}^d)$ if and only if $f \in \mathscr{H}^1_{loc}(\mathbb{R}^d)$ when f is a positive function, the result by Müller follows from a local variant of Theorem 41.

In the div-curl setting, a result similar to Theorem 41 is obtained by the same authors.

Theorem 42 *Let $E \in L^p(\mathbb{R}^d)^d$ and $B \in L^q(\mathbb{R}^d)^d$ where $\frac{1}{p} + \frac{1}{q} = 1$ and $1 < p < \infty$, and assume that* $\operatorname{div} E = 0$ *and* $\operatorname{curl} B = 0$. *Then $E \cdot B$ belongs to $\mathscr{H}^1(\mathbb{R}^d)$.*

The weak continuity result of the compensated compactness Theorem 39 can then be improved using this result. Indeed, in contrast to L^1, the Hardy space $\mathscr{H}^1$ is sequentially closed for its weak-$*$ topology: if in the above result E and B are replaced by L^2-bounded sequences $(E_n)_{n\geq 0}$ and $(B_n)_{n\geq 0}$ which are respectively divergence and curl free, then $E_n \cdot B_n$ is bounded in $\mathscr{H}^1$ and a subsequence converges in the weak* topology towards the products $E \cdot B$ of the L^2 weak limits of E_n and B_n.

Yves Meyer and his collaborators have proposed several approaches for establishing the $\mathscr{H}^1$ boundedness of the above nonlinear quantities. A very insightful one, introduced together with Ronald Coifman and Sylvia Dobyinski [16], is based on the idea that proper *renormalization* procedures may significantly improve the behaviour of nonlinear quantities.

For the product of two functions, one such renormalization emerged from the work by Jean-Michel Bony on microlocal analysis and paradifferential operators [5]. It can be described through the Littlewood–Paley decomposition (106): expanding the product uv into all products $\Delta_j u \Delta_l v$, one extracts the terms for which $|j-l| \leq 1$ that correspond to resonating frequencies. The resulting sum has the form

$$\mathscr{P}(u, v) = \sum_{j\in\mathbb{Z}} \Delta_j u S_{j-2} v + \sum_{j\in\mathbb{Z}} \Delta_j v S_{j-2} u =: \Pi_v u + \Pi_u v, \tag{151}$$

where both terms are called *paraproducts*.

Using the fact that the norm of the Hardy space $\mathscr{H}^1(\mathbb{R}^d)$ can be defined by

$$\|f\|_{\mathscr{H}^1} = \|Sf\|_{L^1}, \quad Sf = \Big(\sum_{j\in\mathbb{Z}} |\Delta j f|^2\Big)^{1/2}, \tag{152}$$

it can be checked that the paraproduct of two functions of $L^p(\mathbb{R}^d)$ and $L^q(\mathbb{R}^d)$ with $\frac{1}{q} + \frac{1}{p} = 1$ and $1 < p < \infty$ belongs to $\mathscr{H}^1(\mathbb{R}^d)$. This is readily applied to the vector fields E and B in Theorem 42, and one is left with estimating the remaining term

$$\mathscr{R}(u, v) = \sum_{|i-j|\leq 1} \Delta_j E \cdot \Delta_i B, \tag{153}$$

that is the source of trouble for general vector fields. Coifman, Dobyinsky and Meyer prove that when $\operatorname{div} E = 0$ and $\operatorname{curl} B = 0$, this term belongs to the Besov space $B_1^{0,1}(\mathbb{R}^d)$ which is even smaller than $\mathscr{H}^1(\mathbb{R}^d)$. Similar renormalization techniques can be performed using wavelet expansions in place of Littlewood–Paley decompositions.

4.3 *Improved Sobolev Embeddings and the Space* BV

Embedding inequalities between Sobolev spaces are of ubiquitous use in the analysis of PDEs. One most classical instance is the following: if $f \in W^{1,p}(\mathbb{R}^d)$ for some $p < d$, then $f \in L^q(\mathbb{R}^d)$ for $\frac{1}{p} - \frac{1}{q} = \frac{1}{d}$, and one has

$$\|f\|_{L^q} \leq C\|\nabla f\|_{L^p}, \tag{154}$$

where $C = C(p, d)$. The above inequality expresses a trade-off: integrability is gained at the prize of a loss in smoothness. The right side in (154) is the norm of the homogeneous Sobolev space $\dot{W}^{1,p}(\mathbb{R}^d)$ which thus also embeds continuously in $L^q(\mathbb{R}^d)$.

The above inequality is invariant under the action of the affine group, that is, by translation and scaling: when applying it to the function $g(x) = f(ax + b)$ in place of f, both sides of (154) are multiplied by the same factor $a^{-d/q}$. On the other hand, it is not invariant by modulation: when applying it to $g(x) = e^{i\omega \cdot x} f(x)$, the left side of (154) is left unchanged since $|g| = |f|$, while the right side is of the order $\|g\| \sim \|\nabla f\|_{L^p} + |\omega|\|f\|_{L^p}$. This reveals that (154) is not sharp for oscillatory functions: $\|g\|_{L^q}$ becomes highly overestimated as $|\omega| \to \infty$.

A remedy to this defect is provided by improved versions of the above Sobolev inequality where the right-side includes a norm that decreases as f oscillates. Intuitively such norms should measure the size of the primitives of f instead of their derivatives. In other words, they should be norms of negative smoothness spaces. More precisely, Patrick Gérard, Yves Meyer and Frédéric Oru establish in [37] the following improvement to (154).

Theorem 43 *Let $f \in \dot{W}^{1,p}(\mathbb{R}^d)$ for some $1 < p < d$, then with $\frac{1}{q} := \frac{1}{p} - \frac{1}{d}$ and $r := \frac{d}{q}$*

$$\|f\|_{L^q} \leq C\|\nabla f\|_{L^p}^{p/q} \|f\|_{\dot{B}_\infty^{-r,\infty}}^{1-p/q}, \tag{155}$$

where $C = C(p, d)$.

The norm of the homogeneous Besov space $\dot{B}_\infty^{-r,\infty}$ can be defined by

$$\|f\|_{\dot{B}_\infty^{-r,\infty}} = \sup_{j \in \mathbb{Z}} 2^{rj} \|S_j f\|_{L^\infty}, \tag{156}$$

where $S_j f$ are the low-pass components in the Littlewood–Paley analysis introduced in (105). It is readily checked that for $g(x) = e^{i\omega \cdot x} f(x)$, one has

$\|g\|_{\dot{B}_\infty^{-r,\infty}} \sim |\omega|^{-r}$ as $|\omega| \to \infty$, and therefore (155) is now robust with respect to modulations. Since $W^{1,q}(\mathbb{R}^d)$ is continously embedded in $\dot{B}_\infty^{-r,\infty}(\mathbb{R}^d)$, the standard Sobolev inequality (154) follows from (155).

Gérard, Meyer and Oru proved a somewhat more general result: if $1 < p < q < \infty$, $r > 0$ and $s = r(\frac{q}{p} - 1)$, one has

$$\|f\|_{L^q} \leq C\|\Lambda^s f\|_{L^p}^{p/q} \|f\|_{\dot{B}_\infty^{-r,\infty}}^{1-p/q}, \tag{157}$$

where $\Lambda := \sqrt{-\Delta}$ is the Riesz operator defined by $\widehat{\Lambda f}(\omega) = |\omega|\hat{f}(\omega)$. In the particular case where $\frac{1}{p} - \frac{1}{q} = \frac{s}{d}$, the above is an improvement on $\|f\|_{L^q} \leq C\|\Lambda^s f\|_{L^p}$ which expresses the embedding of the homogeneous Riesz potential space $\dot{H}^{s,p}(\mathbb{R}^d)$ (that coincides with $\dot{W}^{s,p}(\mathbb{R}^d)$ when s is an integer) into $L^q(\mathbb{R}^d)$.

This last embedding is not compact and the improved inequality (157) plays a critical role in the proof of subsequent results which give a precise description of this lack of compactness in terms of *profiles* with asymptotically disjoint space-scale localization. More precisely, the following result was established by Patrick Gérard [36] in the case $p = 2$ and by Stéphane Jaffard [43] in the more general case $1 < p < \infty$.

Theorem 44 *A sequence $(u_n)_{n\geq 0}$ bounded in $\dot{H}^{s,p}(\mathbb{R}^d)$ can be decomposed, up to a subsequence extraction, according to*

$$u_n = \sum_{l=1}^{L} h_{l,n}^{s-\frac{d}{p}} \phi^l\Big(\frac{\cdot - x_{l,n}}{h_{l,n}}\Big) + r_{n,L} \tag{158}$$

where $(\phi^l)_{l>0}$ is sequence in $\dot{H}^{s,p}(\mathbb{R}^d)$ and $\lim_{L\to\infty}\Big(\limsup_{n\to\infty} \|r_{n,L}\|_{L^q}\Big) = 0$. This decomposition is "asymptotically orthogonal" in the sense that for $k \neq l$

$$|\log(h_{l,n}/h_{k,n})| \to \infty \;\text{ or }\; |x_{l,n} - x_{k,n}|/h_{l,n} \to \infty, \;\text{ as }\; n \to \infty. \tag{159}$$

Profile decompositions of the above type have been used in the study of nonlinear evolution equations such as the critical nonlinear wave equations [2] or the critical nonlinear Schrödinger equation [62].

The improved Sobolev inequalities (157) are in spirit similar to classical inequalities for *interpolation spaces*, of the form

$$\|f\|_Z \leq C\|f\|_X^{1-\theta}\|f\|_Y^{\theta}. \tag{160}$$

Here $Z = [X, Y]_{\theta,\tau}$ is the real interpolation space between X and Y for given parameter $0 < \theta < 1$ and $0 < \tau \leq \infty$, see [4] for an introduction to such spaces. Classical examples are the Gagliardo–Nirenberg inequalities. The proof of (157) however requires a more specific treatment since the interpolation spaces between $\dot{H}^{s,p}(\mathbb{R}^d)$ and $\dot{B}_\infty^{-r,\infty}$ do not have a simple description. The strategy adopted by

Gérard, Meyer and Oru is based on cleverly combining estimates for the Littlewood–Paley blocks of f that result from the finiteness of the right-hand side in (157).

This strategy is however uneffective in the limit case where $p = 1$, for example when considering the embedding between $\dot{W}^{1,1}(\mathbb{R}^d)$ and $L^q(\mathbb{R}^d)$ with $q := \frac{d}{d-1}$. The corresponding inequality

$$\|f\|_{L^q} \leq C\|\nabla f\|_{L^1}^{\frac{1}{q}}\|f\|_{\dot{B}_\infty^{1-d,\infty}}^{\frac{1}{d}}, \tag{161}$$

which only holds when $d \geq 2$, was proved by Yves Meyer and his collaborators in [14] using a different approach. The main ingredient is provided by weak-type estimates for the wavelet coefficients of functions of bounded variations established in [12] by Albert Cohen, Wolfgang Dahmen, Ronald DeVore and Ingrid Daubechies. In the simplest case $d = q = 2$, one such estimate states that the decreasing rearrangement $(d_n)_{n\geq 1}$ of the coefficients d_λ in the wavelet expansion (121) of f satisfies

$$d_n \leq C|f|_{\mathrm{BV}}n^{-1}, \quad n \geq 1. \tag{162}$$

Here $|f|_{\mathrm{BV}}$ is the total variation of f which coincides with $\|\nabla f\|_{L^1}$ if $f \in \dot{W}^{1,1}$. This shows that the weak-ℓ^1 norm of (d_λ) is controlled by $|f|_{\mathrm{BV}}$. Since $\|f\|_{\dot{B}_\infty^{-1,\infty}}$ is equivalent to the ℓ^∞ norm of this sequence, the improved Sobolev inequality follows directly from the fact that the ℓ^2 norm is controlled by the geometric mean of the weak-ℓ^1 and ℓ^∞ norms. An alternate proof of (161), using heat kernel techniques, was provided by Michel Ledoux in [51].

The particular interest of Yves Meyer for the space BV was stimulated by its use in image processing: functions in $\mathrm{BV}(\mathbb{R}^2)$ are allowed to have jump discontinuities on curves of finite length and are therefore a plausible model for images containing objects with sharp edges, in the absence of noise and texture. A prominent model based on this intuition was proposed by Stanley Osher and Leonid Rudin [82]: it consists of splitting the image f into two pieces u and v, by solving a variational problem of the form

$$\min\{\|v\|_{L^2}^2 + \lambda|u|_{\mathrm{BV}} \;:\; f = u + v\}, \tag{163}$$

The function u models the object component of the image, as v captures the textural and noise component. The parameter $\lambda > 0$ tunes the importance given to each of these and can be related to the noise level.

This model is closely related to that proposed by David Mumford and Jayant Shah in which u is searched for as piecewise H^1 with discontinuities on curves of finite length. Another related model, proposed in [10], is obtained when replacing the space BV by slightly smaller Besov space $B_1^{1,1}$. In view of (124), the resulting minimization problem has then a simple expression using wavelet expansions of the form (121): each coefficient u_λ of u individually minimizes $|f_\lambda - u_\lambda|^2 + \lambda|u_\lambda|$, where f_λ is the coefficient of f, and is therefore given by a shrinkage of the form

$$u_\lambda = \max\{0, |f_\lambda| - \lambda/2\} \operatorname{sign}(f_\lambda). \tag{164}$$

Such wavelet thresholding strategies were first advocated and analyzed by David Donoho, Iain Johnstone, Gérard Kerkyacharian and Dominique Picard in statistical estimation [33]. In this approach, the coefficients of the textural part v all satisfy the uniform bound $|v_\lambda| \leq \lambda/2$, which means that the $B_\infty^{-1,\infty}$ norm of this component is controlled by $\lambda/2$.

This remark led Yves Meyer to investigate the negative order smoothness spaces that are involved in the original Osher–Rudin model (163). Here, the natural candidate is the space F of divergences of bounded vector fields, with norm

$$\|v\|_F := \min\{\|w\|_{L^\infty} \ : \ v = \operatorname{div}(w)\}, \tag{165}$$

This space coincides with the dual of $W^{1,1}(\mathbb{R}^2)$ and is slightly smaller than $B_\infty^{-1,\infty}$. Yves Meyer proved that the v component in (163) is naturally controlled in this norm, and used the space F both in order to analyze the qualitative properties this decomposition and to explore alternative strategies aiming at better capturing the textural part, for example by minimizing $\|v\|_F + \lambda |u|_{\mathrm{BV}}$

4.4 Oscillations in Nonlinear Evolution PDEs

One major object of study in nonlinear evolution PDEs is the phenomenon of explosion in finite time of the solution measured in some given norm. This study goes together with understanding which type of initial condition prevent such blow up to occur, and in which function space global existence of solutions can be established.

Yves Meyer developed a research program, built on the intuition that in several relevant cases, *blow up can be prevented by oscillations* in the initial condition, and that such oscillations should be quantified by means of negative order smoothness spaces. This led to a series of breakthrough for PDEs such as the nonlinear heat equation and the Navier–Stokes equations. These two models may be written in general semi-linear format as

$$\frac{\partial u}{\partial t} = \Delta u + N(u), \quad u(t = 0) = u_0. \tag{166}$$

where N is a nonlinear operator. The solution process studied by Yves Meyer is based on the integral formulation

$$u(t) = S(t)u_0 + \int_0^t S(t-s)N(u(s))ds, \tag{167}$$

where $S(t) = e^{\Delta t}$ is the semi-group associated with the linear heat equation $\frac{\partial u}{\partial t} = \Delta u$. In this approach, which is similar in spirit to that used in the proof of the classical Cauchy–Lipschitz theorem for ODEs, the function u is thus viewed as a fixed point of a nonlinear operator $u \mapsto G(u)$ defined by the right side of (167). Solutions $u \in C([0,T], X)$ to such fixed point problems, for some given Banach space X, are referred to as *mild solutions* to (166).

In the case of the nonlinear heat equation

$$\frac{\partial u}{\partial t} = \Delta u + u^3, \quad (x,t) \in \mathbb{R}^3 \times]0,\infty[, \quad u(x,0) = u_0(x), \tag{168}$$

formal multiplication by u and integration over $\mathbb{R}^3$ yields $\frac{d}{dt}\|u\|_{L^2}^2 = -2\|\nabla u\|_{L^2}^2 + \|u\|_{L^4}^4$, which hints that the evolution of the L^2 norm of the solution is governed by a competition between $\|\nabla u\|_{L^2}^2$ and $\|u\|_{L^4}^4$. This can be made more precise by a sufficient condition for blow up which follows from work by Howard Levine [54].

Theorem 45 *If u_0 is a non-trivial smooth and compactly supported function and if*

$$\|\nabla u_0\|_{L^2}^2 \leq \frac{1}{2}\|u_0\|_{L^4}^4, \tag{169}$$

then there exists a finite T_0 such that $\|u(\cdot,t)\|_{L^2}$ is unbounded as $t \to T_0$.

The improved Sobolev inequalities (157) show that (169) will not hold if an appropriate negative smoothness norm $\|u_0\|_X$ is small enough, in particular when u_0 is highly oscillatory. This suggests that there should exist negative smoothness spaces X satisfying the improved inequality

$$\|f\|_{L^4} \leq \|\nabla f\|_{L^2}^{1/2} \|f\|_X^{1/2}, \tag{170}$$

and such that $\|u_0\|_X \leq \eta$ for some small enough $\eta > 0$ implies global existence of a mild solution to (168) in $C([0,\infty[, X)$. In addition, due to the invariance of (168) by rescaling $u_\lambda(x,t) = \lambda u(\lambda x, \lambda^2 t)$, such a norm should satisfy $\|f_\lambda\|_X = \|f\|_X$ for $f_\lambda(x) = \lambda f(\lambda x)$. Otherwise this norm could be made arbitrarily small by a proper rescaling and the condition $\|u_0\|_X \leq \eta$ becomes irrelevant. For Lebesgue spaces, the L^3 norm satisfies this scaling, and existence was established by Fred Weissler [92].

Theorem 46 *There exists $\eta > 0$ such that $\|u_0\|_{L^3} \leq \eta$ implies the existence of a global mild solution $u \in C([0,\infty[, L^3(\mathbb{R}^3))$ to* (168).

The largest negative smoothness space X that satisfies all requirements is the homogeneous Besov space $\dot{B}_\infty^{-1,\infty}(\mathbb{R}^3)$. A global existence result is not known for this space but Yves Meyer established such a result for a smaller negative smoothness space which also satisfies all requirements.

Theorem 47 *Let $X = \dot{B}_6^{-1/2,\infty}(\mathbb{R}^3)$. There exists $\eta > 0$ such that $\|u_0\|_X \le \eta$ implies the existence of a global mild solution $u \in C([0,\infty[, X)$ to* (168).

In the case of the incompressible Navier–Stokes equations

$$\frac{\partial u}{\partial t} = \Delta u - (u \cdot \nabla)u - \nabla p \quad \text{and} \quad \mathrm{div}(u) = 0, \quad (x,t) \in \mathbb{R}^3 \times]0,\infty[, \tag{171}$$

mild solutions are defined by applying to the equation the Leray–Hopf projector onto divergence free vector fields. This projector can be defined from the Riesz transforms $\widehat{R_j f}(\omega) := -i\omega_j|\omega|^{-1}\hat{f}(\omega)$ by

$$\Pi(f_1, f_2, f_3) = (g_1, g_2, g_3), \quad g_j = (R - R_j R)f_j, \quad R := R_1 + R_2 + R_3. \tag{172}$$

This eliminates the pressure p from (171) and one reaches the formulation (167) for the velocity u with $N(u) = -\Pi(u \cdot \nabla)u$. The invariant scaling for u is $\lambda u(\lambda x, \lambda^2 t)$, similar to the nonlinear heat equation, therefore suggesting that $L^3(\mathbb{R}^3)$ is the natural solution space. The study of mild solutions to the Navier–Stokes equations was initiated by Tosio Kato who proved a result of global existence in $u \in C([0,\infty[, L^3(\mathbb{R}^3)^3)$ exactly similar to Theorem 46.

It is worth mentioning however that the situation differs between the two equations as to uniqueness. For the nonlinear heat equation, it is known that uniqueness of solutions cannot be expected in $C([0,\infty[, L^3(\mathbb{R}^3))$, but Fred Weissler could establish it by appending the additional smallness condition

$$\sup_{t>0} t^{1/2}\|u(\cdot,t)\|_{L^\infty} \le \eta. \tag{173}$$

For the Navier–Stokes equation, uniqueness of mild solutions in $C([0,\infty[, L^3(\mathbb{R}^3)^3)$ was proved by Giulia Furioli, Pierre Gilles Lemarié and Elide Terraneo, without any such assumption [35]. One key argument in the proof of this result is an idea introduced by Yves Meyer which says that the fluctuation of the exact solution $u(\cdot,t)$ around the solution $S(t)u_0$ of the linear heat equation can be controlled in a smaller space. A simpler proof, not based on this argument, was later provided by Sylvie Monniaux [78], and Yves Meyer showed that the space $L^3(\mathbb{R}^3)$ could be replaced by a larger space defined as the closure of the test functions $\mathscr{D}(\mathbb{R}^3)$ in the weak space $L^{3,\infty}(\mathbb{R}^3)$. It was also proved by Luis Escauriaza, Gregory Seregin and Vladimír Sverák that such solutions are smooth for positive time [34].

Yves Meyer raised the question of whether a weaker condition of the form $\|u_0\|_X \le \eta$, for some negative smoothness space could ensure global existence of a mild solution to the Navier–Stokes equations. First results in this direction came from the work of Marco Cannone and Fabrice Planchon, in particular the following from [83].

Theorem 48 *For* $3 \leq q < \infty$ *and* $X_q = \dot{B}_q^{-(1-3/q),\infty}(\mathbb{R}^3)^3$, *there exists* $\eta_q > 0$ *such that*

$$\operatorname{div}(u_0) = 0, \quad u_0 \in L^3(\mathbb{R}^3)^3, \quad \text{and} \quad \|u_0\|_{X_q} \leq \eta_q \tag{174}$$

imply the existence of a global solution $u \in C([0,\infty[, L^3(\mathbb{R}^3)^3)$ *to* (171).

In contrast to Kato's global existence theorem, the above result does not require u_0 to be small in L^3, as long as its oscillations ensure smallness in X. The best result along these lines is due to Herbert Koch and Daniel Tataru [50] and requires smallness of the components of u_0 in a space G that contains all the above X_q spaces and which consists of divergences of vector fields from $\mathrm{BMO}(\mathbb{R}^3)^3$. Recall that $\mathrm{BMO}(\mathbb{R}^d)$ is a slightly larger space than $L^\infty(\mathbb{R}^d)$ with norm defined by (70). The space G is thus a close substitute to the space F from (165) used by Meyer in the modeling of texture, and to the Besov space $B_\infty^{-1,\infty}$ that appears in improved Sobolev inequalities. More precisely, one has the embedding chain $F \subset G \subset B_\infty^{-1,\infty}$.

The research program initiated by Yves Meyer therefore reveals that very similar spaces play a critical role for modeling the relevant oscillating patterns of textures in image processing and initial conditions in nonlinear PDEs.

References

1. P. Auscher, S. Hofmann, M. Lacey, A. McIntosh and P. Tchamitchian. The solution of the Kato square root problem for second order elliptic operators on $\mathbb{R}^n$. *Ann. of Math. (2)*, 156, 633–654, 2002.
2. H. Bahouri and P. Gérard. High frequency approximation of solutions to critical nonlinear wave equations. *Amer. J. Math.*, 121, 131–175, 1999.
3. G. Beylkin, R. Coifman and V. Rokhlin. Fast wavelet transforms and numerical algorithms. *Comm. Pure Appl. Math.*, 44, 141–183, 1991.
4. J. Bergh and J. Löfström. *Interpolation spaces*, volume 223 of *Grundlehren der Mathematischen Wissenschaften*, Springer-Verlag, Berlin–New York, 1976.
5. J.-M.Bony. Calcul symbolique et propagation des singularités pour les équations aux dérivées partielles non linéaires. *Ann. Sci. École Norm. Sup. (4)*, 14, 209–246, 1981.
6. A.-P. Calderón. Commutators of singular integral operators. *Proc. Nat. Acad. Sci. U.S.A.*, 53, 1092–1099, 1965.
7. A.-P. Calderón. Algebra of singular integral operators. In *Proc. Sympos. Pure Math., Chicago, Ill., 1966*, pages 18–55, Amer. Math. Soc., Providence, RI, 1967.
8. E. Candès, J. Romberg and T. Tao. Stable signal recovery from incomplete and inaccurate measurements. *Comm. Pure Appl. Math.*, 59, 1207–1223, 2006.
9. E. Candès and T. Tao. Near-optimal signal recovery from random projections: universal encoding strategies? *IEEE Trans. Inform. Theory*, 53, 5406–2425, 2006.
10. A. Chambolle, R. DeVore, N. Lee and B. Lucier. Nonlinear wavelet image processing: variational problems, compression, and noise removal through wavelet shrinkage. *IEEE Trans. Image Process.*, 7, 319–335, 1998.
11. A. Cohen, W. Dahmen and R. DeVore. Adaptive wavelet methods fo elliptic operator equations: Convergence rates. *Math. Comp.*, 70, 27–75, 2000.

12. A. Cohen, W. Dahmen, I. Daubechies and R. DeVore, Harmonic analysis of the space BV. *Rev. Mat. Iberoamericana*, 19, 235–263, 2003.
13. A. Cohen, I. Daubechies and J.C. Feauveau, Biorthogonal bases of compactly supported wavelets, *Comm. Pure Appl. Math.*, 45, 485–560, 1992.
14. A. Cohen, Y. Meyer and F. Oru. Improved Sobolev inequalities. In *Séminaire sur les Équations aux Dérivées Partielles, 1997–1998*, Exp. No. XVI, 16 pp. École Polytech., Palaiseau, 1998.
15. R. Coifman, G. David and Y. Meyer. La solution des conjecture de Calderón. *Adv. in Math.*, 48, 144–148, 1983.
16. R. Coifman, S. Dobyinsky and Y. Meyer. Opérateurs bilinéaires et renormalisation. In *Essays on Fourier analysis in honor of Elias M. Stein*, volume 42 of *Princeton Math. Ser.*, pages 146–161. Princeton Univ. Press, Princeton, NJ, 1995.
17. R. Coifman, P. Jones and S. Semmes. Two elementary proofs of the L^2 boundedness of Cauchy integrals on Lipschitz curves. *J. Amer. Math. Soc. 2*, 3, 553–564, 1989.
18. R. Coifman, P.-L. Lions, Y. Meyer and S. Semmes. Compensated compactness and Hardy spaces. *J. Math. Pures Appl. (9)*, 72, 247–286, 1993.
19. R. Coifman, A. McIntosh and Y. Meyer. L'intégrale de Cauchy définit un opérateur borné sur L^2 pour les courbes lipschitziennes. *Ann. of Math. (2)*, 116, 361–387, 1982.
20. R. Coifman and Y. Meyer. On commutators of singular integrals and bilinear singular integrals. *Trans. Amer. Math. Soc.*, 212, 315–331, 1975.
21. R. Coifman and Y. Meyer. Commutateurs d'intégrales singulières et opérateurs multilinéaires. *Ann. Inst. Fourier*, 28, 177–202, 1978.
22. R. Coifman and Y. Meyer. *Au delà des opérateurs pseudo-différentiels*, volume 57 of *Astérisque*. Société Mathématique de France, Paris, 1978.
23. A. Cordoba. La formule sommatoire de Poisson. *C. R. Acad. Sci. Paris Sér. I Math.*, 306, 373–376, 1984.
24. I. Daubechies. Orthonormal bases of compactly supported wavelets. *Comm. Pure Appl. Math.*, 41, 909–996, 1988.
25. I. Daubechies. *Ten Lectures on Wavelets*, volume 61 of *CBMS-NSF Regional Conference Series in Applied Mathematics*. Society for Industrial and Applied Mathematics (SIAM), Philadelphia, PA, 1992.
26. I. Daubechies, S. Jaffard and J.-L. Journé. A simple Wilson orthonormal basis with exponential decay. *SIAM J. Math. Anal.*, 22, 554–573, 1991.
27. G. David. Unrectifiable 1-sets have vanishing analytic capacity. *Rev. Mat. Iberoamericana*, 14, 369–479, 1998.
28. G. David, J.-L. Journé, and S. Semmes. Opérateurs de Calderón–Zygmund, fonctions para-accrétives et interpolation. *Rev. Mat. Iberoamericana*, 1, 1–56, 1985.
29. R. DeVore. Nonlinear approximation. In *Acta Numerica Vol. 7*, pages 51–150. Cambridge University Press, Cambridge, 1998.
30. J.I. Diaz and Y. Meyer. Poisson summation formulae and the wave equation with a finitely supported measure as initial velocity. *Afr. Diaspora J. Math.*, 20, 1–13, 2017.
31. D. Donoho. Unconditional bases are optimal for data compression and statistical estimation. *Appl. Comput. Harmon. Anal.*, 1, 100–115, 1993.
32. D. Donoho. Compressed Sensing. *IEEE Trans. Inform. Theory*, 52, 1289–1306, 2006.
33. D. Donoho, I. Johnstone, G. Kerkyacharian and D. Picard. Wavelet shrinkage: Asymptotia? With discussion and a reply by the authors. *J. Roy. Statist. Soc. Ser. B*, 57, 301–369, 1995.
34. L. Escauriaza, G.-A. Seregin and V. Sverak. $L_{3,\infty}$ solutions of the Navier–Stokes equations and backward uniqueness. *Russ. Math. Surv.*, 58, 211–250, 2003.
35. G. Furioli, P.-G. Lemarié and E. Terraneo. Sur l'unicité dans $L^3(\mathbb{R}^3)$ des solutions "mild" des équations de Navier–Stokes. *C. R. Acad. Sci. Paris Sér. I Math.*, 325, 1253–1256, 1997.
36. P. Gérard. Description du défaut de compacité de l'injection de Sobolev. *ESAIM Control Optim. Calc. Var.*, 3, 213–233, 1998.
37. P. Gérard, Y. Meyer and F. Oru. Inégalités de Sobolev précisées. In *Séminaire sur les Équations aux Dérivées Partielles, 1996–1997*, Exp. No. IV, 11 pp. École Polytech., Palaiseau, 1997.

38. A.P. Guinand. Concordance and the harmonic analysis of sequences. *Acta Math.*, 101, 235–271, 1959.
39. A. Haraux and S. Jaffard. Pointwise and spectral control of some plate models. *Rev. Mat. Iberoamericana*, 7, 1–24, 1991.
40. A. Haraux and V. Komornik. Oscillations of anharmonic Fourier series and the wave equation. *Rev. Mat. Iberoamericana*, 1, 57–77, 1985.
41. C.S. Herz. Spectral synthesis for the Cantor set. *Proc. Nat. Acad. Sci. U.S.A.*, 42, 42–43, 1956.
42. S. Jaffard. Contrôle interne exact des vibrations d'une plaque rectangulaire. *Portugal. Math.* 47, 423–429, 1990.
43. S. Jaffard. Analysis of the lack of compactness in the critical Sobolev embeddings. *J. Funct. Anal.*, 161, 384—396, 1999.
44. S. Jaffard. On the Frisch–Parisi conjecture. *J. Math. Pures Appl. (9)* 79, 525–552, 2000.
45. S. Jaffard and Y. Meyer. *Wavelet Methods for Pointwise Regularity and Local Oscillations of Functions*, volume 123 of *Mem. Amer. Math. Soc.* Amer. Math. Soc., Providence, RI, 1996.
46. S. Jaffard, Y. Meyer and R. Ryan. *Wavelets. Tools for Science and Technology*. Society for Industrial and Applied Mathematics (SIAM), Philadelphia, PA, 2001.
47. J.-P. Kahane. Sur les fonctions moyenne-périodiques bornées. *Ann. Inst. Fourier*, 293–314, 1957.
48. J.-P. Kahane and R. Salem. *Ensembles Parfaits et Séries Trigonométriques*. Hermann, Paris, 1963.
49. J.C. Lagarias. Meyer's concept of quasicrystal and quasiregular sets. *Comm. Math. Phys.*, 179, 365–376, 1996.
50. H. Koch and D. Tataru, Well-posedness for the Navier–Stokes equations. *Adv. Math.*, 157, 22–35, 2001.
51. M. Ledoux. On improved Sobolev embedding theorems. *Math. Res. Lett.*, 10, 659–669, 2003.
52. N. Lev and A. Olevskii. Quasicrystals and Poisson's summation formula. *Invent. Math.*, 200, 585–606, 2015.
53. N. Lev and A. Olevskii. Quasicrystals with discrete support and spectrum. *Rev. Mat. Iberoamericana*, 32, 1341–1352, 2016.
54. H.A. Levine. Some nonexistence and instability theorems for solutions of formally parabolic equations of the form $Pu_t = Au + \mathscr{F}(u)$. *Arch. Rational Mech. Anal.*, 51, 371–386, 1973.
55. H.J. Landau. Necessary density conditions for sampling and interpolation of certain entire functions. *Acta Math.*, 117, 37–52, 1967.
56. A. McIntosh and Y. Meyer. Algèbres d'opérateurs définis par des intégrales singulières. *C. R. Acad. Sci. Paris Sér. I Math.*, 301, 395–397, 1985.
57. S. Mallat. Multiresolution approximation and wavelet orthonormal bases of $L^2(\mathbb{R})$. *Trans. Amer. Math. Soc.*, 315, 69–88, 1989.
58. S. Mallat. *A Wavelet Tour of Signal Processing*. Academic Press, San Diego, CA, 1998.
59. B. Matei and Y. Meyer. A variant of compressed sensing. *Rev. Mat. Iberoamericana*, 25, 669–692, 2009.
60. B. Matei and Y. Meyer. Simple quasicrystals are sets of stable sampling. *Complex Var. Elliptic Equ.*, 55, 947–964, 2010.
61. M.S. Melnikov and J. Verdera. A geometric proof of the L^2 boundedness of the Cauchy integral on Lipschitz graphs. *Internat. Math. Res. Notices*, 7, 325–331, 1995.
62. F. Merle and L. Vega. Compactness at blow-up time for L_2 solutions of the critical nonlinear Schrödinger equation in 2D. *Internat. Math. Res. Notices*, 8, 399–425, 1998.
63. Y. Meyer. Nombres algébriques, nombres transcendants et équirepartition modulo 1. *Acta Arith.*, 16, 347–350, 1970.
64. Y. Meyer. *Nombres de Pisot, Nombres de Salem et Analyse Harmonique*, volume 117 of *Lecture Notes in Mathematics*, Springer, Berlin–New York, 1970.
65. Y. Meyer. *Algebraic Numbers and Harmonic Analysis*, volume 2 of *North-Holland Mathematical Library*. North-Holland, Amsterdam–London, 1972.
66. Y. Meyer. Nombres premiers et vibrations. In *Séminaire Delange–Pisot–Poitou 1971–1972, Théorie des nombres*, Fasc. 2, Exp. No. 15, 5 pp. Paris, 1973.

67. Y. Meyer. *Trois problèmes sur les séries trigonométriques*, volume 1 of *Astérisque*. Société Mathématique de France, Paris, 1973.
68. Y. Meyer. Adèles et séries trigonométriques spéciales. *Ann. of Math. (2)*, 97, 171–186, 1973.
69. Y. Meyer. Etude d'un modèle mathématique issu du controle des structures spatiales déformables. In *Nonlinear partial differential equations and their applications. Collège de France seminar, Vol. VII (Paris, 1983–1984)*, pages 234–242. Res. Notes in Math., 122, Pitman, Boston, 1985.
70. Y. Meyer. *Wavelets and Operators*, volume 37 of *Cambridge Studies in Advanced Mathematics*. Cambridge University Press, Cambridge, 1992.
71. Y. Meyer. *Wavelets, Vibrations and Scalings*, volume 9 of *CRM Monograph Series*, Amer. Math. Soc., Providence, RI, 1998.
72. Y. Meyer. *Oscillating Patterns In Image Processing and In Some Nonlinear Evolution Equations*, volume 22 of *University Lecture Series*. The fifteenth Dean Jacqueline B. Lewis memorial lectures. Amer. Math. Soc., Providence, RI, 2001.
73. Y. Meyer. La conjecture de Kato (d'après Pascal Auscher, Steve Hofmann, Michael Lacey, John Lewis, Alan McIntosh et Philippe Tchamitchian). *Astérisque*, 290, 193–206, 2003.
74. Y. Meyer. Measures with locally finite support and spectrum. *Proc. Natl. Acad. Sci. USA*, 113, 3152–3158, 2016.
75. Y. Meyer. Irregular sampling of signals and images. To appear in Det Kongelige Norske Videnskabers Selskabs Skrifter, 2018.
76. Y. Meyer and R. Coifman. *Wavelets: Calderón–Zygmund and Multilinear Operators*, volume 48 of *Cambridge Studies in Advanced Mathematics*. Cambridge University Press, Cambridge, 1997.
77. F. Murat. Compacité par compensation. *Ann. Scuola Norm. Sup. Pisa Cl. Sci. (4)*, 5, 489–507, 1978.
78. S. Monniaux. Uniqueness of mild solutions of the Navier–Stokes equation and maximal L^p-regularity. *C. R. Acad. Sci. Paris Sér. I Math.*, 328, 663–668, 1999.
79. S. Müller. Higher integrability of determinants and weak convergence in L^1, *J. Reine Angew. Math.*, 412, 20–34, 1990.
80. A. Olevskii and A. Ulanovskii. A universal sampling of band-limited signals. *C. R. Math. Acad. Sci. Paris* 342, 927–931, 2006.
81. A. Olevskii and A. Ulanovskii. Universal sampling and interpolation of band-limited signals. *Geom. Funct. Anal.*, 18, 1029–1052, 2008.
82. L. Rudin and S. Osher. Total variation based image restoration with free local constraints. *IEEE International Conference on Image Processing*, 1, 31–35, 1994.
83. F. Planchon. Asymptotic behavior of global solutions to the Navier–Stokes equations in $\mathbb{R}^3$. *Rev. Mat. Iberoamericana*, 14, 71–93, 1998.
84. R. Salem and A. Zygmund. Sur un théorème de Piatecki–Shapiro. *C. R. Acad. Sci. Paris*, 40, 2040–2042, 1955.
85. L. Schwartz. Théorie générale des fonctions à moyenne-périodique. *Ann. of Math. (2)*, 857–929, 1947.
86. E. Stein. *Singular Integrals and Differentiability Properties of Functions*, volume 30 of *Princeton Mathematical Series*. Princeton University Press, Princeton, NJ, 1970.
87. D. Shechtman, I.A. Blech, D. Gratias and J.W. Cahn. Metallic phase with long range orientational order and no translational symmetry. *Phys. Rev. Lett.*, 53, 1951–1953, 1984.
88. J.-O. Strömberg. A modified Franklin system and higher order spline systems on $\mathbb{R}^n$ as unconditional bases for Hardy spaces. In *Conference on harmonic analysis in honor of A. Zygmund, Vol. I, II* pages 475–494. Wadsworth Math. Ser., Wadsworth, Belmont, CA, 1983.
89. L. Tartar. Compensated compactness and applications to partial differential equations. In *Nonlinear analysis and mechanics: Heriot-Watt Symposium, Vol. IV*, pages 136–212. Res. Notes in Math., 39, Pitman, Boston, MA, 1979.
90. X. Tolsa. Painlevé's problem and the semi-additivity of analytic capacity. *Acta Math.*, 190, 105–149, 2003.

91. G. Verchota. Layer potentials and regularity for the Dirichlet problem for Laplace's equation in Lipschitz domains. *J. Funct. Anal.*, 59, 572–611, 1984.
92. F. Weissler. Local existence and nonexistence for semilinear parabolic equations in L^p. *Indiana Univ. Math. J.*, 29, 79–102, 1980.

List of Publications for Yves F. Meyer

1966

[1] Le spectre de Wiener. *Studia Math.*, 27:189–201.
[2] Complément à un théorème de Paley. *C. R. Acad. Sci. Paris Sér. A-B*, 262:A281–A282.
[3] Prolongement des multiplicateurs d'idéaux fermés de $L^1(R^n)$. *C. R. Acad. Sci. Paris Sér. A-B*, 262:A744–A745.
[4] (with H.P. Rosenthal). Convexité et ensemble de Ditkin forts. *C. R. Acad. Sci. Paris Sér. A-B*, 262:A1404–A1406.
[5] Multiplicateurs des coefficients de Fourier des fonctions intégrables analytiques. *C. R. Acad. Sci. Paris Sér. A-B*, 263:A385–A387.
[6] Idéaux fermés de L^1 dans lesquels une suite approche l'identité. *Math. Scand.*, 19:219–222.

1967

[7] Isomorphismes entre certaines algèbres de restrictions. *C. R. Acad. Sci. Paris Sér. A-B*, 265:A18–A20.

1968

[8] Isomorphismes entre certaines algèbres de restrictions. *Ann. Inst. Fourier (Grenoble)*, 18(2):73–86.
[9] Endomorphismes des idéaux fermés de $L^1(G)$, classes de Hardy et séries de Fourier lacunaires. *Ann. Sci. École Norm. Sup. (4)*, 1:499–580.
[10] Une caractérisation des nombres de Pisot. *C. R. Acad. Sci. Paris Sér. A-B*, 266:A63–A64.
[11] Problème de l'unicité et de la synthèse en analyse harmonique. *C. R. Acad. Sci. Paris Sér. A-B*, 266:A275–A276.
[12] Nombres de Pisot et analyse harmonique. *C. R. Acad. Sci. Paris Sér. A-B*, 267:A196–A198.

H. Holden, R. Piene (eds.), *The Abel Prize 2013–2017*,
https://doi.org/10.1007/978-3-319-99028-6_31

[13] Spectres des mesures et mesures absolument continues. *Studia Math.*, 30: 87–99.

1969

[14] Synthèse harmonique. In *Séminaire d'Analyse Harmonique (année 1968/1969)*, Exp. No. 1, 7 pp.

[15] À propos de la formule de Poisson. In *Séminaire d'Analyse Harmonique (année 1968/1969)*, Exp. No. 2, 14 pp.

[16] Algèbres de restrictions non isomorphes. *Ann. Inst. Fourier (Grenoble)*, 19(1):117–124, x.

[17] Nombres de Pisot, nombres de Salem et répartition modulo 1. In *Séminaire Delange–Pisot–Poitou: 1968/69, Théorie des Nombres*, Fasc. 1, Exp. 2, 6 pp. Secrétariat mathématique, Paris.

[18] Caractérisation des nombres de Pisot–Vijayaraghavan (P.-V.). In *Séminaire Delange–Pisot–Poitou: 1967/68, Théorie des Nombres*, Fasc. 1, Exp. 8, 4 pp. Secrétariat mathématique, Paris.

[19] Nombres algébrique et répartition modulo 1. *C. R. Acad. Sci. Paris Sér. A-B*, 268:A25–A27.

[20] (with A. Bonami). Propriétés de convergence de certaines séries trigonométriques. *C. R. Acad. Sci. Paris Sér. A-B*, 269:A68–A70.

[21] Les problèmes de l'unicité et de la synthèse harmonique et la répartition modulo 1. *C. R. Acad. Sci. Paris Sér. A-B*, 269:A320–A322.

[22] (with J.-P. Schreiber). Quelques fonctions moyennes-périodiques non bornées. *Ann. Inst. Fourier (Grenoble)*, 19(1):231–236, xi.

[23] Nombres algébriques, nombres transcendants et équiré-partition modulo 1. *Acta Arith.*, 16:347–350.

1970

[24] Les nombres de Pisot et l'analyse harmonique. *Studia Math.*, 34:127–147.

[25] Nombres algébriques et analyse harmonique. *Ann. Sci. École Norm. Sup. (4)*, 3:75–110.

[26] Les nombres de Pisot et la synthèse harmonique. *Ann. Sci. École Norm. Sup. (4)*, 3:235–246.

[27] *Nombres de Pisot, nombres de Salem et analyse harmonique*. Lecture Notes in Mathematics, Vol. 117. Springer-Verlag, Berlin-New York.

1971

[28] Nombres transcendants et répartition modulo 1. *Colloque de Théorie des Nombres (Univ. Bordeaux, Bordeaux, 1969)*, 143–149. Bull. Soc. Math. France, Mém. 25, Soc. Math. France, Paris.

[29] Special trigonometric series. *Journées Arithmétiques Françaises (Univ. Provence, Marseille, 1971)*, Exp. No. 5, 5 pp. Univ. Provence, Marseille.

[30] Ultra-thin symmetric sets and harmonic analysis. In *Thin sets in harmonic analysis (Sem. Inst. Mittag-Leffler, Univ. Stockholm, Stockholm, 1969–1970)*,

161–167, 177–180. Lecture Notes in Pure and Appl. Math., Vol. 2, Dekker, New York.

[31] Pisot numbers and the problem of synthesis. In *Thin sets in harmonic analysis (Sem. Inst. Mittag-Leffler, Univ. Stockholm, Stockholm, 1969–1970)*, 168–176, 177–180. Lecture Notes in Pure and Appl. Math., Vol. 2, Dekker, New York.

[32] Nombres de Pisot et analyse harmonique. In *Actes du Congrès International des Mathématiciens (Nice, 1970)*, Tome 2, 663–665. Gauthier-Villars, Paris.

1972

[33] Séries trigonométriques spéciales et corps quadratiques. Collection of articles honoring the completion by Antoni Zygmund of 50 years of scientific activity, IV. *Studia Math.*, 44:321–333.

[34] Recent advances in spectral synthesis. In *Conference on Harmonic Analysis (Univ. Maryland, College Park, Md., 1971)*, 239–253. Lecture Notes in Math., 266, Springer, Berlin.

[35] Étude asymptotique des vibrations des sphères. In *Séminaire Goulaouic–Schwartz 1971–1972: Équations aux dérivées partielles et analyse fonctionnelle*, Exp. No. 28, 9 pp. École Polytech., Centre de Math., Paris.

[36] *Algebraic numbers and harmonic analysis.* North-Holland Mathematical Library, Vol. 2. North-Holland Publishing Co., Amsterdam-London; American Elsevier Publishing Co., Inc., New York. x+274 pp.

1973

[37] Adèles et séries trigonométriques spéciales. *Ann. of Math. (2)*, 97:171–186.

[38] Théorie L^p des sommes trigonométriques apériodiques. *C. R. Acad. Sci. Paris Sér. A-B*, 277:A15–A17.

[39] Sur un problème de M. Mandelbrojt. In *L'analyse harmonique dans le domaine complexe (Actes Table Ronde Internat., Centre Nat. Recherche Sci., Montpellier, 1972)*, 161–167. Lecture Notes in Math., Vol. 336, Springer, Berlin.

[40] Adèles et séries trigonométriques spéciales. In *Séminaire Delange–Pisot–Poitou (13e année: 1971/72), Théorie des nombres*, Fasc. 1, Exp. No. 11, 16 pp. Secrétariat Mathématique, Paris.

[41] Nombres premiers et vibrations. In *Séminaire Delange–Pisot–Poitou (13e année: 1971/72), Théorie des nombres*, Fasc. 2, Exp. No. 15, 5 pp. Secrétariat Mathématique, Paris.

[42] *Trois problèmes sur les sommes trigonométriques.* Astérisque, 1. Société Mathématique de France, Paris. i+387 pp.

1974

[43] (with F. Gramain). Ensembles de fréquences et fonctions presque périodiques. *Colloq. Math.*, 30:269–275.

[44] Théorie L^p des sommes trigonométriques apériodiques. *Ann. Inst. Fourier (Grenoble)*, 24(4):xi, 189–211 (1975).

[45] Les nombres de Pisot et la synthèse spectrale. In *Journées Arithmétiques (Grenoble, 1973)*, 117–120. Bull. Soc. Math. France Mém., No. 37, Soc. Math. France, Paris.

1975

[46] (with R. R. Coifman). On commutators of singular integrals and bilinear singular integrals. *Trans. Amer. Math. Soc.*, 212:315–331.
[47] (with F. Gramain). Quelques fonctions moyenne-périodiques bornées. *Colloq. Math.*, 33(1):133–137.
[48] (with L. Lesieur, C. Joulain, and J. Lefebvre). *Algèbre générale*. Collection U, Série "Mathématiques". Librairie Armand Colin, Paris.

1976

[49] Remarques sur un théorème de J. Delsarte. *Ann. Inst. Fourier (Grenoble)*, 26(2):xi, 133–152.
[50] Harmonic analysis of mean-periodic functions. In *Studies in harmonic analysis (Proc. Conf., DePaul Univ., Chicago, Ill., 1974)*, 151–160. MAA Stud. Math., Vol. 13, Math. Assoc. Amer., Washington, D.C.
[51] Comportement asymptotique des solutions de certaines équations de convolution. *J. Math. Pures Appl. (9)*, 55(1):69–97.

1977

[52] Quelques problèmes sur les fonctions presque-périodiques. In *Séminaire d'Analyse Harmonique (1976–1977)*, 21–27. Publ. Math. Orsay, No. 77–77.
[53] (with R. Coifman). Opérateurs pseudo-différentiels et théorème de Calderón. In *Séminaire d'Analyse Harmonique (1976–1977)*, 28–40. Publ. Math. Orsay, No. 77–77.

1978

[54] Nouvelles estimations pour les opérateurs pseudo-différentiels. In *Séminaire Goulaouic–Schwartz (1977/1978)*, Exp. No. 10, 6 pp. École Polytech., Palaiseau.
[55] (with R. Coifman). Commutateurs d'intégrales singulières et opérateurs multilinéaires. *Ann. Inst. Fourier (Grenoble)*, 28(3):xi, 177–202.
[56] (with R. R. Coifman). *Au delà des opérateurs pseudo-différentiels*, *Astérisque*, 57. Société Mathématique de France, Paris. i+185 pp.
[57] (with R. Coifman). Commutators of singular integrals. In *Fourier analysis and approximation theory (Proc. Colloq., Budapest, 1976), Vol. I*, 265–273, *Colloq. Math. Soc. János Bolyai*, 19, North-Holland, Amsterdam-New York.
[58] Estimations L^2 pour les opérateurs pseudo-différentiels. In *Séminaire d'Analyse Harmonique (1977/1978)*, 47–53, *Publ. Math. Orsay 78*, 12, Univ. Paris XI, Orsay.
[59] Inégalités L^2 à poids pour les opérateurs différentiels. In *Séminaire d'Analyse Harmonique (1977/1978)*, 54–62, *Publ. Math. Orsay 78*, 12, Univ. Paris XI, Orsay.

[60] (with S. Hartman). Interpolation harmonique sur les compacts. *Colloq. Math.*, 40(2):265–276.

1979

[61] Sharp estimates for pseudodifferential operators. In *Harmonic analysis in Euclidean spaces (Proc. Sympos. Pure Math., Williams Coll., Williamstown, Mass., 1978)*, 105–113, Proc. Sympos. Pure Math., XXXV, Part 2, Amer. Math. Soc., Providence, R.I.
[62] (with R. Coifman). Le théorème de Calderón par les "méthodes de variable réelle". *C. R. Acad. Sci. Paris Sér. A-B*, 289(7):A425–A428.
[63] Produits de Riesz généralisés. In *Seminar on Harmonic Analysis, 1978–1979 (French)*, 38–48, *Publ. Math. Orsay 79*, 7, Univ. Paris XI, Orsay.
[64] (with R. Coifman). Le théorème de Calderón par les "méthodes de variable réelle". In *Seminar on Harmonic Analysis, 1978–1979*, 49–55, *Publ. Math. Orsay 79*, 7, Univ. Paris XI, Orsay.

1980

[65] Les nouvelles intégrales singulières de Calderón. In *Séminaire Bourbaki (1978/79)*, Exp. No. 528, 57–65, *Lecture Notes in Math.*, 770, Springer, Berlin.
[66] (with R. R. Coifman). Fourier analysis of multilinear convolutions, Calderón's theorem, and analysis of Lipschitz curves. In *Euclidean harmonic analysis (Proc. Sem., Univ. Maryland, College Park, Md., 1979)*, 104–122, *Lecture Notes in Math.*, 779, Springer, Berlin.
[67] (with R. R. Coifman). Une généralisation du théorème de Calderón sur l'intégrale de Cauchy. In *Fourier analysis (Proc. Sem., El Escorial, 1979)*, 87–116, *Asoc. Mat. Española*, 1, Asoc. Mat. Española, Madrid.

1981

[68] Multiplication of distributions. In *Mathematical analysis and applications, Part B*, 603–615, *Adv. in Math. Suppl. Stud.*, 7b, Academic Press, New York-London.
[69] Régularité des solutions des équations aux dérivées partielles non linéaires (d'après J.-M. Bony). In *Bourbaki Seminar, Vol. 1979/80*, 293–302, *Lecture Notes in Math.*, 842, Springer, Berlin-New York.
[70] Remarques sur un théorème de J.-M. Bony. Proceedings of the Seminar on Harmonic Analysis (Pisa, 1980). *Rend. Circ. Mat. Palermo (2)*, suppl. 1, 1–20.
[71] Les opérateurs pseudo-différentiels classiques et leurs conjugués par changement de variable. In *Goulaouic–Meyer–Schwartz Seminar, 1980–1981*, Exp. No. X, 17 pp. École Polytech., Palaiseau.

1982

[72] (with M. Gatesoupe). Quelques propriétés de l'algèbre de Fourier du groupe des entiers dyadiques. *Ark. Mat.*, 20(1):147–156.

[73] Nouvelles estimations pour les solutions d'équations aux dérivées partielles non linéaires. In *Goulaouic-Meyer-Schwartz Seminar, 1981/1982*, Exp. No. VI, 13 pp. École Polytech., Palaiseau.

[74] (with R. R. Coifman and A. McIntosh). L'intégrale de Cauchy définit un opérateur borné sur L^2 pour les courbes lipschitziennes. *Ann. of Math. (2)*, 116(2):361–387.

[75] (with R. Coifman and A. McIntosh). The Hilbert transform on Lipschitz curves. In *Miniconference on partial differential equations (Canberra, 1981)*, 26–69, *Proc. Centre Math. Anal. Austral. Nat. Univ.*, 1, Austral. Nat. Univ., Canberra.

1983

[76] (with R. R. Coifman and D. G. Deng). Domaine de la racine carrée de certains opérateurs différentiels accrétifs. *Ann. Inst. Fourier (Grenoble)*, 33(2):x, 123–134.

[77] (with R. R. Coifman and G. David). La solution des conjecture de Calderón. *Adv. in Math.*, 48(2):144–148.

[78] (with G. Bourdaud). Inégalités L^2 précisées pour la classe $S^0_{0,0}$. In *Seminar on Harmonic Analysis, 1981–1982*, 48–58, *Publ. Math. Orsay 83*, 2, Univ. Paris XI, Orsay.

[79] (with R. R. Coifman and E. M. Stein). Un nouvel éspace fonctionnel adapté à l'étude des opérateurs définis par des intégrales singulières. In *Harmonic analysis (Cortona, 1982)*, 1–15, *Lecture Notes in Math.*, 992, Springer, Berlin.

[80] (with R. R. Coifman and A. McIntosh). Estimations L^2 pour les noyaux singuliers. In *Conference on harmonic analysis in honor of Antoni Zygmund, Vol. I, II (Chicago, Ill., 1981)*, 287–294, Wadsworth Math. Ser., Wadsworth, Belmont, CA.

[81] Sur un problème de Michael Herman. In *Conference on harmonic analysis in honor of Antoni Zygmund, Vol. I, II (Chicago, Ill., 1981)*, 726–731, Wadsworth Math. Ser., Wadsworth, Belmont, CA.

[82] (with R. R. Coifman). L'analyse harmonique non linéaire. In *Topics in modern harmonic analysis, Vol. I, II (Turin/Milan, 1982)*, 707–721. Ist. Naz. Alta Mat. Francesco Severi, Rome.

1984

[83] Intégrales singulières, opérateurs multilinéaires, analyse complexe et équations aux derivées partielles. In *Proceedings of the International Congress of Mathematicians, Vol. 1, 2 (Warsaw, 1983)*, 1001–1010, PWN, Warsaw.

[84] Continuité sur les espaces de Hölder et de Sobolev des opérateurs définis par des intégrales singulières. In *Goulaouic-Meyer-Schwartz seminar, 1983–1984*, Exp. No. 1, 12 pp. École Polytech., Palaiseau.

1985

[85] (with R. R. Coifman and E. M. Stein). Some new function spaces and their applications to harmonic analysis. *J. Funct. Anal.*, 62(2):304–335.

[86] (with M. H. Taibleson and G. Weiss). Some functional analytic properties of the spaces B_q generated by blocks. *Indiana Univ. Math. J.*, 34(3):493–515.

[87] (with A. McIntosh). Algèbres d'opérateurs définis par des intégrales singulières. *C. R. Acad. Sci. Paris Sér. I Math.*, 301(8):395–397.

[88] (with R. R. Coifman). Nonlinear harmonic analysis and analytic dependence. In *Pseudodifferential operators and applications (Notre Dame, Ind., 1984)*, 71–78, *Proc. Sympos. Pure Math.*, 43, Amer. Math. Soc., Providence, RI.

[89] Real analysis and operator theory. In *Pseudodifferential operators and applications (Notre Dame, Ind., 1984)*, 219–235, *Proc. Sympos. Pure Math.*, 43, Amer. Math. Soc., Providence, RI.

[90] Le lemme de Cotlar et Stein et la continuité L^2 des opérateurs définis par des intégrales singulières. Colloquium in honor of Laurent Schwartz, Vol. 1 (Palaiseau, 1983). *Astérisque*, 131:115–125.

[91] Les nouveaux opérateurs de Calderón–Zygmund. Colloquium in honor of Laurent Schwartz, Vol. 1 (Palaiseau, 1983). *Astérisque*, 131:237–254.

[92] Minimalité de certains espaces fonctionnels et applications à la théorie des opérateurs. In *Bony–Sjöstrand–Meyer seminar, 1984–1985*, Exp. No. 3, 13 pp. École Polytech., Palaiseau.

[93] (with C. Kenig). Kato's square roots of accretive operators and Cauchy kernels on Lipschitz curves are the same. In *Recent progress in Fourier analysis (El Escorial, 1983)*, 123–143, *North-Holland Math. Stud.*, 111, Notas Mat., 101, North-Holland, Amsterdam.

[94] Continuité sur les espaces de Hölder et de Sobolev des opérateurs définis par des intégrales singulières. In *Recent progress in Fourier analysis (El Escorial, 1983)*, 145–172, *North-Holland Math. Stud.*, 111, Notas Mat., 101, North-Holland, Amsterdam.

[95] Étude d'un modèle mathématique issu du contrôle des structures spatiales déformables. In *Nonlinear partial differential equations and their applications. Collège de France seminar, Vol. VII (Paris, 1983–1984)*, 8–9, 234–242, *Res. Notes in Math.*, 122, Pitman, Boston, MA.

1986

[96] (with R. R. Coifman). A simple proof of a theorem by G. David and J.-L. Journé on singular integral operators. In *Probability theory and harmonic analysis (Cleveland, Ohio, 1983)*, 61–65, *Monogr. Textbooks Pure Appl. Math.*, 98, Dekker, New York.

[97] (with I. Daubechies and A. Grossmann). Painless nonorthogonal expansions. *J. Math. Phys.*, 27(5):1271–1283.

[98] (with R. R. Coifman). Nonlinear harmonic analysis, operator theory and P.D.E. In *Beijing lectures in harmonic analysis (Beijing, 1984)*, 3–45, *Ann. of Math. Stud.*, 112, Princeton Univ. Press, Princeton, NJ.

[99] (with P. G. Lemarié). Ondelettes et bases hilbertiennes. *Rev. Mat. Iberoamericana*, 2(1–2):1–18.

[100] De la recherche pétrolière à la géométrie des espaces de Banach en passant par les paraproduits. In *Séminaire sur les équations aux dérivées partielles, 1985–1986*, Exp. No. I, 11 pp. École Polytech., Palaiseau.

1987

[101] Principe d'incertitude, bases hilbertiennes et algèbres d'opérateurs. Séminaire Bourbaki, Vol. 1985/86. *Astérisque*, 145–146(4):209–223.
[102] Les ondelettes. In *Contributions to nonlinear partial differential equations, Vol. II (Paris, 1985)*, 158–171, *Pitman Res. Notes Math. Ser.*, 155, Longman Sci. Tech., Harlow.
[103] Ondelettes et fonctions splines. In *Séminaire sur les équations aux dérivées partielles 1986–1987*, Exp. No. VI, 18 pp. École Polytech., Palaiseau.
[104] Ondelettes, fonctions splines et analyses graduées. *Rend. Sem. Mat. Univ. Politec. Torino*, 45(1):1–42.

1988

[105] (with G. Bourdaud). Inégalités L^2 précisées pour la classe $S^0_{0,0}$. *Bull. Soc. Math. France*, 116(4):401–412.
[106] Constructions de bases orthonormées d'ondelettes. *Rev. Mat. Iberoamericana*, 4(1):31–39.

1989

[107] (with S. Jaffard). Bases d'ondelettes dans des ouverts de $\mathbf{R}^n$. *J. Math. Pures Appl. (9)*, 68(1):95–108.
[108] Ondelettes, filtres miroirs en quadrature et traitement numérique de l'image. *Gaz. Math.*, (40):31–42.
[109] Wavelets and operators. In *Analysis at Urbana, Vol. I (Urbana, IL, 1986–1987)*, 256–365, *London Math. Soc. Lecture Note Ser.*, 137, Cambridge Univ. Press, Cambridge.
[110] Orthonormal wavelets. In *Wavelets (Marseille, 1987)*, 21–37, Inverse Probl. Theoret. Imaging, Springer, Berlin.
[111] (with R. Coifman, G. David, and S. Semmes). ω-Calderón–Zygmund operators. In *Harmonic analysis and partial differential equations (El Escorial, 1987)*, 132–145, *Lecture Notes in Math.*, 1384, Springer, Berlin.
[112] (with S. Jaffard). Les ondelettes. In *Harmonic analysis and partial differential equations (El Escorial, 1987)*, 182–192, *Lecture Notes in Math.*, 1384, Springer, Berlin.
[113] Orthonormal wavelets. In *IXth International Congress on Mathematical Physics (Swansea, 1988)*, 38–47, Hilger, Bristol.
[114] (with R. Coifman, P.-L. Lions, and S. Semmes). Compacité par compensation et espaces de Hardy. *C. R. Acad. Sci. Paris Sér. I Math.*, 309(18):945–949.

1990

[115] (with R. Coifman, P.-L. Lions, and S. Semmes). Compacité par compensation et espaces de Hardy. In *Séminaire sur les Équations aux Dérivées Partielles, 1989–1990*, Exp. No. XIV, 10 pp. École Polytech., Palaiseau.

[116] Ondelettes, filtres miroirs en quadrature et traitement numérique de l'image. In *Les ondelettes en 1989 (Orsay, 1989)*, 14–25, 196–197, *Lecture Notes in Math.*, 1438, Springer, Berlin.

[117] *Ondelettes et opérateurs. I.* Actualités Mathématiques. Hermann, Paris. xii+215 pp.

[118] *Ondelettes et opérateurs. II.* Actualités Mathématiques. Hermann, Paris. pp. i-xii and 217–384.

[119] Construction de bases orthonormées d'ondelettes. *Colloq. Math.*, 60/61(1):141–149.

1991

[120] (with R. R. Coifman). Remarques sur l'analyse de Fourier à fenêtre. *C. R. Acad. Sci. Paris Sér. I Math.*, 312(3):259–261.

[121] Orthonormal wavelets. In *Frontiers in pure and applied mathematics*, 235–245. North-Holland, Amsterdam.

[122] (with G. Bourdaud). Fonctions qui opèrent sur les espaces de Sobolev. *J. Funct. Anal.*, 97(2):351–360.

[123] (with R. J. DiPerna and P.-L. Lions). L^p regularity of velocity averages. *Ann. Inst. H. Poincaré Anal. Non Linéaire*, 8(3–4):271–287.

[124] Ondelettes sur l'intervalle. *Rev. Mat. Iberoamericana*, 7(2):115–133.

[125] Wavelets and applications. In *Proceedings of the International Congress of Mathematicians, Vol. I, II (Kyoto, 1990)*, 1619–1626. Math. Soc. Japan, Tokyo.

[126] (with R. R. Coifman). *Ondelettes et opérateurs. III.* Actualités Mathématiques. pp. i–xii and 383–538. Hermann, Paris.

1992

[127] *Wavelets and operators.* Translated from the 1990 French original [117] by D. H. Salinger. *Cambridge Studies in Advanced Mathematics*, 37. Cambridge University Press, Cambridge. xvi+224 pp.

[128] (with R. R. Coifman). Analyse de Fourier adaptée à une partition par des cubes de Whitney. *Colloq. Math.*, 63(1):111–117.

[129] (with R. Coifman and V. Wickerhauser). Adapted wave form analysis, wavelet-packets and applications. In *ICIAM 91 (Washington, DC, 1991)*, 41–50. SIAM, Philadelphia, PA.

[130] (with R. R. Coifman and V. Wickerhauser). Wavelet analysis and signal processing. In *Wavelets and their applications*, 153–178. Jones and Bartlett, Boston, MA.

[131] (with R. R. Coifman and V. Wickerhauser). Size properties of wavelet-packets. In *Wavelets and their applications*, 453–470. Jones and Bartlett, Boston, MA.

[132] *Ondelettes et algorithmes concurrents, Actualités Scientifiques et Industrielles*, 1435. Hermann, Paris.

[133] (with S. Dobyinsky). Lemme div-curl et renormalisation. In *Séminaire sur les Équations aux Dérivées Partielles, 1991–1992*, Exp. No. II, 6 pp. École Polytech., Palaiseau.

1993

[134] *Wavelets.* Algorithms & applications. Society for Industrial and Applied Mathematics (SIAM), Philadelphia, PA. xii+133 pp. Translated from the French and with a foreword by Robert D. Ryan.

[135] (with R. Coifman, P.-L. Lions, and S. Semmes). Compensated compactness and Hardy spaces. *J. Math. Pures Appl. (9)*, 72(3):247–286.

[136] (with F. Paiva). Remarques sur la construction des ondelettes orthogonales. *J. Anal. Math.*, 60:227–240.

[137] Wavelets and operators. In *Different perspectives on wavelets (San Antonio, TX, 1993)*, volume 47 of *Proc. Sympos. Appl. Math.*, pages 35–58. Amer. Math. Soc., Providence, RI.

[138] Wavelets: their past and their future. In *Progress in wavelet analysis and applications (Toulouse, 1992)*, 9–18. Frontières, Gif-sur-Yvette.

[139] (with R. R. Coifman, S. Quake, and M. V. Wickerhauser). Signal processing and compression with wavelet packets. In *Progress in wavelet analysis and applications (Toulouse, 1992)*, 77–93. Frontières, Gif-sur-Yvette.

1994

[140] (with M. Cannone and F. Planchon). Solutions auto-similaires des équations de Navier-Stokes. In *Séminaire sur les Équations aux Dérivées Partielles, 1993–1994*, Exp. No. VIII, 12 pp. École Polytech., Palaiseau.

1995

[141] (with R. R. Coifman and S. Dobyinsky). Opérateurs bilinéaires et renormalisation. In *Essays on Fourier analysis in honor of Elias M. Stein (Princeton, NJ, 1991)*, 146–161, *Princeton Math. Ser.*, 42, Princeton Univ. Press, Princeton, NJ.

[142] (with R. R. Coifman and V. Wickerhauser). Numerical harmonic analysis. In *Essays on Fourier analysis in honor of Elias M. Stein (Princeton, NJ, 1991)*, 162–174, *Princeton Math. Ser.*, 42, Princeton Univ. Press, Princeton, NJ.

[143] (with R. R. Coifman). Gaussian bases. *Appl. Comput. Harmon. Anal.*, 2(3):299–302.

[144] Wavelet analysis, local Fourier analysis, and 2-microlocalization. In *Harmonic analysis and operator theory (Caracas, 1994)*, 393–401, *Contemp. Math.*, 189. Amer. Math. Soc., Providence, RI.

[145] (with M. Cannone). Littlewood–Paley decomposition and Navier–Stokes equations. *Methods Appl. Anal.*, 2(3):307–319.

[146] Chirps n-dimensionnels et analyse 2-microlocale. In *Séminaire sur les Équations aux Dérivées Partielles, 1994–1995*, Exp. No. I, 12 pp. École Polytech., Palaiseau.

[147] Quasicrystals, Diophantine approximation and algebraic numbers. In *Beyond quasicrystals (Les Houches, 1994)*, 3–16. Springer, Berlin.

[148] Problèmes de l'unicité, de la synthèse et des isomorphismes en analyse harmonique. In *Séminaire Bourbaki, Vol. 10*, Exp. No. 341, 463–471. Soc. Math. France, Paris.

1996

[149] (with S. Jaffard). Wavelet methods for pointwise regularity and local oscillations of functions. *Mem. Amer. Math. Soc.*, 123(587), x+110.

[150] (with Y.-S. Han). A characterization of Hilbert spaces and the vector-valued Littlewood–Paley theorem. *Methods Appl. Anal.*, 3(2):228–234.

[151] (with L. Auslander). A generalized Poisson summation formula. *Appl. Comput. Harmon. Anal.*, 3(4):372–376.

1997

[152] (with R. Coifman and G. Matviyenko). Modulated Malvar–Wilson bases. *Appl. Comput. Harmon. Anal.*, 4(1):58–61.

[153] (with R. Coifman). *Wavelets*. Calderón-Zygmund and multilinear operators. *Cambridge Studies in Advanced Mathematics*, 48. Cambridge University Press, Cambridge. xx+315 pp. Translated from the 1990 and 1991 French originals ([118], [126]) by David Salinger.

[154] Wavelets and fast numerical algorithms. In *Handbook of numerical analysis, Vol. V*, 639–713, Handb. Numer. Anal., V, North-Holland, Amsterdam.

[155] (with H. Xu). Wavelet analysis and chirps. *Appl. Comput. Harmon. Anal.*, 4(4):366–379.

[156] (with P. Gerard and F. Oru). Inégalités de Sobolev précisées. In *Séminaire sur les Équations aux Dérivées Partielles, 1996–1997*, Exp. No. IV, 11 pp. École Polytech., Palaiseau.

[157] Wavelets, paraproducts, and Navier–Stokes equations. In *Current developments in mathematics, 1996 (Cambridge, MA)*, 105–212, Int. Press, Boston, MA.

1998

[158] *Wavelets, vibrations and scalings*, *CRM Monograph Series*, 9. Amer. Math. Soc., Providence, RI. x+133 pp.

[159] (with K. Daoudi and J. Lévy Véhel). Construction of continuous functions with prescribed local regularity. *Constr. Approx.*, 14(3):349–385.

[160] (with A. Cohen and F. Oru). Improved Sobolev embedding theorem. In *Séminaire sur les Équations aux Dérivées Partielles, 1997–1998*, Exp. No. XVI, 16 pp. École Polytech.,

1999

[161] Large-time behavior and self-similar solutions of some semilinear diffusion equations. In *Harmonic analysis and partial differential equations (Chicago, IL, 1996)*, 241–261, Chicago Lectures in Math., Univ. Chicago Press, Chicago, IL.
[162] (with F. Sellan and M. S. Taqqu). Wavelets, generalized white noise and fractional integration: the synthesis of fractional Brownian motion. *J. Fourier Anal. Appl.*, 5(5):465–494.

2000

[163] Le traitement du signal et l'analyse mathématique. *Ann. Inst. Fourier (Grenoble)*, 50(2):593–632.
[164] (with S. Jaffard). On the pointwise regularity of functions in critical Besov spaces. *J. Funct. Anal.*, 175(2):415–434.
[165] Wavelets and functions with bounded variation from image processing to pure mathematics. *Atti Accad. Naz. Lincei Cl. Sci. Fis. Mat. Natur. Rend. Lincei (9) Mat. Appl.*, Special Issue, 77–105.

2001

[166] (with R. Coifman and F. Geshwind). Noiselets. *Appl. Comput. Harmon. Anal.*, 10(1):27–44.
[167] (with S. Jaffard and R. D. Ryan). *Wavelets*. Society for Industrial and Applied Mathematics (SIAM), Philadelphia, PA. xiv+256 pp. Revised edition.
[168] *Oscillating patterns in image processing and nonlinear evolution equations*, *University Lecture Series*, 22. Amer. Math. Soc., Providence, RI. x+122 pp.
[169] The role of oscillations in some nonlinear problems. In *European Congress of Mathematics, Vol. I (Barcelona, 2000)*, 75–99, *Progr. Math.*, 201, Birkhäuser, Basel.

2002

[170] Why and how does one do mathematical research? *Gac. R. Soc. Mat. Esp.*, 5(1):57–61.
[171] (with L. Brandolese). On the instantaneous spreading for the Navier–Stokes system in the whole space. *ESAIM Control Optim. Calc. Var.*, 8:273–285.

2003

[172] (with T. Rivière). A partial regularity result for a class of stationary Yang–Mills fields in high dimension. *Rev. Mat. Iberoamericana*, 19(1):195–219.
[173] La conjecture de Kato (d'après Pascal Auscher, Steve Hofmann, Michael Lacey, John Lewis, Alan McIntosh et Philippe Tchamitchian). Séminaire Bourbaki, Vol. 2001/2002, Exp. No. 902. *Astérisque*, 290:viii, 193–206.

[174] (with J.-P. Kahane, P. Cartier, V. I. Arnold, J. Coates, D. Zagier, H. Brézis, F. Pham, B. Malgrange, H. Helson, Y. Colin de Verdière, and P.-L. Lions). *Leçons de mathématiques d'aujourd'hui*, Vol. 1, *Le Sel et le Fer*, 4. Cassini, Paris, second edition. xvi+322 pp.

2004

[175] Guy David, lauréat du Prix Servant et du Prix Sunyer i Balaguer. *Gaz. Math.*, 101:84–91.

[176] Jean Leray et la recherche de la vérité. In *Actes des Journées Mathématiques à la Mémoire de Jean Leray*, 1–12, *Sémin. Congr.*, 9, Soc. Math. France, Paris.

2005

[177] Compression des images fixes. *Gaz. Math.*, (103):9–23.

[178] (with E. Wesfreid and V. L. Billat). Multifractal analysis of heartbeat time series in human races. *Appl. Comput. Harmon. Anal.*, 18(3):329–335.

2006

[179] Oscillating patterns in some nonlinear evolution equations. In *Mathematical foundation of turbulent viscous flows*, 101–187, *Lecture Notes in Math.*, 1871, Springer, Berlin.

[180] From wavelets to atoms. In *150 years of mathematics at Washington University in St. Louis*, 105–117, *Contemp. Math.*, 395, Amer. Math. Soc., Providence, RI.

[181] Comment mesurer les surfaces? *Gaz. Math.*, (109):23–36.

[182] (with J. Azéma, C. Dellacherie, C. Doléans-Dade, M. Emery, Y. Le Jan, B. Maisonneuve, J. Neveu, N. Privault, and D. Revuz). Témoignages. In *In memoriam Paul-André Meyer: Séminaire de Probabilités XXXIX*, 35–46, *Lecture Notes in Math.*, 1874, Springer, Berlin.

[183] (with G. Bourdaud). Le calcul fonctionnel sous-linéaire dans les espaces de Besov homogènes. *Rev. Mat. Iberoam.*, 22(2):725–746, loose erratum.

2007

[184] (with A. Haddad). An improvement of Rudin–Osher–Fatemi model. *Appl. Comput. Harmon. Anal.*, 22(3):319–334.

[185] Traitement du signal. *Gaz. Math.*, (112):28–30.

[186] (with J. B. Garnett, T. M. Le, and L. A. Vese). Image decompositions using bounded variation and generalized homogeneous Besov spaces. *Appl. Comput. Harmon. Anal.*, 23(1):25–56.

[187] (with A. Haddad). Variational methods in image processing. In *Perspectives in nonlinear partial differential equations*, 273–295, *Contemp. Math.*, 446, Amer. Math. Soc., Providence, RI.

2008

[188] (with Q. X. Yang). Continuity of Calderón–Zygmund operators on Besov or Triebel–Lizorkin spaces. *Anal. Appl. (Singap.)*, 6(1):51–81.

[189] Complex analysis and operator theory in Alberto Calderón's work. In *Selected papers of Alberto P. Calderón*, 593–606. Amer. Math. Soc., Providence, RI.

[190] (with B. Matei). Quasicrystals are sets of stable sampling. *C. R. Math. Acad. Sci. Paris*, 346(23–24):1235–1238.

2009

[191] (with B. Matei). A variant of compressed sensing. *Rev. Mat. Iberoam.*, 25(2):669–692.

2010

[192] (with M. Lanza de Cristoforis). In honour of Gérard Bourdaud. *Complex Var. Elliptic Equ.*, 55(8–10):691–692.

[193] (with B. Matei). Simple quasicrystals are sets of stable sampling. *Complex Var. Elliptic Equ.*, 55(8–10):947–964.

[194] (with J. Gilles). Properties of BV $- G$ structures + textures decomposition models. Application to road detection in satellite images. *IEEE Trans. Image Process.*, 19(11):2793–2800.

2011

[195] Pattern theory. *Eur. Math. Soc. Newsl.*, (81):54–56.

2012

[196] Quasicrystals, almost periodic patterns, mean-periodic functions and irregular sampling. *Afr. Diaspora J. Math.*, 13(1):1–45. Addendum, *ibid.*, 17(1):65–74, 2014.

[197] Souvenirs de Rauzy. *Gaz. Math.*, (132):91–92.

2014

[198] (with B. Matei and J. Ortega-Cerdà). Stable sampling and Fourier multipliers. *Publ. Mat.*, 58(2):341–351.

2015

[199] Quasicrystals and control theory. In *Analysis and geometry*, 219–229, *Springer Proc. Math. Stat.*, 127, Springer, Cham.

2016

[200] Measures with locally finite support and spectrum. *Proc. Natl. Acad. Sci. USA*, 113(12):3152–3158.

[201] (with E. Chassande-Mottin and S. Jaffard). Des ondelettes pour détecter les ondes gravitationnelles. *Gaz. Math.*, 148:61–64.

2017

[202] (with J. I. Díaz). Poisson summation formulae and the wave equation with a finitely supported measure as initial velocity. *Afr. Diaspora J. Math.*, 20(1): 1–13.

[203] Guinand's measures are almost periodic distributions. *Bull. Hellenic Math. Soc.*, 61:11–20.

[204] Measures with locally finite support and spectrum. *Rev. Mat. Iberoam.*, 33(3):1025–1036.

2018

[205] Mean-periodic functions and irregular sampling. *Transactions of the Royal Norwegian Society of Sciences and Letters*, (1):19 pp.

[206] Global and local estimates on trigonometric sums. *Transactions of the Royal Norwegian Society of Sciences and Letters*, (2):25 pp.

Curriculum Vitae for Yves F. Meyer

Born:	July 19, 1939 in Paris, France
Degrees/education:	Agrégation, École Normale Supérieure, 1960
	PhD, Université de Strasbourg, 1967
Positions:	Teacher, Prytanée National Militaire, 1960–1963
	Teaching Assistant, Université de Strasbourg, 1963–1966
	Professor, Université Paris-Sud, 1966–1980
	Professor, École Polytechnique, 1980–1986
	Professor, Université Paris-Dauphine, 1986–1995
	Senior Researcher, Centre national de la recherche scientifique, 1995–1999
	Professor, École Normale Supérieure de Cachan, 1999–2003
	Professor emeritus, École Normale Supérieure de Cachan, 2004–

H. Holden, R. Piene (eds.), *The Abel Prize 2013–2017*,
https://doi.org/10.1007/978-3-319-99028-6_32

Memberships:	Académie des Sciences, 1993
	American Academy of Arts and Sciences, foreign honorary member, 1994
	American Mathematical Society, Fellow, 2013
	National Academy of Sciences, foreign associate, 2014
	Norwegian Academy of Science and Letters, 2017
	Real Academia de Ciencias Exactas, Físicas y Naturales, foreign member, 2018
Awards and prizes:	Salem Prize, 1970
	Prix Carrière, Académie des Sciences, 1972
	Prix de l'État, Académie des Sciences, 1984
	Carl Friedrich Gauss Prize, 2000
	Abel Prize, 2017
Honorary degrees:	Universidad Autonoma de Madrid, 1997
	École Polytechnique Fédérale de Lausanne, 2017

Part VI
Abel Activities

The Abel Prize. (Photo: Calle Huth/Studio)

(2006) Each year, during the week of the Abel Prize ceremony, the main street in Oslo, Karl Johans gate, is decorated with Abel banners. (Photo: Harald Hanche-Olsen)

(2007) The Abel Laureates Lennart Carleson, Srinivasa Varadhan, and Peter Lax in front of the Abel Monument in the park of the Royal Castle in Oslo. (Photo: Heiko Junge/Scanpix)

(2016) Crown Prince Haakon arriving at the University Aula for the Award Ceremony. (Photo: Yngve Vogt, UiO)

(2003) The Abel Laureates are invited to a private audience at the Royal Palace. King Harald and Queen Sonja receiving Jean-Pierre Serre. (Photo: NTB Scanpix)

(2012) The Award Ceremony in the Aula of the University of Oslo. The wall paintings are by Edvard Munch. (Photo: Erlend Aas/NTB Scanpix)

(2012) The Award Ceremony for Endre Szemerédi in the University Aula. (Photo: Harald Hanche-Olsen)

(2009) Abel Laureate Mikhail Gromov with Minister of research and higher education Tora Aasland. (Photo: Knut Falch/Scanpix)

(2013) King Harald and Abel Laureate Pierre Deligne at the banquet in Akershus Castle. (Photo: Fredrik Varfjell/NTB Scanpix)

(2016) The banquet at Akershus Castle. (Photo: Audun Braastad)

(2015) John Nash giving his Abel Lecture at the University of Oslo. (Photo: Eirik Furu Baardsen)

(2015) Nash and Nirenberg at the Abel lectures in Bergen. (Photo: Anne-Marie Astad)

(2017) From the Abel Lectures at the University of Oslo. (Photo: Harald Hanche-Olsen)

(2017) The Abel Lecturers: Stéphane Mallat, Yves Meyer, Ingrid Daubechies, and Emmanuel Candès. (Photo: Ola Gamst Sæther)

(2012) From the party at the Norwegian Academy of Science and Letters in the evening of the day of the Abel Lectures. (Photo: Harald Hanche-Olsen)

(2017) Nadia Hasnaoui interviews Abel Laureate Yves Meyer at Det Norske Teatret right after the prize ceremony. (Photo: Eirik Furu Baardsen)

(2017) Terrence Tao discussing the work of Abel Laureate Yves Meyer with Eldrid Borgan at the House of Literature in Oslo. (Photo: Eirik Furu Baardsen)

(2005) Lax with children in Bergen. (Photo: NTB/Scanpix)

(2010) Tate with children in Kristiansand. (Photo: UiA/Tor Martin Lien)

(2016) Wiles with children in Kristiansand. (Photo: Anne-Marie Astad)

(2011) Sigbjørn Hals, after having received the Bernt Michael Holmboe Memorial Prize, with Abel Laureate Milnor and Minister of research and higher education Kristin Halvorsen, at the Cathedral School in Oslo. (Photo: Stian Lysberg Solum/Scanpix)

(2017) The recipient of the Bernt Michael Holmboe Memorial Prize, the mathematics teacher Hanan Abdelrahman, with Abel Laureate Yves Meyer. (Photo: Håkon Mosvold Larsen/NTB)

(2014) The winner of the national Niels Henrik Abel competition, Johan Sokrates Wind, Kongsbakken vgs, with Prime Minister Erna Solberg (left) and CEO of Petroleum Geo-Services Jon Erik Reinhardsen. (Photo: Henrik Fjørtoft/NTNU Komm. avd.)

(2016) The winning team of UngeAbel. (Photo: Gro Berg)

(2016) From the final of the national Niels Henrik Abel competition, held at the Norwegian University of Science and Technology. (Photo: Harald Hanche-Olsen)

(2012) The signing of the agreement regarding the Heidelberg Laureate Forum between the Klaus Tschira Foundation and Norwegian Academy of Science and Letters. In the front Andreas Reuter (left) and Klaus Tschira (right). In the back, from left to right, Nils Christian Stenseth, Fabrizio Gagliardi, Ingrid Daubechies, Helge Holden, Øivind Andersen, Detlev Rünger. (Photo: Eirik Furu Baardsen)

(2014) Abel Laureate Sir Michael Atiyah with students at the Heidelberg Laureate Forum. (Photo: Christian Flemming)

(2007) The winner of the 2006 Ramanujan Prize, Sujatha Ramdorai, with Prime Minister Stoltenberg and Abel Laureate Srinivasa Varadhan at the banquet at Akershus Castle. (Photo: Heiko Junge/Scanpix)

(2007) Alf Bjørseth hands over the Abel manuscripts to the director of the National Library, Vigdis Skarstein. (Photo: Heiko Junge/Scanpix)

(2004) The participants of the first Abel symposium. (Photo: Samfoto/Svein Erik Dahl)

(2017) Abel symposium participants on their way to Svolvær. (Photo: Martin Gulbrandsen)

(2011) Signing of an agreement of cooperation between the National Institute for Mathematical Sciences in Kumasi, Ghana, The Norwegian Academy of Science and Letters, and Petroleum Geo-Services. (Photo: Trine Gerlyng)

(2014) The unveiling of the Abel Plaque in Berlin. From the left: Jürg Kramer, Martin Grötschel, Erika Kappel, Sven Erik Svedman og Øivind Andersen. (Photo: Øyvind R. Haugen)

Abel Prize Citations 2003–2012

2003—Jean-Pierre Serre

The Norwegian Academy of Science and Letters has decided to award the Abel Prize for 2003 to **Jean-Pierre Serre**, Collège de France, Paris, France,

> for playing a key role in shaping the modern form of many parts of mathematics, including topology, algebraic geometry and number theory

The first Abel Prize has been awarded to Jean-Pierre Serre, one of the great mathematicians of our time. Serre is an Emeritus Professor at the Collège de France in Paris. He has made profound contributions to the progress of mathematics for over half a century, and continues to do so. Serre's work is of extraordinary breadth, depth and influence. He has played a key role in shaping the modern form of many parts of mathematics, including:

- Topology, which treats the question: what remains the same in geometry even when the length is distorted?

H. Holden, R. Piene (eds.), *The Abel Prize 2013–2017*,
https://doi.org/10.1007/978-3-319-99028-6_33

- Algebraic geometry, which treats the question: what is the geometry of solutions of polynomial equations?
- Number theory, the study of basic properties of numbers. For example prime numbers and the solution of polynomial equations as in Fermat's Last Theorem.

Serre developed revolutionary algebraic methods for studying topology, and in particular studied the transformations between spheres of higher dimensions. He is responsible for a spectacular clarification of the work of the Italian algebraic geometers by introducing and developing the right algebraic machinery for determining when their geometric construction worked. This powerful technique of Serre, with its new language and viewpoint, ushered in a golden age for algebraic geometry.

For the past four decades Serre's magnificent work and vision of number theory have been instrumental in bringing that subject to its current glory. This work connects and extends in many ways the mathematical ideas introduced by Abel, in particular his proof of the impossibility of solving the 5th degree equation by radicals, and his analytic techniques for the study of polynomial equations in two variables. Serre's research has been vital in setting the stage for many of the most celebrated recent breakthroughs, including the proof by Wiles of Fermat's Last Theorem.

Although Serre's effort has been directed to more conceptual mathematics, his contributions have connection to important applications. The practical issues of finding efficient error-correcting codes and of public-key cryptography, both make use of solutions of polynomial equations (specifically over finite fields) and Serre's work has substantially deepened our understanding of this topic.

2004—Sir Michael Francis Atiyah and Isadore M. Singer

The Norwegian Academy of Science and Letters has decided to award the Abel Prize for 2004 to **Sir Michael Francis Atiyah**, University of Edinburgh, and **Isadore M. Singer**, Massachusetts Institute of Technology,

> for their discovery and proof of the index theorem, bringing together topology, geometry and analysis, and their outstanding role in building new bridges between mathematics and theoretical physics

The second Abel Prize has been awarded jointly to Michael Francis Atiyah and Isadore M. Singer. The Atiyah–Singer index theorem is one of the great landmarks of twentieth-century mathematics, influencing profoundly many of the most important later developments in topology, differential geometry and quantum field theory. Its authors, both jointly and individually, have been instrumental in repairing a rift between the worlds of pure mathematics and theoretical particle physics, initiating a cross-fertilization which has been one of the most exciting developments of the last decades.

We describe the world by measuring quantities and forces that vary over time and space. The rules of nature are often expressed by formulas involving their rates of change, so-called differential equations. Such formulas may have an "index", the number of solutions of the formulas minus the number of restrictions which they impose on the values of the quantities being computed. The index theorem calculates this number in terms of the geometry of the surrounding space.

A simple case is illustrated by a famous paradoxical etching of M.C. Escher, "Ascending and Descending", where the people, going uphill all the time, still manage to circle the castle courtyard. The index theorem would have told them this was impossible!

The Atiyah–Singer index theorem was the culmination and crowning achievement of a more than one-hundred-year-old evolution of ideas, from Stokes's theorem, which students learn in calculus classes, to sophisticated modern theories like Hodge's theory of harmonic integrals and Hirzebruch's signature theorem.

The problem solved by the Atiyah–Singer theorem is truly ubiquitous. In the 40 years since its discovery, the theorem has had innumerable applications, first in mathematics and then, beginning in the late 1970s, in theoretical physics: gauge theory, instantons, monopoles, string theory, the theory of anomalies etc.

At first, the applications in physics came as a complete surprise to both the mathematics and physics communities. Now the index theorem has become an integral part of their cultures. Atiyah and Singer, together and individually, have been tireless in their attempts to explain the insights of physicists to mathematicians. At the same time, they brought modern differential geometry and analysis as it applies to quantum field theory to the attention of physicists and suggested new directions in physics itself. This cross-fertilization continues to fruitful for both sciences.

Michael Francis Atiyah and Isadore M. Singer are among the most influential mathematicians of the last century and are still working. With the index theorem they changed the landscape of mathematics. Over a period of 20 years they worked together on the index theorem and its ramifications.

Atiyah and Singer came originally from different fields of mathematics: Atiyah from algebraic geometry and topology, Singer from analysis. Their main contributions in their respective areas are also highly recognized. Atiyah's early work on meromorphic forms on algebraic varieties and his important 1961 paper on Thom complexes are such examples. Atiyah's pioneering work together with Friedrich Hirzebruch on the development of the topological analogue of Grothendieck's K-theory had numerous applications to classical problems of topology and turned out later to be deeply connected with the index theorem.

Singer established the subject of triangular operator algebras (jointly with Richard V. Kadison). Singer's name is also associated with the Ambrose–Singer holonomy theorem and the Ray–Singer torsion invariant. Singer, together with Henry P. McKean, pointed out the deep geometrical information hidden in heat kernels, a discovery that had great impact.

2005—Peter D. Lax

The Norwegian Academy of Science and Letters has decided to award the Abel Prize for 2005 to **Peter D. Lax**, Courant Institute of Mathematical Sciences, New York University, for his groundbreaking contributions to the theory and application of partial differential equations and to the computation of their solutions

Ever since Newton, differential equations have been the basis for the scientific understanding of nature. Linear differential equations, in which cause and effect are directly proportional, are reasonably well understood. The equations that arise in such fields as aerodynamics, meteorology and elasticity are nonlinear and much more complex: their solutions can develop singularities. Think of the shock waves that appear when an airplane breaks the sound barrier.

In the 1950s and 1960s, Lax laid the foundations for the modern theory of nonlinear equations of this type (hyperbolic systems). He constructed explicit solutions, identified classes of especially well-behaved systems, introduced an important notion of entropy, and, with Glimm, made a penetrating study of how solutions behave over a long period of time. In addition, he introduced the widely used Lax–Friedrichs and Lax–Wendroff numerical schemes for computing solutions. His work in this area was important for the further theoretical developments. It has also been extraordinarily fruitful for practical applications, from weather prediction to airplane design.

Another important cornerstone of modern numerical analysis is the "Lax Equivalence Theorem". Inspired by Richtmyer, Lax established with this theorem the conditions under which a numerical implementation gives a valid approximation to the solution of a differential equation. This result brought enormous clarity to the subject.

A system of differential equations is called "integrable" if its solutions are completely characterized by some crucial quantities that do not change in time. A classical example is the spinning top or gyroscope, where these conserved quantities are energy and angular momentum.

Integrable systems have been studied since the nineteenth century and are important in pure as well as applied mathematics. In the late 1960s a revolution occurred when Kruskal and co-workers discovered a new family of examples, which have "soliton" solutions: single-crested waves that maintain their shape as they travel. Lax became fascinated by these mysterious solutions and found a unifying concept for understanding them, rewriting the equations in terms of what are now called "Lax pairs". This developed into an essential tool for the whole field, leading to new constructions of integrable systems and facilitating their study.

Scattering theory is concerned with the change in a wave as it goes around an obstacle. This phenomenon occurs not only for fluids, but also, for instance, in atomic physics (Schrödinger equation). Together with Phillips, Lax developed a broad theory of scattering and described the long-term behaviour of solutions (specifically, the decay of energy). Their work also turned out to be important in fields of mathematics apparently very distant from differential equations, such as number theory. This is an unusual and very beautiful example of a framework built for applied mathematics leading to new insights within pure mathematics.

Peter D. Lax has been described as the most versatile mathematician of his generation. The impressive list above by no means states all of his achievements.

His use of geometric optics to study the propagation of singularities inaugurated the theory of Fourier Integral Operators. With Nirenberg, he derived the definitive Gårding-type estimates for systems of equations. Other celebrated results include the Lax–Milgram lemma and Lax's version of the Phragmén–Lindelöf principle for elliptic equations. Peter D. Lax stands out in joining together pure and applied mathematics, combining a deep understanding of analysis with an extraordinary capacity to find unifying concepts. He has had a profound influence, not only by his research, but also by his writing, his lifelong commitment to education and his generosity to younger mathematicians.

2006—Lennart Carleson

The Norwegian Academy of Science and Letters has decided to award the Abel Prize for 2006 to **Lennart Carleson**, Royal Institute of Technology, Sweden,

> for his profound and seminal contributions to harmonic analysis and the theory of smooth dynamical systems

In 1807, the versatile mathematician, engineer and Egyptologist Jean Baptiste Joseph Fourier made the revolutionary discovery that many phenomena, ranging from the typical profiles describing the propagation of heat through a metal bar to the vibrations of violin strings, can be viewed as sums of simple wave patterns called sines and cosines. Such summations are now called Fourier series. Harmonic analysis is the branch of mathematics that studies these series and similar objects.

For more than 150 years after Fourier's discovery, no adequate formulation and justification was found of his claim that every function equals the sum of its Fourier series. In hindsight this loose statement should be interpreted as regarding every function for which "it is possible to draw the graph", or more precisely, every continuous function. Despite contributions by several mathematicians, the problem remained open.

In 1913 it was formalized by the Russian mathematician Lusin in the form of what became known as Lusin's conjecture. A famous negative result of Kolmogorov in 1926, together with the lack of any progress, made experts believe that it would only be a matter of time before someone constructed a continuous function for which the sum of its Fourier series failed to give the function value anywhere. In 1966, to the surprise of the mathematical community, Carleson broke the decades-long impasse by proving Lusin's conjecture that every square-integrable function, and thus in particular every continuous function, equals the sum of its Fourier series "almost everywhere".

The proof of this result is so difficult that for over 30 years it stood mostly isolated from the rest of harmonic analysis. It is only within the past decade that mathematicians have understood the general theory of operators into which this theorem fits and have started to use his powerful ideas in their own work.

Carleson has made many other fundamental contributions to harmonic analysis, complex analysis, quasi-conformal mappings, and dynamical systems. Standing out among them is his solution of the famous corona problem, so called because it examines structures that become apparent "around" a disk when the disk itself is "obscured", poetically analogous to the corona of the sun seen during an eclipse. In this work he introduced what has become known as Carleson measures, now a fundamental tool of both complex and harmonic analysis.

The influence of Carleson's original work in complex and harmonic analysis does not limit itself to this. For example, the Carleson–Sjölin theorem on Fourier multipliers has become a standard tool in the study of the "Kakeya problem", the prototype of which is the "turning needle problem": how can we turn a needle 180° in a plane, while sweeping as little area as possible? Although the Kakeya problem originated as a toy, the description of the volume swept in the general case turns out to contain important and deep clues about the structure of Euclidean space.

Dynamical systems are mathematical models that seek to describe the behaviour in time of large classes of phenomena, such as those observed in meteorology, financial markets, and many biological systems, from fluctuations in fish populations to epidemiology. Even the simplest dynamical systems can be mathematically surprisingly complex. With Benedicks, Carleson studied the Hénon map, a dynamical system first proposed in 1976 by the astronomer Michel Hénon, a simple system exhibiting the intricacies of weather dynamics and turbulence. This system was generally believed to have a so-called strange attractor, drawn in beautiful detail by computer graphics tools, but poorly understood mathematically. In a great tour de force, Benedicks and Carleson provided the first proof of the existence of this strange attractor in 1991; this development opened the way to a systematic study of this class of dynamical systems.

Carleson's work has forever altered our view of analysis. Not only did he prove extremely hard theorems, but the methods he introduced to prove them have turned out to be as important as the theorems themselves. His unique style is characterized by geometric insight combined with amazing control of the branching complexities of the proofs.

2007—Srinivasa S. R. Varadhan

The Norwegian Academy of Science and Letters has decided to award the Abel Prize for 2007 to **Srinivasa S. R. Varadhan**, Courant Institute of Mathematical Sciences, New York,

> for his fundamental contributions to probability theory and in particular for creating a unified theory of large deviations

Probability theory is the mathematical tool for analyzing situations governed by chance. The law of large numbers, discovered by Jacob Bernoulli in the eighteenth century, shows that the average outcome of a long sequence of coin tosses is usually close to the expected value. Yet the unexpected happens, and the question is: how? The theory of large deviations studies the occurrence of rare events. This subject has concrete applications to fields as diverse as physics, biology, economics, statistics, computer science, and engineering.

The law of large numbers states that the probability of a deviation beyond a given level goes to zero. However, for practical applications, it is crucial to know how fast it vanishes. For example, what capital reserves are needed to keep the probability of default of an insurance company below acceptable levels? In analyzing such actuarial "ruin problems", Harald Cramér discovered in 1937 that standard approximations based on the Central Limit Theorem (as visualized by the bell curve) are actually misleading. He then found the first precise estimates of large deviations for a sequence of independent random variables. It took 30 years before Varadhan discovered the underlying general principles and began to demonstrate their tremendous scope, far beyond the classical setting of independent trials.

In his landmark paper "Asymptotic probabilities and differential equations" in 1966 and his surprising solution of the polaron problem of Euclidean quantum field theory in 1969, Varadhan began to shape a general theory of large deviations that was much more than a quantitative improvement of convergence rates. It addresses a fundamental question: what is the qualitative behaviour of a stochastic system if it deviates from the ergodic behaviour predicted by some law of large numbers or if it arises as a small perturbation of a deterministic system? The key to the answer is a powerful variational principle that describes the unexpected behaviour in terms of a new probabilistic model minimizing a suitable entropy distance to the initial probability measure. In a series of joint papers with Monroe D. Donsker exploring the hierarchy of large deviations in the context of Markov processes, Varadhan demonstrated the relevance and the power of this new approach. A striking application is their solution of a conjecture of Mark Kac concerning large time asymptotics of a tubular neighbourhood of the Brownian motion path, the so-called "Wiener sausage".

Varadhan's theory of large deviations provides a unifying and efficient method for clarifying a rich variety of phenomena arising in complex stochastic systems, in fields as diverse as quantum field theory, statistical physics, population dynamics, econometrics and finance, and traffic engineering. It has also greatly expanded our ability to use computers to simulate and analyze the occurrence of rare events. Over the last four decades, the theory of large deviations has become a cornerstone of modern probability, both pure and applied.

Varadhan has made key contributions in several other areas of probability. In joint work with Daniel W. Stroock, he developed a martingale method for characterizing diffusion processes, such as solutions of stochastic differential equations. This new approach turned out to be an extremely powerful way of constructing new Markov processes, for example infinite-dimensional diffusions arising in population genetics.

Another major theme is the analysis of hydrodynamical limits describing the macroscopic behaviour of very large systems of interacting particles. A first breakthrough came in joint work with Maozheng Guo and George C. Papanicolaou on gradient models. Varadhan went even further by showing how to handle non-gradient models, greatly extending the scope of the theory. His ideas also had a strong influence on the analysis of random walks in a random environment. His name is now attached to the method of "viewing the environment from the travelling particle", one of the few general tools in the field.

Varadhan's work has great conceptual strength and ageless beauty. His ideas have been hugely influential and will continue to stimulate further research for a long time.

2008—John Griggs Thompson and Jacques Tits

The Norwegian Academy of Science and Letters has decided to award the Abel Prize for 2008 to **John Griggs Thompson**, Graduate Research Professor, University of Florida, and **Jacques Tits**, Professor Emeritus, Collège de France, Paris

> for their profound achievements in algebra and in particular for shaping modern group theory

Modern algebra grew out of two ancient traditions in mathematics, the art of solving equations, and the use of symmetry as for example in the patterns of the tiles of the Alhambra. The two came together in late eighteenth century, when it was first conceived that the key to understanding even the simplest equations lies in the symmetries of their solutions. This vision was brilliantly realised by two young mathematicians, Niels Henrik Abel and Évariste Galois, in the early nineteenth century. Eventually it led to the notion of a group, the most powerful way to capture the idea of symmetry. In the twentieth century, the group theoretical approach was a crucial ingredient in the development of modern physics, from the understanding of crystalline symmetries to the formulation of models for fundamental particles and forces.

In mathematics, the idea of a group proved enormously fertile. Groups have striking properties that unite many phenomena in different areas. The most important groups are finite groups, arising for example in the study of permutations, and linear groups, which are made up of symmetries that preserve an underlying geometry. The work of the two laureates has been complementary: John Thompson concentrated on finite groups, while Jacques Tits worked predominantly with linear groups.

Thompson revolutionised the theory of finite groups by proving extraordinarily deep theorems that laid the foundation for the complete classification of finite simple groups, one of the greatest achievements of twentieth century mathematics. Simple groups are atoms from which all finite groups are built. In a major breakthrough, Feit and Thompson proved that every non-elementary simple group has an even number of elements. Later Thompson extended this result to establish a classification of an important kind of finite simple group called an N-group. At this point, the classification project came within reach and was carried to completion by others. Its almost incredible conclusion is that all finite simple groups belong to certain standard families, except for 26 sporadic groups. Thompson and his students played a major role in understanding the fascinating properties of these sporadic groups, including the largest, the so-called Monster.

Tits created a new and highly influential vision of groups as geometric objects. He introduced what is now known as a Tits building, which encodes in geometric terms the algebraic structure of linear groups. The theory of buildings is a central unifying principle with an amazing range of applications, for example to the classification of algebraic and Lie groups as well as finite simple groups, to Kac–Moody groups (used by theoretical physicists), to combinatorial geometry (used in computer science), and to the study of rigidity phenomena in negatively curved spaces. Tits's geometric approach was essential in the study and realisation of the sporadic groups, including the Monster. He also established the celebrated "Tits alternative": every finitely generated linear group is either virtually solvable or contains a copy of the free group on two generators. This result has inspired numerous variations and applications.

The achievements of John Thompson and of Jacques Tits are of extraordinary depth and influence. They complement each other and together form the backbone of modern group theory.

2009—Mikhail Leonidovich Gromov

The Norwegian Academy of Science and Letters has decided to award the Abel Prize for 2009 to **Mikhail Leonidovich Gromov**, Permanent Professor, Institut des Hautes Études Scientifiques, France,

for his revolutionary contributions to geometry

Geometry is one of the oldest fields of mathematics; it has engaged the attention of great mathematicians through the centuries, but has undergone revolutionary change during the last 50 years. Mikhail Gromov has led some of the most important developments, producing profoundly original general ideas, which have resulted in new perspectives on geometry and other areas of mathematics.

Riemannian geometry developed from the study of curved surfaces and their higher dimensional analogues and has found applications, for instance, in the theory of general relativity. Gromov played a decisive role in the creation of modern global Riemannian geometry. His solutions of important problems in global geometry relied on new general concepts, such as the convergence of Riemannian manifolds and a compactness principle, which now bear his name.

Gromov is one of the founders of the field of global symplectic geometry. Holomorphic curves were known to be an important tool in the geometry of complex manifolds. However, the environment of integrable complex structures was too rigid. In a famous paper in 1985, he extended the concept of holomorphic curves to J-holomorphic curves on symplectic manifolds. This led to the theory of Gromov–Witten invariants, which is now an extremely active subject linked to modern quantum field theory. It also led to the creation of symplectic topology and gradually penetrated and transformed many other areas of mathematics.

Gromov's work on groups of polynomial growth introduced ideas that forever changed the way in which a discrete infinite group is viewed. Gromov discovered the geometry of discrete groups and solved several outstanding problems. His geometrical approach rendered complicated combinatorial arguments much more natural and powerful.

Mikhail Gromov is always in pursuit of new questions and is constantly thinking of new ideas for solutions of old problems. He has produced deep and original work throughout his career and remains remarkably creative. The work of Gromov will continue to be a source of inspiration for many future mathematical discoveries.

2010—John Torrence Tate

The Norwegian Academy of Science and Letters has decided to award the Abel Prize for 2010 to **John Torrence Tate**, University of Texas at Austin,

for his vast and lasting impact on the theory of numbers

Beyond the simple arithmetic of $1, 2, 3, \ldots$ lies a complex and intricate world that has challenged some of the finest minds throughout history. This world stretches from the mysteries of the prime numbers to the way we store, transmit, and secure information in modern computers. It is called the theory of numbers. Over the past century it has grown into one of the most elaborate and sophisticated branches of mathematics, interacting profoundly with other areas such as algebraic geometry and the theory of automorphic forms. John Tate is a prime architect of this development.

Tate's 1950 thesis on Fourier analysis in number fields paved the way for the modern theory of automorphic forms and their L-functions. He revolutionized global class field theory with Emil Artin, using novel techniques of group cohomology. With Jonathan Lubin, he recast local class field theory by the ingenious use of formal groups. Tate's invention of rigid analytic spaces spawned the whole field of rigid analytic geometry. He found a p-adic analogue of Hodge theory, now called Hodge-Tate theory, which has blossomed into another central technique of modern algebraic number theory.

A wealth of further essential mathematical ideas and constructions were initiated by Tate, including Tate cohomology, the Tate duality theorem, Barsotti–Tate groups, the Tate motive, the Tate module, Tate's algorithm for elliptic curves, the Néron–Tate height on Mordell–Weil groups of abelian varieties, Mumford–Tate groups, the Tate isogeny theorem and the Honda–Tate theorem for abelian varieties over finite fields, Serre–Tate deformation theory, Tate–Shafarevich groups, and the Sato–Tate conjecture concerning families of elliptic curves. The list goes on and on.

Many of the major lines of research in algebraic number theory and arithmetic geometry are only possible because of the incisive contribution and illuminating insight of John Tate. He has truly left a conspicuous imprint on modern mathematics.

2011—John Willard Milnor

The Norwegian Academy of Science and Letters has decided to award the Abel Prize for 2011 to **John Willard Milnor**, Institute for Mathematical Sciences, Stony Brook University, New York, USA, for pioneering discoveries in topology, geometry and algebra

All of Milnor's works display marks of great research: profound insights, vivid imagination, elements of surprise, and supreme beauty.

Milnor's discovery of exotic smooth spheres in seven dimensions was completely unexpected. It signaled the arrival of differential topology and an explosion of work by a generation of brilliant mathematicians; this explosion has lasted for decades and changed the landscape of mathematics. With Michel Kervaire, Milnor went on to give a complete inventory of all the distinct differentiable structures on spheres of all dimensions; in particular they showed that the 7-dimensional sphere carries exactly 28 distinct differentiable structures. They were among the first to identify the special nature of four-dimensional manifolds, foreshadowing fundamental developments in topology.

Milnor's disproof of the long-standing *Hauptvermutung* overturned expectations about combinatorial topology dating back to Poincaré. Milnor also discovered homeomorphic smooth manifolds with nonisomorphic tangent bundles, for which he developed the theory of microbundles. In three-manifold theory, he proved an elegant unique factorization theorem.

Outside topology, Milnor made significant contributions to differential geometry, algebra, and dynamical systems. In each area Milnor touched upon, his insights and approaches have had a profound impact on subsequent developments.

His monograph on isolated hypersurface singularities is considered the single most influential work in singularity theory; it gave us the Milnor number and the Milnor fibration.

Topologists started to actively use Hopf algebras and coalgebras after the definitive work by Milnor and J. C. Moore. Milnor himself came up with new insights into the structure of the Steenrod algebra (of cohomology operations) using the theory of Hopf algebras. In algebraic K-theory, Milnor introduced the degree two functor; his celebrated conjecture about the functor—eventually proved by Voevodsky—spurred new directions in the study of motives in algebraic geometry. Milnor's introduction of the growth invariant of a group linked combinatorial group theory to geometry, prefiguring Gromov's theory of hyperbolic groups.

More recently, John Milnor turned his attention to dynamical systems in low dimensions. With Thurston, he pioneered "kneading theory" for interval maps, laying down the combinatorial foundations of interval dynamics, creating a focus of intense research for three decades. The Milnor–Thurston conjecture on entropy monotonicity prompted efforts to fully understand dynamics in the real quadratic family, bridging real and complex dynamics in a deep way and triggering exciting advances.

Milnor is a wonderfully gifted expositor of sophisticated mathematics. He has often tackled difficult, cutting-edge subjects, where no account in book form existed. Adding novel insights, he produced a stream of timely yet lasting works of masterly lucidity. Like an inspired musical composer who is also a charismatic performer, John Milnor is both a discoverer and an expositor.

2012—Endre Szemerédi

The Norwegian Academy of Science and Letters has decided to award the Abel Prize for 2012 to **Endre Szemerédi**, Alfréd Rényi Institute of Mathematics, Hungarian Academy of Sciences, Budapest, Hungary, and Department of Computer Science, Rutgers, The State University of New Jersey, USA

> for his fundamental contributions to discrete mathematics and theoretical computer science, and in recognition of the profound and lasting impact of these contributions on additive number theory and ergodic theory

Discrete mathematics is the study of structures such as graphs, sequences, permutations, and geometric configurations. The mathematics of such structures forms the foundation of theoretical computer science and information theory. For instance, communication networks such as the internet can be described and analyzed using the tools of graph theory, and the design of efficient computational algorithms relies crucially on insights from discrete mathematics. The combinatorics of discrete structures is also a major component of many areas of pure mathematics, including number theory, probability, algebra, geometry, and analysis.

Endre Szemerédi has revolutionized discrete mathematics by introducing ingenious and novel techniques, and by solving many fundamental problems. His work has brought combinatorics to the center-stage of mathematics, by revealing its deep connections to such fields as additive number theory, ergodic theory, theoretical computer science, and incidence geometry.

In 1975, Endre Szemerédi first attracted the attention of many mathematicians with his solution of the famous Erdős–Turán conjecture, showing that in any set of integers with positive density, there are arbitrarily long arithmetic progressions. This was a surprise, since even the case of progressions of lengths 3 or 4 had earlier required substantial effort, by Klaus Roth and by Szemerédi himself, respectively. A bigger surprise lay ahead. Szemerédi's proof was a masterpiece of combinatorial reasoning, and was immediately recognized to be of exceptional depth and importance. A key step in the proof, now known as the Szemerédi Regularity Lemma, is a structural classification of large graphs. Over time, this lemma has become a central tool of both graph theory and theoretical computer science, leading to the solution of major problems in property testing, and giving rise to the theory of graph limits.

Still other surprises lay in wait. Beyond its impact on discrete mathematics and additive number theory, Szemerédi's theorem inspired Hillel Furstenberg to develop ergodic theory in new directions. Furstenberg gave a new proof of Szemerédi's theorem by establishing the Multiple Recurrence Theorem in ergodic theory, thereby unexpectedly linking questions in discrete mathematics to the theory of dynamical systems. This fundamental connection led to many further developments, such as the Green–Tao theorem asserting that there are arbitrarily long arithmetic progressions of prime numbers.

Szemerédi has made many additional deep, important, and influential contributions to both discrete mathematics and theoretical computer science.

Examples in discrete mathematics include the Szemerédi–Trotter theorem, the Ajtai–Komlós–Szemerédi semi-random method, the Erdős–Szemerédi sum-product theorem, and the Balog–Szemerédi–Gowers lemma. Examples in theoretical computer science include the Ajtai–Komlós–Szemerédi sorting network, the Fredman–Komlós–Szemerédi hashing scheme, and the Paul–Pippenger–Szemerédi–Trotter theorem separating deterministic and non-deterministic linear time.

Szemerédi's approach to mathematics exemplifies the strong Hungarian problem-solving tradition. Yet, the theoretical impact of his work has been a game-changer.

The Abel Committee

2013

Ragni Piene (University of Oslo, Norway), chair
Noga Alon (Tel Aviv University, Israel)
Stanislav Smirnov (University of Geneva, Switzerland)
Gang Tian (Princeton University, USA, and Beijing University, China)
Terence Tao (University of California at Los Angeles, USA)

2014

Ragni Piene (University of Oslo, Norway), chair
Maria Esteban (Université Paris–Dauphine, France)
Stanislav Smirnov (University of Geneva, Switzerland)
Gang Tian (Princeton University, USA and Beijing University, China)ă
Cédric Villani (Université de Lyon and Institut Henri Poincaré, France)

2015

John Rognes (University of Oslo, Norway), chair
Maria Esteban (Université Paris–Dauphine, France)
Rahul Pandharipande (ETH, Zürich, Switzerland)
Éva Tardos (Cornell University, USA)
Cédric Villani (Université de Lyon and Institut Henri Poincaré, France)

2016

John Rognes (University of Oslo, Norway), chair
Rahul Pandharipande (ETH, Zürich, Switzerland)
Éva Tardos (Cornell University, USA)
Luigi Ambrosio (Scuola Normale Superiore, Pisa, Italy)
Marta Sanz-Solé (Universitat de Barcelona, Spain)

H. Holden, R. Piene (eds.), *The Abel Prize 2013–2017*,
https://doi.org/10.1007/978-3-319-99028-6

2017

John Rognes (University of Oslo, Norway), chair
Luigi Ambrosio (Scuola Normale Superiore, Pisa, Italy)
Marta Sanz-Solé (Universitat de Barcelona, Spain)
Ben J. Green (University of Oxford, UK)
Marie-France Vignéras (Institut de Mathématiques de Jussieu, Paris, France)

The Niels Henrik Abel Board

2013
Helge Holden (chair)
Anne Borg
Kari Gjetrang
Arne Bang Huseby
Hans Munthe-Kaas
Øivind Andersen (observer)

2014
Helge Holden (chair)
Anne Borg
Arne Bang Huseby
Hans Munthe-Kaas
Anne Carine Tanum
Øivind Andersen (observer)

2015
Kristian Ranestad (chair)
Anne Borg
Hans Munthe-Kaas
Einar Rønquist
Anne Carine Tanum
Øivind Andersen (observer)

H. Holden, R. Piene (eds.), *The Abel Prize 2013–2017*,
https://doi.org/10.1007/978-3-319-99028-6

2016

Kristian Ranestad (chair)
Anne Borg
Hans Munthe-Kaas
Einar Rønquist
Anne Carine Tanum
Øivind Andersen (observer)

2017

Kristian Ranestad (chair)
Anne Borg
Hans Munthe-Kaas
Einar Rønquist
Anne Carine Tanum
Øystein Hov (observer)

The Abel Lectures 2013–2017

2013

Pierre Deligne (Institute for Advanced Study, Princeton): *Hidden symmetries of algebraic varieties*
Nick Katz (Princeton University): *Life over finite fields*
Claire Voisin (Université Paris VI): *Mixed Hodge structures and the topology of algebraic varieties*
Ravi Vakil (Stanford University): *Algebraic geometry and the ongoing unification of mathematics* [Science Lecture]

2014

Yakov G. Sinai (Princeton University and The Russian Academy of Sciences): *Now everything has been started? The origin of deterministic chaos*
Gregory A. Margulis (Yale University): *Kolmogorov–Sinai entropy and homogeneous dynamics*
Konstantin Khanin (University of Toronto): *Between mathematics and physics*
Domokos Szász (Budapest University of Technology): *Mathematical billiards and chaos* [Science Lecture]

2015

John F. Nash, Jr. (Princeton University): *An interesting equation*
Louis Nirenberg (New York University): *Some remarks on mathematics*
Camillo De Lellis (Universität Zürich): *Surely you're joking, Mr. Nash?*
Tristan Rivière (ETH, Zürich): *Exploring the unknown, the work of Louis Nirenberg on Partial Differential Equations*[1]
Frank Morgan (Williams College, USA): *Soap bubbles and mathematics* [Science Lecture]

[1]Published in *Notices of the AMS* 63(2):120–125 (2016).

H. Holden, R. Piene (eds.), *The Abel Prize 2013–2017*,
https://doi.org/10.1007/978-3-319-99028-6

2016

Sir Andrew Wiles (Oxford University): *Fermat's Last Theorem: abelian and non-abelian approaches*
Henri Darmon (McGill University, Canada): *Andrew Wiles' marvelous proof*[2]
Manjul Bhargava (Princeton University): *What is the Birch–Swinnerton-Dyer Conjecture, and what is known about it?*
Simon Singh: *From Fermat's Last Theorem to Homer's Last Theorem* [Popular Lecture]

2017

Yves Meyer (École Normale Supérieure Paris-Saclay): *Detection of gravitational waves and time-frequency wavelets*
Stéphane Mallet (École Polytechnique, Paris): *A wavelet zoom to analyze a multi-scale world*
Ingrid Daubechies (Duke University): *Wavelet bases: roots, surprises and applications*
Emmanuel Jean Candès (Stanford University): *Wavelets, sparsity and its consequences*

[2]Published in *Notices of the AMS* 64(3):209–216 (2017), and *Newsletter of the EMS*, Issue 104 (June 2017), 7–13.

The Abel Laureate Presenters 2013–2017

In March each year when the President of the Norwegian Academy of Science and Letters announces the Abel Laureate and the Chair of the Abel Committee states the reasons for the selection, a mathematician presents the work of the Laureate. Below we list the presenters for the period 2013–2017:

2013 (P. Deligne) Timothy Gowers, University of Cambridge
2014 (Y. G. Sinai) Jordan S. Ellenberg, University of Wisconsin
2015 (J. F. Nash, Jr. and L. Nirenberg) Alexander Bellos, London
2016 (A. Wiles) Alexander Bellos, London
2017 (Y. Meyer) Terence Tao, University of California at Berkeley

H. Holden, R. Piene (eds.), *The Abel Prize 2013–2017*,
https://doi.org/10.1007/978-3-319-99028-6

The Interviews with the Abel Laureates

Transcripts of parts of the interviews that Christian Skau, Norwegian University of Science and Technology, (2013–2017), Martin Raussen, Aalborg University, (2013–2016), and Bjørn I. Dundas, University of Bergen, (2017) made with each laureate in connection with the Prize ceremonies, can be found in the following publications:

2013 Pierre Deligne

EMS Newsletter, issue 89 (Sep. 2013) 15–23,
AMS Notices, **61** (2014) 177–185.

2014 Yakov Sinai

EMS Newsletter, issue 93 (Sep. 2014) 12–19,
AMS Notices, **62** (2015) 152–160.

2015 John F. Nash Jr. and Louis Nirenberg

EMS Newsletter, issue 97 (Sep. 2015) 26–31, (Nash),
EMS Newsletter, issue 98 (Dec. 2015) 33–38, (Nirenberg),
AMS Notices, **63** (2016) 135–140, (Nirenberg),
AMS Notices, **63** (2016) 486–491, (Nash).

2016 Sir Andrew J. Wiles

EMS Newsletter, issue 101 (Sep. 2016) 29–38,
AMS Notices, **64** (2017) 198–207.

2017 Yves Meyer

EMS Newsletter, issue 105 (Sep. 2017) 14–22,
AMS Notices, **65** (2018) 520–529.

H. Holden, R. Piene (eds.), *The Abel Prize 2013–2017*,
https://doi.org/10.1007/978-3-319-99028-6

The interviews for the period 2003–2016 have been published in the book M. Raussen and C. Skau (eds.) *Interviews with the Abel Prize Laureates 2003–2016*, European Mathematical Society Publishing House, Zürich, 2017.

Addenda, Errata, and Updates[1]

2003 Jean-Pierre Serre

(i) Publications

2013

[210], [211], [212], and [261] have all been reprinted by Springer as *Oeuvres — Collected Papers, Vols. I–IV*.

2014

[297] Bases normales autoduales et groupes unitaires en caractéristique 2. *Transform. Groups* 19(2):643–698. Erratum *loc. cit.* 20(1):305 (2015).

[298] Henri Cartan 1904–2008. *Bull. Lond. Math. Soc.* 46(1):211–216.

2015

[299] (co-edited with P. Colmez). Correspondance Serre–Tate: Volume I (1956–1973). *Documents Mathématiques* 13, xxviii+448 pp.

[300] (co-edited with P. Colmez). Correspondance Serre–Tate: Volume II (1973–2000). *Documents Mathématiques* 14, xvi+521 pp.

2016

[301] *Finite Groups: An Introduction.* Surveys of Modern Mathematics, Vol. 10. International Press, Somerville, MA; Higher Education Press, Beijing.

[302] Lettre à Armand Borel. In *Frobenius distributions: Lang–Trotter and Sato–Tate conjectures.* Contemp. Math., 663, Amer. Math. Soc., Providence, RI, 1–9.

[1] H. Holden, R. Piene (eds.): *The Abel Prize 2003–2007. The First Five Years*. Springer, Berlin, 2010, and H. Holden, R. Piene (eds.): *The Abel Prize 2008–2012*. Springer, Berlin, 2014.

H. Holden, R. Piene (eds.), *The Abel Prize 2013–2017*,
https://doi.org/10.1007/978-3-319-99028-6

[303] *Collected works of John Tate. Part I (1951–1975)*. Edited by Barry Mazur and Jean-Pierre Serre. American Mathematical Society, Providence, RI. xxvii+716 pp.

[304] *Collected works of John Tate. Part II (1976–2006)*. Edited by Barry Mazur and Jean-Pierre Serre. American Mathematical Society, Providence, RI. xxv+751 pp.

2017

[305] On the mod p reduction of orthogonal representations. *arXiv:1708.00046.*

2018

[306] Cohomological invariants mod 2 of Weyl groups. To appear in *Oberwolfach Reports* 15 (2018). *arXiv:1805.07172.*

(ii) Addendum CV

Honorary degree, Tsing Hua University, 2017.

2004 Sir Michael Atiyah and Isadore M. Singer

(i) Publications by M. Atiyah

2013

[265] published in *Math. Proc. Cambridge Philos. Soc.* 155, no. 1, 13–37.

2014

[266] *Collected works. Vol. 7. (2002–2013)* Oxford University Press, Oxford.

[267] Geometry in 2, 3 and 4 dimensions. In *The Poincaré conjecture*, Clay Math. Proc. 19:1–6.

[268] Solomon Lefschetz and Mexico. The influence of Solomon Lefschetz in geometry and topology. *Contemp. Math.* vol. 621, Amer. Math. Soc.

[269] (with S. Zeki, J. P. Romaya, D. M. T. Benincasa) The experience of mathematical beauty and its neural correlates. *Frontiers in Human Neuroscience*, Vol. 8, Article 68, 1–12.

2015

[270] The pre-history of the European mathematical society. *Eur. Math. Soc. Newsl.* 95:17–18.

[271] (with G. Franchetti and B. J. Schroers). Time evolution in a geometric model of a particle. *J. High Energy Phys.* no. 2, 062, front matter+16 pp.

[272] The art of mathematics. In *Art in the Life of Mathematicians*. A. K. Szemerédi, editor. Amer. Math. Soc., Providence, RI.

2016

[273] Riemann's influence in geometry, analysis and number theory. In *The legacy of Bernhard Riemann after one hundred and fifty years.* Vol. I, 57–67, Adv. Lect. Math. 35.1, Int. Press, Somerville, MA.

[274] The Hirzebruch signature theorem for conical metrics. In *Arbeitstagung Bonn 2013*, 1–15, Progr. Math., 319, Birkhäuser/Springer, Cham.

[275] (with N. S. Manton) Complex Geometry of Nuclei and Atoms. *arXiv 1609.02816*.

[276] The non-existent complex 6-sphere. *arXiv 1610.09366*.

2017

[277] Geometric Models of Helium. *Modern Phys. Lett.* A 32 , no. 14, 1750079, 11 pp.

[278] (with M. Dunajski and L.J. Mason) Twistor theory at fifty: from contour integrals to twistor strings. *Proc. R. Soc. A.* 473, no. 2206, 20170530, 33 pp.

[279] Scalar curvature, flat Borromean rings and the 3-body problem. *arXiv 1709.01539*.

[280] (with M. Marcolli) Anyons in geometric models of matter. *J. High Energ. Phys.* 76, 23 pp.

2018

[281] (with J. Malkoun) The Relativistic Geometry and Dynamics of Electrons. *Found. Phys.* 48, no. 2, 199–208.

[282] William Leonard Edge. *Eur. J. Math.* 4, no. 1, 437–438.

[283] (with C. Zapata-Carratala) K-theory and the Jones polynomial. In *Ludwig Faddeev Memorial Volume: A Life in Mathematical Physics*. M.-L. Ge, A.J. Niemi, K. K. Phua, and L. A. Takhtajan, editors. World Sci. Publ., Hackensack, NJ, pp. 53–57. Also published in *Rev. Math. Phys.* 30, no. 6, 1840002, 5 pp.

[284] Understanding the 6-dimensional sphere. In *Foundations of Mathematics and Physics One Century After Hilbert*. J. Kouneiher editor, Springer, Cham, pp. 129–133.

2019

[285] Arithmetic physics. In *Proceedings of the ICM 2018, Rio de Janeiro, Brazil.* Vol. I. World Scientific. pp. 263–270.

(ii) Addendum CV for Atiyah

Member, Academia Europaea, 1988.

(iii) Articles and books about Atiyah and Singer in connection with the Abel Prize

A. Mukherhjee: Atiyah–Singer Index Theorem. An Introduction. Hindustan Book Agency, 2013.
S. Roberts: Michael Atiyah's Imaginative State of Mind. Quanta Magazine, https://www.quantamagazine.org/read-offline/21348/.

2005 Peter D. Lax

(i) Publications

2013

[215], [216] have been reprinted in Springer Collected Works in Mathematics, Springer, New York, 2013.

2014

[233] (with M. S. Terell) *Calculus with Applications.* Undergraduate Texts in Mathematics. Springer, New York.
[234] (with J.L. Kazdan, A.B. Novikoff, C. D., Hill, A. Jameson, E. V. Swenson, R.B. Shapiro, C. Garabedian, and E. Garabedian) Emily Paul Roesel Garabedian (1927–2010). *Notices Amer. Math. Soc.* 61, no. 3, 244–255.
[235] Numbers and functions: from a classical experimental mathematicians point of view [book review]. *Amer. Math. Monthly* 121, no. 2, 183.

2015

[236] Painting and mathematics. In *Art in the Life of Mathematicians.* A. K. Szemerédi, editor. Amer. Math. Soc., Providence, RI.
[237] (with S. Anastasio, R.D. Douglas, C.Foias, W.M. Ching, M. Davis, M. Sharir, M. Wigler, J.A. Fisher, L. Nirenberg, P.M. Willig, and H. Porta) In memory of Jacob Schwartz. *Notices Amer. Math. Soc.* 62, no. 5, 473–490.

2018

[238] (with M.S. Terrell) *Multivariable Calculus with Applications*. Undergraduate Texts in Mathematics. Springer, Cham, 483 pp.

(iii) A biography of Peter Lax

R. Hersch: *Peter Lax, Mathematician. An Illustrated Memoir*. American Mathematical Society, Providence, 2015.

2006 Lennart Carleson

(ii) Addendum CV

Member, Academia Europaea, 1993.

(iii) Articles about Carleson in connection with the Abel Prize

Lennart Carleson erhält den Abel-Preis. [German] *Mitt. Dtsch. Math.-Ver.* 14 (2006), no. 2, 87–88.
H. Duistermaat: Abel Prize winner 2006; Lennart Carleson: achievements until now. *Nieuw Arch. Wiskd.* (5) 8 (2007), no. 3, 175–177.
A, Vargas: 2006 Abel Prize: Lennart Carleson. [Catalan] *SCM Not.*, no. 21 (2007), 63–65.
J. Ortega-Certà, J. Tatier: Lennart Carleson, 2006 Abel Prize. [Catalan] *Butl. Soc. Catalana Mat.* 22 (2007), no. 2, 153–164, 230 (2008).

2007 S. R. Srinivasa Varadhan

(i) Publications by S. R. S. Varadhan

1988

[68] Reprinted in R. Azencott, M. Freidlin, S.R.S. Varadhan (editors) *Large Deviations at Saint-Flour*, 213–261. Springer, Heidelberg.

2010

[136] was published in *Ann. Probab.* 42(2):649–688 in 2014.

2013

[144] (with S. Sethuraman). Large deviations for the current and tagged particle in $1D$ nearest-neighbor symmetric simple exclusion. *Ann. Probab.* 41(3A):1461–1512.
[145] Entropy, large deviations, and scaling limits. *Comm. Pure Appl. Math.* 66, no. 12, 1914–1932.

2014

[146] (with Y. Kifer). Nonconventional large deviations theorems. *Probab. Theory Related Fields* 158(1–2): 197–224.
[147] (with Y. Kifer). Nonconventional limit theorems in discrete and continuous time via martingales. *Ann. Probab.* 42(2):649–688.

2015

[148] Topics in occupation times and Gaussian free fields. [book review]. *Bull. Amer. Math. Soc. (N.S.)* 152(1):167–169.
[149] (with M. Gorny) Fluctuations of the self-normalized sum in the Curie–Weiss model of SOC. *J. Stat. Phys.* 160, no. 3, 513–518.
[150] Entropy and its many avatars. *J. Math. Soc. Japan* 67, no. 4, 1845–1857.

2016

[151] (with A. Dembo, M. Shkolnikov, O. Zeitouni) Large deviations for diffusions interacting through their ranks. *Comm. Pure Appl. Math.* 69, no. 7, 1259–1313.
[152] *Large Deviations*. Courant Lecture Notes in Mathematics, 27. American Mathematical Society, Providence, RI, 2016. vii+104 pp.
[153] (with C. Mukherjee) Brownian occupation measures, compactness and large deviations. *Ann. Probab.* 44, no. 6, 3934–3964.
[154] Martingale methods for the central limit theorem. In *Rabi N. Bhattacharya–Selected Papers*. M. Denker and E. Waymire, editors. Contemp. Mathematicians, Birkhäuser/Springer, Cham. pp. 131–135.

2017

[155] (with Y. Kifer) Tails of polynomials of random variables and stable limits for nonconventional sums. *J. Stat. Phys.* 166, no. 3–4, 575–608.

2018

[156] (with C. Mukherjee) Identification of the Polaron measure and its central limit theorem. *arXiv:1802.05696.*
[157] (with C. Mukherjee) Strong coupling limit of the Polaron measure and the Pekar process. *arXiv:1806.06865.*

(iii) Articles about Varadhan in connection with the Abel Prize

On the award of the Abel Prize to S. R. S. Varadhan. [Russian] *Teor. Veroyatn. Primen* 52 (2007), no. 3, 417–418; translation in *Theory Probab. Appl.* 52 (2008), no. 3, 371.

2008 John G. Thompson and Jacques Tits

(i) Publications by J. G. Thompson

2012

[99] (with H.L. Montgomery). Geometric properties of the zeta function. *Acta Arith.* 155(4):373–396.

2014

[100] (with P. Balister, B. Bollobás, and Z. Füredi). Minimal symmetric differences of lines in projective planes. *J. Combin. Des.* 22(10):435–451.
[101] (with P. Sin). Some uniserial representations of certain special linear groups. *J. Algebra* 398:448–460.

(i) Publications by J. Tits

2011

[208a] Sur les groupes algébriques affins. Théorèmes fondamentaux de structure. Classification des groupes semisimples et geométries associéees. In *Groups, Lie rings and cohomology theory*, C.I.M.E. Summer School, 20, Springer, Heidelberg. pp. 185–264.

2013

[209] is published in four volumes.

[210] Résumés des cours au Collège de France 1973–2000. In *Documents Mathématiques (Paris)*, 12. Société Mathématique de France, Paris, xii+390 pp.

(iii) Article about Tits in connection with the Abel Prize

G. Rousseau: Les immeubles, une théorie de Jacques Tits, prix Abel 2008. [French] *Gaz. Math.* No. 121 (2009), 47–64.

2009 Mikhail Gromov

(i) Publications by M. Gromov

2013

[131] In a search for a structure, Part 1: On entropy. In *European Congress of Mathematics*, 51–78, Eur. Math. Soc., Zürich.

2014

[132] Plateau–Stein manifolds. *Cent. Eur. J. Math.*, 12:923–951.

[133] Dirac and Plateau billiards in domains with corners. *Cent. Eur. J. Math.*, 12:1109–1156.

2015

[134] Symmetry, probability, entropy: synopsis of the lecture at MAXENT 2014. *Entropy* 17, no. 3, 1273–1277.

[135] Where do we come from? What are we? Where are we going? In *The Poincaré conjecture*, Clay Math. Proc. 19:81–144.

[136] Colorful categories. *Russian Math. Surveys*, 70:591–655. Also available in *Uspekhi Mat. Nauk,* no. 4(424), 3–76 (In Russian).

2016

[137] Introduction John Nash: theorems and ideas. In *Open Problems in Mathematics*. J. F. Nash and M. T. Rassias editors. Springer, pages xi–xiii.

2017

[138] Math currents in the brain. In *Simplicity: ideals of practice in mathematics and the arts*, 107–118, Springer.

[139] (with C. LeBrun, G. Besson, J. Simons, J. Cheeger, J.-F. Bourguignon, D. Sullivan, J. Lafontaine, J. Kazdan, M.-L. Michelsohn, P. Pansu, D. Ebin, and K. Grove) Marcel Berger remembered. *Notices Amer. Math. Soc.* 64, no. 11, 1285–1295.

[140] Geometric, algebraic, and analytic descendants of Nash isometric embedding theorems. *Bull. Amer. Math. Soc. (N.S.)*, 54, no. 2, 173–245.

[141] Morse spectra, homology measures and parametric packing problems. *arXiv 1710.03616.*

2018

[142] Metric inequalities with scalar curvature. *Geom. Funct. Anal.* 28, no. 3, 645–726.

[143] A dozen problems, questions and conjectures about positive scalar curvature. In *Foundations of Mathematics and Physics One Century After Hilbert.* J. Kouneiher editor. Springer, Cham, pp. 135–158.

[144] *Great Circle of Mysteries. Mathematics, the World, the Mind.* Birkhäuser/Springer, Cham, 202 pp. Translated from Russian.

[145] Scalar curvature of manifolds with boundaries: Natural questions and artificial constructions. *arXiv:1811.04311.*

(ii) Addendum CV

Member, Academia Europaea, 1993.

(iii) Articles about Gromov in connection with the Abel Prize

O. Bogopolski: Mikhail Gromov, 2009 Abel Prize. [Catalan] *SCM Not.* No. 27 (2009), 39–40.

F. Forstnerič: The 2009 Abel Prize to Mikhael Gromov. [Slovenian] *Obzornik Mat. Fiz* 57 (2010), no. 2, 41–52.

2010 John T. Tate

(i) Publications

2012

[84] An oft cited letter from Tate to Serre on computing local heights on elliptic curves. *arXiv 1207.5765.*

2015

[85] (co-edited by P. Colmez and J. P. Serre). Correspondance Serre–Tate: Volume I (1956–1973). *Documents Mathématiques* 13, xxviii+448 pp.

[86] (co-edited by P. Colmez and J.P. Serre). Correspondance Serre–Tate: Volume II (1973–2000). *Documents Mathématiques* 14, xvi+521 pp.

2016

[87] (with M. Artin, A. Jackson, D. Mumford) John Alexandre Grothendieck 1928–2014, Part 2. *Notices Amer. Math. Soc.* 63, no. 4, 401–413.

[88] An old letter from J. Tate to B. Dwork, as viewed recently by J. Tate and B. Mazur. *Izv. Ross. Akad. Nauk Ser. Mat.* 80, no. 5, 153–156. Russian translation in *Izv. Math.* 80, no. 5, 954–957.

[89] *Collected works of John Tate. Part I (1951–1975).* Edited by Barry Mazur and Jean-Pierre Serre. American Mathematical Society, Providence, RI. xxvii+716 pp.

[90] *Collected works of John Tate. Part II (1976–2006).* Edited by Barry Mazur and Jean-Pierre Serre. American Mathematical Society, Providence, RI. xxv+751 pp.

[91] Abelian varieties isogenous to a power of an elliptic curve. *arXiv 1602.06237.*

2017

[92] Number theory in the 20th century: Part 1. *Bull. Amer. Math. Soc. (N.S.)* 54, no. 4, 547–550.

2018

[93] (with B.W. Jordan, A.G. Keeton, G. Allan, B. Poonen, E.M. Rains, N. Shepherd-Barron) Abelian varieties isogenous to a power of an elliptic curve. *Compos. Math.* 154, no. 5, 934–959.

(ii) Addendum CV

Fellow, American Mathematical Society, 2013.

(iii) Articles about Tate in connection with the Abel Prize

L. D. Olson: Abel Prize 2010—John Tate. [Norwegian] *Normat* 58 (2010), no. 1, 1–5.

S. Friedlander: Introduction to the Tate issue. *Bull. Amer. Math. Soc. (N.S.)*, 54 (2017), no. 4, 541–543.

2011 John W. Milnor

(i) Publications

2014

[150] Collected papers of John Millnor. VII. Dynamical systems (1984–2012). Edited by A. Bonifant. American Mathematical Society, Providence, RI. xvi+592 pp.

[151] Arithmetic of unicritical polynomial maps. In *Frontiers in complex dynamics*, 15–24. Princeton Math. Ser., 51, Princeton Univ. Press, Princeton, NJ.

2015

[152] Topology through the centuries: low dimensional manifolds. *Bull. Amer. Math. Soc. (N.S.)* 52, no. 4, 545–584.

[153] (with M.A. Kervaire) The Kervaire–Milnor correspondence 1958–1961. *Bull. Amer. Math. Soc. (N.S.)* 52, no. 4, 611–658.

2017

[154] (with A. Bonifant) On real and complex cubic curves. *Enseign. Math.* 63, no. 1, 21–61.

2018

[155] (with A. Bonifant and X. Buff) On antipode preserving cubic maps. *Proc. London Math. Soc.* 116:670–728.

[156] (with A. Bonifant) Group actions, divisors, and plane curves. *arXiv: 1809.05191.*

(ii) Addendum CV

Fellow, American Mathematical Society, 2014.

(iii) Articles about Milnor in connection with the Abel Prize

D. Schleicher: Der Abel-Preis für John W. Milnor. [German] *Mitt. Dtsch. Math.-Ver.* 19 (2011), no. 2, 81–85.

J. Porti: John Milnor: 2011 Abel Prize. [Catalan] *SCM Not.* No. 31 (2011), 32–36.

S. Khare: On Abel Prize 2011 to John Willard Milnor: a brief description of his significant work. *Math. Student* 82 (2013), no. 1–4, 247–279.

S. Friedlander: Introductory comments. *Bull. Amer. Math. Soc. (N.S.)* 52 (2015), no. 4, 543–554.

2012 Endre Szemerédi

(i) Publications by E. Szemerédi:

2012

[185] appeared with the title "The approximate Loebl–Komlós–Sós conjecture and embedding trees in sparse graphs" in *Electron. Res. Announc. Math. Sci.*, (2015) 22:1–11.

2013

[186] Is laziness paying off? ("Absorbing" method). In *Colloquium de Giorgi 2010–2012*, Vol. 4, 17–34.

2015

[187] Arithmetic progressions, different regularity lemmas and removal lemmas. *Commun. Math. Stat.* 3, no. 3, 315–328.

[188] (with J. Hladký, D. Piguet, M. Simonovits, M. Stein) The approximate Loebl–Komlós-Sós conjecture and embedding trees in sparse graphs. *Electron. Res. Announc. Math. Sci.* 22, 1–11.

2016

[187] Erdős's unit distance problem. In *Open problems in mathematics* (eds. J. Nash Jr. and M. Rassias), 459–477. Springer.

[188] Structural approach to subset sum problems. *Found. Comput. Math.* 16, no. 6, 1737–1749.

2017

[187] (with V. Rödl, A. Ruciński, M. Schacht) On the Hamiltonicity of triple systems with high minimum degree. *Ann. Comb.* 21, no. 1, 95–117.

[188] (with J. Hladký, J. Komlós, D. Piguet, M. Simonovits, M. Stein) The approximate Loebl–Komlós-Sós conjecture I: The sparse decomposition. *SIAM J. Discrete Math.* 31, no. 2, 945–982.

[189] (with J. Hladký, J. Komlós, D. Piguet, M. Simonovits, M. Stein) The approximate Loebl–Komlós-Sós conjecture II: The rough structure of LKS graphs. *SIAM J. Discrete Math.* 31, no. 2, 983–1016.

[190] (with J. Hladký, J. Komlós, D. Piguet, M. Simonovits, M. Stein) The approximate Loebl–Komlós-Sós conjecture III: The finer structure of LKS graphs. *SIAM J. Discrete Math.* 31, no. 2, 1017–1071.

[191] (with J. Hladký, J. Komlós, D. Piguet, M. Simonovits, M. Stein) The approximate Loebl–Komlós-Sós conjecture IV: Embedding techniques and the proof of the main result. *SIAM J. Discrete Math.* 31, no. 2, 1072–1148.

(iii) Articles about Szemerédi in connection with the Abel Prize

M. Kang: The 2012 Abel laureate Endre Szemerédi and his celebrated work. *Internationale Mathematische Nachrichten*. Nr. 221, December 2012, pp. 1–19.

M. Beiglböck, R. Winkler: Endre Szemerédi: Ein mathematisches Universum in kombinatorischem Gewande. [German] *Internationale Mathematische Nachrichten*. Nr. 221, December 2012, pp. 21–38.

G. Lugosi, O. Serra: Endre Szemerédi, 2012 Abel Prize. [Spanish] *Gac. R. Soc. Mat. Esp.* 15 (2012), no. 3, 537–559.

G. Lugosi, O. Serra: Endre Szemerédi, 2012 Abel Prize. [Catalan] *Butl. Soc. Catalana Mat.* 28 (2013) no. 1, 87–115.

O. Serra: Endre Szemerédi, 2012 Abel Prize. [Catalan] *SCM Not.*, no. 33 (2013), 42–43.

R. Ramachandran: Endre Szemerédi receives 2012 Abel Prize. (Reproduced from *The Hindu*, March 22, 2012.) *Asia Pac. Math. Newsl.* 2 (2012), no. 2, 37.

V. Lundsgaard Hansen, P. G. Hjorth: Abelpris til informatikkens teoretiske stjerne: Kombinatorik i en klarsynet hjerne. [Danish] *Aktuel Naturvidenskab*, Vol. 6, No. December, 2012, p. 12–15.

Articles about the Abel Prize

A. Quirós Gracián: The Abel Prize. [Spanish] *Gac. R. Soc. Mat. Esp.* 5 (2002), no. 1, 155–161.
J. Echegaray y Eizaguirre: The Newton of the North. [Spanish] *Gac. R. Soc. Mat. Esp.* 5 (2002), no. 1, 162–173.
The Abel Prize. *Austral. Math. Soc. Gaz.* 30 (2003), no. 2, 102–103.
N. Hungerbühler, J. Kramer, Y. Reichelt: The Abel Prize—the new coronation of mathematical careers. *Elem. Math.* 58 (2003), no. 2, 45–48.
K. Piaskowska: The missing Nobel Prize? On the international Abel Prize in mathematics. (Polish) *Wiad. Mat.* 47 (2011), no. 2, 225–234.
F. T. Bruss: Abel in Holland and a few reflections on the Abel Prize. *Eur. Math. Soc. Newsl.* No. 80 (2011), 20–21.
K. G. Helsvig: Ti år med Abelprisen — den manglende Nobelprisen i matematikk? [Norwegian] *Historisk tidsskrift,* Volume 92 (2013), 261–289.
K. G. Helsvig: The Abel Prize: The Missing Nobel in Mathematics. In H. Holden, R. Piene (eds.): *The Abel Prize 2008–2012*, Springer, Berlin, 2014. Also in *Centaurus*, Vol. 56 (2014), 1–30.
M. Křížek, L. Somer, and M. Markl: The First Ten Abel Prizes for Mathematics. [Czech]. *The Union of Czech Mathematicians and Physicists*, Prague, 2013, pp. 88.
F. Zalamea: Major currents of mathematics in the twentieth century. VI. Panorama of the Abel Prizes (2003–2013). [Spanish] *Bol. Mat.* 20 (2013), 53–64.

The Abel Lecture at the International Congress of Mathematicians

2010 (Hyderabad) S.R.S. Varadhan: "Large Deviations" (Published in Proceedings of the International Congress of Mathematicians Hyderabad, India, 2010, Vol. I, pp. 622–639)
2014 (Seoul) J. W. Milnor: "Topology through the centuries: Low dimensional manifolds" (Published in Bull. Amer. Math. Soc. 52(4):545–584 (2015))
2018 (Rio de Janeiro) M. Atiyah: "Arithmetic Physics" (Published in Proceedings of the International Congress of Mathematicians, Rio de Janeiro, Brazil, Vol. I, pp. 263–270)

The Abel Lecture at the European Congress of Mathematics

2016 (Berlin) Endre Szemerédi: "Arithmetic progressions and graph theoretic lemmas"

Printed by Printforce, the Netherlands